Genetically Engineered Mice for Cancer Research

Jeffrey E. Green • Thomas Ried
Editors

Genetically Engineered Mice for Cancer Research

Design, Analysis, Pathways, Validation and Pre-Clinical Testing

Springer

Editors
Jeffrey E. Green
Transgenic Oncogenesis
and Genomics Section
Laboratory of Cancer Biology and Genetics
National Cancer Institute
Bethesda, MD, USA
jegreen@nih.gov

Thomas Ried
Genetics Branch, Center for
Cancer Research, National Institutes
of Health/National Cancer Institute
Bethesda, MD, USA
riedt@mail.nih.gov

ISBN 978-1-4899-9125-6 ISBN 978-0-387-69805-2 (eBook)
DOI 10.1007/978-0-387-69805-2
Springer New York Dordrecht Heidelberg London

Printed on acid-free paper

Springer is part of Springer Science+Business Media (www.springer.com)

Foreword

The extraordinary explosion in our knowledge about cancer over the last 40 years has finally begun to illuminate the black box that defines what we now know to be a large collection of distinct but highly related diseases called cancer. Our knowledge is far from complete-insights, surprises and alternate views still await us as our journey through cancer research continues. Despite this progress, the speed and extent of advances in our ability to prevent and treat cancer is frustrating to all. Much is now discussed about the need to emphasize the translation of our fundamental insights into new approaches to people with cancer. This book and the efforts described are essentially about that translational challenge. There are two critical roadblocks to translate basic research into new cancer interventions:

1. It is essential that we can definitely relate and apply our knowledge gained in the laboratory to the actual human disease as it exists in people with cancer and
2. We need a very robust translational enterprise that links the discovery, development, and testing of safe and effective interventions, such as diagnostics and drugs, to the relevant science

The modern use of mouse models for human cancer is critical to both of these roadblocks. Unfortunately, the very term "mouse models" tends to imply a somewhat limited view of the potential of the mouse for teaching us about cancer.

First of all, mice can get cancer and they can be experimentally manipulated to get cancer. This, in and of itself, provides us with the extraordinary opportunity to study mouse oncology as a way to take manipulable experimental systems way beyond human cancer cells grown on plastic surfaces to study the whole organism context of the complex "organs" called tumors. We have only begun to scratch the surface of the field of experimental mouse oncology, but it represents an enormous opportunity for making our fundamental discoveries much more sophisticated and relevant to aspects of human tumor biology that we have only the most rudimentary insights into. Way too little attention has been given to the value of mouse oncology but, to my mind, it represents the most translatable approach to the next period of cancer biology research that we have.

I find it useful to distinguish "mouse oncology" from "mouse models of human cancer." While they are clearly two sides of the same coin, they take place with different expectations and different criteria of success. Mouse oncology, to my way of thinking, nearly uniquely allows us to ask sophisticated questions about cancer, including:

- The complex cellular and noncellular development, evolution and fates of tumors
- Metastasis, spread, and growth of tumors
- The cellular origin and development of cancer
- Host–tumor interactions, including genetic context, immune and inflammatory controls

"Mouse models," on the other hand, according to this description, provides us with experimental tools that we hope are more predictive settings for studying interventions preclinically to help in our assessment of new human interventions. These studies may include efficacy, pharmacodynamics, dosing schedules, and the development of resistance among other measures. The value of a "model" is in its predictive power as a surrogate of what will happen in people. We are slowly gaining some experience with more sophisticated mouse models of certain human cancers, such as CML that support the hoped for value of these models as potentially reliable paths on the road to successful drug development for humans, but this aspect of the use of mice in cancer research, while the most widely discussed, is just in its infancy.

Both the ability to study mouse oncology for fundamental insights in actual tumor biology and the generation of mouse models for human cancer rely on the continued development of new technologies to generate defined cancers in mice and the technologies to characterize and study cancer in the mouse. These technologies are well described in this volume as are emerging views of the characterization of those tumors at the molecular and other levels of analysis.

What is so attractive to me in this volume is the attention to mouse oncology and the insights to be garnered that is enabled by the study of mouse cancer, including the testing of the some of the most accepted molecular pathways in cancer, and host–tumor interactions.

The volume ends back at the translational potential of the mouse as "model" for predicting human interventions. The jury is still out as to the full translational potential of modern mouse models, as it should be. Despite all of the progress, we are still in the early days of learning how to generate, characterize, and understand mouse cancer and to even assess the similarities and differences with human disease.

Richard Klausner
Director, National Cancer Institute (2001)

Preface

The landmark studies by Gordon and Ruddle in the early 1980s (Gordon et al. 1980; Gordon and Ruddle 1981) demonstrating that the mouse genome could be permanently altered through transgenesis ushered in a remarkable era in biotechnology that has expanded the landscape of molecular biology and functional genomics. Methods to alter the mouse genome have become increasingly sophisticated during the past three decades. Beginning with the initial technique of pronuclear injection of embryos and their transfer into foster recipient female mice, the mouse genome can be modified through homologous recombination to alter a gene in every cell of the animal or only in a subset of targeted cell types at different stages of development. Methods whereby genes or noncoding RNAs can be inducibly expressed or repressed have added the important dimension of more precise temporal control to regulating the expression of transgenes in the animal.

This book provides important overviews regarding the state-of-the art of mouse modeling related to cancer research written by leaders in the field. The generation of genetically engineered mouse models (GEMMs) of cancer has become increasingly refined to the point where genetic lesions identified in human tumors can be introduced alone or in relevant combinations to recapitulate oncogenic processes that drive cancer formation and progression in humans. The challenge now is to harvest the value of these models for preclinical testing of novel therapeutic strategies to ultimately advance the treatment of cancer in patients.

Part I of the book is devoted to methodologies to generate GEMMs. Chapter 1 provides a general overview of the techniques used to manipulate the mouse genome. Chapter 2 provides detailed expositions on the use of the *cre*-lox system to conditionally alter a gene and methods for the inducible expression of a gene. The use of the very powerful recombineering approach is discussed in Chapter 3. This technique allows for the introduction of precise alterations in bacterial artificial chromosomes (BACs) and opens up the ability to manipulate very large segments of chromosomal DNA. This has simplified both the replacement of altered genes back into the germ line through homologous recombination and the generation of constructs that contain the authentic regulatory elements distributed over a genetic locus in order to express genes in a manner that exactly recapitulates endogenous expression of that locus.

Forward genetic screens have proved very powerful in identifying genetic interactions that may lead to tumor progression and the use of insertional mutagenesis is described in Chapter 4. Chapter 5 delineates the application of the novel TVA system to target somatic expression of individual or multiple transgenes that has provided important insights into cooperativity between oncogenic pathways. As described in Chapter 6, the use of the chemical carcinogen ENU has been particularly useful as a mutagen to develop rodent models of cancer.

While hundreds of GEMM cancer models have been developed, it is critical that they be thoroughly evaluated on multiple levels to determine in what ways they do or do not represent subtypes of human cancer. Part II of the book explores how various approaches are used to compare mouse models of cancer with human tumors. Morphologic and biomarker studies remain the most important method for diagnosing, staging, and predicting outcome for human patients. Much effort has gone into performing cross-species pathology analyses between human tumors and their counterparts arising in GEMMs. These important comparisons are highlighted in Chapter 7. As high-throughput genomic studies have demonstrated, human cancers arising in the same organ that have similar morphologic appearances may be quite different on a molecular level. Clearly, no single mouse model will represent multiple subtypes of human cancer, but particular GEMMs may be excellent models for a certain subtype of cancer. Identifying such models, therefore, is in keeping with the concept of "personalized medicine," and will be key in their use for understanding key biologic distinctions between tumor subtypes and for developing new therapeutics. Relevant mouse models are now being evaluated using high-throughput genomic approaches and are being compared to similar studies performed on human tumors. Chapter 8 discusses how this has been performed using methods to determine copy number alterations in the genomes of GEMMs. Chapter 9 summarizes advanced molecular cytogenetic techniques with a special emphasis on their use in visualizing chromosomal translocations- the hallmark of hematological malignancies- and the study of mechanisms by which they arise. Gene expression profiling has provided important insights into how particular GEMMs may cluster together with particular subtypes of human cancer and is reviewed in Chapter 10. As expounded in Chapter 11, advances in *in vivo* imaging modalities have greatly advanced the ability to follow tumor progression in a living animal in real time and to determine how tumors respond to particular therapies without sacrificing the animal. These technologies parallel many aspects of how tumors are assessed and followed in human patients, making such studies in mice highly translational to the clinic.

Part III of the book provides several important examples of how mouse modeling has shed new insights into molecular mechanisms and biologic processes that are fundamental to tumor development. An important example of how normal differentiation programs are related to tumor development is discussed in Chapter 13. Chapter 14 presents an overview of how the functions of p53 and pRB have been dissected using mouse models and how this relates to loss of function of these key tumor suppressor genes in many human cancers. Several genes and pathways have been identified as being involved in human colorectal cancers and have been manipulated in GEM models to generate gastrointestinal tumors. The variety of such models and the pathways they represent are presented in Chapter 15. This is followed by Chapters 16 and 17 that describe how modeling in the mouse has shed light on the involvement of the Src family members and signaling pathway and maspin expression on regulating metastatic tumor progression. While GEMMs have been developed primarily to manipulate the genome at the DNA level, epigenetic regulation of the genome is increasingly recognized as a major determinant of development, differentiation, and oncogenesis. Understanding the epigenome has been advanced by important studies in mouse models as highlighted in Chapter 18.

The transforming growth factor β (TGFβ) family is composed of a large and complex set of ligands and receptors whose roles in normal tissue homeostasis and tumor formation have been a great challenge to understand. Chapter 19 describes how various approaches in GEMMs have provided important insights into the function of this family of genes in cancer biology. Much of the complexity of how the TGFβ family operates is due to its cross talk between multiple cell types. The vital interplay between epithelial tumor cells and their neighboring stromal components is now recognized as fundamental to the development and progression of cancer. How GEMMs have contributed to our knowledge about stromal–epithelial cross talk that influences the development of tumors is presented in Chapter 20. The critical role of the immune system in participating in cross talk with tumor cells is expanded in Chapter 21.

Part IV of the book focuses on how GEM models are being exploited to improve cancer prevention and preclinical testing of novel therapeutic approaches. Unfortunately, the use of GEMMs for drug development has been seriously hampered by intellectual property issues related to the patents, which were awarded to the development of the "*Oncomouse*" (Stewart et al. 1984). However, academia, biotechnology companies, and the pharmaceutical industries have recognized the value of these models for preclinical applications and in some cases, the preclinical studies in GEMMs have motivated the development of clinical trials in patients. There is a growing trend to utilize many of these models for pre-clinical testing, although their value for predicting response in patients remains to be shown in many cases. Chapter 22 provides a concise historical perspective of preclinical testing using non-GEM animal models, human cell lines, and xenografts up to the recent use of GEMMs for testing drug therapies. Applying genomic approaches to identify new drug targets, particularly in the context of specific genetic alterations in a tumor, is discussed in Chapter 23. An important use of GEMMs has been to test approaches for cancer prevention. As described in Chapter 24, prevention trials in GEMMs have led to important understandings of how some agents work and to new clinical trials. Interesting approaches using inducible oncogene systems in GEMMs led to the concept of "oncogene addition," as discussed in Chapter 25. This has underscored the hope that by identifying a pathway or pathways that are the "Achilles heel" of tumor survival, genes critical for tumor maintenance can be functionally identified and targeted for therapeutic intervention.

Chapters 26 and 27 provide excellent examples of how GEMMs have been utilized for testing novel therapies for tumors of the CNS and hematological malignancies. A perspective on how the pharmaceutical industry envisions the incorporation of GEMMs into research and drug development is discussed in Chapter 28.

This book covers many important topics related to the generation, validation, and use of GEM models for advancing our knowledge of the molecular biology of cancer and how GEMMs may be used for translational research. Nonetheless, we realize that many important topics and the outstanding work of many other investigators who have shaped the field of mouse modeling could not be included in this book due to space constraints. We are most grateful to all of those who have contributed to this effort which we feel will further educate students, teachers, investigators, and mouse modelers about the tremendous value gleaned from GEM models utilized in cancer research. We are convinced that the comprehensive and systematic analysis of sophisticated mouse models will provide critical information which ultimately will benefit cancer patients.

Many of the advances described in this book were greatly influenced by the support of the National Cancer Institute for the Mouse Models of Human Cancer Consortium (MMHCC) that was initiated by Dr. Richard Klausner when he was director of the NCI. Many of the advances in cancer modeling and the infrastructure to support this enterprise are the results of the MMHCC initiative through the efforts of the Program Director, Cheryl Marks and Division Director, Dinah Singer. The scientific cancer and mouse modeling community are indebted to their foresight and dedicated efforts.

Bethesda, MD, USA Jeffrey E. Green
Bethesda, MD, USA Thomas Ried

References

Gordon JW, Scangost GA, Plotkin DJ, Barbosaf JA, Ruddle FH (1980) Genetic transformation of mouse embryos by microinjection of purified DNA. Proc Natl Acad Sci U S A 77:7380–7384

Gordon JW, Ruddle FH (1981) Integration and stable germ line transmission of genes injected into mouse pronuclei. Science 214:1244–1246

Stewart TA, Pattengale PK, Leder P (1984) Spontaneous mammary adenocarcinomas in transgenic mice that carry and express MTV/myc fusion genes. Cell 38:627–637

Contents

Contributors

Michelle Ahalt Developmental Therapeutics Program, NCI-Frederick, 1050 Boyles St, Building 1052/239, Frederick, MD 21702, USA

Donna G. Albertson Department of Laboratory Medicine and UCSF Helen Diller Family Comprehensive Cancer Center, Cancer Research Institute, University of California San Francisco, San Francisco, CA 94143-0808, USA

Sergio Alcoser Developmental Therapeutics Program, NCI-Frederick, 1050 Boyles St, Building 1052/239, Frederick, MD 21702, USA

James V. Alvarez Department of Cancer Biology, Raymond and Ruth Perelman School of Medicine at the University of Pennsylvania, Philadelphia, PA 19104, USA

Abramson Family Cancer Research Institute, Raymond and Ruth Perelman School of Medicine at the University of Pennsylvania, Philadelphia, PA 19104-6160, USA

Laura Attardi Department of Radiation and Cancer Biology, Stanford University School of Medicine, Stanford, CA 94305, USA

Alice Hawley Berger Cancer Genetics Program, Beth Israel Deaconess Cancer Center, Departments of Medicine and Pathology, Beth Israel Deaconess Medical Center, Harvard Medical School, Boston, MA, USA

Weill Graduate School of Medical Sciences, Cornell University, New York, NY, USA

Neil A. Bhowmick Department of Medicine, Uro-Oncology Research Program, Cedars-Sinai Medical Center, 8750 Beverly Boulevard, Atrium 103, Los Angeles, CA 90048, USA

Powel H. Brown Department of Clinical Cancer Prevention, The University of Texas MD Anderson Cancer Center, Houston, TX 77030, USA

Eletha Carbajal Program in Cancer Biology and Genetics, Memorial Sloan Kettering Cancer Center, New York, NY 10021, USA

Robert D. Cardiff Department of Pathology and Laboratory Medicine, Center for Comparative Medicine, Center for Genomic Pathology, University of California, Davis, CA 95616, USA

Kenneth C. Carter Noble Life Sciences, Inc., 22 Firstfield Road, Gaithersburg, MD 20878, USA

Lewis A. Chodosh Department of Cancer Biology, Raymond and Ruth Perelman School of Medicine at the University of Pennsylvania, Philadelphia, PA 19104, USA

Department of Cell and Developmental Biology, Raymond and Ruth Perelman School of Medicine at the University of Pennsylvania, Philadelphia, PA 19104, USA

Department of Medicine, Raymond and Ruth Perelman School of Medicine at the University of Pennsylvania, Philadelphia, PA 19104, USA

Abramson Family Cancer Research Institute, Raymond and Ruth Perelman School of Medicine at the University of Pennsylvania, Philadelphia, PA 19104-6160, USA

Christopher H. Contag Molecular Imaging Program at Stanford, Stanford School of Medicine, Stanford, CA, USA

Department of Pediatrics, Stanford School of Medicine, Stanford, CA 94305, USA

Department of Microbiology and Immunology, Stanford School of Medicine, Stanford, CA 94305, USA

Lisa M. Coussens Department of Pathology, University of California, San Francisco, 513 Parnassus Ave. HSW-450C, San Francisco, CA 94143, USA

Helen Diller Family Comprehensive Cancer Center, University of California, San Francisco, 513 Parnassus Ave. HSW-450C, San Francisco, CA 94143-0502, USA

Edwin Cuppen Hubrecht Institute for Developmental Biology and Stem Cell Research, Section Functional Genomics and Bioinformatics, Uppsalalaan 8, Utrecht 3584, CT, The Netherlands

Karin E. de Visser Department of Molecular Biology, The Netherlands Cancer Institute, Plesmanlaan 121, CX 1066, Amsterdam, The Netherlands

Chu-Xia Deng Genetics of Development and Disease Branch, National Institute of Diabetes, Digestive and Kidney Diseases, National Institutes of Health, 10/9N105, 10 Center Drive, Bethesda, MD 20892, USA

Michael J. Difilippantonio Division of Cancer Treatment and Diagnosis, National Cancer Institute, NIH, 31 Center Drive, Suite 3A44, Bethesda, MD 20892, USA

Thomas Doetschman Department of Cellular and Molecular Medicine, University of Arizona, Tucson, AZ, USA

BIO5 Institute, University of Arizona, Tucson, AZ, USA

Reinhard Ebner National Cancer Institute, Section of Cancer Genomics, National Institutes of Health, Bethesda, MD 20892, USA

Manel Esteller Cancer Epigenetics Group, Spanish National Cancer Research Centre (CNIO), Madrid, Spain

Cancer Epigenetics and Biology Program (PEBC), Catalan Institute of Oncology (ICO), Barcelona, Catalonia, Spain

Yi Feng Department of Cancer Biology, Raymond and Ruth Perelman School of Medicine at the University of Pennsylvania, Philadelphia, PA 19104, USA

Abramson Family Cancer Research Institute, Raymond and Ruth Perelman School of Medicine at the University of Pennsylvania, Philadelphia, PA 19104-6160, USA

Andrea Ferris HIV Drug Resistance Program, National Cancer Institute-Frederick, Frederick, MD 21702, USA

Omar E. Franco Department of Urologic Surgery, Vanderbilt University Medical Center, Nashville, TN 37232-2765, USA

Andrew O. Giacomelli Department of Biochemistry, McMaster University, ON, Canada

Michael N. Gould McArdle Laboratory for Cancer Research, Department of Oncology, University of Wisconsin – Madison, 1400 University Avenue, Madison, WI 53706, USA

Jeffrey E. Green Transgenic Oncogenesis and Genomics Section, Laboratory of Cancer Biology and Genetics, National Cancer Institute, Building 37, Room 4054, 37 Convent Drive., Bethesda, MD 20892, USA

Robin M. Hallett Department of Biochemistry, McMaster University, ON, Canada

John A. Hassell Department of Biochemistry, McMaster University, ON, Canada

Centre for Functional Genomics, McMaster University, 1200 Main Street West, ON, L8N 3Z5 Canada

Simon W. Hayward Department of Urologic Surgery, Department of Cancer Biology, Vanderbilt-Ingram Cancer Center, Vanderbilt University Medical Center, Nashville, TN 37232-2765, USA

Department of Urologic Surgery, Vanderbilt University Medical Center, A-1302 Medical Center North, 1161 21st Avenue South, Nashville, TN 37232-2765, USA

Michael T. Hemann The Koch Institute for Integrative Cancer Research at MIT, 700 Main Street, Cambridge, MA 02139, USA

Eric C. Holland Program in Cancer Biology and Genetics, Memorial Sloan Kettering Cancer Center, New York, NY 10021, USA

Melinda Hollingshead Developmental Therapeutics Program, NCI-Frederick, 1050 Boyles St, Building 1052/239, Frederick, MD 21702, USA

Stephen K. Horrigan Noble Life Sciences, Inc., 22 Firstfield Road, Gaithersburg, MD 20878, USA

Stephen H. Hughes HIV Drug Resistance Program, National Cancer Institute-Frederick, Frederick, MD 21702, USA

Ken Hung Department of Gastroenterology, Tufts Medical Center, Boston, MA 02111, USA

Richard Klausner The Column Group, 1700 Owen st., Suite 500, San Francisco, CA 94158, USA

Hosein Kouros-Mehr Genentech, 1 DNA Way, MS-60, South San Francisco, CA 94080, USA

Melanie Kucherlapati Division of Genetics, Department of Medicine, Brigham and Women's Hospital, Harvard Medical School, Boston, MA 02115, USA

Raju Kucherlapati Division of Genetics, Department of Medicine, Brigham and Women's Hospital, Harvard Medical School, Boston, MA 02115, USA

Mari Kuraguchi Department of Medical Oncology, Belfer Institute for Applied Cancer Science, Dana Farber Cancer Institute, Boston, MA 02115, USA

David A. Largaespada The Department of Genetics, Cell Biology and Development, The Center for Genome Engineering, Masonic Cancer Center, The University of Minnesota, Twin Cities; 6-160 Jackson Hall; 321 Church St. S.E., Minneapolis, MN 55455, USA

Brian C. Lewis Program in Gene Function and Expression, University of Massachusetts Medical Center, Worcester, MA 01605, USA

Yi Li Lester and Sue Smith Breast Center and Department of Molecular and Cell Biology, Baylor College of Medicine, Houston, TX 77030, USA

Richard Marcotte Ontario Cancer Institute, University of Toronto, Toronto, ON M5G 1L7, Canada

Aleksandra M. Michalowski Transgenic Oncogenesis and Genomics Section, Laboratory of Cancer Biology and Genetics, Center for Cancer Research, National Cancer Institute, Bethesda, MD, USA

Willam J. Muller Goodman Cancer Center, 1160 Pine Ave., Montreal, QC H3A 1A3, Canada

Sandra Orsulic Women's Cancer Research Institute, Cedars-Sinai Medical Center, Los Angeles, CA 90048, USA

Pier Paolo Pandolfi Cancer Genetics Program, Beth Israel Deaconess Cancer Center, Departments of Medicine and Pathology, Beth Israel Deaconess Medical Center, Harvard Medical School, Boston, MA, USA

Weill Graduate School of Medical Sciences, Cornell University, New York, NY, USA

Subha Philip Mouse Cancer Genetics Program, Center for Cancer Research, NCI-Frederick, Building 560, Room 32-31C, 1050 Boyles Street, Frederick, MD 21702, USA

Veronica R. Placencio Departments of Cancer Biology, Urologic Surgery, Vanderbilt-Ingram Cancer Center, Vanderbilt University, Nashville, TN 37232-2765, USA

Jeniffer A. Prescher Molecular Imaging Program at Stanford, Stanford School of Medicine, Stanford, CA, USA

Lauren Reinke Department of Molecular Pharmacology and Biological Chemistry, Northwestern University Feinberg School of Medicine, Robert H. Lurie Comprehensive Cancer Center, 303 E. Superior Street, Chicago, IL 60611, USA

Thomas Ried Genetics Branch, Center for Cancer Research, National Institutes of Health/National Cancer Institute, 50 South Drive Room 1408, Bethesda, MD 20892, USA

Cecilia Rosales Cancer Epigenetics Group, Spanish National Cancer Research Centre (CNIO), Madrid, Spain

Cancer Epigenetics and Biology Program (PEBC), Catalan Institute of Oncology (ICO), Barcelona, Catalonia, Spain

Julien Sage Department of Pediatries and Genetics, Stanford University, Stanford, CA 94305, USA

L. Philip Sanford BIO5 Institute, University of Arizona, Tucson, AZ, USA

Shyam K. Sharan Mouse Cancer Genetics Program, Center for Cancer Research, National Cancer Institute at Frederick, Building 560, Room 32-31C, 1050 Boyles Street, Frederick, MD 21702, USA

Bart M.G. Smits McArdle Laboratory for Cancer Research, Department of Oncology, University of Wisconsin – Madison, 1400 University Avenue, Madison, Rm 506A, WI 53706, USA

Douglas W. Strand Department of Urologic Surgery, Vanderbilt University Medical Center, Nashville, TN 37232-2765, USA

Jeffrey W. Strovel Noble Life Sciences, Inc., 22 Firstfield Road, Gaithersburg, MD 20878, USA

Terry Van Dyke Mouse Cancer Genetics Program, National Cancer Institute at Frederick, 1050 Boyles Street, Building 560, Room 32-32, Frederick, MD 21702, USA

Chunyu Wang Department of Clinical Cancer Prevention, The University of Texas MD Anderson Cancer Center, Houston, TX 77030, USA

Bart O. Williams Molecular Medicine and Virology Group, Van Andel Research Institute, Grand Rapids, MI 49503, USA

Elizabeth S. Yeh Department of Cancer Biology, Raymond and Ruth Perelman School of Medicine at the University of Pennsylvania, Philadelphia, PA 19104, USA

Abramson Family Cancer Research Institute, Raymond and Ruth Perelman School of Medicine at the University of Pennsylvania, Philadelphia, PA 19104-6160, USA

Ming Zhang Department of Molecular Pharmacology and Biological Chemistry, Northwestern University Feinberg School of Medicine, Robert H. Lurie Comprehensive Cancer Center, 303 E. Superior Street, Chicago, IL 60611, USA

Min Zhu Transgenic Oncogenesis and Genomics Section, Laboratory of Cancer Biology and Genetics, Center for Cancer Research, National Cancer Institute, Bethesda, MD, USA

Chapter 1
Overview of Designing Genetically Engineered Mouse (GEM) Models

Thomas Doetschman and L. Philip Sanford

1.1 Introduction

It is very important to spend time and effort on vector design considerations when planning to make a GEM. The vector designer will ask what information is desired from the genetically modified animal, and an engineering scheme will be devised. It is strongly recommended that the investigator consult experienced GEM vector producers with all the information that is desired from the GEM. The investigator will be apprised of the feasibility of each design consideration, and usually learns of additional design elements that may expand the information that can be obtained from the GEM and that can in turn expand the overall research yield. Our experience is that the extra time, effort, and care that is put into the coordination of GEM design with research objectives saves much time and effort in the long run. In addition, we have found that careful GEM design consideration will greatly improve the success of GEM production. In this chapter, we discuss gene targeting design considerations that should be made before initiating production of the engineered mouse strain.

1.2 Has the GEM Already Been Made?

The time, effort, and expense of GEM production make it imperative that the investigator first determine whether his/her "favorite gene" has been knocked out or otherwise modified by another research group in a way that is applicable to the research objectives. Our GEM

T. Doetschman (✉)
Department of Cellular and Molecular Medicine, University of Arizona, Tucson, AZ, USA

BIO5 Institute, University of Arizona, Tucson, AZ, USA
e-mail: tdoetsch@u.arizona.edu

L.P. Sanford
BIO5 Institute, University of Arizona, Tucson, AZ, USA
e-mail: psanford@email.arizona.edu

J.E. Green and T. Ried (eds.), *Genetically Engineered Mice for Cancer Research: Design, Analysis, Pathways, Validation and Pre-Clinical Testing*,
DOI 10.1007/978-0-387-69805-2_1,

production service facility has experienced instances where in the investigator had failed to look for a pre-existing GEM only to find later that the GEM already existed. Queries to colleagues can also turn up GEMs that have yet to be published or that are in the production phase. The near universal availability of PubMed searches make this possible if a publication has occurred. As an intermediate step one can often find whether an ES cell line with the targeted gene is already available. The advent of the NIH's Knockout Mouse Project (KOMP) and other gene targeting repositories have made locating gene-targeted ES cells quite easy (http://www.komp.org/).

1.3 Considerations in Choosing ES Cell Genetic Background

The variety of ES cell genetic backgrounds for gene targeting has greatly expanded in recent years. While many of the early experiments were done with ES cells derived from a 129 background (rev. by Elizabeth Simpson (Simpson et al. 1997)), other genetic backgrounds are now available including C56L/6J, C57BL/6N, C57BL/6 TyrC-, Balb/C, CH3/HeN, DBA/2N, FVB/N, NZW, 129S6, 129X1, and many F1 ES cell lines (Millipore, Jackson Labs, Hyclone, Open Biosystems and Taconic). For many years mouse geneticists have advocated the use of inbred mouse lines to assist in phenotype standardization and penetrance. This push gained much traction with the NIH's KOMP and the efforts of the German Gene Trap Consortium (GGTC), North American Conditional Mouse Mutagenesis project (NorComm), and the International Gene Trap Consortium (IGTC). As a consequence of these efforts the C57BL/6 mouse has become the de facto standard mouse background for gene targeting.

Most early GEM producers used a 129 substrain of ES cell and bred the chimeras to C57BL/6 or Black Swiss females resulting in mixed strain GEMs. An advantage of working with mixed genetic backgrounds is that although the degree of penetrance may be less and variability in expressivity may be greater on a mixed than inbred background, the range of phenotypes is also likely to be greater. In addition, the phenotypes are more likely to be of physiological relevance because phenotypes that could occur only in an epistatic background may not represent an important gene function at the level of the animal in the wild or of the species. Therefore, by maintaining mixed background GEM strains as "Advanced Intercross Lines" so that the lines remain in a "partially outbred" fashion so as to avoid inadvertent inbreeding and subsequent loss of phenotype penetrance, we can more thoroughly assess the multiple functions of a gene and be more assured that the phenotypes we see are physiologically important (Doetschman 1999). The variable penetrance that can occur on mixed genetic backgrounds (rev. in Kallapur et al. 1999; Sanford et al. 2001) may better reflect the phenotypic variability found in the outbred human population. Finally, a practical advantage is that outbred mice tend to be easier to maintain than the inbred lines which often have smaller average litter sizes.

The advantages of investigating GEMs on inbred genetic backgrounds are also compelling. Penetrance of phenotype is high and there is little variability in expressivity of the phenotype(s). An obvious advantage is that the number of mice needed for most studies is considerably reduced, thereby reducing the time and cost of experiments.

The investigator must keep in mind, however, that moving a targeted allele from one background to another can dramatically alter phenotype (see Hide et al. 2002; Almind et al. 2003; Jiang et al. 2005; McLin and Steward 2006 for phenotypic changes seen when moving a targeted allele to another background from C57BL/6). Finally, high penetrance and invariable expressivity of phenotype allows one to better detect nongenetic influences on phenotype such as environment and nutrition. In summary, the investigator must take into account the genetic/environmental context in which the desired information is to be obtained from the GEM, and that will dictate the choice of genetic background to be used. Both strain dependency of phenotype and modifier genes are discussed in detail in later chapters.

1.4 Considerations for Simple Gene Knockout Vector Design

The most common gene targeting scheme has been the simple knockout (Fig. 1.1). Typically, this approach entails the replacement of a vital gene coding sequence with a selection cassette. The advantage of this approach is that the production of the null allele is usually fairly straight forward and there are repositories of ready to use targeted ES cell lines that contain this type of gene manipulation or a related gene inactivation called a gene trap. The disadvantages of this type of gene targeting are often unpredictable but may include the following: a gene thought to be expressed only in adult tissues turns out to have an essential role during embryogenesis (Huang et al. 1993, 1995; Cossee et al. 2000; Den et al. 2006). Consequently, if one desires to investigate an adult phenotype, it would be worthwhile to know the embryonic and fetal expression of the gene in order to determine whether any reposited ES lines will be adequate for the research objectives.

Another problem that can be encountered when designing simple gene knockouts is the inadvertent production of a dominant negative allele (Bhattacharyya et al. 2002; Moll and Slade 2004) or possibly a gain of function allele. A dominant negative allele can lead to haploinsufficiency and thus impair normal function of the targeted mouse line. This is a possibility with gene products that bind to or form ologomers. Another potential problem lies in the disruption or deletion of control elements located in an intron, and intronic control elements can occasionally affect distant gene loci. A more detailed discussion of these issues is presented below in the section discussing the production of conditional alleles. Lastly, a number of gene targeting vectors use the phosphoglycerate kinase (*pgk*) promoter in either their selection or reporter cassettes. This promoter is problematic and should be avoided or at the very least removed from the ES cells prior to making the mouse as there are a number of cases detailing the ability of the *pgk* promoter to deregulate the expression of genes located on the same or opposite DNA strand. Such spurious gene expression can easily alter the phenotype of the gene knockout under study and is of particular concern when small gene modifications, such as the introduction of a single base pair polymorphism, are expected to have a subtle phenotype. Regional gene disregulation can be very misleading due to direct or downstream effects that provide a spurious phenotype. (Pham et al. 1996; Seidl et al. 1998; Sun and Storb 2001).

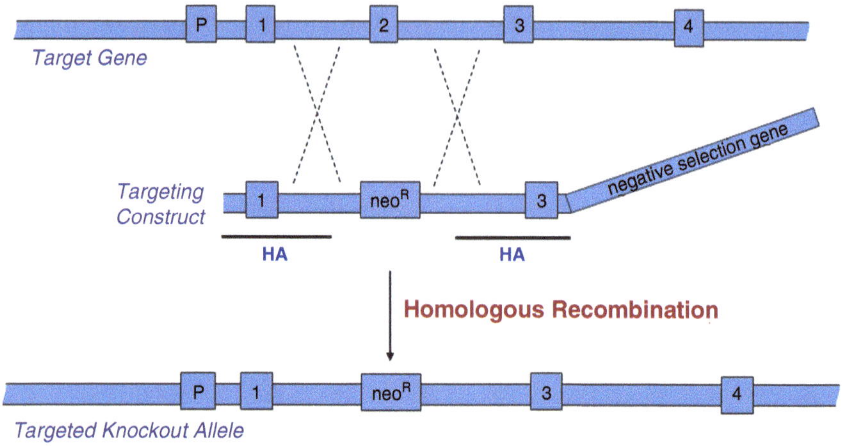

Fig. 1.1 Design considerations for conventional gene knockout. Four exon gene is converted to a dysfunctional gene by replacement of the 2nd exon with a *neo* selectable marker gene which can be either an intact gene or a gene designed for enrichment by being either promoterless or poly (A) addition signal free. A negative selectable marker gene, the Herpes simplex virus thymidine kinase gene is the most commonly used, is placed outside one of the arms of homology (HA) for enrichment. If DNA integration occurs by homologous recombination, the negative selectable marker gene will not be incorporated as it lies outside the region of homology. Other negative selector genes capable of higher expression levels in a mammalian cell may be useful. Homologous arms (HA) as a rule of thumb should be no shorter than ~2 kb and no greater than ~8 kb. The region of the target gene to be removed should be no more than ~20 kb, though longer regions have been deleted, but usually with much lower targeting efficiency. Placement of the positive selectable marker gene must take into consideration the intronic sequences that will be eliminated and their potential for gene regulation. The gene product of the targeted allele will depend on several factors, including whether the positive selectable marker gene has promoter and poly (A), and whether the deleted structural region (exon 2 in the figure) has a number of bp that is a multiple of 3, such that splicing from exon 1 to 3 would keep or destroy coding frame for the remainder of the targeted allele

1.5 Knockin-SNPs

There are several approaches for this type of gene targeting: "hit and run" (Hasty et al. 1991a), "in and out" (Valancius and Smithies 1991), "tag and exchange" (Askew et al. 1993), and "DNA oligonucleotide" (Dekker et al. 2003, 1988). These methods result in a subtly targeted allele with no residual selectable marker genes. Introduction of the exogenously applied *flp* and *cre* recombination systems (Jung et al. 1993; Kuhn et al. 1995; Rajewsky et al. 1996) allow for the removal of selectable marker genes with only the *frt* or *loxP* recombination recognition sites remaining in the subtly mutated allele.

Vector design considerations for the subtle mutation approach are the following. (1) The mouse already has the human SNP in its gene, or another mouse strain has the SNP. (2) There are promoter and intronic SNPs that cause human disease. If these nonstructural SNPs are part of a collection of sites such as transcription factor binding sites, the spacing of the human, and mouse motifs may well be very different (discussed more fully in Sect. 1.16). (3) If no phenotype is seen from the SNP, this may be due strictly to structural differences, or to physiological differences between man and mouse (Hooper et al. 1987).

1.6 Conditional Gene Modifications

Most designs now use some form of conditional system (rev. by Torres and Kühn 1997). These are very useful and popular, yet they bring with them many design considerations that must be taken into account. A generic conditional gene ablation design is shown in Fig. 1.2. Since the conditional allele involves introduction of recombination recognition sites into introns, analysis of the intronic sequences to be disrupted or deleted must be thorough. For example, the floxing of exon 6 in the TGFβ2 gene produced a null allele (Doetschman, unpublished observations). After miRNAs were discovered, it was determined that this null allele was probably due to the juxtaposition of the *loxP* site with a miRNA sequence. Scanning for RNAi motifs can now be done using http://microrna.sanger.ac.uk/sequences/search.shtml. Another intronic feature to be avoided is repetitive elements which can interfere with recombination. These can be identified using Repeat Masker. Other intronic features to be avoided are inter-species conserved sequences, close proximity to splice donor (<100 bp) and acceptor (<300 bp) sites, transcription factor binding sites, and genes on the other strand of DNA. We also avoid placing *loxP* sites farther than 3 kb from the selection cassette because the incidence of inclusion of the distal *loxP* site often drops precipitously with increasing distance, a situation that is locus-specific (Vazquez et al. 1998 and Sanford, unpublished observations).

1.7 Reversible Gene Targeting

Sometimes the science will be best served by the ability to induce a gene ablation or alteration and then to be able to reverse that initial genetic event. We call this a "reversible" gene targeting. Reversible gene targeting is very valuable in discerning whether a single gene causes a disease syndrome. The demonstration that the disease model is completely reversible further strengthens the assertion that the gene in question is solely responsible for the genetic condition under study. The simplest form of this design utilizes a genetic element that is flanked by two pairs of asymmetric site-specific recombinase sites. Typically, one pair would be *loxP* sites such as *lox66* and *lox71* and the other pain would be *frt* sites such as *RE* and *LE* (rev. in Branda and Dymecki 2004). These asymmetric recombinase sites act like one way switches. That is, they allow the genetic element to be inverted only once. This is because the recombinase sites are incompatibly modified by the recombinase. For example, the *cre* recombinase can be used to invert a promoter or exon into a non-functional orientation, and subsequent *flp*-mediated recombination can reverse this process.

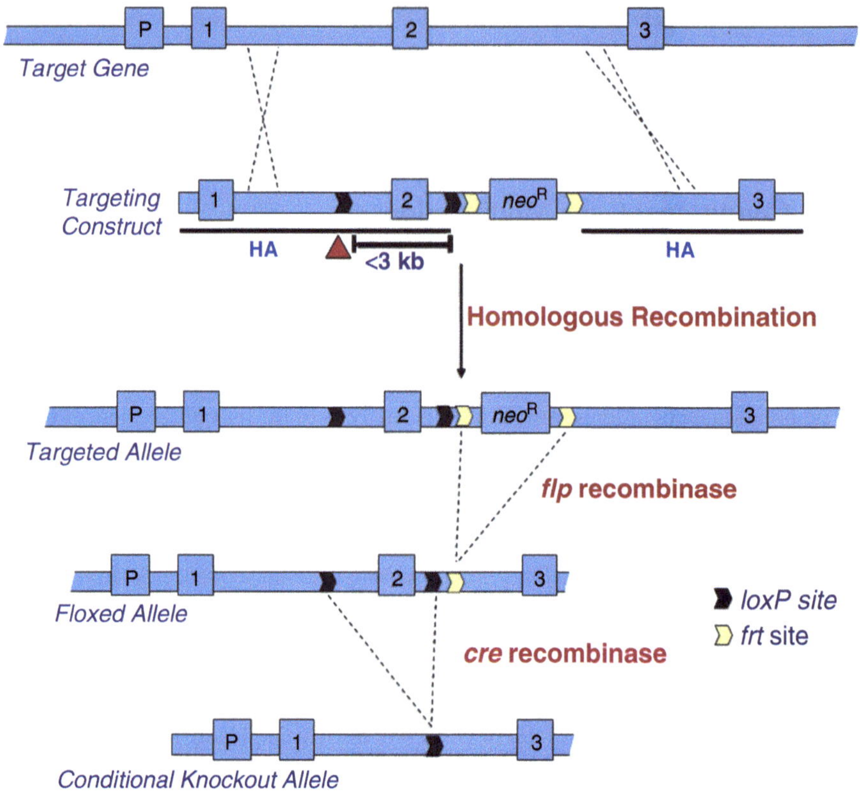

Fig. 1.2 Design considerations for conditional allele. A ≥3 exon gene is converted to a conditional knockout allele in which exon 2 can be conditionally removed when the mouse is genetically combined with a mice carrying a tissue-specific or inducible recombinase transgene. Besides the design considerations depicted in the previous figure such as length of homologous arms (HA) and marker gene(s) characteristics and placement, additional considerations involve placement of recombinase recognition sequences (*loxP*, *black directional bar*: *frt*, *yellow directional bar*) and homologous arm discontinuities (*red directional triangle*). Intronic placement of the *loxP*-site-flanked *neo* resistance gene must not interfere with splicing information (>100 bp from splice donor and >300 bp from splice acceptor), repetitive sequences, miRNAs, transcription factor binding sites, interspecies conserved sequences and genes reading from the opposite DNA strand, and there should be ≤3 kb between the farthest recombination recognition site and the selection cassette

1.8 Isogenic DNA

Since gene targeting requires homologous recombination between the targeting vector and the ES cell genome, it was only logical that some of the earliest gene targeting experiments solely utilized isogenic DNA (Doetschman et al. 1987, 1988; Valancius and Smithies 1991; Hasty et al. 1991a, rev. in van Deursen and Wieringa 1992; Doetschman 1994; Zhou et al. 2001). A systematic analysis of the efficiency of isogenic versus nonisogenic DNA demonstrated that the use of isogenic DNA in the targeting vector can dramatically increase targeting efficiency (Deng and Capecchi 1992).

1.9 Length of Homologous Arms

It is reasonable to assume that the longer the length of homology between the targeting vector and the target gene, the more likely the two DNA molecules will find each other. Also, there is probably a minimum length of homology required by the recombination machinery to carry out homologous recombination. Bradley (Hasty et al. 1991b) and Capecchi (Deng and Capecchi 1992) demonstrated the direct relationship between length of homology and gene targeting efficiency. Very short arms of homology can support homologous recombination, but not always with complete fidelity. That the recombination machinery requires a minimum length of homology was suggested by Smithies in an experiment in which small deletions of 14 and 37 bp were found adjacent to the end of homology when the length of homology was only 32 bp (Doetschman et al. 1988). With respect to very long arms of homology, no systematic study has been made, but it is theoretically possible that because longer homologous arms will likely contain repetitive sequences, the latter may increase the number of genomic targets, thereby decreasing the targeting efficiency. An early study by John Wilson (Zheng and Wilson 1990) found that the targeting efficiency (number of targeted alleles per cell) was the same regardless of whether the cell had two targets (endogenous *DHFR* genes) or 800 targets (amplified *DHFR* genes). Hence, if a homologous arm contains a sequence that is repeated a thousand times in the genome, they may all become targets, thereby decreasing the targeting efficiency at the desired locus. In general, it is recommended that a total of 6–8 kb of homology be used in a targeting construct, with a minimum of 2 kb in one of the arms of homology, and no more than 10 kb in the long arm.

1.10 Distance Between Homologous Arms

It is generally accepted that the greater the distance between homologous arms, the poorer the targeting efficiency. There is little in the way of systematic studies in the literature to substantiate this, in part, because this may be locus specific. As a rule of thumb, however, it is thought that efficiency drops off with greater than an ~20 kb deletion (19.2 kb, Zhang et al. 1994).

1.11 Heterologous Elements Within Homologous Arms

Complex targeting schemes in which homology discontinuities are to be introduced such as diagnostic restriction sites, small polymorphic sequences, nonselectable marker genes, or *frt* or *loxP* recombination recognition sites (see Fig. 1.2) often result in site-specific targeting without incorporation of the sequences causing the discontinuity. In an early gene targeting study by Smithies, the long arm of homology was interrupted by a nonhomologous restriction site which led to some targeted clones with and others without the restriction site (Doetschman et al. 1987), indicating that the sequence at which the recombination "crossover" occurs is unpredictable. In another example, the separation of a green fluorescent protein (GFP) reporter from a *neo* selection cassette by as little as 200 bp of homologous sequence nearly eliminated the chances of getting GFP into the otherwise targeted allele. It appeared that the nonselectable or heterologous sequences promoted the crossover between the selection cassette and the second element (Sanford, unpublished observations). Consequently, complex targeting constructs with discontinuities in the homologous arms must be carefully sequenced to identify the clones with all of the required discontinuities intact.

1.12 Positive Selection

A positive selectable marker gene is needed to identify the small percentage of electroporated colonies that have undergone homologous recombination (refer to Fig. 1.1). It requires a promoter and poly(A) addition signal. One scheme for improving the ratio of targeted to selected colonies (often referred to as enrichment) is to use a promoterless positive selector gene which is based on the idea that the selector gene will be expressed only if inserted in the correct orientation relative to the promoter of an endogenous gene, thereby reducing the number of integrants capable of expressing the selector gene (Doetschman et al. 1988). Another scheme relies on the targeted gene providing the poly(A) addition signal for the positive selector gene. Here, enrichment is based upon the idea that integration in intergenic region or in the wrong orientation within a gene would prevent its expression (Thomas and Capecchi 1987). Such experiments have yielded modest enrichments estimated from 19- to 120-fold and 2- to 5-fold, respectively. Other considerations with respect to the positive selector gene are its potential for incorporation into the structural components of the gene product, and whether it disrupts reading frame of the target gene during splicing. For example, a cryptic splice acceptor site has been identified in the popularly used neomycin resistance cassette (Moens et al. 1992).

1.13 Negative Selection

Negative selection in ES cells was introduced by Capecchi (Mansour et al. 1988) as a means to select against nonhomologous insertions of the targeting sequences, the idea being that if the integration were homologous, the negative selection gene would be lost if it were placed outside of the regions of genomic homology. Enrichments afforded by negative selection have ranged from 4- to 2,000-fold.

A possible reason for this wide range of enrichment may lie in the fact that the strength of negative selection should match the difficulty of getting the desired genetic rearrangement. Recently, we were working on a large deletion. After analyzing >2,000 negative selection-resistant ES cell clones from which no targeted clones were identified, we redesigned the gene targeting vector utilizing a much stronger negative selection scheme. This resulted in a 2 log smaller number of negatively selected clones, of which nearly half were targeted.

1.14 Pretesting Diagnostic Procedures

It is highly recommended that before generating the targeting construct, positive control constructs be generated. Many investigators have lost much time and effort by having to redesign the diagnostic tools for identifying targeted ES cells and the offspring of the germline chimeras, or worse, having to redesign the targeting constructs and redo the targeting experiments. All too frequently, a diagnostic tool that works on paper, or even at the plasmid level, does not work in the presence of the proper amounts of genomic DNA that is prepared as it will be from the GEM mice. Conversely, all diagnostic approaches must be tested for false positives or false negatives in genomic DNA. Our experience in producing GEMs as a service to other investigators is that when we began making the constructs ourselves and verifying them for effective diagnostics, rather than using the constructs or ES cells produced by our clients, our GEM production success rate went from about 70% to nearly 100% (Sanford, unpublished data).

1.15 Recombineering Systems

Bacterial-based recombination engineering (Muyrers et al. 1999) uses genetic elements flanked by relatively short 50 bp regions of sequence homology coupled with inducible viral DNA recombinases to reliably modify DNA from plasmids, bacterial artificial chromosomes (BACs), and the *E. coli* chromosome. The sequence of the homology regions can be chosen freely and therefore any position on a target molecule can be specifically altered. Since this method uses sequence homology to locate the genetic rearrangements, it overcomes the traditional restriction endonuclease site limitations of conventional cloning and gene manipulation (Copeland et al. 2001; Muyrers et al. 2001; Court et al. 2002).

Of particular attractiveness is the high efficiency at which these recombination reactions occur allowing serial recombineering reactions without checking for rare product intermediates after each manipulation as is necessary with conventional molecular biology approaches. Through a series of marker selections and counter selections multiple modifications can be rapidly processed. Bacterial colonies can be screened for the desired product after several reactions since the background of unwanted reactions is so low at each operation (Wang et al, 2006). These strengths have resulted in the creation of large-scale recombineering projects that are designed to produce large numbers of BAC-based transgenes and to make gene targeting vectors for large-scale functional genomics programs (see the KOMP: http://www.komp.org/, ECOMM: http://www.eucomm.org/, and related programs).

1.16 BAC Transgenics Versus Subtle Mutations

When testing the effects of a human susceptibility polymorphism in the mouse, the question arises whether to introduce the polymorphism as a human transgene, preferably in a BAC, into an endogenous gene knockout background or whether to introduce the human polymorphism into the endogenous mouse gene. The advantages of the former would be that the entire human gene product would be introduced, hopefully bringing its regulatory regions with it and it would be easier to do. However, the disadvantages would be that the regulatory regions may have some incompatibilities with the mouse regulatory molecules, it would be difficult to demonstrate that no other subtle mutations occurred in the BAC during production of the transgenic mouse, and BAC copy number and insertion-site mutations which could affect subtle phenotypic differences might occur. Sequence analysis of the regulatory information surrounding the polymorphisms would be informative, but not conclusive. For example, if all of the regulatory information such as DNA binding elements is identical but the spacing between them is slightly different, is the spacing necessary because the mouse's regulatory molecules have structures that require the differential spacing, or does the spacing reflect differences in regulation? If it is the former, then the polymorphism should be put into the mouse gene; if it is the latter, then a BAC transgene might be better. Improvements in the technology to introduce subtle mutations into the mouse genome might improve our willingness to try the latter approach.

1.17 Summary

Since the technologies for generating GEMs have now become quite sophisticated, ranging in extent from SNP knockins to chromosomal engineering, and varying in expression capabilities from combinations of conditional, inducible, and reversible systems, it is clear that vector design be matched with overall experimental design and research objectives. This impressive repertoire of possibilities may best be utilized through consultation with genetic engineers in order to optimize the usefulness of the GEM. Here, we have delineated some of the experimental and vector design considerations that should be taken into account to efficiently produce a GEM that can maximize the research utility of the mouse.

1.18 Brief Historical Perspective

1.18.1 Development of GEM Field

Transgenic mice: Early attempts to introduce exogenous genes into mice involved SV40 viral DNA microinjection into preimplantation blastocysts resulting in adult animals with viral DNA sequences (Jaenisch and Mintz 1974), and germline transmission was first attained by MMLV infection of preimplantation embryos (Jaenisch 1976). Pronuclear microinjection of SV40 DNA sequences resulted in transgene sequences in most organs of resulting newborn mice (Gordon et al. 1980). Transgene expression was achieved by inclusion of an intron in a β-globin transgene (Wagner et al. 1981), and copy-dependent β-globin transgene expression was achieved by inclusion of a distant *cis*-acting locus control region (Grosveld et al. 1987). YAC (Lamb et al. 1993) and BAC (Antoch et al. 1997) transgenics were used to delineate the multigene regions responsible for trisomy and for rescuing loss of a chromosomal region, respectively. Transgenic animal technology and its importance in general and for cancer in particular have been reviewed (Jaenisch 1988; Hanahan 1989).

Embryonic stem cells and gene targeting: Spontaneous teratomas in an inbred strain of mice (Stevens and Little 1954) were shown to harbor pluripotent cells that supported transplantable teratomas (Stevens 1960). Teratocarcinoma cell lines were derived from the teratomas (Evans 1972), and their *in vivo* pluripotency was demonstrated by blastocyst injection but with little germline efficiency (Brinster 1974; Papaioannou et al. 1975). ES cells were isolated directly from the inner cell mass of mouse embryos (Evans and Kaufman 1981; Martin 1981), were subsequently shown to colonize the germline at practical frequencies (Bradley et al. 1984), and were shown to be capable of transmitting transgenes to the germline (Robertson et al. 1986) in a generationally stable fashion (Gossler et al. 1986). Gene targeting was achieved in ES cells through homologous recombination (Doetschman et al. 1987; Thomas and Capecchi 1987), and those ES cells could be used to stably transmit the targeted gene into a new strain of mouse (Koller et al. 1989; Thompson et al. 1989). Finally, it was demonstrated that genes not expressed in ES cells could be targeted at the same frequency as expressed genes (Johnson et al. 1989). Reviews on the early developments in gene targeting are available (Mansour 1990; Koller and Smithies 1992).

These transgenic and gene-targeted mouse studies have provided the technical foundation for the development of GEMs that have led to a wave of second generation GEMs with more complex genetic combinations and more highly controlled regulation of tumor suppressor genes and oncogenes, rev. in (Tuveson and Jacks 2002; Jackson-Grusby 2002; Jonkers and Berns 2002; Van Dyke and Jacks 2002). GEMs are now being used in a preclinical setting to identify and validate novel diagnostic (Shaw et al. 2005; Hung et al. 2009) and therapeutic (Giraudo et al. 2004; Daniel et al. 2005; Carver and Pandolfi 2006) approaches for the detection and treatment of human cancer.

References

Almind K, Kulkarni RN, Lannon SM, Kahn CR (2003) Identification of interactive loci linked to insulin and leptin in mice with genetic insulin resistance. Diabetes 52:1535–1543

Antoch MP, Song EJ, Chang AM, Vitaterna MH, Zhao Y, Wilsbacher LD, Sangoram AM, King DP, Pinto LH, Takahashi JS (1997) Functional identification of the mouse circadian Clock gene by transgenic BAC rescue. Cell 89:655–667

Askew GR, Doetschman T, Lingrel JB (1993) Site-directed point mutations in embryonic stem cells: a gene-targeting tag-and-exchange strategy. Mol Cell Biol 13:4115–4124

Bhattacharyya R, Bhaumik M, Raju TS, Stanley P (2002) Truncated, inactive N-acetylglucosaminyltransferase III (GlcNAc-TIII) induces neurological and other traits absent in mice that lack GlcNAc-TIII. J Biol Chem 277:26300–26309

Bradley A, Evans M, Kaufman MH, Robertson E (1984) Formation of germ-line chimaeras from embryo-derived teratocarcinoma cell lines. Nature 309:255–256

Branda CS, Dymecki SM (2004) Talking about a revolution: the impact of site-specific recombinases on genetic analyses in mice. Dev Cell 6:7–28

Brinster RL (1974) The effect of cells transferred into the mouse blastocyst on subsequent development. J Exp Med 140:1049–1056

Carver BS, Pandolfi PP (2006) Mouse modeling in oncologic preclinical and translational research. Clin Cancer Res 12:5305–5311

Copeland NG, Jenkins NA, Court DL (2001) Recombineering: a powerful new tool for mouse functional genomics. Nat Rev Genet 2:769–779

Cossee M, Puccio H, Gansmuller A, Koutnikova H, Dierich A, LeMeur M, Fischbeck K, Dolle P, Koenig M (2000) Inactivation of the Friedreich ataxia mouse gene leads to early embryonic lethality without iron accumulation. Hum Mol Genet 9:1219–1226

Court DL, Sawitzke JA, Thomason LC (2002) Genetic engineering using homologous recombination. Annu Rev Genet 36:361–388

Daniel D, Chiu C, Giraudo E, Inoue M, Mizzen LA, Chu NR, Hanahan D (2005) CD4+ T cell-mediated antigen-specific immunotherapy in a mouse model of cervical cancer. Cancer Res 65:2018–2025

Dekker M, Brouwers C, te-Riele H (2003) Targeted gene modification in mismatch-repair-deficient embryonic stem cells by single-stranded DNA oligonucleotides. Nucleic Acids Res 31:e27

Den Z, Cheng X, Merched-Sauvage M, Koch PJ (2006) Desmocollin 3 is required for pre-implantation development of the mouse embryo. J Cell Sci 119:482–489

Deng C, Capecchi MR (1992) Reexamination of gene targeting frequency as a function of the extent of homology between the targeting vector and the target locus. Mol Cell Biol 12:3365–3371

Doetschman T, Gregg RG, Maeda N, Hooper ML, Melton DW, Thompson S, Smithies O (1987) Targetted correction of a mutant HPRT gene in mouse embryonic stem cells. Nature 330:576–578

Doetschman T, Maeda N, Smithies O (1988) Targeted mutation of the Hprt gene in mouse embryonic stem cells. Proc Natl Acad Sci USA 85:8583–8587

Doetschman T (1994) Gene transfer in embryonic stem cells. In: Pinkert CA (ed) Transgenic animal technology: a laboratory handbook. Academic, New York, pp 115–146

Doetschman T (1999) Interpretation of phenotype in genetically engineered mice. Lab Anim Sci 49:137–143

Evans MJ (1972) The isolation and properties of a clonal tissue culture strain of pluripotent mouse teratoma cells. J Embryol Exp Morphol 28:163–176

Evans MJ, Kaufman MH (1981) Establishment in culture of pluripotential cells from mouse embryos. Nature 292:154–156

Giraudo E, Inoue M, Hanahan D (2004) An amino-bisphosphonate targets MMP-9-expressing macrophages and angiogenesis to impair cervical carcinogenesis. J Clin Invest 114:623–633

Gordon JW, Scangos GA, Plotkin DJ, Barbosa JA, Ruddle FH (1980) Genetic transformation of mouse embryos by microinjection of purified DNA. Proc Natl Acad Sci USA 77:7380–7384
Gossler A, Doetschman T, Korn R, Serfling E, Kemler R (1986) Transgenesis by means of blastocyst-derived embryonic stem cell lines. Proc Natl Acad Sci USA 83:9065–9069
Grosveld F, van Assendelft GB, Greaves DR, Kollias G (1987) Position-independent, high-level expression of the human beta-globin gene in transgenic mice. Cell 51:975–985
Hanahan D (1989) Transgenic mice as probes into complex systems. Science 246:1265–1275
Hasty P, Ramirez-Solis R, Krumlauf R, Bradley A (1991a) Introduction of a subtle mutation into the Hox-2.6 locus in embryonic stem cells. Nature 350:243–246
Hasty P, Rivera-Perez J, Bradley A (1991b) The length of homology required for gene targeting in embryonic stem cells. Mol Cell Biol 11:5586–5591
Hide T, Hatakeyama J, Kimura-Yoshida C, Tian E, Takeda N, Ushio Y, Shiroishi T, Aizawa S, Matsuo I (2002) Genetic modifiers of otocephalic phenotypes in Otx2 heterozygous mutant mice. Development 129:4347–4357
Hooper M, Hardy K, Handyside A, Hunter S, Monk M (1987) HPRT-deficient (Lesch-Nyhan) mouse embryos derived from germline colonization by cultured cells. Nature 326:292–295
Huang LS, Voyiaziakis E, Markenson DF, Sokol KA, Hayek T, Breslow JL (1995) apo B gene knockout in mice results in embryonic lethality in homozygotes and neural tube defects, male infertility, and reduced HDL cholesterol ester and apo A-I transport rates in heterozygotes. J Clin Invest 96:2152–2161
Huang PL, Dawson TM, Bredt DS, Snyder SH, Fishman MC (1993) Targeted disruption of the neuronal nitric oxide synthase gene. Cell 75:1273–1286
Hung KE, Faca V, Song K, Sarracino DA, Richard LG, Krastins B, Forrester S, Porter A, Kunin A, Mahmood U, Haab BB, Hanash SM, Kucherlapati R (2009) Comprehensive proteome analysis of an Apc mouse model uncovers proteins associated with intestinal tumorigenesis. Cancer Prev Res 2:224–233
Jackson-Grusby L (2002) Modeling cancer in mice. Oncogene 21:5504–5514
Jaenisch R (1976) Germ line integration and Mendelian transmission of the exogenous Moloney leukemia virus. Proc Natl Acad Sci USA 73:1260–1264
Jaenisch R (1988) Transgenic animals. Science 240:1468–1474
Jaenisch R, Mintz B (1974) Simian virus 40 DNA sequences in DNA of healthy adult mice derived from preimplantation blastocysts injected with viral DNA. Proc Natl Acad Sci USA 71:1250–1254
Jiang W, Anderson MS, Bronson R, Mathis D, Benoist C (2005) Modifier loci condition autoimmunity provoked by Aire deficiency. J Exp Med 202:805–815
Johnson RS, Sheng M, Greenberg ME, Kolodner RD, Papaioannou VE, Spiegelman BM (1989) Targeting of nonexpressed genes in embryonic stem cells via homologous recombination. Science 245:1234–1236
Jonkers J, Berns A (2002) Conditional mouse models of sporadic cancer. Nat Rev Cancer 2:251–265
Jung S, Rajewsky K, Radbruch A (1993) Shutdown of class switch recombination by deletion of a switch region control element. Science 259:984–987
Kallapur S, Ormsby I, Doetschman T (1999) Strain dependency of TGFbeta1 function during embryogenesis. Mol Reprod Dev 52:341–349
Koller BH, Hagemann LJ, Doetschman T, Hagaman JR, Huang S, Williams PJ, First NL, Maeda N, Smithies O (1989) Germ-line transmission of a planned alteration made in a hypoxanthine phosphoribosyltransferase gene by homologous recombination in embryonic stem cells. Proc Natl Acad Sci USA 86:8927–8931
Koller BH, Smithies O (1992) Altering genes in animals by gene targeting. Annu Rev Immunol 10:705–730
Kuhn R, Schwenk F, Aguet M, Rajewsky K (1995) Inducible gene targeting in mice. Science 269:1427–1429
Lamb BT, Sisodia SS, Lawler AM, Slunt HH, Kitt CA, Kearns WG, Pearson PL, Price DL, Gearhart JD (1993) Introduction and expression of the 400 kilobase amyloid precursor protein gene in transgenic mice. Nat Genet 5:22–30

Mansour SL, Thomas KR, Capecchi MR (1988) Disruption of the proto-oncogene int-2 in mouse embryo-derived stem cells: a general strategy for targeting mutations to non-selectable genes. Nature 336:348–352

Mansour SL (1990) Gene targeting in murine embryonic stem cells: introduction of specific alterations into the mammalian genome. Genet Anal Tech Appl 7:219–227

Martin GR (1981) Isolation of a pluripotent cell line from early mouse embryos cultured in medium conditioned by teratocarcinoma stem cells. Proc Natl Acad Sci USA 78:7634–7638

McLin JP, Steward O (2006) Comparison of seizure phenotype and neurodegeneration induced by systemic kainic acid in inbred, outbred, and hybrid mouse strains. Eur J Neurosci 24:2191–2202

Moens CB, Auerback AB, Conlon RA, Joyner AL, Rossant J (1992) A targeted mutation reveals a role for N-myc in branching morphogenesis in the embryonic mouse lung. Genes Develop 6:691–704

Moll UM, Slade N (2004) p63 and p73: roles in development and tumor formation. Mol Cancer Res 2:371–386

Muyrers JP, Zhang Y, Stewart AF (2001) Techniques: Recombinogenic engineering – new options for cloning and manipulating DNA. Trends Biochem Sci 26:325–331

Muyrers JP, Zhang Y, Testa G, Stewart AF (1999) Rapid modification of bacterial artificial chromosomes by ET-recombination. Nucleic Acids Res 27:1555–1557

Papaioannou VE, McBurney MW, Gardner RL, Evans MJ (1975) Fate of teratocarcinoma cells injected into early mouse embryos. Nature 258:70–73

Pham CT, MacIvor DM, Hug BA, Heusel JW, Ley TJ (1996) Long-range disruption of gene expression by a selectable marker cassette. Proc Natl Acad Sci USA 93:13090–13095

Rajewsky K, Gu H, Kuhn R, Betz UA, Muller W, Roes J, Schwenk F (1996) Conditional gene targeting. J Clin Invest 98:600–603

Robertson E, Bradley A, Kuehn M, Evans M (1986) Germ-line transmission of genes introduced into cultured pluripotential cells by retroviral vector. Nature 323:445–448

Sanford LP, Kallapur S, Ormsby I, Doetschman T (2001) Influence of genetic background on knockout mouse phenotypes. Methods Mol Biol 158:217–225

Seidl KJ, Bottaro A, Vo A, Zhang J, Davidson L, Alt FW (1998) An expressed neo(r) cassette provides required functions of the 1gamma2b exon for class switching. Int Immunol 10:1683–1692

Shaw AT, Kirsch DG, Jacks T (2005) Future of early detection of lung cancer: the role of mouse models. Clin Cancer Res 11:4999s–5003s

Simpson EM, Linder CC, Sargent EE, Davisson MT, Mobraaten LE, Sharp JJ (1997) Genetic variation among 129 substrains and its importance for targeted mutagenesis in mice. Nat Genet 16:19–27

Stevens LC (1960) Embryonic potency of embryoid bodies derived from a transplantable testicular teratoma of the mouse. Dev Biol 2:285–297

Stevens LC, Little CC (1954) Spontaneous testicular teratomas in an inbred strain of mice. Proc Natl Acad Sci USA 40:1080–1087

Sun T, Storb U (2001) Insertion of phosphoglycerine kinase (PGK)-neo 5′ of Jlambda1 dramatically enhances VJlambda1 rearrangement. J Exp Med 193:699–712

Thomas KR, Capecchi MR (1987) Site-directed mutagenesis by gene targeting in mouse embryo-derived stem cells. Cell 51:503–512

Thompson S, Clarke AR, Pow AM, Hooper ML, Melton DW (1989) Germ line transmission and expression of a corrected HPRT gene produced by gene targeting in embryonic stem cells. Cell 56:313–321

Torres RM, Kühn R (1997) Laboratory protocols for conditional gene targeting. Oxford University Press, Oxford

Tuveson DA, Jacks T (2002) Technologically advanced cancer modeling in mice. Curr Opin Genet Dev 12:105–110

Valancius V, Smithies O (1991) Testing an "in-out" targeting procedure for making subtle genomic modifications in mouse embryonic stem cells. Mol Cell Biol 11:1402–1408

van Deursen J, Wieringa B (1992) Targeting of the creatine kinase M gene in embryonic stem cells using isogenic and nonisogenic vectors. Nucleic Acids Res 20:3815–3820

Van Dyke T, Jacks T (2002) Cancer modeling in the modern era: progress and challenges. Cell 108:135–144

Vazquez JC, Nogues C, Rucker EB, Piedrahita JA (1998) Factors affecting the efficiency of introducing precise genetic changes in ES cells by homologous recombination: tag-and-exchange versus the Cre-loxp system. Transgenic Res 7:181–193

Wagner TE, Hoppe PC, Jollick JD, Scholl DR, Hodinka RL, Gault JB (1981) Microinjection of a rabbit beta-globin gene into zygotes and its subsequent expression in adult mice and their offspring. Proc Natl Acad Sci USA 78:6376–6380

Wang J, Sarov M, Rientjes J, Fu J, Hollak H, Kranz H, Xie W, Stewart AF, Zhang Y (2006) An improved recombineering approach by adding RecA to lambda Red recombination. Mol Biotechnol 32:43–53

Zhang H, Hasty P, Bradley A (1994) Targeting frequency for deletion vectors in embryonic stem cells. Mol Cell Biol 14:2404–2410

Zheng H, Wilson JH (1990) Gene targeting in normal and amplified cell lines. Nature 344:170–173

Zhou L, Rowley DL, Mi QS, Sefcovic N, Matthes HW, Kieffer BL, Donovan DM (2001) Murine inter-strain polymorphisms alter gene targeting frequencies at the mu opioid receptor locus in embryonic stem cells. Mamm Genome 12:772–778

Chapter 2
The Use of Cre–loxP Technology and Inducible Systems to Generate Mouse Models of Cancer

Chu-Xia Deng

2.1 Introduction

Drs. Mario R. Capecchi, Martin J. Evans, and Oliver Smithies received the 2007 Nobel Prize in Physiology or Medicine for their pioneering work in introducing specific gene modifications in mice by the use of embryonic stem (ES) cells (Deng 2007). This technology, commonly referred to as gene targeting or knockout, is based on homologous recombination between DNA sequences residing in the chromosome and newly introduced DNA to mutate genes of interest in the mouse genome (Capecchi 1989). Gene targeting has proven to be a powerful means for precise manipulation of the mammalian genome, which has generated thousands of mutant mouse strains. Studies of these mutant mice have yielded enormously useful information in virtually all fields of biological and biomedical sciences. Indeed, gene targeting can theoretically be used to generate mutant mice for all genes in the near future. However, many genes are indispensable for embryonic and/or early postnatal development. In such cases, germline mutations of these genes often result in embryonic, neonatal, or preadult lethality, preventing further studies of their functions in later stages of development and tumorigenesis (Weinstein et al. 2000; Deng 2002b; Coumoul and Deng 2003; Friedberg and Meira 2006).

In the past decade, the Cre–loxP technology, combined with inducible systems, has been used to overcome embryonic and early postnatal lethality (Le and Sauer 2000; Nagy 2000). Many tumor suppressor genes and oncogenes have been mutated or activated in a spatial and temporal manner, making it possible for studying their function in a way that would otherwise not be possible. This chapter discusses

C.-X. Deng (✉)
Genetics of Development and Disease Branch, National Institute of Diabetes, Digestive and Kidney Diseases, National Institutes of Health, 10/9N105, 10 Center Drive, Bethesda, MD 20892, USA
e-mail: chuxiad@bdg10.niddk.nih.gov

J.E. Green and T. Ried (eds.), *Genetically Engineered Mice for Cancer Research: Design, Analysis, Pathways, Validation and Pre-Clinical Testing*,
DOI 10.1007/978-0-387-69805-2_2,

details for designing and generating mice carrying conditional loss or gain of function mutations, and strategies for tissue-specific Cre–loxP-mediated recombination. Advances of several major inducible systems and their applications to cancer research are also discussed.

2.2 Cre–LoxP System

The Cre–loxP site-specific recombination system of Coliphase P1 is particularly simple and well characterized (Argos et al. 1986; Sternberg et al. 1986; Sauer and Henderson 1988). *Cre* (*c*yclization *re*combination) gene encodes a 38-kDa site-specific DNA recombinase, called Cre, which recognizes 34-bp sites, loxP (*lo*cus of *X*-over of *P*1), and catalyzes both intra and intermolecular recombination between two loxP sites (Fig. 2.1). The loxP site consists of an 8-bp nonpalindromic core region flanked by two 13-bp inverted repeats (Fig. 2.1a). Cre–loxP-mediated recombination between two directly repeated loxP sites excises all DNA sequences located within the two sites as a covalently closed circle (Fig. 2.1b). Because Cre–loxP-mediated recombination occurs at high efficiency and it does not require any other host factors, except for its substrate, i.e., DNA, it has been widely used in a variety of experimental model systems. In most cases, loxP sites are placed in the same chromosome in direct repeat position so that the intervening

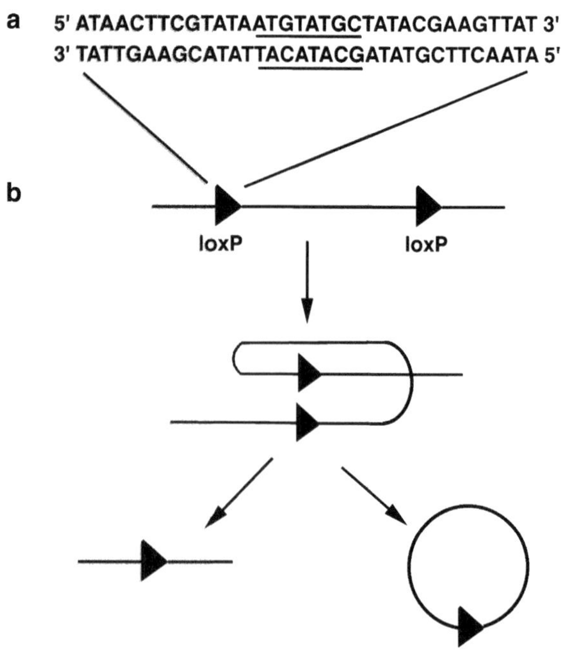

Fig. 2.1 Schematic representation of Cre–loxP-mediated recombination. (**a**) The loxP site consists of an 8-bp nonpalindromic core region (*underlined*) flanked by two 13-bp inverted repeats. (**b**). Cre–loxP-mediated recombination between two directly repeated loxP sites generates a linear product containing one loxP and a covalently closed circle containing excised DNA sequence located between two loxP sites

DNA sequence can be deleted. The loxP sites can also be placed in different chromosomes to promote recombination between different chromosomes, and placed in an inverted position in the same chromosome to create a switch to inactivate and activate genes of interest.

2.3 Cre–LoxP-Mediated Gene Inactivation

2.3.1 Generation of a Conditional Mutant Allele in Mice

The first step in the Cre–loxP-mediated gene inactivation is to generate a targeting vector for the gene of interest. The vector can be constructed by using multiple established procedures that were described in detail elsewhere (Zhang et al. 2002; Deng and Xu 2004; Iiizumi et al. 2006). Using the *Smad4* gene as an example, a replacement type targeting vector, commonly used for co-transfer of a selectable marker and a nonselectable marker (Deng et al. 1993) is discussed (Fig. 2.2). Such a vector contains a neomycin (*neo*) gene for positive selection and a thymidine kinase (*tk*) gene for negative selection (Mansour et al. 1988) (Fig. 2.2a). The *neo* gene is flanked with two loxP sites and is inserted into intron 8, and the third loxP site is placed in intron 7 of the *Smad4* gene. Thus, exon 8 of *Smad4* gene is flanked by loxP sites (floxed) and can be deleted upon Cre–loxP-mediated recombination (Fig. 2.2b). After introducing such a 3-loxP gene-targeting construct into ES cells, the cells containing predicted homologous recombination are identified by Southern blots and/or PCR (Fig. 2.2c), and injected into blastocysts for germline transmission by standard techniques.

2.3.2 Deletion of the Neo Gene from a Conditional Mutant Allele

It has been shown that the presence of the *neo* gene in an intron frequently affects endogenous gene expression and results in the reduction or complete inactivation of the floxed genes (Hirotsune et al. 1998; Chen et al. 1999; Iwata et al. 2000; Rucker et al. 2000); (Xu et al. 2001b). Thus, it is important to be able to remove the *neo* gene from targeted loci whenever it is necessary. The *neo* gene, if it is flanked by loxP, can be removed using several methods either in ES cells or mutant mice. The removal of the *neo* gene in ES cells by transient Cre expression has been used successfully in generating conditional knockouts (Gu et al. 1994). Although it is a quick way to delete the *neo* gene, it requires additional modification of ES cells and it may compromise totipotency and increase the difficulty of obtaining germline transmission. On the other hand, the presence of *neo* in an intron of a gene does not always generate obvious effects and sometime it can even create serial hypomorphic alleles that are useful for studying the function of genes of interest (Hirotsune et al. 1998; Chen et al. 1999; Iwata et al. 2000; Rucker et al. 2000); (Xu et al. 2001b).

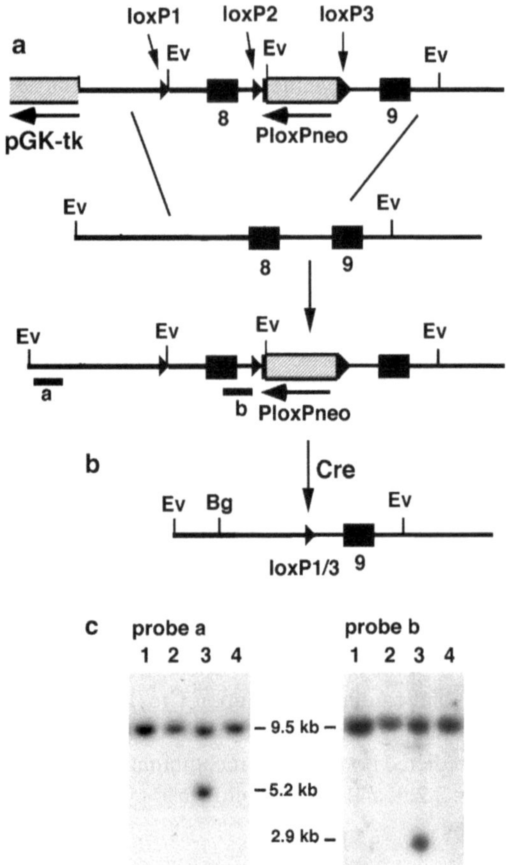

Fig. 2.2 Introduction of loxP sites into the *Smad4* locus. (**a**) A targeting vector that contains a loxP in the intron 7 and a *ploxPneo* in intron 8. Through a double cross event, the vector introduces all three loxP into the *Smad4* locus. (**b**) Cre–loxP-mediated complete recombination can delete all DNA sequence between loxP 1 and 3. (**c**) Targeted events were identified by Southern blot analysis of Ev (*Eco*RV)-digested genomic DNAs with a 5′ flanking probe (probe a). The wild-type clones only show a fragment of 9.5 kb and the targeted clones showed an additional fragment of 5.5 kb due to the introduction of an *Eco*RV site. The *Eco*RV-digested genomic DNA was also blotted using an internal probe (probe b) to verify the presence of the *ploxPneo* gene. In this case, the targeted ES clones showed a 3-kb fragment in additional to the wild-type fragment of 9.5 kb

In such cases, it is beneficial to keep the *neo* gene in ES cells, and remove it later in mutant mice after its physiological impact is assessed.

Currently, four approaches have been developed in case the *neo* gene needs to be removed from the conditional knockout allele in mice. Xu et al. described two approaches to delete *ploxPneo* from mice. The first approach is to cross the mice containing the 3-loxP mutant allele with the EIIa-Cre transgenic mice (Lakso et al. 1996), and the second one is to microinject the Cre expression construct into the pronucleus of fertilized eggs (Xu et al. 2001b). The third method removes the floxed

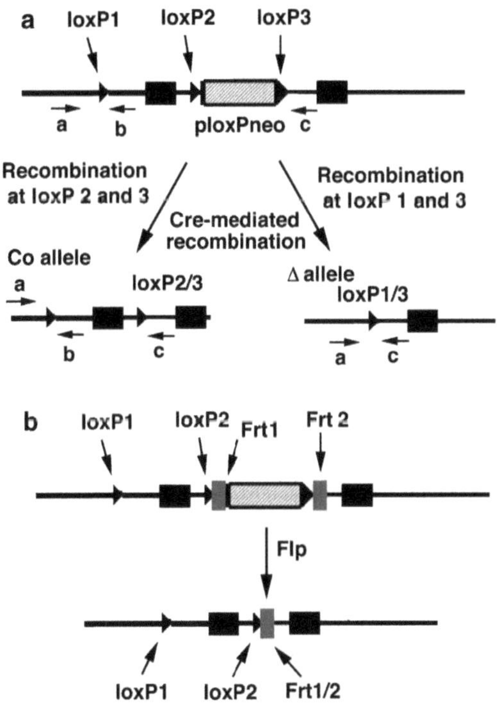

Fig. 2.3 Strategies for the removal of the *neo* gene from the conditional knockout allele. (**a**) Removal of *ploxPneo* gene from a 3-loxP knockout allele. Cre-mediated recombination between loxP 2 and 3 deletes the *ploxPneo* while the recombination between loxP 1 and 3 deletes all DNA sequences between loxP 1 and 3. Different recombination events can be detected by PCR analysis using primers a, b, and c. If this is performed in mice, it could generate conditional mutant mice and null mice carrying the delta allele at the same time. (**b**) Removal of a Frt-floxed *neo* gene through the expression of Flp recombinase. Because the loxP 2 is placed outside the Frt1, the Flip/Frt-mediated recombination only deletes the Frt-foxed *neo*, generating a loxP floxed allele, which can be used for conditional knockout

neo gene by infecting 16-cell stage morulae with the recombinant Cre adenovirus (Kaartinen and Nagy 2001). All these approaches are based on the fact that Cre-mediated recombination is normally incomplete, and the allele without the *neo* cassette can be identified by PCR analysis using different sets of primers in the offspring (Fig. 2.3a). To avoid screening for the incomplete recombination product generated by Cre/loxP, Meyers et al. (1998) reported a method using a combined Cre/loxP and Flp/Frt system to excise the *neo* gene in the germline of the adult mouse (Meyers et al. 1998) (Fig. 2.3b). The Flp/Frt site-specific recombination system was initially found in yeast and it works efficiently in *Drosophila* and in mammalian cells (Golic and Lindquist 1989; O'Gorman et al. 1991). In this approach, the *neo* gene is flanked by a combined loxP/Frt site on one side and an Frt on the other side. An advantage is that deletion of *neo* by Flp recombinase does not affect the loxP-flanked fragment.

Table 2.1 Comparison of advantages and disadvantages of several approaches for deletion of the *neo* gene

Methods to remove floxed *neo* gene	Advantages	Disadvantages
Transit expression of Cre in ES cells	Quick, technically easy	Additional modifications may compromise pluripotency of ES cells
Cross with EIIa-Cre transgenic mice	Avoids ES manipulation and removes the *neo* gene in mice with a high reliability	Requires two rounds of animal mating, i.e., first with the EIIa-Cre mice and the second with wild-type mice to separate alleles carrying different Cre/loxP-mediated recombination. The screening for incomplete Cre/loxP-mediated recombination can be time consuming
Oocyte injection	Direct injection of a Cre expression plasmid into oocyte. The amount of input Cre can be adjusted to increase efficiency of removing the *neo*	Requires two rounds of mating with wild type mice to obtain oocytes and separate alleles carrying different Cre/loxP-mediated recombination. In addition, it requires experience in microinjection, embryo manipulation, and implantation
Infecting morulae with recombinant Cre adenovirus	High efficiency of Cre adenovirus to infect morulae, which may delete floxed neo with high efficiency	Two rounds of mating with wild-type mice to obtain morulae and separate alleles carrying different Cre/loxP-mediated recombination. In addition, it requires experience in adenovirus production, embryo manipulation, and implantation
Combination of Cre/loxP–Flp/Frt	Deletion of the *neo* is independent of Cre/loxP system. It is a straightforward screen for the complete recombination product	Screening large number of offspring is expected due to a low efficiency of the Flp/Frt system in mouse

However, it was found that the efficiency of the Flp/Frt system in mouse is much lower than the Cre/loxP system (Meyers et al. 1998), which requires screening of a relatively large number of animals to obtain the correct allele. A summary of advantages and disadvantages of these approaches is listed in Table 2.1.

2.3.3 Tissue-Specific Conditional Knockout Mice

Once mice carrying conditional knockout alleles of genes are created, the mutant mice can be crossed with mice carrying Cre that is controlled by desired promoters to achieve targeted gene knockout in a spatial–temporal fashion. Numerous transgenic mice carrying tissue-specific and/or inducible Cre expression have been generated

Table 2.2 Transgenic mouse lines carrying inducible Cre

Promoters	Organs/tissues	References
NMDA-type glutamate receptor subunit gene	Cerebellar granule cell-specific and inducible expression inducible by antiprogestins	Tsujita et al. (1999)
SM22	Temporally controlled somatic mutagenesis in smooth muscle – tamoxifen inducible	Kuhbandner et al. (2000)
E_{μ}/P_{SV40}	B cell – tamoxifen inducible	Schwenk et al. (1998)
Transthyretin	Fetal and adult liver – tamoxifen inducible	Tannour-Louet et al. (2002)
Alpha1-antitrypsin	Hepatocyte – tamoxifen inducible	Imai et al. (2000)
Keratin 14	Epidermis – RU486 inducible	Berton et al. (2000)
WAP	Mammary gland – Tet-inducible	Utomo et al. (1999)
CMV	Ubiquitous – tamoxifen inducible	Hayashi and McMahon (2002)
Wnt	Embryonic neural tube – tamoxifen inducible	Danielian et al. (1998)
An interferon-responsive promoter	Liver and nearly complete in lymphocytes – interferon inducible	Kuhn et al. (1995)
Hsp70	Ubiquitous – heat shock inducible	Dietrich et al. (2000)
CamKIIalpha	Olfactory lobe, cortex, striatum, hippocampus and Purkinje cells – doxycycline inducible	Lindeberg et al. (2002)

(Nagy and Mar 2001, http://www.mshri.on.ca/nagy/Cre-pub.html, and Table 2.2). Many of these mice have been used to knock out tumor suppressor genes, including adenomatous polyposis coli (APC) (Clarke 2005), breast cancer-associated gene 1 (BRCA1) (Xu et al. 1999a), breast cancer-associated gene 2 (BRCA2) (Jonkers et al. 2001), Neurofibromatosis type one (NF1) (Gitler et al. 2004), p53 (Jonkers et al. 2001), phosphatase and tensin homolog deleted on chromosome 10 (PTEN) (Li et al. 2002), retinoblastoma (RB) (Ruiz et al. 2006), SMAD4 (Li et al. 2003), and transforming growth factor beta (TGF-beta)-type II receptor (Ijichi et al. 2006). These studies provide valuable information regarding functions of these genes in tumor initiation and progression. The progresses achieved using SMAD4 and BRCA1 conditional knockout mice are briefly reviewed below.

2.3.3.1 Cre–loxP-Mediated Knockout of SMAD4 in Multiple Tissues

SMAD4 serves as a common mediator of the TGF-beta superfamily that comprises over 40 growth and differentiation factors, including members in the subfamily of TGF-beta, activin, inhibin, and bone morphogenetic protein, which play numerous important functions in diverse developmental processes by regulating proliferation, differentiation, and apoptosis (Heldin et al. 1997; Massague 1998; Derynck et al. 2001;

Pollard 2001; Wakefield et al. 2001). In humans, *SMAD4* is a well-known tumor suppressor gene, and its mutations are frequently detected in pancreatic cancer, stomach cancer, liver cancer, and colon cancer (Hahn et al. 1996a, b; Nagatake et al. 1996; Schutte et al. 1996; Maesawa et al. 1997; Friedl et al. 1999). Germline mutations of *SMAD4* also contribute to familial juvenile polyposis, an autosomal dominant disorder characterized by predisposition to hamartomatous polyps and gastrointestinal cancer (Howe et al. 1998).

In mice, loss of SMAD4 results in lethality at embryonic (E) days 6–7 due to impaired extraembryonic membrane formation and decreased epiblast proliferation (Sirard et al. 1998; Yang et al. 1998). Because SMAD4 serves as a common mediator for the TGF-beta superfamily, SMAD4 conditional mutant mice generated by using the Cre–loxP approach (Yang et al. 2002; Bardeesy et al. 2006) should serve as a valuable tool for studying TGF-beta/SMAD4 signaling during postnatal development and tumorigenesis.

Currently, conditional knockout of SMAD4 has been performed in many organs/tissues, and tumorigenesis was observed in the mammary gland (Li et al. 2003), skin (Qiao et al. 2006), forestomach (Teng et al. 2006), liver (Yang et al. 2005; Xu et al. 2006), and pancreas (Bardeesy et al. 2006; Izeradjene et al. 2007; Kojima et al. 2007). Despite the finding that SMAD4 is mutated in about 60% of pancreatic ductal adenocarcinoma (PDAC) (Hahn et al. 1996a, b), SMAD4 deletion alone in the pancreas does not induce tumor formation (Bardeesy et al. 2006; Izeradjene et al. 2007; Kojima et al. 2007). Loss of SMAD4 also does not interfere with pancreas development and physiologic functions. However, when combined with an activated K-ras (G12D) allele, SMAD4 deficiency enabled rapid development of a distinct class of tumors resembling intraductal papillary mucinous neoplasia (MCN), a precursor to PDAC in humans. Progression of MCNs in both mice and humans is accompanied by loss of heterozygosity of p53 or p16 (Izeradjene et al. 2007). These data suggest that the invasive PDACs in humans and mice share similar overall mutational spectra, and the loss of Smad4 is a later event in pancreatic tumorigenesis.

Similarly, knockout of SMAD4 in the liver alone by albumin promoter-driven Cre (*Smad4*$^{Co/Co}$;*Alb-Cre*) does not cause developmental defects and tumor formation (Wang et al. 2005). Instead, it leads to the surprising finding that liver-specific knockout of SMAD4 causes iron overload in multiple organs, most pronounced in liver, kidney, and pancreas. The phenotypes of mutant mice resemble those found in hereditary hemochromatosis, a common genetic disorder among Caucasians (Pietrangelo 2006; Beutler 2007). Further studies indicate that the absence of SMAD4 results in marked decreased expression of hepcidin in the liver. Hepcidin is produced predominantly by the liver, although a number of other organs, such as lung and heart, also express it at much lower levels (Leong and Lonnerdal 2004). Prohepcidin is then cleaved to form the mature form, a 25 aa peptide, which is secreted into the circulation, and transported to duodenum and intestine, where it negatively regulates iron absorption in crypt cells and/or villous enterocytes. The absence of SMAD4 reduced production of hepatic hepcidin, leading to an increased expression of genes involved in intestinal iron absorption, including Dcytb, DMT1, and ferroportin (Wang et al. 2005). These data uncover a novel role of TGF-beta/

SMAD4 in regulating hepcidin expression and thus intestinal iron transport and iron homeostasis.

The lack of cancer formation in the liver suggests that SMAD4 deficiency alone is not enough to cause malignant transformation. However, it was found that the liver of *Smad4*$^{Co/Co}$*;Alb-Cre* mice exhibited increased expression of the PTEN tumor suppressor, which is mutated in a wide range of human cancers (Sansal and Sellers 2004). These data suggest that the increased expression of PTEN could inhibit the effect of SMAD4 deficiency on tumor induction. To test this, Xu et al. introduced a conditional mutation of PTEN (Groszer et al. 2001) into *Smad4*$^{Co/Co}$*;Alb-Cre* mice to knockout PTEN and SMAD4 simultaneously (Xu et al. 2006). In the PTEN and SMAD4 double mutant (*Smad4*$^{Co/Co}$*;Pten*$^{Co/Co}$*;Alb-Cre*) mice, hyperplastic foci emerged exclusively from bile ducts at 2 months of age (Fig. 2.4a–d). The hyperplastic foci progressed through multiple stages, including hyperplasia, dysplasia, carcinoma *in situ*, and eventually well-established cholangiocarcinoma (CC) in all animals at 4–7 months of age (Fig. 2.4e, f).

Because the endogenous albumin promoter is only expressed in hepatocytes but not in bile ducts (Yakar et al. 1999), it was surprising that the tumors derived exclusively from bile ducts. To investigate this, the *Alb-Cre* mice were mated with transgenic mice bearing a Rosa-26 reporter mouse [β-galactocidase expression upon Cre–LoxP-mediated recombination (Soriano 1999)]. β-Galactosidase positive cells were initially detected in both bile ducts and hepatocytes in the liver in a stochastic fashion in E15.5 embryos (Fig. 2.4g, h), and spread to a majority of hepatocytes and bile duct epithelial cells at P30 (Fig. 2.4i). These data suggest that the bile duct is more sensitive to tumorigenesis induced by deficiency of both PTEN and SMAD4 than hepatocytes in mice.

Further analysis indicated that CC formation follows a multistep progression of histopathological changes that are associated with significant alterations, including high levels of phosphorylated AKT, FOXO1, GSK-3β, mTOR, and ERK, and increased levels of cyclin D1, β-catenin, and c-Myc. CC accounts for about 15% of total liver cancer cases in the world with significant variations from country to country, and is associated with poor prognosis; most patients die soon after diagnosis (Taylor-Robinson et al. 2001; Okuda et al. 2002; Olnes and Erlich 2004; Sirica 2005). Studies on human CC also revealed similar alterations, including p53, p16, p27, p57, SMAD4, and increased levels of β-catenin, cyclin D1, ERK, Ras, AKT, and c-Myc (Sugimachi et al. 2001a, b; Ito et al. 2002; Kang et al. 2002; Wu et al. 2004; Sirica 2005). These findings elucidate a common mechanism between human and mouse CC formation and thus provide an animal model for the discovery of drugs for the treatment of CC.

2.3.3.2 Cre–loxP-Mediated Knockout of BRCA1 in Breast Cancer Research

Breast cancer is the leading cause of cancer incidence affecting approximately one in nine women in Western countries (Alberg and Helzlsouer 1997; Paterson 1998; Alberg et al. 1999; Kerr and Ashworth 2001; Nathanson and Weber 2001). Familial breast cancer is responsible for about 5–10% of total breast cancer cases caused by

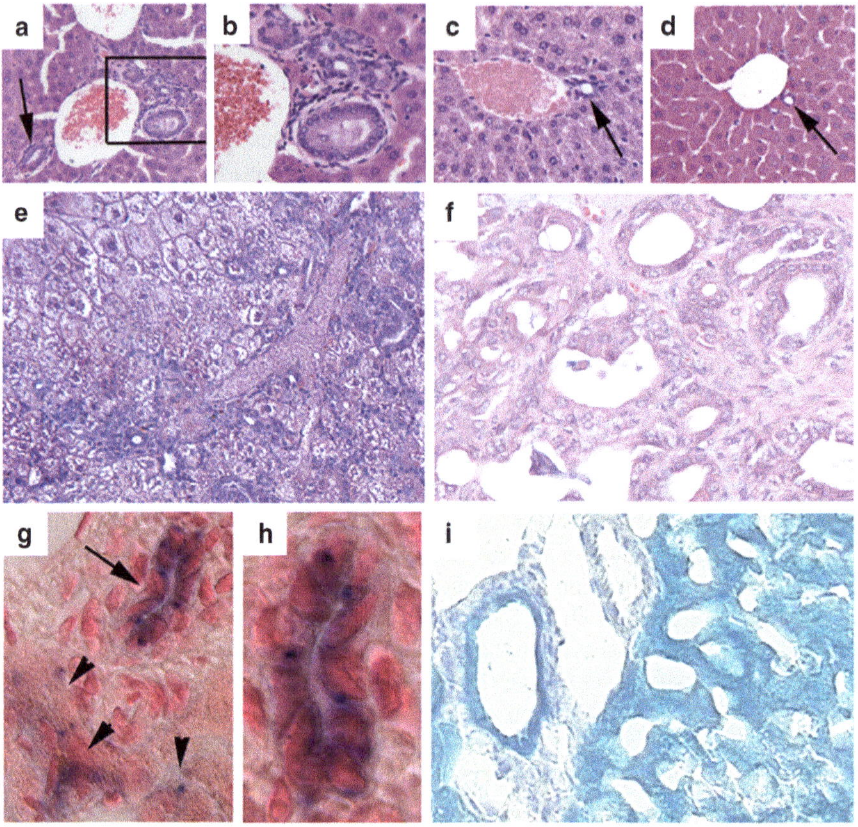

Fig. 2.4 Targeted disruption of SMAD4 and PTEN results in cholangiocarcinoma in the liver. (**a–d**) Histologic analysis of livers isolated from 2 months old *Smad4*$^{Co/Co}$;*Pten*$^{Co/Co}$;*Alb-Cre* (**a**, **b**), *Smad4*$^{Co/Co}$*Alb-Cre* (**c**) and wild-type (**d**) mice. *Arrows* point to bile ducts. (**b**) is the *boxed area* in (**a**). (**e**) An H&E liver section showing significantly increased bile duct branching in the liver of a 3-month-old *Smad4*$^{Co/Co}$;*Pten*$^{Co/Co}$;*Alb-Cre* mouse. (**f**) A well-developed CC found in *Smad4*$^{Co/Co}$; *Pten*$^{Co/Co}$;*Alb-Cre* liver. (**a–c**) Albumin-Cre activity assayed by using Rosa-26 reporter mice at P15 (**g**, **h**), and P30 (**i**)

mutations of BRCA1 and BRCA2, and other unidentified tumor suppressor genes (Alberg and Helzlsouer 1997; Paterson 1998; Kerr and Ashworth 2001; Nathanson and Weber 2001). Germline mutations of BRCA1 have been found to contribute to about 45% of the familial breast cancer cases and about 90% of the familial breast and ovarian cancer (Alberg and Helzlsouer 1997; Paterson 1998). BRCA1 was mapped in 1990 and was subsequently cloned in 1994 (Hall et al. 1990; Miki et al. 1994). Germline mutations in *BRCA1* have been detected in approximately half of familial breast cancer cases and most cases of combined familial breast/ovarian cancers (Alberg and Helzlsouer 1997; Paterson 1998). BRCA1 mutation carriers have a 50–80% risk of developing breast cancer by the age of 70 (Easton et al. 1995; Struewing et al. 1997; Ford et al. 1998).

In mice, loss of function mutation of BRCA1 generated by gene targeting is not compatible with embryonic development. Most mutant mice carrying various mutations died during gestation displaying growth retardation and apoptosis (Gowen et al. 1996; Hakem et al. 1996; Liu et al. 1996; Ludwig et al. 1997; Shen et al. 1998; Xu et al. 2001c). Studies on these mice demonstrated that BRCA1-deficiency resulted in defective DNA damage repair, abnormal centrosome duplication, impaired homologous recombination, defective cell cycle checkpoint, growth retardation, increased apoptosis, and genetic instability (Deng 2002a, 2006; Deng and Wang 2003). To overcome the early lethality and create animal models for BRCA1-associated hereditary breast cancer, several mutant mice carrying conditional knockout BRCA1 have been generated (Xu et al. 1999a; Mak et al. 2000; Liu et al. 2007). A most commonly used model of BRCA1 conditional mutant mice carries floxed exon 11 of the BRCA1 gene (Xu et al. 1999a), and the mutant mice are crossed with transgenic mice carrying either MMTV-Cre or WAP-Cre (Wagner et al. 1997) to specifically delete the BRCA1 in mammary epithelial cells. Analysis of these BRCA1 conditional mutant mice ($Brca1^{Co/Co}$*;MMTV-Cre* and $Brca1^{Co/Co}$*;MMTV-Cre*) revealed abnormal ductal and alveolar development of mutant mammary glands. There was also significantly increased apoptosis of epithelial cells, suggesting that cell death triggered by the loss of BRCA1 may be a primary cause for the abnormalities in branch morphogenesis. Despite these abnormalities, about 25% of BRCA1 conditional mutant mice developed mammary tumors when they were on average 18 months of age (Xu et al. 1999a). Further studies revealed that BRCA1 plays an important role in DNA damage repair and multiple cell cycle checkpoints (Xu et al. 1999b, 2001a, 2003; Weaver et al. 2002; Wang et al. 2004). The absence of BRCA1 results in genetic instability, which activates the tumor suppressor p53, leading to apoptosis. Consistent with this, disruption of p53 in BRCA1 mutant mice attenuates apoptosis and accelerates tumor formation (Brodie et al. 2001; Xu et al. 2001c). Recent studies revealed that increased insulin/IGF signaling (Shukla et al. 2006), activation of estrogen/ER-alpha signaling (Li et al. 2007; Jones et al. 2008), and increased expression of angiogenic factors, including angiopoietin-1 (Furuta et al. 2006) also facilitate breast cancer formation in BRCA1-deficient mice.

2.4 Cre–loxP-Mediated Gene Activation

Another important application of the Cre–loxP system in cancer research is to achieve gene activation. Many human cancers are caused by activation of numerous oncogenes; for example, activating mutations of the *RAS* oncogene are found in approximately one-third of all human cancers (Bos 1989; Khosravi-Far and Der 1994). Much of our knowledge on oncogenic signaling and its influence on tumor formation came from mouse models carrying activated oncogenes. Using K-ras as an example, the general strategy used for the generation of mutant mice by the Cre–LoxP technology is discussed below.

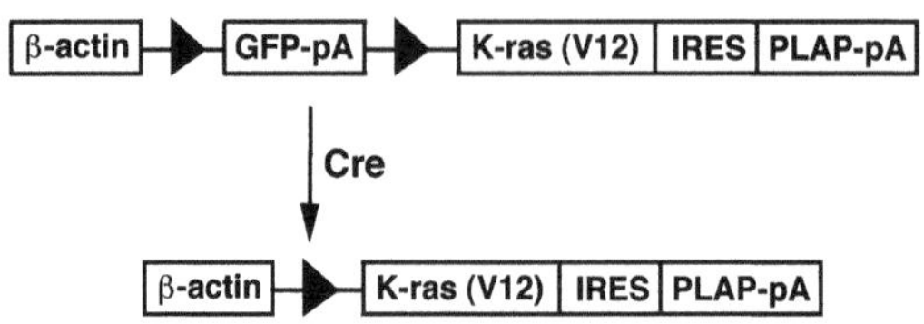

Fig. 2.5 Schematic representation of the conditional K-ras^{G12V} construct. The conditional K-ras^{G12V} transgene is driven by a broadly active beta-actin promoter, followed by a floxed-GFP, and then a K-ras^{G12V} cDNA combined with a PLAP expression construct through IRES (internal ribosomal entry site). Without Cre recombinase, GFP mRNA is expressed and the K-ras^{G12V} oncogene remains silent. After Cre-mediated deletion of the floxed-GFP, the K-ras^{G12V} oncogene is placed directly under control of the beta-actin promoter. The K-ras^{G12V} oncogene is transcribed together with the PLAP cDNA. The expression of alkaline phosphatase can serve as a marker for Cre–loxP-mediated recombination (modified from Meuwissen et al. 2001)

2.4.1 Activation of Oncogenes Using the Cre–loxP Technology

The Ras gene family contains three genes, K-ras, N-ras, and H-ras. Activation of *KRAS*, which occurs more frequently than that of the other two members, is found in many different types of human tumors, including adenocarcinomas of the pancreas (90%), colon (50%), and lung (30%) (Rodenhuis et al. 1988; Mills et al. 1995; Huncharek et al. 1999). Meuwissen et al. (2001) made a mouse model carrying an activated K-Ras (K-ras^{G12V}) mutation that specifically targets lung epithelial cells (Meuwissen et al. 2001). As shown in Fig. 2.5, the conditional K-ras^{G12V} transgene contained a broadly active beta-actin promoter, followed by a GFP (green fluorescence protein) expression cassette flanked by two loxP sites (floxed-GFP), and then a K-ras^{G12V} cDNA combined with a PLAP (human placenta-like alkaline phosphatase) expression construct. The floxed-GFP not only works as an indicator for the presence of the transgene but more importantly, it also serves to prevent expression of K-ras^{G12V}. Thus, the activated K-ras can only be expressed upon the removal of the block through Cre–LoxP-mediated recombination (Fig. 2.5). In this study, the researchers directly injected adenoviruses carrying Cre recombinase (Ad-Cre) intratracheally to K-ras^{G12V} transgenic mice to activate the K-ras in lung epithelial cells. This gave rise to rapid onset of pulmonary adenocarcinomas with 100% incidence 9–13 weeks postinjection. The tumor lesions also shared many features with human non-small cell lung cancer. These data demonstrate that sporadic expression of the activated K-Ras oncogene is sufficient to elicit lung tumorigenesis, which mimics human lung cancer.

Ad-Cre was also directly injected into the pancreatic ducts and acini through the common bile duct of K-ras transgenic mice to induce pancreatic cancer (Ueda et al. 2006). Alternatively, the K-ras oncogene can also be activated by breeding K-ras^{G12V} transgenic mice with mice carrying temporal–spatial-regulated Cre expression in organs/tissues of interest, such as the intestine (Luo et al. 2007). Similar approaches have also been used to express some other oncogenes in order to study their functions for tumor formation (Jager et al. 2004).

2.4.2 Activation of Tumor Suppressor Genes Using the Cre–LoxP Technology

Caner development is often associated with the inactivation of tumor suppressor genes. For example, loss of function mutation of the tumor suppressor p53 is found in approximately 50% of all human cancers (Morgan and Kastan 1997). However, it is unclear whether sustained inactivation of p53 is required for tumor maintenance. To investigate this, a reactivatable p53 knockout allele (p53-LSL) was generated using the Cre–loxP strategy (Ventura et al. 2007). In this case, transcription of p53 is shut off by a floxed blocker that is inserted in intron 1 of the gene. The p53-LSL mice were crossed with mice carrying a Cre recombinase–estrogen-receptor-T2 (Cre–ERT2) allele targeted to the ubiquitously expressed ROSA26 locus. The temporally controlled p53 reactivation *in vivo* can be achieved by tamoxifen administration, which allows the Cre recombinase to translocate from the cytoplasm to the nucleus (Indra et al. 1999), thus permitting the recombination of genomic loxP sites. The data showed that deletion of the blocker restored endogenous p53 expression and resulted in regression of autochthonous lymphomas and sarcomas in mice without affecting normal tissues (Ventura et al. 2007). The p53 restoration primarily induced apoptosis in lymphomas, while in sarcomas it primarily suppressed cell growth with features of cellular senescence. Cre–loxP-mediated bax gene activation was also used to reduce growth rate and increase sensitivity to chemotherapeutic agents in human gastric cancer cells and cervical carcinoma (Komatsu et al. 2000; Huh et al. 2001). This study serves as an example that a therapeutic effect can be achieved by the activation of tumor suppressor genes.

2.5 Conclusion and Future Directions

The Cre–loxP technology, combined with inducible systems, has been widely used to generate animal models for spatial and temporal regulated gene activation and inactivation. Studies of these mutant mice not only advance our knowledge of functions of numerous tumor suppressor genes and oncogenes, but also provide enormously useful information in virtually all areas of cancer biology oncology. It is anticipated that more animal models carrying spatial–temporal inducible systems will be generated in the near future. Using these animals, studies should be directed toward the detection of specific tumor signature profiles, and oncogenic signaling pathways that may be associated with certain tumor suppressors and oncogenes during tumorigenesis and tumor progression. Studies should also be designed to reveal extensive interactions between different genes and their relationship with genetic background modifiers and nongenetic factors (i.e., hormones). Animals can also serve as models for early tumor diagnosis, chemoprevention, and gene therapy studies, including the targeted delivery of drugs and tissue-specific activation of tumor suppressor genes to inhibit cancer growth and metastasis. Furthermore, the Cre–loxP inducible system combined with RNA interference (RNAi) technology has been used in mice to knockdown endogenous genes with high efficiency (Chang et al. 2004; Ventura et al. 2004; Coumoul and Deng 2006; Coumoul et al. 2005; Shukla et al. 2007a). Of note, a recent study performed in a mouse model for human FGFR2-related craniosynostosis indicates that mutant alleles bearing point mutations can be specifically targeted using RNAi technology with high efficiency without affecting wild-type mRNA levels (Shukla et al. 2007b). Because many human cancers are caused by point mutations of oncogenes, this data points to the future direction of using the Cre–loxP mediated RNAi inducible system for the therapeutic treatment of cancers that are caused by dominant mutations while allowing normal expression of wild-type alleles.

Acknowledgments I thank Dr. John T. Lahusen for the critical reading of the manuscript. This research was supported by the Intramural Research Program of the National Institute of Diabetes, Digestive and Kidney Diseases, National Institutes of Health, USA.

References

Alberg AJ, Helzlsouer KJ (1997) Epidemiology, prevention, and early detection of breast cancer. Curr Opin Oncol 9:505–511

Alberg AJ, Lam AP, Helzlsouer KJ (1999) Epidemiology, prevention, and early detection of breast cancer. Curr Opin Oncol 11:435–441

Argos P, Landy A, Abremski K, Egan JB, Haggard-Ljungquist E, Hoess RH, Kahn ML, Kalionis B, Narayana SV, Pierson LS 3rd et al (1986) The integrase family of site-specific recombinases: regional similarities and global diversity. EMBO J 5:433–440

Bardeesy N, Cheng KH, Berger JH, Chu GC, Pahler J, Olson P, Hezel AF, Horner J, Lauwers GY, Hanahan D, DePinho RA (2006) Smad4 is dispensable for normal pancreas development yet critical in progression and tumor biology of pancreas cancer. Genes Dev 20:3130–3146

Berton TR, Wang XJ, Zhou Z, Kellendonk C, Schutz G, Tsai S, Roop DR (2000) Characterization of an inducible, epidermal-specific knockout system: differential expression of lacZ in different Cre reporter mouse strains. Genesis 26:160–161

Beutler E (2007) Iron storage disease: facts, fiction and progress. Blood Cells Mol Dis 39:140–147

Bos JL (1989) ras oncogenes in human cancer: a review. Cancer Res 49:4682–4689

Brodie SG, Xu X, Qiao W, Li WM, Cao L, Deng CX (2001) Multiple genetic changes are associated with mammary tumorigenesis in Brca1 conditional knockout mice. Oncogene 20:7514–7523

Capecchi MR (1989) Altering the genome by homologous recombination [Review]. Science 244:1288–1292

Chang HS, Lin CH, Chen YC, Yu WC (2004) Using siRNA technique to generate transgenic animals with spatiotemporal and conditional gene knockdown. Am J Pathol 165:1535–1541

Chen L, Adar R, Yang X, Monsonego EO, Li C, Hauschka PV, Yayon A, Deng CX (1999) Gly369Cys mutation in mouse FGFR3 causes achondroplasia by affecting both chondrogenesis and osteogenesis. J Clin Invest 104:1517–1525

Clarke AR (2005) Studying the consequences of immediate loss of gene function in the intestine: APC. Biochem Soc Trans 33:665–666

Coumoul X, Deng CX (2003) Roles of FGF receptors in mammalian development and congenital diseases. Birth Defects Res C Embryo Today 69:286–304

Coumoul X, Deng CX (2006) RNAi in mice: a promising approach to decipher gene functions in vivo. Biochimie 88(6):637–643

Coumoul X, Shukla V, Li C, Wang RH, Deng CX (2005) Conditional knockdown of Fgfr2 in mice using Cre-LoxP induced RNA interference. Nucleic Acids Res 33:e102

Danielian PS, Muccino D, Rowitch DH, Michael SK, McMahon AP (1998) Modification of gene activity in mouse embryos in utero by a tamoxifen-inducible form of Cre recombinase. Curr Biol 8:1323–1326

Deng CX (2002a) Roles of BRCA1 in centrosome duplication. Oncogene 21:6222–6227

Deng CX (2002b) Tumor formation in Brca1 conditional mutant mice. Environ Mol Mutagen 39:171–177

Deng CX (2006) BRCA1: cell cycle checkpoint, genetic instability, DNA damage response, and cancer evolution. Nucleic Acids Res 34:1416–1426

Deng C (2007) In celebration of Dr Mario R. Capecchi's Nobel Prize. Int J Biol Sci 3:417–419

Deng CX, Wang RH (2003) Roles of BRCA1 in DNA damage repair: a link between development and cancer. Hum Mol Genet 12:R113–R123

Deng CX, Xu X (2004) Generation and analysis of Brca1 conditional knockout mice. Methods Mol Biol 280:185–200

Deng C, Thomas KR, Capecchi MR (1993) Location of crossovers during gene targeting with insertion and replacement vectors. Mol Cell Biol 13:2134–2140

Derynck R, Akhurst RJ, Balmain A (2001) TGF-beta signaling in tumor suppression and cancer progression. Nat Genet 29:117–129

Dietrich P, Dragatsis I, Xuan S, Zeitlin S, Efstratiadis A (2000) Conditional mutagenesis in mice with heat shock promoter-driven cre transgenes. Mamm Genome 11:196–205

Easton DF, Ford D, Bishop DT (1995) Breast and ovarian cancer incidence in BRCA1-mutation carriers. Breast Cancer Linkage Consortium. Am J Hum Genet 56:265–271

Ford D, Easton DF, Stratton M, Narod S, Goldgar D, Devilee P, Bishop DT, Weber B, Lenoir G, Chang-Claude J, Sobol H, Teare MD, Struewing J, Arason A, Scherneck S, Peto J, Rebbeck TR, Tonin P, Neuhausen S, Barkardottir R, Eyfjord J, Lynch H, Ponder BA, Gayther SA, Zelada-Hedman M et al (1998) Genetic heterogeneity and penetrance analysis of the BRCA1 and BRCA2 genes in breast cancer families. The Breast Cancer Linkage Consortium. Am J Hum Genet 62:676–689

Friedberg EC, Meira LB (2006) Database of mouse strains carrying targeted mutations in genes affecting biological responses to DNA damage Version 7. DNA Repair (Amst) 5:189–209

Friedl W, Kruse R, Uhlhaas S, Stolte M, Schartmann B, Keller KM, Jungck M, Stern M, Loff S, Back W, Propping P, Jenne DE (1999) Frequent 4-bp deletion in exon 9 of the SMAD4/MADH4 gene in familial juvenile polyposis patients. Genes Chromosomes Cancer 25:403–406

Furuta S, Wang JM, Wei S, Jeng YM, Jiang X, Gu B, Chen PL, Lee EY, Lee WH (2006) Removal of BRCA1/CtIP/ZBRK1 repressor complex on ANG1 promoter leads to accelerated mammary tumor growth contributed by prominent vasculature. Cancer Cell 10:13–24

Gitler AD, Kong Y, Choi JK, Zhu Y, Pear WS, Epstein JA (2004) Tie2-Cre-induced inactivation of a conditional mutant Nf1 allele in mouse results in a myeloproliferative disorder that models juvenile myelomonocytic leukemia. Pediatr Res 55:581–584

Golic KG, Lindquist S (1989) The FLP recombinase of yeast catalyzes site-specific recombination in the Drosophila genome. Cell 59:499–509

Gowen LC, Johnson BL, Latour AM, Sulik KK, Koller BH (1996) Brca1 deficiency results in early embryonic lethality characterized by neuroepithelial abnormalities. Nat Genet 12:191–194

Groszer M, Erickson R, Scripture-Adams DD, Lesche R, Trumpp A, Zack JA, Kornblum HI, Liu X, Wu H (2001) Negative regulation of neural stem/progenitor cell proliferation by the Pten tumor suppressor gene in vivo. Science 294:2186–2189

Gu H, Marth JD, Orban PC, Mossmann H, Rajewsky K (1994) Deletion of a DNA polymerase beta gene segment in T cells using cell type-specific gene targeting [see comments]. Science 265:103–106

Hahn SA, Hoque AT, Moskaluk CA, da Costa LT, Schutte M, Rozenblum E, Seymour AB, Weinstein CL, Yeo CJ, Hruban RH, Kern SE (1996a) Homozygous deletion map at 18q21.1 in pancreatic cancer. Cancer Res 56:490–494

Hahn SA, Schutte M, Hoque AT, Moskaluk CA, da Costa LT, Rozenblum E, Weinstein CL, Fischer A, Yeo CJ, Hruban RH, Kern SE (1996b) DPC4, a candidate tumor suppressor gene at human chromosome 18q21.1 [see comments]. Science 271:350–353

Hakem R, de la Pompa JL, Sirard C, Mo R, Woo M, Hakem A, Wakeham A, Potter J, Reitmair A, Billia F, Firpo E, Hui CC, Roberts J, Rossant J, Mak TW (1996) The tumor suppressor gene Brca1 is required for embryonic cellular proliferation in the mouse. Cell 85:1009–1023

Hall JM, Lee MK, Newman B, Morrow JE, Anderson LA, Huey B, King MC (1990) Linkage of early-onset familial breast cancer to chromosome 17q21. Science 250:1684–1689

Hayashi S, McMahon AP (2002) Efficient recombination in diverse tissues by a tamoxifen-inducible form of Cre: a tool for temporally regulated gene activation/inactivation in the mouse. Dev Biol 244:305–318

Heldin CH, Miyazono K, ten Dijke P (1997) TGF-beta signalling from cell membrane to nucleus through SMAD proteins. Nature 390:465–471

Hirotsune S, Fleck MW, Gambello MJ, Bix GJ, Chen A, Clark GD, Ledbetter DH, McBain CJ, Wynshaw-Boris A (1998) Graded reduction of Pafah1b1 (Lis1) activity results in neuronal migration defects and early embryonic lethality. Nat Genet 19:333–339

Howe JR, Roth S, Ringold JC, Summers RW, Jarvinen HJ, Sistonen P, Tomlinson IP, Houlston RS, Bevan S, Mitros FA, Stone EM, Aaltonen LA (1998) Mutations in the SMAD4/DPC4 gene in juvenile polyposis [see comments]. Science 280:1086–1088

Huh WK, Gomez-Navarro J, Arafat WO, Xiang J, Mahasreshti PJ, Alvarez RD, Barnes MN, Curiel DT (2001) Bax-induced apoptosis as a novel gene therapy approach for carcinoma of the cervix. Gynecol Oncol 83:370–377

Huncharek M, Muscat J, Geschwind JF (1999) K-ras oncogene mutation as a prognostic marker in non-small cell lung cancer: a combined analysis of 881 cases. Carcinogenesis 20:1507–1510

Iiizumi S, Nomura Y, So S, Uegaki K, Aoki K, Shibahara K, Adachi N, Koyama H (2006) Simple one-week method to construct gene-targeting vectors: application to production of human knockout cell lines. Biotechniques 41:311–316

Ijichi H, Chytil A, Gorska AE, Aakre ME, Fujitani Y, Fujitani S, Wright CV, Moses HL (2006) Aggressive pancreatic ductal adenocarcinoma in mice caused by pancreas-specific blockade of transforming growth factor-beta signaling in cooperation with active Kras expression. Genes Dev 20:3147–3160

Imai T, Chambon P, Metzger D (2000) Inducible site-specific somatic mutagenesis in mouse hepatocytes. Genesis 26:147–148

Indra AK, Warot X, Brocard J, Bornert JM, Xiao JH, Chambon P, Metzger D (1999) Temporally-controlled site-specific mutagenesis in the basal layer of the epidermis: comparison of the recombinase activity of the tamoxifen-inducible Cre-ER(T) and Cre-ER(T2) recombinases. Nucleic Acids Res 27:4324–4327

Ito Y, Takeda T, Sasaki Y, Sakon M, Yamada T, Ishiguro S, Imaoka S, Tsujimoto M, Monden M, Matsuura N (2002) Expression of p57/Kip2 protein in extrahepatic bile duct carcinoma and intrahepatic cholangiocellular carcinoma. Liver 22:145–149

Iwata T, Chen L, Li C, Ovchinnikov DA, Behringer RR, Francomano CA, Deng CX (2000) A neonatal lethal mutation in FGFR3 uncouples proliferation and differentiation of growth plate chondrocytes in embryos. Hum Mol Genet 9:1603–1613

Izeradjene K, Combs C, Best M, Gopinathan A, Wagner A, Grady WM, Deng CX, Hruban RH, Adsay NV, Tuveson DA, Hingorani SR (2007) Kras(G12D) and Smad4/Dpc4 haploinsufficiency cooperate to induce mucinous cystic neoplasms and invasive adenocarcinoma of the pancreas. Cancer Cell 11:229–243

Jager R, Maurer J, Jacob A, Schorle H (2004) Cell type-specific conditional regulation of the c-myc proto-oncogene by combining Cre/loxP recombination and tamoxifen-mediated activation. Genesis 38:145–150

Jones LP, Tilli MT, Assefnia S, Torre K, Halama ED, Parrish A, Rosen EM, Furth PA (2008) Activation of estrogen signaling pathways collaborates with loss of Brca1 to promote development of ERalpha-negative and ERalpha-positive mammary preneoplasia and cancer. Oncogene 27:794–802

Jonkers J, Meuwissen R, van der Gulden H, Peterse H, van der Valk M, Berns A (2001) Synergistic tumor suppressor activity of BRCA2 and p53 in a conditional mouse model for breast cancer. Nat Genet 29:418–425

Kaartinen V, Nagy A (2001) Removal of the floxed neo gene from a conditional knockout allele by the adenoviral Cre recombinase in vivo. Genesis 31:126–129

Kang YK, Kim WH, Jang JJ (2002) Expression of G1-S modulators (p53, p16, p27, cyclin D1, Rb) and Smad4/Dpc4 in intrahepatic cholangiocarcinoma. Hum Pathol 33:877–883

Kerr P, Ashworth A (2001) New complexities for BRCA1 and BRCA2. Curr Biol 11:R668–R676

Khosravi-Far R, Der CJ (1994) The Ras signal transduction pathway. Cancer Metastasis Rev 13:67–89

Kojima K, Vickers SM, Adsay NV, Jhala NC, Kim HG, Schoeb TR, Grizzle WE, Klug CA (2007) Inactivation of Smad4 accelerates Kras(G12D)-mediated pancreatic neoplasia. Cancer Res 67:8121–8130

Komatsu K, Suzuki S, Shimosegawa T, Miyazaki JI, Toyota T (2000) Cre-loxP-mediated bax gene activation reduces growth rate and increases sensitivity to chemotherapeutic agents in human gastric cancer cells. Cancer Gene Ther 7:885–892

Kuhbandner S, Brummer S, Metzger D, Chambon P, Hofmann F, Feil R (2000) Temporally controlled somatic mutagenesis in smooth muscle. Genesis 28:15–22

Kuhn R, Schwenk F, Aguet M, Rajewsky K (1995) Inducible gene targeting in mice. Science 269:1427–1429

Lakso M, Pichel JG, Gorman JR, Sauer B, Okamoto Y, Lee E, Alt FW, Westphal H (1996) Efficient in vivo manipulation of mouse genomic sequences at the zygote stage. Proc Natl Acad Sci USA 93:5860–5865

Le Y, Sauer B (2000) Conditional gene knockout using cre recombinase [In Process Citation]. Methods Mol Biol 136:477–485

Leong WI, Lonnerdal B (2004) Hepcidin, the recently identified peptide that appears to regulate iron absorption. J Nutr 134:1–4

Li G, Robinson GW, Lesche R, Martinez-Diaz H, Jiang Z, Rozengurt N, Wagner KU, Wu DC, Lane TF, Liu X, Hennighausen L, Wu H (2002) Conditional loss of PTEN leads to precocious development and neoplasia in the mammary gland. Development 129:4159–4170

Li W, Qiao W, Chen L, Xu X, Yang X, Li D, Li C, Brodie SG, Meguid MM, Hennighausen L, Deng CX (2003) Squamous cell carcinoma and mammary abscess formation through squamous metaplasia in Smad4/Dpc4 conditional knockout mice. Development 130:6143–6153

Li W, Xiao C, Vonderhaar BK, Deng CX (2007) A role of estrogen/ERalpha signaling in BRCA1-associated tissue-specific tumor formation. Oncogene 26:7204–7212

Lindeberg J, Mattsson R, Ebendal T (2002) Timing the doxycycline yields different patterns of genomic recombination in brain neurons with a new inducible Cre transgene. J Neurosci Res 68:248–253

Liu CY, Flesken-Nikitin A, Li S, Zeng Y, Lee WH (1996) Inactivation of the mouse Brca1 gene leads to failure in the morphogenesis of the egg cylinder in early postimplantation development. Genes Dev 10:1835–1843

Liu X, Holstege H, van der Gulden H, Treur-Mulder M, Zevenhoven J, Velds A, Kerkhoven RM, van Vliet MH, Wessels LF, Peterse JL, Berns A, Jonkers J (2007) Somatic loss of BRCA1 and p53 in mice induces mammary tumors with features of human BRCA1-mutated basal-like breast cancer. Proc Natl Acad Sci USA 104:12111–12116

Ludwig T, Chapman DL, Papaioannou VE, Efstratiadis A (1997) Targeted mutations of breast cancer susceptibility gene homologs in mice: lethal phenotypes of Brca1, Brca2, Brca1/Brca2, Brca1/p53, and Brca2/p53 nullizygous embryos. Genes Dev 11:1226–1241

Luo F, Brooks DG, Ye H, Hamoudi R, Poulogiannis G, Patek CE, Winton DJ, Arends MJ (2007) Conditional expression of mutated K-ras accelerates intestinal tumorigenesis in Msh2-deficient mice. Oncogene 26:4415–4427

Maesawa C, Tamura G, Nishizuka S, Iwaya T, Ogasawara S, Ishida K, Sakata K, Sato N, Ikeda K, Kimura Y, Saito K, Satodate R (1997) MAD-related genes on 18q21.1, Smad2 and Smad4, are altered infrequently in esophageal squamous cell carcinoma. Jpn J Cancer Res 88:340–343

Mak TW, Hakem A, McPherson JP, Shehabeldin A, Zablocki E, Migon E, Duncan GS, Bouchard D, Wakeham A, Cheung A, Karaskova J, Sarosi I, Squire J, Marth J, Hakem R (2000) Brcal required for T cell lineage development but not TCR loci rearrangement. Nat Immunol 1:77–82

Mansour SL, Thomas KR, Capecchi MR (1988) Disruption of the proto-oncogene int-2 in mouse embryo-derived stem cells: a general strategy for targeting mutations to non-selectable genes. Nature 336:348–352

Massague J (1998) TGF-beta signal transduction. Annu Rev Biochem 67:753–791

Meuwissen R, Linn SC, van der Valk M, Mooi WJ, Berns A (2001) Mouse model for lung tumorigenesis through Cre/lox controlled sporadic activation of the K-Ras oncogene. Oncogene 20:6551–6558

Meyers EN, Lewandoski M, Martin GR (1998) An Fgf8 mutant allelic series generated by Cre- and Flp-mediated recombination. Nat Genet 18:136–141

Miki Y, Swensen J, Shattuck-Eidens D, Futreal PA, Harshman K, Tavtigian S, Liu Q, Cochran C, Bennett LM, Ding W et al (1994) A strong candidate for the breast and ovarian cancer susceptibility gene BRCA1. Science 266:66–71

Mills NE, Fishman CL, Scholes J, Anderson SE, Rom WN, Jacobson DR (1995) Detection of K-ras oncogene mutations in bronchoalveolar lavage fluid for lung cancer diagnosis. J Natl Cancer Inst 87:1056–1060

Morgan SE, Kastan MB (1997) p53 and ATM: cell cycle, cell death, and cancer. Adv Cancer Res 71:1–25

Nagatake M, Takagi Y, Osada H, Uchida K, Mitsudomi T, Saji S, Shimokata K, Takahashi T, Takahashi T (1996) Somatic in vivo alterations of the DPC4 gene at 18q21 in human lung cancers. Cancer Res 56:2718–2720

Nagy A (2000) Cre recombinase: the universal reagent for genome tailoring. Genesis 26:99–109

Nagy A, Mar L (2001) Creation and use of a Cre recombinase transgenic database. Methods Mol Biol 158:95–106

Nathanson KL, Weber BL (2001) "Other" breast cancer susceptibility genes: searching for more holy grail. Hum Mol Genet 10:715–720

O'Gorman S, Fox DT, Wahl GM (1991) Recombinase-mediated gene activation and site-specific integration in mammalian cells. Science 251:1351–1355

Okuda K, Nakanuma Y, Miyazaki M (2002) Cholangiocarcinoma: recent progress. Part 2: molecular pathology and treatment. J Gastroenterol Hepatol 17:1056–1063

Olnes MJ, Erlich R (2004) A review and update on cholangiocarcinoma. Oncology 66:167–179

Paterson JW (1998) BRCA1: a review of structure and putative functions. Dis Markers 13:261–274

Pietrangelo A (2006) Hereditary hemochromatosis. Annu Rev Nutr 26:251–270

Pollard JW (2001) Tumour-stromal interactions. Transforming growth factor-beta isoforms and hepatocyte growth factor/scatter factor in mammary gland ductal morphogenesis. Breast Cancer Res 3:230–237

Qiao W, Li AG, Owens P, Xu X, Wang XJ, Deng CX (2006) Hair follicle defects and squamous cell carcinoma formation in Smad4 conditional knockout mouse skin. Oncogene 25:207–217

Rodenhuis S, Slebos RJ, Boot AJ, Evers SG, Mooi WJ, Wagenaar SS, van Bodegom PC, Bos JL (1988) Incidence and possible clinical significance of K-ras oncogene activation in adenocarcinoma of the human lung. Cancer Res 48:5738–5741

Rucker EB 3rd, Dierisseau P, Wagner KU, Garrett L, Wynshaw-Boris A, Flaws JA, Hennighausen L (2000) Bcl-x and Bax regulate mouse primordial germ cell survival and apoptosis during embryogenesis [In Process Citation]. Mol Endocrinol 14:1038–1052

Ruiz S, Santos M, Paramio JM (2006) Is the loss of pRb essential for the mouse skin carcinogenesis? Cell Cycle 5:625–629

Sansal I, Sellers WR (2004) The biology and clinical relevance of the PTEN tumor suppressor pathway. J Clin Oncol 22:2954–2963

Sauer B, Henderson N (1988) Site-specific DNA recombination in mammalian cells by the Cre recombinase of bacteriophage P1. Proc Natl Acad Sci USA 85:5166–5170

Schutte M, Hruban RH, Hedrick L, Cho KR, Nadasdy GM, Weinstein CL, Bova GS, Isaacs WB, Cairns P, Nawroz H, Sidransky D, Casero RA Jr, Meltzer PS, Hahn SA, Kern SE (1996) DPC4 gene in various tumor types. Cancer Res 56:2527–2530

Schwenk F, Kuhn R, Angrand PO, Rajewsky K, Stewart AF (1998) Temporally and spatially regulated somatic mutagenesis in mice. Nucleic Acids Res 26:1427–1432

Shen SX, Weaver Z, Xu X, Li C, Weinstein M, Chen L, Guan XY, Ried T, Deng CX (1998) A targeted disruption of the murine Brca1 gene causes gamma-irradiation hypersensitivity and genetic instability. Oncogene 17:3115–3124

Shukla V, Coumoul X, Cao L, Wang R, Xiao C, Xu X, Ando S, Yakar S, LeRoith D, Deng D (2006) Absence of the full-length BRCA1 leads to increased expression of IGF signaling axis members. Cancer Res 66:7151–7157

Shukla V, Coumoul X, Deng CX (2007a) RNAi-based conditional gene knockdown in mice using a U6 promoter driven vector. Int J Biol Sci 3:91–99

Shukla V, Coumoul X, Wang RH, Kim HS, Deng CX (2007b) RNA interference and inhibition of MEK-ERK signaling prevent abnormal skeletal phenotypes in a mouse model of craniosynostosis. Nat Genet 39:1145–1150

Sirard C, de la Pompa JL, Elia A, Itie A, Mirtsos C, Cheung A, Hahn S, Wakeham A, Schwartz L, Kern SE, Rossant J, Mak TW (1998) The tumor suppressor gene Smad4/Dpc4 is required for gastrulation and later for anterior development of the mouse embryo. Genes Dev 12:107 119

Sirica AE (2005) Cholangiocarcinoma: molecular targeting strategies for chemoprevention and therapy. Hepatology 41:5–15

Soriano P (1999) Generalized lacZ expression with the ROSA26 Cre reporter strain. Nat Genet 21:70–71

Sternberg N, Sauer B, Hoess R, Abremski K (1986) Bacteriophage P1 cre gene and its regulatory region. Evidence for multiple promoters and for regulation by DNA methylation. J Mol Biol 187:197–212

Struewing JP, Hartge P, Wacholder S, Baker SM, Berlin M, McAdams M, Timmerman MM, Brody LC, Tucker MA (1997) The risk of cancer associated with specific mutations of BRCA1 and BRCA2 among Ashkenazi Jews [see comments]. N Engl J Med 336:1401–1408

Sugimachi K, Aishima S, Taguchi K, Tanaka S, Shimada M, Kajiyama K, Tsuneyoshi M (2001a) The role of overexpression and gene amplification of cyclin D1 in intrahepatic cholangiocarcinoma. J Hepatol 35:74–79

Sugimachi K, Taguchi K, Aishima S, Tanaka S, Shimada M, Kajiyama K, Tsuneyoshi M (2001b) Altered expression of beta-catenin without genetic mutation in intrahepatic cholangiocarcinoma. Mod Pathol 14:900–905

Tannour-Louet M, Porteu A, Vaulont S, Kahn A, Vasseur-Cognet M (2002) A tamoxifen-inducible chimeric Cre recombinase specifically effective in the fetal and adult mouse liver. Hepatology 35:1072–1081

Taylor-Robinson SD, Toledano MB, Arora S, Keegan TJ, Hargreaves S, Beck A, Khan SA, Elliott P, Thomas HC (2001) Increase in mortality rates from intrahepatic cholangiocarcinoma in England and Wales 1968–1998. Gut 48:816–820

Teng Y, Sun AN, Pan XC, Yang G, Yang LL, Wang MR, Yang X (2006) Synergistic function of Smad4 and PTEN in suppressing forestomach squamous cell carcinoma in the mouse. Cancer Res 66:6972–6981

Tsujita M, Mori H, Watanabe M, Suzuki M, Miyazaki J, Mishina M (1999) Cerebellar granule cell-specific and inducible expression of Cre recombinase in the mouse. J Neurosci 19:10318–10323

Ueda S, Fukamachi K, Matsuoka Y, Takasuka N, Takeshita F, Naito A, Iigo M, Alexander DB, Moore MA, Saito I, Ochiya T, Tsuda H (2006) Ductal origin of pancreatic adenocarcinomas induced by conditional activation of a human Ha-ras oncogene in rat pancreas. Carcinogenesis 27:2497–2510

Utomo AR, Nikitin AY, Lee WH (1999) Temporal, spatial, and cell type-specific control of Cre-mediated DNA recombination in transgenic mice. Nat Biotechnol 17:1091–1096

Ventura A, Meissner A, Dillon CP, McManus M, Sharp PA, Van Parijs L, Jaenisch R, Jacks T (2004) Cre-lox-regulated conditional RNA interference from transgenes. Proc Natl Acad Sci USA 101:10380–10385

Ventura A, Kirsch DG, McLaughlin ME, Tuveson DA, Grimm J, Lintault L, Newman J, Reczek EE, Weissleder R, Jacks T (2007) Restoration of p53 function leads to tumour regression in vivo. Nature 445:661–665

Wagner KU, Wall RJ, St-Onge L, Gruss P, Wynshaw-Boris A, Garrett L, Li M, Furth PA, Hennighausen L (1997) Cre-mediated gene deletion in the mammary gland. Nucleic Acids Res 25:4323–4330

Wakefield LM, Piek E, Bottinger EP (2001) TGF-beta signaling in mammary gland development and tumorigenesis. J Mammary Gland Biol Neoplasia 6:67–82

Wang RH, Yu H, Deng CX (2004) A requirement for breast-cancer-associated gene 1 (BRCA1) in the spindle checkpoint. Proc Natl Acad Sci USA 101:17108–17113

Wang RH, Li C, Xu X, Zheng Y, Xiao C, Zerfas P, Cooperman S, Eckhaus M, Rouault T, Mishra L, Deng CX (2005) A role of SMAD4 in iron metabolism through the positive regulation of hepcidin expression. Cell Metab 2:399–409

Weaver Z, Montagna C, Xu X, Howard T, Gadina M, Brodie SG, Deng CX, Ried T (2002) Mammary tumors in mice conditionally mutant for Brca1 exhibit gross genomic instability and centrosome amplification yet display a recurring distribution of genomic imbalances that is similar to human breast cancer. Oncogene 21:5097–5107

Weinstein M, Yang X, Deng C (2000) Functions of mammalian smad genes as revealed by targeted gene disruption in mice [In Process Citation]. Cytokine Growth Factor Rev 11:49–58

Wu T, Leng J, Han C, Demetris AJ (2004) The cyclooxygenase-2 inhibitor celecoxib blocks phosphorylation of Akt and induces apoptosis in human cholangiocarcinoma cells. Mol Cancer Ther 3:299–307

Xu X, Wagner KU, Larson D, Weaver Z, Li C, Ried T, Hennighausen L, Wynshaw-Boris A, Deng CX (1999a) Conditional mutation of Brca1 in mammary epithelial cells results in blunted ductal morphogenesis and tumour formation [see comments]. Nat Genet 22:37–43

Xu X, Weaver Z, Linke SP, Li C, Gotay J, Wang XW, Harris CC, Ried T, Deng CX (1999b) Centrosome amplification and a defective G2-M cell cycle checkpoint induce genetic instability in BRCA1 exon 11 isoform-deficient cells. Mol Cell 3:389–395

Xu B, Kim S, Kastan MB (2001a) Involvement of Brca1 in S-phase and G(2)-phase checkpoints after ionizing irradiation. Mol Cell Biol 21:3445–3450

Xu X, Li C, Garrett-Beal L, Larson D, Wynshaw-Boris A, Deng CX (2001b) Direct removal in the mouse of a floxed neo gene from a three-loxP conditional knockout allele by two novel approaches. Genesis 30:1–6

Xu X, Qiao W, Linke SP, Cao L, Li WM, Furth PA, Harris CC, Deng CX (2001c) Genetic interactions between tumor suppressors Brca1 and p53 in apoptosis, cell cycle and tumorigenesis. Nat Genet 28:266–271

Xu X, Aprelikova O, Moens P, Deng CX, Furth PA (2003) Impaired meiotic DNA-damage repair and lack of crossing-over during spermatogenesis in BRCA1 full-length isoform deficient mice. Development 130:2001–2012

Xu X, Kobayashi S, Qiao W, Li C, Xiao C, Radaeva S, Stiles B, Wang R, Ohara N, Yoshino T, LeRoith D, Torbenson MS, Gores GJ, Wu H, Gao B, Deng C (2006) Induction of intrahepatic cholangiocellular carcinoma by liver specific disruption of Smad4 and Pten in mice. J Clin Invest 116(7):1843–1852

Yakar S, Liu JL, Stannard B, Butler A, Accili D, Sauer B, LeRoith D (1999) Normal growth and development in the absence of hepatic insulin-like growth factor I. Proc Natl Acad Sci USA 96:7324–7329

Yang X, Li C, Xu X, Deng C (1998) The tumor suppressor SMAD4/DPC4 is essential for epiblast proliferation and mesoderm induction in mice. Proc Natl Acad Sci USA 95:3667–3672

Yang X, Li C, Herrera PL, Deng CX (2002) Generation of Smad4/Dpc4 conditional knockout mice. Genesis 32:80–81

Yang L, Mao C, Teng Y, Li W, Zhang J, Cheng X, Li X, Han X, Xia Z, Deng H, Yang X (2005) Targeted disruption of Smad4 in mouse epidermis results in failure of hair follicle cycling and formation of skin tumors. Cancer Res 65:8671–8678

Zhang P, Li MZ, Elledge SJ (2002) Towards genetic genome projects: genomic library screening and gene-targeting vector construction in a single step. Nat Genet 30:31–39

Chapter 3
Using Recombineering Technology to Create Genetically Engineered Mouse Models

Subha Philip and Shyam K. Sharan

3.1 Introduction

The mouse is an attractive model system for studying human diseases, because of its close physiologic and genetic resemblance to humans. Mice also naturally develop conditions similar to human diseases including diabetes, cardiovascular diseases, and cancer. Earlier, mouse models of diseases were developed by selective breeding of naturally occurring mutants with the desired phenotype. Advances in genetic mapping techniques, along with chemically or radiation-induced mutagenesis techniques, have led to the creation of a large reservoir of potential models of human diseases (Bedell et al. 1997; Doolittle et al. 1996).

In the last 30 years, the development of innovative techniques in molecular and stem cell biology has made it possible to manipulate the mouse genome by introducing foreign DNA, by its pronuclear injection into oocytes or by gene targeting in mouse embryonic stem (ES) cells (see Chapter 2; Capecchi 1989; Hogan et al. 1994; Melton 1994). Gene targeting by homologous recombination in ES cells remains the most successful and effective approach for the functional analysis of genes in mice (Bradley et al. 1998). It has been widely used to generate loss-of-function mutations in genes to create mouse models of human diseases. Mouse models that involve the ectopic expression of genes rely on transgenic technology. Over the years, our ability to engineer the mouse genome by gene targeting or by the introduction of transgenes has improved considerably, allowing us to generate better models of human diseases (Bockamp et al. 2002).

Since the sequence of the mouse genome became available, the challenge has been to perform a functional analysis of the genome and to understand the physiological

S. Philip • S.K. Sharan (✉)
Mouse Cancer Genetics Program, Center for Cancer Research, National Cancer Institute at Frederick, Building 560, Room 32-31C, 1050 Boyles Street, Frederick, MD 21702, USA
e-mail: sharans@mail.nih.gov

J.E. Green and T. Ried (eds.), *Genetically Engineered Mice for Cancer Research: Design, Analysis, Pathways, Validation and Pre-Clinical Testing*,
DOI 10.1007/978-0-387-69805-2_3,

significance of various genes (van der Weyden et al. 2002). The functional analysis has been greatly aided by the availability of bacterial artificial chromosomes (BACs). BACs are *Escherichia coli* F factor-based vectors capable of maintaining cloned DNA fragments with an average insert size of 150–200 kb. They can accommodate most eukaryotic genes along with their regulatory elements (Shizuya et al. 1992). This allows the expression of the gene present in the BAC insert to be close to endogenous levels, because its *in vivo* regulatory machinery is kept largely intact. This is in contrast to transgenic mice generated with small DNA fragments because the regulatory sequences of genes can be several kilo bases away from the coding region (Scholz et al. 1997; Valarche et al. 1997), which makes it very difficult to express the transgenes at physiological levels. BACs have allowed us to overcome this limitation to a great extent. In addition, the ease of their handling and the ability to purify them by standard plasmid extraction techniques have made BACs the vector of choice for functional studies in many organisms besides the mouse.

BACs can also be used to generate targeting constructs to disrupt genes in ES cells by homologous recombination, for making knockout mouse models (Liu et al. 2003; for detailed protocol see Wu et al. 2008). Gene targeting can also be used to modify a gene by introducing a subtle alteration in its coding sequence, or to introduce a reporter gene or protein tags to study its expression profile or perform biochemical characterizations of the protein. Another advantage of using BACs is the ability to introduce multiple modifications at different loci that lie far apart using single targeting construct. This would be very time-consuming with conventional gene-targeting techniques, requiring multiple rounds of gene targeting in ES cells when a large gene is altered at multiple loci. Together, these considerations have provided the basis for using BACs to make targeting constructs to generate conditional alleles in ES cells (Liu et al. 2003).

Although the advantage of using BACs to stably harbor large inserts is considerable, finding methods to manipulate the huge BAC insert has been very challenging. Fortunately, recombineering (recombination-based genetic engineering) technology has simplified the manipulations of BACs and revolutionized the way genetically engineered mice are generated. Recombineering allows the manipulation of DNA, not by the conventional restriction enzyme-based methods, but by utilizing short regions of homology to insert, delete, or alter any DNA fragment by recombination. It obviates the need to have conveniently placed unique restriction sites in the cloning vectors and genomic DNA. The only requirement for recombineering is the sequence information of the region to be manipulated. The length of homology required for efficient recombination in *E. coli* can be as short as 35 bases. Such short regions of homology can be introduced into the targeting constructs by polymerase chain reaction (PCR)-based methods using chimeric primers that contain these homology sequences (Fig. 3.1). Development of the recombineering technology is one of the key recent advances in molecular biology. In this chapter, we describe the recombineering technology and discuss its various applications to manipulation of the mouse genome.

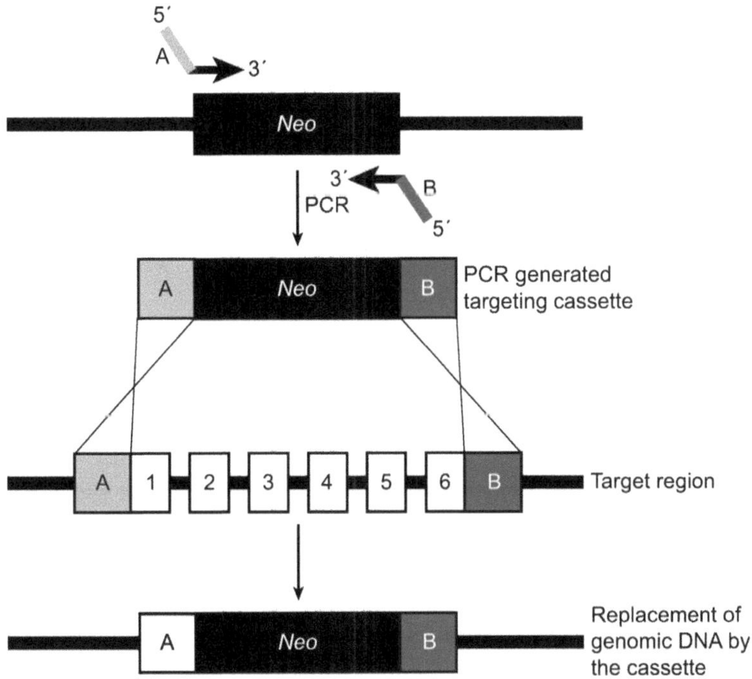

Fig. 3.1 Inserting a selectable marker in the target gene by recombineering. A targeting construct containing 50 bases of homology at both ends to the target region is generated by PCR. Chimeric primers that contain the homology arms (A, B) at the 5′ ends and sequence for priming the amplification of the drug-resistance cassette (e.g., *Kanamycin/Neomycin-resistance* gene, *Neo*) at the 3′ ends are used for PCR amplification. The targeting cassette is introduced into recombination-induced cells where homologous recombination replaces the target gene with the drug-resistance cassette. Recombinant clones can be selected for resistance to antibiotics (e.g., Kanamycin). This figure shows the replacement of the target gene by a drug-resistance cassette but this approach can be used to insert nonselectable sequences such as *Cre*, *EGFP*, or *loxP* by placing such sequences in tandem with the drug-resistance cassette. This approach can also be used to generate subtle alterations such as small insertions, deletions, or point mutations in the target gene by using the two-step selection–counter-selection method

3.2 *In Vivo* Cloning in *E. coli*

The use of homologous recombination to insert linear double-stranded DNA into the genome was first shown in yeast (Baudin et al. 1993; Lafontaine and Tollervey 1996; Yamamoto et al. 1992). Linear DNA introduced into *E. coli* is normally degraded by the RecBCD exonuclease (Wackernagel 1973), so *E. coli* RecBCD$^-$ strains, which lack the linear-double strand DNA exonuclease activity, were among the first *in vivo* cloning systems developed in *E. coli*. This system has been used to clone PCR products with matching terminal sequences into plasmid vectors (Jasin and Schimmel 1984). The major disadvantage of this system is that it is restricted to

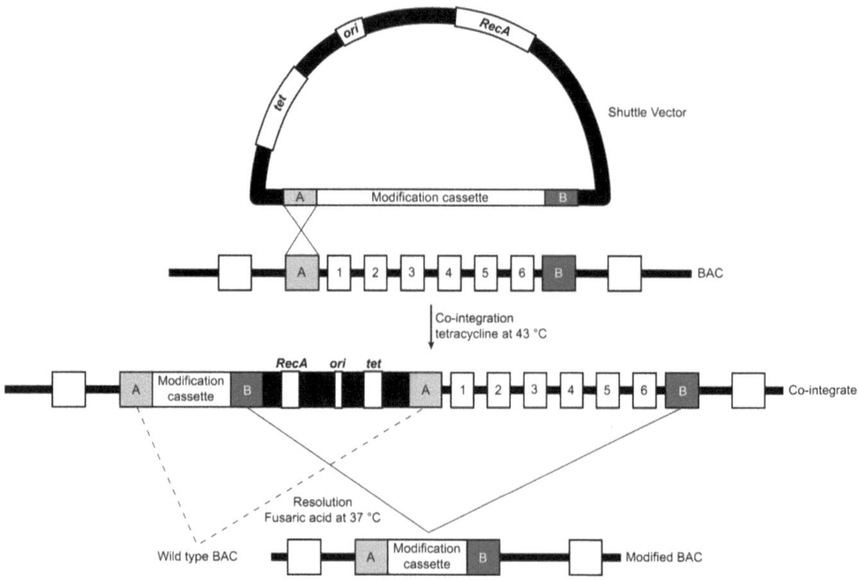

Fig. 3.2 Target gene modification by RecA-mediated homologous recombination. The RecA-mediated BAC modification approach utilizes a shuttle vector that carries the *RecA* gene to provide the recombination function, *Tetracycline-resistance* gene (*tet*) and temperature-sensitive origin of replication (*ori*). A DNA fragment containing the two homology arms (A and B) flanking the modification to be introduced (e.g., EGFP/Cre) is cloned into the vector. In the first step, recombination between the BAC and the shuttle vector through one of the homology arms (A) generates the co-integrant, which can be selected in the presence of tetracycline at 43°C. In the second step, known as resolution, some of the BACs undergo a second recombination through one set of the homology arms, which excises the shuttle vector. Depending on the homology arm used for resolution it generates either the wild-type BAC (if the same homology arm (A) is used for co-integration and resolution) or a recombinant BAC with precise replacement of the target gene by the modification cassette (if the resolution occurs through a different homology arm (B)). The recombinant BACs can be selected in the presence of fusaric acid for loss of the *tet* gene

a specific strain, in which the recombination pathway is constitutively active, which can cause rearrangements in large constructs, like BAC vectors.

Unlike linear DNA, circular DNA is not targeted by the *E. coli* exonucleases; therefore, plasmid-based shuttle vectors have been designed for recombination. The first method for engineering BACs utilized a plasmid with a temperature-sensitive origin of replication to transiently express the *RecA* gene in the BAC host strain, DH10B, which is recombination deficient (*recA*$^{-}$) (Yang et al. 1997). This paradigm allows the replication of the plasmid only at the permissive temperature (30°C), and loss of the plasmid occurs at the restrictive temperature (43°C). This targeting plasmid also carries a tetracycline (*Tet*)-resistance cassette for selection. The homology arms and the modification cassettes are cloned into the plasmid, and introduced into BAC-containing cells. In this technique, the desired recombinants are obtained in two steps (Fig. 3.2). In the first, RecA-mediated recombination at one of the homology arms results in the formation of a cointegrate, which can be selected at 43°C. Then, a fraction of the cointegrates undergo a second recombination event, known as the resolution step, during which the *Tet* and *RecA* genes are deleted from the BAC. Resolved BACs can be identified by loss of tetracycline-resistance gene by selecting for resistance to lipophilic chelating agents, such as fusaric acid. Depending on which homology arm is used during resolution, the process either regenerates the original BAC or the intended modified BAC.

Although the RecA function is only introduced temporarily in this method, it is still present in the cells long enough to cause unwanted rearrangements in the BACs. Another drawback of this system is that it relies on traditional cloning, using restriction enzymes and ligases to create the targeting cassettes. In spite of these limitations, this method has been used successfully to manipulate BACs (Heintz 2001).

3.3 Recombineering Using Bacteriophage Recombination Genes

Soon after the development of the RecA-based BAC engineering method, bacteriophage-encoded homologous recombination functions were utilized for *in vivo* genetic manipulations. The first method devised was based on homologous recombination mediated by the recE and recT proteins encoded by the RAC prophage. The *recE* gene product has 5′–3′ exonuclease activity, and the *recT* gene product is a single strand DNA-binding protein that promotes annealing. This method is referred to as the ET-cloning system. Zhang et al. (1998) showed that the *recE*- and *recT*-encoded proteins can be used to modify genomic DNA by PCR-generated linear dsDNA flanked by short regions of homology (42-bp) (Zhang et al. 1998). A plasmid-based system (pBAD-ETγ) was developed to allow ET cloning in $recBCD^{+}$ *E. coli* strains. In the pBAD-ETγ plasmid, the *recE* gene is under the control of the arabinose-inducible pBAD promoter, and the *recT* gene is under the control of the constitutive EM7 promoter. The plasmid also contains the bacteriophage-λ *gam* gene, controlled by the Tn5 promoter. The λ *gam* gene encodes a RecBCD exonuclease inhibitor that prevents the degradation of linear targeting DNA constructs (Murphy 1991). In this system, recombination is induced by activating recE expression by treatment with arabinose (Zhang et al. 1998).

Similar to the Rac prophage genes, the λ-red homologous recombination system consists of two genes, *red α*, also called *exo* and *red β*, also called *bet*. These two genes are functionally analogous to *recE* and *recT*. The λ-red homologous recombination system also requires *gam* function. Exo has 5′–3′ exonuclease activity that generates ssDNA overhangs; beta binds to these overhangs and stimulates annealing to complementary strands. The pBAD-ETγ equivalent in the Red system is pBAD-αβγ (Muyrers et al. 1999). Because phage-encoded proteins are located on the plasmid, it can be introduced into any *E. coli* strain. However, these systems have two main disadvantages: toxicity caused by the constitutively active gam protein and leaky expression of the recombination function.

3.4 Recombineering Using Defective λ-Prophage

To provide better control over the expression of the λ recombination genes, a system utilizing a defective λ-prophage was developed (Yu et al. 2000). In this system, exo, bet, and gam are expressed from a stably integrated λ-prophage, and their expression is under the control of the temperature-sensitive *λ-cI857* repressor, which is functionally inactive at 42°C. When the repressor is active, at 32°C, the genes responsible for the recombination function are suppressed, but shifting the culture to 42°C alleviates this suppression (Yu et al. 2000). To facilitate BAC engineering, the defective λ-prophage was introduced into a BAC host strain of *E. coli*, DH10B, to yield the *E. coli* strain DY380. Other strains that express additional genes, like Cre (EL250) and Flp (EL350), have been derived from DY380 (Lee et al. 2001). Subsequently, Warming et al. generated SW102, a derivative of the DY380 strain, which lacks the galactokinase gene (*galK*) of the galactose operon (Warming et al. 2005). Since *galK* can serve as both a positive and negative selection marker, this strain has been useful for engineering subtle mutations or for introducing nonselectable sequences into the BAC DNA.

In addition to *E. coli*-based host strains containing defective λ-prophage, mobile BAC DNA recombineering systems are now available that can be introduced into any bacterial strain. Such mobile elements include the mini-λ (Court et al. 2003) and the plasmid-based pSIM vector systems (Datta et al. 2006). Both the mini-λ and the pSIM plasmids provide the same endogenous phage-controlling elements and Red recombination functions described above for the defective prophage strains. Mini-λ is a nonreplicating circular phage DNA that integrates into the bacterial chromosome by site-specific recombination. After its integration, the mini-λ is stable and replicates as part of the host chromosome. Mini-λ can be readily excised to cure the cells of the phage DNA. The excised mini-λ DNA circle can also be purified from the bacterial cells using a standard plasmid purification protocol. The pSIM vectors consist of the elements of the prophage necessary for recombineering carried on a pSC101 plasmid derivative that has temperature-sensitive DNA replication; this plasmid has a low copy number, and the temperature-sensitive replicon permits the loss of the plasmid after recombineering is complete. Various derivatives of pSIMs are available with a variety of selectable antibiotic-resistance markers. Unlike the prophage and the mini-λ, pSIM vectors require drug selection for their stable maintenance.

A new means of delivering the Red system is a λ phage carrying a tetracycline-resistance gene (λTetR). This phage system combines the advantages of stability (integrates into the bacterial chromosomal DNA) as well as mobility. The λTetR can be easily prepared as a high-titer phage lysate and used to transform the desired host by infection. The efficiency of lysogeny is greater than that of plasmid transformation, and once introduced into *E. coli*, the prophage can be stably maintained without any drug selection (Chan et al. 2007).

The recombineering systems described above can be used for a wide range of applications, and the choice of system depends upon the target DNA (for detailed protocol see Sharan et al. 2009). For example, when a plasmid DNA has to be engineered, bacterial strains (e.g., DY380 or SW102) expressing the recombination genes can be used. When a BAC DNA is to be engineered, one of the mobile systems can be introduced into the BAC-containing bacterial cells, which helps to eliminate any potential rearrangement that may occur during BAC transformation.

3.5 Generating Knockout Mouse Models

The generation of knockout mouse models is one of the most frequently used applications of recombineering. Conventional methods for generating gene-targeting constructs rely on mouse genomic libraries in λ-phage or cosmids. These protocols include time-consuming steps, such as generating a restriction map of the insert, subcloning the homology arms, and inserting positive and negative selection markers. Furthermore, although the exons to be deleted are selected based on their functional importance, in most cases the presence of appropriate restriction enzyme sites is a critical determining factor. With the sequencing of the mouse genome and the availability of sequences from the ends of BAC inserts from multiple genomic libraries, the method for generating targeting constructs has changed significantly, and the use of BACs and recombineering to generate knock-out constructs makes it relatively easy to include larger homology arms, which clearly enhances the efficiency of gene targeting in ES cells (Liu et al. 2003). Desirable BAC clones containing a gene of interest can now be selected from the genome browser websites (http://www.genome.ucsc.edu/cgi-bin/hgGateway and http://www.ensembl.org/Mus_musculus/). Once the BAC clone is obtained, a targeting vector can be generated by recombineering in two steps: (1) retrieval or subcloning of the genomic region (Fig. 3.3) containing the two homology arms using a plasmid-based vector and (2) insertion of a positive selection marker for selection in ES cells (Fig. 3.1).

3.5.1 Subcloning or Retrieving Genomic DNA from BAC

DNA fragments can be subcloned from BACs into a linear plasmid vector backbone by recombineering, without the use of restriction enzymes or DNA ligases. The DNA used for such retrieval requires a selectable marker such as the Ampicillin- or Kanamycin-resistance gene flanked by about 50 bases of homology to the 5′ and 3′ ends of the region to be subcloned, and an origin of replication. Gap repair of the linear plasmid DNA by recombination with the target DNA circularizes the plasmid and allows for selection by the appropriate antibiotics (Fig. 3.3).

Retrieval vectors are generated by PCR using two primers and plasmid DNA as the template. A high- or low-copy plasmid backbone can be used, depending on the length of the fragment to be subcloned. A high copy-number plasmid, such as pBluescript, can be used for fragments up to 20–25 kb, but a low copy-number vector,

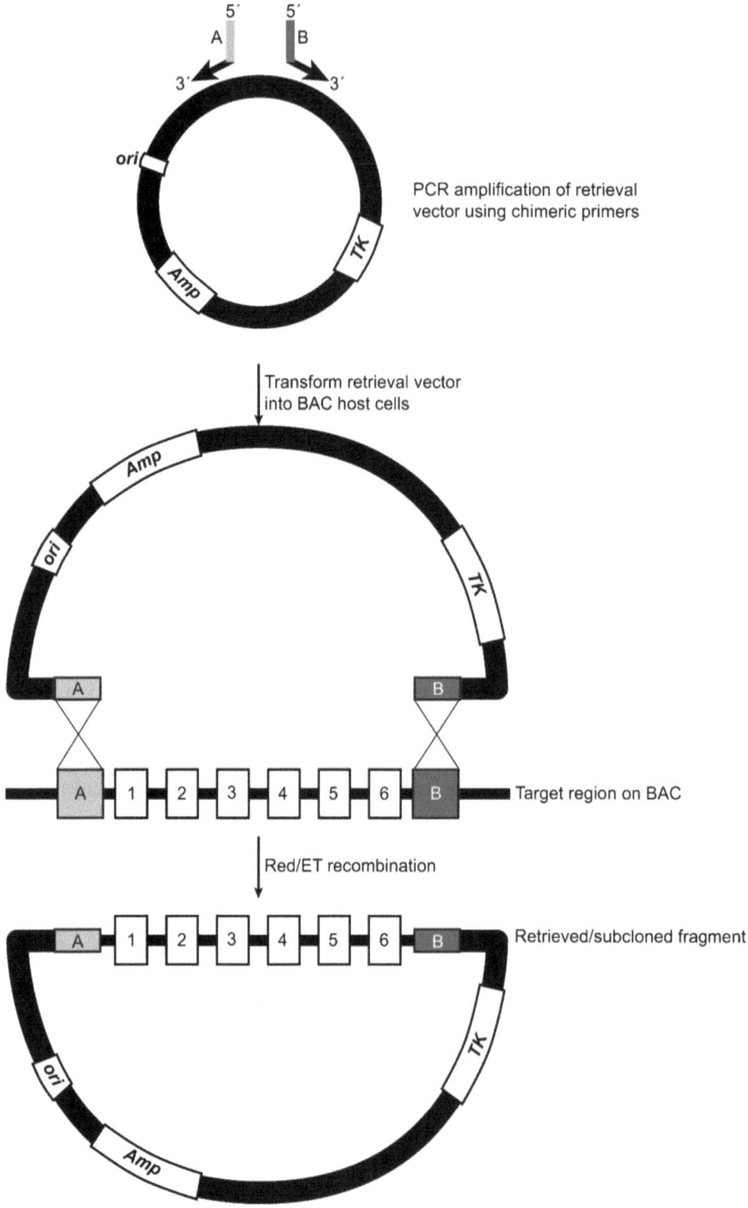

Fig. 3.3 Retrieval or subcloning a genomic fragment from the BAC insert by gap repair. Chimeric primers containing 50 bases of homology (A, B) to target region on their 5′ end and sequence for PCR priming on their 3′ end are used to amplify a linear retrieval construct. The plasmid template used to generate the retrieval construct carries an origin of replication (*ori*) and an antibiotic-resistance marker (e.g., ampicillin-resistance gene, *Amp*). In addition, it contains a negative selection marker (e.g., *Thymidine Kinase, TK*) to be used to select against ES cell clones that do not undergo homologous recombination. When transformed into recombination competent BAC host cells, homologous recombination between the homology arms on the linear retrieval vector and the BAC DNA generates a circular plasmid containing the target region by gap repair. The bacterial cells containing the subcloned fragment can be selected for resistance to Ampicillin

such as pBR322, can be used to subclone fragments as large as 80 kb (Lee et al. 2001). The retrieval plasmid construct can be designed to also contain a negative selection maker (e.g., the gene for *Thymidine Kinase* [*TK*] or Diphtheria toxin [*DT*]), which can be used to ablate ES cells that do not undergo homologous recombination (Liu et al. 2003).

3.5.2 Inserting Selectable Markers

Once the genomic fragment containing the homology arms is subcloned or retrieved, the next step is to delete the exons required to disrupt the gene in ES cells and replace them with a selectable marker. This can be achieved by recombineering using selectable markers that are under the control of a eukaryotic promoter (e.g., the phosphoglycerate kinase-1 *(PGK)* promoter, for the selection in ES cells) as well as a prokaryotic promoter (e.g., *Tn5, Tn10*, or *EM7*, for expression in bacterial cells). Markers that allow resistance to antibiotics such as neomycin, blasticidine, and hygromycin are functional in both bacterial and mammalian cells. Such selectable markers can be introduced by generating targeting constructs that contain the selectable marker flanked by 50 bases with homology to the target sites that are introduced by the PCR-based method described above (Fig. 3.1).

3.6 Conditional Gene Modifications

In many cases, knockout mice die in utero or early postnatally because the ablated gene is essential for embryonic development, which makes it impossible to use this system to study the function of such genes in adult tissues. To circumvent the problem of early lethality and to control the gene disruption temporally and spatially, the conditional knockout approach is commonly used (see Chapter 2; Lewandoski 2001). A conditional knockout allele is generated by inserting a *loxP* or *FRT* sequence at the two ends of a gene or its critical exons. *loxP* and *FRT* are 34-bp DNA sequences that are recognized by a site-specific recombinase, Cre or Flp, respectively. When two *loxP* or *FRT* sites are placed in the same orientation, their recombination (induced by their respective recombinase) deletes the sequence between them. Therefore, by crossing mice carrying a conditional knockout allele of a gene to transgenic mice expressing Cre or Flp under the control of a tissue-specific promoter, a tissue-specific knockout model can be generated (Branda and Dymecki 2004). The conventional preparation of targeting constructs for conditional alleles can be unwieldy and time consuming. New recombineering-based approaches have simplified this procedure. By using selectable markers that are flanked by *loxP* or *FRT* or both, these sites can be inserted easily into the mouse genome (Liu et al. 2003).

3.7 High-Throughput Recombineering

Although genomic fragments retrieved from BACs can be used to generate gene-targeting constructs, the entire BAC insert can also be used directly for homologous recombination in ES cells. Because of the large regions of homology used, this process is extremely efficient (Valenzuela et al. 2003) and has led to the development of a high-throughput and largely automated process, called VelociGene. This method utilizes recombineering to replace the gene of interest with a reporter cassette in the BAC, which not only disrupts the gene but also allows high-resolution expression profiling of the targeted gene in mice.

Although the recombination efficiency is very high, the identification of the targeted clones cannot be performed by routine Southern or PCR-based methods when this system is used, because of the long flanking arms of the BAC targeting vectors. In conventional gene targeting, Southern or PCR-based assays are used with either a probe (for Southern analysis) or by choosing one of the PCR primers from a region outside the homology arms. When the BAC system is used, correctly targeted ES cell clones are identified by quantitative PCR-based methods to detect the loss of the native allele, using gene-specific primers from just inside the deletion points or by fluorescence *in situ* hybridization (FISH) using two different probes, one specific for the native gene and the other specific for the replacing reporter gene (Valenzuela et al. 2003).

Recently, Chan et al. developed a recombineering-based system for generating conventional conditional gene-targeting vectors in a 96-well format (Chan et al. 2007). The high transduction efficiency of the λ-lysogen was used to deliver the recombination function (i.e., *exo, bet, gam*) in BAC-containing DH10B cells. The strategy for the conditional knockout includes the introduction of a selectable marker at the 5′ end of the region to be deleted that is subsequently replaced by another cassette containing an *FRT*-flanked LacZ reporter and a *loxP* site. A second *loxP* site was targeted to the 3′ end of the region to be deleted. The recombineering steps were modified to adapt to the 96-well format. This included the use of a chemical transformation method to introduce the targeting constructs into the *E. coli*. The primers for generating the two targeting cassettes and the retrieval vector were designed by software for genome-wide gene-targeting strategies. The intended deletion region in the genes ranged between 3 and 5 kb. Using this high-throughput method, the authors simultaneously generated gene-targeting constructs for 94 genes on mouse chromosome 11.

3.8 Multiple Alterations

In addition to their high targeting efficiency, multiple simultaneous modifications can be made in the target gene via BAC targeting vectors (Fig. 3.4). Testa et al. generated a BAC targeting construct that simultaneously modified the human mixed

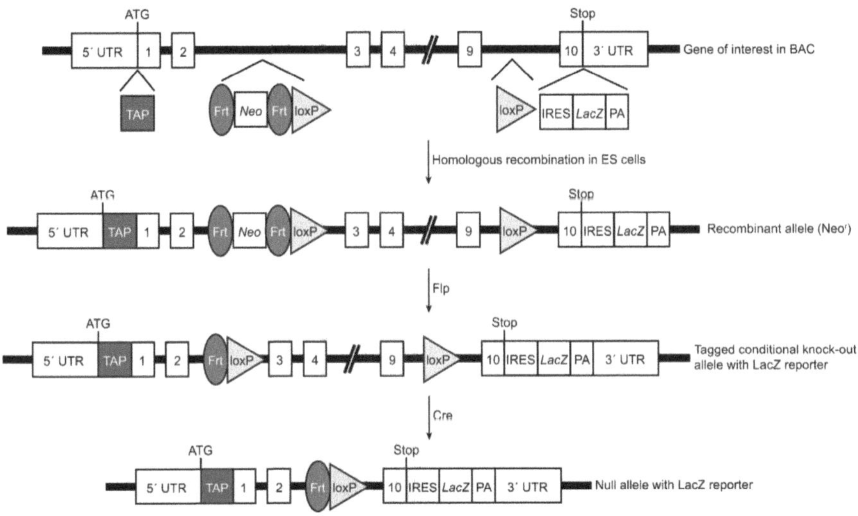

Fig. 3.4 Generation of a multipurpose allele of a target gene using recombineering. A BAC carrying the target gene can be modified by sequential rounds of recombineering to introduce multiple modifications at different regions lying far apart. Represented here is an example that shows the introduction of a N-terminal tandem affinity purification (TAP) tag, a positive selection marker (e.g., *Neo*), a *loxP* site into an intron at the 5′ of the region to be deleted, another loxP site at the 3′ end of the region to be deleted and a reporter cassette (e.g., *LacZ*) with an internal ribosomal entry site (IRES) after the stop codon in the last exon. When the modified BAC is used to target the endogenous gene by homologous recombination in ES cells it can generate a multipurpose allele of the target gene in mice. The allele can serve as a conditional knockout allele to study the function of the gene. The N-terminal tag and the reporter gene are useful for studying the localization and protein interactions of the target gene product. In addition the *Neo*-resistance cassette in one of the introns contains a splice acceptor site and may generate a hypomorphic allele. Such hypomorphic alleles have been shown to be useful for identifying novel functions of the gene that are masked by the null allele

lineage leukemia (MLL) gene at two sites 43 kb apart to generate a MLL translocation model (Testa et al. 2003). The authors introduced a tandem affinity purified (TAP) protein tag at the N terminus and an internal ribosome entry site (IRES) followed by a β-galactosidase–neomycin fusion gene into intron 11. Multiple rounds of recombineering in *E. coli* were performed to engineer the complex targeting construct. The first cassette introduced was the TAP tag, which was placed just after the ATG start site along with a hygromycin-resistance gene flanked by two *FRT* sites. Subsequently, a second cassette containing a splice-acceptor site, an IRES, and a β-galactosidase neomycin fusion gene flanked by two *loxP* sites was introduced. After homologous recombination in ES cells, three different alleles (a knockout allele, a tagged wild-type allele, and a tagged hypomorphic allele) were generated in mice (Testa et al. 2003).

3.9 Generating Transgenic Constructs

In addition to engineering knockout mouse models, recombineering is also being used to make transgenic mouse lines. This is particularly useful when BACs are used to generate the transgenic construct, because variability in the expression pattern among different founder lines is less frequently observed, compared with the significant variability seen with smaller DNA fragments. This feature has made BACs a powerful system for generating transgenic mice. Furthermore, the presence of most *cis*-regulatory elements in a large BAC insert has made BAC technology useful for the functional analysis of genes and for generating reporter lines. The use of BACs to generate transgenic mice was first described in genetic complementation studies in which BACs were used to rescue the phenotype of a mouse *Clock* mutation (King et al. 1997). Since then, BACs have frequently been used to complement mouse mutations in positional cloning experiments (Matesic et al. 2001; Means et al. 2001; Probst et al. 1998; Wakabayashi et al. 1997; Wilson et al. 2000).

3.10 Recombineering-Based Methods for Generating Subtle Alterations

Hypomorphic alleles are valuable tools for the functional dissection of a gene. Such alleles provide functional information that may not be apparent when the entire protein is rendered nonfunctional. In organisms such as bacteria, yeast, flies, and worms, multiple alleles are often generated, because it is relatively easy to screen for desirable mutants after random mutagenesis (Clifford and Schupbach 1994; Lamb et al. 1998). In mice, there are fewer examples of multiple allelic series of a gene, in part because of the time required to generate even a single mutation by homologous recombination in ES cells (Huang et al. 1998; Klinghoffer et al. 2002; Meyers et al. 1998; Nagy et al. 1998; Suh et al. 2002). In an alternative approach, *N*-ethyl-*N*-nitrosourea (ENU), a chemical mutagen, was used to create allelic series for the *Smad2* and *Smad4* genes in ES cells (Chen et al. 2000; Vivian et al. 2002). Although this is an efficient method for generating an allelic series by screening a library of mutagenized ES cells, the alleles are randomly generated. Recombineering technology allows specific subtle alterations to be generated in any gene.

3.10.1 A Selection–Counter Selection Method to Generate Subtle Alterations in BACs

To generate subtle alterations by recombineering, a two-step procedure involving selection and counter selection can be used (Ellis et al. 2001; Muyrers et al. 2000). In the first step, a targeting cassette containing a positive selection marker such as the Kanamycin (*neo*)-resistance gene and a counter-selection marker such as *sacB* is targeted to the region where a mutation is to be generated. Recombinant clones are selected for resistance to Kanamycin. In the second step, another cassette is targeted to the same region to replace the *neo-sacB* cassette. The second cassette is designed to carry the desired mutation in the genomic sequence, which can include point mutations, small insertions, or deletions. The bacterial cells containing the correctly targeted BACs are selected for loss of the *sacB* gene in media containing 7% sucrose, which is toxic to cells expressing *sacB* (Blomfield et al. 1991).

The galactokinase gene (*galK*) has also been used for both selection and counter selection (Warming et al. 2005). This approach makes use of the recombineering strain SW102 in which *galK* of the galactose operon is deleted. The *galK* function can be added in *trans* to the target region, restoring the ability to grow on minimal media with galactose as the only carbon source. In the second step, a desired mutation is targeted to the region, and *galK* function can be selected against by resistance to 2-deoxy galactose (DOG), which is phosphorylated by *galK* into a toxic intermediate. The *galK* strain is reported to generate less background noise from spontaneous mutations in the counter selection marker, compared with the other schemes. Similarly, thymidylate synthase A (*thyA*) can be used in dual selection cassettes in cells containing a *thyA* deletion (Wong et al. 2005).

3.10.2 Modifications Using Single-Stranded Oligonucleotides

A second approach to generating subtle alterations in BACs that does not involve a selectable marker uses single-stranded (ss) oligonucleotides as targeting cassettes for recombineering. The efficiency of recombineering using ssDNA is dramatically higher than when double-stranded (ds) DNA is used (Ellis et al. 2001; Yu et al. 2000). Furthermore, while the *E. coli* recombinase genes *exo, bet*, and *gam* are required for dsDNA recombination, only *bet* is required for ssDNA recombineering. A 30-mer ss oligonucleotide can generate recombinants, but the use of 70-mer ss oligonucleotides is reported to be up to fivefold more efficient (Ellis et al. 2001). The targeting oligonucleotides can correspond to either of the two strands. However, one strand recombines more frequently than the other, and this preferred strand is also the lagging strand during DNA replication. This has been explained by a model in which transient regions of ssDNA are created in the lagging strand during replication, facilitating beta-mediated annealing of the oligonucleotides at the target site (Copeland et al. 2001; Kuzminov 1995).

Recombineering with oligonucleotides in *E. coli* has been used successfully to generate single base changes, small deletions, and the insertion of a Flag tag in the murine homolog of the human breast cancer susceptibility gene *BRCA2* (Swaminathan et al. 2001). The oligonucleotide includes the mutated sequence flanked by short homology arms of 35–70 bases on either side. The recombined BACs are identified by a PCR screening strategy termed, “mismatch amplification mutation assay-PCR” (MAMA PCR), in which the recombinant BAC is selectively amplified (Cha et al. 1992;

Swaminathan et al. 2001). Detection of the recombined BAC by MAMA PCR is sometimes not possible because of the nonspecific amplification of a mismatched PCR product when the mutation involves a deletion, insertion, or alteration of a base that is identical to one of the flanking bases. A two-step hit-and-fix method has been described that overcomes this limitation (Yang and Sharan 2003). First, 6–20 nucleotides are changed around the desired region and then the altered nucleotides are reverted to the original sequence except for the desired subtle change. Since several nucleotides are changed in both steps, specific detection primers can be designed for PCR-based or colony hybridization-based screening methods.

3.11 Humanized Mouse Models of Human Diseases

Success with BAC complementation studies and the engineering of subtle mutations in BACs have led to the development of humanized mouse models of human diseases. For example, a BAC clone containing the human *BRCA1* gene was shown to rescue the embryonic lethal phenotype of a targeted mutation of the murine *Brca1* gene, indicating that the human gene could functionally complement the mouse gene. *Brca1* mutations known to cause human cancer were then engineered into BACs by recombineering, to study their functional significance in mice that were deficient in endogenous *Brca1* (Yang et al. 2003). One of the humanized *BRCA1* mouse models revealed that a missense mutation in codon 64, which was predicted to change a conserved cysteine residue to glycine in the RING domain (Castilla et al. 1994), resulted in a functionally null protein due to aberrant splicing. In another study, a humanized mouse model of glaucoma was generated by using a BAC containing the disease-causing mutation Tyr437His in the human *Myocillin* gene (Senatorov et al. 2006). Similarly, a 5-bp deletion in exon 6 of the human *NBS1* gene, which is observed in 95% of Nijmegen Breakage Syndrome (NBS) patients, was introduced by BAC recombineering to make a mouse model of this disease (Difilippantonio et al. 2005). In both cases, the humanized mouse models recapitulated the pathological changes observed in patients and allowed functional analysis of these mutations.

3.12 Inserting Nonselectable Cassettes

As described above, there are several strategies for introducing nonselectable cassettes into targeting vectors, which can then be used for a wide array of applications. This method is used to introduce *loxP* sites to flank a region to be deleted in conditional knockout experiments. It can also be used to introduce epitope tags (Flag, myc, HA) or protein tags, such as green fluorescent protein (GFP), to be expressed as a fusion protein with a gene of interest. Such transgenic mice are valuable for the biochemical characterization of a particular protein product or for protein localization studies.

A fluorescent reporter can also be used for staining specific cell types and *in vivo* imaging. In addition, recombineering-based BAC transgenic mouse models have proven to be extremely valuable for tissue-specific gene knockout.

3.12.1 Generating Cre/Flp Lines for Conditional Knockout Mouse Models

The successful deletion of a gene of interest in specific tissues relies a great deal on the targeted expression of the Cre or Flp recombinase. When Cre or Flp expression in transgenic mice is driven by a portion of the promoter region of any gene, its expression has been useful but not very reliable. Consequently, many lines have to be screened to identify those that express the transgene as expected. This occurs because we do not yet fully understand the regulatory elements of most genes. In the absence of such detailed knowledge, it is more convenient to use BACs to generate transgenic mice by inserting the Cre or Flp cDNAs at the 5′ or 3′ end of genes by using the selection-counter selection technique or using a floxed selection cassette in tandem with the transgene. Several Cre lines for conditional gene targeting have been generated using this approach. For example, Cre recombinase was inserted in-frame downstream from and under the control of a progesterone receptor promoter within a BAC transgene to generate the PR-BAC^{icre} line. In this line, Cre was expressed only in cell lines expressing the progesterone receptor (Mukherjee et al. 2006). Similarly, a CamkII a^{icre} line was generated using a BAC modification for brain-specific gene targeting. The expression pattern and levels of transgene expression resembled those of the endogenous gene (Casanova et al. 2001).

3.12.2 Generating Reporter Lines

Understanding the expression pattern of a gene can provide important clues to its function. Protein tagging in mammalian cells is usually performed by using cDNA-based transgenes that lack the endogenous regulatory sequences and are driven by nonspecific promoters. The ability to precisely manipulate BACs by recombineering to generate recombinant tagged proteins ensures that the expression pattern and levels resemble those of the endogenous gene, which is useful for characterizing the protein expression patterns and interaction networks.

Poser et al. (2008) have described the use of recombineering in the 96-well format for the high-throughput generation of tagged BAC-based constructs (Poser et al. 2008). The usefulness of this technique for studying protein localization and protein–protein interactions on a genome-wide scale has been shown in yeast. Studies using GFP as a localization marker and tandem affinity tag-based complex purification have helped to generate a comprehensive picture of the core proteome and a localization map of this relatively simple model system (Gavin et al. 2002; Ghaemmaghami et al. 2003; Krogan et al. 2006). The approach for tagged BAC-based transgenesis in the mammalian system has been termed, "BAC transgeneOmics." The authors used a modified localization and affinity purification (LAP) tag, which combines the use of EGFP for LAP, the PreScission protein cleavage site for native elution, and the S-peptide for a second affinity purification. This tag was targeted to the N-terminal coding sequence with a floxed selection cassette and to the C-terminal coding sequence with an IRES-neomycin gene following the tag sequence. The authors successfully used these approaches to generate a large number of transgenes and demonstrated their application to the analysis of protein localization and protein–protein interactions, and to the mapping of a protein's DNA-binding sites (Poser et al. 2008).

A large-scale effort to define the gene expression profile of the central nervous system, the Gene expression nervous system atlas (GENSAT) BAC transgenic project, has been reported by Gong and colleagues (Gong et al. 2003). Using the RecA-based BAC engineering method, the coding sequences of genes present in the BACs were systematically replaced by the EGFP reporter gene for the large-scale generation of EGFP-BAC transgenic lines. Using these lines, the expression of different genes can be localized to specific cell types in the brain, and followed over time to monitor expression during development. BAC fluorescent reporter constructs are also useful for studying gene expression in live cells and for cell lineage tracking. Hsiao et al. (2008) generated a BAC reporter ES cell line for the *NKX2-5* gene, a marker of early cardiomyocyte differentiation. Upon *in vitro* differentiation, EmGFP-positive beating cardiomyocytes were observed. Fluorescence-activated cell sorting was used to isolate the NKX2-5 positive cells. Furthermore, in the chimeric embryos generated from these ES cells, EmGFP was localized to the developing heart tube and foregut (Hsiao et al. 2008).

3.13 Concluding Remarks

We have described here the advances in recombineering-based genetic engineering technology that can lead to the rapid generation of genetically engineered mice bearing a variety of alterations in many genes of interest. The ability to insert, delete, or alter genes easily by recombineering, combined with the accurate transcriptional regulation of the transgene present in the BACs has proven to be a valuable tool for functional genomics. The technology has been improved and streamlined to the extent that it is now being used for several high-throughput applications on a genome-wide scale. These strategies will generate comprehensive BAC resources for conditional gene knockout models and epitope-tagged or reporter constructs covering the whole genome. Revealing the function and expression profile of genes as well as finding their regulatory and interacting partners by using these BAC resources will help generate even better models of human diseases in the near future.

Acknowledgments The authors thank Drs. Kajal Biswas, Suhwan Chang, Lino Tessarollo, and Sergey Kuznetsov for critical review of the manuscript and Tammy Schroyer of the Publication Department for the illustrations. This research was supported by the Center for Cancer Research, National Cancer Institute, Department of Human and Health Services.

References

Baudin A, Ozier-Kalogeropoulos O, Denouel A, Lacroute F, Cullin C (1993) A simple and efficient method for direct gene deletion in Saccharomyces cerevisiae. Nucleic Acids Res 21:3329–3330

Bedell MA, Largaespada DA, Jenkins NA, Copeland NG (1997) Mouse models of human disease. Part II: recent progress and future directions. Genes Dev 11:11–43

Blomfield IC, Vaughn V, Rest RF, Eisenstein BI (1991) Allelic exchange in Escherichia coli using the Bacillus subtilis sacB gene and a temperature-sensitive pSC101 replicon. Mol Microbiol 5:1447–1457

Bockamp E, Maringer M, Spangenberg C, Fees S, Fraser S, Eshkind L, Oesch F, Zabel B (2002) Of mice and models: improved animal models for biomedical research. Physiol Genomics 11:115–132

Bradley A, Zheng B, Liu P (1998) Thirteen years of manipulating the mouse genome: a personal history. Int J Dev Biol 42:943–950

Branda CS, Dymecki SM (2004) Talking about a revolution: the impact of site-specific recombinases on genetic analyses in mice. Dev Cell 6:7–28

Capecchi MR (1989) The new mouse genetics: altering the genome by gene targeting. Trends Genet 5:70–76

Casanova E, Fehsenfeld S, Mantamadiotis T, Lemberger T, Greiner E, Stewart AF, Schutz G (2001) A CamKIIalpha iCre BAC allows brain-specific gene inactivation. Genesis 31:37–42

Castilla LH, Couch FJ, Erdos MR, Hoskins KF, Calzone K, Garber JE, Boyd J, Lubin MB, Deshano ML, Brody LC et al (1994) Mutations in the BRCA1 gene in families with early-onset breast and ovarian cancer. Nat Genet 8:387–391

Cha RS, Zarbl H, Keohavong P, Thilly WG (1992) Mismatch amplification mutation assay (MAMA): application to the c-H-ras gene. PCR Methods Appl 2:14–20

Chan W, Costantino N, Li R, Lee SC, Su Q, Melvin D, Court DL, Liu P (2007) A recombineering based approach for high-throughput conditional knockout targeting vector construction. Nucleic Acids Res 35:e64

Chen Y, Yee D, Dains K, Chatterjee A, Cavalcoli J, Schneider E, Om J, Woychik RP, Magnuson T (2000) Genotype-based screen for ENU-induced mutations in mouse embryonic stem cells. Nat Genet 24:314–317

Clifford R, Schupbach T (1994) Molecular analysis of the Drosophila EGF receptor homolog reveals that several genetically defined classes of alleles cluster in subdomains of the receptor protein. Genetics 137:531–550

Copeland NG, Jenkins NA, Court DL (2001) Recombineering: a powerful new tool for mouse functional genomics. Nat Rev Genet 2:769–779

Court DL, Swaminathan S, Yu D, Wilson H, Baker T, Bubunenko M, Sawitzke J, Sharan SK (2003) Mini-lambda: a tractable system for chromosome and BAC engineering. Gene 315:63–69

Datta S, Costantino N, Court DL (2006) A set of recombineering plasmids for gram-negative bacteria. Gene 379:109–115

Difilippantonio S, Celeste A, Fernandez-Capetillo O, Chen HT, Reina San Martin B, Van Laethem F, Yang YP, Petukhova GV, Eckhaus M, Feigenbaum L, Manova K, Kruhlak M, Camerini-Otero RD, Sharan S, Nussenzweig M, Nussenzweig A (2005) Role of Nbs1 in the activation of the Atm kinase revealed in humanized mouse models. Nat Cell Biol 7:675–685

Doolittle DP, Davisson MT, Guidi JN, Green MC (1996) Catalog of mutant genes and polymorphic loci. In: Lyon SRMF, Brown SDM (eds) Genetic variants and strains of the laboratory mouse. Oxford University Press, Oxford, pp 17–854
Ellis HM, Yu D, DiTizio T, Court DL (2001) High efficiency mutagenesis, repair, and engineering of chromosomal DNA using single-stranded oligonucleotides. Proc Natl Acad Sci USA 98:6742–6746
Gavin AC, Bosche M, Krause R, Grandi P, Marzioch M, Bauer A, Schultz J, Rick JM, Michon AM, Cruciat CM, Remor M, Hofert C, Schelder M, Brajenovic M, Ruffner H, Merino A, Klein K, Hudak M, Dickson D, Rudi T, Gnau V, Bauch A, Bastuck S, Huhse B, Leutwein C, Heurtier MA, Copley RR, Edelmann A, Querfurth E, Rybin V, Drewes G, Raida M, Bouwmeester T, Bork P, Seraphin B, Kuster B, Neubauer G, Superti-Furga G (2002) Functional organization of the yeast proteome by systematic analysis of protein complexes. Nature 415:141–147
Ghaemmaghami S, Huh WK, Bower K, Howson RW, Belle A, Dephoure N, O'Shea EK, Weissman JS (2003) Global analysis of protein expression in yeast. Nature 425:737–741
Gong S, Zheng C, Doughty ML, Losos K, Didkovsky N, Schambra UB, Nowak NJ, Joyner A, Leblanc G, Hatten ME, Heintz N (2003) A gene expression atlas of the central nervous system based on bacterial artificial chromosomes. Nature 425:917–925
Heintz N (2001) BAC to the future: the use of bac transgenic mice for neuroscience research. Nat Rev Neurosci 2:861–870
Hogan B, Beddington R, Costantini F, Lacy E (1994) Manipulating the mouse embryo. Cold Spring Harbor Laboratory Press, Cold Spring Harbor
Hsiao EC, Yoshinaga Y, Nguyen TD, Musone SL, Kim JE, Swinton P, Espineda I, Manalac C, deJong PJ, Conklin BR (2008) Marking embryonic stem cells with a 2A self-cleaving peptide: a NKX2-5 emerald GFP BAC reporter. PLoS One 3:e2532
Huang JD, Mermall V, Strobel MC, Russell LB, Mooseker MS, Copeland NG, Jenkins NA (1998) Molecular genetic dissection of mouse unconventional myosin-VA: tail region mutations. Genetics 148:1963–1972
Jasin M, Schimmel P (1984) Deletion of an essential gene in Escherichia coli by site-specific recombination with linear DNA fragments. J Bacteriol 159:783–786
King DP, Zhao Y, Sangoram AM, Wilsbacher LD, Tanaka M, Antoch MP, Steeves TD, Vitaterna MH, Kornhauser JM, Lowrey PL, Turek FW, Takahashi JS (1997) Positional cloning of the mouse circadian clock gene. Cell 89:641–653
Klinghoffer RA, Hamilton TG, Hoch R, Soriano P (2002) An allelic series at the PDGFalphaR locus indicates unequal contributions of distinct signaling pathways during development. Dev Cell 2:103–113
Krogan NJ, Cagney G, Yu H, Zhong G, Guo X, Ignatchenko A, Li J, Pu S, Datta N, Tikuisis AP, Punna T, Peregrin-Alvarez JM, Shales M, Zhang X, Davey M, Robinson MD, Paccanaro A, Bray JE, Sheung A, Beattie B, Richards DP, Canadien V, Lalev A, Mena F, Wong P, Starostine A, Canete MM, Vlasblom J, Wu S, Orsi C, Collins SR, Chandran S, Haw R, Rilstone JJ, Gandi K, Thompson NJ, Musso G, St Onge P, Ghanny S, Lam MH, Butland G, Altaf-Ul AM, Kanaya S, Shilatifard A, O'Shea E, Weissman JS, Ingles CJ, Hughes TR, Parkinson J, Gerstein M, Wodak SJ, Emili A, Greenblatt JF (2006) Global landscape of protein complexes in the yeast Saccharomyces cerevisiae. Nature 440:637–643
Kuzminov A (1995) Collapse and repair of replication forks in Escherichia coli. Mol Microbiol 16:373–384
Lafontaine D, Tollervey D (1996) One-step PCR mediated strategy for the construction of conditionally expressed and epitope tagged yeast proteins. Nucleic Acids Res 24:3469–3471
Lamb RS, Ward RE, Schweizer L, Fehon RG (1998) Drosophila coracle, a member of the protein 4.1 superfamily, has essential structural functions in the septate junctions and developmental functions in embryonic and adult epithelial cells. Mol Biol Cell 9:3505–3519
Lee EC, Yu D, Martinez de Velasco J, Tessarollo L, Swing DA, Court DL, Jenkins NA, Copeland NG (2001) A highly efficient Escherichia coli-based chromosome engineering system adapted for recombinogenic targeting and subcloning of BAC DNA. Genomics 73:56–65

Lewandoski M (2001) Conditional control of gene expression in the mouse. Nat Rev Genet 2:743–755

Liu P, Jenkins NA, Copeland NG (2003) A highly efficient recombineering-based method for generating conditional knockout mutations. Genome Res 13:476–484

Matesic LE, Yip R, Reuss AE, Swing DA, O'Sullivan TN, Fletcher CF, Copeland NG, Jenkins NA (2001) Mutations in Mlph, encoding a member of the Rab effector family, cause the melanosome transport defects observed in leaden mice. Proc Natl Acad Sci USA 98:10238–10243

Means GD, Boyd Y, Willis CR, Derry JM (2001) Transgenic rescue of the tattered phenotype by using a BAC encoding Ebp. Mamm Genome 12:323–325

Melton DW (1994) Gene targeting in the mouse. Bioessays 16:633–638

Meyers EN, Lewandoski M, Martin GR (1998) An Fgf8 mutant allelic series generated by Cre- and Flp-mediated recombination. Nat Genet 18:136–141

Mukherjee A, Soyal SM, Wheeler DA, Fernandez-Valdivia R, Nguyen J, DeMayo FJ, Lydon JP (2006) Targeting iCre expression to murine progesterone receptor cell-lineages using bacterial artificial chromosome transgenesis. Genesis 44:601–610

Murphy KC (1991) Lambda Gam protein inhibits the helicase and chi-stimulated recombination activities of Escherichia coli RecBCD enzyme. J Bacteriol 173:5808–5821

Muyrers JP, Zhang Y, Benes V, Testa G, Ansorge W, Stewart AF (2000) Point mutation of bacterial artificial chromosomes by ET recombination. EMBO Rep 1:239–243

Muyrers JP, Zhang Y, Testa G, Stewart AF (1999) Rapid modification of bacterial artificial chromosomes by ET-recombination. Nucleic Acids Res 27:1555–1557

Nagy A, Moens C, Ivanyi E, Pawling J, Gertsenstein M, Hadjantonakis AK, Pirity M, Rossant J (1998) Dissecting the role of N-myc in development using a single targeting vector to generate a series of alleles. Curr Biol 8:661–664

Poser I, Sarov M, Hutchins JR, Heriche JK, Toyoda Y, Pozniakovsky A, Weigl D, Nitzsche A, Hegemann B, Bird AW, Pelletier L, Kittler R, Hua S, Naumann R, Augsburg M, Sykora MM, Hofemeister H, Zhang Y, Nasmyth K, White KP, Dietzel S, Mechtler K, Durbin R, Stewart AF, Peters JM, Buchholz F, Hyman AA (2008) BAC TransgeneOmics: a high-throughput method for exploration of protein function in mammals. Nat Methods 5:409–415

Probst FJ, Fridell RA, Raphael Y, Saunders TL, Wang A, Liang Y, Morell RJ, Touchman JW, Lyons RH, Noben-Trauth K, Friedman TB, Camper SA (1998) Correction of deafness in shaker-2 mice by an unconventional myosin in a BAC transgene. Science 280:1444–1447

Scholz H, Bossone SA, Cohen HT, Akella U, Strauss WM, Sukhatme VP (1997) A far upstream cis-element is required for Wilms' tumor-1 (WT1) gene expression in renal cell culture. J Biol Chem 272:32836–32846

Senatorov V, Malyukova I, Fariss R, Wawrousek EF, Swaminathan S, Sharan SK, Tomarev S (2006) Expression of mutated mouse myocilin induces open-angle glaucoma in transgenic mice. J Neurosci 26:11903–11914

Sharan SK, Thomason LC, Kuznetsov SG, Court DL (2009) Recombineering: a homologous recombination based method of genetic engineering. Nat Protoc 4(2):206–223

Shizuya H, Birren B, Kim UJ, Mancino V, Slepak T, Tachiiri Y, Simon M (1992) Cloning and stable maintenance of 300-kilobase-pair fragments of human DNA in Escherichia coli using an F-factor-based vector. Proc Natl Acad Sci USA 89:8794–8797

Suh H, Gage PJ, Drouin J, Camper SA (2002) Pitx2 is required at multiple stages of pituitary organogenesis: pituitary primordium formation and cell specification. Development 129:329–337

Swaminathan S, Ellis HM, Waters LS, Yu D, Lee EC, Court DL, Sharan SK (2001) Rapid engineering of bacterial artificial chromosomes using oligonucleotides. Genesis 29:14–21

Testa G, Zhang Y, Vintersten K, Benes V, Pijnappel WW, Chambers I, Smith AJ, Smith AG, Stewart AF (2003) Engineering the mouse genome with bacterial artificial chromosomes to create multipurpose alleles. Nat Biotechnol 21:443–447

Valarche I, de Graaff W, Deschamps J (1997) A 3′ remote control region is a candidate to modulate Hoxb-8 expression boundaries. Int J Dev Biol 41:705–714

Valenzuela DM, Murphy AJ, Frendewey D, Gale NW, Economides AN, Auerbach W, Poueymirou WT, Adams NC, Rojas J, Yasenchak J, Chernomorsky R, Boucher M, Elsasser AL, Esau L, Zheng J, Griffiths JA, Wang X, Su H, Xue Y, Dominguez MG, Noguera I, Torres R, Macdonald LE, Stewart AF, DeChiara TM, Yancopoulos GD (2003) High-throughput engineering of the mouse genome coupled with high-resolution expression analysis. Nat Biotechnol 21:652–659

van der Weyden L, Adams DJ, Bradley A (2002) Tools for targeted manipulation of the mouse genome. Physiol Genomics 11:133–164

Vivian JL, Chen Y, Yee D, Schneider E, Magnuson T (2002) An allelic series of mutations in Smad2 and Smad4 identified in a genotype-based screen of N-ethyl-N-nitrosourea-mutagenized mouse embryonic stem cells. Proc Natl Acad Sci USA 99:15542–15547

Wackernagel W (1973) Genetic transformation in E. coli: the inhibitory role of the recBC DNase. Biochem Biophys Res Commun 51:306–311

Wakabayashi Y, Kikkawa Y, Matsumoto Y, Shinbo T, Kosugi S, Chou D, Furuya M, Jishage K, Noda T, Yonekawa H, Kominami R (1997) Genetic and physical delineation of the region of the mouse deafness mutation shaker-2. Biochem Biophys Res Commun 234:107–110

Warming S, Costantino N, Court DL, Jenkins NA, Copeland NG (2005) Simple and highly efficient BAC recombineering using galK selection. Nucleic Acids Res 33:e36

Wilson SM, Yip R, Swing DA, O'Sullivan TN, Zhang Y, Novak EK, Swank RT, Russell LB, Copeland NG, Jenkins NA (2000) A mutation in Rab27a causes the vesicle transport defects observed in ashen mice. Proc Natl Acad Sci USA 97:7933–7938

Wong QN, Ng VC, Lin MC, Kung HF, Chan D, Huang JD (2005) Efficient and seamless DNA recombineering using a thymidylate synthase A selection system in Escherichia coli. Nucleic Acids Res 33:e59

Wu S, Ying G, Wu Q, Capecchi MR (2008) A protocol for constructing gene targeting vectors: generating knockout mice for the cadherin family and beyond. Nat Protoc 3:1056–1076

Yamamoto T, Moerschell RP, Wakem LP, Ferguson D, Sherman F (1992) Parameters affecting the frequencies of transformation and co-transformation with synthetic oligonucleotides in yeast. Yeast 8:935–948

Yang XW, Model P, Heintz N (1997) Homologous recombination based modification in Escherichia coli and germline transmission in transgenic mice of a bacterial artificial chromosome. Nat Biotechnol 15:859–865

Yang Y, Sharan SK (2003) A simple two-step, 'hit and fix' method to generate subtle mutations in BACs using short denatured PCR fragments. Nucleic Acids Res 31:e80

Yang Y, Swaminathan S, Martin BK, Sharan SK (2003) Aberrant splicing induced by missense mutations in BRCA1: clues from a humanized mouse model. Hum Mol Genet 12:2121–2131

Yu D, Ellis HM, Lee EC, Jenkins NA, Copeland NG, Court DL (2000) An efficient recombination system for chromosome engineering in Escherichia coli. Proc Natl Acad Sci USA 97:5978–5983

Zhang Y, Buchholz F, Muyrers JP, Stewart AF (1998) A new logic for DNA engineering using recombination in Escherichia coli. Nat Genet 20:123–128

Chapter 4
Insertional Mutagenesis for Generating Mouse Models of Cancer

David A. Largaespada

4.1 Introduction

Despite over 25 years of research on the topic, a PubMed search revealed only 168 manuscripts, 24 of which are reviews, on the topic of "mouse," "cancer," and "insertional mutagenesis." This should not be taken as an indication that insertional mutagenesis has not been instrumental in cancer research. In fact, thousands of other manuscripts on the genes that were discovered as a result of insertional mutagenesis-based forward genetic screens have been published. Many of the most important human cancer genes under study today were identified and implicated in cancer because they were frequent sites of proviral insertion mutation in a mouse model. This includes the mouse homologue of the *TRP53* tumor suppressor gene, which was identified at a recurrent site of proviral insertion mutation in Friend Murine Leukemia Virus (F-MuLV)-induced erythroleukemias in mice (Johnson and Benchimol 1992). The chicken and mouse homologues of the *CMYC* oncogene were found to be frequently altered by proviral insertion in retrovirally induced cancer cells in those species also (Hayward et al. 1981; Payne et al. 1982; Varmus 1983). These examples indicate the important role that somatic insertional mutagenesis has played in the history of cancer research.

If enough cells, or more properly cell clones, suffer enough random mutations cancer will ensue. The fundamental approach to modeling cancer in mice via insertional mutagenesis is to expose enough cells to enough insertion mutations for cancer to ensue. The same concept applies to the use of other mutagens also. The source of these random mutations is important for several reasons. First of all, we can identify sources of dangerous exposures that could cause cancer in people. Secondly, we

---

D.A. Largaespada (✉)
The Department of Genetics, Cell Biology and Development, The Center for Genome Engineering, Masonic Cancer Center, The University of Minnesota,
Twin Cities; 6-160 Jackson Hall; 321 Church St. S.E., Minneapolis, MN 55455, USA
e-mail: larga002@umn.edu

J.E. Green and T. Ried (eds.), *Genetically Engineered Mice for Cancer Research: Design, Analysis, Pathways, Validation and Pre-Clinical Testing*,
DOI 10.1007/978-0-387-69805-2_4,

could use the mutations associated with a given type of DNA damage to identify the genes that actually cause cancer. There are a variety of ways that DNA can be damaged in a cell. This includes endogenous processes like cytosine deamination as well as exposure to exogenous agents, such as mutagenic chemicals and radiation. Using chemical treatment and radiation, it has been possible to model cancer in the mouse and much has been learned about the metabolism of cancer causing agents, the mechanisms by which DNA damage occurs or by which other cellular effects are induced, and the threshold for certain biological outcomes. However, these approaches alone have not yielded a great deal of information about what types or how many cancer genes exist or what their identities are. Instead, this type of information came mainly from the study of virally induced cancer in laboratory animals (for reviews see Gross 1978; van Lohuizen and Berns 1990; Kung et al. 1991; Jonkers and Berns 1996; Mikkers and Berns 2003; Uren et al. 2005). Generally, these models were "found," or noticed serendipitously, and then studied to the great benefit of cancer genetics.

More recently, investigators including those in my own group, have taken concepts learned from the study of the slow transforming retroviruses and applied these to produce very deliberately "created" models of cancer caused by insertional mutagenesis. These models are based on the use of DNA transposable elements introduced as transgenes in mice (Collier et al. 2005; Dupuy et al. 2005). The elements of the system are introduced simply by breeding mice together. The result is a much more flexible system that avoids the barriers that limit the use of retroviral vectors for tumor induction to just a few organs and tissues. There are now several DNA and RNA transposable element systems that have been shown to be active in mouse cells *in vivo* (for reviews see Largaespada 2003; Miskey et al. 2005). The most well-studied transposable element system for use as a genetic tool in the mouse is the *Sleeping Beauty* (SB) transposon system. Many reviews have now been written on its use as a genetic tool in the mouse (Ivics et al. 2004; Izsvak and Ivics 2004). The SB is a member of the Tc1/Mariner family of DNA transposable elements that transpose in a "cut-and-paste" manner and integrate at "TA" dinucleotides. SB was originally identified as a long-dormant transposable element and its relevant transposase in the genomes of salmonid fish. Directed mutagenesis was used to correct mutations that silenced the activity of the transposase (Ivics et al. 1997). Just as with retroviruses, the DNA transposon provides a molecular tag that allows the identification of cancer genes at recurrently mutated sites in cancer cells. However, the SB system, and in the future other transposon systems, provides a much more flexible system for inducing somatic mutagenesis in transgenic mice than is provided by viruses.

Now, the human and mouse genomes, as well as many others, have been sequenced in their entirety. Most protein-encoding genes have likely been identified as well as many genes that encode functional RNAs, such as the microRNAs. Coupled with this knowledge are new technologies for massively parallel DNA sequencing, gene copy number and mRNA expression arrays, and high-throughput proteomics approaches for the analysis of human and mouse tumors. Given this state of affairs one might ask whether it is worth the time and expense to generate and

Table 4.1 Reasons for using insertional mutagenesis to generate mouse models of cancer

1.	An insertional mutagen can provide mutations that cooperate with incomplete carcinogens to cause cancer.
2.	Strongly cooperating pairs of cancer genes can be discovered using insertional mutagenesis.
3.	A cancer gene initially identified in an insertional mutagenesis screen may subsequently be found to be a human cancer gene also.
4.	Insertional mutagenesis in mice offers the potential to perform screens for very specific cancer relevant phenotypes, such as metastasis and treatment resistance.
5.	Some human tumors are likely to be caused in part by insertional mutagenesis.

study mouse models of cancer that are induced by insertional mutagenesis. There are several reasons why it is still worthwhile creating and studying models like these for cancer research. These are described below and summarized in Table 4.1.

An insertional mutagen can provide mutations that cooperate with incomplete carcinogens to cause cancer. It is a common occurrence for transgenic or knockout mice that overexpress a single oncogene or lack a single tumor suppressor gene to not develop rapid, highly penetrant cancer. This is presumably because single mutations must cooperate with other mutations for full-blown cancer to occur. One way around this is to combine more than one transgene or knockout in the same mouse and test one pair of mutations at a time. By using an insertional mutagen in combination with an incomplete mutagen one gets the chance to find many of the potential cooperating genes all in one experiment – rather than by testing them one by one. This approach has been used by many groups in the past for MuLV, MMTV, and transposable elements (Shackleford et al. 1993; van der Lugt et al. 1995; Jonkers et al. 1997; Collier et al. 2005).

In some cases, very strongly cooperating mutations have been discovered using insertional mutagenesis, revealing an unexpected association between two gene products. For example, BXH-2 strain acute myeloid leukemias (AML) with activating *Hoxa9* gene MuLV proviral insertion mutations nearly always also have proviral insertions activating *Meis1* (Nakamura et al. 1996). Later research showed that these two genes indeed cooperate in AML development and that Meis1 serves as a nuclear chaperone for Hoxa9 and binds to DNA sequences in a complex with Hoxa9 (Kroon et al. 1998; Mercader et al. 1999; Shen et al. 1999; Thorsteinsdottir et al. 2001). Thus, the genetic interaction led to the discovery of an underlying biochemical association. *HOXA9* and *MEIS1* are now recognized as important co-transforming genes downstream of human AML fusion oncoproteins involving the *MLL* gene (Lawrence et al. 1999). Such pairs of cooperating oncogenes may reveal "weak points" for attack of a given cancer causing protein. Many more have been recently described using a systematic analysis of MuLV integration sites (de Ridder et al. 2006).

It has often been the case, that a cancer gene initially identified in an insertional mutagenesis screen is subsequently revealed to be a human cancer gene also. Large-scale resequencing and other approaches on human tumor DNA have revealed many, many "passenger" mutations not thought to actually cause tumors but reflect the background error rate, exposure to carcinogenic agents, and loss of replicative fidelity associated with cancer development (Sjoblom et al. 2006; Wood et al. 2007).

The actual "driver" mutations that cause tumor development are present among many, many "passenger" mutations and the two can be difficult to distinguish. It should be informative to compare lists of genes mutated in human tumors with those also altered by insertion mutation in forward, unbiased screens to identify the most likely "driver" mutations.

As research using insertional mutagenesis progresses, it is hoped that insight into very specific cancer relevant phenotypes can be gained. Most mouse models of cancer have been utilized to study the genetics of tumor initiation, i.e., what can cause a tumor to appear? Less well studied are the factors that cause tumor progression to the stage of widely dispersed metastatic disease. Mouse models are even less utilized for studies of treatment responsiveness. Mouse models of cancer induced by insertional mutagenesis continue to evolve as the tumor develops and indeed as it responds to therapy. The evolution of the tumor is theoretically tractable by studying the insertion mutations.

A final reason for studying mouse models of cancer induced via insertional mutagenesis is that some human tumors are likely to be caused in part by the same mechanism. It has been proposed that insertion of the hepatitis B virus genome in human hepatocellular carcinoma is nonrandom and that specific cellular genes are affected by insertion so as to contribute to human cancer (Murakami et al. 2005; Tamori et al. 2005a, b; Pang et al. 2006). Certain retroviruses, the human immunodeficiency virus (HIV) and human T lymphotropic leukemia virus I (HTLV I) specifically, have been shown to affect the expression of candidate cancer genes at proviral insertion sites in lymphoid malignancies (Nakamura et al. 1994; Kubota et al. 1996). The human genome contains active retrotransposable elements (Ostertag and Kazazian 2001; Brouha et al. 2003; de Parseval and Heidmann 2005; Dewannieux and Heidmann 2005). Although not yet clearly implicated, these elements may induce cancer causing insertion mutations in rare somatic cells. Finally, human gene therapies that use integrating vectors, such as retroviruses and lentiviruses, can induce cancer as a side effect of insertional mutagenesis (Hacein-Bey-Abina et al. 2003; Baum et al. 2004; Dave et al. 2004). Thus, mouse models of cancer induced by insertional mutagens may inform human cancer development in many ways.

4.2 Viral Models of Cancer Induced by Insertional Mutagenesis

In general terms, there are two types of cancer causing viruses, the DNA and the RNA tumor viruses. The DNA tumor viruses encode proteins of viral origin that bind to and affect the activity of endogenous cellular proteins that regulate cell growth, thus predisposing infected cells to tumorigenic conversion. The RNA viruses that cause cancer in mice (as well as in chickens, turkeys, cats, and some other species) are retroviruses that fall into two general types: the "slow transforming" retroviruses and the "acute transforming" retroviruses (Maeda et al. 2008). Both types of viruses were discovered by isolating cell-free, tumor lysate filtrates that turned out to contain the viral agent that could cause cancer in a newly inoculated host animal (for a history of this fascinating subject, see Wold and Green 1979).

Acute transforming retroviruses. The acute transforming retroviruses are generally replication defective and rely on a replication competent "helper virus" to spread from cell to cell. These acute transforming retroviruses were found to have viral oncogenes, called v-oncs, that were actually derived from endogenous cellular oncogenes, called c-oncs, during the infection of some rare cell by a replication competent virus. The v-onc, often truncated and otherwise mutated from the c-onc version, is expressed from the retroviral long terminal repeat (LTR) and causes rapid cellular transformation *in vitro* and quickly induces tumors *in vivo* in susceptible hosts. The discovery and study of these v-oncs was critical to early cancer research, for instance providing evidence for the gene-based model of cancer development, for the discovery of tyrosine kinases and their enrichment among cancer genes, and many other things.

Slow transforming retroviruses. In contrast, the slow transforming retroviruses were found not to cause transformation of cells in culture and did not rapidly induce cancer in susceptible hosts. Instead, these slow transforming retroviruses were found to take months or even longer than a year to induce cancer after a prolonged period of chronic infection of the host animal. As it turned out, these retroviruses induce cancer by acting as insertional mutagens (Uren et al. 2005). That is, they mutate chromosomes and genes via the insertion of a foreign piece of DNA. In this case, that foreign DNA is the provirus of the retrovirus. During a normal life cycle of the retrovirus, the RNA genome of the viral particle, once inside an infected cell, is reverse transcribed into a double-stranded DNA copy that then traffics to the nucleus and is integrated into the host cell genome via a virally encoded protein called the integrase. This integrated provirus remains in the chromosome and serves as the substrate for transcription of new copies of the viral RNA genome, as well as spliced mRNAs that produce viral proteins needed to assemble an infectious viral particle. The reverse transcription process generates two identical repeated sequences at the ends of the provirus called the LTRs. These LTRs contain the enhancer/promoter sequences required to initiate these RNA transcripts using the endogenous RNA polymerase II enzyme and other host transcription factors. The LTRs also contain sequences for directing the cell's polyadenylation machinery to the 3′ end of the viral transcripts. The provirus also contains splice donor and splice acceptor sequences used to produce the spliced messages required to produce certain of the viral proteins. Finally, the integrated provirus encodes the proteins used to assemble a viral particle itself. All of the sequence elements can affect an endogenous gene when a provirus is integrated near or within that gene. This topic has been reviewed before (Berns 1988; Mikkers and Berns 2003; Uren et al. 2005), but the main mechanisms by which an integrated provirus can affect an endogenous gene have to do with the LTRs.

The enhancer sequences with the LTR are potent and can affect the frequency of RNA transcript initiation at nearby endogenous promoters and so activate a proto-oncogene by "enhancement" (Nakamura et al. 1996). The LTR of the integrated provirus may initiate a transcript that is fused with sequences from an endogenous gene and thus a proto-oncogene could be activated by "promoter substitution" as occurs with the *Evi1* and other oncogenes (Mukhopadhyaya and Wolff 1992; Hirai et al. 2001). This process may involve splicing from the viral splice donor to endogenous splice acceptors. In this way, an N-terminally truncated protein can be produced. In some cases, cryptic splicing from within retroviral gene leads to the production of a fusion protein consisting of viral protein sequences fused to cellular proto-oncogene protein sequences (Mukhopadhyaya and Wolff 1992). The provirus can also generate a truncated fusion transcript by integrating near the 3′ end of the cellular proto-oncogene message resulting in the removal of 3′ untranslated region (UTR) sequences that destabilize the normal mRNA so that the fusion mRNA is longer lived and protein levels are higher than normal (Selten et al. 1985; Du et al. 2005b). Similarly, it may be possible for an integrated provirus to result in the production of a C-terminally truncated protein (Beinke et al. 2003). If this involved gene is a tumor suppressor gene, then a variety of insertion mutations might destroy the normal function of the gene from insertions in the promoter, an exon, or an intron (Largaespada et al. 1995). The polyadenylation signal in the LTR can result in premature transcript termination (van Lohuizen et al. 1989). If the gene is a classical tumor suppressor gene, then both copies of the gene must be inactivated for cancer to occur. In the tumor, this may result from independent insertion mutations in each allele (Largaespada et al. 1995) or from an insertion mutation in one allele and another form of a mutation in the other allele. Also likely is the existence of insertion mutations into genes that are haploinsufficient for tumor suppression (Cook and McCaw 2000), in which case insertional inactivation of just one allele could be sufficient to promote tumor development.

The most significant advantage of these retrovirally induced cancer models is that the cancer genes that are mutated during tumorigenesis are "molecularly tagged" by the integrated virus. Thus, it is possible to identify these genes simply by cloning and sequencing the proviral integration from tumor genomic DNA (Uren et al. 2005). On a global scale, there are many, many potential integration positions for the provirus in an infected cell. However, only some of the integration positions are near or within the right gene and in the right place and orientation to affect that gene so as to cause cancer. We would expect, therefore, that proviral integrations within normal, non-transformed infected cells would not be enriched for specific sites of the genome, whereas the proviral insertions from tumor genomic DNA induced by a slow transforming retrovirus would show a bias toward specific genes. These biased integration patterns are recognized by the recovery of proviral integration events from multiple independent sites within a single region at a rate higher than what would be expected by chance. These regions of the genome are called common insertion sites (CIS) and prior to completion of the genome were referred to using an acronym that stemmed from the type of experiment that had been done to discover them. For instance, *Fli-1*, meant Friend Leukemia Integration Site 1, because it was a CIS identified in leukemias induced by the Friend Murine Leukemia Virus (Ben-David et al. 1990, 1991). In some instances, the CIS name was then used to designate the gene at that location, as was the case with *Fli-1*. Similarly, the *Meis1* transcription factor gene has been named after the CIS at this locus called Murine Ecotropic Integration Site 1 in AML induced in BXH-2 strain mice by an endogenous ecotropic MuLV

(Moskow et al. 1995). In other cases, the CIS designation does not correspond to the gene affected by proviral insertions at the CIS. Indeed, since it is often unclear what gene is affected by proviral insertion mutations within a given region it is probably best to designate CIS identifiers and not to name genes after CIS designators. The proviral insertion mutations at a given CIS seem to be able to affect just one, or multiple genes at a given locus (Hansen and Justice 1999), or could even affect different genes in different cell types (Sauvageau et al. 2008). Sometimes, proviral insertions seem to affect genes some distance away (60 kb or more) from the integration site, even if other genes exist between the integration site and the affected gene (Lazo et al. 1990; Bartholomew and Ihle 1991; Hansen and Justice 1999). Desperately needed in this work is a clear evaluation of the typical affect of proviral insertion events on genes in a region using large-scale mRNA expression analyses available via microarray hybridization.

Murine leukemia viruses and mouse mammary tumor virsuses. In the mouse, there are two types of retroviruses that have been characterized extensively for their ability to induce cancer. The mouse mammary tumor viruses (MMTVs) are in the betaretroviridea genus and induce mammary tumors generally after transmission from viremic females to female offspring (reviewed in Callahan and Smith 2000). The MuLV are of the gammaretroviridea genus and have three genes that encode proteins required to form an infectious viral particle (reviewed in Gardner 1978). These are the *gag*, *pol*, and *env* genes. The MuLV are typically ~9 kb in length and produce both a full length viral RNA, two copies of which is packaged into the viral particle, and a spliced mRNA encoding the *env* protein. As the name implies the MuLV cause mainly hematopoietic cancer, i.e., leukemia. Depending on the specific virus used and the strain background of the mice infected, various types of leukemia can occur. Indeed, in many cases, several types of leukemia can occur after infection with the same virus in the same strain of mice. The same virus can cause one type of leukemia in one strain of mice and a different type in another strain of mice. The type of disease induced is in part controlled by the activity of the MuLV LTR, which may be more active in some hematopoietic cell types than in others (Bosze et al. 1986; Ishimoto et al. 1987; Nishigaki et al. 2002). The use of the MuLV for cancer modeling is complicated by the fact that in some cases the virus used is normally passed from mother to offspring via the mother's milk or during pregnancy (Bedigian et al. 1993), and in other cases the virus must be introduced via injection into neonatal mice by the experimentalist. Some strains of mice carry infectious MuLV proviral copies in their genomes and the provirus is passed vertically via the germ line (Herr and Gilbert 1983; Mucenski et al. 1986, 1988; Lee and Eicher 1990). Also, mice carry endogenous genes that can restrict viral infection and/or replication of certain viruses. The most prominent among these is the *Fv1* gene, which exists in two forms in laboratory mice; the $Fv1^b$ allele restricts B-tropic MuLV and the $Fv1^n$ allele restricts N-tropic MuLV (Silver and Fredrickson 1983). Therefore, care must be taken in choosing the right virus and the right strain background so as to ensure that viremia can be established in infected mice and so that the desired form of leukemia is induced. Most of the major forms of human leukemia are available as MuLV-induced models in mice, including B- and T lymphoid leukemias (B-ALL, T-ALL) and lymphomas and acute myeloid leukemia (AML). Certain types of human leukemia do not seem to have a MuLV-induced counterpart, however, such as acute megakaryoblastic leukemia and NK cell leukemia. It should be noted, however, that the murine models of MuLV-induced leukemia have generally not been carefully examined to determine which human leukemia they most closely match at a phenotypic level. Instead, morphology and the presence of clonal immunoglobulin receptor (Ig) or T cell receptor (TcR) gene rearrangements have been used to characterize these leukemias (Gilbert et al. 1993). These approaches are hardly complete or definitive. For example, it is not unusual for murine myeloid leukemias to harbor clonal Ig or TcR gene rearrangements (Ru et al. 1993). The National Cancer Institute (NCI) sponsored Mouse Models of Human Cancer Consortium (MMHCC) has helped to motivate better phenotypic characterization of many mouse models of cancer, including definitive rules and approaches for classifying murine leukemia models (Kogan et al. 2002; Morse et al. 2002). However, an approach based on mRNA expression signatures, which has been informative in other mouse models of human cancer (Herschkowitz et al. 2007), is likely to be very useful for the classification of murine leukemia and identification of the closest human counterparts.

MMTV induces various types of mammary tumors in mice after establishing a chronic infection. MMTV has been reviewed extensively (Callahan and Smith 2000). MMTV can infect lymphocytes and thus be passed from mother to offspring via the mother's milk and milk lymphocytes. The MMTV has not only three structural genes like MuLV (i.e., *gag*, *pol*, and *env*), but also produces a regulatory protein called Rem (Indik et al. 2005; Mertz et al. 2005), and a superantigenic protein that activate T cells polyclonally (reviewed in Acha-Orbea and MacDonald 1995). An enhancer within the LTR is strongly estrogen responsive and is activated during puberty, thus resulting in the production of virus that infect and actively replicate in mammary epithelial tissues (Glover and Darbre 1989; Slater et al. 1989). Indeed, the MMTV LTR has been useful for driving transgene expression in mice (Sinn et al. 1987). Chronic infection results in clonal outgrowth of malignant cells having acquired the right suite of insertion mutations to result in tumor development. MMTV-induced mammary tumors generally do not metastasize and are thus less aggressive than many human mammary tumors. Among the genes identified by MMTV insertion mutation in mouse models, growth factor genes are very notable. These presumably act by creating an autocrine mitogenic signaling loop. Among the genes identified at CIS are *Wnt* family genes (such as *Wnt1*), *Notch* family receptors, and fibroblast growth factor (*Fgf*) genes, such as *Fgf3* and *Fgf4* (Shackleford et al. 1993; Callahan and Smith 2000, 2008). More recently, large-scale cloning and sequencing of MMTV insertions has revealed many new CIS and associated candidate genes – some of which are also altered at the sequence of expression level in human breast cancer (Theodorou et al. 2007). These studies have revealed the great potential of the MMTV for identifying human cancer genes and pathways.

The success of the MMTV and the MuLV for modeling human cancer and identifying cancer genes and pathways has stimulated a great deal of interest in creating similar models for other types of malignancy. The idea of insertional mutagenesis should work for many forms of cancer in mice: create enough insertion mutations in enough cells in the right genetic context and cancer should develop as a result of insertional activation and/or inactivation of genes. As mentioned above, the MuLV and MMTV were essentially "found" models of cancer. Although the receptor for the ecotropic Env protein of MuLV is expressed very widely in mice, the level of MuLV infection in most tissues may be very low. This is in part because the activity of the enhancer/promoter in the LTRs is tuned to one specific cell type or a few cell types. Also, these viruses cannot productively infect nondiving cells (Miller et al. 1990). With the exception of some epithelial cell types and cells of the hematopoietic system the vast majority of cells in an adult mammal are quiescent. Some tissues are dividing rapidly, but physical barriers could prevent their infection by MuLV. For example, the basement membrane on one side and a thick mucosal surface could protect cells lining the gastrointestinal tract from efficient infection by MuLV. Despite these obstacles, it should be possible to infect certain rapidly dividing tissues with high titer retroviral preparations to induce cancer in mice, particularly if they are rendered susceptibly to cancer in another way. One notable success in this endeavor has been to use an MuLV-based vector expressing the platelet-derived growth factor beta (*Pdgfb*) gene to induce glioma in mice. The vector can be injected intracranially into newborn C57BL/6 mice, whose brains still harbor rapidly dividing cell populations, the result of which is glioma development after 14–29 weeks (Uhrbom et al. 1998). The overexpression of the Pdgfb protein contributes to glioma growth, but the scientists involved also discovered that specific loci were recurrently mutated by proviral insertion (Johansson et al. 2004). That is, they discovered many CIS and associated genes. The retroviral vector thus both delivered a mitogenic growth factor and also served as an insertional mutagen to provide cooperating mutations. The identified CIS-associated genes are highly likely to contribute to glioma development and progression in cooperation with *Pdgfb*. Indeed, one CIS-associated gene, *Sox10*, has been shown to cooperate with *Pdgfb* in gliomagenesis in a mouse model (Ferletta et al. 2007). A similar approach is probably applicable to other forms of cancer in mice. One could imagine designing retroviral vectors expressing the right growth factor and infecting cells pre- or perinatally during a time of rapid cell growth the result of which would be tumor outgrowth. This concept could be expanded to include cells infected in culture that might then be implanted into host mice. Several reports describe the acquisition of features of transformation in cultured cell populations via infection with an MuLV or infection with an MuLV vector (Du et al. 2005a; Tanaka et al. 2008). This could become an ideal way to do forward genetic screens in primary immortalized human cells since they have been shown to be transformable with specific cancer genes and capable of tumor formation in immunodeficient mice (Elenbaas et al. 2001; Hahn et al. 2002; Lundberg et al. 2002).

4.3 Transposon-Based Models of Cancer Induced by Insertional Mutagenesis

Engineered transposable elements of various kinds can provide a way to circumvent some of the limitations of retroviruses for modeling cancer in mice via insertional mutagenesis (reviewed in Starr and Largaespada 2005; Weiser and Justice 2005; York 2005). The main advantage is that these transposable element systems are delivered via standard transgenesis in mice, using pronuclear injection or homologous recombination in mouse embryonic stem cells. The transposable elements do not have an extracellular stage of their replicative cycle. Since the elements of the system are present in every cell of the mouse, mutagenesis can be made very efficient. The barriers to retroviral infection, such as cell division and tissue barriers, are also not present. This means that mutagenesis can be theoretically achieved in any specific cell type. Since roughly 90% of cancer morbidity and mortality in the world is due to carcinomas, it is highly desirable to create insertional mutagenesis systems that can induce all the most common tumors of epithelial origin and transposons make this possible in mice. Finally, since the complications of retroviral RNA packaging and replication are not present, there is relatively more flexibility available for custom design of the internal components of the mutagenic DNA vector. Indeed, for the DNA transposons, the transposon vector is separated from the transposase, which mobilizes the transposase in trans. The transposon vector needs to be flanked by special inverted terminal repeats, but other than that, the sequences between them can be any sequences that are desired. Therefore, the insertional mutagen can be custom designed to bias toward specific subsets of genes. It might be possible to create vectors that would bias toward tumor suppressor gene inactivation, or just activation of proto-oncogenes, or work only in specific tissues and cell types.

At present, there are several transposable element systems that have been shown to function in transgenic mice. This includes the "cut-and-paste" or DNA transposons, such as *Sleeping Beauty* (SB) (Dupuy et al. 2001; Fischer et al. 2001; Horie et al. 2001), *PiggyBac* (Ding et al. 2005; Wu et al. 2007), *Minos* (Zagoraiou et al. 2001), *Tol2* (Balciunas et al. 2006) and others (reviewed in Largaespada 2003). Also available are the retrotransposons, such as LINE1-based vectors, from human (Ostertag et al. 2002) and mouse (An et al. 2006). Most of the published work has demonstrated that these systems can be used to promote transposition in germ line cells to create heritable mutations. In some cases, these vector systems have been shown to be active for gene delivery in embryos or adult somatic cells *in vivo*. It seems likely that several of these transposon systems will be useful for somatic mutagenesis and cancer gene discovery in mice. However, at present, the *Sleeping Beauty* transposon system is by far the best studied and the only system that has been used for gene transfer in the mouse germ line and soma as well as for insertional mutagenesis in the mouse germ line and soma.

Sleeping Beauty. Briefly, SB is a member of the Tc1/Mariner family of DNA transposable elements that transpose in a "cut-and-paste" manner and integrate at "TA" dinucleotides. SB was originally identified as a long-dormant transposable element and its relevant transposase in the genomes of salmonid fish. Directed mutagenesis

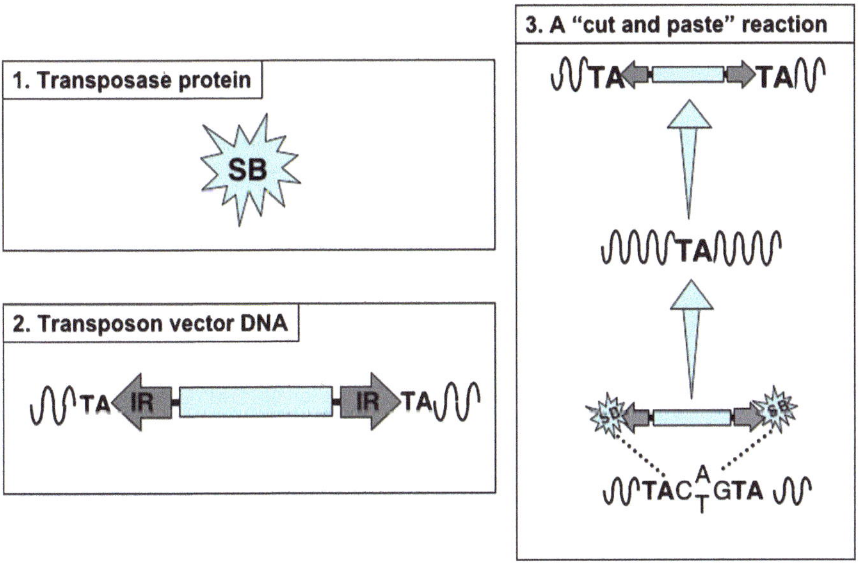

Fig. 4.1 Sleeping Beauty is a two-part transposon system. (1) The transposase protein, which is supplied in trans as mRNA or a DNA transgene. (2) The transposon vector in which sequences of interest are flanked by inverted terminal repeats (IR). (3) In cells that contain transposon vector and express the transposase, a "cut-and-paste" transposition reaction occurs in which transposon excision from the donor locus is followed by reinsertion at a new site at a "TA" dinucleotide

was used to correct mutations that silenced the activity of the transposase (Ivics et al. 1997). SB is a two-component system, composed of the transposase transgene and a transposon vector, flanked by special inverted terminal repeats or IRs (Fig. 4.1). When the transposon vector is present in a cell expressing the transposase transgene, the "cut-and-paste" transposition reaction occurs. The SB transposase enzyme recognizes specific binding sites within the IRs, excises and then integrates the transposon elsewhere at a TA dinucleotide. The excision site is repaired by the host cell machinery leaving behind a 3 bp footprint, either CAG or CTG (Ivics et al. 1997). The transposition reaction can occur from a transfected plasmid to another plasmid or to a chromosome in a transfected cell (Liu et al. 2005). Similarly, an SB transposon vector can jump from one place on a chromosome to another in transgenic mouse germ line and somatic cells (Dupuy et al. 2001; Collier et al. 2005; Dupuy et al. 2005). One interesting and important feature of SB transposition is the called local hopping (Carlson et al. 2003; Horie et al. 2003). This is the tendency of transposons to insert close to the donor locus on the same chromosome when they undergo a transposition reaction. This means that transposon mutagenesis using SB and other transposons that share this feature, is biased to the chromosome and region that carry the transposon vector. We also discovered that transposon mobilization can result in local deletions and other rearrangements near the donor locus in the germ line and soma (Geurts et al. 2006a). As described below, these facts mean that special consideration needs to be made when defining CIS and interpreting the results of SB mutagenesis experiments.

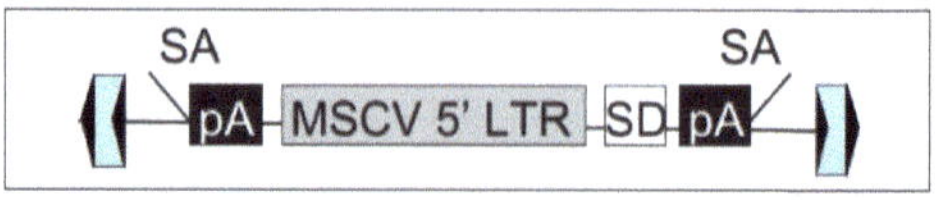

Fig. 4.2 The T2/onc transposon vector. This vector contains elements designed to elicit either transcriptional activation (MSCV 5′ LTR and splice donor) or inactivation (splice acceptors [SA] and polyadenylation signals [pA])

SB insertional mutagenesis-induced models of cancer in the mouse. This topic has been the subject of many reviews in recent years. Although originally shown to be capable of accelerating sarcoma in *p19Arf*$^{-/-}$ mice (Collier et al. 2005), or inducing leukemia in otherwise wild-type mice (Dupuy et al. 2005), our more recent unpublished results show that SB can induce a variety of types of cancer in mice, including carcinomas. We have shown that SB transposon vectors can be mobilized in the soma of transgenic mice allowing forward genetic screens for cancer genes involved in sarcoma and lymphoma/leukemia to be performed in living mice (Collier et al. 2005; Dupuy et al. 2005). We hypothesized that an SB transposon, designed to mimic the ability of a retroviral element to cause both gene loss- and gain-of-function mutations, could be used to "tag" cancer genes (Fig. 4.2). We created a transposon, T2/Onc, which contains splice acceptors (SA) followed by polyadenylation (pA) signals in both orientations. These elements are designed to intercept upstream splice donors and cause premature transcript truncation. Between the two SAs are sequences from the 5′LTR of the murine stem cell virus (MSCV), which contain strong promoter and enhancer elements that are methylation-resistant and active in stem cells (Hawley et al. 1994; Lu et al. 1996; Cherry et al. 2000). Immediately downstream of the LTR is a splice donor (SD) for splicing of a transcript initiated from the LTR into a neighboring gene. The T2/Onc transposon is thus specialized to identify both tumor suppressors and oncogenes. However, the T2/Onc vector could certainly be altered to include a different stringer promoter for instance. Also, variable is the site of T2/Onc transposon array integration and the number of copies integrated. In initial experiments, two different T2/Onc transgenes were used, both of which harbor about 25 copies, one of which is on mouse chromosome 1 and one of which is on mouse chromosome 15. When T2/Onc transgenic mice, carrying either the chromosome 1 or 15 array, were combined with a CAGGS-SB10 transposase transgenic line, in a *p19Arf*$^{-/-}$ background, sarcomagenesis was greatly accelerated compared to *p19Arf*$^{-/-}$ mice (Collier et al. 2005). Accelerated sarcomagenesis was due to T2/Onc insertional mutagenesis. Activation of the *Braf* gene by transposon insertions was common, but many other candidate cancer genes were recurrently mutated by T2/Onc insertion in this study. We later discovered that CAGGS-SB10 transgenic mice express primarily in mesenchymal cells of the mouse perhaps accounting for the ability of the SB system in this study to accelerate sarcomagenesis. A second study, done in collaboration with Dr. Neal Copeland and Dr. Nancy Jenkins, showed that T2/Onc2 transgenic mice crossed to Rosa26-SB11 transgenic mice resulted in leukemiagenesis on an otherwise wild-type background. *Notch1* and other known leukemia genes were identified at CIS in these experiments. In these experiments, a slightly altered form of T2/Onc was used with a longer splice acceptor, which was designated T2/Onc2. Three different T2/Onc2 lines are available integrated on three different chromosomes and which are very high copy lines with anywhere from 100 to 300 copies. Interestingly, although leukemia occurred in all doubly transgenic mice that were aged, 50–75% of doubly transgenic mice died during embryonic development (Dupuy et al. 2005). I hypothesize that the high levels of transposition in these mice resulted in frequent chromosomal deletions and other alterations, similar to those we observed in the germ line (Geurts et al. 2006), early enough in gestation to prevent normal viable development. There are ways to prevent the appearance of embryonic lethality in SB screens: reduce the copy number of T2/Onc or express SB11 tissue-specifically.

We have used the Rosa26-SB11 transgene with T2/Onc low copy lines to generate a large number of leukemias, sarcomas, and other types of tumors in wildtype, *p19Arf*$^{-/-}$ and *Blm*$^{-/-}$ backgrounds recently. The use of the low copy T2/Onc lines with Rosa26-SB11 prevented the embryonic lethality that was observed using the high copy T2/Onc2 transgenic lines. Besides altering the genetic background of the mice in these screens, we have begun to explore tissue-specific mutagenesis with SB to develop informative models of various forms of human solid tumors. Using a conditionally expressed SB11 transposase transgene, it is possible to restrict SB mutagenesis to just tissues expressing the Cre recombinase and avoid the highly penetrant lymphomagenesis that would otherwise occur. Our recent data prove that tissue-specific SB mutagenesis can generate cancer or preneoplasia in mice in a variety of tissues, including in the liver, gastrointestinal tract, brain (glioma and medulloblastoma), and the prostate. As with the other SB-induced or accelerated tumors, amplification and sequencing of the transposon vectors revealed CIS and CIS-associated genes many of which were known human cancer genes. The number of useful projects that can be conceived using SB is very large since most forms of human cancer have not yet been modeled using an insertional mutagenesis approach. Moreover, by altering the strain background, for example using predisposed mutant backgrounds, one might recover a different spectrum of mutant cooperating genes.

4.4 Important Considerations in Creating and Interpreting Results from Mouse Models of Cancer Based on Insertional Mutagenesis

Below, I describe a series of important issues that relate to generating, studying, and interpreting the results from a mouse model of cancer induced via insertional mutagenesis. These issues are important for both viral and transposon-based models. There are also nuances specific for one or the other type of model. It is clear that much work remains to be done to address some of these issues.

Size of the experiment. In all insertional mutagenesis experiments, the more tumors that are generated the more tumor-specific, somatically acquired insertion mutations there are that can be studied and thus the more potential cancer genes there are that can be identified. Obviously, if a CIS is involved in only a relatively small percentage of tumors, then it could only be discovered to be a recurrent insertion site (i.e., a CIS) if a relatively large number of independent tumors are studied. It is expensive to generate and age 100 or more mice for the period of time it takes for a tumor to develop. In many of these models, tumors do not appear until 6 or even 12 months of age. In practice, we therefore try to generate at least 60 experimental class animals, hoping for 60+ tumors. In some of the SB-induced tumor models, more than one tumor clone develops per mouse (unpublished data). Since each tumor clone can be studied independently, it is possible to get away with fewer mice, and still analyze as many independent tumors.

Methods for sequencing insertions. In the past, MuLV and MMTV proviral insertions were often cloned one by one from a given tumor. This was usually done by making a lambda library from each tumor and isolating genomic clones that contain a provirus (Buchberg et al. 1990). Restriction enzyme mapping was followed by limited sequence analysis and derivation of a Southern blot probe that could be used both to genetically map the proviral insertion and to detect proviral insertions in the same region in independent leukemia isolates. Techniques, such as "zoo blotting" and "exon trapping," were used to discover genes in CIS laborious defined this way (van Ooyen et al. 1985; Valk et al. 1997). The use of inverse PCR or linker-mediated PCR-based methods and the sequencing of the mouse genome entirely changed the approach for identification of individual proviral (and later transposon) insertions. A variety of linker-mediated and inverse PCR-based approaches for amplifying proviral and transposon cellular DNA junction fragments have been published (Li et al. 1999; Joosten et al. 2002; Yin and Largaespada 2007). A very clear linker-mediated PCR (LM-PCR)-based method for cloning SB transposon insertions has been published by my group (Largaespada and Collier 2008). The amplified junction fragments are generated by tumor genomic DNA restriction enzyme digestion, linker ligation, and then two rounds of PCR. The amplified junction fragments are often run on an agarose gel, with controls, to verify that the procedure worked. These PCR amplicons can be shotgun cloned in a plasmid vector and sequenced in groups of 96 clones. However, for some particular insertions some restriction enzyme junction fragments are too large to be amplified efficiently. Moreover, most linker-mediated PCR methods employ a second restriction enzyme digestion prior to PCR amplification to prevent amplification of transposons that have not been mobilized from the donor concatemer in the case of SB, or to prevent amplification of internal proviral sequences in the case of MuLV or MMTV. If this second restriction enzyme digestion happens to cleave the linker-ligated junction fragment for some proviral or transposon insertions, then it too would not be amplified. Thus, for the most possible insertions that are present to be recovered one requires more than one independent LM-PCR reaction per sample. Ideally, one should use multiple restriction enzymes and make attempts to clone insertions from the right hand and left hand side of transposon insertions and proviral insertions. The advent of high-throughput "pyrosequencing" or "single molecule" sequencing approaches now means that many, many individual transposon or proviral junction fragments can be sequenced for a reasonable cost (e.g., Vandenbussche et al. 2008). It is fairly straightforward to modify the secondary PCR primers used LM-PCR reactions to incorporate sequences required for the pyrosequencing reaction as well as short barcodes. These barcodes can be used to determine which insertion sites came from which tumor samples after sequencing the PCR products *en masse*.

What are the best criteria for definition of significant CIS? One major complexity in the analysis of CIS is the criteria used to designate the statistically significant amount of proviral clustering required that justifies designation as a CIS and indicate follow-up studies. The underlying assumption in studies such as these is that in the absence of selection the insertion events would be entirely randomly distributed within the genome. Therefore, any statistically significant deviation from a random profile of insertions can be designated a CIS. In the past, very arbitrary rules were developed for designation as a CIS. Some CIS were studied based only on two clustered insertions. Many such CIS are probably "background noise." This is because we now know that the assumption of random proviral insertion in the absence of selection is false. Most retro and lentiviruses that have been studied show a nonrandom genomic insertion pattern (Wu et al. 2003; Wu and Burgess 2004). MuLV shows a strong preference for insertions near the 5′ end of actively transcribed genes (Wu et al. 2003). Thus, the statistical justification for a CIS must take this preference into consideration. One group has developed a model that uses a kernel density estimate for MuLV's insertion preference near 5′ ends of genes to develop a statistical test that can be used to evaluate real proviral insertion site data (de Ridder et al. 2006). Another group cloned MuLV insertions in HeLa cells in culture and determined the number of CIS identified under these presumably nonselected conditions (Wu et al. 2006). The authors concluded that as many as three fourth of the CIS in the literature are statistically unsound and could represent clusters based purely on chance and not on selection.

A general method for defining statistically significant CIS in any given situation has not been agreed upon. However, a proposal can be outlined as follows. First, one should perform a modified Monte Carlo simulation, altered so as to accommodate the known biases of the insertional mutagen in question. For the total number of insertions obtained from a given experiment set, the expected value (E) at 1 for CIS consisting of clusters of 3, 4, 5, 6, or more insertions. You can then define window sizes in kilobase pairs for clusters of 3, 4, 5, 6, or more insertions that would yield $E=1$. This defines CIS window criteria for a given insertional mutagenesis experiment, where only 1 CIS would be expected to occur by chance for clusters of 3, 4, 5, 6, or more insertions. Defining significant clusters of 7, 8, or more insertions usually is not helpful because there is probably no biologically significant meaning to the very large CIS (100+ kb or very much greater than the average size of a gene) that might be found if this were done.

The second step is to obtain real control data for insertions under no selection. To be certain that such CIS are indeed likely to occur once or fewer times, obtain a large number of unselected insertion mutations using the mutagen of choice for this experiment. The number of control insertions obtained in this way should ideally be similar in size to the number obtained from the tumors. This control set of insertions provides a way to test the simulation to determine if it reflects the real number of background CIS that could be expected to occur by chance. This has not typically been done in MuLV or MMTV mutagenesis experiments, but given the relative ease of amplifying and sequencing proviral insertions it should be a priority now.

One implication of the nonrandom distribution of unselected MuLV insertions is that MuLV-based genetic screens do not equally mutagenize the entire genome. Instead, there are many "hot spots" and likely there are very many "cold spots" not easily subjected to insertion mutations. This emphasizes the view that the use of multiple insertional mutagens is the most ideal way to identify the most possible genes that could be involved in a given type of cancer. Transposons certainly have their own bias for insertion preference. SB insertions have to occur at TA dinucleotides and the sequences immediately adjacent to the TA influence the likelihood of insertion at a given TA (Geurts et al. 2006b). Despite this, SB insertion is more or less random on a genome level scale showing little preference to or away from genes (Yant et al. 2005). The most profound source of bias in an SB screen is the "local hopping" phenomenon mentioned above. Because SB transposons land within 10–20 Mbp of the donor locus about 50–80% of the time when mobilized from a chromosome, mutagenesis is biased toward this local region (Carlson et al. 2003; Horie et al. 2003). In fact, the entire chromosome that harbors the donor locus is over-represented by insertions after SB transposition. As mentioned above, local hopping also creates a situation, whereby local deletions and inversions occur, often involving hundreds of kb of sequence adjacent to the donor concatemer (Geurts et al. 2006a). There are several implications of these facts for mutagenesis screens using SB. The first is that CIS is identified near the donor locus simply due to local hopping. Also, if the T2/Onc donor concatemer is close to certain specific cancer genes, then they may be usually mutated by local hopping and influence the genetic pathways, or even the tumor types, that are obtained using that particular T2/Onc transgenic line. It is possible that tumor suppressor genes near the donor T2/Onc locus would be deleted by local deletions, such as those observed in germ line mutagenesis projects (Geurts et al. 2006a). This is a caveat to the use of transposons for cancer gene identification. Part of the cancer causing effects of transposition could be due to genome rearrangements catalyzed by transposition reactions. Nevertheless, our preliminary data suggests that SB-induced tumors do not show global gene copy number changes (unpublished observations).

In order to eliminate CIS due to local hopping, we have taken the approach of using at least two different donor concatemer T2/Onc or T2/Onc2 transgenic lines for each mutagenesis experiment we do. Once the transposon insertion sites have been obtained from all the resultant tumors, we eliminate all same chromosome insertions from the total list of insertions. Only then are CIS sought using the approach described above. This approach allows CIS on donor chromosomes to be identified only if they are recurrently mutated by transposition from another chromosome. This is a conservative approach and probably results in the elimination

Table 4.2 Useful online resources for insertional mutagenesis projects

Resource	URL	Notes
Retrovirus-tagged cancer gene database	http://rtcgd.abcc.ncifcrf.gov/	This is a database of published SB transposon and retroviral insertions and associated CIS, designated by the contributors. It is not comprehensive
Catalogue of somatic mutations in cancer	http://www.sanger.ac.uk/genetics/CGP/cosmic/	This is a continually updated list of known cancer genes and all published somatic mutations described in cancer cells. It is searchable by gene or tumor type
iMAPPER	http://www.sanger.ac.uk/cgi-bin/teams/team113/imapper.cgi	This is an online tool for large-scale analysis of insertional mutagenesis tag sequences against vertebrate genomes

of some true CIS. It is possible to use kernel density estimates to find significant "peaks" of T2/Onc insertions even on the same chromosome (Collier et al. 2005a, b). The generation of an unselected set of transposon insertions can provide an important control for comparison to same chromosome insertions from SB-induced tumors. Obviously, any CIS identified in both sets of data would be highly suspect for an involvement in cancer development.

Informatics analysis of insertion sites. Once CIS have been identified one must define the correct CIS-associated gene. It is crucial to decide which gene or genes are the targets for the selective pressure that led the observation of recurrent insertion mutation at each CIS. Various groups have now developed automated programs for mapping and annotation of the genomic regions near large numbers of insertion mutations (Roberg-Perez et al. 2003). In general, the gene whose 5′ end is closest to the midpoint of the insertions that make up a given CIS is considered the likeliest cancer gene. However, there are many exceptions to this rule (Himmel et al. 2002), and in some cases it is thought that multiple genes near one CIS are making a positive contribution to tumor development (Hansen and Justice 1999). In general, a list of CIS is generated in which one gene per CIS is picked out based on proximity alone. One can imagine generating an expanded list of genes based on proximity that includes more than one gene per CIS. Such a list of genes near CIS associated with AML induced by MuLV indeed shows enrichment for genes dysregulated or differentially regulated in human AML (Erkeland et al. 2006). At present, no convincing standards can be put forth for the correct definition of the gene or genes altered at each of a long list of CIS. Global gene expression arrays, perhaps also exon arrays, may make it possible to define the usual changes associated with proviral or transposon insertions clustered at CIS.

Dr. Keiko Agaki, at the National Cancer Institute, has developed and maintains a database of transposon and retroviral insertions obtained from forward genetic screens for cancer in mice (Akagi et al. 2004). The database includes CIS defined by the investigators who deposited these sequences and can be accessed at: http://rtcgd.abcc.ncifcrf.gov/. The database is searchable using keywords, such as gene names or tumor types. Other useful databases for the analysis of insertion site data can be found in Table 4.2.

The CIS list is the result of a great amount of time, money, and resources dedicated to one project. It is the distillate of a large number of sequenced insertion sites obtained after carefully observing many mice and collecting and analyzing many tumors. Having gone to great efforts to obtain this CIS list, it is illogical to simply scan it and do a few PubMed searches on the gene names that ring a bell and then proceed to study the few that seem like they would make a good story. Thus, the question: what are the most appropriate methods for validation and study of a list of CIS-associated genes?

What are appropriate methods for the validation and the study of genes at CIS loci? After one has completed a forward genetic screen based on insertional mutagenesis, defined CIS and CIS-associated genes a difficult process begins of validating and studying the genes and genetic pathways affected. The appropriate follow-up studies for individual CIS consist of many things, but some specific early studies are certainly warranted. One might also consider the overall approach that should be used. On the one hand, individual genes at CIS can be studied one by one. On the other hand, one could argue the primary strength of the forward genetic screens is to discover pathways important for cancer initiation or progression rather than individual genes. There are various software tools that can be used to determine if genes near CIS from a forward genetic screen are enriched for those that participate in a given biological process or established pathway. For example, Ingenuity Pathway Analysis™ has been used to analyze genes at CIS (Erkeland et al. 2006; Touw and Erkeland 2007). Gene set enrichment analysis (GSEA) could be used to show that genes at CIS from insertional mutagenesis screens are significantly enriched for those altered in various types of human cancer (Subramanian et al. 2005; Erkeland et al. 2006; Touw and Erkeland 2007). A major outcome of these types of analyses should be the definition of sets of cooperating pathways involved in a given neoplastic process.

As for studying individual genes, there are two primary questions that should be addressed before more extensive studies are proposed. The first of these is to obtain evidence that a given insertion mutation is a clonal event, i.e., present in nearly every cell that makes up the tumor mass. Many, indeed most, of the insertions that are recovered by LM-PCR are probably present in only a subclone of a tumor mass. This is true for both transposon and retrovirally induced tumors because as the tumors grow neither transposition nor new proviral integration events cease to occur. Since PCR is used to amplify transposons or proviral junction fragments with cellular genomic DNA, even rare insertion clones can be obtained. The usual method for detecting clonal insertion mutations is to use Southern blotting (for e.g., see Buchberg et al. 1990). But if tumor DNA is very limited, then a 3-primer PCR reaction can be done to verify that an insertion is present in most of the cells that make up the mass of the tumor (Collier et al. 2005). In most cases, the tumor consists not only of tumor cells, but also of large numbers of reactive normal cells from the host, including stromal, endothelial, and hematological cells. Thus, the insertion event is likely to be detected in somewhat less than every genome's worth of genomic DNA, even if it is present in every neoplastic cell.

A second important follow-up experiment is to show that the insertion mutations actually affect the expression of the gene in question in tumors that have the relevant clonal insertion mutation at that CIS. Typically, a northern blot or quantitative reverse transcriptase PCR (Q-RT-PCR) reaction is carried out on tumor RNAs to demonstrate an increase or decrease in the level of expression has occurred in tumors with an insertion mutation compared to tumors without such an insertion and normal tissue controls (Valk et al. 1999). It is ideal to compare levels of gene expression to that seen in both tumors without insertion mutations and in normal nontransformed tissue. The more tumor RNAs that can be assayed the better since some tumors may have alterations in the level of a CIS-associated gene due to upstream genetic or epigenetic alterations rather than an insertion mutation. Some insertion mutations may not alter the level of the gene but instead result in the production of a truncated fusion transcript. For this reason, the position and orientation of the insertion mutation must be taken into account. Vector-specific primers can be designed and combined with gene-specific primers to determine if a fusion transcript is produced (Collier et al. 2005; Dupuy et al. 2005). This mechanism of proto-oncogene activation appears to be common in SB induced models of cancer. For example, the *Braf* gene in SB-accelerated sarcomas and the *Notch1* gene SB-induced T cell leukemias are activated due to insertions into very specific introns leading to the production of N-terminally truncated activated dominant oncoproteins (Collier et al. 2005; Dupuy et al. 2005). In these cases, RT-PCR reactions and/or northern blotting revealed the production of fusion transcripts initiated in the SB T2/Onc vectors. Ideally then, a readout of exon abundance and levels of alternately spliced mRNAs for all CIS-associated genes would be obtained in tumors with insertion mutations compared to the controls mentioned above. It may be practical to use whole genome approaches to assay the relative level of all exons for all genes using RNA from tumors induced in an insertional mutagenesis screen in the future. This could be done using massively parallel sequencing platforms (Morin et al. 2008), or using microarray hybridization platforms available now (Xing et al. 2007). This approach would have the benefit of generating data on the effects of insertion mutations on CIS-associated genes and of generating an overall global gene expression profile that could be compared to gene expression profiles found by studying human tumors. Taking this idea a step further, matching gene expression profiles could indicate that the CIS-associated genes, or pathways they define, are altered in the subset of human tumors that is most similar.

It seems likely that for some mouse models of cancer created by insertional mutagenesis, there is phenotypic heterogeneity similar to that seen in human tumors. This genetic heterogeneity would most likely be caused by the variety and combination of specific genes affected by insertion mutation at specific CIS. However, this idea has not made its way into utilization of any mouse model of cancer caused by insertional mutagenesis. But it is worth reiterating that mouse models of cancer caused by insertional mutagenesis are really multiple models of cancer in one model. The genetic heterogeneity present between differing tumors should be exploited to understand how tumor genotype affects tumor phenotypes of clinical relevance. This is perhaps the most exciting future possibility for research on models of cancer in mice created by insertional mutagenesis.

4.5 Conclusions and Future Progress

The advent of transposon mutagenesis in somatic cells of transgenic mice has allowed forward genetic screens for carcinoma and other solid tumors to be carried out for the first time. A renaissance in these kinds of experiments can be predicted based on this discovery. The solid tumors have been suggested to be less well understood, at a genetic level, than the hematological malignancies. Hopefully this and other new approaches will change this situation. Insertional mutagenesis approaches may also allow certain malignancies to be more easily modeled in the mouse. Insertional mutagenesis may help define complementation groups of interacting genes that can synergize in cancer development. Perhaps the most interesting future development will be the implementation of well-defined genetic screens designed to gain insight into specific phenotypes of clinical relevance.

Insertional mutagenesis has been shown to provide a useful means to model cancer in mice. New cancer genes and pathways have been revealed and insight into specific human cancer genes has been obtained. In the future, I imagine that the approach will be utilized to screen for traits that seem to matter most in determining each patient's outcome. In general, the presence or absence of invasion and metastasis is considered the hallmark of malignancy and the very feature of cancer that kills. It should be possible, using SB or another transposon, to select for and define genes and genetic pathways that control these processes *in vivo*. Another important cancer trait that confounds attempts to cure these diseases is the acquisition of therapy resistance. Indeed, the development of resistance is likely to be encountered for new "molecularly targeted" therapies, just as it is when traditional therapies are used. Despite this, the factors that control the likelihood of resistance development and the mechanisms by which they occur are still very mysterious. It might be possible to develop *in vivo* selection schemes that could result in treatment resistance induced by insertional mutagenesis, which could be fruitfully employed for finding CIS that caused the resistance. Taken together, these cancer phenotypes represent a wonderful opportunity for insertional mutagenesis to help understand the clinical behavior of human tumors. It is likely that many other phenotypes await the application of insertional mutagenesis by creative future scientists also.

Acknowledgments I apologize to colleagues whose work I could not discuss here due to space limitations. I thank the members of the Largaespada laboratory and the Center for Genome Engineering for constant support and helpful discussions. Research in the Largaespada laboratory is supported by the National Institutes of Health (R01 CA113636-01A1 and UO1 CA84221), the American Cancer Society (RPG LIB-106632), and the Leukemia and Lymphoma Society of America (LLS 7019).

References

Acha-Orbea H, MacDonald HR (1995) Superantigens of mouse mammary tumor virus. Annu Rev Immunol 13:459–486

Akagi K, Suzuki T et al (2004) RTCGD: retroviral tagged cancer gene database. Nucleic Acids Res 32(Database issue): D523–D527

An W, Han JS et al (2006) Active retrotransposition by a synthetic L1 element in mice. Proc Natl Acad Sci USA 103(49):18662–18667

Balciunas D, Wangensteen KJ et al (2006) Harnessing a high cargo-capacity transposon for genetic applications in vertebrates. PLoS Genet 2(11):e169

Bartholomew C, Ihle JN (1991) Retroviral insertions 90 kilobases proximal to the Evi-1 myeloid transforming gene activate transcription from the normal promoter. Mol Cell Biol 11(4):1820–1828

Baum C, von Kalle C et al (2004) Chance or necessity? Insertional mutagenesis in gene therapy and its consequences. Mol Ther 9(1):5–13

Bedigian HG, Shepel LA et al (1993) Transplacental transmission of a leukemogenic murine leukemia virus. J Virol 67(10):6105–6109

Beinke S, Deka J et al (2003) NF-kappaB1 p105 negatively regulates TPL-2 MEK kinase activity. Mol Cell Biol 23(14):4739–4752

Ben-David Y, Giddens EB et al (1990) Identification and mapping of a common proviral integration site Fli-1 in erythroleukemia cells induced by Friend murine leukemia virus. Proc Natl Acad Sci USA 87(4):1332–1336

Ben-David Y, Giddens EB et al (1991) Erythroleukemia induction by Friend murine leukemia virus: insertional activation of a new member of the ets gene family, Fli-1, closely linked to c-ets-1. Genes Dev 5(6):908–918

Berns A (1988) Provirus tagging as an instrument to identify oncogenes and to establish synergism between oncogenes. Arch Virol 102(1–2):1–18

Bosze Z, Thiesen HJ et al (1986) A transcriptional enhancer with specificity for erythroid cells is located in the long terminal repeat of the Friend murine leukemia virus. EMBO J 5(7):1615–1623

Brouha B, Schustak J et al (2003) Hot L1s account for the bulk of retrotransposition in the human population. Proc Natl Acad Sci USA 100(9):5280–5285

Buchberg AM, Bedigian HG et al (1990) Evi-2, a common integartion site involved in murine myeloid leukemogenesis. Mol Cell Biol 10:4658–4666

Callahan R, Smith GH (2000) MMTV-induced mammary tumorigenesis: gene discovery, progression to malignancy and cellular pathways. Oncogene 19(8):992–1001

Callahan R, Smith GH (2008) J Mammary Gland Biol Neoplasia 13(3):309–321

Carlson CM, Dupuy AJ et al (2003) Transposon mutagenesis of the mouse germline. Genetics 165(1):243–256

Cherry SR, Biniszkiewicz D et al (2000) Retroviral expression in embryonic stem cells and hematopoietic stem cells. Mol Cell Biol 20(20):7419–7426

Collier LS, Carlson CM et al (2005) Cancer gene discovery in solid tumours using transposon-based somatic mutagenesis in the mouse. Nature 436(7048):272–276

Cook WD, McCaw BJ (2000) Accommodating haploinsufficient tumor suppressor genes in Knudson's model. Oncogene 19(30):3434–3438

Dave UP, Jenkins NA et al (2004) Gene therapy insertional mutagenesis insights. Science 303(5656):333

de Parseval N, Heidmann T (2005) Human endogenous retroviruses: from infectious elements to human genes. Cytogenet Genome Res 110(1–4):318–332

de Ridder J, Uren A et al (2006) Detecting statistically significant common insertion sites in retroviral insertional mutagenesis screens. PLoS Comput Biol 2(12):e166

Dewannieux M, Heidmann T (2005) LINEs, SINEs and processed pseudogenes: parasitic strategies for genome modeling. Cytogenet Genome Res 110(1–4):35–48
Ding S, Wu X et al (2005) Efficient transposition of the piggyBac (PB) transposon in mammalian cells and mice. Cell 122(3):473–483
Du Y, Jenkins NA et al (2005a) Insertional mutagenesis identifies genes that promote the immortalization of primary bone marrow progenitor cells. Blood 106(12):3932–3939
Du Y, Spence SE et al (2005b) Cooperating cancer gene identification via oncogenic retrovirus-induced insertional mutagenesis. Blood 106(7):2498–2505
Dupuy AJ, Akagi K et al (2005) Mammalian mutagenesis using a highly mobile somatic Sleeping Beauty transposon system. Nature 436(7048):221–226
Dupuy AJ, Fritz S et al (2001) Transposition and gene disruption in the male germline of the mouse. Genesis 30(2):82–88
Elenbaas B, Spirio L et al (2001) Human breast cancer cells generated by oncogenic transformation of primary mammary epithelial cells. Genes Dev 15(1):50–65
Erkeland SJ, Verhaak RG et al (2006) Significance of murine retroviral mutagenesis for identification of disease genes in human acute myeloid leukemia. Cancer Res 66(2):622–626
Ferletta M, Uhrbom L et al (2007) Sox10 has a broad expression pattern in gliomas and enhances platelet-derived growth factor-B-induced gliomagenesis. Mol Cancer Res 5(9):891–897
Fischer SE, Wienholds E et al (2001) Regulated transposition of a fish transposon in the mouse germ line. Proc Natl Acad Sci USA 98(12):6759–6764
Gardner MB (1978) Type C viruses of wild mice: characterization and natural history of amphotropic, ecotropic, and xenotropic MuLv. Curr Top Microbiol Immunol 79:215–259
Geurts AM, Collier LS et al (2006a) Gene mutations and genomic rearrangements in the mouse as a result of transposon mobilization from chromosomal concatemers. PLoS Genet 2(9):e156
Geurts AM, Hackett CS et al (2006b) Structure-based prediction of insertion-site preferences of transposons into chromosomes. Nucleic Acids Res 34(9):2803–2811
Gilbert DJ, Neumann PE et al (1993) Susceptibility of AKXD recombinant inbred mouse strains to lymphomas. J Virol 67(4):2083–2090
Glover JF, Darbre PD (1989) Multihormone regulation of MMTV-LTR in transfected T-47-D human breast cancer cells. J Steroid Biochem 32(3):357–363
Gross L (1978) Viral etiology of cancer and leukemia: a look into the past, present and future – G.H.A. Clowes Memorial Lecture. Cancer Res 38(3):485–493
Hacein-Bey-Abina S, Von Kalle C et al (2003) LMO2-associated clonal T cell proliferation in two patients after gene therapy for SCID-X1. Science 302(5644):415–419
Hahn WC, Dessain SK et al (2002) Enumeration of the simian virus 40 early region elements necessary for human cell transformation. Mol Cell Biol 22(7):2111–2123
Hansen GM, Justice MJ (1999) Activation of Hex and mEg5 by retroviral insertion may contribute to mouse B-cell leukemia. Oncogene 18(47):6531–6539
Hawley RG, Lieu FH et al (1994) Versatile retroviral vectors for potential use in gene therapy. Gene Ther 1(2):136–138
Hayward WS, Neel BG et al (1981) Activation of a cellular onc gene by promoter insertion in ALV-induced lymphoid leukosis. Nature 290(5806):475–480
Herr W, Gilbert W (1983) Somatically acquired recombinant murine leukemia proviruses in thymic leukemias of AKR/J mice. J Virol 46(1):70–82
Herschkowitz JI, Simin K et al (2007) Identification of conserved gene expression features between murine mammary carcinoma models and human breast tumors. Genome Biol 8(5):R76
Himmel KL, Bi F et al (2002) Activation of Clg, a novel Dbl family guanine nucleotide exchange factor gene, by proviral insertion at Evi24, a common integration site in B cell and myeloid Leukemia. J Biol Chem 11:11
Hirai H, Izutsu K et al (2001) Oncogenic mechanisms of Evi-1 protein. Cancer Chemother Pharmacol 48(Suppl 1):S35–S40
Horie K, Kuroiwa A et al (2001) Efficient chromosomal transposition of a Tc1/mariner-like transposon Sleeping Beauty in mice. Proc Natl Acad Sci USA 98(16):9191–9196

Horie K, Yusa K et al (2003) Characterization of Sleeping Beauty transposition and its application to genetic screening in mice. Mol Cell Biol 23(24):9189–9207
Indik S, Gunzburg WH et al (2005) A novel, mouse mammary tumor virus encoded protein with Rev-like properties. Virology 337(1):1–6
Ishimoto A, Takimoto M et al (1987) Sequences responsible for erythroid and lymphoid leukemia in the long terminal repeats of Friend-mink cell focus-forming and Moloney murine leukemia viruses. J Virol 61(6):1861–1866
Ivics Z, Hackett PB et al (1997) Molecular reconstruction of Sleeping Beauty, a Tc1-like transposon from fish, and its transposition in human cells. Cell 91(4):501–510
Ivics Z, Kaufman CD et al (2004) The Sleeping Beauty transposable element: evolution, regulation and genetic applications. Curr Issues Mol Biol 6(1):43–55
Izsvak Z, Ivics Z (2004) Sleeping beauty transposition: biology and applications for molecular therapy. Mol Ther 9(2):147–156
Johansson FK, Brodd J et al (2004) Identification of candidate cancer-causing genes in mouse brain tumors by retroviral tagging. Proc Natl Acad Sci USA 101(31):11334–11337
Johnson P, Benchimol S (1992) Friend virus induced murine erythroleukaemia: the p53 locus. Cancer Surv 12:137–151
Jonkers J, Berns A (1996) Retroviral insertional mutagenesis as a strategy to identify cancer genes. Biochim Biophys Acta 1287(1):29–57
Jonkers J, Korswagen HC et al (1997) Activation of a novel proto-oncogene, Frat1, contributes to progression of mouse T-cell lymphomas. EMBO J 16(3):441–450
Joosten M, Vankan-Berkhoudt Y et al (2002) Large-scale identification of novel potential disease loci in mouse leukemia applying an improved strategy for cloning common virus integration sites. Oncogene 21(47):7247–7255
Kogan SC, Ward JM et al (2002) Bethesda proposals for classification of nonlymphoid hematopoietic neoplasms in mice. Blood 100(1):238–245
Kroon E, Krosl J et al (1998) Hoxa9 transforms primary bone marrow cells through specific collaboration with Meis1a but not Pbx1b. EMBO J 17(13):3714–3725
Kubota S, Siomi H et al (1996) Cis/trans-activation of the interleukin-9 receptor gene in an HTLV-I-transformed human lymphocytic cell. Oncogene 12(7):1441–1447
Kung HJ, Boerkoel C et al (1991) Retroviral mutagenesis of cellular oncogenes: a review with insights into the mechanisms of insertional activation. Curr Top Microbiol Immunol 171:1–25
Largaespada DA (2003) Generating and manipulating transgenic animals using transposable elements. Reprod Biol Endocrinol 1(1):80
Largaespada DA, Collier LS (2008) Transposon-mediated mutagenesis in somatic cells: identification of transposon-genomic DNA junctions. Methods Mol Biol 435:95–108
Largaespada DA, Shaughnessy JD Jr et al (1995) Retroviral integration at the Evi-2 locus in BXH-2 myeloid leukemia cell lines disrupts Nf1 expression without changes in steady-state Ras-GTP levels. J Virol 69(8):5095–5102
Lawrence HJ, Rozenfeld S et al (1999) Frequent co-expression of the HOXA9 and MEIS1 homeobox genes in human myeloid leukemias. Leukemia 13(12):1993–1999
Lazo PA, Lee JS et al (1990) Long-distance activation of the Myc protooncogene by provirus insertion in Mlvi-1 or Mlvi-4 in rat T-cell lymphomas. Proc Natl Acad Sci USA 87(1):170–173
Lee BK, Eicher EM (1990) Segregation patterns of endogenous mouse mammary tumor viruses in five recombinant inbred strain sets. J Virol 64(9):4568–4572
Li J, Shen H et al (1999) Leukaemia disease genes: large-scale cloning and pathway predictions [see comments]. Nat Genet 23(3):348–353
Liu G, Geurts AM et al (2005) Target-site preferences of Sleeping Beauty transposons. J Mol Biol 346(1):161–173
Lu M, Zhang N et al (1996) Retrovirus-mediated gene expression in hematopoietic cells correlates inversely with growth factor stimulation. Hum Gene Ther 7(18):2263–2271
Lundberg AS, Randell SH et al (2002) Immortalization and transformation of primary human airway epithelial cells by gene transfer. Oncogene 21(29):4577–4586

Maeda N, Fan H et al (2008) Oncogenesis by retroviruses: old and new paradigms. Rev Med Virol 18(6):387–405
Mercader N, Leonardo E et al (1999) Conserved regulation of proximodistal limb axis development by Meis1/Hth. Nature 402(6760):425–429
Mertz JA, Simper MS et al (2005) Mouse mammary tumor virus encodes a self-regulatory RNA export protein and is a complex retrovirus. J Virol 79(23):14737–14747
Mikkers H, Berns A (2003) Retroviral insertional mutagenesis: tagging cancer pathways. Adv Cancer Res 88:53–99
Miller DG, Adam MA et al (1990) Gene transfer by retrovirus vectors occurs only in cells that are actively replicating at the time of infection. Mol Cell Biol 10(8):4239–4242
Miskey C, Izsvak Z et al (2005) DNA transposons in vertebrate functional genomics. Cell Mol Life Sci 62(6):629–641
Morin R, Bainbridge M et al (2008) Profiling the HeLa S3 transcriptome using randomly primed cDNA and massively parallel short-read sequencing. Biotechniques 45(1):81–94
Morse HC 3rd, Anver MR et al (2002) Bethesda proposals for classification of lymphoid neoplasms in mice. Blood 100(1):246–258
Moskow JJ, Bullrich F et al (1995) Meis1, a PBX1-related homeobox gene involved in myeloid leukemia in BXH-2 mice. Mol Cell Biol 15(10):5434–5443
Mucenski ML, Bedigian HG et al (1988) Comparative molecular genetic analysis of lymphomas from six inbred mouse strains. J Virol 62(3):839–846
Mucenski ML, Taylor BA et al (1986) AKXD recombinant inbred strains: models for studying the molecular genetic basis of murine lymphomas. Mol Cell Biol 6(12):4236–4243
Mukhopadhyaya R, Wolff L (1992) New sites of proviral integration associated with murine promonocytic leukemias and evidence for alternate modes of c-myb activation [published erratum appears in J Virol 1993 May;67(5):2960]. J Virol 66(10):6035–6044
Murakami Y, Saigo K et al (2005) Large scaled analysis of hepatitis B virus (HBV) DNA integration in HBV related hepatocellular carcinomas. Gut 54(8):1162–1168
Nakamura T, Largaespada DA et al (1996) Cooperative activation of Hoxa and Pbx1-related genes in murine myeloid leukaemias. Nat Genet 12(2):149–153
Nakamura Y, Moriuchi R et al (1994) Altered expression of a novel cellular gene as a consequence of integration of human T cell lymphotropic virus type 1. J Gen Virol 75(Pt 10):2625–2633
Nishigaki K, Hanson C et al (2002) Analysis of the disease potential of a recombinant retrovirus containing Friend murine leukemia virus sequences and a unique long terminal repeat from feline leukemia virus. J Virol 76(3):1527–1532
Ostertag EM, DeBerardinis RJ et al (2002) A mouse model of human L1 retrotransposition. Nat Genet 32(4):655–660
Ostertag EM, Kazazian HH Jr (2001) Biology of mammalian L1 retrotransposons. Annu Rev Genet 35:501–538
Pang R, Tse E et al (2006) Molecular pathways in hepatocellular carcinoma. Cancer Lett 240(2):157–169, Epub 2005 Oct 17
Payne GS, Bishop JM et al (1982) Multiple arrangements of viral DNA and an activated host oncogene in bursal lymphomas. Nature 295(5846):209–214
Roberg-Perez K, Carlson CM et al (2003) MTID: a database of Sleeping Beauty transposon insertions in mice. Nucleic Acids Res 31(1):78–81
Ru M, Shustik C et al (1993) Graffi murine leukemia virus: molecular cloning and characterization of the myeloid leukemia-inducing agent. J Virol 67(8):4722–4731
Sauvageau M, Miller M et al (2008) Quantitative expression profiling guided by common retroviral insertion sites reveals novel and cell type specific cancer genes in leukemia. Blood 111(2):790–799
Selten G, Cuypers HT et al (1985) Proviral activation of the putative oncogene Pim-1 in MuLV induced T-cell lymphomas. EMBO J 4(7):1793–1798
Shackleford GM, MacArthur CA et al (1993) Mouse mammary tumor virus infection accelerates mammary carcinogenesis in Wnt-1 transgenic mice by insertional activation of int-2/Fgf-3 and hst/Fgf-4. Proc Natl Acad Sci USA 90(2):740–744

Shen WF, Rozenfeld S et al (1999) HOXA9 forms triple complexes with PBX2 and MEIS1 in myeloid cells. Mol Cell Biol 19(4):3051–3061
Silver JE, Fredrickson TN (1983) Susceptibility to Friend helper virus leukemias in CXB recombinant inbred mice. J Exp Med 158(5):1693–1702
Sinn E, Muller W et al (1987) Coexpression of MMTV/v-Ha-ras and MMTV/c-myc genes in transgenic mice: synergistic action of oncogenes in vivo. Cell 49(4):465–475
Sjoblom T, Jones S et al (2006) The consensus coding sequences of human breast and colorectal cancers. Science 314(5797):268–274, Epub 2006 Sep 7
Slater EP, Posseckert G et al (1989) Binding of steroid receptors to the HREs of mouse mammary tumor virus, chicken and xenopus vitellogenin and rabbit uteroglobin genes: correlation with induction. J Steroid Biochem 34(1–6):11–16
Starr TK, Largaespada DA (2005) Cancer gene discovery using the Sleeping Beauty transposon. Cell Cycle 4(12):1744–1748
Subramanian A, Tamayo P et al (2005) Gene set enrichment analysis: a knowledge-based approach for interpreting genome-wide expression profiles. Proc Natl Acad Sci USA 102(43):15545–15550
Tamori A, Nishiguchi S et al (2005a) Hepatitis B virus DNA integration in hepatocellular carcinoma after interferon-induced disappearance of hepatitis C virus. Am J Gastroenterol 100(8):1748–1753
Tamori A, Yamanishi Y et al (2005b) Alteration of gene expression in human hepatocellular carcinoma with integrated hepatitis B virus DNA. Clin Cancer Res 11(16):5821–5826
Tanaka M, Jin G et al (2008) Identification of candidate cooperative genes of the Apc mutation in transformation of the colon epithelial cell by retroviral insertional mutagenesis. Cancer Sci 99(5):979–985
Theodorou V, Kimm MA et al (2007) MMTV insertional mutagenesis identifies genes, gene families and pathways involved in mammary cancer. Nat Genet 39(6):759–769
Thorsteinsdottir U, Kroon E et al (2001) Defining roles for HOX and MEIS1 genes in induction of acute myeloid leukemia. Mol Cell Biol 21(1):224–234
Touw IP, Erkeland SJ (2007) Retroviral insertion mutagenesis in mice as a comparative oncogenomics tool to identify disease genes in human leukemia. Mol Ther 15(1):13–19
Uhrbom L, Hesselager G et al (1998) Induction of brain tumors in mice using a recombinant platelet-derived growth factor B-chain retrovirus. Cancer Res 58(23):5275–5279
Uren AG, Kool J et al (2005) Retroviral insertional mutagenesis: past, present and future. Oncogene 24(52):7656–7672
Valk PJ, Hol S et al (1997) The genes encoding the peripheral cannabinoid receptor and alpha-L-fucosidase are located near a newly identified common virus integration site, Evi11. J Virol 71(9):6796–6804
Valk PJ, Vankan Y et al (1999) Retroviral insertions in Evi12, a novel common virus integration site upstream of Tra1/Grp94, frequently coincide with insertions in the gene encoding the peripheral cannabinoid receptor Cnr2. J Virol 73(5):3595–3602
van der Lugt NM, Domen J et al (1995) Proviral tagging in E mu-myc transgenic mice lacking the Pim-1 proto-oncogene leads to compensatory activation of Pim-2. EMBO J 14(11):2536–2544
van Lohuizen M, Berns A (1990) Tumorigenesis by slow-transforming retroviruses – an update. Biochim Biophys Acta 1032(2–3):213–235
van Lohuizen M, Breuer M et al (1989) N-myc is frequently activated by proviral insertion in MuLV-induced T cell lymphomas. EMBO J 8(1):133–136
van Ooyen A, Kwee V et al (1985) The nucleotide sequence of the human int-1 mammary oncogene; evolutionary conservation of coding and non-coding sequences. EMBO J 4(11):2905–2909
Vandenbussche M, Janssen A et al (2008) Generation of a 3D indexed Petunia insertion database for reverse genetics. Plant J 54(6):1105–1114
Varmus HE (1983) Using retroviruses as insertional mutagens to identify cellular oncogenes. Prog Clin Biol Res 119:23–35
Weiser KC, Justice MJ (2005) Cancer biology: Sleeping Beauty awakens. Nature 436(7048): 184–186
Wold WS, Green M (1979) Historic milestones in cancer virology. Semin Oncol 6(4):461–478

Wood LD, Parsons DW et al (2007) The genomic landscapes of human breast and colorectal cancers. Science 318(5853):1108–1113
Wu S, Ying G et al (2007) Toward simpler and faster genome-wide mutagenesis in mice. Nat Genet 39(7):922–930
Wu X, Burgess SM (2004) Integration target site selection for retroviruses and transposable elements. Cell Mol Life Sci 61(19–20):2588–2596
Wu X, Li Y et al (2003) Transcription start regions in the human genome are favored targets for MLV integration. Science 300(5626):1749–1751
Wu X, Luke BT et al (2006) Redefining the common insertion site. Virology 344(2):292–295
Xing Y, Ouyang Z et al (2007) Assessing the conservation of mammalian gene expression using high-density exon arrays. Mol Biol Evol 24(6):1283–1285
Yant SR, Wu X et al (2005) High-resolution genome-wide mapping of transposon integration in mammals. Mol Cell Biol 25(6):2085–2094
Yin B, Largaespada DA (2007) PCR-based procedures to isolate insertion sites of DNA elements. Biotechniques 43(1):79–84
York A (2005) Sleeping beauty offers new method to find cancer genes. Lancet Oncol 6(8):545
Zagoraiou L, Drabek D et al (2001) In vivo transposition of Minos, a Drosophila mobile element, in mammalian tissues. Proc Natl Acad Sci USA 98(20):11474–11478

Chapter 5
The RCAS/TVA Somatic Gene Transfer Method in Modeling Human Cancer

Yi Li, Andrea Ferris, Brian C. Lewis, Sandra Orsulic, Bart O. Williams, Eric C. Holland, and Stephen H. Hughes

5.1 Introduction

Human cancer usually arises from a small number of somatic cells that have gained one or a few critical genetic mutations. This clonal evolution takes place in a field of normal tissue. Elucidating the molecular and cellular events of this process may be assisted by models in which transforming genetic alterations are generated in a small number of cells in a tissue that has fully developed. Among the models is viral insertional mutagenesis, which has been used to identify causal genetic lesions and to learn about their effects on carcinogenesis (Nusse and Varmus 1982; Theodorou et al. 2007 and Chapter 4 in this book). However, viral mutagenesis is not suitable for

Y. Li (✉)
Lester and Sue Smith Breast Center and Department of Molecular and Cell Biology,
Baylor College of Medicine, Houston, TX 77030, USA
e-mail: liyi@bcm.edu

A. Ferris • S.H. Hughes
HIV Drug Resistance Program, National Cancer Institute-Frederick,
Frederick, MD 21702, USA

B.C. Lewis
Program in Gene Function and Expression, University of Massachusetts
Medical Center, Worcester, MA 01605, USA

S. Orsulic
Women's Cancer Research Institute, Cedars-Sinai Medical Center,
Los Angeles, CA 90048, USA

B.O. Williams
Molecular Medicine and Virology Group, Van Andel Research Institute,
Grand Rapids, MI 49503, USA

E.C. Holland
Program in Cancer Biology and Genetics, Memorial Sloan Kettering
Cancer Center, New York, NY 10021, USA

J.E. Green and T. Ried (eds.), *Genetically Engineered Mice for Cancer Research: Design, Analysis, Pathways, Validation and Pre-Clinical Testing*,
DOI 10.1007/978-0-387-69805-2_5, © Springer Science+Business Media, LLC 2012

testing the transforming potential of specific genes. In 1994, Federspiel et al. reported that exogenous genes could be selectively introduced into myoblasts of transgenic mice using a vector that is a member of the avian leukosis virus subgroup A (ALV-A). The myoblasts in these mice were made susceptible to infection by transgenic expression of *tva* from the *α-actin* promoter. The technology was adapted to facilitate the introduction of oncogenes into the brain of neonatal transgenic mice expressing of an avian viral receptor (*tva*) from glial cell-specific promoters (Holland et al. 1998, 2000b). Subsequently, this method of somatic gene transfer has been used to create tumor models in several other tissues, including the ovary (Orsulic et al. 2002), pancreas (Du et al. 2007; Lewis et al. 2003; Seidler et al. 2008), liver (Lewis et al. 2005), mammary gland (Du et al. 2006), vascular endothelium (Montaner et al. 2003; Sausville et al. 2008; Vervoort et al. 2008), and others (Fu et al. 2005; Murphy et al. 2003; Pao et al. 2003). In this chapter, we (1) introduce the TVA receptor, the ALV life cycle, and RCAS vectors; (2) discuss the use of the TVA method for introducing genes into mammalian cells *in vivo*; (3) summarize special features involved in using the TVA method in selected tissue types; and (4) describe practical protocols and tips in the use of this technology.

5.2 Overview of TVA and RCAS

The *tva* gene was originally cloned using a gene transfer approach to identify chicken DNA fragments that conferred susceptibility to infection by ALV-A on mammalian cells (Young et al. 1993). The peptide sequence of the *tva* gene product (TVA) is homologous to the ligand-binding repeat of the family of low-density lipoprotein receptors (LDLRs) (Bates et al. 1993), but the physiologic functions of TVA remain unknown. TVA is translated from an alternatively spliced mRNA and is present in both GPI (glycosylphosphatidylinositol)-linked and transmembrane forms. *tva*800 (800-bp in length) encodes the GPI-linked receptor while *tva*950 (950-bp in length) encodes the transmembrane isoform. Three other ALV receptors have also been identified. TVB is the receptor for ALV-B, -D, and -E, and is related to the tumor necrosis factor receptor family of proteins (Adkins et al. 2000; Brojatsch et al. 1996). chNHE1 (Na^+/H^+ exchanger type 1) is the receptor for ALV-J (Chai and Bates 2006). The receptor for ALV-C is a member of the immunoglobulin superfamily (Elleder et al. 2005).

ALV is a member of the avian sarcoma-leukosis virus (ASLV) family of retroviruses (Petropoulos and Hughes 1991), which also includes Rous sarcoma virus (RSV) and its derivative, the *R*eplication *C*ompetent *A*SLV-LTR with a *S*plice acceptor (RCAS) avian retroviral vectors (Hughes et al. 1987). ALV is replication competent-and causes disease slowly, usually by inserting near a protooncogene in the host genome and activating it. On the other hand, acutely transforming retroviruses have already acquired oncogenes from the host cell and have lost part or all of at least one essential viral gene. As a consequence, these viruses are defective and require a helper to replicate. However, when RSV arose from ALV, by acquiring

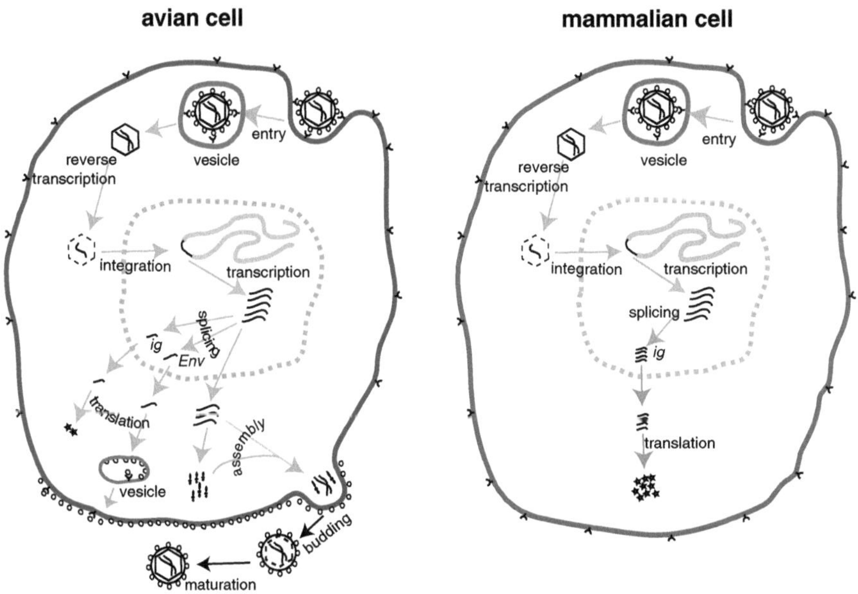

Fig. 5.1 Infection of avian versus mammalian cells with an RCASBP(A) vector that also carries an inserted gene (*ig*), such as an oncogene. For avian cells, the virus recognizes TVA on the cell surface, and is endocytosed to form a vesicle. The viral core is released from the vesicle into the cytoplasm, a process that requires the acidification of the vesicle (Mothes et al. 2000). Reverse-transcription converts the diploid single-stranded viral genome into a double-stranded DNA. This DNA genome, with the help of its associated viral proteins, migrates to the nucleus and integrates into the host genome. RCAS vectors can infect nondividing cells at reduced efficiency, which means that the viral DNA can enter the nucleus without the breakdown of the nuclear membrane that occurs during mitosis. The integrated genome (provirus) is treated like a host gene and is transcribed by host RNA polymerase II. Both spliced and unspliced RNAs are exported into cytoplasm. In the RCAS vectors, both Env and ig proteins are translated from spliced *env* and *ig* mRNAs, respectively. If an RCAN vector with an internal promoter is used, then a separate message is produced from the internal promoter. Note that the Env proteins are produced in the ER-Golgi and then transferred into the cytoplasmic membrane. Newly produced Env in an infected cell can bind to TVA and prevent new infection (viral interference). Gag and Gag-Pol polyproteins are made from the unspliced mRNA, and form a complex with the unspliced mRNA (viral genome) at the cell membrane. These nascent virions bud from the cell, and mature into infectious particles after auto-proteolytic processing of the polyproteins. In most mammalian cells, only the *ig* mRNA is produced in significant quantities; consequently, RCAS infection leads to the preferential production of the product of the inserted gene. Because Env is not made in significant quantities in most mammalian cells, TVA is not blocked in infected cells; thus, infected mammalian cells remain susceptible to new infections

the cellular oncogene *src*, it retained all of the sequences needed for replication, and it is the only known acutely transforming retrovirus that is replication competent.

Like other retroviruses, ASLVs efficiently infect only those cells that produce the receptor corresponding to the envelope glycoprotein on the virus. Following the binding of viral envelope glycoproteins to the specific receptor protein on the surface of susceptible cells, the viral envelope fuses with the host membrane, introducing the virion core into the cytoplasm (Fig. 5.1). Subsequently, the diploid single-stranded RNA genome in the virion is converted into a linear double-stranded DNA by the reverse transcriptase associated with the virion. The resulting viral DNA-protein complex (which also includes integrase) enters the nucleus. The viral DNA is then integrated into the host genome by the viral integrase. Integration occurs at many sites in the host chromatin, but is not entirely random (Barr et al. 2005). The integrated viral DNA (provirus) now functions as part of the host genome and is transcribed into RNA by host RNA polymerase II. A portion of the RNA molecules are spliced and exported from the nucleus to produce the viral envelope, Env. Unspliced RNAs are exported from the nucleus and function as genomic RNA as well as mRNA for two polyproteins, Gag and Gag-Pol. At the cytoplasmic membrane, the polyproteins and viral RNA form nascent virions, which bud from cells, taking along part of cellular membrane and Env. Newly budded virions contain unprocessed Gag and Gag-Pol polyproteins at a ratio of approximately 20 to 1. The polyproteins in the nascent virions are cleaved by the viral protease (a process called maturation) to produce infectious viruses. In most retroviruses, processing of the Gag polyprotein gives rise to the viral structural proteins (matrix, capsid, and nucleocapsid); Gag-Pol gives rise to the viral enzymes (protease, reverse transcriptase, and integrase). It is not known whether the Gag portion of Gag-Pol also gives rise to a small amount of the viral structural proteins. ASLVs differ from other retroviruses in that the protease is part of Gag.

The details of the construction of the RCAS vectors are complex (Hughes et al. 1987), but the approach used to generate the RCAS vectors was straightforward: The *src* oncogene in RSV was replaced with a unique ClaI site so that exogenous genes can be cloned into the vector. The vector can establish a productive infection in avian cells; these cells are used to produce viral stocks. Additional information about the vector construction can be obtained from Hughes et al. (1987) and from the RCAS Web site: http://home.ncifcrf.gov/hivdrp/RCAS/.

Several RCAS-based vectors are available (Hughes et al. 1987; Petropoulos and Hughes 1991). *R*eplication *C*ompetent *A*SLV-LTR *N*o splice acceptor (RCAN) lacks the splice acceptor in RSV and RCAS. Without the splice acceptor, inserts are not expressed from the viral LTR so that an internal promoter is usually cloned along with an exogenous gene in the RCAN vectors. This modification both allows cell type-specific expression, and permits different levels of expression of the inserted gene based on the exogenous promoter used. The level of expression of an inserted gene can also be controlled by choosing LTRs with promoters that have different levels of expression. For instance, RCOS and RCON were made by substituting the LTR of the endogenous virus RAV-0 for the LTR in RCAS/RCAN. Because the LTR promoters in RCOS and RCON are relatively weak, the vectors are difficult to grow and are not generally used for producing viral stocks for experiments in mammalian cells (Hughes 2004). In contrast, substituting the polymerase region from the Bryan high-titer strain of RSV (Hanafusa et al. 1963) enhances the titer of the original RCAS/RCAN vectors by five- to tenfold, and there is a concomitant increase in the level of expression of the inserted gene from the LTR promoter (via the spliced message.) The *B*ryan *P*olymerase-containing RCAS/RCAN vectors are designated RCASBP and RCANBP (Petropoulos and Hughes 1991), respectively. RCASBP(A) and RCANBP(A) (replication-competent ALV LTR with or without a splice acceptor, Bryan polymerase, subgroup A, respectively; Fig. 5.2) are commonly

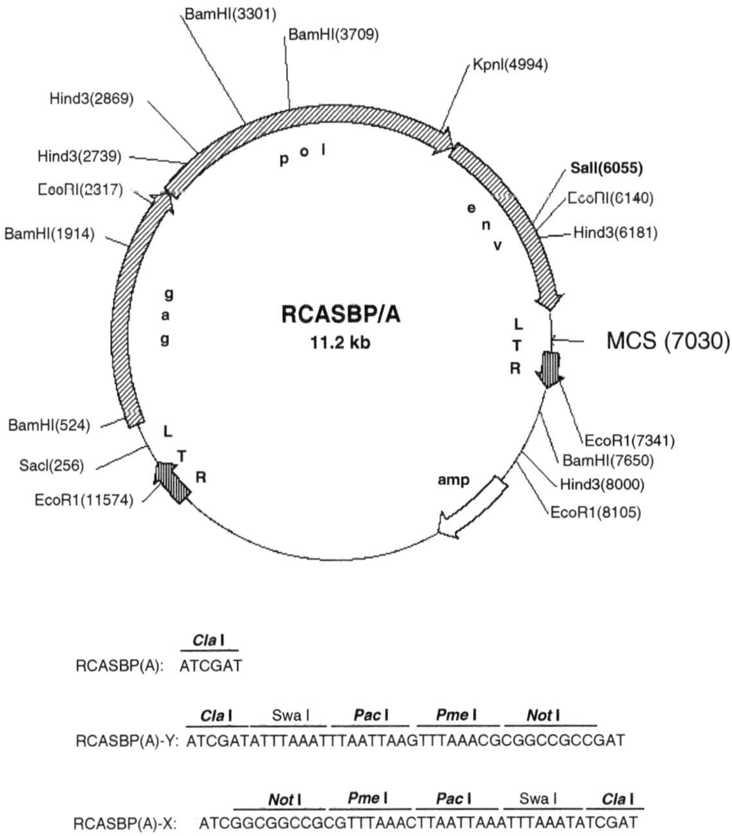

Fig. 5.2 Map of RCASBP(A) and its derivatives RCASBP(A)-X and RCASBP(A)-Y. Unique sites are in bold. There is another *Cla* I site (not labeled), which is *dam*–methylated in Dam+ bacteria, and is thus not usually cleavable. The positions of *Gag*, *Pol*, and *Env* are indicated (not to scale). The sequence of multiple cloning sites (MCS) is shown for each vector

used to produce viral vector stocks that can infect mammalian cells expressing the TVA receptor as a transgene. These vectors are often referred to collectively as RCAS in literature on animal models using the TVA method. Because the differences between RCAS and RCASBP are significant, the correct nomenclature should be used in any description of specific experiments. An immortalized chicken fibroblast cell line, DF-1, is commonly used to propagate RCAS virus (Himly et al. 1998; Kim et al. 2001; Schaefer-Klein et al. 1998). This line is available from ATCC (CRL-12203). A titer of approximately 10^7 to 10^8 can be easily obtained in DF-1 culture medium with RCASBP(A) or RCANBP(A).

Most of the genes that have been successfully expressed using RCASBP(A) are less than 3 kb in size. This is because the overall size of retroviral genomes is limited. The exact limiting factors are unknown, although they are probably related to the volume of the folded RNA in the virion, rather than the linear length of the genome. However, there is not a sharp, well-defined cutoff in size beyond which the vector will not work. Viruses without inserts generally replicate more efficiently than those with inserts, and any RCAS vector that has an insert will tend to lose the insert through recombination if the virus is allowed to replicate. If the inserts in a standard RCAS vector are larger than 2.5 kb, it is often difficult or impossible to obtain stocks in which a useful fraction of the virus retains the insert. Inserts smaller than 2 kb are usually stable.

In order to accommodate a larger insert, vectors that have part of the viral structural genes removed have been developed. BBAN lacks *env* and can accommodate an insert as large as 4.5 kb (Boerkoel et al. 1993). Infectious virus is made in cells that express the missing *env* in trans through stable expression or transient transfection. However, it is generally much more difficult to prepare high titer stocks with defective vectors. (Lentiviral vectors pseudotyped with Env A can be used to make vectors with even larger inserts; this approach will be discussed later.)

5.3 Ectopic *tva* Expression Mediates RCAS Entry into Mammalian Cells

Mammalian cells lack receptors for ASLV and are thus resistant to infection by ASLV viruses and RCAS vectors. However, ectopic expression of *tva* can render mammalian cells susceptible to entry of ALSV-A or RCAS(A) (Bates et al. 1993; Young et al. 1993). In mammalian cells, the virus can complete the first half of its life cycle normally – following entry the RNA genome of the virus is copied into DNA and integrated into the host genome – but the infected mammalian cells do not produce infectious viruses. Although we do not know what prevents viral replication, there appear to be problems in the synthesis and splicing of the viral RNA and in the transport and assembly of the viral proteins. In contrast to normal cellular genes, retroviruses must give rise to both spliced and unspliced RNAs, and both must be successfully exported from the nucleus. A proper balance between spliced and unspliced RNAs is required for the cell to make the viral proteins and genomic RNA in the proper amounts to allow the assembly of infectious virions. In at least some mammalian cells infected with RCAS vectors, it appears that the balance is shifted away from the synthesis of unspliced RNA to the production of spliced RNAs (Nasioulas et al. 1995). In a mouse infected with an RCAS vector, this can be helpful because it increases the synthesis of the protein of interest and reduces harmful effects associated with the synthesis of the viral proteins. In at least some cases, the production of a foreign protein in mice can lead to the elimination of cells that are productively infected with RCAS vectors (Pinto et al. 2000).

The RCAS LTR promoter is often less efficient in mammalian cells than in avian cells (unpublished observations). The LTR promoter seems to be more efficient in human HEK293 cells than in murine fibroblasts, such as NIH3T3 cells, but whether the LTR is generally stronger in primate than in murine cells is unknown. In mice, the promoter strength probably varies depending on the cell type (Zheng and Hughes 1999).

Exogenous genes can be expressed from an internal promoter using the RCANBP(A) vectors. The use of an internal promoter is especially important if the goal is to achieve high level expression in cell types in which the LTR promoter is weakly expressed. Because the total length of RCAN vectors is still limited, the internal promoter must be relatively small to allow room for insertion of the exogenous gene.

RCASBP(A) and RCANBP(A) are the preferred vectors for *in vivo* gene expression because they give rise to high titer viral stocks (in excess of 10^7). The vectors can easily be concentrated to produce stocks with titers as high as 10^9 IUs per milliliter. Both viral suspension and viral producer cells (DF-1) can be used to deliver virus into TVA^+ cells in mice. After injection of a viral suspension in mice, free viral particles can diffuse relatively quickly, enter the circulation, and probably die within a day or two. On the other hand, producer cells stay at the injection site and can continuously produce virus until they are destroyed (presumably by the immune system of the host), usually within a week after injection. Therefore, the injection of producer cells may lead to higher numbers of infected cells at the injection site (Federspiel et al. 1994).

As an alternative to using ectopic expression of *tva* to allow ASLV vectors to infect mammalian cells, RCAS vectors have been made that express envelope glycoproteins that recognize receptors that are normally present on mammalian cells. For instance, RCAS vectors have been developed in which the avian envelope glycoprotein was replaced by either the ecotropic or the amphotropic envelope from MLV (Barsov and Hughes 1996; Chang et al. 2005; Koo et al. 2004). The ecotropic RCAS vectors infect only murine cells while the amphotropic vectors infect murine and other mammalian cells. RCASBPM2C(797–8) carries a modified version of the MLV amphotropic envelope. This vector is replication-competent in avian cells and infects mammalian and avian cells permissive for amphotropic MLV infection. In contrast, RCASBP(eco) infection is restricted to murine cells or cells engineered to express the ecotropic receptor. This vector can establish a productive infection and grows to titers of 10^5 to 10^6 in an avian cell line based on DF-1, DFJ8, that expresses the murine ecotropic receptor (Barsov et al. 2001). These modified vectors can be used to infect all strains of mice, not just those that express the *tva* transgene, which reduces the cost and complexity of the animal husbandry. However, because these vectors infect murine cells with little tissue or cell type specificity, a major advantage of the TVA system is lost.

5.4 The Use of the RCAS/TVA Method to Introduce Genes into Mice for Cancer Modeling

Federspiel et al. (1994) tested whether ectopic expression of *tva* in mice could mediate RCAS infection and the stable introduction of exogenous genes into a specific subset of somatic mouse cells *in vivo*. They created transgenic mice expressing

Top 10 Reasons for Using RCAS Vectors in Modeling Cancer

1. Express oncogene in a small number of cells within a field of normal cells *in vivo*.
2. Express oncogene at a desirable tissue developmental time.
3. Continue to express a test gene even after a change in cell fate.
4. Carry DNA encoding shRNA against a tumor suppressor.
5. Express Cre for conditional deletion of a floxed tumor suppressor.
6. Carry rtTA for inducible expression.
7. Do not spread to other animals in the same facility.
8. Do not infect humans.
9. Can readily generate high titer viruses.
10. Can test many genes in a single TVA transgenic line.

tva from the *α-actin* promoter. Intramuscular injection of RCASBP(A) expressing the gene encoding alkaline phosphatase (AP) into neonatal *tva* transgenic mice led to expression of virally encoded AP in the injected limb. This report established that *tva* expressed as a transgene in mice can be used to mediate viral infection, thus providing a route for genetic manipulation of selected mouse cells *in vivo*.

Holland and Varmus tested whether this method could be used to create cancer models in mice. They created transgenic mice expressing *tva* from the glial fibrillary acidic protein (GFAP) promoter. Intracranial co-injection of these mice with RCASBP(A)-*bFGF* and RCASBP(A)-*AP* caused the infected cells (detected by staining for AP) to migrate away from the needle track (Holland and Varmus 1998). Although no tumors were found in the initial study, they later demonstrated that an RCASBP(A)-delivered oncogene encoding an activated version of epidermal growth factor receptor (EGFR) could cause glioma-like lesions in GFAP-*tva* transgenic mice that were predisposed to cancer through the germ line loss of INK4a/ARF (Holland et al. 1998). Similar lesions were generated by simultaneous infection of GFAP-tva mice that did not harbor any cancer-predisposing germ line mutation with RCASBP(A) vectors expressing both *EGRF* and *CDK4*. Further work showed that co-infection of GFAP-*tva* mice with RCASBP(A) vectors expressing both activated *Ras* and *Akt* could induce glioblastomas (Holland et al. 2000a).

Since the initial success of RCAS/TVA mice for modeling brain tumors, the system has been used for cancer studies in several other organs and tissue types, for studying normal development, and for tracing cell lineages (Doetsch et al. 1999; Dunn et al. 2001, 2000; Fults et al. 2002; Gaur et al. 2001; Hou et al. 2004; Murphy et al. 2003; Murphy and Leavitt 1999). Recently, a knock-in mouse line was developed in which the expression of *tva* is controlled by Cre-loxP. This system makes it possible to use the available tissue-specific Cre lines to control the temporal and spatial expression of *tva* (Seidler et al. 2008).

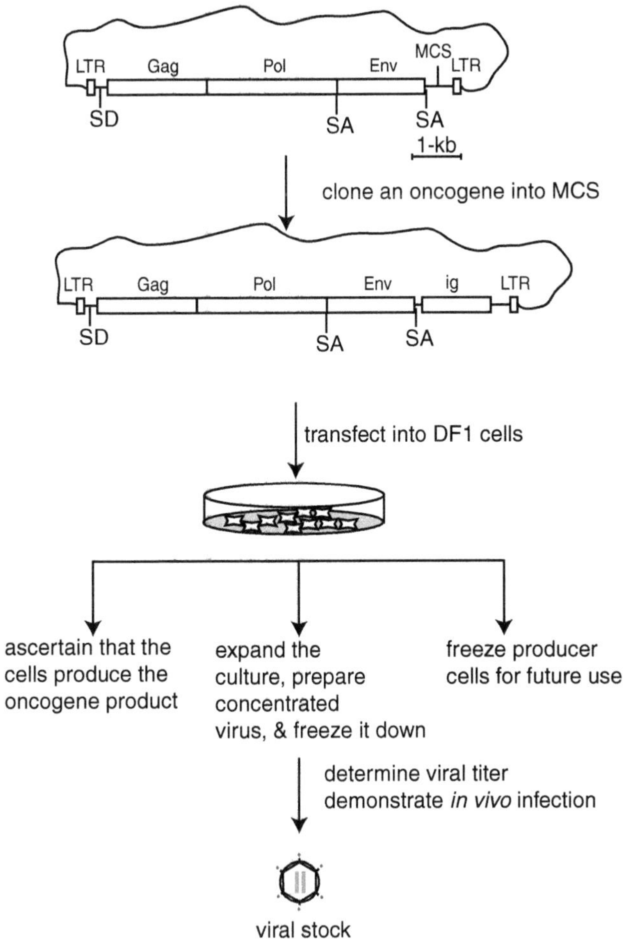

Fig. 5.3 Flow chart for generating and validating RCASBP(A) viruses. The vector backbone (*curvy line*, not drawn to scale) is lost upon viral integration. Splice donor (SD), splice acceptors (SAs), and inserted gene (ig) are indicated

Organ-specific applications in cancer modeling are discussed in the next section, but the following summarizes some general applications of the TVA method in cancer modeling (Fig. 5.3). An oncogene introduced by RCAS is expressed from the LTR promoter in the provirus. Even if there is a change in cell fate due to the expression of the exogenous gene, the oncogene continues to be expressed. This is usually not the case in conventional transgenic models, where transgene expression is linked to a promoter that specifies cell type-specific expression so that the transgene may be turned off by cell differentiation or dedifferentiation. This difference may explain some of the histopathological differences between tumors induced by an oncogene expressed by RCAS vs. by the same oncogene expressed as a transgene (Du et al. 2006; Lewis et al. 2003; Morton et al. 2008).

As initially shown in glioma modeling, RCAS vectors are useful for introducing an oncogene into cancer-predisposed mice to study collaboration between the RCAS-delivered genetic lesion and the germline mutation, or to simultaneously introduce two oncogenes for the test of synergy. For example, multiple genetic events were modeled by infecting ovarian cells ex vivo with multiple viruses and then transplanting the cells back into mice (Orsulic et al. 2002). Furthermore, when RCASBP(A)-*ErbB2* was introduced into TVA+ cells in a mammary gland that also carried a *Wnt-1* oncogene, mammary tumors appeared much faster than when either RCASBP(A)-*ErbB2* or the *Wnt-1* transgene was used alone (Du et al. 2006). In addition, RCASBP(A) expressing shRNA against *p53* accelerated pancreatic tumor progression in mice that carried a *K-Ras(G12D)* mutant allele (Seidler et al. 2008). These approaches for testing oncogenic collaboration are a cost-effective alternative to conventional mouse models, which usually require extensive breeding of transgenic and knockout animals to study synergistic interactions. This approach is especially valuable for exposing weak transforming potential of a test gene, and for identifying its collaborating partners without the need for an elaborate breeding program. However, the co-infection method is not particularly useful for testing the potential suppressor function of a gene, because only a minor fraction of cells in a target tissue is usually infected by more than one of the admixed vectors. Tumors may emerge without a detectable reduction of latency from the cells that were not infected by the second vector carrying the potential tumor suppressor. This problem can be avoided by expressing the potential suppressor, together with other genes, in the same vector (see below).

RCAS vectors can also be used to delineate genetic factors regulating tumor progression. For example, transgenic mice expressing SV40 T antigen from the rat insulin promoter (RIP-TAG) progress to pancreatic islet cancer through defined histopathological changes; however, the molecular events underlying these alterations were not well-defined. By creating bi-transgenic mice carrying both RIP-*TAG* and RIP-*tva*, Du et al. (2007) showed that RCASBP(A)-mediated introduction of a dominant-negative mutant of *E-cadherin 1* or *BCL-xL* into either hyperplastic or early dysplastic lesions in these bi-transgenic mice accelerated islet tumor formation and promoted islet tumor cell invasion and lymph node metastasis. This *in vivo* observation led to the discovery that BCL-xL can suppress E-cadherin 1, remodel cytoskeleton, and stimulate migration and invasion, possibly through an interaction with myosin Va.

In productively infected avian cells, the Env protein produced by a newly integrated virus occupies the cognate receptor on the cell surface and prevents further infection, a process called receptor interference or superinfection resistance. If two RCAS viruses carrying different genes are provided simultaneously, both can enter the cell before superinfection resistance is established. Because mammalian cells express low levels of Env, receptor interference may not pose a significant problem for subsequent infection, potentially allowing sequential introduction of oncogenes using RCAS vectors. However, in most organs, sequential infection is technically difficult because the infected cells usually comprise only a small fraction of cells in a target tissue. Directing the second virus to the cells previously infected with the first virus can be challenging. It may be more practical to co-inject a test oncogenic virus with a virus producing a tetracycline-activated transcriptional activator (tTA) into *tva* mice that harbor an inducible oncogene so that doxycycline can be used to turn on the second oncogene under the control of tetO (described below).

Oncogenic collaboration can also be tested by cloning two genes into a single RCAS vector. An IRES can be used to express the second gene, though the expression levels of both genes appear reduced when this technique is used. Expressing two genes within the same vector has also been used to investigate tumor suppressor function by testing whether the expression of one gene interferes with the transforming potential of another gene (Wolf et al. 2003).

Expression of *rtTA* using RCAS vectors allows inducible expression of a transgene expressed from the doxycycline-inducible promoter (tetO). Transplantation of fibroblasts from tetO-*K-RasG12D/β-actin* bi-transgenic mice infected ex vivo with RCASBP(A)-*rtTA-IRES-EGFP* resulted in sarcomas when doxycycline was present as the inducer. Upon withdrawal of doxycycline, the tumors underwent apoptosis (Pao et al. 2003). Subsequently, *in vivo* injection of RCASBP(A)-*rtTA* was found to be useful for a glioblastoma tumor maintenance study in nestin-*tva* mice (Holmen and Williams 2005). In addition, Cre can also be introduced using an RCAS vector to delete tumor suppressor genes in *tva* transgenic mice that harbor a tumor suppressor gene flanked by loxP sites (Hu et al. 2005).

For reasons discussed earlier, viral proteins that derive from Gag, Gag-Pol, and Env are produced at very low levels in many mammalian cells, although there are mammalian cell lines, such as 293 and D-17, that do express significant amounts of viral proteins (Nasioulas et al. 1995; Ferris, unpublished observations). The poor production of RCAS viral proteins in most murine cells predicts that RCAS vectors are unlikely to cause an immune response in mice (Pinto et al. 2000). Of course, any gene cloned in a RCAS vector may provoke an immune response if the encoded protein product is immunogenic. This means that a human ortholog introduced by RCAS may encode a protein that is immunogenic in mouse cells. Certain oncogenic mutations in the coding sequence may lead to the generation of a new epitope that could elicit an immune response as well. Indeed, this somatic expression system could be useful for studying CTL responses to naturally-occurring mutations in carcinogenesis. The clearance of cells producing a virally encoded immunogenic protein can be relatively rapid (a few weeks) while the severity of the response depends on the protein and on the cell type(s) that are infected (Pinto et al. 2000).

The expression of genes in RCAS proviruses may be lost due to a poorly understood phenomenon called gene silencing. Over time, infected cells that initially expressed a gene inserted into a retroviral vector may stop expressing the gene. Genes expressed from both LTR promoters and internal promoters are subject to silencing. In cultured cells, MLV-based and ASLV-based vectors appear to be more susceptible to silencing than lentiviral vectors (Katz et al. 2007). In cultured cells, silencing can be relatively well-defined experimentally, both because there is no CTL response and because there are treatments that can reactivate the expression of genes carried by silenced proviruses. It is less clear how frequently gene silencing occurs in animal models. In muscle, RCAS silencing is not a major problem for a period as long as several months (Federspiel et al. 1994). In the mammary gland, gene products can be detected 18 months following infection (unpublished observation). However, it is not clear whether silencing has occurred in some of the infected cells. In addition, infected cells may have a finite life span. If the gene of interest does not improve the survival of the infected cells, they may be lost over time. Therefore, with time, both the number of cells infected *in vivo* or transplanted into mice after ex vivo infection may decline, and of the infected cells that remain, some may lose the ability to express the inserted gene.

It is generally accepted that the preintegration complexes (PICs) produced by infection with murine leukemia virus (MLV), and with vectors that derive from MLV, cannot enter the intact nucleus and that these vectors do not infect nondividing cells. On the other hand, RCAS vectors can infect nondividing cells in culture, though with reduced efficiency; the PICs of RCAS and ALSVs appear to be capable of entering the intact nucleus (Hatziioannou and Goff 2001; Katz et al. 2002). The reduction in the efficiency of infection in nondividing cells has not been carefully quantified, and there is little or no data on the ability of the RCAS vectors to infect nondividing cells in mice.

In contrast, lentiviruses and lentivirus-derived vectors infect nondividing cells reasonably well, although the efficiency is still reduced compared to dividing cells. Modified lentiviral vectors whose *env* gene has been replaced by the ALV(A) *env* have been reported to infect nondividing cultured cells more efficiently than RCAS (Lewis et al. 2001; Pizzato et al. 2008). Lentiviral vectors have also been used to infect mammary cells in *tva* transgenic mice (Siwko et al. 2008). A benefit of using lentiviral vectors is that relatively large inserts can be expressed. However, unlike RCAS, lentiviral vectors do not carry viral genes for replication so that the virus must be generated by cotransfecting the vector DNA with other plasmids that supply Gag, Pol, and ALV Env(A). Such cotransfections usually lead to much lower titers than are routinely achieved with the replication-competent RCAS vectors.

Finally, it should be noted that following integration, RCAS and other retroviral vectors may alter expression of host genes, predisposing the infected cells to cancer. RCAS can integrate into or near a protooncogene, activating the oncogene in one of several ways. If RCAS is inserted immediately upstream of the transcriptional start site of a protooncogene, the viral LTR can increase the expression of the protooncogene. If the provirus is inserted in an intron or near the protooncogene, the enhancer elements in the LTR can cause increased expression of the protooncogene. Insertion into the 3′UTR may cause premature termination of transcription, making the mRNA either more or less stable. Intragenic integration can disrupt the function of a tumor suppressor gene, but this usually does not have deleterious effects in a diploid cell, unless the remaining intact locus is haplo-insufficient or the truncated gene product gains dominant-negative functions. Because microRNA has important regulatory functions in cancer, RCAS insertion may deregulate microRNA, predisposing host cells to cancer. However, RCAS vectors have not been reported to induce tumors by insertional mutagenesis in mammals, in contrast to what has been seen with some other vectors, including MLV, MMTV, and transposons (see Chapter 2). This may be due to the small number of cells that are infected when RCAS vectors are used, which would reduce the chance that a proviral insertion would lead to the deregulation of protooncogenes or tumor suppressor genes. Experiments in which RCAS was used to infect hematopoietic stem cells in monkeys suggest that, at least in this model, the RCAS vectors may have a lower potential for insertional activation of oncogenes than MLV vectors (Hu et al. 2007, 2008).

5.5 Organ-Specific Cancer Modeling Using the TVA Technology

As mentioned earlier, the TVA method has been used to study carcinogenesis in multiple tissues. While the general principle is the same for all tissues, there are tissue-specific issues in the application of this technique. The following narrative summarizes organ-specific applications of this method with focus on unique considerations for each organ.

5.5.1 Brain Cancer

Many genetic alterations have been reported in brain tumors. A number of these oncogenic lesions have been studied in *tva* transgenic mice, including *EGFR*, *Akt*, *Ras*, *PDGF*, *sonic hedgehog*, *β-catenin*, *Sox10*, *p21*, and *IGF* (Ferletta et al. 2007; Fults et al. 2002; Holland et al. 2000a, 1998; Liu et al. 2007; Momota et al. 2008; Rao et al. 2004). While *PDGF* alone appears to be able to induce tumors (Dai et al. 2001), the other oncogenes require collaborating genetic lesions. As has already been discussed, oncogene collaboration has been investigated by mixing two viruses at the time of injection. A second oncogene may affect tumor latency as well as the histopathology of the resulting tumors (Dai et al. 2005; Momota et al. 2008; Uhrbom et al. 2002). The combination of different oncogenes and *tva* lines has resulted in models for many subtypes of brain tumors, greatly expanding the pool of preclinical models for therapeutic testing (Hambardzumyan et al. 2008). In addition, glioma maintenance has been investigated by co-infection with RCASBP(A)-*rtTA* and the RCASBP(A) vector that carried an oncogene under the transcriptional control of the tet-responsive element (TRE) (Holmen and Williams 2005).

The normal human brain comprises glia, neurons, vascular endothelium, and meningeal cells. Glial cells are the predominant cell type, and are further subgrouped into astrocytes, oligodendrocytes, microglia, and ependymal cells. Gliomas are the most common type of brain tumor in adults. The more aggressive gliomas are also called glioblastomas or glioblastoma multiforme (GBM). Gliomas can be further subtyped into astrocytomas and oligodendrogliomas based on their histological features, but it is not clear whether they arise from normal astrocytes and oligodendrocytes, respectively, from just one of these cell types, or from a common progenitor cell. Brain tumors were the first cancer to be modeled using the RCAS/TVA system (Holland and Varmus 1998). Two transgenic mouse strains have been used for studying brain tumors: GFAP-*tva* and nestin-*tva*. RCAS viruses are generally delivered into the brains of these mice by intracranial injection of DF-1 cells producing the RCAS vector. Neonatal mice were initially used, but adult mice are also susceptible to infection.

GFAP encodes an intermediate filament protein and is expressed primarily in astrocytes postnatally (Holland and Varmus 1998). Direct intracranial injection of DF-1 cells producing RCAS vectors into neonatal GFAP-*tva* transgenic mice results in infection of the *tva*-expressing cells. This mouse strain was used to show that infection with an RCASBP(A) virus expressing *bFGF* induced extensive migration of glial cells. Subsequently, this mouse model has been used to generate tumors and to study how oncogenes modulate dedifferentiation of astrocytes (Lassman et al. 2004) or preferentially transform undifferentiated astrocytes. *Nestin* also encodes an intermediate filament protein, and is expressed in the progenitor population within the central nervous system (Lendahl et al. 1990). In general, Nestin-*tva* mice are more prone to the development of tumors upon RCASBP(A)-facilitated viral transduction than are GFAP-*tva* mice. For example, a combination of RCASBP(A)-*Akt* and RCASBP(A)-*K-Ras* induced tumors in 26% of the injected nestin-*tva* mice, but in none of the infected GFAP-*tva* mice (Holland et al. 2000a).

5.5.2 Breast Cancer

The breast epithelium forms during the late stage of embryogenesis through invagination of the embryonic epidermis, and develops into a tree-like structure that comprises an inner layer of epithelial cells and an outer layer of myoepithelial cells. This ductal tree is wrapped by a thin layer of basement membrane and is embedded in a stroma composed mostly of adipocytes. Many genetic lesions have already been found to play a causal role in breast cancer, but many that have been implicated still need to be tested. How these genetic alterations collaborate to transform breast cells is still not clear. Further complicating the matter is that the exact cellular origin of breast cancer has not been identified. Possibilities include differentiated breast epithelial cells, progenitor cells, and stem cells. Several transgenic lines have been made that express *tva* in different subsets of mammary cells, including lines using the MMTV promoter (Du et al. 2006), *WAP* (YL, unpublished), *keratin 5* (Orsulic et al. 2002), *keratin 19* (Morton et al. 2007; Siwko et al. 2008), and *keratin 6* (Bu et al. 2011). The MMTV promoter is active in an undefined subset of mammary epithelial cells, and the reported MMTV-*tva* MA line expresses *tva* in approximately half of the mammary epithelial cells in adult mice (Du et al. 2006; Siwko et al. 2008). *Keratin 19* is ubiquitously expressed in all mammary epithelial cells at all stages of mammary development. *WAP* is expressed in highly differentiated mammary epithelial cells, but may also be expressed in some progenitor cells (Boulanger et al. 2005; Li et al. 2007; Matulka et al. 2007; Robinson et al. 1995). *Keratin 6* is expressed in mammary progenitor cells but not stem cells (Bu et al. 2011).

While efficient infection can be achieved for many organs by injecting virus into the peritoneum or into the organ, infection of mammary epithelial cells is inefficient if the virus is injected intraperitoneally or directly into the exposed mammary fat pad. Intraperitoneal injection of RCASBP(A)-*PyMT* into MMTV-*tva* never led to tumors in a pilot study (YL, unpublished). Direct injection of RCASBP(A)-*PyMT* into exposed mammary glands in 7 MMTV-*tva* transgenic mice resulted in only two tumors after 1 year. On the other hand, though injection of virus via the lactiferous duct (Nguyen et al. 2000) is a challenging technique, especially in black mice, this route of viral delivery led to improved infection of approximately 0.3% of mammary cells in pubertal MMTV-*tva* transgenic mice (Du et al. 2006). Using this method, injection of 10^7 IU RCASBP(A)-*PyMT* in 10 μl caused tumors in all injected mice in approximately 1 month (Du et al. 2006). Injection of RCASBP(A) carrying cellular oncogenes such as *ErbB2* has also been shown to induce mammary tumors (Du et al. 2006; Reddy et al. 2010).

The rate of proliferation of epithelial cells in a developing ductal tree is much higher during puberty than in the adult mammary gland. However, the tree is much larger and denser in adults, so both the total number of epithelial cells and the number of proliferating epithelial cells are higher in adult mice. Consequently, delivery of the same amount of virus usually results in more infected cells in young adult mice (12-week-old) than in pubertal mice. Since lactiferous ducts are clogged in older mice (>25 weeks), injection of virus by this route is problematic in older mice. It is very difficult to inject virus into prepubescent mice because the ductal tree has not developed.

Co-injection of two viruses leads to co-infection in the mammary epithelium in MMTV-*tva* transgenic mice, although co-infection appears to be infrequent (less than 10% of the infected cells) (Dong and YL, unpublished). Oncogenic collaboration can be tested by injecting two admixed viruses. Sequential infection has not been tested in mammary glands. It is unlikely that this will be a useful approach for introducing oncogenes into previously infected cells in the mammary epithelium because the ductal tree is large and only a small volume can be delivered at a time, making it very difficult to retarget the same cells in subsequent infections.

Since the great majority (>95%) of mammary cells are not in the cell cycle in adult mammary glands, the use of pseudotyped lentiviral vectors may improve the infection efficiency in adult mammary gland. Siwko et al. (2008) have shown that Env(A)-pseudotyped lentivirus carrying *PyMT* can induce mammary tumors following injection into adult keratin 19-*tva* mice, but the relatively low titers of these vectors did not produce an infection rate better than that seen with RCAS viruses.

It has been speculated that stem cells may be the primary cell of origin for breast cancer (Clarke and Fuller 2006; Wicha et al. 2006). However, transgenic mice that express *tva* selectively in stem cells have not been reported. Selective expression of *tva* in the mammary stem cell population is especially challenging because a promoter for the selective expression in mouse mammary stem cells has not been found. The use of a combination of markers can enrich for stem cells (Shackleton et al. 2006; Stingl et al. 2006), but it is technically difficult to restrict *tva* expression to a subset of cells by the use of more than one promoter. In addition, stem cells are rare, and even if they are made the only susceptible target cells that can be infected by RCAS vectors, it still may be difficult to use these vectors to introduce oncogenes into stem cells for cancer modeling.

5.5.3 Liver Cancer

Primary liver cancers, predominantly hepatocellular carcinoma (HCC), are a leading cause of cancer-related mortalities. The poor survival rate of patients afflicted with this malignancy is primarily due to the development of HCC within the context of chronic hepatitis and liver cirrhosis, which significantly impairs normal liver function. In addition, a significant fraction of HCC patients develop metastases of the extra-hepatic segment of the portal vein and the lungs. Although the involvement of hepatitis and liver cirrhosis is well-documented, the molecular mechanisms that underlie the development of liver cancer and the progression of benign tumors to a malignant state are poorly understood. Important genetic alterations in HCC include mutated *TP53* and *INK4A/ARF*, amplified *c-Myc*, mutated *CTNNB1* encoding a stabilized β-catenin, and mutated *PI3KCA* encoding an activated p110 subunit of PI3K (Buendia 2000; Lee et al. 2005).

Lewis and colleagues made transgenic mice expressing *tva* under the control of the albumin promoter and enhancer, and used direct injection of RCASBP(A) virus into liver parenchyma in newborn mice to achieve infection. RCASBP(A)-*PyMT* caused liver tumors in approximately 65% of animals (Lewis et al. 2005). Although these tumors could exceed 1 cm in diameter, they rarely metastasized. Germline or liver-specific deletion of the *Trp53* tumor suppressor locus did not increase tumor development, but facilitated the development of pulmonary metastases in RCASBP(A)-*PyMT*-infected mice (Chen et al. 2007; Lewis et al. 2005). These findings are consistent with the proposal that the loss of *p53* is a late genetic event in HCC and underscore the role of *p53* in the progression of this malignancy (Buendia 2000). By introducing RCASBP(A) viruses encoding PyMT mutants impaired in the stimulation of specific downstream signaling pathways, Lewis et al. (2005) identified the PI3K/Akt/m-TOR signaling axis as a key mediator of the metastatic phenotype, a finding corroborated in subsequent analysis of migration and invasion properties of HCC cell lines *in vitro* (Chen et al. 2007).

While deletion of the *Ink4a/Arf* locus did not cooperate with *PyMT* in the formation of metastatic tumors, concomitant deletion of the *Trp53* and *Ink4a/Arf* tumor suppressor loci further accelerated the onset of metastatic disease, possibly by enhancing the invasive activity of tumor cells (Chen et al. 2007). Further, these studies led to the observation that the Arf tumor suppressor restrained HCC cell invasion in a p53-independent manner (Chen et al. 2008).

5.5.4 Ovarian Cancer

Human ovarian carcinomas are thought to arise from the epithelium that covers the surface of the ovary. Many genetic alterations have been reported, including *c-myc*, *K-ras*, *Akt*, and *p53*. In the absence of a promoter that is expressed exclusively in the epithelial cells on the surface of the ovary, Orsulic et al. (2002) used K5-*tva* transgenic mice that express *tva* from the *keratin 5* promoter. Although *tva* is expressed in various epithelial cell types in the K5-*tva* transgenic mice, the surface epithelium is the only TVA-positive cell type in the ovaries of these mice. Using this line of mice, Orsulic et al. (2002) developed the first mouse model for ovarian carcinoma that recapitulates human ovarian carcinoma development and progression.

In this model, the ovary-specific gene delivery was achieved by isolating the ovaries from the keratin 5-*tva* mice and infecting them ex vivo with RCASBP(A) vectors. A few days later, the infected ovarian cells were implanted orthotopically into nude or immunocompetent mice. While implantation of ovarian cells co-infected by RCASBP(A)-*Akt*, RCASBP(A)-*Ras*, and RCASBP(A)-*c-Myc* did not lead to tumors in 6 months, implantation of ovarian cells from p53-null *tva* mice that were infected by any two of these three RCASBP(A) viruses caused palpable tumors in 3–6 weeks. These findings showed that *p53* mutations have an essential role in the transformation of these ovarian cells with these three oncogenes. Remarkably, the mice developed ovarian tumors that phenocopy human ovarian papillary carcinoma and resemble the development and metastatic spread seen in human patients. The initial tumorigenic growth was confined to the implanted ovary, followed by spread to adjacent tissues, accumulation of ascites, and finally metastatic growth on the surfaces of intraperitoneal organs with a special affinity for the omentum and the mesentery.

Because these tumors developed very rapidly, it is reasonable to conclude that these combinations of genetic alterations were sufficient to transform ovarian epithelial cells. This genetically defined mouse model provides an opportunity to study genotype–phenotype correlations that may lead to a better understanding of the contributions of individual genetic alterations to tumor progression and metastasis (Miao et al. 2007; Xing et al. 2006). Mouse ovarian tumors that are induced with different combinations of defined genetic alterations make it possible to develop models for cooperation and cross-talk between redundant biochemical pathways, which appear to be the main reasons for the failure of therapeutic agents that are designed to interfere with a specific molecular pathway. For example, this model was successfully used to determine the molecular mechanisms of ovarian tumor resistance to mTOR-targeted therapy (Xing and Orsulic 2005a; Xing and Orsulic 2005b). A more detailed understanding of the biochemical pathways that are responsible for cancer maintenance and progression in this genetically defined mouse model may be valuable for the future design of effective targeted therapies of human ovarian cancer. The similarities in metastatic behavior to human ovarian cancer make this model particularly useful for developing and testing therapeutic approaches aimed at the advanced stages of human ovarian cancer.

5.5.5 Pancreatic Cancer

The human pancreas comprises primarily pancreatic exocrine cells (acinar cells) and their associated ducts. The islets of Langerhans are small clusters of endocrine cells embedded within this largely exocrine tissue. Pancreatic tumors display the histologic features of these major cell types. However, more than 90% of pancreatic tumors exhibit ductal character, and the vast majority of these are ductal adenocarcinomas. The current dogma is that the various pancreatic tumor types are derived from the transformation of their normal counterparts. However, the actual cells of origin for pancreatic cancers have not been identified.

Lewis and colleagues (2003) established a mouse model for pancreatic tumorigenesis by expressing *tva* under the control of the *elastase* promoter, which is expressed in mature pancreatic acinar cells as well as some progenitor cells (Chiang and Melton 2003). Intraperitoneal injection of concentrated RCAS viruses into neonatal mice was used as the route of viral delivery. Using this approach, RCASBP(A)-*PyMT*-induced both ductal and acinar precursor lesions with a latency period between 8 and 14 months (Lewis et al. 2003; Morton et al. 2008). If RCASBP(A)-*PyMT* was delivered to elastase-*tva* mice that were also null for the *Ink4a/Arf* locus, both acinar and ductal carcinomas arose, and the tumor cells produced the transcription factor Pdx1, a marker for early pancreas progenitors, along with the endocrine marker synaptophysin (Lewis et al. 2003; Morton et al. 2008). Metastases could not be detected in these mice, most likely due to early euthanasia that was required because other lesions (lymphomas, sarcomas) unrelated to the pancreatic phenotype were caused by the null *Ink4a/Arf* alleles. Delivery of RCASBP(A)-*PyMT* to elastase-*tva* mice with pancreas-specific deletion of the *Trp53* tumor suppressor locus led to the development of metastases, most frequently to the liver, consistent with features of the human disease (Morton et al. 2008). In contrast to *PyMT*, delivery of RCASBP(A)-*c-Myc* to elastase-*tva* transgenic mouse deleted for the *Ink4a/Arf* locus led to the development of pancreatic endocrine neoplasms that stained positive for insulin and endocrine-restricted transcription factors, such as Isl1 and Nkx2.2 (Lewis et al. 2003). These studies provide an insight on the relationships of the cancer cell of origin, the initiating oncogene, and the cancer phenotype. It is likely that elastase-positive pancreas progenitors can be stimulated to differentiate along the ductal axis by signaling pathways activated by PyMT, and along the endocrine axis by c-Myc activation. Indeed, activating mutations in K-Ras, which is an important downstream mediator of PyMT, occur in greater than 90% of pancreatic ductal adenocarcinomas (Hezel et al. 2006).

In recently published work, Seidler et al. (2008) described a new knock-in mouse line with Cre-induced *tva* expression from the Rosa26 locus. By crossing this mouse strain with the ptf1a-*cre* mouse line, they demonstrated that postnatal delivery of RCASBP(A) viruses encoding K-RasG12D stimulated the formation of pancreatic cancer precursor lesions which occasionally progressed to adenocarcinoma. They further showed that TVA-mediated infection of mice engineered to express *K-Ras(G12D)* in pancreatic progenitor cells with RCASBP(A) virus carrying a short hairpin RNA targeting *p53*, resulted in accelerated progression to invasive adenocarcinomas. Importantly, by using different *Cre* drivers, this *tva* line can be used to model other malignancies as well.

Using the RIP to drive the expression of *tva* in mice, Du et al. (2007) reported selective infection of islet cells. Using this line to study islet tumorigenesis in a background of RIP-*TAG* transgenic expression, dominant-negative E-cadherin or Bcl-xL can accelerate the formation of invasive and metastatic islet tumors initiated by RIP-*TAG*. This model exemplifies the use of somatic TVA models as tools for discovering genetic lesions controlling cancer progression, especially invasion and metastasis.

5.5.6 Other Cancers

Intraperitoneal injection of neonatal mice transgenic for β-actin-*tva* with RCASBP(A)-*PyMT* or RCASBP(A)-*Neu* induced hemangiomas and hemangiosarcomas in multiple sites in a few weeks to a few months (YL, unpublished). RCASBP(A)-*PyMT* also induced hemangiomas in mice expressing *tva* under the control of the stem cell leukemia gene (*scl*) promoter and SCL +19 enhancer (Sausville et al. 2008). By expressing *tva* selectively in endothelial cells using the promoter for TIE2, Montaner et al. (2003) showed that a herpesvirus gene encoding a G protein-coupled receptor (vGPCR) can induce endothelial malignancies that resemble human Kaposi sarcomas. Furthermore, the same group showed that Akt can induce hemangiomas in *Pten*-deficient mice (Sodhi et al. 2004).

Several laboratories have initiated studies aimed at using the RCAS/TVA system to gain insights into the mechanisms underlying melanocyte differentiation and the development of melanoma. Dunn et al. (2000) developed a mouse strain expressing *tva* under the control of the promoter for the dopachrome tautomerase (DCT, also known as tyrosinase-related protein 2). Using neural tube explant cultures, members of the Wnt signaling pathway were found to control both the expansion and cell fate of neural crest-derived melanocytes (Dunn et al. 2000, 2005). Transplantation of mouse melanocytes transduced ex vivo with MLV *env*-pseudotyped RCASBP(A) expressing different *Ras* isoforms led to better understanding of the specificity of Ras isoforms in melanocyte tumorigenesis (Whitwam et al. 2007). More recently, VanBrocklin et al. (2010) found that in DCT-*tva* mice that also harbored loxP sites flanking *INK4a/ARF*, melanomas could be induced within a few months after a subcutaneous injection of RCASBP(A) producing N-Ras^{Q61R} and Cre.

5.6 Practical Protocols and Tips for Using TVA-Mediated Gene Transfer Method

5.6.1 Selection of a tva cDNA for Transgenic Expression

Ectopic expression of either *tva*800 or *tva*950 sensitizes all cells tested so far for infection by RCAS. It is not known whether one TVA isoform may be better than the other in mediating viral entry or higher expression. *tva*800 is the one used by most investigators while *tva950* has also been reported to mediate RCAS infection *in vivo* (Murphy et al. 2003; Murphy and Leavitt 1999). Of note, the level of *tva* expression does not seem to correlate with the sensitivity to infection by RCAS. For *tva*800, it has been suggested that too high a level of expression might interfere with infection due to the potential competition for viral particles by soluble receptor that may be released from cell surface (Federspiel et al. 1994).

5.6.2 Cloning Exogenous Genes into RCAS Vectors

1. Clone the gene of interest into a selected RCAS plasmid. RCASBP(A) and derivatives (Fig. 5.2) are usually preferred.
2. Purify the plasmid DNA using a method that will yield pure DNA (cesium chloride gradients, Qiagen Maxiprep, etc.).
3. Confirm the plasmid identity again by restriction digestions with several enzymes or by sequencing.
4. Store the DNA at –20°C.

5.6.2.1 Tips

RCASBP(A) has only one cloning site (*Cla* I). (The other *Cla* I site in this vector is methylated when the plasmids are propagated in Dam^+ bacteria.) Directly cloning exogenous genes into such a large vector using *Cla* I can be inconvenient. Adaptor plasmids have been developed to facilitate the insertion of genes into the vectors (Hughes et al. 1987). Furthermore, these vectors have been modified to include more cloning sites. For example, RCASBP(A)-X contains *Not* I, *Pme* I, *Pac* I, and *Swa* I in addition to *Cla* I, and RCASBP(A)-Y contains the same sites in the reverse orientation (Fig. 5.2). RCASBP(A) has also been modified to allow Gateway-cloning (Dunn et al. 2000; Loftus et al. 2001), and this modified vector is available from Addgene. Gateway cloning is especially useful when many inserts need to be tested or when these inserts are already in Gateway compatible vectors.

The insert size for RCASBP(A) is limited to approximately 2.5 kb. Avoid inserting sequences that cause transcriptional termination or polyadenylation or sequences that have splice sites (unless splicing is a desired outcome).

Use only filtered pipet tips to dispense viral vector DNA to avoid cross-contamination. If a less transforming viral plasmid is even slightly contaminated by a more potent one, with time in culture, the more potent virus will overtake the less potent virus, since DF-1 cells harboring the contaminating virus will usually proliferate faster. Furthermore, in mice, the more transforming virus will likely mask any phenotype that may be associated with the weakly transforming virus.

5.6.3 Generation of DF-1 Cells Producing RCAS Viruses

Note: The following protocol is for generating virus from a single viral construct. Care should be exercised to avoid cross-contamination when multiple viruses need to be generated at the same time.

1. Maintain DF-1 chicken fibroblasts in growth medium (DMEM with high glucose, 10% fetal bovine serum, 2 mM L-glutamine, 10 units/ml penicillin, 10 μg/ml streptomycin) at 39°C (or 37°C) and 5% CO_2.
2. One day before transfection, pass DF-1 cells into three 60 mm tissue culture dishes so that they will be approximately 30% confluent at the time of transfection.
3. Transfect one dish with the experimental plasmid using Superfect (Qiagen) following the manufacturer's protocol (calcium phosphate and lipofectamine reagents can also be used). It is desirable to include a parallel transfection with the RCASBP(A)-*GFP* control plasmid so that both efficiency of transfection and possible contamination of virus in DF-1 culture can be evaluated. Caution should be taken in preparing and transferring viral plasmid DNA. The use of filtered tips is strongly recommended to avoid cross-contamination between viral plasmids. Keep the third dish as an untransfected control.
4. Expand the culture 24–48 h later, by passing the cells at a ratio of 1:6.
5. Examine the RCASBP(A)-*GFP*-transfected dish under an inverted fluorescence microscope to estimate the efficiency of transfection.
6. Continue passing cells for 5–7 days until all of the cells are presumably infected. At this time, the RCASBP(A)-*GFP*-transfected dishes should be near 100% GFP^+, and other dishes should not have any GFP^+ cells. Discard all dishes if the RCASBP(A)-*GFP* dishes are not near 100% GFP^+ or if GFP is found in non-RCASBP(A)-*GFP* transfected dishes.
7. At approximately 7 days posttransfection, freeze down cells from two 10-cm dishes carrying the virus of interest in 10% DMSO and 20% fetal bovine serum. For long-term storage, store the frozen cells in a liquid nitrogen storage tank.

8. At 7–10 days posttransfection, ascertain that the producer cells are stably producing the protein from the gene cloned in RCAS. Use Western blotting to validate the size of the protein, and immunofluorescent staining for validation of the proper cellular location of the protein and to confirm complete viral spread. Include the control dishes as well.

5.6.3.1 Tips

DF-1 cells are now widely used to grow RCAS viruses. Historically, chicken embryo fibroblasts (CEFs) were used to grow ALSV and RCAS vectors. However, CEFs have a finite lifespan and they have to be harvested frequently from fertilized eggs from strains of chickens that express *tva*. Furthermore, unless the CEFs are prepared from specific pathogen-free eggs, there is a very real possibility that the embryos could be infected with exogenous ASLVs, which can interfere with the replication of an RCAS vector and/or recombine with the vector. To make matters worse, most chickens have in their genomes one or more endogenous viruses that are closely related to the ASLVs, and by extension, to RCAS, and these can also recombine with the RCAS vectors. A strain of chickens (called EV-0) was developed that lacks endogenous viruses closely related to ALSV. Vector stocks can be safely made in cells from EV-0 birds. DF-1 cells were developed from CEFs made from this line (Himly et al. 1998; Kim et al. 2001; Schaefer-Klein et al. 1998), and can be purchased from ATCC (http://www.atcc.org/ATCCAdvancedCatalogSearch/ProductDetails/tabid/452/Default.aspx?ATCCNum=CRL-12203&Template=cellBiology).

If the protein of interest is undetectable or is produced at very low levels or in less than 50% of the DF-1 cells 7 or more days after transfection, the virus stock is not very useful. There are several obvious possibilities. An insert may cause recombination (due to size or repetitive sequences). The insert may encode a protein harmful to the producer cells, which could make it difficult or impossible to prepare a viral stock that expresses the insert because cells that harbor viral variants that have lost the insert outgrow the cells that express the insert. RCASBP(A)-*Cre* and RCASBP(A)-*TGFβ* both appear to be harmful to DF-1 cells, and are difficult to produce in high titers (see below for titer determination).

Chicken fibroblasts are traditionally grown at 39°C. DF-1 cells also grow well and produce high titer RCAS viruses at 37°C, though a careful comparison of viral yields at these two temperatures has not been done. The ATCC-recommended medium is very similar to what is described here, and can also support high titer production of RCAS viruses in DF-1 cells.

5.6.4 Virus Collection and Concentration

1. Expand the producer cells in the medium as stated above until there are about 20 15-cm dishes.

2. When the monolayer reaches confluence, remove excess medium to leave only approximately 12 ml for each dish. (The use of minimum amounts of medium gives more concentrated virus.)
3. Collect the supernatant once 24 h later (when the medium should have turned slightly yellowish), add fresh medium (12 ml/dish), and collect virus again daily for up to 7 days.
4. Concentrate the fresh supernatant each day. Remove cell debris by a low speed centrifugation (500 × *g*, 10 min, 4°C). Transfer the supernatant to sterile disposable ultracentrifuge tubes (30 ml), and spin tubes at 26,000 rpm (50,000 × *g*) for 1.5 h at 4°C in an ultracentrifuge. Remove all except 300 μl to obtain 100-fold or more concentration. Resuspend the pellet by vortexing for 2 min at a medium speed. Alternatively, cell culture supernatant that has been cleared of debris can be concentrated up to 200-fold using the Vivaspins20 centrifugal concentrator (Vivascience Ltd.), which has a molecular weight cutoff of 10,000 and can accommodate up to 15 ml in each tube.
5. Freeze the virus in aliquots at −80°C.

5.6.4.1 Tips

Virus producer cells should not be passed for more than 3 weeks before viral collection to avoid cross-contamination and a drop in titer, which often occurs because cells that produce high levels of virus often grow more slowly than those that make lower levels of virus. Although the viral genome is relatively stable in fully infected cells, passage of the virus often results in the loss of the inserted gene. Following multiple rounds of infection, culture supernatants may have many mutant viruses, but due to viral interference, only a limited number of new infections occur in fully infected producer cells. Thus, viral mutations can be minimized by passing/propagating the producer cells. If more viruses are needed for viral production, either prepare a fresh stock of cells by transfection or thaw frozen producer cells. If frozen stocks are exhausted, it is advisable to make new infected cells by transfecting plasmid DNA.

Ultracentrifugation increases titers relative to the unconcentrated stock; however, there is usually a slight loss in the viral yield since it is difficult to resuspend all the viral particles. Viral loss can be minimized using the Vivaspin columns, but some media components are also concentrated if this method is used.

Aliquoting the virus is important, since there is typically a loss of up to one log of viral titer in one cycle of freezing and thawing. Store aliquoted virus collected from each day into one row in the box. In this way, virus of the same titer can be obtained across different collection times by mixing viruses from all tubes in one column. Upon thawing, virus should be kept on ice and used for infection of cultured cells or for injection into *tva* mice as quickly as possible. We have noticed reduced rates of infection of the mammary gland if the virus was kept on ice for more than a few hours.

Handle dishes for one virus at a time. When working in the hood, allowing at least 5 minutes before working with another virus may reduce the chance of cross-contamination.

5.6.5 Viral Titer Determination

1. Viral titers can be determined by limiting dilution on DF-1 cells or mammalian cells stably expressing *tva* (e.g., NIH3T3-*tva* or 293T-*tva*). If DF-1 cells are used, fresh viral supernatants should be filtered using a 0.8 μm filter to avoid contamination by detached DF-1 producer cells that can be present in the viral supernatant.
2. One day before the limiting dilution assay, pass cells to 60 mm tissue culture dishes so that the cells will be approximately 30% confluent at the time of infection.
3. Make a series of tenfold dilutions of the viral supernatant (from 10^0 to 10^{11}) in growth medium.
4. Add 1 ml of each dilution to a 60 mm TC dish containing DF-1 cells; make duplicate plates for each dilution point.
5. Incubate for 3 h at 37°C.
6. Add 2 ml of growth medium to each dish; allow cells to grow for 4–7 days. For DF-1 cells, pass cells as needed to keep them in logarithmic growth so that the entire dish will be infected.
7. Assay for gene expression by an immunofluorescence assay (DF-1 or mammalian cells) or by Western blotting (DF-1 cells only). ELISA and real-time PCR can also be used.

5.6.5.1 Tips

If more precise titer determination is necessary, sequentially dilute the virus twofold instead of tenfold before adding into the culture dishes.

Fresh supernatants from RCASBP(A)-infected DF-1 cells usually have a titer of 10^7 to 10^8 on avian cells, but the measured titers on mammalian cells are usually a few-fold lower. This is probably due to lower activities of the RCAS LTR in mammalian cells.

Acknowledgments We thank Drs. Harold Varmus, Sheri Holmen, Vidya Sinha, and Gary Chamness for critical comments on this manuscript. This research was supported in part by National Institutes of Health R01 CA113869 and CA124820 (to YL) and the Intramural Research Program of the NIH, National Cancer Institute, Center for Cancer Research (to SHH).

References

Adkins HB, Brojatsch J, Young JA (2000) Identification and characterization of a shared TNFR-related receptor for subgroup B, D, and E avian leukosis viruses reveal cysteine residues required specifically for subgroup E viral entry. J Virol 74:3572–3578

Barr SD, Leipzig J, Shinn P, Ecker JR, Bushman FD (2005) Integration targeting by avian sarcoma-leukosis virus and human immunodeficiency virus in the chicken genome. J Virol 79:12035–12044

Barsov EV, Hughes SH (1996) Gene transfer into mammalian cells by a Rous sarcoma virus-based retroviral vector with the host range of the amphotropic murine leukemia virus. J Virol 70:3922–3929

Barsov EV, Payne WS, and Hughes SH (2001) Adaptation of chimeric retroviruses in vitro and in vivo: isolation of avian retroviral vectors with extended host range. J Virol 75:4973–4983

Bates P, Young JA, Varmus HE (1993) A receptor for subgroup A Rous sarcoma virus is related to the low density lipoprotein receptor. Cell 74:1043–1051

Boerkoel CF, Federspiel MJ, Salter DW, Payne W, Crittenden LB, Kung HJ, Hughes SH (1993) A new defective retroviral vector system based on the Bryan strain of Rous sarcoma virus. Virology 195:669–679

Boulanger CA, Wagner KU, Smith GH (2005) Parity-induced mouse mammary epithelial cells are pluripotent, self-renewing and sensitive to TGF-beta1 expression. Oncogene 24(4):552–560

Brojatsch J, Naughton J, Rolls MM, Zingler K, Young JA (1996) CAR1, a TNFR-related protein, is a cellular receptor for cytopathic avian leukosis-sarcoma viruses and mediates apoptosis. Cell 87:845–855

Buendia MA (2000) Genetics of hepatocellular carcinoma. Semin Cancer Biol 10:185–200

Bu W, Chen J, Morrison GD, Huang S, Creighton CJ, Huang J, Chamness GC, Hilsenbeck SG, Roop DR, Leavitt AD (2011) Keratin 6a marks mammary bipotential progenitor cells that can give rise to a unique tumor model resembling human normal-like breast cancer. Oncogene Epub May 2, 2011

Chai N, Bates P (2006) Na+/H+ exchanger type 1 is a receptor for pathogenic subgroup J avian leukosis virus. Proc Natl Acad Sci USA 103:5531–5536

Chang KW, Barsov EV, Ferris AL, Hughes SH (2005) Mutations of a residue within the polyproline-rich region of Env alter the replication rate and level of cytopathic effects in chimeric avian retroviral vectors. J Virol 79:10258–10267

Chen YW, Klimstra DS, Mongeau ME, Tatem JL, Boyartchuk V, Lewis BC (2007) Loss of p53 and Ink4a/Arf cooperate in a cell autonomous fashion to induce metastasis of hepatocellular carcinoma cells. Cancer Res 67:7589–7596

Chen YW, Paliwal S, Draheim K, Grossman SR, Lewis BC (2008) p19Arf inhibits the invasion of hepatocellular carcinoma cells by binding to C-terminal binding protein. Cancer Res 68:476–482

Chiang MK, Melton DA (2003) Single-cell transcript analysis of pancreas development. Dev Cell 4:383–393

Clarke MF, Fuller M (2006) Stem cells and cancer: two faces of eve. Cell 124:1111–1115

Dai C, Celestino JC, Okada Y, Louis DN, Fuller GN, Holland EC (2001) PDGF autocrine stimulation dedifferentiates cultured astrocytes and induces oligodendrogliomas and oligoastrocytomas from neural progenitors and astrocytes in vivo. Genes Dev 15:1913–1925

Dai C, Lyustikman Y, Shih A, Hu X, Fuller GN, Rosenblum M, Holland EC (2005) The characteristics of astrocytomas and oligodendrogliomas are caused by two distinct and interchangeable signaling formats. Neoplasia 7:397–406

Doetsch F, Caille I, Lim DA, Garcia-Verdugo JM, Alvarez-Buylla A (1999) Subventricular zone astrocytes are neural stem cells in the adult mammalian brain. Cell 97:703–716

Du YC, Lewis BC, Hanahan D, Varmus H (2007) Assessing tumor progression factors by somatic gene transfer into a mouse model: Bcl-xL promotes islet tumor cell invasion. PLoS Biol 5:2255–2269

Du Z, Podsypanina K, Huang H, McGrath A, Toneff MJ, Bogoslovskaia E, Zhang X, Moraes RC, Fluck MM, Allred DC et al (2006) Introduction of oncogenes into mammary glands in vivo with an avian retroviral vector initiates and promotes carcinogenesis in mouse models. Proc Natl Acad Sci USA 103:17396–17401

Dunn KJ, Brady M, Ochsenbauer-Jambor C, Snyder S, Incao A, Pavan WJ (2005) WNT1 and WNT3a promote expansion of melanocytes through distinct modes of action. Pigment Cell Res 18:167–180

Dunn KJ, Incao A, Watkins-Chow D, Li Y, Pavan WJ (2001) In utero complementation of a neural crest-derived melanocyte defect using cell directed gene transfer. Genesis 30:70–76
Dunn KJ, Williams BO, Li Y, Pavan WJ (2000) Neural crest-directed gene transfer demonstrates Wnt1 role in melanocyte expansion and differentiation during mouse development. Proc Natl Acad Sci USA 97:10050–10055
Elleder D, Stepanets V, Melder DC, Senigl F, Geryk J, Pajer P, Plachy J, Hejnar J, Svoboda J, Federspiel MJ (2005) The receptor for the subgroup C avian sarcoma and leukosis viruses, Tvc, is related to mammalian butyrophilins, members of the immunoglobulin superfamily. J Virol 79:10408–10419
Federspiel MJ, Bates P, Young JA, Varmus HE, Hughes SH (1994) A system for tissue-specific gene targeting: transgenic mice susceptible to subgroup A avian leukosis virus-based retroviral vectors. Proc Natl Acad Sci USA 91:11241–11245
Ferletta M, Uhrbom L, Olofsson T, Ponten F, Westermark B (2007) Sox10 has a broad expression pattern in gliomas and enhances platelet-derived growth factor-B-induced gliomagenesis. Mol Cancer Res 5:891–897
Fu SL, Huang YJ, Liang FP, Huang YF, Chuang CF, Wang SW, Yao JW (2005) Malignant transformation of an epithelial cell by v-Src via tv-a-mediated retroviral infection: a new cell model for studying carcinogenesis. Biochem Biophys Res Commun 338:830–838
Fults D, Pedone C, Dai C, Holland EC (2002) MYC expression promotes the proliferation of neural progenitor cells in culture and in vivo. Neoplasia 4:32–39
Gaur M, Murphy GJ, deSauvage FJ, Leavitt AD (2001) Characterization of Mpl mutants using primary megakaryocyte-lineage cells from mpl(–/–) mice: a new system for Mpl structure-function studies. Blood 97:1653–1661
Hambardzumyan D, Becher OJ, Rosenblum MK, Pandolfi PP, Manova-Todorova K, Holland EC (2008) PI3K pathway regulates survival of cancer stem cells residing in the perivascular niche following radiation in medulloblastoma in vivo. Genes Dev 22:436–448
Hanafusa H, Hanafusa T, Rubin H (1963) The defectiveness of Rous sarcoma virus. Proc Natl Acad Sci USA 49:572–580
Hatziioannou T, Goff SP (2001) Infection of nondividing cells by Rous sarcoma virus. J Virol 75:9526–9531
Hezel AF, Kimmelman AC, Stanger BZ, Bardeesy N, Depinho RA (2006) Genetics and biology of pancreatic ductal adenocarcinoma. Genes Dev 20:1218–1249
Himly M, Foster DN, Bottoli I, Iacovoni JS, Vogt PK (1998) The DF-1 chicken fibroblast cell line: transformation induced by diverse oncogenes and cell death resulting from infection by avian leukosis viruses. Virology 248:295–304
Holland EC, Celestino J, Dai C, Schaefer L, Sawaya RE, Fuller GN (2000a) Combined activation of Ras and Akt in neural progenitors induces glioblastoma formation in mice. Nat Genet 25:55–57
Holland EC, Hively WP, DePinho RA, Varmus HE (1998) A constitutively active epidermal growth factor receptor cooperates with disruption of G1 cell-cycle arrest pathways to induce glioma-like lesions in mice [In Process Citation]. Genes Dev 12:3675–3685
Holland EC, Li Y, Celestino J, Dai C, Schaefer L, Sawaya RA, Fuller GN (2000b) Astrocytes give rise to oligodendrogliomas and astrocytomas after gene transfer of polyoma virus middle T antigen in vivo. Am J Pathol 157:1031–1037
Holland EC, Varmus HE (1998) Basic fibroblast growth factor induces cell migration and proliferation after glia-specific gene transfer in mice. Proc Natl Acad Sci USA 95:1218–1223
Holmen SL, Williams BO (2005) Essential role for Ras signaling in glioblastoma maintenance. Cancer Res 65:8250–8255
Hou L, Loftus SK, Incao A, Chen A, Pavan WJ (2004) Complementation of melanocyte development in SOX10 mutant neural crest using lineage-directed gene transfer. Dev Dyn 229:54–62
Hu J, Ferris A, Larochelle A, Krouse AE, Metzger ME, Donahue RE, Hughes SH, Dunbar CE (2007) Transduction of rhesus macaque hematopoietic stem and progenitor cells with avian sarcoma and leukosis virus vectors. Hum Gene Ther 18:691–700
Hu J, Renaud G, Ferris A, Hendrie PC, Donahue RE, Hughes SH, Wolfsberg TG, Russell DW, Dunbar CE (2008) Reduced genotoxicity of avian sarcoma leukosis virus vectors in rhesus

long-term repopulating cells compared to standard murine retrovirus vectors. Mol Ther 16(9):1617–1623
Hu X, Pandolfi PP, Li Y, Koutcher JA, Rosenblum M, Holland EC (2005) mTOR promotes survival and astrocytic characteristics induced by Pten/AKT signaling in glioblastoma. Neoplasia 7:356–368
Hughes SH (2004) The RCAS vector system. Folia Biol (Praha) 50:107–119
Hughes SH, Greenhouse JJ, Petropoulos CJ, Sutrave P (1987) Adaptor plasmids simplify the insertion of foreign DNA into helper-independent retroviral vectors. J Virol 61:3004–3012
Katz RA, Greger JG, Darby K, Boimel P, Rall GF, Skalka AM (2002) Transduction of interphase cells by avian sarcoma virus. J Virol 76:5422–5434
Katz RA, Jack-Scott E, Narezkina A, Palagin I, Boimel P, Kulkosky J, Nicolas E, Greger JG, Skalka AM (2007) High-frequency epigenetic repression and silencing of retroviruses can be antagonized by histone deacetylase inhibitors and transcriptional activators, but uniform reactivation in cell clones is restricted by additional mechanisms. J Virol 81:2592–2604
Kim H, You S, Kim IJ, Farris J, Foster LK, Foster DN (2001) Increased mitochondrial-encoded gene transcription in immortal DF-1 cells. Exp Cell Res 265:339–347
Koo BC, Kwon MS, Choi BR, Lee HT, Choi HJ, Kim JH, Kim NH, Jeon I, Chang W, Kim T (2004) Retrovirus-mediated gene transfer and expression of EGFP in chicken. Mol Reprod Dev 68:429–434
Lassman AB, Dai C, Fuller GN, Vickers AJ, Holland EC (2004) Overexpression of c-MYC promotes an undifferentiated phenotype in cultured astrocytes and allows elevated Ras and Akt signaling to induce gliomas from GFAP-expressing cells in mice. Neuron Glia Biol 1:157–163
Lee JW, Soung YH, Kim SY, Lee HW, Park WS, Nam SW, Kim SH, Lee JY, Yoo NJ, Lee SH (2005) PIK3CA gene is frequently mutated in breast carcinomas and hepatocellular carcinomas. Oncogene 24:1477–1480
Lendahl U, Zimmerman LB, McKay RD (1990) CNS stem cells express a new class of intermediate filament protein. Cell 60:585–595
Lewis BC, Chinnasamy N, Morgan RA, Varmus HE (2001) Development of an avian leukosis-sarcoma virus subgroup A pseudotyped lentiviral vector. J Virol 75:9339–9344
Lewis BC, Klimstra DS, Socci ND, Xu S, Koutcher JA, Varmus HE (2005) The absence of p53 promotes metastasis in a novel somatic mouse model for hepatocellular carcinoma. Mol Cell Biol 25:1228–1237
Lewis BC, Klimstra DS, Varmus HE (2003) The c-myc and PyMT oncogenes induce different tumor types in a somatic mouse model for pancreatic cancer. Genes Dev 17:3127–3138
Li Z, Tognon CE, Godinho FJ, Yasaitis L, Hock H, Herschkowitz JI, Lannon CL, Cho E, Kim SJ, Bronson RT et al (2007) ETV6-NTRK3 fusion oncogene initiates breast cancer from committed mammary progenitors via activation of AP1 complex. Cancer Cell 12:542–558
Liu Y, Yeh N, Zhu XH, Leversha M, Cordon-Cardo C, Ghossein R, Singh B, Holland E, Koff A (2007) Somatic cell type specific gene transfer reveals a tumor-promoting function for p21(Waf1/Cip1). EMBO J 26:4683–4693
Loftus SK, Larson DM, Watkins-Chow D, Church DM, Pavan WJ (2001) Generation of RCAS vectors useful for functional genomic analyses. DNA Res 8:221–226
Matulka LA, Triplett AA, Wagner KU (2007) Parity-induced mammary epithelial cells are multipotent and express cell surface markers associated with stem cells. Dev Biol 303:29–44
Miao J, Wang Z, Provencher H, Muir B, Dahiya S, Carney E, Leong CO, Sgroi DC, and Orsulic S (2007) HOXB13 promotes ovarian cancer progression. Proc Natl Acad Sci USA 104:17093–17098
Momota H, Shih AH, Edgar MA, Holland EC (2008) c-Myc and beta-catenin cooperate with loss of p53 to generate multiple members of the primitive neuroectodermal tumor family in mice. Oncogene 27(32):4392–4401
Montaner S, Sodhi A, Molinolo A, Bugge TH, Sawai ET, He Y, Li Y, Ray PE, Gutkind JS (2003) Endothelial infection with KSHV genes in vivo reveals that vGPCR initiates Kaposi's sarcomagenesis and can promote the tumorigenic potential of viral latent genes. Cancer Cell 3:23–36
Morton JP, Klimstra DS, Mongeau ME, Lewis BC (2008) Trp53 deletion stimulates the formation of metastatic pancreatic tumors. Am J Pathol 172:1081–1087

Morton JP, Mongeau ME, Klimstra DS, Morris JP, Lee YC, Kawaguchi Y, Wright CV, Hebrok M, Lewis BC (2007) Sonic hedgehog acts at multiple stages during pancreatic tumorigenesis. Proc Natl Acad Sci USA 104:5103–5108

Mothes W, Boerger AL, Narayan S, Cunningham JM, Young JA (2000) Retroviral entry mediated by receptor priming and low pH triggering of an envelope glycoprotein. Cell 103:679–689

Murphy GJ, Gottgens B, Vegiopoulos A, Sanchez MJ, Leavitt AD, Watson SP, Green AR, Frampton J (2003) Manipulation of mouse hematopoietic progenitors by specific retroviral infection. J Biol Chem 278:43556–43563

Murphy GJ, Leavitt AD (1999) A model for studying megakaryocyte development and biology. Proc Natl Acad Sci USA 96:3065–3070

Nasioulas G, Hughes SH, Felber BK, Whitcomb JM (1995) Production of avian leukosis virus particles in mammalian cells can be mediated by the interaction of the human immunodeficiency virus protein Rev and the Rev-responsive element. Proc Natl Acad Sci USA 92:11940–11944

Nguyen D, Beeman N, Lewis MT, Schaack J, Neville MC (2000) Intraductal injection into the mouse mammary gland. In: Ip MM, Asch BB (eds) Methods in mammay gland biology and breast cancer research. Kluwer, New York, pp 259–270

Nusse R, Varmus HE (1982) Many tumors induced by the mouse mammary tumor virus contain a provirus integrated in the same region of the host genome. Cell 31:99–109

Orsulic S, Li Y, Soslow RA, Vitale-Cross LA, Gutkind JS, Varmus HE (2002) Induction of ovarian cancer by defined multiple genetic changes in a mouse model system. Cancer Cell 1:53–62

Pao W, Klimstra DS, Fisher GH, Varmus HE (2003) Use of avian retroviral vectors to introduce transcriptional regulators into mammalian cells for analyses of tumor maintenance. Proc Natl Acad Sci USA 100:8764–8769

Petropoulos CJ, Hughes SH (1991) Replication-competent retrovirus vectors for the transfer and expression of gene cassettes in avian cells. J Virol 65:3728–3737

Pinto VB, Prasad S, Yewdell J, Bennink J, Hughes SH (2000) Restricting expression prolongs expression of foreign genes introduced into animals by retroviruses. J Virol 74:10202–10206

Pizzato M, Popova E, Gottlinger HG (2008) Nef can enhance the infectivity of receptor-pseudotyped human immunodeficiency virus type 1 particles. J Virol 82(21):10811–10819

Rao G, Pedone CA, Del Valle L, Reiss K, Holland EC, Fults DW (2004) Sonic hedgehog and insulin-like growth factor signaling synergize to induce medulloblastoma formation from nestin-expressing neural progenitors in mice. Oncogene 23:6156–6162

Reddy JP, Peddibhotla S, Bu W, Zhao J, Haricharan S, Du YC, Podsypanina K, Rosen JM, Donehower LA, and Li Y (2010) Defining the ATM-mediated barrier to tumorigenesis in somatic mammary cells following ErbB2 activation. Proceedings of the National Academy of Sciences of the United States of America 107:3728–3733

Robinson GW, McKnight RA, Smith GH, Hennighausen L (1995) Mammary epithelial cells undergo secretory differentiation in cycling virgins but require pregnancy for the establishment of terminal differentiation. Development 121:2079–2090

Robinson JP, Vanbrocklin MW, Lastwika KJ, McKinney AJ, Brandner S, and Holmen SL (2010) Activated MEK cooperates with Ink4a/Arf loss or Akt activation to induce gliomas in vivo. Oncogene.

Sausville J, Molinolo AA, Cheng X, Frampton J, Takebe N, Gutkind JS, Feldman RA (2008) RCAS/SCL-TVA animal model allows targeted delivery of polyoma middle T oncogene to vascular endothelial progenitors in vivo and results in hemangioma development. Clin Cancer Res 14:3948–3955

Schaefer-Klein J, Givol I, Barsov EV, Whitcomb JM, VanBrocklin M, Foster DN, Federspiel MJ, Hughes SH (1998) The EV-O-derived cell line DF-1 supports the efficient replication of avian leukosis-sarcoma viruses and vectors. Virology 248:305–311

Seidler B, Schmidt A, Mayr U, Nakhai H, Schmid RM, Schneider G, Saur D (2008) A Cre-loxP-based mouse model for conditional somatic gene expression and knockdown in vivo by using avian retroviral vectors. Proc Natl Acad Sci USA 105:10137–10142

Shackleton M, Vaillant F, Simpson KJ, Stingl J, Smyth GK, Asselin-Labat ML, Wu L, Lindeman GJ, Visvader JE (2006) Generation of a functional mammary gland from a single stem cell. Nature 439:84–88

Siwko S, Bu W, Gutierrez C, Lewis BC, Jechlinger M, Schaffhausen B, Li Y (2008) Lentivirus-mediated oncogene introduction into mammary cells in vivo induces tumors. Neoplasia 11:653–662

Sodhi A, Montaner S, Patel V, Gomez-Roman JJ, Li Y, Sausville EA, Sawai ET, Gutkind JS (2004) Akt plays a central role in sarcomagenesis induced by Kaposi's sarcoma herpesvirus-encoded G protein-coupled receptor. Proc Natl Acad Sci USA 101:4821–4826

Stingl J, Eirew P, Ricketson I, Shackleton M, Vaillant F, Choi D, Li HI, Eaves CJ (2006) Purification and unique properties of mammary epithelial stem cells. Nature 439:993–997

Theodorou V, Kimm MA, Boer M, Wessels L, Theelen W, Jonkers J, Hilkens J (2007) MMTV insertional mutagenesis identifies genes, gene families and pathways involved in mammary cancer. Nat Genet 39:759–769

Uhrbom L, Dai C, Celestino JC, Rosenblum MK, Fuller GN, Holland EC (2002) Ink4a-Arf loss cooperates with KRas activation in astrocytes and neural progenitors to generate glioblastomas of various morphologies depending on activated Akt. Cancer Res 62:5551–5558

VanBrocklin MW, Robinson JP, Lastwika KJ, Khoury JD, and Holmen SL (2010) Targeted delivery of NRASQ61R and Cre-recombinase to post-natal melanocytes induces melanoma in Ink4a/Arflox/lox mice. Pigment Cell Melanoma Res 23:531–541

Vervoort VS, Lu M, Valencia F, Lesperance J, Breier G, Oshima R, Pasquale EB (2008) A novel Flk1-TVA transgenic mouse model for gene delivery to angiogenic vasculature. Transgenic Res 17:403–415

Whitwam T, Vanbrocklin MW, Russo ME, Haak PT, Bilgili D, Resau JH, Koo HM, Holmen SL (2007) Differential oncogenic potential of activated RAS isoforms in melanocytes. Oncogene 26:4563–4570

Wicha MS, Liu S, Dontu G (2006) Cancer stem cells: an old idea – a paradigm shift. Cancer Res 66:1883–1890

Wolf RM, Draghi N, Liang X, Dai C, Uhrbom L, Eklof C, Westermark B, Holland EC, Resh MD (2003) p190RhoGAP can act to inhibit PDGF-induced gliomas in mice: a putative tumor suppressor encoded on human chromosome 19q13.3. Genes Dev 17:476–487

Xing D, and Orsulic S (2005a) A genetically defined mouse ovarian carcinoma model for the molecular characterization of pathway-targeted therapy and tumor resistance. Proc Natl Acad Sci USA 102:6936–6941

Xing D, and Orsulic S (2005b) Modeling resistance to pathway-targeted therapy in ovarian cancer. Cell Cycle 4:1004–1006

Xing D, and Orsulic S (2006) A mouse model for the molecular characterization of brca1-associated ovarian carcinoma. Cancer Res 66:8949–8953

Young JA, Bates P, Varmus HE (1993) Isolation of a chicken gene that confers susceptibility to infection by subgroup A avian leukosis and sarcoma viruses. J Virol 67:1811–1816

Zheng XH, Hughes SH (1999) An avian sarcoma/leukosis virus-based gene trap vector for mammalian cells. J Virol 73:6946–6952

Chapter 6
Target-Selected ENU Mutagenesis to Develop Cancer Models in the Rat

Bart M.G. Smits, Edwin Cuppen, and Michael N. Gould

6.1 Introduction to Rat Genetics

The study of mammalian model systems is an indispensable tool to gain insight in human health and disease. In the search for the contribution of gene function to human traits and diseases, comparative genomic approaches have proven to be very powerful. Nowadays, comparative genomic approaches are greatly facilitated by the availability of near-complete and draft genome sequences of many mammals, including human (Lander et al. 2001; Venter et al. 2001), chimpanzee (Chimp Genome Consortium 2005), dog (Lindblad-Toh et al. 2005), rat (Gibbs et al. 2004), mouse (Waterston et al. 2002), and more. However, for practical purposes rodents are the preferred model organisms in the area of mammalian genetics, owing to their reproductive aptitude and small size.

In the early 1900s, rodent genetics started with rediscovering Mendel's law in animals. For both mice and rats certain coat color characteristics were found to inherit in Mendelian ratios (reviewed in Lindsey 1979; Paigen 2003). Soon after that, the first rat (PA) and mouse (DBA) inbred strains were established, which provided the foundation for the creation of thousands of inbred rodent strains that exist to date to mimic specific characteristics of human health and disease (http://www.informatics.jax.org/menus/strain_menu.shtml; http://www.rgd.mcw.edu/strains). Despite their equally early start, however, the mouse became the model organism of

B.M.G. Smits • M.N. Gould (✉)
McArdle Laboratory for Cancer Research, Department of Oncology,
University of Wisconsin – Madison, 1400 University Avenue, Rm 506A,
Madison, WI 53706, USA
e-mail: gould@oncology.wisc.edu

E. Cuppen
Hubrecht Institute for Developmental Biology and Stem Cell Research, Section Functional Genomics and Bioinformatics, Uppsalalaan 8, Utrecht 3584, CT, The Netherlands

J.E. Green and T. Ried (eds.), *Genetically Engineered Mice for Cancer Research: Design, Analysis, Pathways, Validation and Pre-Clinical Testing*,
DOI 10.1007/978-0-387-69805-2_6,

choice for geneticists. The rat has been favored for physiological and nutritional studies, probably owing to its bigger size.

In the last 30 years, mouse geneticists have made tremendous advances in the development of novel genetic techniques to manipulate the mouse germ line. Genome manipulation in a random fashion was already reported as early as 1979 using chemical mutagenesis of the male germ line (Russell et al. 1979). In 1980 and 1981, the first mouse germ line transgenics were published (Brinster et al. 1981; Costantini and Lacy 1981; Gordon et al. 1980; Harbers et al. 1981; Wagner et al. 1981) that were produced via zygotic pronuclear injections of naked DNA. The proceeding with the biggest impact on mammalian genetics came in the late 1980s, when the first gene-targeted mice were reported that were produced via homologous recombination in embryonic stem (ES) cells (Koller et al. 1989; Thomas and Capecchi 1990; Thompson et al. 1989). The major advantage of this technology has been that any genetic construct can be "knocked in" at a precisely selected position in the genome. With this discovery the mouse genetic toolbox was well-equipped, although many variations on these technologies have been developed over the last two decades (reviewed in Capecchi 2005).

Meanwhile, rat genetic research has mainly centered on positional cloning of Quantitative Trait Loci (QTL) (Jacob and Kwitek 2002). Literally, hundreds of inbred strains have been specifically bred to reflect particular aspects of human physiology, such as blood glucose levels, blood pressure levels, and cancer predisposition. By means of QTL mapping, many traits have been anchored to the genome. Employing positional cloning using congenic rat lines has in many cases reduced the size of the interval, sometimes even to single gene level (Lazar et al. 2005; Aitman et al. 2008). Nevertheless, it requires a tremendous amount of animals and dedication of the researchers to positionally clone each QTL to single gene level. Bearing in mind that phenotypes may be attenuated or diluted due to locus heterogeneity and epistasis (Glazier et al. 2002), positional cloning in many cases could have its limitations. Additionally, several susceptibility loci have been localized to noncoding portions of the genome (e.g., Samuelson et al. 2007), indicating that not only gene function, but also gene regulation could underlie a phenotype defined by a QTL interval.

It becomes increasingly evident that the genetic mechanisms harnessed by QTL are tremendously complex. Many efforts are required to dissect a QTL's genetic landscape that includes coding genes, noncoding RNAs (e.g., large noncoding RNAs, miRNAs), and regulatory genetic elements (e.g., enhancers, repressors, insulators). To examine the contribution of a genetic element's function to a QTL, reverse genetics is a broadly accepted approach. In a reverse genetic or gene-driven approach, a genetic element of unknown function is manipulated and the consequences of this alteration to the phenotype are studied.

Rat genetic tools and genomic resources are almost as abundant as those for the mouse (Lazar et al. 2005; Smits and Cuppen 2006). For example, the rat has a near-complete genome sequence (Gibbs et al. 2004), genome-wide polymorphism maps (Guryev et al. 2004; Kwitek et al. 2004; Zimdahl et al. 2004), and transgenic technology (Mullins et al. 2002) available. Yet one technology has been missing, namely the ability to target-specific sequences by homologous recombination in ES cells. Thus far, the rat has failed to produce pluripotent ES cells that are capable of forming all tissues, including the germ line in a chimeric embryo. Until recently, this has hampered rat researchers to knockout genes of interest. To bypass this technical hurdle, the Gould lab (McArdle Laboratory for Cancer Research, University of Wisconsin – Madison, USA) and the Cuppen lab (Hubrecht Institute for Developmental Biology and Stem Cell Research, Utrecht, The Netherlands) independently developed an alternative gene-inactivation strategy based on target-selected chemical mutagenesis with *N*-ethyl-*N*-nitrosourea (ENU). Among various other genes, four tumor suppressor genes have been successfully targeted, resulting in valuable novel model systems to assess certain aspects of hereditary cancer. This chapter reviews the establishment of ENU mutagenesis-based knockout technology in the rat and the characterization of the phenotypes of four tumor suppressor gene knockout models.

6.2 Rat ENU Mutagenesis-Based Knockout Technology

Target-selected ENU mutagenesis approaches have a simple outline, illustrated in Fig. 6.1. It starts with mutagenizing the male germ line, which in the rat occurs via intraperitoneal injection of ENU. This treatment results in random point mutations, which will be passed on to the F1 generation by mating mutagenized males with untreated females. DNA samples are extracted from a tissue sample of all F1 animals and screened for induced mutations in selected target genes, using high-throughput PCR-based screening technology. Occasionally, an induced mutation introduces a premature stop codon resulting in a truncated open reading frame (ORF). The heterozygous F1 animal is identified and the mutation is crossed to homozygosity. A series of biochemical and/or functional experiments need to be carried out to verify complete absence of the targeted protein.

Technical guidelines to initiate a rat ENU mutagenesis experiment have recently been described in detail (Smits et al. 2008). Here, we review the development of rat ENU mutagenesis-based knockout technology by the Gould lab and Cuppen lab.

6.2.1 ENU Mutagenesis of the Rat

ENU is considered the most potent mammalian stem cell mutagen (Justice et al. 2000). When injected in male rats, the alkylating action of ENU is thought to induce DNA adducts in the spermatogonial stem cells (van Zeeland et al. 1990). Through proliferation during spermatogenesis, some of the DNA adducts could cause incorporation of the wrong nucleotide during DNA replication, ultimately leading to fixation of a mutation in the descending sperm cells (Noveroske et al. 2000).

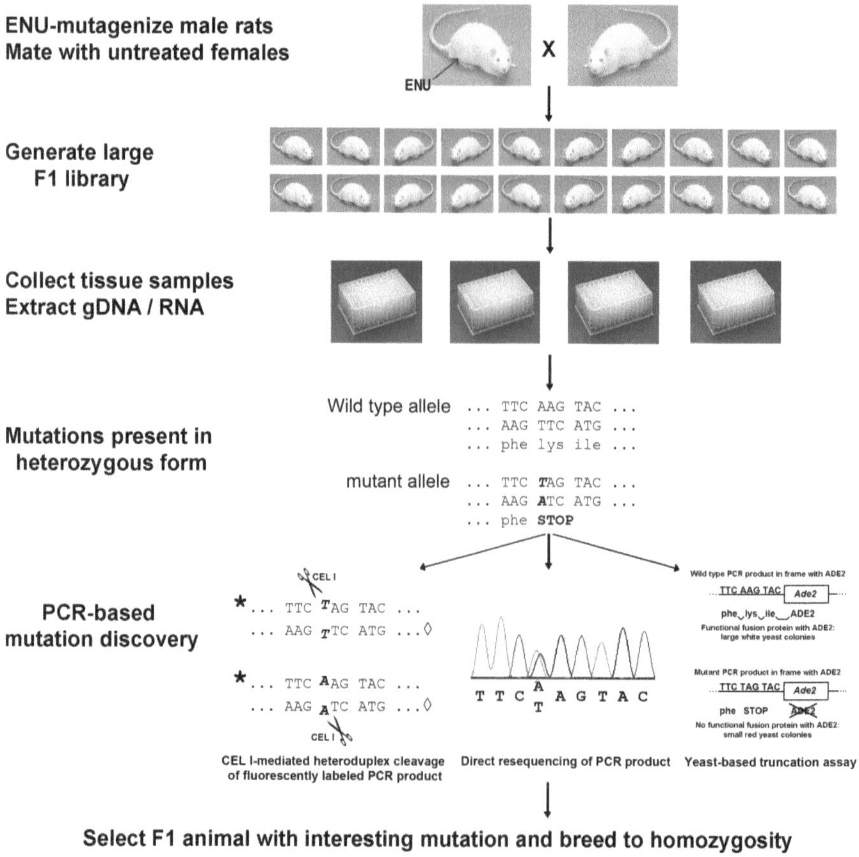

Fig. 6.1 Schematic overview of rat ENU-mutagenesis-driven knockout technology. Male rats are treated with ENU and mated to untreated females to generate a large cohort of F1 animals. From each F1 animal a tissue sample is collected from which DNA or RNA is extracted. Each F1 animal carries many random heterozygous mutations, some of which are located in the coding region or splice site of a gene to introduce a premature stop codon. Exons of genes of interest undergo inspection for heterozygous mutations in all collected DNA/RNA samples (see text for details). Animals carrying mutations in genes of interest are selected and bred to homozygosity

The final sperm population of a mutagenized male contains numerous fixed mutations. Since induction of mutations by ENU mutagenesis is a stochastic process, multiple males need to be mutagenized in order to generate a large F1 library of mutant animals that is needed to increase the chance that a particular gene of interest will be hit (Fig. 6.1).

The hit rate of ENU, i.e., the mutation frequency is strongly depending on the dosage. It has been determined in the mouse that the relationship between dosage and hit rate is linear above a certain threshold (Russell et al. 1982b). In addition, it has now been determined for mice and rats that dose repetition of ENU administration has a positive effect on the hit rate (Russell et al. 1982a; Hitotsumachi et al.

1985; Zan et al. 2003). However, too high doses could cause sterility (Zan et al. 2003; Smits et al. 2004), or sometimes even lethality. Too low doses result in less efficient mutagenesis and subsequently lower mutation rates. The dosage of ENU that can be administered to the rat is strongly strain dependent (Zan et al. 2003; Smits et al. 2006b, 2008), just like it is for the mouse (Justice et al. 2000).

It has been reported in mice and rats that ENU mutagenesis could cause a temporary reduction of fertility by killing off mature sperm cells in the testes (Justice et al. 2000; Zan et al. 2003; Smits et al. 2004). After a full round of spermatogenesis, which takes around 9–10 weeks in the rat, fertility rates should be back to normal, and mutations should have become fixed in fresh population of sperm. The optimal dose of ENU for a certain rat strain can be estimated as the dose at which around 50% of the treated males are fertile at 10 weeks of age. This is important to be able to effectively build a large F1 library. It turns out that effective ENU mutagenesis in the rat is characterized by a dip in pup production around 8 weeks after treatment. Suboptimal doses of ENU do not have this characteristic (Smits et al. 2008). The optimal doses of ENU for strains Brown Norway (BN), Fisher F344/Crl, Fisher F344/NHsd, Sprague Dawley (SD), and Wistar are 3×20, 3×40, 2×60, 2×60, and 3×40 milligrams per kilogram bodyweight, respectively. Other strains have been found to tolerate much lower doses and the optimal dose has not been determined yet (reviewed in Smits et al. 2008).

6.2.2 Discovery of Induced Mutations

ENU mutagenesis results in the introduction of single base pair changes (and sporadically small deletions) randomly distributed across the genome. In coding regions of genes, such mutations could lead to the introduction of a premature stop codon directly or via a frame-shift mutation. Frame shifts could be caused by a small deletion or an aberrant splicing event, in the case a splice donor or acceptor site was hit. In addition to stop codons, mutations could also alter an amino acid of a protein. Amino-acid changing mutations could potentially result in a reduction of protein function (hypomorph), or a gain of protein function (hypermorph) and therefore generate an allelic series. An allelic series of mutations could accelerate functional characterization of the genetic element. For example, while the knockout allele of a gene could result in a dramatic life-compromising developmental phenotype, a reduction-of-function allele could provide information about its function in later stages of life or about its contribution to a disease. In addition, amino-acid changing mutations could provide information on the protein domain they reside in. Finally, mutations could also be silent.

The discovery of newly induced mutations is tailored toward exons of genes of interest. Many mutation discovery platforms have been developed over the last decade. In the rat ENU mutagenesis projects three technologies have been applied, namely CEL I-mediated heteroduplex cleavage (Smits et al. 2004), high-throughput resequencing (Smits et al. 2006b), and a yeast-based truncation assay (Zan et al. 2003).

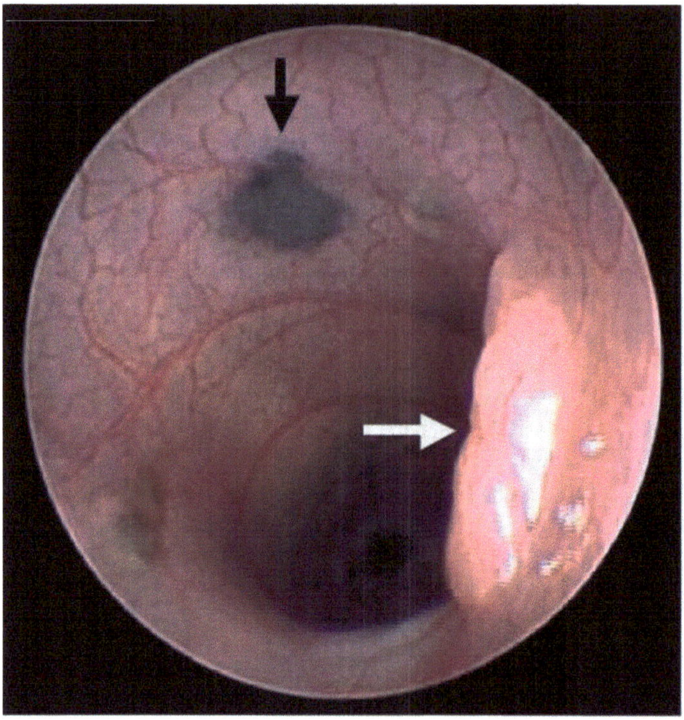

Fig. 6.2 Pirc colonic tumors (*white arrow*) are identified by endoscopy and marked with India ink tattoo (*black arrow*) placed in the submucosa of the adjacent normal epithelium. Tattoos have persisted more than 90 days and are likely to persist during the life span of the animal. Picture kindly contributed by Dr. James Amos-Landgraf of the McArdle Laboratory for Cancer Research (University of Wisconsin – Madison)

Each methodology has its own characteristics. A mutation discovery method should be selected, depending on the type of mutations and the amount of genes a researcher is interested in (Smits et al. 2008).

The yeast-based truncation assay developed by the Gould lab, could be applied when only full knockouts are desired (Zan et al. 2003). The method relies on cloning of a large exon or a large piece of cDNA in frame with a yeast reporter gene in a vector that is simultaneously shuttled into yeast cells. Yeast cells that harbor the functional chimeric construct will grow into large white colonies on a plate. Yeast cells that get a truncated version will result in smaller red colonies on a plate. A heterozygous mutant having a truncation of the ORF of the targeted gene is easily recognizable by a plate with half large white colonies and half small red colonies (Fig. 6.2). This method does not identify amino acid changing mutations, but focuses on truncations that have the highest probably of resulting in a full knockout situation. Since it requires significant manual operation, but also low investment and running costs, the method is suitable for laboratories that would like to have a rat knockout for a few genes. For routine production of rat mutants and knockouts, other, more automated methods have been developed.

The CEL I-based heteroduplex cleavage methodology is commonly used for mutation detection after chemical mutagenesis (Colbert et al. 2001; Wienholds et al. 2003; Smits et al. 2004). Briefly, an exon of interest is amplified by PCR using fluorescently labeled primers. The labeled product is then denatured and reannealed. In case a mutation was present, a heteroduplex will form that can be specifically cleaved by the CEL I endonuclease. These cleavage products are visualized as extra dots on a slap gel image, produced by a LI-COR DNA analyzer. The DNA of animals having "extra dots" in the exon of interest is resequenced to reveal the nature of the mutation. This method has the potency of identifying all mutations, including premature stop codons, splice-site mutations, and amino acid changing mutations. The assay can be run in a semiautomated fashion in small and larger scale experiments. The costs are rather low, especially owing to pooling of samples. The CEL I method scores moderately on robustness of data quality and accuracy.

A very accurate mutation discovery strategy is dideoxy sequencing. Fully automated, scaled-up versions of high-throughput resequencing technology have been applied to mutagenized libraries of rats, zebrafish, and *Caenorhabditis elegans* (Wienholds et al. 2002; Smits et al. 2006b; Cuppen et al. 2007). This strategy is only cost-effective if routine production of mutants is desired, as investment costs are considerably high. Using high-throughput resequencing on a library of mutant F1 rats, the Cuppen group was able to deduce the exact molecular mutation rates for two strains. These are 1 in 1.76×10^6 and 1 in 1.24×10^6 base pairs for F344/Crl and Wistar, respectively (Smits et al. 2006b), which is comparable to the mutation rates recently found for certain mouse strains (Concepcion et al. 2004; Quwailid et al. 2004). However, higher ENU-induced mutation rates have been reported for the genome of a vertebrate species. For example, the mutation rate of zebrafish could rise to approximately 1 in 2×10^5 base pairs (Wienholds et al. 2003).

Recently, a fourth mutation discovery technology was applied to a (cryopreserved) library of mutant F1 animals. This technology, named Mut-POWER, is based on Mu-transposition into heteroduplex positions (Mashimo et al. 2008).

6.2.3 Follow Up After Identification of an Interesting Mutant

If an interesting mutation is found, the corresponding F1 animal has to be bred to homozygosity. Once homozygous, the absence or functional alteration of the protein product needs to be demonstrated prior to phenotypic characterization of the knockout animal. While phenotyping a mutant animal, it should be kept in mind that potentially confounding background mutations inevitably introduced by ENU mutagenesis, could be present. To reduce the amount of background mutations, the mutation of interest must be backcrossed to the wild-type genetic background. Theoretically, the amount of background mutations should diminish approximately twofold with every backcross. However, if some background mutations are close to the mutation of interest, they have a high probability of being inherited together with the mutation of interest. For the rat, we estimated based on a mutation rate of 1 in 1.25 Mb and inspired by simulations for the mouse chromosomes (Keays et al. 2006), that after ten backcrosses DNA segments linked to the mutated site of interest could still contain 8–45? background mutations, depending on chromosome size (Smits et al. 2008). Although the chance that one of these linked background mutations happens to modify the phenotype under study is extremely small, precaution is required when analyzing homozygous mutants. Therefore, nonmutant littermates should always be included as a control group. In every cross, the homozygous genotype to phenotype relationship must be 1:1. If a confounding linked mutation is suspected, generating a second mutant or a phenocopy using a RNA interference-based approach could be considered.

6.3 ENU-Induced Rat Models for Hereditary Cancer

The rat model organism is widely used in cancer research. Many experiments rely on chemical carcinogenesis, xenografts, or other means of artificial cancer induction. Although examples of spontaneous dominant Mendelian inherited forms of cancer in the rat exist (Eker and Mossige 1961), models for hereditary cancer are relatively scarce. In search for development of novel hereditary cancer rat models, the Gould lab and Cuppen lab independently explored ENU mutagenesis as an alternative gene knockout strategy. This technology has only recently yielded the first mutant rat models. Here, we describe the initial characterization of four rat knockout models for the tumor suppressor genes *APC*, *BRCA1*, *BRCA2*, and *MSH6*. The most important characteristics are summarized in Table 6.1.

6.3.1 Adenomatous Polyposis Coli

In 2007, the Dove and Gould laboratories (McArdle Lab, University of Wisconsin – Madison) collaboratively published the generation and characterization of a rat model for familial colon cancer (Amos-Landgraf et al. 2007). This ENU-induced model carries a knockout allele of the *Apc* gene. The dominantly inherited human condition in which adenomatous polyposis coli (APC)-ablating heterozygous mutations lead to numerous colonic polyps is termed familial adenomatous polyposis (FAP). In fact, the majority of all human colon tumors (familial and sporadic) carry *APC* alleles that abolish protein function (Groden et al. 1991; Kinzler et al. 1991; Nagase and Nakamura 1993). Studies using mouse knockouts have provided major understanding of APC's role in cancer etiology.

In the early 1990s, the Dove lab identified the underlying mutation of the mouse mutant Min (multiple intestinal neoplasia). This mutant turned out to carry a truncating point mutation in the *Apc* gene, which results in tens of polyps in the small intestine (Moser et al. 1990). Additional knockout alleles of *Apc* in the mouse resulted in histologically similar small intestinal polyp adenomas, although multiplicity differed dramatically (Fodde et al. 1994; Oshima et al. 1995). The neoplasia occurrence of

Table 6.1 Properties of target-selected ENU mutagenesis-induced rat knockout models for hereditary cancer

Model	Mutation	Median survival and tumor phenotype description
Apc	K1137X	Homozygous embryonic lethal, heterozygous survival ~11 months for males, >17 months for females. At 4 months, extensive colonic and intestinal polyposis in heterozygotes was observed
Brca1	Splice site Δexon22	Homozygous embryonic lethal, heterozygous survival normal (>18 months). No increased predisposition to cancer in heterozygous animals was observed
Brca2	Y1359X	Homozygous survival ~14–15 months. Homozygous animals are sterile, smaller and develop multiple tumors in various organs. Mammary cancer incidence was not elevated compared to control littermates. Heterozygous survival normal (>18 months) and no increased predisposition to cancer animals was observed
Msh6	L306X	Homozygous survival ~14 months. At 9 months, homozygous animals develop multiple tumors, of which T- and B-cell lymphoblastic lymphomas were the most predominant

Please note that all viable rat models survive longer compared to complementary mouse knockout models. See text for details

Apc knockout mice could only be shifted toward primarily colonic in the presence of a null allele in *Cdx2* or *Smad3* (Aoki et al. 2003; Sodir et al. 2006), although this genetic system clearly has its limitations in terms of resembling the genotype of human FAP patients. An animal model that would recapitulate human APC deficiency-derived colonic carcinogenesis more directly is highly preferred.

The desired ENU-induced truncating mutation in the rat *Apc* gene was found by screening 1,360 progeny of mutagenized male founder animals, using the yeast-based truncation assay (Amos-Landgraf et al. 2007). The mutation was induced in the F344/NTac inbred rat strain. Resequencing of the entire coding portion of *Apc* revealed a single base pair change, introducing a premature stop codon (K1137X). Since *Apc* is an essential gene, lack of functional protein in the homozygous state cannot be demonstrated. Nevertheless, the heterozygous phenotype is fully penetrant. After 4 months of age, heterozygous animals develop extensive polyposis in the colon and microadenomas are found in both colon and small intestine. This mutant rat kindred was termed Pirc for polyposis in the rat colon. Importantly, mutant rats developed macroadenomas in the colon and small intestine in a 1:1 ratio, which contrasts the Min mouse model and closely resembles human FAP patients (Amos-Landgraf et al. 2007). In addition, colonic adenoma multiplicity can be significantly boosted using ENU treatment in the Pirc model system, if higher statistical power is needed.

Besides these phenotypic advantages over the existing mouse models, the long-lived Pirc model provides unique opportunities for *in vivo* tumor imaging, using microCT technology or classical endoscopy (Amos-Landgraf et al. 2007). The most appealing application in this respect would be *in vivo* monitoring of individual adenomas' behavior (progression, regression, etc.) upon pharmacological intervention or preventive therapy. Recently, individual adenomas have been successfully tagged *in vivo* using microtatooing on healthy tissue next to the tumor (Amos-Landgraf and Dove, personal communication; Fig. 6.2), which allows for tracking of the adenoma even though its morphology may change drastically upon treatment.

The Pirc model also represents interesting genetic features for further study. It is believed that an early transforming event in the majority of colonic tumors is the loss of the wild-type *Apc* allele, which was also confirmed in colonic and small intestinal adenomas of Pirc rats. Yet, the rat metacentric karyotype allows for distinguishing between whole chromosome loss, somatic recombination, or somatic local deletion as the mechanism through which LOH (loss-of-heterozygosity) at mutated site in the *Apc* gene occurs. Somatic recombination (and deletion) involves only one arm of the chromosome. Since the mouse karyotype is acrocentric, examination of both sides of the centromere is unavailable. To test if the entire wild-type chromosome is lost in tumors compared to normal adjacent tissue, Amos-Landgraf et al. (2007) made use of polymorphic sites on both arms of the chromosome, obtained by mating Pirc animals of the F344 strain to animals of the Wistar-Furth (WF) strain. None of the tumors analyzed showed loss of both arms, suggesting that whole chromosome loss is a rare event in APC deficiency-initiated carcinogenesis in this model. The majority of tumors showed LOH on one arm, often extending over 10 Mb of *Apc*, indicating that LOH in these cases was caused by somatic recombination or deletion.

Another genetic aspect of the Pirc model that is subjected to further investigations is the identification of modifiers of *Apc* polyposis. It has been observed that the rat *Apc* knockout allele in different inbred genetic backgrounds results in altered multiplicity. More importantly, one genetic background entirely shifts adenoma formation to the colon (Amos-Landgraf et al., manuscript in prep). This situation in the Min mouse could only be achieved by adding a knockout allele of *Smad3* (Sodir et al. 2006).

In combination with experimental advantages the rat model organism provides by its greater size, the Pirc model is a valuable addition to the plethora of animal models of *APC* polyposis.

6.3.2 *Brca1/Brca2*

The first knockout alleles generated in the rat worldwide were *Brca1* and *Brca2* (Zan et al. 2003). *BRCA1* and *BRCA2* are tumor suppressor genes that both function in DNA repair. In the mid-1990s, *BRCA1* and *BRCA2* were found to be predisposition genes to human breast cancer by positional cloning and linkage analysis (Hall et al. 1990; Castilla et al. 1994; Friedman et al. 1994; Miki et al. 1994; Wooster et al. 1995). Disease-causing variants not only confer high risk of breast cancer (10–20-fold increase), but also elevate risk to ovarian and other cancers (Breast Cancer Linkage Consortium 1999; Thompson and Easton 2002). Similar to FAP patients, *BRCA1* and *BRCA2* carriers are thought to develop tumors due to loss of the wild-type allele.

Mouse models lacking *brca1* or *brca2* are embryonic lethal (reviewed in Evers and Jonkers 2006). Disappointingly, in heterozygous state none of these mutants increases tumor incidence. If *brca1* or *brca2* was targeted via cre recombination in the mammary gland, it required mammary-specific loss of *Tp53* to boost mammary carcinogenesis (Brodie et al. 2001; Jonkers et al. 2001; McCarthy et al. 2007). The tumors in the mammary-specific (K14-cre) *brca1 Tp53* knockout model were described as solid in growth and highly proliferative with anaplastic nuclei. The majority expressed basal keratin, and were ER/PR/Erbb2 negative, which resembles a poor prognosis-group of human breast cancers (McCarthy et al. 2007). Similarly, the *brca2 Tp53* mammary-specific (K14-Cre) knockout situation generated mammary carcinomas with very high efficiency (Jonkers et al. 2001).

Generally, rat mammary tumors accurately recapitulate human breast tumors with respect to responsiveness to hormones and initiation in the mammary duct (Russo et al. 1990; Nandi et al. 1995). In search for an improved genetic model for human breast cancer, the Gould lab published the generation of knockout alleles for *Brca1* and *Brca2* using ENU mutagenesis in the SD rat strain (Zan et al. 2003). Using the yeast-based truncation assay, a splice acceptor site mutation and a premature stop codon were identified in *Brca1* and *Brca2*, respectively. Identical to mouse *brca1* knockouts, heterozygous *Brca1* rat intercrosses fail to produce homozygous pups, implicating embryonic lethality (Smits et al. 2006a). Interestingly, homozygous *Brca2* rat knockouts are viable, in contrast to their mouse counterparts (Zan et al. 2003). The *Brca2* homozygotes, however, appear to be significantly smaller than their heterozygous littermates, and they are infertile due to impaired gonad development (Cotroneo et al. 2007). The rat *Brca2* knockouts showed a shortened lifespan as a result of the formation of a variety of tumors, of which osteosarcomas were the most frequent (Cotroneo et al. 2007). The formation of mammary tumors was not increased over control animals. Possibly, the perturbed hormonal environment (due to ovarian dysfunction) hampered mammary carcinogenesis. To overcome this situation, the mutation was transferred to an inbred genetic background (WF) in order to allow for mammary gland transplant studies. Even in a hormonally normal setting, transplanted Brca2 lacking mammary glands fail to grow carcinomas, although accelerated growth was observed, as defined by a higher degree of branching and lobularity compared to transplanted control mammary glands (Smits et al. 2006a).

Finally, heterozygous *Brca1* and *Brca2* rat knockouts were monitored over prolonged time periods to record their putative increased mammary cancer predisposition. Comparable to heterozygous mouse *brca1* and *brca2* knockouts, no increased tumor incidence was observed, even when treated with the chemical carcinogens DMBA and NMU (Smits et al. 2006a). The development of *Brca1* and *Brca2* knockout rats into genetic mammary carcinogenesis models has not yet succeeded. Nevertheless, the viability of the *Brca2* homozygous knockout state and the subsequent formation of cataracts in these animals have afforded the possibility to study Brca2 function in a tissue-specific manner in an animal model system.

6.3.3 *Msh6*

One of the knockout alleles identified in a large target-selected ENU mutagenesis-driven rat reverse genetics project was a premature stop codon in the *Msh6* gene (Smits et al. 2006b). The MSH6 protein is a member of the mismatch repair (MMR) protein family. Like for other MMR proteins, it is believed that MSH6 deficiency contributes to a cell's transformation by accumulation of mutations and/or by preventing an appropriate apoptotic response to the DNA damage (Kinzler and Vogelstein 1996; Fishel 2001; Yang et al. 2004). Deleterious mutations in the *MSH6* gene are the underlying genetic defect for <10% of HNPCC (Hereditary Non-Polyposis Colorectal Cancer) or Lynch syndrome cases (Akiyama et al. 1997; Miyaki et al. 1997; Wijnen et al. 1999; Berends et al. 2002), which is a dominantly inherited cancer predisposition syndrome (Lynch and Smyrk 1996). Again, tumor development is most probably initiated by LOH likely associated with the genetic instability caused by MSH6 deficiency. Recently, characterization of the rat *Msh6* knockout model revealed an interesting spectrum of tumors, not identical to complementary mouse models and with high similarity to the tumor spectrum of atypical HNPCC patients (van Boxtel et al. 2008).

The knockout allele was identified by resequencing in a library of progeny from mutagenized male Wistar rats. The mutation changes a leucine to a premature stop codon (L306X) in the beginning of the DNA MMR domain. Homozygous mutant rats showed no detectable protein on a Western blot. In addition, these rats show a mutator phenotype, characterized by microsatellite instability and accumulation of point mutations in the germ line. These observations indicate that the rat *Msh6* mutant is a full knockout (van Boxtel et al. 2008).

Insight in the contribution of MSH6 malfunctioning to cancer etiology is predominantly obtained from mouse knockout studies. Homozygous *msh6* knockout mice have a median survival of 6–10 months with tumors arising as early as 3 months of age (Edelmann et al. 1997; de Wind et al. 1999). The tumor spectrum of *msh6*$^{-/-}$ mice consists primarily of invasive T and B-cell lymphomas and GI tract tumors (Edelmann et al. 1997). Liver, lung, skin, fibrous cell, and endothelial cell tumors have also been observed.

The most frequently occurring tumor types in *Msh6*$^{-/-}$ rats are also T and B-cell lymphomas (lymphoblastic). Other tumors observed in *Msh6*$^{-/-}$ rats include squamous cell carcinoma, leiomyocarcinoma, Leydig cell tumor, and adeno(carcino)ma in the stomach, uterus, testis, and mammary gland, respectively (van Boxtel et al. 2008). Strikingly, four of seven female *Msh6*$^{-/-}$ rats developed uterus cancer of which three suffered from endometrial cancer, a type of uterus cancer seen in female *msh6*$^{-/-}$ mice at a much lower frequency (3 in 22 in only one report; de Wind et al. 1999). Atypical HNPCC patients bearing mutations in *MSH6* show a reduced penetrance of colorectal cancer and prevalence of endometrial cancer (Kolodner et al. 1999; Wijnen et al. 1999; Wu et al. 1999; Wagner et al. 2001). In addition, *MSH6* has been found mutated relatively frequently in an unselected series of endometrial cancers (Goodfellow et al. 2003). The rat *Msh6* knockout model provides an excellent opportunity to study genetic instability-related endometrial carcinogenesis in an animal model system that is accessible to *in vivo* endoscopic procedures and irradiation treatment.

The most striking differences between Msh6 deficient mice and rats were the delayed appearance of tumors and prolonged survival in the rat model. The *Msh6*$^{-/-}$ rats have a median survival of ~14 months with tumors arising around 9 months, compared to 6–11 months of median survival and tumor appearance around 3 months of age in the mouse model. This potentially provides an extra experimental window in a MMR affected mammalian carcinogenesis model.

6.4 The Impact of Rat Knockout Technology

It is evident that rat genetic research made strong progression over the last decades (Lazar et al. 2005; Smits and Cuppen 2006; Aitman et al. 2008). At present, geneticists are increasingly interested in understanding the genetics underlying complex traits, such as hypertension, diabetes, and cancer susceptibility, for which the rat model system has been utilized extensively. However, while rat genetic research is gaining momentum, one important methodology has been missing until recently, namely the ability to knockout a gene of interest. Embryonic stem cell-based technology remains elusive for the rat. In fact, gene-targeting technology by homologous recombination in ES cells is strictly limited to the mouse. To circumvent the need for ES cells, we successfully explored target-selected ENU mutagenesis as a means to produce knockouts in the rat (Zan et al. 2003; Smits et al. 2004, 2006b). This strategy has been successfully developed for other model organisms, such as zebrafish (Wienholds et al. 2002), *Arabidopsis* (McCallum et al. 2000), *C. elegans* (Jansen et al. 1997), *Drosophila* (Bentley et al. 2000), and also in the mouse (Beier 2000; Coghill et al. 2002).

Originally, ENU mutagenesis in the mouse has been developed primarily to boost mutation rates in order to allow for large-scale phenotype-driven screens (Hrabe de Angelis et al. 2000; Nolan et al. 2000). Deduced from a limited number of gene-based mouse ENU mutagenesis screens, the ENU-induced mutation rates are comparable between the rat and the mouse (Coghill et al. 2002; Concepcion et al. 2004; Quwailid et al. 2004), namely roughly 1 in 1.25 million base pairs (Smits et al. 2006b). Taking into account that the rat model organism could be suited for the study of a different variety of phenotypes, large-scale phenotypic ENU mutagenesis screens in the rat is not a mission impossible anymore. In addition, genome-wide genetic variation, which is essential for genetic mapping of induced phenotypes, is becoming available in high resolution. A genome-wide SNP mapping panel was already used to identify the molecular lesion in the *Myo7a* gene underlying the ENU-induced tornado phenotype in the rat, illustrating that the approach is viable (Smits et al. 2005).

Ideally, rat knockout technology would allow for a knockout of every gene of interest. It has been estimated that roughly 50,000 F1 animals would have to be generated to have a 96% chance to find an induced premature stop codon in the average rat gene of 1,300 bp, based on a mutation rate of 1 in 1.2 Mb and the three most frequently occurring mutations induced by ENU mutagenesis (Smits et al. 2006b). Hence, larger than average genes are more suited to be targeted. For example, the Gould lab screened libraries of 1,965, 788, and 1,360 F1 rats to find a truncation mutation in the relatively large ORFs of *Brca1*, *Brca2*, and *Apc*, respectively. In such cases, it would be perfectly suitable to follow a rolling circle model for F1 library production. ENU-treated males are continuously bred at a pace that allows for continuous screening of the progeny without accumulation of F1 animals. F1 animals having an interesting mutation are kept and all other animals are discarded after screening of the target genes.

An attractive alternative has recently been published, which is to create a library of cryopreserved sperm from all male F1 rats (Mashimo et al. 2008). The gene of interest is screened in the corresponding DNA samples and the heterozygous animal is rederived from the frozen sperm sample that contains the desired mutation. It does require intracytoplasmatic sperm injection (ICSI) to fertilize oocytes with mutant sperm prior to placement of the embryo in the uterus, which is a specialized procedure not suitable for every laboratory. Other examples of successful rederivation of rats from cryopreserved sperm via artificial (intrauterine) insemination exist (Nakatsukasa et al. 2001, 2003), but this method is hardly transferable to other laboratories.

Recently, other means of generating knockout rats have been explored. The most prominent technology is Sleeping Beauty (SB) transposon mutagenesis (Kitada et al. 2007; Lu et al. 2007; see Chapter 4). Although the first characterized knockout has not yet appeared in publication, several groups have announced to have identified multiple knockout alleles (Rat Genomics and Models Meeting 2007, Cold Spring Harbor, NY). Another promising technology that would allow for homologous gene targeting, similar to mouse ES cell-based technology, is to generate cloned rats via somatic cell nuclear transfer. Although homologous recombination has not yet been shown for the rat nuclear transfer technology, cloned rats have already been produced (Zhou et al. 2003).

The rat ENU mutagenesis programs initiated by the Gould lab, the Cuppen lab, the PhysGen program (Medical College of Wisconsin – Milwaukee, WI; http://www.pga.mcw.edu), and the Serikawa lab (Kyoto University, Kyoto, Japan) have already yielded many interesting induced mutants. The newly produced rat knockout models for the serotonin transporter (*Sert*) (Homberg et al. 2007) and the four widely studied tumor suppressor genes reviewed here (*Apc*, *Brca1*, *Brca2*, and *Msh6*) (Amos-Landgraf et al. 2007; Cotroneo et al. 2007; van Boxtel et al. 2008) have just made their appearance in the scientific community. The size of the rat allows for novel research angles inapplicable in existing models, for example, the colonic imaging options in the Pirc model (Amos-Landgraf et al. 2007). Another advantageous common feature among the *Apc*, *Brca2*, and *Msh6* knockouts is the longer lifespan and the capacity to bear larger tumor burden compared to existing knockout models in the mouse enabling longitudinal studies of tumor development and treatment. Finally, the phenotypes have various other unique features, which make these novel rat models a valuable addition to the existing complementary murine models.

Acknowledgments We are grateful to Dr. James Amos-Landgraf (University of Wisconsin – Madison) for kindly providing an endoscopic picture of a rat colonic adenoma marked with a tattoo. We are grateful to Mr. Ruben van Boxtel (Hubrecht Institute – Utrecht, Netherlands) for critically reading the manuscript. BMGS has been supported by a Postdoctoral Award from the Era of Hope Breast Cancer Research Program of the Department of Defense.

References

Aitman TJ, Critser JK, Cuppen E, Dominiczak A, Fernandez-Suarez XM, Flint J, Gauguier D, Geurts AM, Gould M, Harris PC (2008) Nat Genet 40(5):516–22. Review

Akiyama Y, Sato H, Yamada T, Nagasaki H, Tsuchiya A, Abe R, Yuasa Y (1997) Germ-line mutation of the hMSH6/GTBP gene in an atypical hereditary nonpolyposis colorectal cancer kindred. Cancer Res 57:3920–3923

Amos-Landgraf JM, Kwong LN, Kendziorski CM, Reichelderfer M, Torrealba J, Weichert J, Haag JD, Chen KS, Waller JL, Gould MN et al (2007) A target-selected Apc-mutant rat kindred enhances the modeling of familial human colon cancer. Proc Natl Acad Sci USA 104:4036–4041

Aoki K, Tamai Y, Horiike S, Oshima M, Taketo MM (2003) Colonic polyposis caused by mTOR-mediated chromosomal instability in Apc+/Delta716 Cdx2+/– compound mutant mice. Nat Genet 35:323–330

Beier DR (2000) Sequence-based analysis of mutagenized mice. Mamm Genome 11:594–597

Bentley A, MacLennan B, Calvo J, Dearolf CR (2000) Targeted recovery of mutations in Drosophila. Genetics 156:1169–1173

Berends MJ, Wu Y, Sijmons RH, Mensink RG, van der Sluis T, Hordijk-Hos JM, de Vries EG, Hollema H, Karrenbeld A, Buys CH et al (2002) Molecular and clinical characteristics of MSH6 variants: an analysis of 25 index carriers of a germline variant. Am J Hum Genet 70:26–37

Brinster RL, Chen HY, Trumbauer M, Senear AW, Warren R, Palmiter RD (1981) Somatic expression of herpes thymidine kinase in mice following injection of a fusion gene into eggs. Cell 27:223–231

Brodie SG, Xu X, Li C, Kuo A, Leder P, Deng CX (2001) Inactivation of p53 tumor suppressor gene acts synergistically with c-neu oncogene in salivary gland tumorigenesis. Oncogene 20:1445–1454

Capecchi MR (2005) Gene targeting in mice: functional analysis of the mammalian genome for the twenty-first century. Nat Rev Genet 6:507–512

Castilla LH, Couch FJ, Erdos MR, Hoskins KF, Calzone K, Garber JE, Boyd J, Lubin MB, Deshano ML, Brody LC et al (1994) Mutations in the BRCA1 gene in families with early onset breast and ovarian cancer. Nat Genet 8:387–391

Chimp Genome Consortium (2005) Initial sequence of the chimpanzee genome and comparison with the human genome. Nature 437:69–87

Coghill EL, Hugill A, Parkinson N, Davison C, Glenister P, Clements S, Hunter J, Cox RD, Brown SD (2002) A gene-driven approach to the identification of ENU mutants in the mouse. Nat Genet 30:255–256

Colbert T, Till BJ, Tompa R, Reynolds S, Steine MN, Yeung AT, McCallum CM, Comai L, Henikoff S (2001) High-throughput screening for induced point mutations. Plant Physiol 126:480–484

Concepcion D, Seburn KL, Wen G, Frankel WN, Hamilton BA (2004) Mutation rate and predicted phenotypic target sizes in ethylnitrosourea-treated mice. Genetics 168:953–959

Costantini F, Lacy E (1981) Introduction of a rabbit beta-globin gene into the mouse germ line. Nature 294:92–94

Cotroneo MS, Haag JD, Zan Y, Lopez CC, Thuwajit P, Petukhova GV, Camerini-Otero RD, Gendron-Fitzpatrick A, Griep AE, Murphy CJ et al (2007) Characterizing a rat Brca2 knockout model. Oncogene 26:1626–1635

Cuppen E, Gort E, Hazendonk E, Mudde J, van de Belt J, Nijman IJ, Guryev V, Plasterk RH (2007) Efficient target-selected mutagenesis in Caenorhabditis elegans: toward a knockout for every gene. Genome Res 17:649–658
de Wind N, Dekker M, Claij N, Jansen L, van Klink Y, Radman M, Riggins G, van der Valk M, Van't Wout K, te Riele H (1999) HNPCC-like cancer predisposition in mice through simultaneous loss of Msh3 and Msh6 mismatch-repair protein functions. Nat Genet 23:359–362
Edelmann W, Yang K, Umar A, Heyer J, Lau K, Fan K, Liedtke W, Cohen PE, Kane MF, Lipford JR et al (1997) Mutation in the mismatch repair gene Msh6 causes cancer susceptibility. Cell 91:467–477
Eker R, Mossige J (1961) A dominant gene for renal adenomas in the Rat. Nature 189:858–859
Evers B, Jonkers J (2006) Mouse models of BRCA1 and BRCA2 deficiency: past lessons, current understanding and future prospects. Oncogene 25:5885–5897
Fishel R (2001) The selection for mismatch repair defects in hereditary nonpolyposis colorectal cancer: revising the mutator hypothesis. Cancer Res 61:7369–7374
Fodde R, Edelmann W, Yang K, van Leeuwen C, Carlson C, Renault B, Breukel C, Alt E, Lipkin M, Khan PM et al (1994) A targeted chain-termination mutation in the mouse Apc gene results in multiple intestinal tumors. Proc Natl Acad Sci USA 91:8969–8973
Friedman LS, Ostermeyer EA, Szabo CI, Dowd P, Lynch ED, Rowell SE, King MC (1994) Confirmation of BRCA1 by analysis of germline mutations linked to breast and ovarian cancer in ten families. Nat Genet 8:399–404
Gibbs RAGM, Metzker WML, Muzny DM, Sodergren EJ, Scherer S, Scott G, Steffen D, Worley KC, Burch PE et al (2004) Genome sequence of the Brown Norway rat yields insights into mammalian evolution. Nature 428:493–521
Glazier AM, Nadeau JH, Aitman TJ (2002) Finding genes that underlie complex traits. Science 298:2345–2349
Goodfellow PJ, Buttin BM, Herzog TJ, Rader JS, Gibb RK, Swisher E, Look K, Walls KC, Fan MY, Mutch DG (2003) Prevalence of defective DNA mismatch repair and MSH6 mutation in an unselected series of endometrial cancers. Proc Natl Acad Sci USA 100:5908–5913
Gordon JW, Scangos GA, Plotkin DJ, Barbosa JA, Ruddle FH (1980) Genetic transformation of mouse embryos by microinjection of purified DNA. Proc Natl Acad Sci USA 77:7380–7384
Groden J, Thliveris A, Samowitz W, Carlson M, Gelbert L, Albertsen H, Joslyn G, Stevens J, Spirio L, Robertson M et al (1991) Identification and characterization of the familial adenomatous polyposis coli gene. Cell 66:589–600
Guryev V, Berezikov E, Malik R, Plasterk RH, Cuppen E (2004) Single nucleotide polymorphisms associated with rat expressed sequences. Genome Res 14:1438–1443
Hall JM, Lee MK, Newman B, Morrow JE, Anderson LA, Huey B, King MC (1990) Linkage of early onset familial breast cancer to chromosome 17q21. Science 250:1684–1689
Harbers K, Jahner D, Jaenisch R (1981) Microinjection of cloned retroviral genomes into mouse zygotes: integration and expression in the animal. Nature 293:540–542
Hitotsumachi S, Carpenter DA, Russell WL (1985) Dose-repetition increases the mutagenic effectiveness of *N*-ethyl-*N*-nitrosourea in mouse spermatogonia. Proc Natl Acad Sci USA 82:6619–6621
Homberg JR, Olivier JD, Smits BM, Mul JD, Mudde J, Verheul M, Nieuwenhuizen OF, Cools AR, Ronken E, Cremers T et al (2007) Characterization of the serotonin transporter knockout rat: a selective change in the functioning of the serotonergic system. Neuroscience 146:1662–1676
Hrabe de Angelis MH, Flaswinkel H, Fuchs H, Rathkolb B, Soewarto D, Marschall S, Heffner S, Pargent W, Wuensch K, Jung M et al (2000) Genome-wide, large-scale production of mutant mice by ENU mutagenesis. Nat Genet 25:444–447
Jacob HJ, Kwitek AE (2002) Rat genetics: attaching physiology and pharmacology to the genome. Nat Rev Genet 3:33–42
Jansen G, Hazendonk E, Thijssen KL, Plasterk RH (1997) Reverse genetics by chemical mutagenesis in Caenorhabditis elegans. Nat Genet 17:119–121
Jonkers J, Meuwissen R, van der Gulden H, Peterse H, van der Valk M, Berns A (2001) Synergistic tumor suppressor activity of BRCA2 and p53 in a conditional mouse model for breast cancer. Nat Genet 29:418–425

Justice MJ, Carpenter DA, Favor J, Neuhauser-Klaus A, Hrabe de Angelis M, Soewarto D, Moser A, Cordes S, Miller D, Chapman V et al (2000) Effects of ENU dosage on mouse strains. Mamm Genome 11:484–488

Keays DA, Clark TG, Flint J (2006) Estimating the number of coding mutations in genotypic- and phenotypic-driven *N*-ethyl-*N*-nitrosourea (ENU) screens. Mamm Genome 17:230–238

Kinzler KW, Nilbert MC, Su LK, Vogelstein B, Bryan TM, Levy DB, Smith KJ, Preisinger AC, Hedge P, McKechnie D et al (1991) Identification of FAP locus genes from chromosome 5q21. Science 253:661–665

Kinzler KW, Vogelstein B (1996) Lessons from hereditary colorectal cancer. Cell 87:159–170

Kitada K, Ishishita S, Tosaka K, Takahashi R, Ueda M, Keng VW, Horie K, Takeda J (2007) Transposon-tagged mutagenesis in the rat. Nat Methods 4:131–133

Koller BH, Hagemann LJ, Doetschman T, Hagaman JR, Huang S, Williams PJ, First NL, Maeda N, Smithies O (1989) Germ-line transmission of a planned alteration made in a hypoxanthine phosphoribosyltransferase gene by homologous recombination in embryonic stem cells. Proc Natl Acad Sci USA 86:8927–8931

Kolodner RD, Tytell JD, Schmeits JL, Kane MF, Gupta RD, Weger J, Wahlberg S, Fox EA, Peel D, Ziogas A et al (1999) Germ-line msh6 mutations in colorectal cancer families. Cancer Res 59:5068–5074

Kwitek AE, Gullings-Handley J, Yu J, Carlos DC, Orlebeke K, Nie J, Eckert J, Lemke A, Andrae JW, Bromberg S et al (2004) High-density rat radiation hybrid maps containing over 24,000 SSLPs, genes, and ESTs provide a direct link to the rat genome sequence. Genome Res 14:750–757

Lander ESLM, Birren LB, Nusbaum C, Zody MC, Baldwin J, Devon K, Dewar K, Doyle M, FitzHugh W et al (2001) Initial sequencing and analysis of the human genome. Nature 409:860–921

Lazar J, Moreno C, Jacob HJ, Kwitek AE (2005) Impact of genomics on research in the rat. Genome Res 15:1717–1728

Lindblad-Toh KCM, Mikkelsen WTS, Karlsson EK, Jaffe DB, Kamal M, Clamp M, Chang JL, Kulbokas EJ 3rd, Zody MC et al (2005) Genome sequence, comparative analysis and haplotype structure of the domestic dog. Nature 438:803–819

Lindsey JR (1979) Historical foundations. Academic Press, New York

Lu B, Geurts AM, Poirier C, Petit DC, Harrison W, Overbeek PA, Bishop CE (2007) Generation of rat mutants using a coat color-tagged sleeping beauty transposon system. Mamm Genome 18:338–346

Lynch HT, Smyrk T (1996) Hereditary nonpolyposis colorectal cancer (Lynch syndrome). An updated review Cancer 78:1149–1167

Mashimo T, Yanagihara K, Tokuda S, Voigt B, Takizawa A, Nakajima R, Kato M, Hirabayashi M, Kuramoto T, Serikawa T. (2008) An ENinduced mutant archive for gene targeting in rats. Nat Genet 40(5):514–5. No abstract available

McCallum CM, Comai L, Greene EA, Henikoff S (2000) Targeted screening for induced mutations. Nat Biotechnol 18:455–457

McCarthy A, Savage K, Gabriel A, Naceur C, Reis-Filho JS, Ashworth A (2007) A mouse model of basal-like breast carcinoma with metaplastic elements. J Pathol 211:389–398

Miki Y, Swensen J, Shattuck-Eidens D, Futreal PA, Harshman K, Tavtigian S, Liu Q, Cochran C, Bennett LM, Ding W et al (1994) A strong candidate for the breast and ovarian cancer susceptibility gene BRCA1. Science 266:66–71

Miyaki M, Konishi M, Tanaka K, Kikuchi-Yanoshita R, Muraoka M, Yasuno M, Igari T, Koike M, Chiba M, Mori T (1997) Germline mutation of MSH6 as the cause of hereditary nonpolyposis colorectal cancer. Nat Genet 17:271–272

Moser AR, Pitot HC, Dove WF (1990) A dominant mutation that predisposes to multiple intestinal neoplasia in the mouse. Science 247:322–324

Mullins LJ, Brooker G, Mullins JJ (2002) Transgenesis in the rat. Methods Mol Biol 180:255–270

Nagase H, Nakamura Y (1993) Mutations of the APC (adenomatous polyposis coli) gene. Hum Mutat 2:425–434

Nakatsukasa E, Inomata T, Ikeda T, Shino M, Kashiwazaki N (2001) Generation of live rat offspring by intrauterine insemination with epididymal spermatozoa cryopreserved at −196 degrees C. Reproduction 122:463–467

Nakatsukasa E, Kashiwazaki N, Takizawa A, Shino M, Kitada K, Serikawa T, Hakamata Y, Kobayashi E, Takahashi R, Ueda M et al (2003) Cryopreservation of spermatozoa from closed colonies, and inbred, spontaneous mutant, and transgenic strains of rats. Comp Med 53:639–641

Nandi S, Guzman RC, Yang J (1995) Hormones and mammary carcinogenesis in mice, rats, and humans: a unifying hypothesis. Proc Natl Acad Sci USA 92:3650–3657

Nolan PM, Peters J, Strivens M, Rogers D, Hagan J, Spurr N, Gray IC, Vizor L, Brooker D, Whitehill E et al (2000) A systematic, genome-wide, phenotype-driven mutagenesis programme for gene function studies in the mouse. Nat Genet 25:440–443

Noveroske JK, Weber JS, Justice MJ (2000) The mutagenic action of *N*-ethyl-*N*-nitrosourea in the mouse. Mamm Genome 11:478–483

Oshima M, Oshima H, Kitagawa K, Kobayashi M, Itakura C, Taketo M (1995) Loss of Apc heterozygosity and abnormal tissue building in nascent intestinal polyps in mice carrying a truncated Apc gene. Proc Natl Acad Sci USA 92:4482–4486

Paigen K (2003) One hundred years of mouse genetics: an intellectual history. I. The classical period (1902–1980). Genetics 163:1–7

Quwailid MM, Hugill A, Dear N, Vizor L, Wells S, Horner E, Fuller S, Weedon J, McMath H, Woodman P et al (2004) A gene-driven ENU-based approach to generating an allelic series in any gene. Mamm Genome 15:585–591

Russell WL, Hunsicker PR, Carpenter DA, Cornett CV, Guinn GM (1982a) Effect of dose fractionation on the ethylnitrosourea induction of specific-locus mutations in mouse spermatogonia. Proc Natl Acad Sci USA 79:3592–3593

Russell WL, Hunsicker PR, Raymer GD, Steele MH, Stelzner KF, Thompson HM (1982b) Dose–response curve for ethylnitrosourea-induced specific-locus mutations in mouse spermatogonia. Proc Natl Acad Sci USA 79:3589–3591

Russell WL, Kelly EM, Hunsicker PR, Bangham JW, Maddux SC, Phipps EL (1979) Specific-locus test shows ethylnitrosourea to be the most potent mutagen in the mouse. Proc Natl Acad Sci USA 76:5818–5819

Russo J, Gusterson BA, Rogers AE, Russo IH, Wellings SR, van Zwieten MJ (1990) Comparative study of human and rat mammary tumorigenesis. Lab Invest 62:244–278

Samuelson DJ, Hesselson SE, Aperavich BA, Zan Y, Haag JD, Trentham-Dietz A, Hampton JM, Mau B, Chen KS, Baynes C et al (2007) Rat Mcs5a is a compound quantitative trait locus with orthologous human loci that associate with breast cancer risk. Proc Natl Acad Sci USA 104:6299–6304

Smits BM, Cotroneo MS, Haag JD, Gould MN (2006a) Genetically engineered rat models for breast cancer. Breast Dis 28:53–61

Smits BM, Cuppen E (2006) Rat genetics: the next episode. Trends Genet 22:232–240

Smits BM, Haag JD, Gould MN, Cuppen E (2008) Rat knockout and mutant models. In: Conn PM (ed) Sourcebook of models for biomedical research. Humana Press, Totowa, New Jersey, pp 171–178

Smits BM, Mudde J, Plasterk RH, Cuppen E (2004) Target-selected mutagenesis of the rat. Genomics 83:332–334

Smits BM, Mudde JB, van de Belt J, Verheul M, Olivier J, Homberg J, Guryev V, Cools AR, Ellenbroek BA, Plasterk RH et al (2006b) Generation of gene knockouts and mutant models in the laboratory rat by ENU-driven target-selected mutagenesis. Pharmacogenet Genomics 16:159–169

Smits BM, Peters TA, Mul JD, Croes HJ, Fransen JA, Beynon AJ, Guryev V, Plasterk RH, Cuppen E (2005) Identification of a rat model for usher syndrome type 1B by *N*-ethyl-*N*-nitrosourea mutagenesis-driven forward genetics. Genetics 170:1887–1896

Sodir NM, Chen X, Park R, Nickel AE, Conti PS, Moats R, Bading JR, Shibata D, Laird PW (2006) Smad3 deficiency promotes tumorigenesis in the distal colon of ApcMin/+ mice. Cancer Res 66:8430–8438

The Breast Cancer Linkage Consortium (1999) Cancer risks in BRCA2 mutation carriers. J Natl Cancer Inst 91:1310–1316

Thomas KR, Capecchi MR (1990) Targeted disruption of the murine int-1 proto-oncogene resulting in severe abnormalities in midbrain and cerebellar development. Nature 346:847–850

Thompson D, Easton DF (2002) Cancer Incidence in BRCA1 mutation carriers. J Natl Cancer Inst 94:1358–1365

Thompson S, Clarke AR, Pow AM, Hooper ML, Melton DW (1989) Germ line transmission and expression of a corrected HPRT gene produced by gene targeting in embryonic stem cells. Cell 56:313–321

van Boxtel R, Toonen P, van Roekel H, Verheul M, Smits BM, Korving J, de Bruin A, Cuppen E (2008) Lack of DNA mismatch repair protein MSH6 in the rat results in hereditary non-polyposis colorectal cancer-like tumorigenesis. Carcinogenesis 29(6).1290–7

van Zeeland AA, de Groot A, Neuhauser-Klaus A (1990) DNA adduct formation in mouse testis by ethylating agents: a comparison with germ-cell mutagenesis. Mutat Res 231:55–62

Venter JCMD, Myers AEW, Li PW, Mural RJ, Sutton GG, Smith HO, Yandell M, Evans CA, Holt RA et al (2001) The sequence of the human genome. Science 291:1304–1351

Wagner A, Hendriks Y, Meijers-Heijboer EJ, de Leeuw WJ, Morreau H, Hofstra R, Tops C, Bik E, Brocker-Vriends AH, van Der Meer C et al (2001) Atypical HNPCC owing to MSH6 germline mutations: analysis of a large Dutch pedigree. J Med Genet 38:318–322

Wagner TE, Hoppe PC, Jollick JD, Scholl DR, Hodinka RL, Gault JB (1981) Microinjection of a rabbit beta-globin gene into zygotes and its subsequent expression in adult mice and their offspring. Proc Natl Acad Sci USA 78:6376–6380

Waterston RHK, Birney Lindblad-Toh E, Rogers J, Abril JF, Agarwal P, Agarwala R, Ainscough R, Alexandersson M, An P et al (2002) Initial sequencing and comparative analysis of the mouse genome. Nature 420:520–562

Wienholds E, Schulte-Merker S, Walderich B, Plasterk RH (2002) Target-selected inactivation of the zebrafish rag1 gene. Science 297:99–102

Wienholds E, van Eeden F, Kosters M, Mudde J, Plasterk RH, Cuppen E (2003) Efficient target-selected mutagenesis in zebrafish. Genome Res 13:2700–2707

Wijnen J, de Leeuw W, Vasen H, van der Klift H, Moller P, Stormorken A, Meijers-Heijboer H, Lindhout D, Menko F, Vossen S et al (1999) Familial endometrial cancer in female carriers of MSH6 germline mutations. Nat Genet 23:142–144

Wooster R, Bignell G, Lancaster J, Swift S, Seal S, Mangion J, Collins N, Gregory S, Gumbs C, Micklem G (1995) Identification of the breast cancer susceptibility gene BRCA2. Nature 378:789–792

Wu Y, Berends MJ, Mensink RG, Kempinga C, Sijmons RH, van Der Zee AG, Hollema H, Kleibeuker JH, Buys CH, Hofstra RM (1999) Association of hereditary nonpolyposis colorectal cancer-related tumors displaying low microsatellite instability with MSH6 germline mutations. Am J Hum Genet 65:1291–1298

Yang G, Scherer SJ, Shell SS, Yang K, Kim M, Lipkin M, Kucherlapati R, Kolodner RD, Edelmann W (2004) Dominant effects of an Msh6 missense mutation on DNA repair and cancer susceptibility. Cancer Cell 6:139–150

Zan Y, Haag JD, Chen KS, Shepel LA, Wigington D, Wang YR, Hu R, Lopez-Guajardo CC, Brose HL, Porter KI et al (2003) Production of knockout rats using ENU mutagenesis and a yeast-based screening assay. Nat Biotechnol 21:645–651

Zhou Q, Renard JP, Le Friec G, Brochard V, Beaujean N, Cherifi Y, Fraichard A, Cozzi J (2003) Generation of fertile cloned rats by regulating oocyte activation. Science 302:1179

Zimdahl H, Nyakatura G, Brandt P, Schulz H, Hummel O, Fartmann B, Brett D, Droege M, Monti J, Lee YA et al (2004) A SNP map of the rat genome generated from cDNA sequences. Science 303:807

Chapter 7
The Tumor Pathology of Genetically Engineered Mice: Genomic Pathology

Robert D. Cardiff

7.1 Introduction

This is the genomic era. Remarkably powerful technologies bring us new insights into the molecular mechanisms of disease with the concomitant promise that those insights will result in molecule-based cures. Medical oncology has been and will be the major beneficiary of genomics. Genomics also brings a renewed interest in comparative pathobiology because the sequencing of mammalian genomes emphasizes the remarkable similarity between all animals and, in particular, in mammalian species. The genes related to a given disease in humans result in the same disease in laboratory animals. One gene results in one disease that can be cured by one medicine. The mantra "One Gene, One Disease, One Medicine" has been officially endorsed by various national and international societies as "One Health" (Conrad et al. 2009; Kahn et al. 2009; Kaplan et al. 2009; Karesh and Cook 2009; Cardiff et al. 2008a, b; Cardiff 2007).

The genetically engineered mouse is the animal that is used to phenotype each of our 20,000+/– genes (Anonymous 2007; Austin 2004). The mouse is used to fulfill the Koch's Postulates of modern science (Daley 1993). Modern oncology is already a major beneficiary of genetic engineering. Most of this book's readers are well-aware of these facts and are almost certainly already using genetic engineering to determine the function of one gene or another. Most of you want to know whether the neoplasm created by your genetic manipulation "looks like" a comparable human tumor. Rendering this opinion requires either that you are an accomplished comparative pathologist or that you consult with someone who is. The pathologist is a person

R.D. Cardiff (✉)
Department of Pathology and Laboratory Medicine, Center for Comparative Medicine, Center for Genomic Pathology, University of California, Davis, CA 95616, USA
e-mail: rdcardiff@ucdavis.edu

J.E. Green and T. Ried (eds.), *Genetically Engineered Mice for Cancer Research: Design, Analysis, Pathways, Validation and Pre-Clinical Testing*,
DOI 10.1007/978-0-387-69805-2_7,

with particular skill of "looking at" tumors through a microscope to "validate" your tumor phenotype (Cardiff et al. 2004). While most of you do not have the training required for microscopic validation, you need to appreciate what your expert can tell you about your tumor to appreciate this part of the excitement of genomics.

Examination of tumors in this genomic era has led to new concepts concerning structure and function that are discussed in the context of four principles (Cardiff et al. 2006a). This chapter describe how the introduction of specific molecular changes into the cell's genome creates unique microscopic morphological tumor phenotypes in almost all mouse organ systems (Cardiff 1996; Cardiff et al. 1991). These unique tumors have characteristic, "signature" phenotypes that can be recognized in tumors from different organ systems (Figs. 7.1–7.7). The tumors arising from genetic alteration of a given molecular pathway share morphological characteristics belonging to that signal transduction pathway (Fig. 7.6). Tumors initiated by a given genetic change undergo neoplastic progression that is similar to that observed in the so-called spontaneous tumors of mice, humans, and other animals (Figs. 7.8–7.10). In crosses of one oncogenic GEM with another GEM bearing other oncogenic transgenes, one transgene dominates the morphological changes (Cardiff et al. 1991). The dominant transgenes form a recognizable molecular-structural hierarchy. In contrast, tumor-suppressor genes generally produce a mixture of tumor phenotypes when they are silenced or knocked out (Fig. 7.7). The initiating oncogene dominates the phenotype when paired with tumor-suppressor knockouts or with another oncogene (Fig. 7.7). Host modifiers, such as genetic background, age, gender, and immunological status, rarely have a significant effect on tumor morphology but do influence the biology of the tumor.

This chapter also provides a more standard short review discussing and illustrating the systemic pathology that most students require for their research. The "Systemic Pathology" section is designed to illustrate the more common genetically engineered systems in major organ systems. It helps address questions about comparisons between the newest tumor model and the more established tumor models in the major organ systems. The major proportion of this chapter, however, is devoted to the broader perspective of the general effects of genetic manipulation on tumorigenesis, genomic pathology.

7.1.1 Comparative Pathology

Pathology is used for validation of genetically engineered mice (GEM) models of human disease (Cardiff et al. 2004; Green et al. 2002). Validation differs from verification in that verification by a pathologist certifies that the process is neoplastic but does not necessarily compare the characteristics of the neoplasm with the human disease. Verification is diagnosis driven and confirms hypotheses by naming the processes. In contrast, validation is discovery-driven study of the disease process comparing and contrasting the characteristics in the two species leading to discovery-driven formulation of hypotheses. Validation occurs at many different levels. With GEM tumor biology, the engineering represents a validation step for each engineered gene.

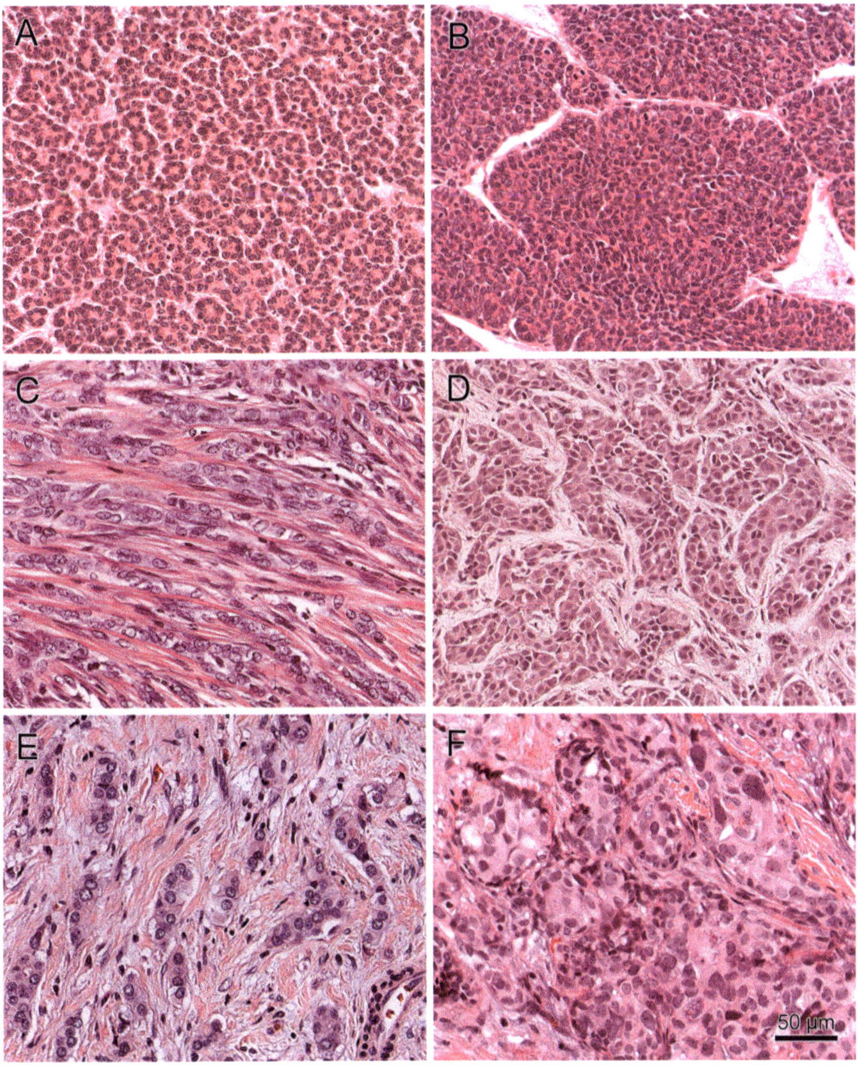

Fig. 7.1 Mouse-Human Comparisons. (**a** and **b**) illustrate two patterns of classical MMTV-induced mouse mammary tumors. (**c** and **d**) are examples of patterns obtained from GEM mice. (**e** and **f**) are examples of comparable human tumors. (**a**) Microacinar pattern which was described as Type A by Thelma Dunn. (**b**) Solid Type B pattern of Dunn. (**c**) Mammary tumor in a GEM with *e-Cadherin* knocked out. Note the dense connective tissue and tumor cells in linear arrays. (**d**) Mammary tumor in a GEM with high levels of expression of *ErbB2*. The growth pattern is of irregular cords and nests of cells. (**e**) Human lobular carcinoma of the breast associated with silencing of *e-Cadherin*. Note similar growth pattern as the mouse shown in (**c**). (**f**) Human breast cancer with amplification of *ErbB2*. Compare with the pattern from the mouse in (**d**). Scale bar for all images in (**f**)

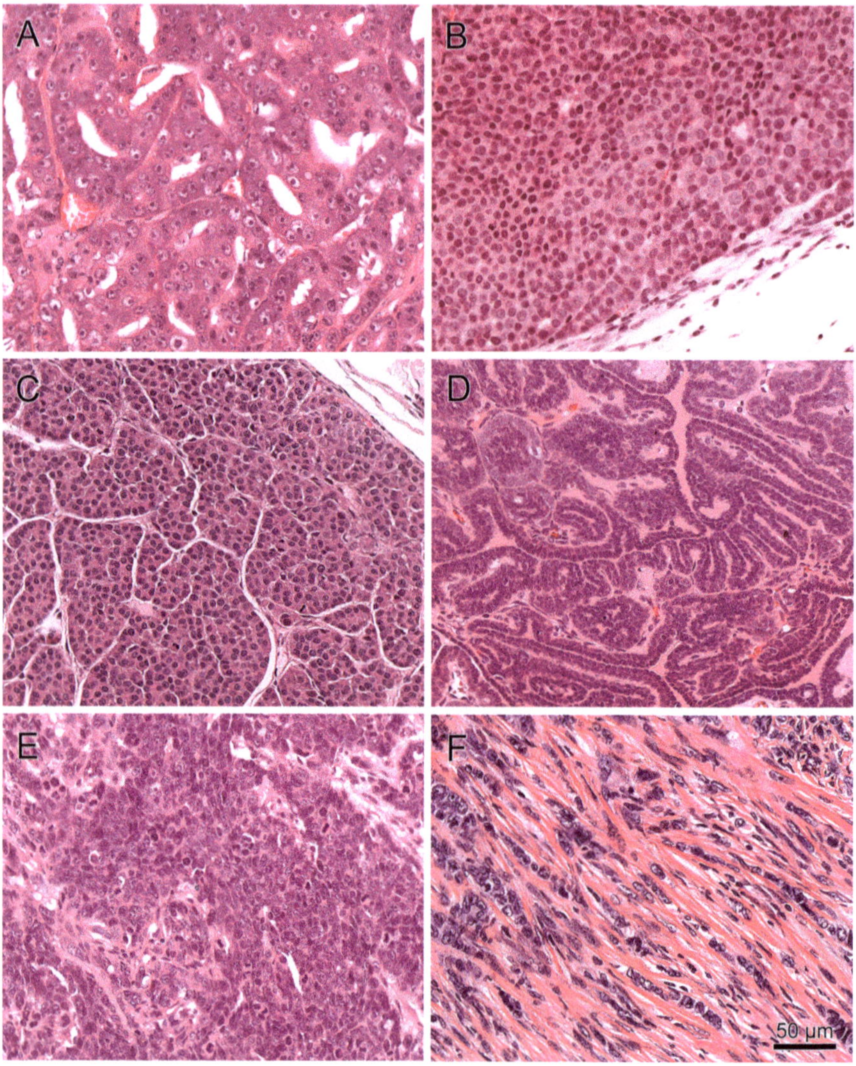

Fig. 7.2 Mouse Mammary Tumors. (**a**) Mammary tumor in a Tg(*Myc*) mouse showing a glandular pattern with amphophilic staining cells containing large nuclei and very prominent large nucleoli. (**b**) Mammary tumor from a Tg(*ErbB2*) mouse showing a solid nodular growth pattern with larger, lighter staining peripheral cells. (**c**) Mammary tumor from a Tg(*Ras*) mouse. The tumor cells contain small nuclei and grow in solid nests. Compare the nuclear size with those in (**a** and **b**). (**d**) Mammary tumor in a Tg(*Wnt1*) mouse showing a complex, branching ductal architecture. There are two distinct cell layers, the outer layer contains myoepithelial cells. (**e**) Mammary tumor from a Tg(*C(3)-SV40-Tag*) mouse. The tumor cells contain small nuclei and scanty cytoplasm. (**f**) Mammary tumor in a Tm(*e-Cadherin* $^{-/-}$) mouse with linear rows of cells in dense connective tissue. Scale bar for all images in (**f**)

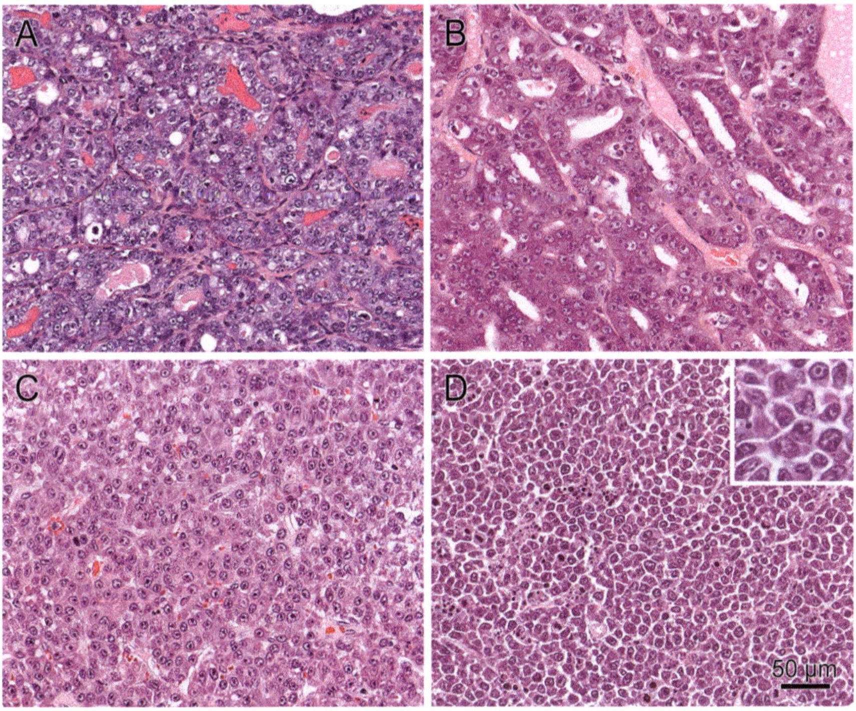

Fig. 7.3 *MYC*. Various patterns seen in different tissues with *Myc*-associated tumors. (**a**) Prostate. (**b**) Mammary gland. (**c**) Liver. (**d**) Lymph node. Whatever the tissue, the tumor cells contain large nuclei with prominent nuclei and amphophilic cytoplasm. The inset in (**d**) shows lymphoma cells at higher magnification. Scale bar for all images other than the inset in (**d**) is shown in (**f**)

Does the gene in question create a specific malignant neoplasm when overexpressed or ablated in the target tissue? The answer in the case of many genes, such as Myc and Neu, is, unequivocally, yes. But does genetic engineering result in a tumor that looks like the human cancer? How far can the similarity between human and mouse be carried? What does the data tell us?

Using modifications of the anatomic foundational ontology as a guideline, GEM are clearly "valid" models in most instances and to certain levels (Cardiff et al. 2004). That is, the overexpression of a gene, such as Myc, causes similar-appearing neoplasms in both human and man arising in the same organs or the same cells. This is the basis for "One Medicine" in the genomic era (Cardiff et al. 2008a). For example, the Myc- and neu-induced tumors of the mouse mammary gland look like some human breast cancers. The specific examples provided in this chapter show the difference between the standard, spontaneous mouse mammary tumor virus (MMTV)-induced mammary tumors (Fig. 7.1a, b) and GEM-related mammary tumors (Fig. 7.1c, d).

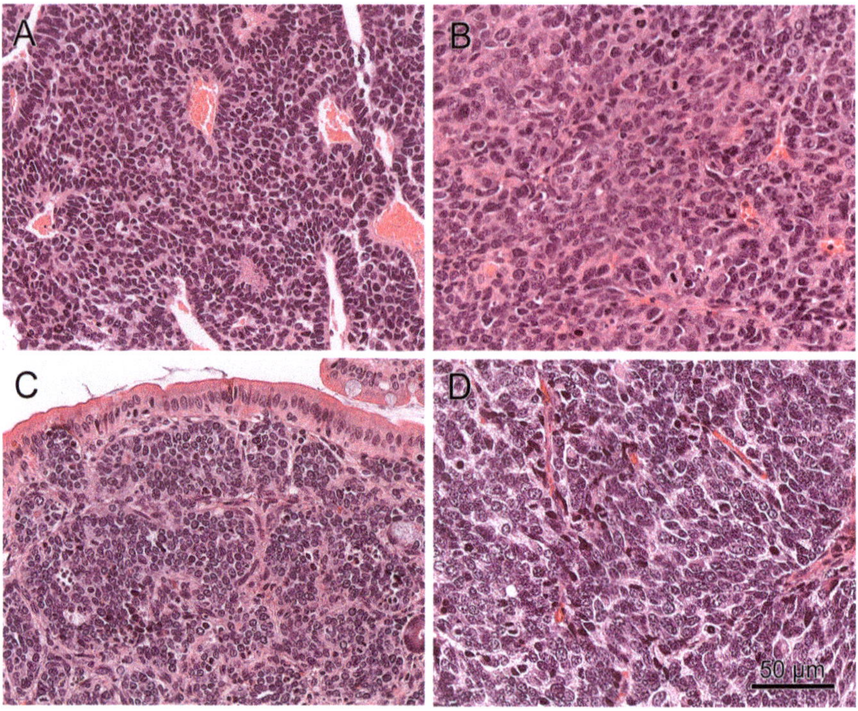

Fig. 7.4 *SV40-Tag* Various patterns of *SV-40Tag*-associated tumors. (**a**) Brain. (**b**) Lung. (**c**) Intestine. (**d**) Prostate. All tumors demonstrate cells with scanty cytoplasm and "pink" stroma. Scale bar for all is in (**d**)

The GEM tumors associated with silencing of the e-cadherin gene (Fig. 7.1c) or the overexpression of ErbB2 (Fig. 7.1d) closely resemble the human lobular breast cancers with silenced e-cadherin (Fig. 7.1e) or amplification of ErbB2 (Fig. 7.1f). Numerous other examples of morphological similarities between mouse and human tumors have been provided in almost all organ systems (Cardiff et al. 1977, 1992, 2000a, b, 2001, 2004; Cardiff 1996, 2001, 2003; Cardiff and Wellings 1999; Cardiff and Munn 1995; Borowsky et al. 2003; Holland 2004).

Therefore, GEM, by these criteria, are certainly "valid" models of human disease. The gene, tissue, and microscopic anatomy are similar. However, the modern investigator needs an even more detailed validation. What about host response, and the tumor stroma or the vasculature? What about the tumor biology?

The biology of the tumor is the single most important factor in assessing the lesions identified microscopically. Ethical and legal boundaries compel the medical oncologist/scientist to rely on the inferences of "guilt by association," rather than experimental evidence, in their attempts to relate function and structure. In contrast, the comparative oncologist has the opportunity to experimentally verify the empirical inferences. Thus, each inference can be tested using rigorous experimental biology. The test by transplantation is but one such approach available to the GEM pathologist and should be used much more often.

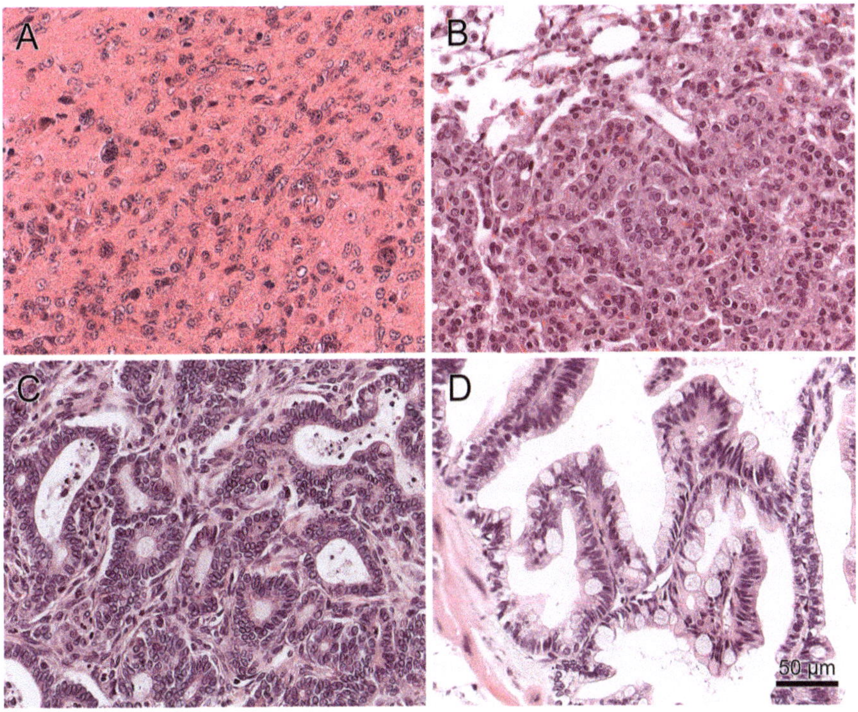

Fig. 7.5 *Ras* is not a dominant gene with a notable pattern in all tissues. The resulting tumors demonstrate a variety of growth patterns. (**a**) Brain (Oligodendroglioma). (**b**) Lung (Papillary adenoma). (**c**) Gut (Adenocarcinoma). (**d**) Prostate (Intestinal metaplasia). Scale bar for all images in (**d**)

7.1.1.1 Digital Imaging

Excellent reviews outline the six essential categories of genetic aberrations and the numerous factors that influence the neoplastic process (Hanahan and Weinberg 2000; Hager and Hanahan 1999). A checklist can be devised to remind us to examine all of the pertinent features, since each feature is important. However, a checklist alone is not adequate in the absence of comparisons with other models and other species. There is no substitute for a broad experience base. The issue is how to share and compare. The sharing of experience can now be accomplished using digital imaging.

In recent years, the digital capture, compression, storage, and display of images of the whole slide have become possible. Whole slide imaging (WSI) technology provides images that can be reviewed from remote locations. The remote viewer can "look at" the critical information in the context of the entire slide. WSI has revolutionized the capture, storage, retrieval, and display of microscopic images (Cardiff et al. 2004). The images are now presented over the Internet allowing the viewer to scan any part of the whole slide image at any magnification. These slides are available as archives for comparative pathology (http://imagearchive.compmed.ucdavis.edu/).

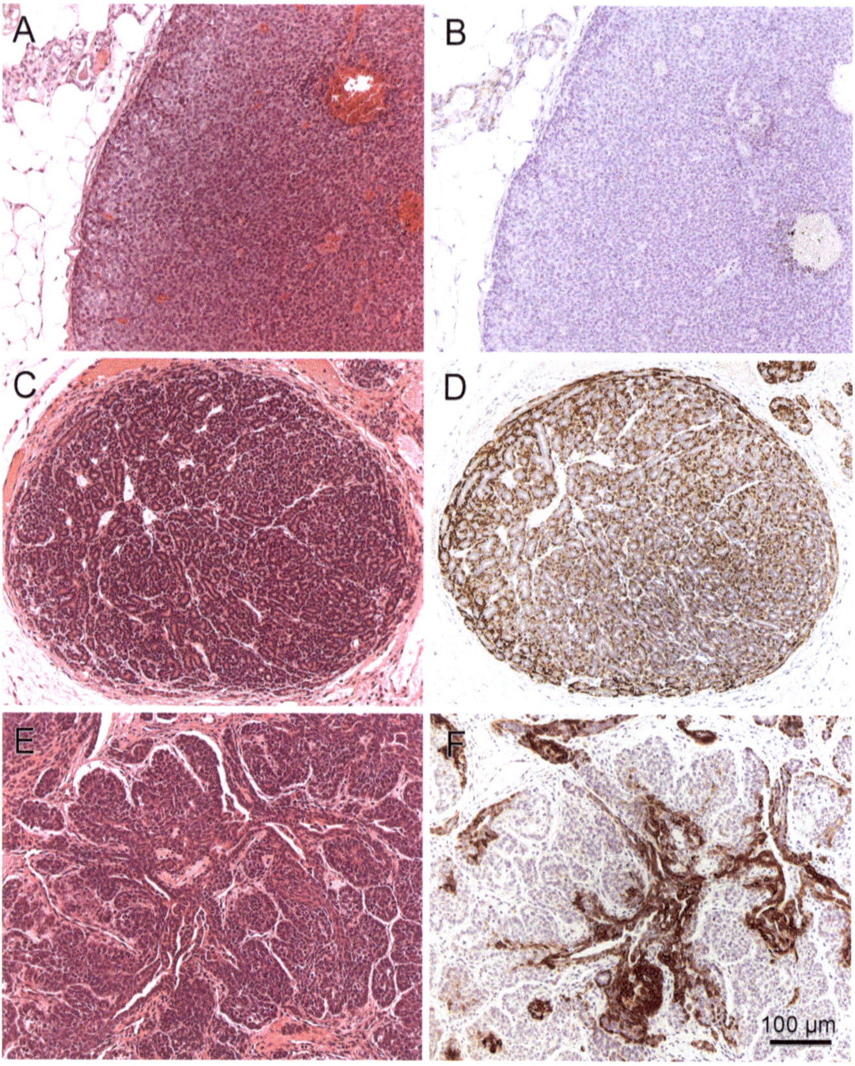

Fig. 7.6 Pathway Pathology. Transgenic mice with *ErbB* and *Wnt* pathway tumors exhibit different patterns of growth. *ErbB* tumors tend to be solid, without glandular, ductular, or squamous differentiation. They do not stain with cytokeratin-14 or other myoepithelial markers. *Wnt1* pathway tumors exhibit many different growth patterns. (**a**) Hematoxylin and eosin stained tumor from an *ErbB2* mouse. Notice the solid grown pattern. (**b**) Anti-cytokeratin-14 stain of a similar *ErbB2* tumor with no staining for myoepithelial or squamous elements. (**c**) Hematoxylin and eosin stain of a *Wnt1* tumor showing a microacinar growth pattern. (**d**) Anti-cytokeratin-14 staining of a similar microacinar *Wnt1* tumor. Myoepithelial cells show strong staining. (**e**) Hematoxylin and eosin stain of a *Wnt1* tumor with solid nests and branching ducts. (**f**) Anti-cytokeratin-14 staining. The branching ducts show strong positive staining. Scale bar for all is in (**f**)

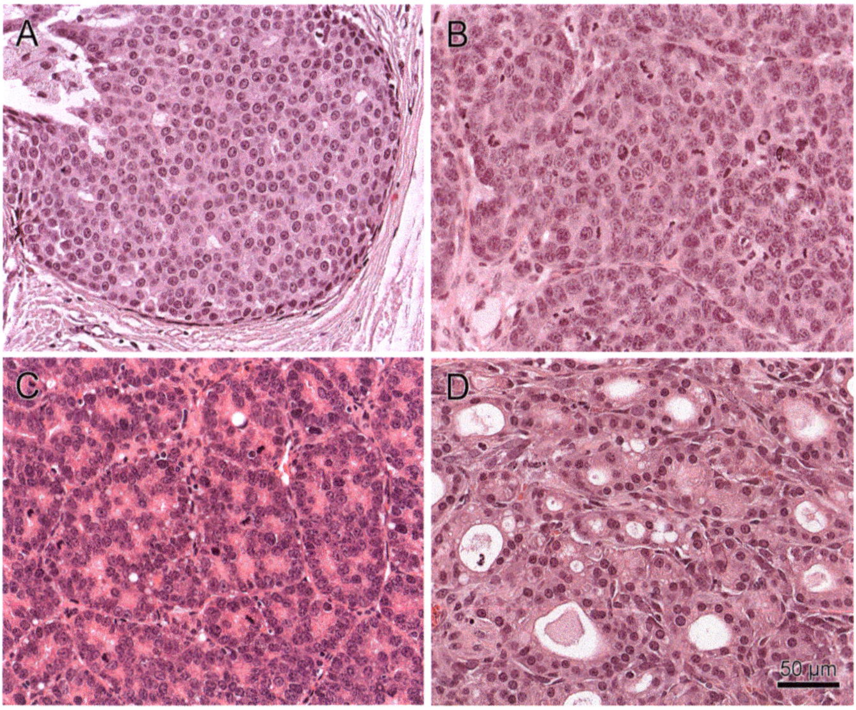

Fig. 7.7 Tumor Suppressor Genes. GEM tumors with tumor suppressor genes exhibit a variety of morphologies. (**a**) Mammary tumor from a Tg(*ErbB2*) mouse shows a solid growth pattern with uniform cells and nuclei. (**b**) When the GEM contains both Tg(*ErbB2*) and Tg(*Trp53*$^{+/-}$), the cells are larger and contain pleomorphic nuclei and numerous mitoses. (**c**) Tumor from a Tm(*Trp53*$^{-/-}$) mouse plus another unknown complementary oncogene exhibits a different glandular growth pattern. (**d**) Prostatic tumor from a Tm(*Pten*$^{+/-}$) mouse demonstrates back-to-back glands and nuclei with prominent nucleoli. Scale bar for all is in (**d**)

7.1.1.2 Morphometrics

Validation of GEM models as representing different attributes of human disease raises another interesting issue. Perhaps, validation is best performed by more objective measurements. While the skill of the expert pathologist equipped with exquisite descriptive language cannot be replaced by the computer, many of the current quandaries and controversies are resolved as descriptive phrases, such as "increasing nuclear pleomorphism," and are represented mathematically. How reproducible is the pathology among different experts? Opinion is replaced by measurement.

Captured, archived digital images can be retrieved and morphometrically analyzed. As morphometric analysis of digital images improves, the vast archives of digital whole slide images are subjected to analysis. This permits quantification of

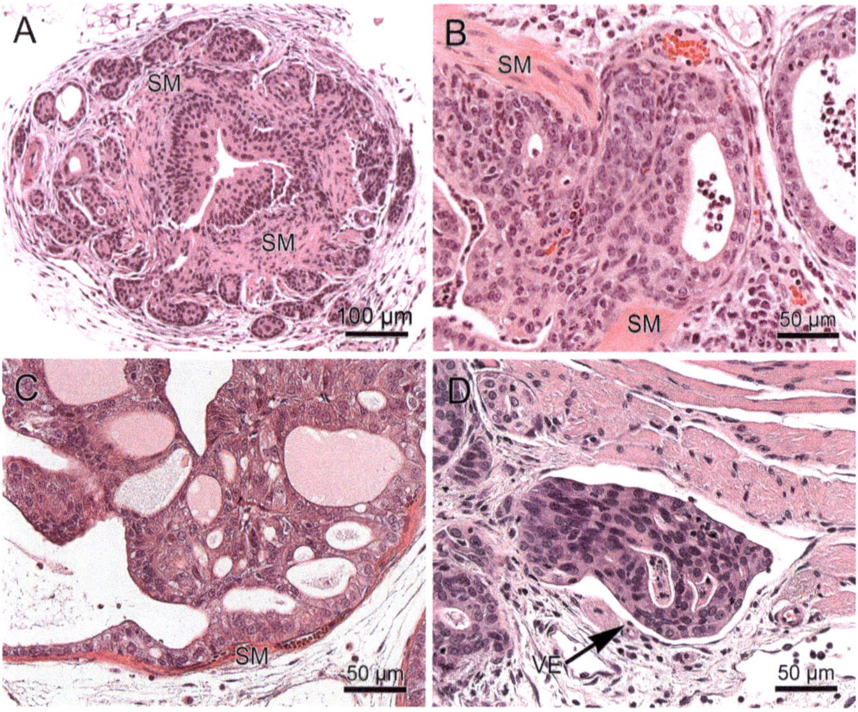

Fig. 7.8 Neoplastic Processes. Herniation is a process in which mechanical forces can disrupt structural integrity and produce abnormal histology. (**a**) Shows herniation of the lining epithelium through the smooth muscle (SM) of a mouse ureter. Groups of epithelial cells are exterior to the smooth muscle, giving the appearance of neoplastic invasion. (**b**), (**c**), and (**d**) show different stages of the neoplastic process. (**b**) Contains microinvasion of a group of cells which have penetrated through the smooth muscle (SM) layer of mouse prostate and into the surrounding stroma. (**c**) Shows *in situ* carcinoma (PIN) of mouse dorsolateral prostate. At the lower left is a small area of relatively normal prostatic epithelium, and a larger area within the prostate gland containing a cribriform pattern of proliferating cells (PIN III). The proliferation has not disrupted the smooth muscle (SM) layer of the gland. (**d**) Contains a small embolus of tumor cells which have penetrated into a vascular channel of mouse prostate. The intact vascular endothelium (VE) is easily visible. Scale bars are as indicated for each image

subtle characteristics, such as neoangiogenesis. More important, the investigator is able to normalize characteristic on the basis of other attributes, such as nuclear size. These data permit multivariant analysis to reveal novel relationships.

However, digital imaging and comparative image analysis requires rigorous preprocessing of the histological samples. If samples from different laboratories are to be compared, they must be prepared using the same fixative and the same dehydration and embedding techniques and materials. Otherwise, any real or imagined difference in the sample might be an artifact of preprocessing. Several reviews are available that emphasize the need to control the preprocessing and processing variables (Espina et al. 2005, 2007, 2008; Gulmann et al. 2005).

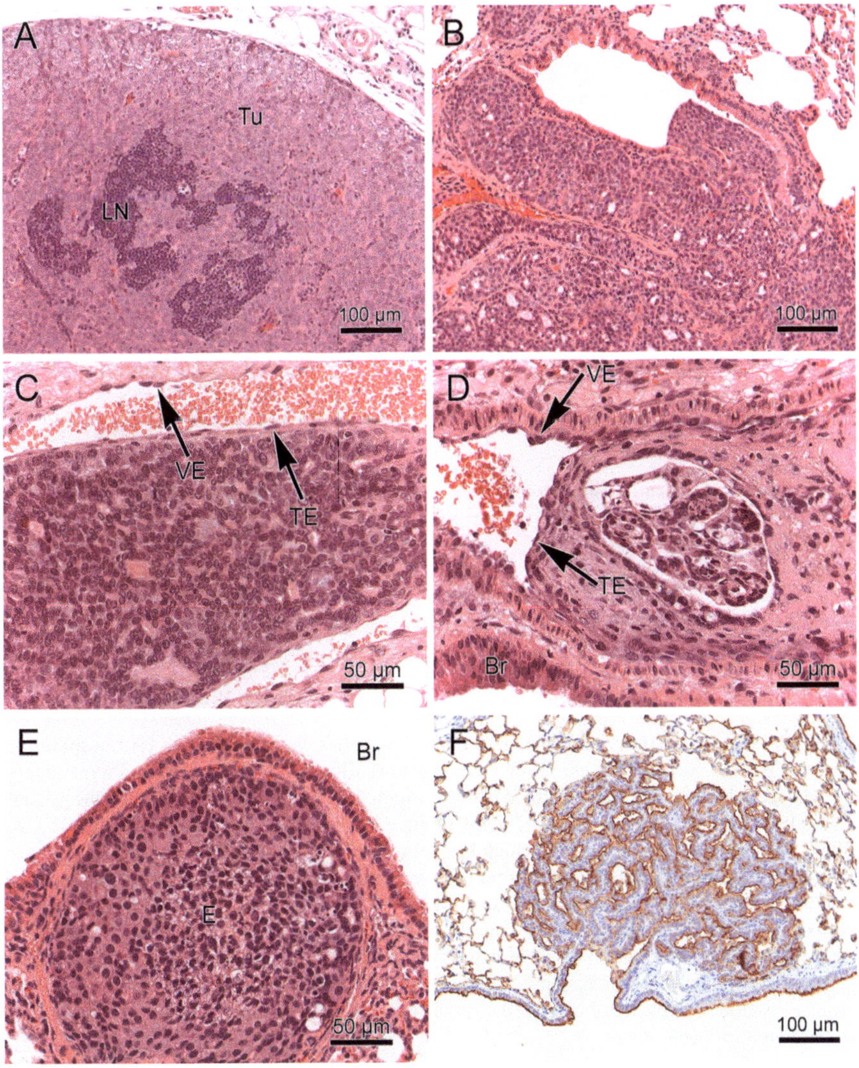

Fig. 7.9 Metastatic and Non-Metastatic Processes. In the mouse, metastases are pulmonary and intravascular. (**a**) Shows cells from a local mammary tumor (Tu) which have invaded and virtually replaced the normal cells of a lymph node. Only a small remnant of remaining lymph node cells (LN) is visible in the center of the image. (**b**) Shows metastatic mammary tumor cells within and next to a bronchiole in mouse lung. (**c**) A large tumor embolus of metastatic mammary carcinoma is present in the center of the image. The embolus is surrounded by tumor endothelium (TE) and is within a blood vessel lined by vascular endothelium (VE). (**d**) Another example of a metastatic mammary tumor embolus within a blood vessel in mouse lung. The tumor endothelium (TE) and vascular endothelium (VE) are readily visible. (**e**) A tumor embolus (**E**) is within a sub-bronchial (Br) blood vessel. Notice the vascular wall is intact. (**f**) Muc-1 staining readily identifies the papillary pattern of a primary lung papillary broncioalveolar adenoma. Scale bars are as indicated for each image

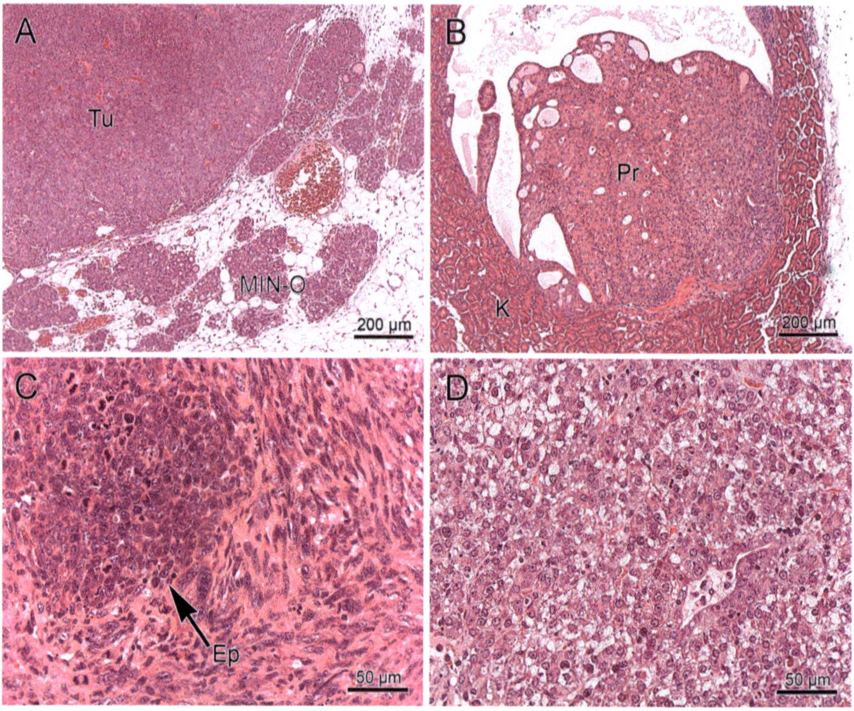

Fig. 7.10 Transplantation and Induced Tumors. There are several techniques for transplanting tumor cells into various organs. Three techniques are illustrated here. (**a**) Shows a mammary intraepithelial neoplasia-outgrowth (MIN-O) which has been transplanted into the gland-cleared fat pad of a mouse. A large solid tumor (Tu) has formed from the outgrowth. (**b**) Shows mouse prostatic tissue (Pr) which has been transplanted under the capsule of a mouse kidney (K). (**c** and **d**) illustrate recurrent tumors from GEM mice with doxycycline-induced tumors. (**c**) Epithelial-mesenchymal-transition tumor from a doxycycline-induced Tg(*Wnt1*) mouse. Most of the tumor cells are spindle-shaped, but occasional nests of solid epithelial cells (Ep) are also present. When *Wnt* expression was de-induced, the tumor grew. (**d**) When hepatoblastoma cells are transplanted subdermally into a *Myc* mouse, repression of *Myc* will induce the tumor to partially regress, but persist as cells resembling normal liver. Scale bars as indicated for each image

7.1.1.3 Conditions that Affect Tumorigenesis

Before beginning the details of comparative genomic pathology, the reader must be aware of the various host conditions that may influence the occurrence and interpretation of tumors. For example, several research programs use GEM to search for genetic modifiers (Nagase et al. 1999; Balmain and Nagase 1998; Dragani 2003). Pathologic analysis is also expected to detect subtle phenomena, perhaps, distant from the target tumor, that might reveal important insights. The microbiological flora could be important in interpreting the biology of the neoplastic process but is rarely considered in the natural history of neoplastic disease.

Most inbred laboratory mice were originally developed to study cancer and, thus, selected for the tumors they developed. All strains develop tumors with aging. For example, FVB mice are widely used for transgenesis. Spontaneous pulmonary adenomas occur in many elderly FVB females (Mahler et al. 1996). Since the mouse lung is the site of most metastases, the adenomas are frequently misinterpreted as metastases (Fig. 7.9f). Spontaneous adenomas are papillary and often involve the bronchiolar mucosa, differentiating them from intravascular tumor emboli.

Spontaneous pituitary adenomas also occur in elderly FVB/N females and are not fully appreciated (Mahler et al. 1996; Wakefield et al. 2003). These tumors are all prolactinomas that stimulate mammary lobuloalveolar hyperplasia and an increase in mammary adenocarcinomas (Wakefield et al. 2003; Nieto et al. 2003; Radaelli et al. 2009a). Therefore, age-matched controls and samples of the pituitaries are necessary when working with FVB mice more than 1 year old.

The 129 strain ES cells are used to create knockouts. Mouse ovarian tumors are commonly teratomas or granulosa cell tumors that are particularly common in the strains of 129 mice (Stevens 1980; Galvez et al. 2004). Further, the genetic contamination of the parental 129 strain requires that the exact variant 129 strain is recorded.

The GEM has other interesting, but frequently overlooked, pathology. For example, some examples of prostatic hyperplasia are due to interstitial cell (Leydig cell) hyperplasias. Dilation and engorgement of seminal vesicle can be extensive and even unilateral in some wild-type males and are common age-related pathologies in C57BL/6 mice. Enlarged, engorged seminal vesicles have also been associated with hyperplasia of the ampullary glands and not the prostatic glands (Donjacour et al. 1998). In short, know the background pathology of your mice.

7.2 Genomic Pathology

This section discusses pathology using the four established principles of comparative tumor biology (Cardiff et al. 2006a).

Principle 1: Oncogenic events initiating tumorigenesis profoundly influence the morphological phenotype of the tumor.
Principle 2: Tumorigenesis is influenced by tissue context.
Principle 3: Neoplastic progression is a multistep process associated with sequential morphological changes.
Principle 4: Mouse tumors mimic many aspects of human cancers (Figs. 7.1–7.7).

The biology of spontaneous or experimentally induced murine tumors is profoundly influenced by endogenous, oncogenic retroviruses, age, hormonal milieu, genetic background, and exogenous factors, such as exogenous retroviruses and chemical carcinogens. Therefore, the biologist needs to have an appreciation of the types of "spontaneous" and carcinogen-induced tumors previously described in mice.

Carcinogen-induced and viral-induced tumors of the lymphocytes, myelocytes, liver, bowel, and connective tissue may resemble some types of human cancers. However, the microscopic appearances of spontaneous, solid epithelial tumors of mice rarely “look like” human cancers. For instance, the traditional, “spontaneous” MMTV-induced tumors do not look like human breast cancers (Fig. 7.1) (Cardiff and Wellings 1999). However, GEM tumors are microscopically different from anything previously seen in experimental carcinogenesis and often closely mimic cancers occurring in humans.

For example, the virus-induced tumor types, spontaneous mouse mammary tumors, typically have an organized myoepithelium that is not found in human breast cancer (Cardiff et al. 2000a) (Fig. 7.6d, f). Medically trained surgical pathologists regard these murine tumors as benign because of the organized myoepithelium and expansile margins. However, up to 60% of these tumors in mice are metastatic (Cardiff et al. 2006a).

All strains of laboratory mice develop leukemia but some strains, such as AKR, were specifically selected and bred for MuLV-induced leukemia (Kogan et al. 2002). While leukemias are generally lymphocytic or myeloid, a number of neoplastic, hematopoietic cell types can be recognized and classified (Morse et al. 2002).

The “spontaneous” mouse tumors are all associated with tumor viruses. “Spontaneous” mammary tumors arise from activation of Wnt, Notch, or Fgf protooncogenes by insertion activation of the MMTV proviral DNA (van Leeuwen and Nusse 1995). The process has been experimentally reproduced using the MMTV LTR to promote the same genes (Cardiff 1996) resulting in identical microscopic phenotypes that conform to the Dunn classification (Figs. 7.1a, b and 7.2d).

The sarcomas are also retrovirus-induced lesions. Several types of sarcoma viruses have been isolated from different mouse strains and, like the chicken sarcoma viruses, involve the transduction of cellular genes, such as fos (Verma and Graham 1987) and mos, by exogenous leukemia retroviruses (Yew and Strobel 1993). Other sarcomas are the result of leukemia virus transduction of genes from other species, such as rats, cats, and gibbon apes (Snyder and Theilen 1969; Gardner et al. 1970, 1971; Kawakami et al. 1973; Kumar et al. 1990; Kirsten and Mayer 1969; Naharro et al. 1983).

The “buyer beware” because not all spindle cell tumors in mice are “sarcomas.” The mouse is prone to developing epithelial-mesenchymal-transition (EMT) tumors (Radaelli et al. 2009b). Many of the “spindle cell” tumors previously called sarcomas are not mesenchymal in origin but are a transition of mouse epithelium into a mesenchymal spindle cell phenotype (Fig. 7.10c) (Landesman-Bollag et al. 2001; White et al. 2001). These EMT-type tumors are generally transgene-induced which no longer express the transgene (Landesman-Bollag et al. 2001; White et al. 2001). These tumors have increased expression of the transcription factors: twist, snail, or slug. They can be recognized by dual staining for low-molecular-weight keratin and vimentin or smooth muscle actin. The reader should be forewarned that EMT is a well-known phenomenon in cultured mouse cells and their explants (Radaelli et al. 2009a, b).

In contrast to spontaneous tumors, tumors produced in GEM using genes from human cancers produce cancers that are remarkable mimics of their human counterparts (Cardiff 1996, 2001, 2003, 2000a; Cardiff and Wellings 1999; Cardiff and Muller 1993) (Fig. 7.1). The induced dysregulation of most oncogenes leads to tumors with a unique or signature phenotypes (Cardiff et al. 1991, 1992; Cardiff 1988; Cardiff and Aguilar-Cordova 1988). Signature phenotypes were initially recognized in the mammary gland as gene-specific phenotypes for Tg(Myc) (Figs. 7.2a and 7.3b), Tg(Neu) (Fig. 7.2b), and Tg(Ras) (Fig. 7.2c) (Cardiff et al. 1991). While more GEM modeling has been done in the mammary gland resulting in a broader experience in mammary biology than most other organ systems (Figs. 7.1c,d and 7.3–7.5), signature phenotypes are observed with the same transgenes in other organ systems.

The initial mammary transgenic tumors recognized with gene-specific phenotypes were Ras, Neu, and Myc (Cardiff et al. 1991). These three transgenes produce distinct histological phenotypes, closely resemble human breast cancer, and are distinct from spontaneous mouse tumors related to Wnt, Notch, and Fgfr activation. Crosses with mice with combinations of two or three of the transgenes result in acceleration of mammary tumorigenesis (Cardiff et al. 1991). The phenotypes resulting from the crosses are not unique but dominated by one initiating oncogene. Molecular profiling has confirmed that the Tg(Myc), Tg(SV40 Tag), and Tg(ErbB2) tumors form distinctly separate molecular and morphologic clusters (Desai et al. 2002; Fargiano et al. 2003).

The Myc oncogene is a dominant oncogenic transgene that is overexpressed in many human cancers (Schmidt 1999, 2004). The Tg(Myc) tumor phenotype in GEM is characterized by large cells with large pleomorphic nuclei with a coarse-clumped chromatin and prominent nucleoli (Fig. 7.3) (Cardiff et al. 1991). The cytoplasm is amphophilic, implying abundant RNA. These mammary tumors generally have a glandular pattern. Similar morphological patterns are found in Tg(Myc)-induced tumors of the mammary gland (Figs. 7.2a and 7.3b), liver (Fig. 7.3c), lung, and lymphomas (Figs. 7.3d and 7.10).

The signature Tg(Ras) tumor is different from the large blue cell phenotype of Tg(Myc)-based tumors (Cardiff et al. 1991). The signature Tg(Ras) tumor has relatively small cells with oval to round nuclei and a delicate chromatin pattern with bright red cytoplasm (Figs. 7.2c and 7.5). The Tg(Ras) mammary tumors tend to have a papillary pattern that is similar to transitional cell carcinomas of the human bladder.

The Tg(Neu) phenotype is a third distinctive type and is composed of cells intermediate in size that have large, round nuclei with a delicate chromatin pattern and a small nucleolus (Fig. 7.2b) (Cardiff et al. 1991). The cytoplasm is a lighter pink than the Tg(Ras) tumor cell cytoplasm. The Tg(Neu) mammary tumors tend to form solid, nodular, expansile masses. The nodules are frequently "zonal" with increasing differentiation toward the center (Deckard-Janatpour et al. 1997; DiGiovanna et al. 1998).

Tg(SV40 Tag) is another dominant oncogenic transgene that has been expressed in many different organ systems behind various promoter systems (Figs. 7.2e and 7.4). The SV40 Tag is a viral gene whose product binds to and inactivates the tumor-suppressor genes P53 and Rb1 (Ali and DeCaprio 2001; Furth 1998). Therefore, Tag is a powerful antitumor-suppressor protein (Ali and DeCaprio 2001). One mutational variant, T^{121}, is designed to bind Rb1 but not P53 (Saenz Robles et al. 1994).

Tg(SV40 Tag) tumor phenotype is an invasive, metastatic tumor that is generally a neuroendocrine tumor (Shappell et al. 2004; Powell et al. 2003; Evangelou et al. 2004). The neuroendocrine tumors have pleomorphic nuclei, with a typical "salt and pepper" chromatin pattern and a fibrillar, "stromal" neuropil (Fig. 7.4). The tumors are undifferentiated but form gland-like "pseudorosettes." The diagnosis can be confirmed with antisynaptophysin, chromagranin, or antilow-molecular-weight keratin stains, such as anti-K8 or anti-K18. The antikeratin stains show a paranuclear granule characteristic of neuroendocrine tumors in both humans and mice.

Tg(SV40 Tag)-related neuroendocrine tumors are found in the prostate (Fig. 7.4d), lung (Fig. 7.4b), pancreas, and gut (Fig. 7.4c) (Hager and Hanahan 1999; Shappell et al. 2004; Powell et al. 2003; Evangelou et al. 2004; Kasper et al. 1998; Masumori et al. 2001; Nikitin et al. 2004; Borowsky et al. 2004).

Two signal transcription pathways have distinctive tumor phenotypes: Tg(Wnt1) and Tg(ErbB2)-related tumors (Figs. 7.2 and 7.6) (Rosner et al. 2002) (Desai). The Tg(ErbB2) tumors have their characteristic pattern and cytology. The Tg(Wnt1) tumors, in contrast, show the classical phenotypes of the MMTV-induced tumors described by Thelma Dunn (Figs. 7.1 and 7.2d) (Sass and Dunn 1979). Interestingly, a number of genes in the Wnt pathway share one or more of the characteristics of these tumors.

It is significant that the oncogenic transgenes belonging to the same signal transduction pathways share morphological characteristics (Fig. 7.6) (Rosner et al. 2002) because common features in any tumor group may provide a clue to the underlying genetic aberration. Molecular expression profiling demonstrates that Tg(Ras), Tg(PyV-mT), and Tg(ErbB2) form gene expression clusters distinct from the Tg(Myc) and Tg(SV40Tag) tumors (Desai et al. 2002; Fargiano et al. 2003). Genetic crosses of Tg(Myc), Tg(Wnt), or Tg(ErbB2) GEM with cyclin D1 (Ccnd1) knockout mice (Tm(Ccnd1)) demonstrate that tumorigenesis is abrogated in the Tg(ErbB2), Tm($Ccnd1^{-/-}$) null mice but not in the Tm($Ccnd1^{-/-}$) null mice crossed with Tg(Myc) or Tg(Wnt) GEM (Yu et al. 2001). Thus, these distinctive patterns need to be recognized.

The Tg(ErbB2/Ras) pathway mammary tumors have solid, nodular growth patterns that clearly originate in the mammary side buds (Fig. 7.6a, b). Tumor-bearing animals belonging to this group include Tg(ErbB2), Tg(Neu), Tg(PyV-mT), Tg(Ras), and Tg(Igf1R) (Rosner et al. 2002).

In contrast, the Tg(Wnt/Fgf) family of mammary tumors have more complex growth patterns that begin with a ductal dysmorphogenesis (Fig. 7.6e, f) (Rosner et al. 2002). The terminal ends of the ducts differentiate into microacinar (type A of Dunn) (Fig. 7.6c, d), adenosquamous, solid embryoid (type B of Dunn), or pilar patterns (Fig. 7.6e, f) (Rosner et al. 2002). These tumors have been associated with a large number of models, including Tg(Wnt) (Cardiff et al. 2000a, b, 2001; Rosner et al. 2002; Li et al. 2003), Tg(Fgf), Min mice (C57BL/6J-Apc^{Min}/J) (Moser et al. 2001), Tg(CTNNB1) (beta-catenin)(Miyoshi et al. 2002a), Tg glycogen synthetase kinase1

(GSK1), and Tg casein kinase 2a (CK2a) (Landesman-Bollag et al. 2001; Rosner et al. 2002; Miyoshi et al. 2002b). Various transgenes produce tumors resembling different subsets of the Tg(Wnt) tumors and MMTV-induced tumors. For example, the MIN mice treated with chemical carcinogens develop almost solely "pilar" tumors with very distinctive production of extracellular hard keratin characteristic of hair (Seldin PMID: 16263698) (Rosner et al. 2002). The related Tg(CTNNB1) (beta-catenin) mice also produce pilar tumors (Miyoshi et al. 2002a). In contrast, Tg(Wnt1) tumors are predominantly microacinar or ductal (Figs. 7.2d and 7.6c, e), (Rosner et al. 2002) with prominent, well-organized myoepithelium (Fig. 7.6d, f).

Pten, Nkx3.1, Akt1, Fgf, and similar oncogenes or tumor-suppressor genes result in premalignant lesions and adenocarcinomas that are indistinguishable from each other (Figs. 7.7d and 7.8c) (Kim et al. 2002; Song et al. 2002; Freeman et al. 2003; Abate-Shen et al. 2003). Even when crossed with functional knockout Tg(T^{121}) mice, Tm($Pten^{-/-}$) animals develop a subset of lesions with an unusual Akt-signature papillary lesion (Hill et al. 2005).

The Pten/Akt prostate cancer has cells with large oval nuclei with a relatively open chromatin and distinct, large nucleolus. The cytoplasm is abundant and pale pink. This pattern differs from the SV40 Tag type of prostate cancer cell that has smaller nuclei with more diffuse hyperchromatic chromatin and relatively scanty cytoplasm.

7.2.1 The Tumor Biology

The biology of these tumors is important in your interpretation. Some important considerations include the observations that:

1. Some genes have a stronger influence than others on the microscopic structure of the tumor.
2. Expression levels of the oncogenic transgenes influence the microscopic structure of the tumor.
3. Mutational changes in the initiating oncogenic transgene may change the fundamental microscopic structure of the resulting tumor.
4. The site of oncogene insertion has little or no effect on the microscopic structure of the resulting tumor.
5. The microscopic structure of tumors initiated by weak oncogenes can be influenced by tissue context.
6. The tissue and other factors influence the tumor biology but not the tumor morphology.
7. The microscopic structure of the neoplasm is not affected by the background strain.
8. Neoplastic progression is a multistep process associated with sequential morphological changes.

These and other aspects of tumor pathobiology are discussed in detail elsewhere (Cardiff et al. 2006a; Cardiff 1996, 2003).

7.2.2 *Mouse Tumors Mimic Human Cancers*

Genetic engineering has provided a new basis for understanding the correlation between structure and function. This correlation also provides a basis for comparing human and murine tumors leading to a new aspect of comparative tumor pathology. The new genomic pathology does not change the fundamental premises of tumor biology, but does challenge everyone to examine the morphological organization of the tumor cell in a different way. The new pathology is based on molecular events that provide unique insights into their structural consequences. The specific molecular event clearly results in specific microscopic tumor phenotypes.

7.3 Systems Pathology

Pathology is traditionally presented as systems pathology. The important principles and examples have been presented. However, students are eager to understand these principles in the context of their specific model and their specific organ system. Although this is an important perspective, the world of genetic engineering is so dynamic that any discussion or description of all of the tumors in an organ system is doomed to be obsolete by the time the publication appears. The discussion provided here includes an introduction to the standard models in the organ system with illustrations from human pathology.

7.4 General Principles

Our emphasis is on neoplasms of the mouse. Most neoplasms appear as discrete masses. The sole exception to this rule is the leukemias and lymphomas that tend to present as diffuse lesions causing enlargement of the involved organ (organomegaly). On the other hand, not all space-occupying masses are neoplastic. Abscesses, granulomas, and infarcts are also discrete focal entities. Abscesses are focal areas of acute inflammation with central necrosis generally filled with pus. Granulomas are focal areas of chronic inflammation with fibrosis and monocytes with multinucleated giant cells. Infarcts are discrete areas of ischemic necrosis that follow the organ's blood supply. Each has distinct gross and microscopic features that need to be understood, compared, and contrasted to neoplasms.

Neoplastic masses in the mouse are mostly, but not always, expansile with pushing margins. In contrast to most human malignancies, they rarely have infiltrative, invasive margins. Even malignant neoplasms of the mouse that have a high metastatic potential may not appear as invasive as malignant human tumors. Necrosis occurs in the larger tumor masses as they outgrow their blood supply. Hemorrhage and necrosis are basic gross indicators of malignancy.

The evaluation of biology of the neoplasm requires an understanding of its interactions with the host. In fact, the most important biology is frequently found at the host–tumor interface. Therefore, it is extremely important to note features, such as blood supply and inflammation, in the tissue around the tumor itself. In contrast to the need of the molecular biologist to have as pure a sample of the tumor as possible, the pathologist generally needs to examine the host–tumor interface and wants a generous sample of the tissue adjacent to the tumor. In general, pathologists take an area of host tissue equal to the area or volume of the tumor.

Mouse malignancies have a tendency to metastasize to the lung. In comparison to humans which frequently have local lymph node involvement, the lymphatic drainage of mice is limited and rarely contains lymph node metastasis. Bone metastases is also rare in the mouse but can be created experimentally. Metastasis in most textbook cartoons is represented as invasion of blood vessels by individual malignant cells. Empirically, metastases to the mouse lung are usually seen in large clusters of cells that are trapped in relatively large vessels (Siegel et al. 2003). Many of these clusters have two layers of endothelium (Fig. 7.10). This phenomenon is thought to be caused by noninvasive, intravascular metastasis in which large nests of tumor cells are pinched off with their endothelium intact and are carried to the lungs, where they lodge with producing two layers (Sugino et al. 2002). This interpretation is supported by the observation of such tumor emboli in vessels supplying the neoplasm (Fig. 7.9).

Grading and staging systems that are so useful in medical oncology have had significant predictive or prognostic value in the mouse. Mouse tumors that appear benign to the medical pathologist are frequently metastatic. Mouse tumors, such as EMT tumors, are locally aggressive but do not appear to be metastatic.

7.5 Mammary Pathology (Cardiff 2003) (Fig. 7.11)

Pathology: Most mouse mammary tumors are easily recognized as palpable masses deep to the skin. They can infiltrate the dermis to adhere to the skin and ulcerate through the epidermis if neglected. On gross examination of the exposed tumor, large feeder vessels are generally present and the tumors may be pale, pink, or black depending on the degree of hemorrhagic necrosis. In general, the mouse mammary tumors are fleshy and softer than the classical hard, firm, human breast cancers. Most mouse tumors have pushing borders. Most mouse model systems produce multiple mammary tumors.

The route of metastasis is also different with the mouse displaying a hematogenous spread to the lung. In contrast, the human breast cancers have regional lymph node involvement with preferential spread to bone, brain, and liver.

The precancers of mice appear as small, focal atypias that are difficult to see with the naked eye. Under the dissecting scope, these early lesions can be identified by the slight pink blush of their abnormal vasculature or, in some mice, a slight yellow discoloration. These lesions are generally flat in the plane of the

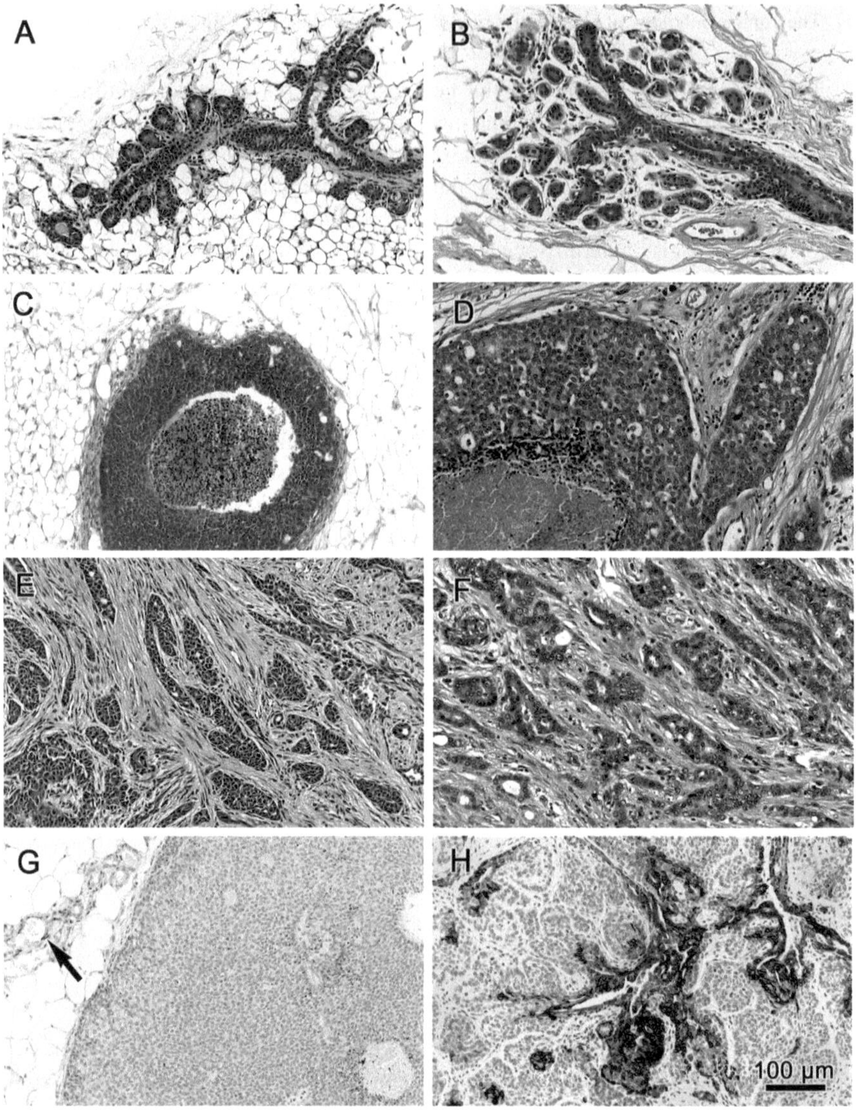

Fig. 7.11 Comparative pathology of mouse and human mammary neoplasia. Histology of mouse and human mammary gland with hematoxylin and eosin stains (**a–f**) and cytokeratin-14 immunohistochemistry (**g–h**). In the top six panels, mouse is on the left (**a**, **c**, **e**) with human on the right (**b**, **d**, **f**). (**a**) Mouse normal virgin mammary gland near nipple showing duct branching and budding. (**b**) Human normal resting lobule. (**c**) Mouse MIN in a Tg(C(3)1-SV40 Tag) showing central necrosis within a dysplastic intraductular proliferation. (**d**) Human DCIS with luminal necrosis and ductular extension. (**e**) Mouse invasive mammary carcinoma with a scirrhous-like growth pattern in a Tet-inducible Tg(cNeu). (**f**) Human invasive mammary carcinoma, typical scirrhous, "ductal" pattern. (**g**) Immunocytochemical stain for cytokeratin-14 of a mouse Tg(MMTVLTR-ERBB2) mammary tumor with a solid growth pattern, showing positive immunostaining only in the adjacent normal ducts (*arrow*). The tumor cells do not exhibit myoepithelial differentiation as detected by cytokeratin-14 staining. (**h**) Mouse Tg(MMTVLTR-WNT) mammary tumor showing cytokeratin-14-positive cells within branching ducts terminating in immunonegative solid nodules. The 100-μm scale bar (**h**) is applicable to all panels

mammary fat pad. They expand but do not invade through the fascia surrounding the mammary fat pad.

Proof of biological potential is traditionally provided by the test by transplantation (Cardiff et al. 1977, 2006b; Park et al. 2002). By operational definition, growth of malignant neoplasm is not restricted to the mammary fat pad and expands in all directions in orthotopic transplantation. In contrast, precancers are limited to growth in the mammary fat pad and are readily growth inhibited by normal mammary gland.

As discussed above, and elsewhere, the histology of mammary neoplasms is quite variable. The specific phenotype is determined by the pathway and the initiating oncogene (Rosner et al. 2002). With rare exception, the phenotype is consistent with an epithelial neoplasm and ranges from solid to glandular. However, in contrast to the human counterpart, the degree of tissue, cytological, or glandular differentiation does not predict biological outcome. In short, pathological grading is not useful in predicting biological behavior.

Since the early 1900s, mouse mammary tumorigenesis has been the primary model for experimental tumor pathology (Cardiff and Kenney 2007). Mice were specifically inbred to study mammary carcinogenesis. As a result, more is known about spontaneous mouse mammary tumors than any other organ system. GEM with mammary tumors were among the earliest developed using transgenesis (Cardiff and Kenney 2007).

Over 100 GEM models of human breast cancer currently exist. As discussed above, mammary tumors have characteristic or unique gene-specific, "signature" phenotypes that readily can be identified microscopically. The principle that genotype predicts phenotype can be applied to other GEM and extended to include entire molecular pathways, referred to as pathway pathology. The wnt and the ErbB signal transduction pathways have been the most thoroughly dissected using targeted transgenesis.

A majority of breast cancer models have been induced using the MMTV-LTR promoter. Other promoters, such as C3(1), WAP, and BLG, also have been used. These promoters introduce a slightly different biology to the system. More recently, inducible promoters, such as Tamoxifin and Doxycycline, have been used. The tumors retain the phenotype characteristic of the oncogene. Inserting the transgene behind the native promoter or "knocking in" an oncogenic variant at the normal gene locus has resulted in tumors that closely resemble the oncogene induced by the MMTV-LTR. These tumors also resemble the phenotypes that develop in association with other promoters.

The key question is always: Do GEM model human breast cancer? Genes that cause cancer of the human mammary epithelium cause cancer of the mouse mammary epithelium (verification). Some cases have a remarkable morphological similarity of the cancers from the two species. For example, human lobular carcinoma with loss of e-cadherin is precisely modeled by the e-cadherin mouse knockout (Derksen et al. 2006).

Mice and humans, however, are different species with different biologies. The biology of their tumors is also different. In contrast to 70% of human breast cancers that are ER+, most mouse models of breast cancer are hormone independent and lack detectable estrogen receptors. This attribute alone does not make them triple negative, basal phenotype tumors. Recently, several publications have appeared suggesting the presence of ER^+ and PR^+ tumors. However, examination of these tumors generally reveals that less than 10% of the cell population is ER positive. Further, most studies do not include ovarian ablation to determine whether such tumors are hormone dependent. We have recently seen a model that produces ER^+ and PR^+ ovarian-dependant tumors.

7.6 Prostate Pathology (Borowsky et al. 2004) (Fig. 7.12)

Gross pathology: Since mice rarely develop spontaneous prostate cancers, any prostate tumor arising in the context of genetic modification is directly related to the inserted or knockout genes. Mouse prostatic tumors are generally detected as advanced lesions at necropsy. Most tumors are first discovered when quite large and advanced. Since the normal rodent prostate is composed of delicate fern-like structures, any discoloration or mass is suspect. They may be hemorrhagic when detected. The earlier precancerous prostatic intraepithelial neoplasia (PIN) lesions are more difficult to find and may appear as a slight bulge in the gland. PIN lesions can be detected using indirect lighting and a dissecting microscope. The seminal vesicles in the SV40 Tag models also develop relatively benign tumors that should not be mistaken as malignant.

Microscopic pathology: Three phenotypes of malignant tumors have been described in the mouse prostate. Poorly differentiated neuroendocrine tumors are characteristic of SV40 Tag models. Small, back-to-back glands are found in a myc model (Ellwood-Yen et al. 2003). The AKT-related models develop a larger cell-type neoplasm which invades in clusters (Park et al. 2002). Regional lymph node and bone and pulmonary metastases have all been described.

A number of human prostatic cancer models have now been developed. The mouse prostate is significantly different from human, being divided into separate recognizable lobes, the anterior (or coagulating gland), the ventral, and the dorsolateral. The human prostate is divided into zones with peripheral, central, and transitional zones but is structurally a solid organ. Human prostate carcinoma is known to arise in the peripheral zone. No definitive relationship between the mouse prostate lobes and the zones of the human prostate has been identified. Both human and mouse prostate glands are characterized by cuboidal and columnar epithelium.

Fig. 7.12 (continued) (*left center*). Smaller, uninvolved glands are seen (*right and periphery*). (**e**) Mouse neuroendocrine carcinoma in "TRAMP" model showing neuroendocrine rosette formation. Tumor was synaptophysin positive (not shown), confirming the neuroendocrine differentiation. (**f**) Human neuroendocrine carcinoma, *solid areas*, and residual glandular differentiation. Tumor was also synaptophysin positive (not shown), confirming neuroendocrine differentiation. (**g**) Mouse prostate adenocarcinoma with small acinar pattern and reactive stroma in the "LADY" model. (**h**) Human prostate adenocarcinoma with a microacinar growth pattern (Gleason 3 pattern) and cytology showing prominent central nucleoli. The 100-μm scale bar (**h**) applicable to all panels

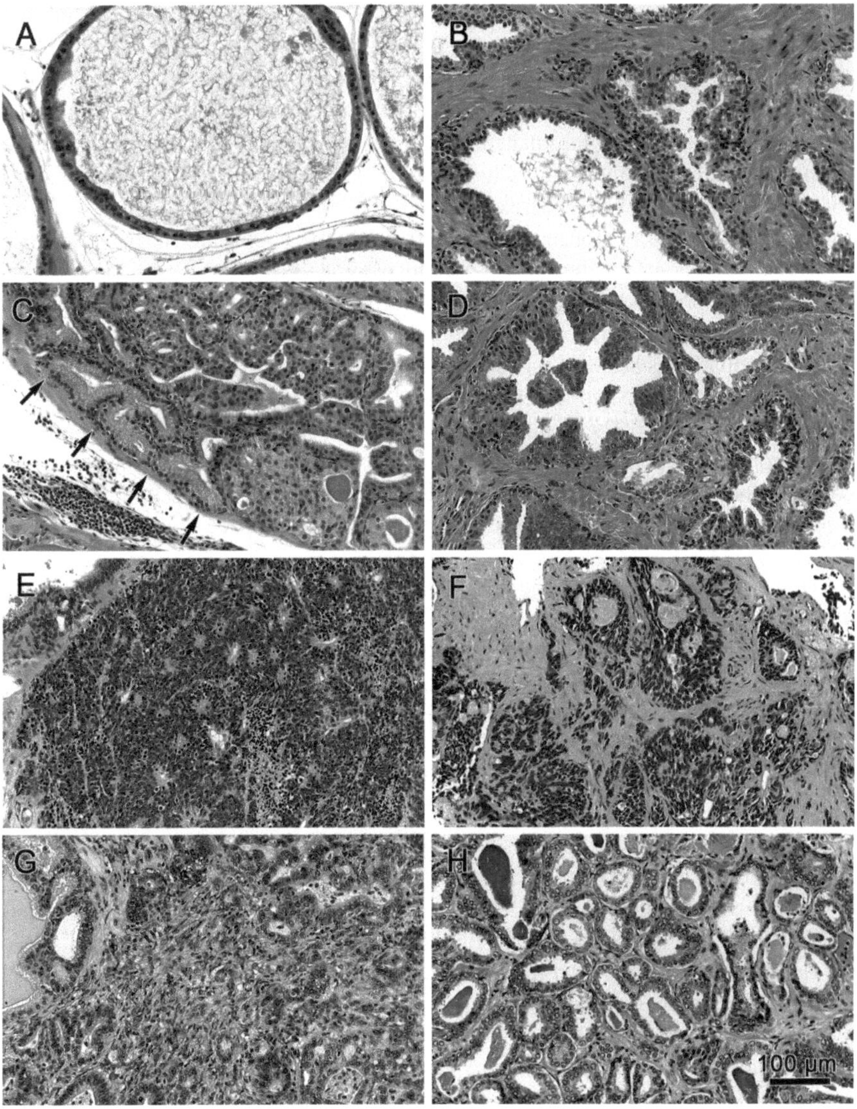

Fig. 7.12 Comparative pathology of mouse and human prostate neoplasia. Histology of mouse and human prostate gland with H&E stains. Mouse is in (**a**), (**c**), (**e**), and (**g**). The human is in (**b**), (**d**), (**f**), and (**h**). (**a**) Mouse normal dorsolateral prostate gland showing simple cuboidal epithelium, inconspicuous basal cells, and a thin muscular stroma surrounded by loose-connective tissue. (**b**) Human normal prostate: Two glands illustrate the variability usually seen in the human prostate from more simple cuboidal (*right*) to more columnar epithelium (*left*). (**c**) Mouse prostatic intraepithelial neoplasia (mPIN III) in a Tm(Nkx3.1$^{(-/+)}$ × Pten $^{(-/+)}$) with foci of transformation seen as multiple cell layers with altered cytology, and loss of polarity adjacent to a normal area evident as a single layer of epithelium (*arrows point to normal area*). (**d**) Human high-grade prostatic intraepithelial neoplasia (HGPIN) with multiple cell layers showing loss of cell polarity and atypical nuclear features with prominent single central nucleoli involving the large gland space

Both species have a thin but distinct layer of basal cells surrounding luminal cells.

In contrast to humans, hyperplasia does not occur in the normal aging mouse. Some of the GEM prostate models, such as SV40 Tag-related models, do develop a diffuse hyperplasia of the luminal epithelium. The differentiation of hyperplasia and neoplasia in GEM is important (Chiaverotti et al. 2008). When the lesions arise as focal (or multifocal) and progressive proliferations, the intraepithelial glandular foci can be defined as mouse prostatic intraepithelial neoplasia (mPIN) (Shappell et al. 2004). In contrast, the diffuse hyperplasias seen in some SV40-Tag mice do not appear to progress to cancer and should be regarded as atypical hyperplasias (Couto and Cardiff 2008).

The comparisons between mouse and human prostate are important in interpretation of pathogenesis (Shappell et al. 2004). The human prostate gland is embedded in a muscular stroma. The mouse prostate lobes are ensheathed by a thin fibromuscular stroma and a peritoneal reflection. They are free to expand within the peritoneal cavity. When the human prostate enlarges, it compresses the urethra. When the mouse prostate enlarges, it bulges into the surrounding space and the abdominal cavity.

Thus far, three distinct prostate phenotypes have been seen in GEM models. The most extensively studied phenotype is derived from Tg(SV40 Tag). While multiple variants are available, they are all associated with a diffuse hyperplasia of the luminal epithelium, formation of large "botyroides" polyps and poorly differentiated tumors with a high rate of metastasis. The invasive carcinomas are neuroendocrine tumors with a lineage separate from the luminal hyperplasias (Nikitin et al. 2007; Zhou et al. 2007). However, infrequent, true adenocarcinomas can be found.

The second phenotype includes a number of GEM tumor types with activation of the AKT pathway (Park et al. 2002). AKT is the nexus of the major signal transduction pathway activated in these models. The morphological lesions of activated AKT, whether in the Nkx3.1 or PTEN knockouts or the Tg(AKT), all have close cytological similarity with large cells, abundant pale cytoplasm, and large pleomorphic nuclei with open chromatin.

Some of the non-SV40 T antigen models progress to invasive and metastatic carcinoma. They include a number of models designed to mimic molecular events seen in human prostate cancer, such as the Nkx3.1 (–/+), PTEN (–/+), a "conditional" knockout – floxed RXRalpha × probasin-Cre, and transgenic probasin (ARR2PB)-FGF8b. They effectively mimic common losses of heterozygosity or growth factor overexpression seen in human CaP. They consistently produce metastatic carcinoma but with prolonged latencies (Shappell et al. 2004; Powell et al. 2003; Abate-Shen and Shen 2000). The myc model also produces an almost exact morphological mimic of human prostate cancer (Ellwood-Yen et al. 2003).

Several ras- and myc-driven models have been developed (Ellwood-Yen et al. 2003; Scherl et al. 2004). One ras model was noteworthy in that it featured an intestinal metaplasia (Scherl et al. 2004). However, most ras and myc modeling has failed to produce consistent neoplastic transformation.

7.7 Hematopathology (Fig. 7.13)

Leukemias and lymphomas are the most common spontaneous malignancies of mice. Many strains of laboratory mice were specifically selected for the study of "spontaneous" leukemias and lymphomas. Of course, these diseases are triggered by the murine leukemia viruses. They have served as important models for understanding normal and malignant hematopoiesis. In the GEM era, genes-associated human leukemias, when placed into the mouse, have recreated the human disease with remarkable accuracy and are being used for drug development and preclinical trials (Carver and Pandolfi 2006).

Gross pathology: Animals with leukemias or lymphomas are usually clinically ill. The diseases result in organomegally affecting multiple organs. The spleen, liver, lymph nodes, and thymus become enlarged. The specific disease and the strain background result in differing patterns of involvement. Cut surfaces of the organs generally have a mottled white appearance but can also be quite hemorrhagic and red.

Mouse lymphomas differ significantly from their human counterparts. For example, the most common spontaneous lymphomas in mice do not form "follicle-like" tumors, but the immunophenotyping shows that the mouse lesions are similar to the human follicular B-cell lymphoma (FBL). The diffuse large B-cell lymphomas (DLBCLs), common in human, are rare in mice. The mouse FBL has mixed populations of B cells and is classified with DLBCL as an FBL. Thus, the classifications are difficult to understand and do not readily mimic the morphology of human disease.

Interestingly, GEM develop lymphomas that are rarely found in laboratory mouse strains. Examples include splenic marginal-zone lymphomas in p53 null mice (Fig. 7.1d, e), Burkitt lymphomas associated with a translocated MYC-transgene (Fig. 7.1f), and prolymphocytic lymphomas in TCL1-transgenic mice. B-natural killer-cell lymphomas have been found in some mouse strains and not reported in humans.

The classification and characterization of these lesions now require detailed cytological, immunological, and molecular study (Kogan et al. 2002; Morse et al. 2002). They can be divided into nonlymphoid (Kogan et al. 2002) and lymphoid neoplasms (Morse et al. 2002).

7.8 Nonlymphoid Neoplasms (Fig. 7.14)

Mouse leukemias were originally classified by histopathologic diagnosis. These diseases were classified in mice, including granulocytic leukemia, erythroid leukemia, histiocytic sarcoma, and mast cell tumor (Kogan et al. 2002). The Mouse Models of Human Cancers Consortium of the National Cancer Institute (the USA) developed new proposals for classifying nonlymphoid hematopoietic neoplasms of mice to include those now found in GEM (Kogan et al. 2002).

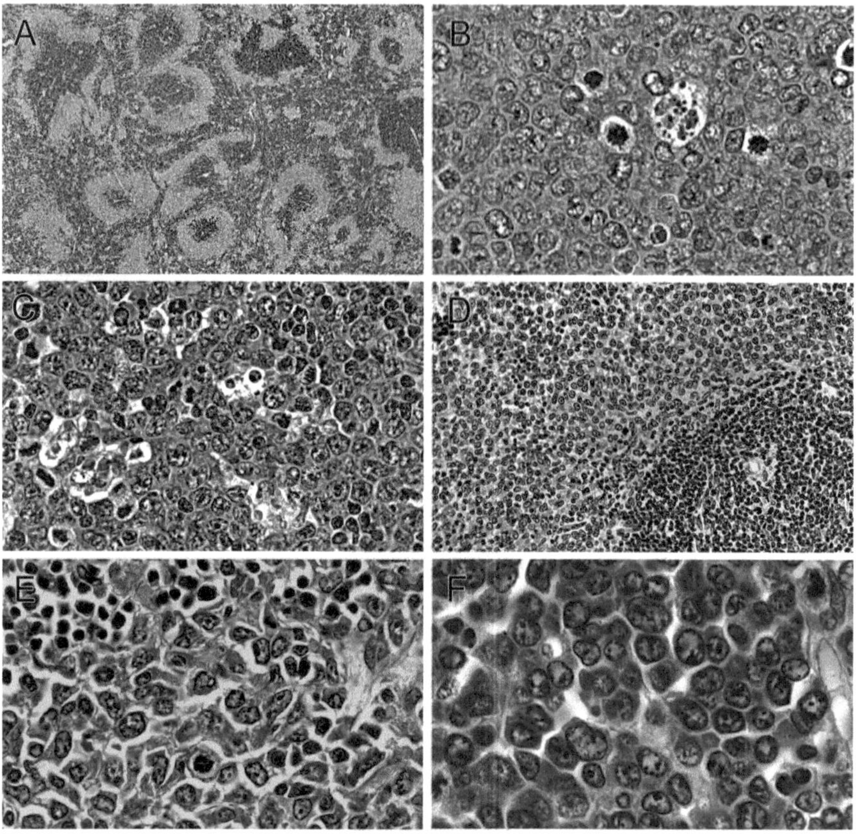

Fig. 7.13 Images illustrating some of the stains and the typical cytological features of several large and small cell variant lymphomas. (**a**) Immunohistochemistry showing TdT expression in a TG(p53$^{-/-}$) with TdT-positive metastatic thymic T-cell lymphoma to the splenic white and red pulp and negative marginal-zone lymphoma. (**b**) Diffuse large B-cell lymphoma. (**c**) Thymic T-cell lymphoblastic lymphoma. (**d**) Early splenic, marginal-zone lymphoma with normal white pulp on right side in a p53 null mouse. (**e**) Splenic, marginal-zone lymphoma invading red pulp in a p53 null mouse. Note pleomorphic cells with abundant cytoplasm. (**f**) Burkitt's B-cell lymphoblastic lymphoma in a transgenic mouse (images courtesy of Dr. J.M. Ward)

Mice are normally myeloproliferative "factories." The marrow and spleen respond to infections and stress with expansion and increased production of myeloid elements. Thus, splenic enlargement most frequently indicates reactive change rather than neoplastic disease. This biology has made it difficult to ascertain whether the mouse has "myelodysplastic syndromes (MDS)." What is the appropriate use of the terms "leukemia" and "myeloproliferative disease (MPD)" in mice? The nonlymphoid, hematopoietic "neoplasms" of mice have been divided into four categories: (1) nonlymphoid, hematopoietic sarcomas, (2) nonlymphoid leukemias, (3) myeloid dysplasias, and (4) myeloid proliferations (nonreactive).

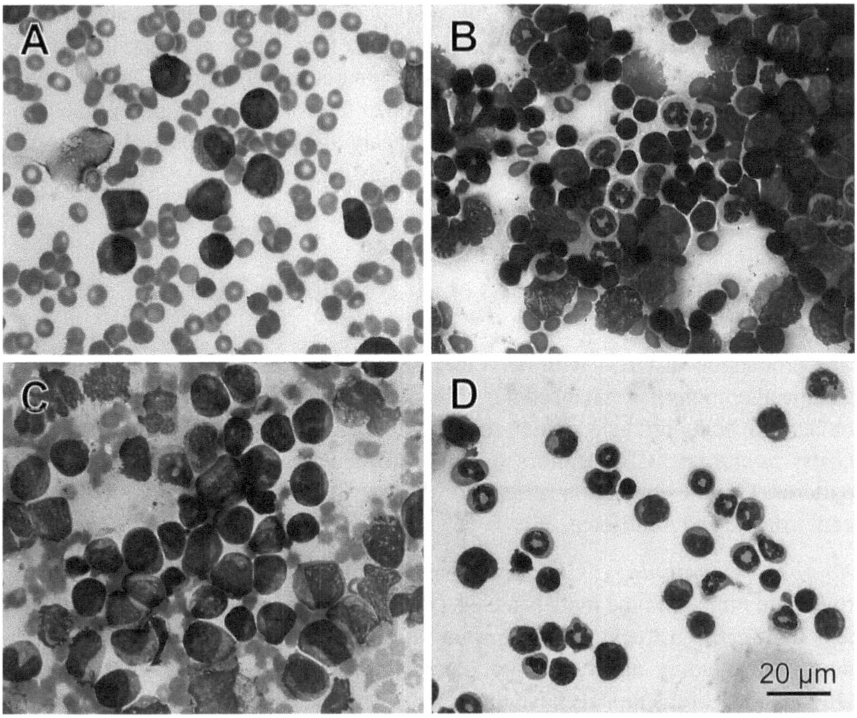

Fig. 7.14 Cytology of human and mouse myeloid neoplasms. Wright's Giemsa stain. Red blood cells and debris from broken cells are seen in most panels. (**a**) Bone marrow smear of human patient with acute promyelocytic leukemia. Leukemic promyelocytes have open chromatin and numerous azurophilic cytoplasmic granules. Note cell with abnormal, rod-shaped primary granules (Auer rods), characteristic of human acute promyelocytic leukemia. (**b**) Bone marrow smear of a normal FVB/N mouse showing a heterogeneous mixture of nucleated blood cells, including neutrophilic cells with pale cytoplasm and ring-shaped to irregular nuclei with dense chromatin, dark staining round erythroid progenitors, small lymphocytes with pale purple round nuclei and high nuclear:cytoplasmic ratios, and immature cells with round nuclei and basophilic cytoplasm. (**c**) Bone marrow smear of leukemia in a mouse expressing a *PML–RARA* transgene that is associated with human acute promyelocytic leukemia. Normal heterogeneous mixture of marrow cells is replaced by leukemic immature forms/blasts with high nuclear:cytoplasmic ratios, open chromatin, and cytoplasm containing azurophilic granules. (**d**) Myeloproliferative disease in mouse with *Nf1*$^{-/-}$ bone marrow; cytospin of bone marrow. Cells are predominantly neutrophilic, accompanied by immature myelomonocytic cells. Few small round cells, including lymphocytes and residual erythroid cells, are present. Bar = 20 μm (images and legend courtesy of Dr. S. Kogan, UCSF)

This terminology is designed to permit more accurate comparisons to human illness while recognizing the similarities and distinct pathologies of mice and humans (Kahn et al. 2009). In contrast to humans, mice develop aggressive leukemias composed of well-differentiated granulocytic/monocytic cells. Thus, the "grading" of mouse leukemias and other tumors is not useful for evaluating biological progression.

Classification of nonlymphoid leukemias with diagnostic criteria can be found elsewhere (Kogan et al. 2002). The classification of these lesions generally requires more detailed analysis and an experienced comparative pathologist. The investigator needs to recognize the disease and organize the collection of appropriate samples. For example, any case with splenomegaly should be suspect. The samples should include spleen, liver, marrow, and peripheral blood.

7.9 Neuropathology (Weiss et al. 2002) (Fig. 7.15)

The production of GEM with nervous system tumors using biologically relevant genetic alterations has enhanced the study of CNS tumors. An increasing number of models are being produced which reproduce models of almost all human tumors. *Gross pathology*: CNS tumors in mice are detected by abnormal behavioral patterns. Gross examination reveals focal expansile or infiltrative foci that can be hemorrhagic and multifocal.

Microscopic pathology: CNS neoplasms in mice are morphologically similar to human tumors. Some models are exact molecular and morphological phenocopies. High-grade astrocytic tumors show the characteristic "pseudopalisading" of nuclei surrounding areas of tumor necrosis exactly like human glioblastomas. Oligodendroglioma models have tumor cells with the "perinuclear haloes" characteristic of human oligodendrogliomas. Intrinsic GEM tumors have patterns of tumor cell infiltration characteristic of most human gliomas.

Immunohistochemical and ultrastructual methods are useful for characterizing CNS tumors. Neuroepithelial tumors, such as GEM gliomas and GEM medulloblastomas, are GFAP, S-100 protein, synaptophysin, and NeuN positive. Biomarkers for CNS tumors include nestin, neurofilament protein, and vimentin. Immunohistochemistry using S-100 protein, neurofilament protein, and the 75-kDa, low-affinity nerve growth factor receptor (LNGFR) has been used to identify peripheral nerve. Biomarkers of peripheral nervous system tumors also include collagen type IV, epithelial membrane antigen (EMA), and laminin.

7.10 Gastrointestinal Pathology (Boivin et al. 2003) (Figs. 7.16 and 7.17)

Gross pathology: The numerous models of human GI cancer all have gross intestinal polyps or sessile plaques. In contrast to human cancers which are usually in the stomach and colon, the small bowel appears to be the favored target in mice. Since mice are coprophagic, hindgut formentors, the topographical differences might be expected rather than surprising. The lesions in GEM are generally multifocal and can be observed grossly. The multifocality creates obstructions and may explain the lack of metastases.

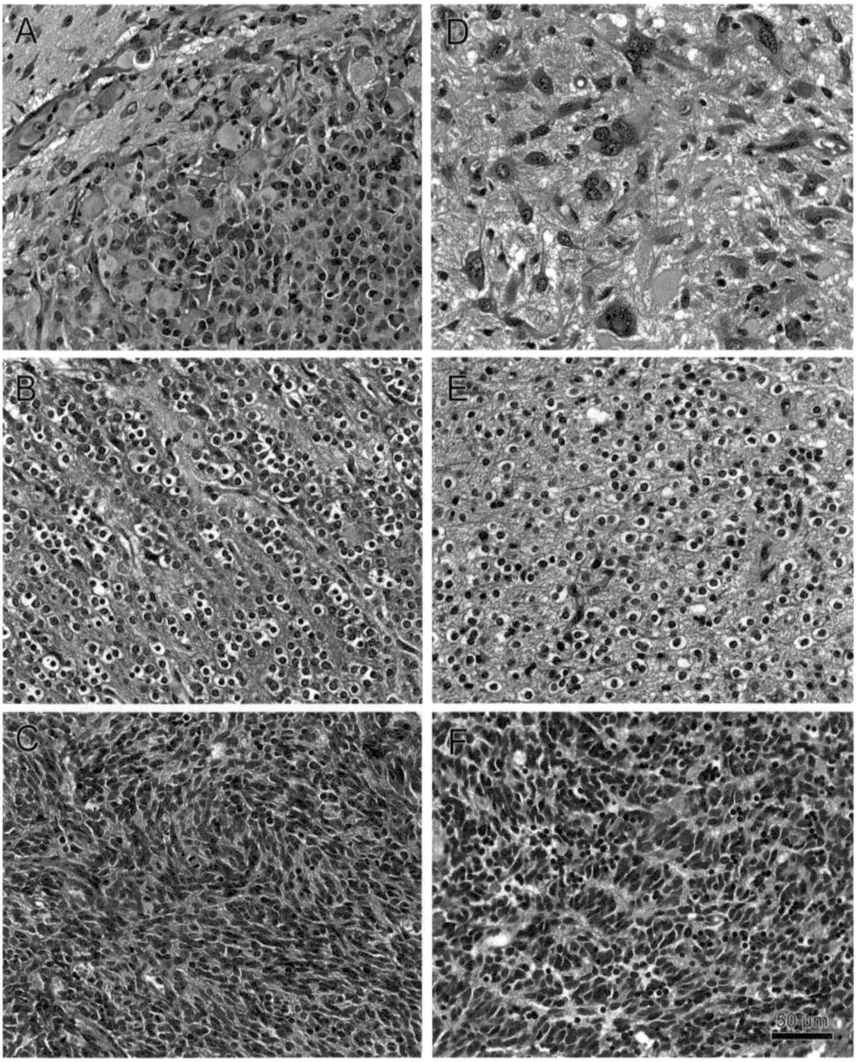

Fig. 7.15 Comparative pathology of mouse and human nervous system neoplasia. Histology of mouse and human tumors stained with H&E. Mouse is in (**a**–**c**). The human is in (**d**–**f**). (**a**) Brain showing a mouse astrocytoma, high grade (glioblastoma) with prominent "gemistocytic" features consisting of anaplastic glial cells with abundant eosinophilic cytoplasm. Infiltrative margins (*top left*), vascular proliferation (*prominently mid and bottom right*), and zones of tumor necrosis (not shown in figure) characterize this tumor. The tumor is from a nestin tv-a (retrovirus receptor)-transgenic mouse with RCAS retrovirus infection carrying *ras* and *Akt* oncogenes (**d**) Human glioblastoma with "gemistocytic" features. (**b**) Brain showing a mouse oligodendroglioma with characteristic clear cytoplasm and uniform cell morphology infiltrating brain parenchyma. This tumor is from a nestin tv-a-transgenic/Ink4a (–/–) mouse with RCAS retrovirus infection carrying PDGF (**e**) Human oligodendroglioma, grade II, showing similar characteristic tumor cell cytology. (**c**) Mouse medulloblastoma from a mouse heterozygous for Patched (*ptc* –/+). (**f**) Human medulloblastoma showing typical hyperchromatic cytology and neural architectural features. The 50-μm scale bar (**f**, lower right) applicable to all panels

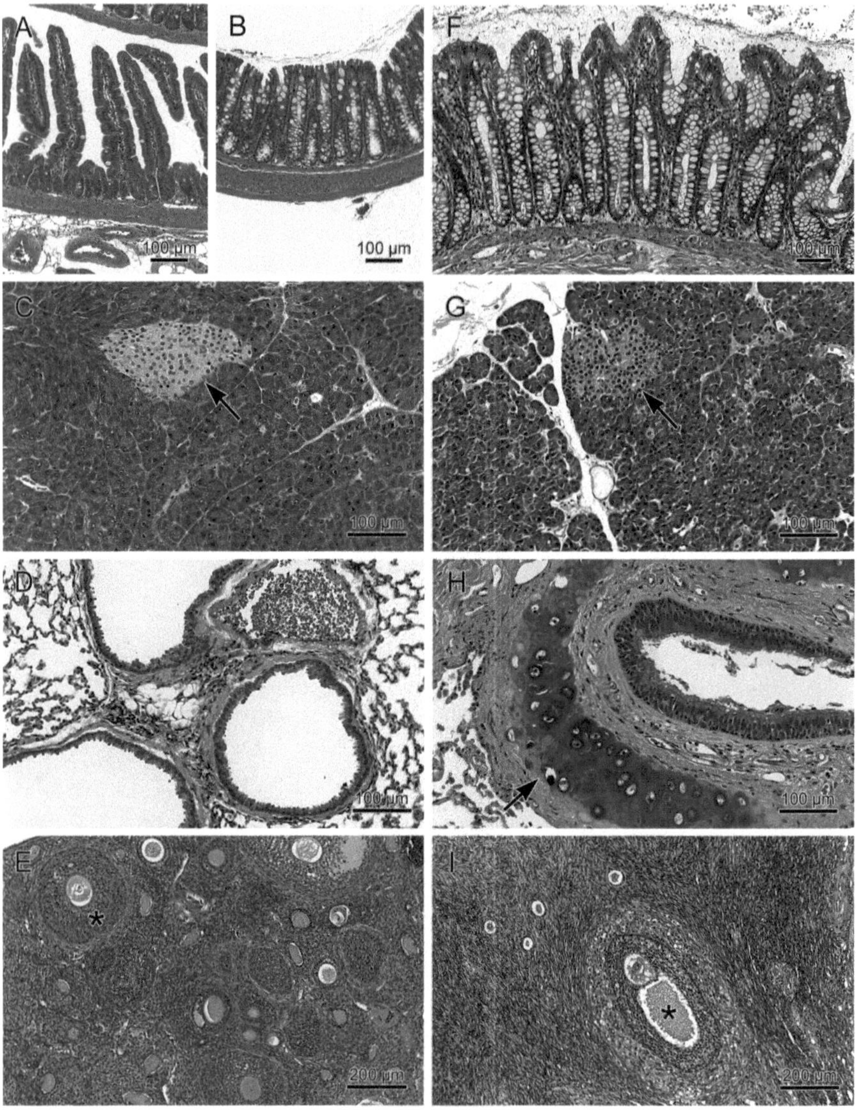

Fig. 7.16 Comparative histology of normal mouse and human tissues. Digital photomicrographs of histologic sections of normal mouse and human intestine, pancreas, lung, and ovary stained with H&E. The magnification is the same within each tissue type, illustrating the relative size of mouse and human cells. Mouse tissue is on the left (**a**–**e**) with human tissue on right (**f**–**i**). (**a**) Mouse small intestine and (**b**) mouse large bowel. (**f**) Human large bowel. (**c**) Mouse pancreas with (**g**) human pancreas. *Arrows* point to islets of Langerhans in each tissue. (**d**) Mouse lung and (**h**) human lung. The mouse bronchi lack the cartilage seen in human (*arrow*). Note the multitude of developing ova (*asterisk*) in mouse ovary (**e**) compared with human (**i**). *Scale bars* as indicated for each panel

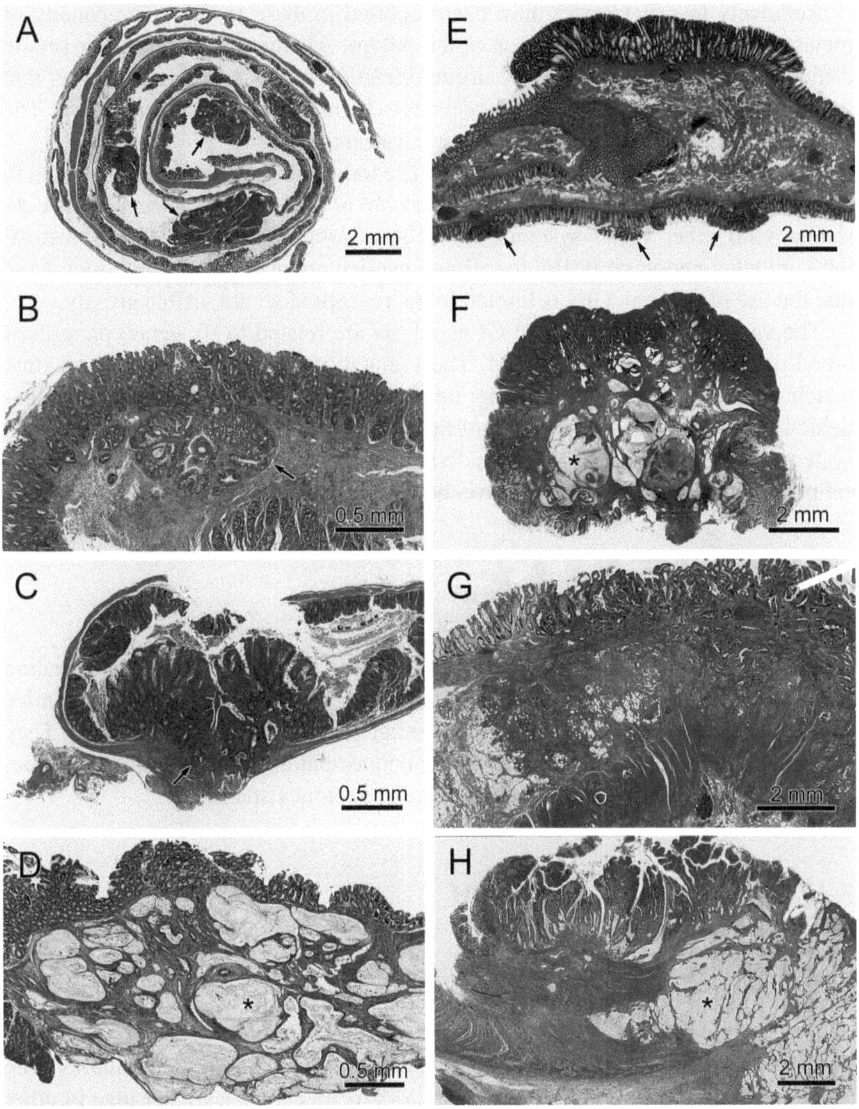

Fig. 7.17 Comparative pathology of mouse and human intestinal neoplasia. Digital photomicrographs of histologic sections of mouse and human small and large bowel stained with H&E. Mouse tissue is in (**a**–**d**) and human is in (**e**–**h**). (**a**) Swiss roll of mouse intestine with multiple broad-based adenomatous polyps (*arrows*) in an AOM-treated mouse. (**e**) Section of human large bowel from a patient with familial adenomatous polyposis and a large sessile adenomatous polyp (*at top*) as well as several smaller tubular adenomas (*arrows*). (**b**) Large bowel of a Tm(*MSH*$^{-/-}$) showing pseudo-invasion beneath the muscularis mucosae by a focus of normal epithelium (*arrow*). (**f**) Entrapped mucin (*asterisk*) in a human polyp which has undergone torsion. (**c**) Invasive adenocarcinoma of cecum in a Tm(TGFb1$^{-/-}$×P21$^{-/-}$). *Arrow* indicates invasion through muscle layers. (**g**) Invasive adenocarcinoma in human colon. (**d**) Mucinous carcinoma in a *Rag2 TGFb1* mouse showing extensive pools of mucin (*asterisk*). (**h**) Human mucinous carcinoma with pools of mucin within and penetrating through the muscle layers (*asterisk*). *Scale bars* as indicated for each panel

Relatively few metastases have been reported in these models. The paucity of metastases has made interpretation of the polyps, whether pedunculated or sessile difficult. Another major difficulty in interpretation of these mouse lesions is that hyperplastic intestinal mucosa of mice tends to herniate through the muscularis. The herniation has been misinterpreted as invasive carcinoma.

As a result, descriptive terms are used. The terminology is similar to that used in human but departs even further in consideration of potential precancers. As is consistent with other organ systems, the mouse terminology uses gastrointestinal intraepithelial neoplasia (GIN) for all noninvasive atypical microscopic foci. Note that the use of the term GIN is limited to microscopic foci not visible grossly.

The various mouse models of GI neoplasia are related to six genotypic groups based on the construct of the GEM. These mutations include Wnt signaling, mismatch repair genes, TGFβ signaling, immune deficiency with colitis, carcinogen-treated mice, and others which do not fit the above categories. Unlike other organ systems, these gene-based categories do not appear to have distinctive morphological phenotypes. Each group does have, however, distinctive biological properties.

7.10.1 Wnt Signaling Pathway

Several mouse models now carry mutations in genes associated with the Wnt signaling pathway. In particular, they involve mutations in β-catenin and Apc. These mice develop adenomas/polyps localized in the small intestine (Oshima et al. 1995). They are primarily pedunculated (polypoid) adenomas that arise in the mucosa without inflammation. Progression to carcinoma is rare or nonexistent.

7.10.2 Mismatch Repair GEM

DNA replication errors, genetic recombination, and chemical modification result in DNA mismatches. MMR genes encode proteins that correct DNA mismatches. The models include Mlh1, Pms1, Pms2, Msh2, Msh3, and Msh6. Tm(Mlh1$^{-/-}$), Tm(Msh3$^{-/-}$), and Tm(Msh6$^{-/-}$) mice also develop small intestinal tumors (see Chapter 15). However, the ratio of carcinomas to adenomas is higher than in other models (Edelmann et al. 2000). These lesions are predominantly pedunculated tumors. Msh2$^{-/-}$ mice also develop intestinal adenomas and carcinomas in the duodenum and jejunum. The Msh1$^{-/-}$ tumors are typically plaque lesions.

7.10.3 TGFβ Signaling Pathway

TGFβ signaling suppresses intestinal tumorigenesis (Engle et al. 1999). Combinations of Tgfβ1 (+/+, +/–, and –/–) with Rag2$^{-/-}$ develop cecal and colonic neoplasms.

All double-knockout mice, Tgfβ1$^{-/-}$ Rag2$^{-/-}$ mice, develop tumors. The tumors begin as sessile adenomas and carcinomas that may be observed at 3–6 months. The microflora may be involved in their development as Tgfβ1$^{-/-}$ Rag2$^{-/-}$ mice derived in a germ-free environment do not develop neoplasms. Smad4$^{+/-}$ mice also develop polyps in the duodenum and stomach. The duodenal polyps are sessile and consist of cystic deep glands with overlying hyperplastic mucosa. Most of the gastric polyps in this model appear to be hamartomas, but dysplastic glands and signet ring cell carcinoma have also been observed.

7.10.4 Immune-Deficient Mouse Models

Immune-deficient knockout mouse models (Il-10, Il-2, GNAI2, and Tcrα) develop inflammatory lesions of the large bowel, rectal prolapse, and proliferative lesions that sometimes progress to adenocarcinomas. Rederivation in bacteria- and virus-free environments ablates tumorigenesis in these immunodeficient models.

7.10.5 Endocrine Pancreas Tumors (Galvez et al. 2004) (Fig. 7.18)

Gross pathology: The gross pathology and distribution of islet cell tumors is similar in the various models. Islet cell tumors can arise in all regions of the pancreas. Exocrine and endocrine tumors cannot always be distinguished on the basis of gross examination. Microscopic confirmation is required. Islet cell tumors typically contain little fibrous tissue and are softer than many other exocrine tumors. Islet cell tumors are highly vascular, providing a characteristic pink color to the tumors. Islet cell tumors in mice range from 1 to 10 mm in diameter. Since the smaller tumors are not grossly detectable, accurate assessments of the tumors require microscopic examination.

Microscopic pathology: Well-differentiated islet cell tumors are recognized with routine hematoxylin and eosin staining because they closely resemble normal islets. Anaplastic islet cell tumors are more difficult to distinguish from poorly differentiated exocrine tumors. Well-differentiated tumor cells have moderate amounts of pale eosinophilic, granular cytoplasm, and uniform round to oval nuclei. The cells are ordinarily very uniform in an islet cell tumor.

Normal islets may be difficult to distinguish from large normal islets, islet cell hyperplasia, and small islet cell. The presence of intraislet fibrosis, which distorts the shape of the islet and produces a lobular appearance, is very suggestive of hyperplasia. On the other hand, adenomas tend to be spherical and contain little fibrous tissue. Islet cell hyperplasia could be a preneoplastic stage in the development of islet cell tumors.

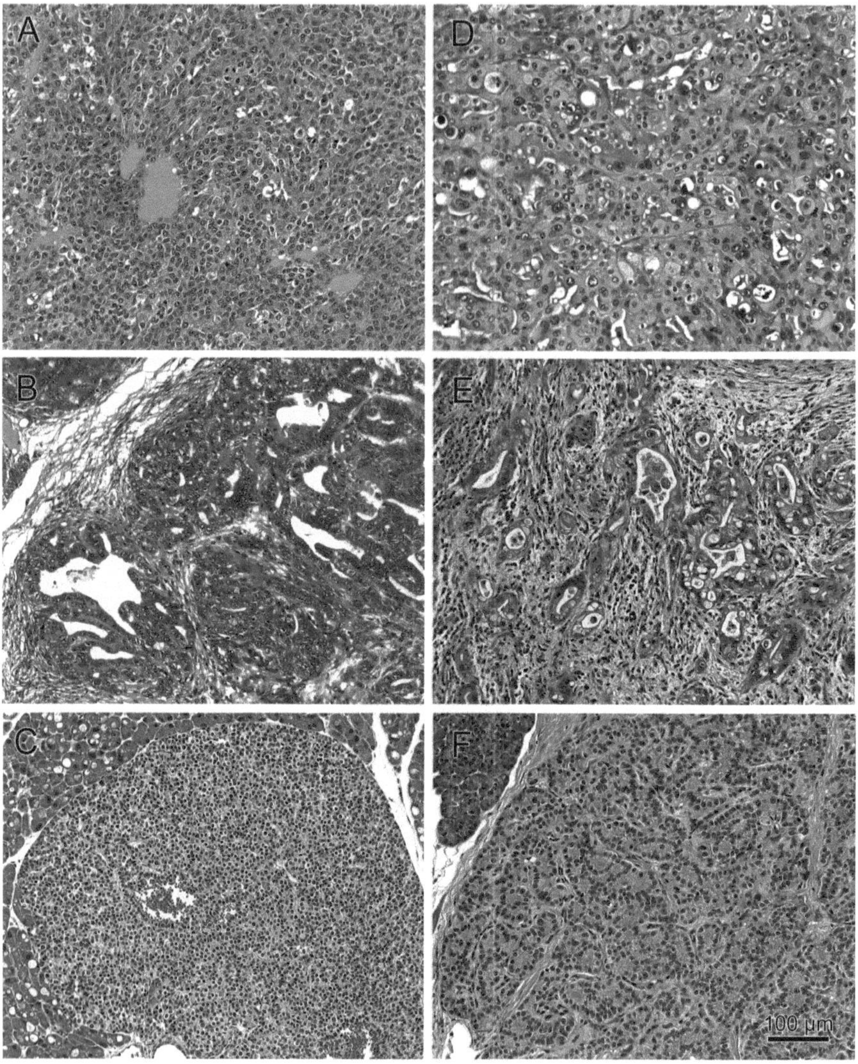

Fig. 7.18 Comparative pathology of mouse and human pancreatic neoplasia. Digital photomicrographs of histologic sections of endocrine and exocrine pancreatic neoplasms stained with H&E. Mouse tissue is on the left (**a**, **b**), with human tissue on the right (**c**, **d**). (**a**) Mouse exocrine pancreatic adenocarcinoma in a polyomavirus middle T mouse (image generously provided by Robert Oshima (Cecena et al. 2006)). (**c**) Typical appearance of a human poorly differentiated exocrine pancreatic adenocarcinoma. (**b**) Section of mouse islet cell tumor (insulinoma) in a rat insulin II promoter-transgenic mouse. (**d**) Section of a human islet cell tumor (insulinoma). The 100-μm scale bar (**d**) applicable to all panels

Because normal islets have a trabecular arrangement separated by capillaries, a more solid pattern of cells in a small tumor is suggestive of an adenomatous neoplasm rather than a large, "normal" or hyperplastic islet. Although the tumors generally form solid sheets, some larger islet cell tumors may have an accentuated trabecular pattern with neoplastic cells aligned in rows or ribbons along thin-walled sinusoidal blood vessels. Hemorrhages, hemorrhagic cysts, and groups of iron-laden macrophages are commonly observed within islet tumors.

Islet cell tumors are subclassified as adenomas and adenocarcinomas (Galvez et al. 2004). However, the presence of metastasis is the only reliable criterion for the diagnosis of a well-differentiated islet cell carcinoma. Metastasis is usually associated with larger tumors but can arise from small, well-differentiated tumors. The liver is usually the initial site of distant metastasis. In the absence of metastasis, the diagnosis of islet cell adenocarcinoma can be made only in the presence of extensive local invasion, invasion of a blood vessel, or anaplasia.

7.10.6 Exocrine Pancreas (Galvez et al. 2004) (Figs. 7.18 and 7.19)

Tumors of the exocrine pancreas, though rare in laboratory mice, are now produced with a variety of constructs (Hruban et al. 2006; Hingorani et al. 2003; Cecena et al. 2006). The constructs target the exocrine cells and produce a mixture of adenomas and adenocarcinomas. Gross examination reveals numerous large masses. The adenomas tend to be fleshy and pink. The adenocarcinomas, as suggested above tend to have significant fibrosis and a hard, firm consistency.

7.11 Pulmonary Pathology (Nikitin et al. 2004) (Fig. 7.20)

7.11.1 Gross Pathology

Spontaneous tumors of the lung are well-known in several mouse strains and are related to ras-associated genes. These tumors are typically bronchioalveolar adenomas(BAAs)/adenocarcinomas. They appear as single or multiple masses in the lung that can be appreciated on gross examination or by using lung whole mounts.

7.11.2 Microscopic Pathology

The BAAs typically form papillary tumors, but they can also have a lepidinal pattern which spreads along the alveolar walls. With a little diligence, a direct connection

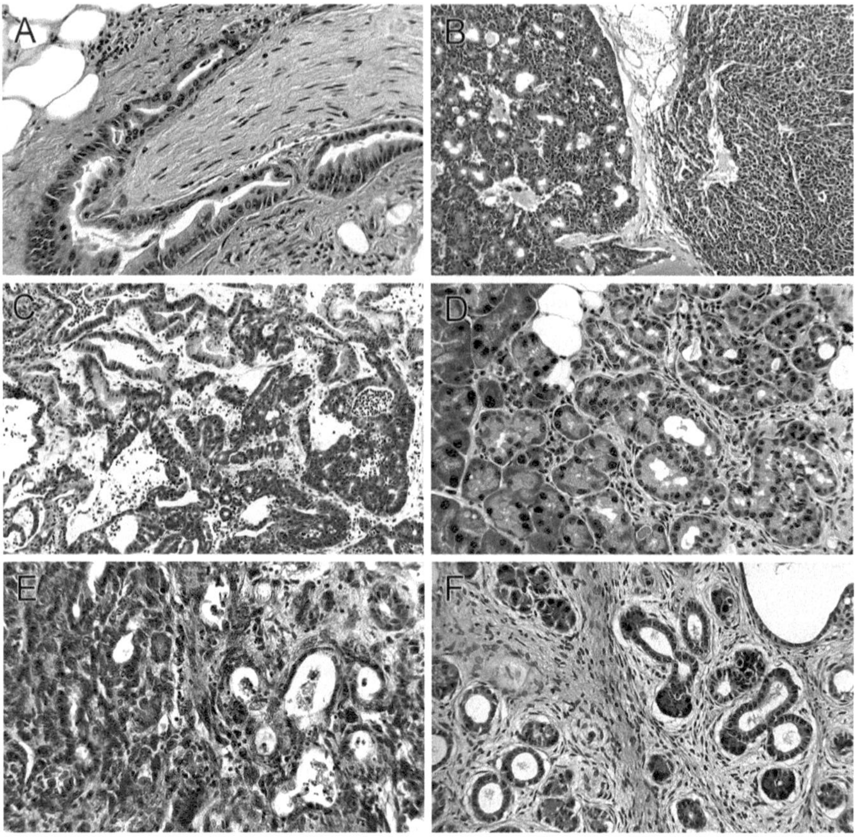

Fig. 7.19 Comparisons of human and mouse pancreatic cancers. These H&E-stained slides compare a human pancreatic adenocarcinoma (A) with various mouse models. (**a**) Human ductal adenocarcinoma of the pancreas with perineural invasion. (**b**) Two adjacent acinar cell carcinomas from Tg(EL-1-Tag) mouse pancreas. The carcinoma on the left is moderately differentiated. The carcinoma on the right is poorly differentiated, but acinar tubules are identifiable. (**c**) Ductal metaplasia (*upper left*) in a poorly differentiated acinar cell carcinoma (*lower right*) from the pancreas of a Tg(EL-1-myc) mouse. (**d**) Ductal metaplasia (*right*) of dysplastic acinar cells (*left*) in the pancreas from a Tg(EL-1-Kras) mouse. Slide courtesy of Eric Sandgren. (**e**) Acinar cell carcinoma with focal ductal metaplasia from pancreas of a Tg(EL-1-TGF∝-p53$^{-/-}$) mouse. Slide courtesy of Roland Schmid. (**f**) Pancreatic lobule with ductal metaplasia and fibrosis from a 24-week-old Tg(MT-TGF∝) mouse. Slide courtesy of Steven Leach. H&E

between the normal bronchial epithelium and the tumor can be demonstrated. These tumors are often mistaken for metastases. Since they occur in strains, such as FVB, that are used for genetic engineering, one needs to be familiar with these tumors. There are no clear-cut criteria to distinguish between benign and malignant tumors. Toxicological pathologists tend to use size as a criteria rather than microscopic appearance.

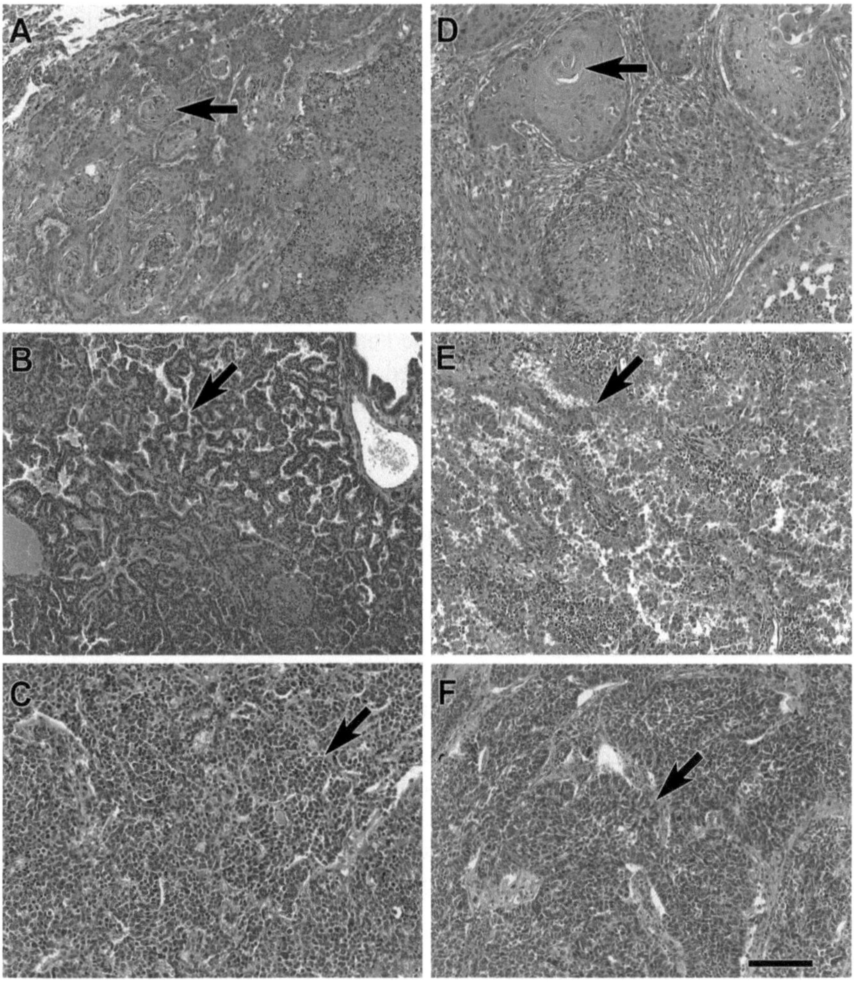

Fig. 7.20 Comparative pathology of mouse and human lung neoplasms. Digital images of H&E-stained sections of several types of lung tumors. Mouse sections are on the left (**a**–**c**), with human on the right (**d**–**f**). (**a**) Mouse keratinizing squamous cell carcinoma induced by intratracheal administration of 3-methylcholanthrene. (**d**) Human keratinizing squamous cell carcinoma. In both (**a**) and (**d**), squamous differentiation is demonstrated by formation of keratin pearls (*arrows*). (**b**) Papillary adenocarcinoma induced in the mouse with STOP^{flox}K-rasG12D knock-in upon administration with AdCre. (**e**) Human papillary adenocarcinoma. In both (**b**) and (**e**), tumor cells form a papillary pattern with central fibrovascular cores (*arrows*). (**c**) Neuroendocrine carcinoma in a mouse with Tm($p53^{floxP/floxP}Rb1^{floxP/floxP}$) conditional alleles after intratracheal administration of AdCre. (**f**) Human small cell lung carcinoma. The mouse (**c**) shows hyperchromatic cells arranged in small nests (*arrow*). The human (**f**) is characterized by small hyperchromatic cells forming palisades (*arrow*). Hematoxylin and eosin. Calibration bar for all panels is 100 μm

Significantly, the GEM models do not include the squamous metaplasia of the bronchial mucosa found as a precursor in human lung cancers. In fact, no models of squamous cell carcinoma of GEM lungs have been reported. Numerous attempts to recapitulate the smoking-related injuries in non-GEM models have produced few squamous neoplasms.

Genetic engineering using the oncogenic ras transgene appears to recreate the papillary bronchioalveolar adenoma of spontaneous origin. These ras-derived lesions can usually be distinguished from spontaneous tumors in that the GEM have multiple tumors at necropsy.

Genetic engineering has produced other forms of lung cancer. The neuroendocrine tumors produced with SV40 Tag or double knockout of Rb and p53 are striking because they produce a poorly differentiated, nonlarge cell carcinoma with neuroendocrine features that have never been seen in wild-type mice.

7.12 Dermatopathology (Borowsky et al. 2004) (Fig. 7.21)

Mouse models of skin cancer are also accurate mimics of human skin cancer that recapitulate the complex tumor–stromal interactions, angiogenesis, and the multistep progression of human skin cancers. Mouse models of malignant melanoma, squamous cell carcinoma, and basal cell carcinoma are available for study.

Malignant melanoma: GEM models of melanoma with transgene expression under the control of the melanocyte-specific tyrosinase promoter include the SV40 early region and activated alleles of H-Ras. Chemical carcinogenesis has increased the rates of melanomas in some knockout mice by using 7,12-dimethylbenz(a)anthracene (DMBA) as an initiator and 12-O-tetradecanoylphorbol-13-acetate (TPA) as a promoter.

The biology of the mouse melanoma is different from human melanoma. For example, early human melanoma is characterized by the spread of atypical melanocytes within the epidermis called “Pagetoid spread.” Pagetoid spread is rarely seen in mice. However, overexpression of hepatocyte growth factor with perinatal ultraviolet radiation can result in Pagetoid spread (Fig. 7.4c, d). GEM models of melanoma have far fewer metastases than human melanomas.

Squamous cell carcinoma: Human risk factors for squamous cell carcinoma include excessive ultraviolet radiation exposure, immune suppression, human papillomavirus (HPV) infection, certain chronic dermatoses, and topical arsenic exposure.

Chemical carcinogen-induced squamous lesions in mice and rats have been extensively studied. Exophytic squamous lesions arise that resemble the relatively benign human lesions (keratoacanthoma-like). These lesions, however, can progress to carcinoma, with invasion of the polyp stalk and the underlying tissue.

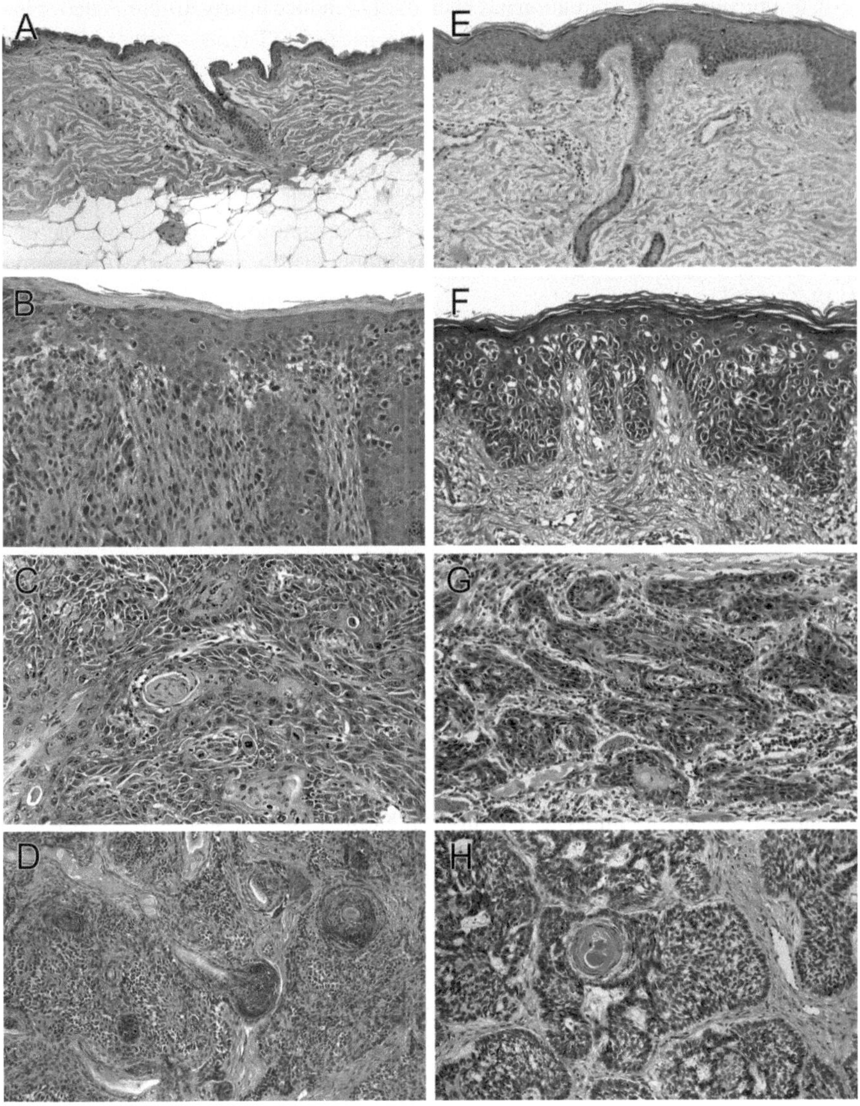

Fig. 7.21 Comparison of mouse and human skin and skin tumors. (**a**) Normal mouse skin from the trunk with epidermis that ranges from two to four cells in thickness and a hair follicle with attached sebaceous glands and piloarrector muscle. (**b**) Normal human skin from the back with epidermis six to ten cells in thickness, thicker dermis, and an eccrine duct that connects with the overlying epidermis. (**c**) Mouse model of malignant melanoma exhibiting prominent intra-epidermal spread produced by perinatal UV irradiation of Tg(metallothionein-hepatocyte growth factor) mice. (**d**) Human malignant melanoma *in situ* exhibiting prominent intraepidermal spread. (**e**) Mouse model of squamous cell carcinoma (moderately differentiated) arising in Tg(keratin 14-HPV early region) mouse. (**f**) Human squamous cell carcinoma, moderately differentiated. (**g**) Basaloid follicular tumor arising in a Tg(Patched$^{+/-}$) mouse. Some aspects of the tumor are similar to human basal cell carcinoma, but the finding of several nearly formed hair follicles in the lesion would be an unusual finding. (**h**) Human basal cell carcinoma, micronodular type

Chronic ultraviolet irradiation has been used to induce poorly differentiated spindle cell squamous cell carcinomas. These lesions are EMT tumors that can be identified with dual-IHC stains for vimentin and keratin.

Mouse models using cytokeratin promoters, such as keratin 5 or keratin 14, have a predilection for forming squamous cell carcinomas. The model precancerous squamous lesions show squamous proliferation with atypia that resembles squamous cell carcinoma *in situ*. GEM models, with the expression of HPV16 early region, mimic the multistep progression in squamous cell carcinoma, including epidermal dysplasia, squamous cell carcinoma *in situ*, and invasive carcinoma (Fig. 7.21e, f).

Basal cell and adnexal tumors: Hedgehog pathway mutations result in a diverse array of adnexal neoplasms resembling some forms of human basal cell carcinoma. The tumors have basaloid cytology and often some evidence of primitive hair follicle differentiation (Fig. 7.21g, h). Tumors in $Ptch^{+/-}$ mice and K5-Gli-1 also have some features of human basal cell carcinoma. In the mouse, they are classified as "complex follicular germinative neoplasms" called "basaloid follicular neoplasms" with the addition of further descriptors to fit the individual cases.

7.13 GYN Pathology

7.13.1 Ovary (Galvez et al. 2004) (Fig. 7.22)

Spontaneous ovarian tumors are rare in the majority of mouse strains, and most neoplasms are derived from sex cord-stromal elements not the surface epithelium as in human. However, some models have been developed (Flesken-Nikitin et al. 2003).

7.13.2 Cervix (Arbeit 2003) (Fig. 7.23)

Cervical cancer has been directly linked to genital infection by HPV, and HPV DNA has been detected in nearly all cervical cancers (Arbeit 1996). Persistent viral disease is associated with cervical neoplasia and invasive cervical cancers. Several studies support a functional role for the E6 and E7 oncogenes in tumorigenesis. Ubiquitous or targeted expression of E6 and E7 has been shown to form benign tumors or cancers in transgenic mice (Elson et al. 2000). Expression of the E6 and E7 oncogenes in premalignant dysplastic lesions, and in cervix cancers (Eckert et al. 2000), is inferential support for a functional role for these viral oncogenes in human carcinogenesis.

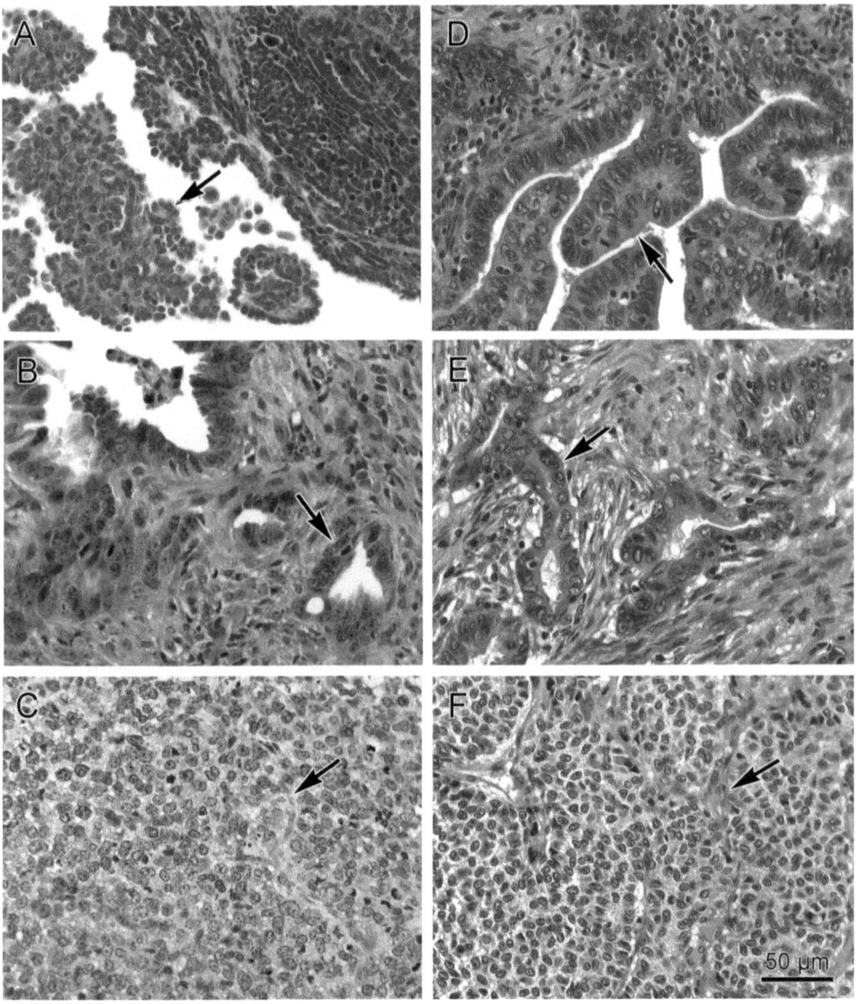

Fig. 7.22 Comparative pathology of mouse and human ovarian neoplasms. (**a**) Poorly differentiated carcinoma in Tg(MISIIRTag) mouse. Neoplastic cells form papillary structures present at the surface of the ovary and in the intrabursal space (*arrow*). (**d**) Human serous papillary adenocarcinoma. Note papillae formation by atypical cells (*arrow*). (**b**) Poorly differentiated adenocarcinoma induced in the Tm($p53^{floxP/floxP}Rb1^{floxP/floxP}$) mouse after AdCre-mediated *inactivation* of p53 and Rb1. Note location of tubules (*arrow*) in fibrous stroma similar to those in (**e**). (**e**) Human poorly differentiated serous adenocarcinoma. The tumor is composed of tubular glands (*arrow*) in fibrous stroma. (**c**) Granulosa cell tumor in Tg(OSP-1Tag) mouse. Cells and few capillaries (*arrow*) are arranged in a diffuse pattern. (**f**) Human granulosa cell tumor. Cells contain moderate amounts of cytoplasm with oval nuclei and form nests separated by narrow perivascular trabeculae (*arrow*). Calibration bar for all panels = 50 μm

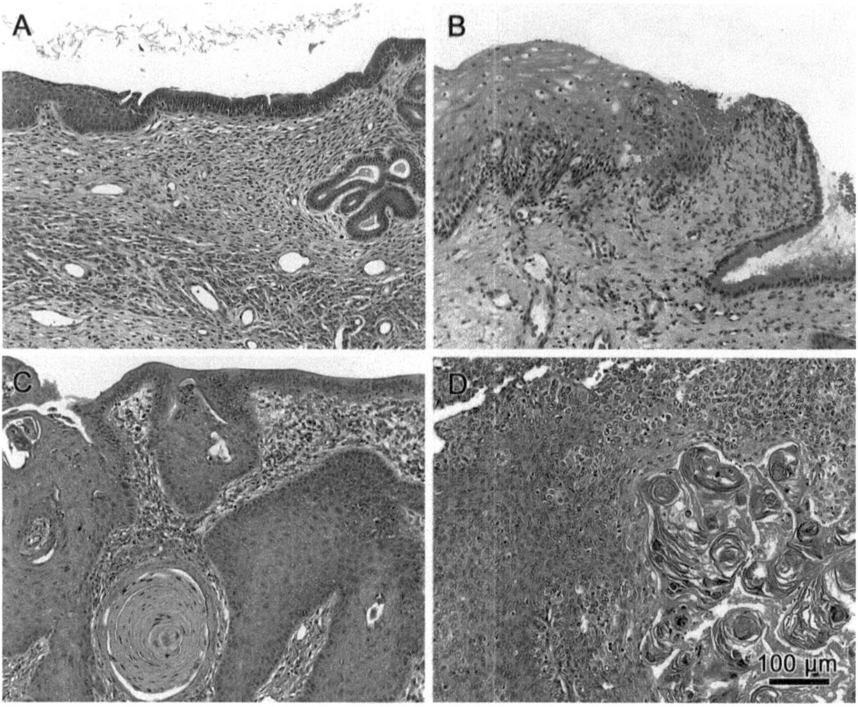

Fig. 7.23 Comparative pathology of mouse and human cervix neoplasia. Histology of mouse and human cervix with H&E staining. Mouse is (**a**) and (**c**) and human is (**b** and **d**). (**a**) Mouse normal cervix transition zone showing squamous (*left*) in transition to columnar (*right*) epithelium overlaying smooth muscular stroma with endocervical glands (*right*). (**b**) Human normal cervix transition zone showing squamous (*left*) and columnar (*right*) epithelium. The Luminal columnar epithelium invaginates to form the endocervical glands. (**c**) Mouse squamous carcinoma in an estrogen stimulated TG(K14-HPV16) mouse. Squamous epithelium is markedly thickened on the surface and invades the reactive stroma forming focal keratin "pearls" (*bottom, mid-left*). (**d**) Human cervical carcinoma showing broad sheets of dysplastic squamous epithelium with entrapped keratin clusters. All panels are at identical magnification with the 100-μm scale bar (**d**) applicable to all panels

Several transgenic mouse models of HPV disease have been developed targeting HPV oncogene expression to ocular lens (Munger 2002). Each of these models exhibit dysregulation of proliferation and altered cellular differentiation. These mice also progress to full-blown malignancy (Munger 2002).

Models of cervical carcinogenesis utilizing K14-HPV16-transgenic mice have been developed (zur Hausen 2002). The mice transgenic for the HPV16 early region under control of the human keratin-14 promoter express HPV16 E6 and E7 in the basal squamous epithelial cells. They model HPV16 viral persistence in basal squamous epithelial cells and are potentially requisite for cervical carcinogenic progression. The mice are treated with high-dose 17-beta-estradiol for up to 6 months and undergo a multistep progression through to invasive malignancy in the vagina and cervix (Elson et al. 2000). Low doses, however, induce squamous carcinogenesis localized to the cervical transformation zone (zur Hausen 2002).

Acknowledgments Dr. Cardiff wrote and takes sole responsibility for the text. The author is indebted to Drs. A.D. Borowsky, J.M. Ward, M. Bosenberg, J. Longfellow, J. Arbeit, A.Y. Nikitin, S. Kogan, J.J. Galvez, K. Aldape, and others who provided the outlines for the systemic pathology. We appreciate the willingness of investigators who shared unpublished data and images with us. This work has been supported, in part, by grants and contracts R01 CA089140, 22XS037A, U01 CA10102, and U01 CA084294 from the National Cancer Institute and National Centers for Research Resources (UR42-RR1495) from the NIH.

References

Abate-Shen C, Shen MM (2000) Molecular genetics of prostate cancer. Genes Dev 14(19):2410–2434

Abate-Shen C, Banach-Petrosky WA, Sun X, Economides KD, Desai N, Gregg JP et al (2003) Nkx3.1; Pten mutant mice develop invasive prostate adenocarcinoma and lymph node metastases. Cancer Res 63(14):3886–3890

Ali SH, DeCaprio JA (2001) Cellular transformation by SV40 large T antigen: interaction with host proteins. Semin Cancer Biol 11(1):15–23

Anonymous (2007) Mutant mice galore. Nature 446(7135):469–470

Arbeit JM (1996) Transgenic models of epidermal neoplasia and multistage carcinogenesis. Cancer Surv 26:7–34

Arbeit JM (2003) Mouse models of cervical cancer. Comp Med 53(3):256–258

Austin CP (2004) The knock out mouse project. Nat Genet 36(2):921–924 [Commentary]

Balmain A, Nagase H (1998) Cancer resistance genes in mice: models for the study of tumour modifiers. Trends Genet 14(4):139–144

Boivin GP, Washington K, Yang K, Ward JM, Pretlow TP, Russell R et al (2003) Pathology of mouse models of intestinal cancer: consensus report and recommendations. Gastroenterology 124(3):762–777

Borowsky A, Galvez J, Munn R, Cardiff R (2003) Comparative pathology of mouse models of human cancers. Comp Med 53(3):248–249

Borowsky AD, Munn RJ, Galvez JJ, Cardiff RD, Ward JM, Morse HC 3rd et al (2004) Mouse models of human cancers (part 3). Comp Med 54(3):258–270

Cardiff RD (1988) Cellular and molecular aspects of neoplastic progression in the mammary gland. Eur J Cancer Clin Oncol 24(1):15–20

Cardiff RD (1996) The biology of mammary transgenes: five rules. J Mammary Gland Biol Neoplasia 1(1):61–73

Cardiff RD (2001) Validity of mouse mammary tumour models for human breast cancer: comparative pathology. Microsc Res Tech 52(2):224–230

Cardiff RD (2003) Mouse models of human breast cancer. Comp Med 53(3):250–253

Cardiff RD (2007) Pathologists needed to cope with mutant mice. Nature 447(7144):528

Cardiff RD, Aguilar-Cordova E (1988) Protoneoplasia revisited: the molecular biology of mouse mammary hyperplasia. Anticancer Res 8(5):925–933

Cardiff RD, Kenney N (2007) Mouse mammary tumor biology: a short history. Adv Cancer Res 98:53–116

Cardiff RD, Muller WJ (1993) Transgenic mouse models of mammary tumorigenesis. Cancer Surv 16:97–113

Cardiff RD, Munn RJ (1995) Comparative pathology of mammary tumorigenesis in transgenic mice. Cancer Lett 90(1):13–19
Cardiff RD, Wellings SR (1999) The comparative pathology of human and mouse mammary glands. J Mammary Gland Biol Neoplasia 4(1):105–122
Cardiff RD, Wellings SR, Faulkin LJ (1977) Biology of breast preneoplasia. Cancer 39 (6 Suppl):2734–2746
Cardiff RD, Sinn E, Muller W, Leder P (1991) Transgenic oncogene mice. Tumor phenotype predicts genotype. Am J Pathol 139(3):495–501
Cardiff RD, Ornitz D, Lee F, Moreadith R, Sinn E, Muller W et al (1992) Mammary morphogenesis and oncogenes. In: Breast cancer: progress in biology, clinical management and prevention. pp 41–55
Cardiff RD, Anver MR, Gusterson BA, Hennighausen L, Jensen RA, Merino MJ et al (2000a) The mammary pathology of genetically engineered mice: the consensus report and recommendations from the Annapolis meeting. Oncogene 19(8):968–988
Cardiff RD, Wagner U, Henninghausen L (2000b) Mammary cancer in humans and mice: a tutorial for comparative pathology. The CD-ROM. J Mammary Gland Biol Neoplasia 5(2):243–244
Cardiff RD, Wagner U, Henninghausen L (2001) Mammary cancer in humans and mice: a tutorial for comparative pathology. Vet Pathol 38(4):357–358
Cardiff RD, Rosner A, Hogarth MA, Galvez JJ, Borowsky AD, Gregg JP (2004) Validation: the new challenge for pathology. Toxicol Pathol 32(Suppl 1):31–39
Cardiff RD, Munn RJ, Galvez JJ (2006a) The tumor pathology of genetically engineered mice: a new approach to molecular pathology. In: Fox JG, Davisson MT, Quimby FW, Barthold SW, Newcomer CE, Smith AL (eds) The mouse in biomedical research: experimental biology and oncology, 2nd edn. Elsevier, New York, pp 581–622
Cardiff RD, Anver MR, Boivin GP, Bosenberg MW, Maronpot RR, Molinolo AA et al (2006b) Precancer in mice: animal models used to understand, prevent, and treat human precancers. Toxicol Pathol 34(6):699–707
Cardiff RD, Ward JM, Barthold SW (2008a) 'One medicine-one pathology': are veterinary and human pathology prepared? Lab Invest 88(1):18–26
Cardiff RD, Ward JM, Barthold SW (2008b) 'One medicine–one pathology': are veterinary and human pathology prepared? Lab Invest 88(1):18–26
Carver BS, Pandolfi PP (2006) Mouse modeling in oncologic preclinical and translational research. Clin Cancer Res 12(18):5305–5311
Cecena G, Wen F, Cardiff RD, Oshima RG (2006) Differential sensitivity of mouse epithelial tissues to the polyomavirus middle T oncogene. Am J Pathol 168(1):310–320
Chiaverotti T, Couto SS, Donjacour A, Mao JH, Nagase H, Cardiff RD et al (2008) Dissociation of epithelial and neuroendocrine carcinoma lineages in the transgenic adenocarcinoma of mouse prostate model of prostate cancer. Am J Pathol 172(1):236–246
Conrad PA, Mazet JA, Clifford D, Scott C, Wilkes M (2009) Evolution of a transdisciplinary "One Medicine-One Health" approach to global health education at the University of California, Davis. Prev Vet Med 92(4):268–274
Couto SS, Cardiff RD (2008) The genomic revolution and endocrine pathology. Endocr Pathol 19(3):139–147
Daley GQ (1993) Animal models of BCR/ABL-induced leukemias. Leuk Lymphoma 11(Suppl 1): 57–60
Deckard-Janatpour K, Muller WJ, Chodosh LA, Gardner HP, Marquis ST, Coffey RJ et al (1997) Differential expression of the neu transgene in murine mammary tissues. Int J Oncol 11:235–244
Derksen PW, Liu X, Saridin F, van der Gulden H, Zevenhoven J, Evers B et al (2006) Somatic inactivation of E-cadherin and p53 in mice leads to metastatic lobular mammary carcinoma through induction of anoikis resistance and angiogenesis. Cancer Cell 10(5):437–449
Desai KV, Xiao N, Wang W, Gangi L, Greene J, Powell JI et al (2002) Initiating oncogenic event determines gene-expression patterns of human breast cancer models. Proc Natl Acad Sci USA 99(10):6967–6972

DiGiovanna MP, Lerman MA, Coffey RJ, Muller WJ, Cardiff RD, Stern DF (1998) Active signaling by Neu in transgenic mice. Oncogene 17(14):1877–1884
Donjacour AA, Thomson AA, Cunha GR (1998) Enlargement of the ampullary gland and seminal vesicle, but not the prostate in int-2/Fgf-3 transgenic mice. Differentiation 62(5):227–237
Dragani TA (2003) 10 years of mouse cancer modifier loci: human relevance. Cancer Res 63(12):3011–3018
Dunn TB (1953) Morphology of mammary tumors in mice. In: Homburger F, Fishman WH (eds) The physiopathology of cancer. Cassel, London, pp 123–148
Eckert RL, Crish JF, Balasubramanian S, Rorke EA (2000) Transgenic animal models of human papillomavirus-dependent disease. (Review). Int J Oncol 16(5):853–870
Edelmann W, Umar A, Yang K, Heyer J, Kucherlapati M, Lia M et al (2000) The DNA mismatch repair genes Msh3 and Msh6 cooperate in intestinal tumor suppression. Cancer Res 60(4): 803–807
Ellwood-Yen K, Graeber TG, Wongvipat J, Iruela-Arispe ML, Zhang J, Matusik R et al (2003) Myc-driven murine prostate cancer shares molecular features with human prostate tumors. Cancer Cell 4(3):223–238
Elson DA, Riley RR, Lacey A, Thordarson G, Talamantes FJ, Arbeit JM (2000) Sensitivity of the cervical transformation zone to estrogen-induced squamous carcinogenesis. Cancer Res 60(5):1267–1275
Engle SJ, Hoying JB, Boivin GP, Ormsby I, Gartside PS, Doetschman T (1999) Transforming growth factor beta1 suppresses nonmetastatic colon cancer at an early stage of tumorigenesis. Cancer Res 59(14):3379–3386
Espina V, Geho D, Mehta AI, Petricoin EF 3rd, Liotta LA, Rosenblatt KP (2005) Pathology of the future: molecular profiling for targeted therapy. Cancer Invest 23(1):36–46
Espina V, Heiby M, Pierobon M, Liotta LA (2007) Laser capture microdissection technology. Expert Rev Mol Diagn 7(5):647–657
Espina V, Edmiston KH, Heiby M, Pierobon M, Sciro M, Merritt B et al (2008) A portrait of tissue phosphoprotein stability in the clinical tissue procurement process. Mol Cell Proteomics 7(10):1998–2018
Evangelou AI, Winter SF, Huss WJ, Bok RA, Greenberg NM (2004) Steroid hormones, polypeptide growth factors, hormone refractory prostate cancer, and the neuroendocrine phenotype. J Cell Biochem 91(4):671–683
Fargiano AA, Desai KV, Green JE (2003) Interrogating mouse mammary cancer models: insights from gene expression profiling. J Mammary Gland Biol Neoplasia 8(3):321–334
Flesken-Nikitin A, Choi KC, Eng JP, Shmidt EN, Nikitin AY (2003) Induction of carcinogenesis by concurrent inactivation of p53 and Rb1 in the mouse ovarian surface epithelium. Cancer Res 63(13):3459–3463
Freeman DJ, Li AG, Wei G, Li HH, Kertesz N, Lesche R et al (2003) PTEN tumor suppressor regulates p53 protein levels and activity through phosphatase-dependent and -independent mechanisms. Cancer Cell 3(2):117–130
Furth PA (1998) SV40 rodent tumour models as paradigms of human disease: transgenic mouse models. Dev Biol Stand 94:281–287
Galvez JJ, Cardiff RD, Munn RJ, Borowsky AD, Boivin GP, Groden J et al (2004) Mouse models of human cancers (Part 2). Comp Med 54(1):13–28
Gardner MB, Rongey RW, Arnstein P, Estes JD, Sarma P, Huebner RJ et al (1970) Experimental transmission of feline fibrosarcoma to cats and dogs. Nature 226(5248):807–809
Gardner MB, Arnstein P, Johnson E, Rongey RW, Charman HP, Huebner RJ (1971) Feline sarcoma virus tumor induction in cats and dogs. J Am Vet Med Assoc 158(6 Suppl 2):1046–1053
Green JE, Cardiff R, Hennighausen L, Wakefield L, Wagner U, Lee E et al (2002) Validation of transgenic mammary cancer models: goals of the NCI Mouse Models of Human Cancer Consortium and the mammary cancer CD-ROM. Transgenic Res 11(6):635–636
Gulmann C, Espina V, Petricoin E 3rd, Longo DL, Santi M, Knutsen T et al (2005) Proteomic analysis of apoptotic pathways reveals prognostic factors in follicular lymphoma. Clin Cancer Res 11(16):5847–5855

Hager JH, Hanahan D (1999) Tumor cells utilize multiple pathways to down-modulate apoptosis. Lessons from a mouse model of islet cell carcinogenesis. Ann N Y Acad Sci 887:150–163

Hanahan D, Weinberg RA (2000) The hallmarks of cancer. Cell 100(1):57–70

Hill R, Song Y, Cardiff RD, Van Dyke T (2005) Heterogeneous tumor evolution initiated by loss of pRb function in a preclinical prostate cancer model. Cancer Res 65(22):10243–10254

Hingorani SR, Petricoin EF, Maitra A, Rajapakse V, King C, Jacobetz MA et al (2003) Preinvasive and invasive ductal pancreatic cancer and its early detection in the mouse. Cancer Cell 4(6):437–450

Holland EC (ed) (2004) Mouse models of human cancer. Wiley, Hoboken, NJ, pp 15–30

Hruban RH, Adsay NV, Albores-Saavedra J, Anver MR, Biankin AV, Boivin GP et al (2006) Pathology of genetically engineered mouse models of pancreatic exocrine cancer: consensus report and recommendations. Cancer Res 66(1):95–106

Kahn RE, Clouser DF, Richt JA (2009) Emerging infections: a tribute to the one medicine, one health concept. Zoonoses Public Health 56(6–7):407–428

Kaplan B, Kahn LH, Monath TP, Woodall J (2009) 'ONE HEALTH' and parasitology. Parasit Vectors 2(1):36

Karesh WB, Cook RA (2009) One world–one health. Clin Med 9(3):259–260

Kasper S, Sheppard PC, Yan Y, Pettigrew N, Borowsky AD, Prins GS et al (1998) Development, progression, and androgen-dependence of prostate tumors in probasin-large T antigen transgenic mice: a model for prostate cancer. Lab Invest 78(3):319–333

Kawakami TG, Buckley P, Huff S, McKain D, Fielding H (1973) A comparative study in vitro of a simian virus isolated from spontaneous woolly monkey fibrosarcoma and of a known feline fibrosarcoma virus. Bibl Haematol 39:236–243

Kim MJ, Cardiff RD, Desai N, Banach-Petrosky WA, Parsons R, Shen MM et al (2002) Cooperativity of Nkx3.1 and Pten loss of function in a mouse model of prostate carcinogenesis. Proc Natl Acad Sci USA 19:19

Kirsten WH, Mayer LA (1969) Malignant lymphomas of extrathymic origin induced in rats by murine erythroblastosis virus. J Natl Cancer Inst 43(3):735–746

Kogan SC, Ward JM, Anver MR, Berman JJ, Brayton C, Cardiff RD et al (2002) Bethesda proposals for classification of nonlymphoid hematopoietic neoplasms in mice. Blood 100(1):238–245

Kumar R, Sukumar S, Barbacid M (1990) Activation of ras oncogenes preceding the onset of neoplasia. Science 248(4959):1101–1104

Landesman-Bollag E, Romieu-Mourez R, Song DH, Sonenshein GE, Cardiff RD, Seldin DC (2001) Protein kinase CK2 in mammary gland tumorigenesis. Oncogene 20(25):3247–3257

Li Y, Welm B, Podsypanina K, Huang S, Chamorro M, Zhang X et al (2003) Evidence that transgenes encoding components of the Wnt signaling pathway preferentially induce mammary cancers from progenitor cells. Proc Natl Acad Sci USA 100(26):15853–15858

Mahler JF, Stokes W, Mann PC, Takaoka M, Maronpot RR (1996) Spontaneous lesions in aging FVB/N mice. Toxicol Pathol 24(6):710–716

Masumori N, Thomas TZ, Chaurand P, Case T, Paul M, Kasper S et al (2001) A probasin-large T antigen transgenic mouse line develops prostate adenocarcinoma and neuroendocrine carcinoma with metastatic potential. Cancer Res 61(5):2239–2249

Miyoshi K, Shillingford JM, Le Provost F, Gounari F, Bronson R, von Boehmer H et al (2002a) Activation of beta-catenin signaling in differentiated mammary secretory cells induces transdifferentiation into epidermis and squamous metaplasias. Proc Natl Acad Sci USA 99(1):219–224

Miyoshi K, Rosner A, Nozawa M, Byrd C, Morgan F, Landesman-Bollag E et al (2002b) Activation of different Wnt/beta-catenin signaling components in mammary epithelium induces transdifferentiation and the formation of pilar tumors. Oncogene 21(36):5548–5556

Morse HC 3rd, Anver MR, Fredrickson TN, Haines DC, Harris AW, Harris NL et al (2002) Bethesda proposals for classification of lymphoid neoplasms in mice. Blood 100(1):246–258

Moser AR, Hegge LF, Cardiff RD (2001) Genetic background affects susceptibility to mammary hyperplasias and carcinomas in Apc(min)/+ mice. Cancer Res 61(8):3480–3485

Munger K (2002) The role of human papillomaviruses in human cancers. Front Biosci 7:d641–d649

Nagase H, Mao JH, Balmain A (1999) A subset of skin tumor modifier loci determines survival time of tumor-bearing mice. Proc Natl Acad Sci USA 96(26):15032–15037

Naharro G, Tronick SR, Rasheed S, Gardner MB, Aaronson SA, Robbins KC (1983) Molecular cloning of integrated Gardner-Rasheed feline sarcoma virus: genetic structure of its cell-derived sequence differs from that of other tyrosine kinase-coding onc genes. J Virol 47(3):611–619
Nieto AI, Shyamala G, Galvez JJ, Cardiff RD (2003) Persistent hyperplasia in FVB/N. Comp Med 53(4):361–366
Nikitin AY, Alcaraz A, Anver MR, Bronson RT, Cardiff RD, Dixon D et al (2004) Classification of proliferative pulmonary lesions of the mouse: recommendations of the mouse models of human cancers consortium. Cancer Res 64(7):2307–2316
Nikitin AY, Matoso A, Roy-Burman P (2007) Prostate stem cells and cancer. Histol Histopathol 22(9):1043–1049
Oshima M, Oshima H, Kitagawa K, Kobayashi M, Itakura C, Taketo M (1995) Loss of Apc heterozygosity and abnormal tissue building in nascent intestinal polyps in mice carrying a truncated Apc gene. Proc Natl Acad Sci USA 92(10):4482–4486
Park JH, Walls JE, Galvez JJ, Kim M, Abate-Shen C, Shen MM et al (2002) Prostatic intraepithelial neoplasia in genetically engineered mice. Am J Pathol 161(2):727–735
Powell WC, Cardiff RD, Cohen MB, Miller GJ, Roy-Burman P (2003) Mouse strains for prostate tumorigenesis based on genes altered in human prostate cancer. Curr Drug Targets 4(3):263–279
Radaelli E, Arnold A, Papanikolaou A, Garcia-Fernandez RA, Mattiello S, Scanziani E et al (2009a) Mammary tumor phenotypes in wild-type aging female FVB/N mice with pituitary prolactinomas. Vet Pathol 46(4):736–745
Radaelli E, Damonte P, Cardiff RD (2009b) Epithelial-mesenchymal transition in mouse mammary tumorigenesis. Future Oncol 5(8):1113–1127
Rosner A, Miyoshi K, Landesman-Bollag E, Xu X, Seldin DC, Moser AR et al (2002) Pathway pathology: histological differences between ErbB/Ras and Wnt pathway transgenic mammary tumors. Am J Pathol 161(3):1087–1097
Saenz Robles MT, Symonds H, Chen J, Van Dyke T (1994) Induction versus progression of brain tumor development: differential functions for the pRB- and p53-targeting domains of simian virus 40T antigen. Mol Cell Biol 14(4):2686–2698
Sass B, Dunn TB (1979) Classification of mouse mammary tumors in Dunn's miscellaneous group including recently reported types. J Natl Cancer Inst 62(5):1287–1293
Scherl A, Li JF, Cardiff RD, Schreiber-Agus N (2004) Prostatic intraepithelial neoplasia and intestinal metaplasia in prostates of probasin-RAS transgenic mice. Prostate 59(4):448–459
Schmidt EV (1999) The role of c-myc in cellular growth control. Oncogene 18(19):2988–2996
Schmidt EV (2004) The role of c-myc in regulation of translation initiation. Oncogene 23(18):3217–3221
Shappell SB, Thomas GV, Roberts RL, Herbert R, Ittmann MM, Rubin MA et al (2004) Prostate pathology of genetically engineered mice: definitions and classification. The consensus report from the Bar Harbor meeting of the Mouse Models of Human Cancer Consortium Prostate Pathology Committee. Cancer Res 64(6):2270–2305
Siegel PM, Shu W, Cardiff RD, Muller WJ, Massague J (2003) Transforming growth factor beta signaling impairs Neu-induced mammary tumorigenesis while promoting pulmonary metastasis. Proc Natl Acad Sci USA 100(14):8430–8435
Snyder SP, Theilen GH (1969) Transmissible feline fibrosarcoma. Nature 221(185):1074–1075
Song Z, Wu X, Powell WC, Cardiff RD, Cohen MB, Tin RT et al (2002) Fibroblast growth factor 8 isoform B overexpression in prostate epithelium: a new mouse model for prostatic intraepithelial neoplasia. Cancer Res 62(17):5096–5105
Stevens LC (1980) Teratocarcinogenesis and spontaneous parthenogenesis in mice. Results Probl Cell Differ 11:265–274
Sugino T, Kusakabe T, Hoshi N, Yamaguchi T, Kawaguchi T, Goodison S et al (2002) An invasion-independent pathway of blood-borne metastasis: a new murine mammary tumor model. Am J Pathol 160(6):1973–1980
van Leeuwen F, Nusse R (1995) Oncogene activation and oncogene cooperation in MMTV-induced mouse mammary cancer. Semin Cancer Biol 6(3):127–133
Verma IM, Graham WR (1987) The fos oncogene. Adv Cancer Res 49:29–52

Wakefield LM, Thordarson G, Anver M, Cardiff RD (2003) Spontaneous pituitary abnormalities and mammary hyperplasias in FVB/NCr mice: implications for mouse modeling. Comp Med 53(4):348–356
Weiss WA, Israel M, Cobbs C, Holland E, James CD, Louis DN et al (2002) Neuropathology of genetically engineered mice: consensus report and recommendations from an international forum. Oncogene 21(49):7453–7463
White DE, Cardiff RD, Dedhar S, Muller WJ (2001) Mammary epithelial-specific expression of the integrin-linked kinase (ILK) results in the induction of mammary gland hyperplasias and tumors in transgenic mice. Oncogene 20(48):7064–7072
Yew N, Strobel M, Vande Woude GF (1993) Mos and the cell cycle: the molecular basis of the transformed phenotype. Curr Opin Genet Dev 3(1):19–25
Yu Q, Geng Y, Sicinski P (2001) Specific protection against breast cancers by cyclin D1 ablation. Nature 411(6841):1017–1021
Zhou Z, Flesken-Nikitin A, Nikitin AY (2007) Prostate cancer associated with p53 and Rb deficiency arises from the stem/progenitor cell-enriched proximal region of prostatic ducts. Cancer Res 67(12):5683–5690
zur Hausen H (2002) Papillomaviruses and cancer: from basic studies to clinical application. Nat Rev Cancer 2(5):342–350

Chapter 8
Genomic DNA Copy Number Alterations in Mouse Cancer Models and Human Cancer

Donna G. Albertson

Cancer development is associated with the acquisition of genomic and epigenetic alterations. These changes may be brought about at the genomic level in a variety of ways, including altered karyotypes, point mutations, and epigenetic mechanisms. Cytogenetic and more recently molecular and array-based analytic methods have found great variation in the numbers and types of chromosome level alterations present in human tumors (Mitelman et al. 2008). Recurrent alterations involving chromosomal copy number changes have been observed frequently, particularly in solid tumors. Reduced or enhanced expression of one or more genes mapping to such regions is expected to provide the selection for maintenance of the aberrations. Thus, recurrent chromosomal aberrations provide a route to discovery of genes or pathways that contribute to development and progression of tumors. Nevertheless, uncovering the driver genes for recurrent chromosomal aberrations remains challenging because the aberrations may span large regions of the genome and contain many genes. Here, we illustrate with examples how mouse cancer models recapitulate genomic alterations observed in human cancer, and how cross-species comparisons have facilitated the identification of oncogenes and tumor suppressors, as well as polymorphic variants contributing to individual cancer susceptibility.

8.1 Genomic DNA Copy Number Alterations

Genomic aberrations resulting in copy number alterations may involve gains or losses of whole chromosomes (aneuploidy) or parts of chromosomes (deletions, duplications, nonreciprocal translocations) or focal amplifications, defined as copy

D.G. Albertson (✉)
Department of Laboratory Medicine and UCSF Helen Diller Family
Comprehensive Cancer Center, Cancer Research Institute, University of California
San Francisco, San Francisco, CA 94143-0808, USA
e-mail: albertson@cc.ucsf.edu

J.E. Green and T. Ried (eds.), *Genetically Engineered Mice for Cancer Research: Design, Analysis, Pathways, Validation and Pre-Clinical Testing*,
DOI 10.1007/978-0-387-69805-2_8, © Springer Science+Business Media, LLC 2012

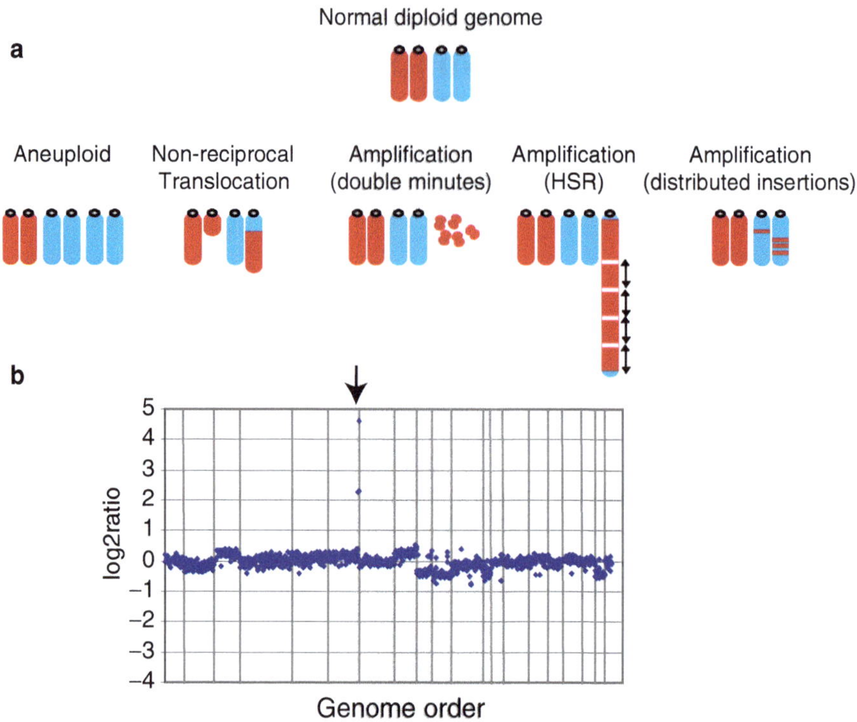

Fig. 8.1 Genomic alterations. (**a**) A normal diploid karyotype represented by two pairs of chromosomes may undergo chromosome rearrangements that result in copy number aberrations, including gains or losses of whole chromosomes (aneuploidy), nonreciprocal translocations in which as shown here, one half-translocation may be retained and the other lost, and amplifications. The amplified DNA may be present as extra chromosomal copies (double minutes), tandem or inverted repeats often recognized cytologically as homogeneously staining regions (HSRs), or extra DNA copies may be inserted at various locations across the genome. (**b**) Copy number profile of a mammary tumor from a C3(1)/SV40 Tag transgenic mouse. Hybridization was carried out to an array of BAC clones and shown is the copy number ratio at each clone ordered according to position in the genome from chromosome 1 to X. Both low level gains and losses and the recurrent amplification of *Kras* on the distal end of chromosome 6 (*arrow*) are present

number increases of restricted chromosomal regions (Fig. 8.1a). Models for amplification formation require an initiating double-strand break in cells lacking robust checkpoints, and subsequent rounds of genomic rearrangement leading to increased copy number of a restricted region of the genome (Albertson 2006). While chromosomal copy number changes resulting from nonreciprocal translocations appear to be relatively stable in model systems, regions of amplification are unstable (Snijders et al. 2008) and are maintained in the genome by selection. Thus, amplicons are unlikely to be passenger aberrations, highlighting their utility for the identification of genes or pathways of functional importance to the tumor at the time when the sample was obtained.

8.2 Comparative Genomic Hybridization Reveals Genome-Wide Copy Number Aberrations

Cytogenetics provided early evidence of recurrent chromosomal aberrations in cancer, beginning with the identification of the Philadelphia chromosome in chronic myelogenous leukemia. Detection of recurrent aberrations in solid tumors was more difficult by cytogenetics, but a substantial number were found. The development of comparative genomic hybridization (CGH) provided the first genome-wide analytical tool for interrogating tumor genomes for copy number alterations (Kallioniemi et al. 1992). As originally described, CGH involved hybridization of differentially labeled test (tumor) and reference genomes to metaphase chromosomes, which provided the map of the genome with cytogenetic resolution. The initial application of CGH immediately revealed previously unappreciated aberrations in human tumor genomes, including amplification at chromosome 20q13. Positional cloning and functional analysis lead to the identification of *ZNF217* as the driver gene for this recurrent amplicon in breast cancer (Collins et al. 1998). Subsequently, the availability of genomic clones and genome sequence facilitated implementation of array-based platforms for the analysis of mammalian genomes (Fig. 8.1b) (Solinas-Toldo et al. 1997; Pinkel et al. 1998). Chromosomal and array CGH studies of numerous tumor types have revealed specific spectra of aberrations associated with a particular tumor tissue type or subtype, and distinguished genomic profiles associated with disease progression or response to therapy.

8.3 Early Applications of CGH to the Analysis of Mouse Mammary Models of Human Cancer Revealed Similarities and Differences with Human Tumors

Application of chromosomal CGH to the analysis of murine tumors was more challenging due to the small size and difficulty in identifying individual mouse chromosomes (Shi et al. 1997). Nevertheless, the analysis of several mouse models in which mammary tumors are induced by disruption of signaling pathways, tumor suppressor gene deficiency and overexpression of oncogenes under control of the MMTV LTR or endogenous promoters revealed both similarities and differences with human breast tumors (Liu et al. 1998; Weber et al. 1998; Weaver et al. 1999; Montagna et al. 2002, 2003). In all models, recurrent copy number changes were observed (Fig. 8.2), indicating that secondary genomic alterations were required for tumorigenesis. Notably, the recurrent aberrations often occurred at frequencies >50% of cases.

Mammary cell lines derived from primary tumors induced in a model with a mutated activated *HER2/neu* gene under its endogenous promoter when analyzed by spectral karyotyping and CGH revealed recurrent deletions of chromosome 4

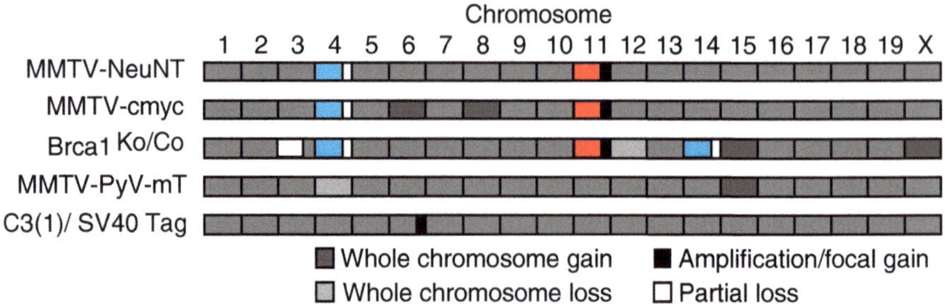

Fig. 8.2 Summary of genomic aberrations reported in mouse mammary tumors (see text for details). Chromosomes are shown as *gray* boxes with recurrent gains or losses of whole chromosomes in *red* and *blue*, respectively. Regions of partial losses of proximal or distal chromosome arms are indicated in *white* and focal gains or amplifications in *black*

(7/11 cases) and genomic amplification of *HER2/neu* (6/12 cases) (Montagna et al. 2002). A similar analysis of cell lines derived from mammary gland tumors from mice that overexpress *Myc* under the control of the MMTV-LTR promoter also found complete or partial loss of chromosome 4 (6/8 cases), and less frequently (3/8 cases) gains of chromosome 6 and partial gains of chromosomes 8 and 11 (Weaver et al. 1999). Mammary tumors arising in mice conditionally mutant for *Brca1* (*Brca1$^{Ko/Co}$ Wap-Cre* or *MMTV-Cre*) were the most genomically unstable, yet recurrent copy number changes were observed, and included gain of distal chromosome 11, partial gain of chromosome 15, loss of distal chromosome 14, loss of chromosome 4, and gain of the X chromosome. The minimal regions of gain on chromosomes 11 and 15 were centered on bands 11D-E and 15D2-D3, respectively (Weaver et al. 2002). Tumors induced in mice expressing the polyoma virus middle T antigen (MMTV-PyV-mT transgenic mice) were also analyzed (Montagna et al. 2003). Expression of middle T antigen deregulates growth factor signaling pathways resulting in stimulation of the mitogen-activated protein kinase cascade. Consistent genomic alterations in cell lines derived from mammary tumors in this model included frequent losses of chromosome 4 (10/26 cases), gain of chromosome 15 (10/26 cases), and copy number increases or amplification on chromosome 11 (17/26 cases). In two cases, the amplicons were present as double minute chromosomes. The gene, *Sept9* was identified as the likely driver gene because it was overexpressed when amplified and also overexpressed frequently in mammary tumors from other mouse models as well as human breast tumor cell lines (Montagna et al. 2003).

It is clear from Fig. 8.2 that tumors from these models share recurrent genomic alterations of chromosomes 4, 11, and 15. Distal chromosome 4 is syntenic with human chromosome 1p32-p36, a region frequently lost in human tumors. The region of mouse chromosome 4 is also syntenic with 9p, including *CDKN2A*, a gene frequently lost in tumors cell lines. Since these genomic analyses were carried out on cell lines, it is possible that selection for the loss of *CDKN2A* promoted the chromosome 4 losses during establishment of the cell cultures. Similar losses of chromosome 4, however, were found in a recent array CGH analysis of *HER2/neu*- and polyoma virus middle T antigen-based tumor models using DNA extracted from frozen tissue (Hodgson et al. 2005), indicating that chromosome 4 loss is likely to be a characteristic of the tumor genomes and not a genomic aberration selected by passage in cell culture. Distal mouse chromosome 11 (bands 11D-E) is syntenic to human chromosome 17q11-qter. Two regions of amplification have been mapped to this location in breast tumors. The proximal amplicon includes *ERBB2* at 17q12, while the distal one contains 11 overexpressed genes in highly amplified tumors, including *FAM33A, DHX40, CLTC, PTRH2, TMEM49, TUBD1, RPS6KB1, ABC1, USP32, APPBP2*, and *PPM1D* (Parssinen et al. 2007). *SEPT9*, identified as the candidate driver gene for amplification of chromosome 11D-E in mammary tumors arising in MMTV-PyV-mT antigen transgenic mice (Montagna et al. 2003), maps distal to the minimal 17q23 amplicon in human breast cancer. Thus, although overexpression of *SEPT9* in human and mouse tumors is frequent, the amplicons arising in mammary tumors from MMTV-PyV-mT transgenic mice appear not to be orthologous to the 17q23 amplicon present in sporadic human breast tumors. On the other hand, the common region of gain on mouse chromosome 15 (bands D2–D3) in tumors from *Brca1* deficient and MMTV-PyV-mT transgenic mice is syntenic to 8q24, a region that is frequently gained or amplified in both sporadic and *BRCA1* familial human breast tumors. A likely candidate oncogene driving gain or amplification of this region in both mouse and human tumors is *MYC*.

In contrast to the four mammary tumor models discussed above, mice carrying the C3(1)/SV40 Tag transgene, which interferes with *Trp53* and *Rb1* function, developed tumors that consistently amplified a region of distal mouse chromosome 6, including *Kras*. Amplification was associated with tumor progression, increasing in frequency from 12.5% of tumors at 4 months to 68% of tumors at 6 months (Liu et al. 1998). Although *KRAS* is not amplified in human breast cancer, elevated RAS protein levels are observed and are associated with lymph node metastasis. Thus, functional similarities between the C3(1)/SV40 Tag mouse model and human cancer exist, but the tumors appear to be promoted by different genetic routes.

8.4 Identification of Oncogenes by Cross-Species Comparisons

Human tumor genomes have many genomic alterations that potentially can guide the identification of tumor suppressor genes and oncogenes. As discussed above, amplicons are particularly attractive for the identification of oncogenes, since they are unstable and without selection would disappear. These observations suggest that focusing cancer gene discovery efforts on these regions is likely to reveal functionally important genes. Nevertheless, analysis of amplicons in human tumors may be challenging, because they often span large regions containing many genes and are highly complex with variable levels of amplification across the region. Some human tumor types, however, offer unique opportunities, because they characteristically amplify genes with high frequency (Bastian et al. 2000) or, as in the case of oral squamous cell carcinoma (SCC), the tumor genomes characteristically display recurrent amplicons, which may be rare (occurring in <5% of cases), but narrow (e.g., <3 Mb), which facilitates the identification of driver oncogenes (Snijders et al. 2005). The genomes of tumors arising in mouse models, on the other hand, generally display few copy number changes, often consisting of whole chromosome changes and in some cases high level amplifications, as exemplified by the spectra of copy number changes in mammary tumors shown in Fig. 8.2.

The advent of array CGH which employs the currently available extensive molecular knowledge of both human and murine genomes has enabled mapping of copy number alterations to the genome sequence of these species and thereby the precise identification of recurrent copy number changes occurring in syntenic regions. Thus, the standard paradigm for positional cloning of candidate oncogenes, i.e., requiring that the gene be overexpressed when it is amplified, as well as sometimes when not, can be enriched by adding cross-species comparisons to evaluate the well-defined set of genes common to the aberrations in tumors from both species. This approach is illustrated by studies of a mouse model of hepatocellular carcinoma (HCC). Precisely mapped syntenic amplicons were identified that were conserved between species (Zender et al. 2006). A focal amplicon at 9qA1 was identified in mouse tumors induced when *Trp53* null hepatoblasts co-expressing *Myc* were transplanted back into a mouse. The amplicon is syntenic with a previously identified 11q22 amplicon that occurs at low frequency in a number of human cancers, including human HCC (Zender et al. 2006), pancreas (Bashyam et al. 2005; Gysin et al. 2005), esophageal (Imoto et al. 2001), lung (Dai et al. 2003), and oral SCC (Snijders et al. 2005). Whereas, *BIRC2* and/or *BIRC3*, two genes in the minimal region of common amplification had been identified as candidate driver genes for amplification in pancreas, lung, and esophageal cancer, studies in oral SCC had ruled out *BIRC3* and instead suggested that *YAP1* and *BIRC2* were likely to be the driver oncogenes (Snijders et al. 2005). Using the genetically modified mouse hepatoblasts, it was possible to show that both *BIRC2* and *YAP1* have oncogenic properties and further that the two genes collaborate to promote tumorigenesis (Zender et al. 2006). Thus, while it is not possible to predict whether a particular tumor model will have a propensity to acquire focal amplicons, their presence in mouse models facilitates the identification of oncogenes as in human cancers. An advantage offered by the mouse models is the possibility of carrying out further functional studies in the same type of cells that induced tumor formation. An analogous approach has been used to identify *NEDD9* as a metastasis promoting gene in melanoma (Kim et al. 2006). Thus, cross-species positional cloning and functional analyses, which are now readily performed using current genomics resources and technologies, as well as increasingly sophisticated mouse tumor models, are likely to facilitate evaluation and validation of the functional importance of the catalogue of potential driver cancer genes identified by high-throughput genomic analyses of tumors.

8.5 Enhancing Genomic Instability in Mouse Models

The simplicity of the genomic alterations arising in mouse tumor models has both advantages and disadvantages for the identification of cancer genes. On the one hand, chromosomal aberrations are simple, facilitating analysis. On the other hand, there are generally few genomic copy number alterations in mouse tumors compared to human tumors. Thus, the models do not recapitulate the genomic diversity of human tumors.

One possible cause of enhanced genomic instability in human tumors is the decrease in telomere length with age, resulting in cells with chromosomes with dysfunctional short telomeres prone to chromosomal rearrangements. The telomeres of mice, on the other hand, are considerably longer so that telomere attrition is less likely to contribute to tumor formation. To investigate the role of telomere dysfunction in tumor genome instability and tumor formation, mice lacking the telomerase RNA component (*Terc−/−*) were generated and chromosomal instability and tumor formation compared in early and late generation animals (Blasco et al. 1997). In *Trp53* mutant mice, these studies revealed a change in tumor spectrum with age, including colon, skin, and mammary tumors, and enhanced tumor genomic instability (Artandi et al. 2000). Telomere dysfunction also appears to be associated with aneuploidy in human tumors. In human breast tumors, an association of significantly shorter telomere lengths with aneuploidy and metastasis have been reported (Griffith et al. 1999), and a recent investigation of gene expression changes associated with genomic instability in breast cancer found altered expression of genes involved in telomere dysfunction and shorter telomeres to be associated with amplification (Fridlyand et al. 2006).

In the presence of intact *Trp53*-dependent checkpoints, on the other hand, telomere dysfunction is associated with increased numbers of early stage neoplastic lesions, but a decrease in later high-grade lesions, suggesting that tumor progression is suppressed in the *Terc−/−* mice and that reactivation of telomerase is required to restore sufficient telomerase function for tumor cell viability (Chin et al. 1999; Greenberg et al. 1999; Gonzalez-Suarez et al. 2000; Rudolph et al. 2001; Farazi et al. 2003). Studies of telomerase activity in breast cancer and benign breast lesions are consistent with the mouse model studies. For example, while telomerase activity could be detected in the majority of breast cancers, it was not found in benign lesions, such as ductal hyperplasia and atypical ductal hyperplasia, although telomerase activity was detected in 59% of ductal carcinoma *in situ* cases (Poremba et al. 1998).

In addition to telomere dysfunction, defective DNA repair and impaired DNA damage checkpoints, including *Atm* deficiency or dominant negative *Trp53* mutations have been reported to destabilize the genome of murine tumors. A triple mutant murine lymphoma model combining *Trp53*, *Atm* and telomere deficiency developed tumors with numerous copy number changes, including focal amplifications and deletions (Maser et al. 2007). A comparison with human T-ALL revealed that 18 of the 160 aberrations identified in the mouse model were syntenic with regions of copy number aberration in human tumors. Targeted sequencing together with CGH revealed frequent inactivation of *Fbxw7* and *Pten* in the mouse lymphoma model and human T-ALL. Moreover, regions of aberration were shared with human tumors from different tissues, such as pancreas, lung, and colon. These studies highlight the synteny conserved between the genomes of human tumors and those arising in mice with engineered genomic instability. They not only show that the species share biological processes leading to tumorigenesis, but also demonstrate the utility of such mouse models for prioritizing evaluation of genomic alterations identified in different human tumor types.

8.6 Cross-species Comparisons Facilitate Identification of Cancer Susceptibility Genes

The genetic differences between mouse strains can be utilized in interspecific crosses to identify loci that control various tumor phenotypes and genes that interact to modify the effect of another gene. Interspecific crosses and backcrosses between two divergent strains often use tumor resistant mice, such as *Mus spretus* and sensitive inbred lines of *Mus musculus*. The genomes of tumors arising in such mice can be interrogated for recurrent regions of loss of heterozygosity (LOH) in order to identify genetic events required for tumor development. In this way, a skin tumor susceptibility locus on mouse chromosome 2 was identified by crosses between outbred *M. spretus* and inbred lines of *M. musculus*, which allowed both linkage analysis and haplotyping to be used to define the region (Ewart-Toland et al. 2003). Subsequent comparison to human breast tumors in which the syntenic region at 20q13.2 is often amplified, focused attention on three genes *Znf217*, *Cyp24a1*, and *Stk6* (human gene, *STK15*). Expression analysis ruled out *Cyp24a1* and identified *Stk6* as the best candidate susceptibility locus because the *M. musculus Stk6* allele was preferentially overexpressed and amplified in F1 tumor cell lines. While no polymorphism distinguishing *M. spretus* and *M. musculus* could be identified, statistically significant preferential amplification of the *STK15* 91A allele was detected in human colon cancer. Moreover, the high-risk Ile31 STK15 isoform promoted enhanced growth and transformation compared to the low-risk Phe31 form. The identification of combinations of common variants that modify susceptibility to sporadic human cancers requires thousands of cases and controls for loci with low relative risk. Cross-species strategies incorporating information from mouse models and human cancer may be a general approach that facilitates the identification of other low penetrance human susceptibility genes (Ruivenkamp et al. 2002; Ewart-Toland et al. 2003).

8.7 Future Directions

The completion of genome sequences for human and mouse have enhanced the capability to make cross-species comparisons and to use information gained from studies of tumor genomes from both species to rapidly identify candidate oncogenes and tumor suppressors mapping to regions of recurrent aberrations. It is likely that second generation sequencing and SNP genotyping platforms will allow even higher resolution comparison of these tumor genomes, thereby bringing cross-species genome-wide analysis of genomic alterations to the nucleotide level.

References

Albertson DG (2006) Gene amplification in cancer. Trends Genet 22:447–455

Artandi SE, Chang S, Lee SL, Alson S, Gottlieb GJ, Chin L, DePinho RA (2000) Telomere dysfunction promotes non-reciprocal translocations and epithelial cancers in mice. Nature 406:641–645

Bashyam MD, Bair R, Kim YH, Wang P, Hernandez-Boussard T, Karikari CA, Tibshirani R, Maitra A, Pollack JR (2005) Array-based comparative genomic hybridization identifies localized DNA amplifications and homozygous deletions in pancreatic cancer. Neoplasia 7:556–562

Bastian BC, Kashani-Sabet M, Hamm H, Godfrey T, Moore DH 2nd, Brocker EB, LeBoit PE, Pinkel D (2000) Gene amplifications characterize acral melanoma and permit the detection of occult tumor cells in the surrounding skin. Cancer Res 60:1968–1973

Blasco MA, Lee HW, Hande MP, Samper E, Lansdorp PM, DePinho RA, Greider CW (1997) Telomere shortening and tumor formation by mouse cells lacking telomerase rna. Cell 91:25–34

Chin L, Artandi SE, Shen Q, Tam A, Lee SL, Gottlieb GJ, Greider CW, DePinho RA (1999) P53 deficiency rescues the adverse effects of telomere loss and cooperates with telomere dysfunction to accelerate carcinogenesis. Cell 97:527–538

Collins C, Rommens JM, Kowbel D, Godfrey T, Tanner M, Hwang SI, Polikoff D, Nonet G, Cochran J, Myambo K, Jay KE, Froula J, Cloutier T, Kuo WL, Yaswen P, Dairkee S, Giovanola J, Hutchinson GB, Isola J, Kallioniemi OP, Palazzolo M, Martin C, Ericsson C, Pinkel D, Albertson D, Li WB, Gray JW (1998) Positional cloning of znf217 and nabc1: Genes amplified at 20q13.2 and overexpressed in breast carcinoma. Proc Natl Acad Sci USA 95:8703–8708

Dai Z, Zhu WG, Morrison CD, Brena RM, Smiraglia DJ, Raval A, Wu YZ, Rush LJ, Ross P, Molina JR, Otterson GA, Plass C (2003) A comprehensive search for DNA amplification in lung cancer identifies inhibitors of apoptosis ciap1 and ciap2 as candidate oncogenes. Hum Mol Genet 12:791–801

Ewart-Toland A, Briassouli P, de Koning JP, Mao JH, Yuan J, Chan F, MacCarthy-Morrogh L, Ponder BA, Nagase H, Burn J, Ball S, Almeida M, Linardopoulos S, Balmain A (2003) Identification of stk6/stk15 as a candidate low-penetrance tumor-susceptibility gene in mouse and human. Nat Genet 34:403–412

Farazi PA, Glickman J, Jiang S, Yu A, Rudolph KL, DePinho RA (2003) Differential impact of telomere dysfunction on initiation and progression of hepatocellular carcinoma. Cancer Res 63:5021–5027

Fridlyand J, Snijders AM, Ylstra B, Li H, Olshen A, Segraves R, Dairkee S, Tokuyasu T, Ljung BM, Jain AN, McLennan J, Ziegler J, Chin K, Devries S, Feiler H, Gray JW, Waldman F, Pinkel D, Albertson DG (2006) Breast tumor copy number aberration phenotypes and genomic instability. BMC Cancer 6:96

Gonzalez-Suarez E, Samper E, Flores JM, Blasco MA (2000) Telomerase-deficient mice with short telomeres are resistant to skin tumorigenesis. Nat Genet 26:114–117

Greenberg RA, Chin L, Femino A, Lee KH, Gottlieb GJ, Singer RH, Greider CW, DePinho RA (1999) Short dysfunctional telomeres impair tumorigenesis in the ink4a(delta2/3) cancer-prone mouse. Cell 97:515–525

Griffith JK, Bryant JE, Fordyce CA, Gilliland FD, Joste NE, Moyzis RK (1999) Reduced telomere DNA content is correlated with genomic instability and metastasis in invasive human breast carcinoma. Breast Cancer Res Treat 54:59–64

Gysin S, Rickert P, Kastury K, McMahon M (2005) Analysis of genomic DNA alterations and mrna expression patterns in a panel of human pancreatic cancer cell lines. Genes Chromosomes Cancer 44:37–51

Hodgson JG, Malek T, Bornstein S, Hariono S, Ginzinger DG, Muller WJ, Gray JW (2005) Copy number aberrations in mouse breast tumors reveal loci and genes important in tumorigenic receptor tyrosine kinase signaling. Cancer Res 65:9695–9704

Imoto I, Yang ZQ, Pimkhaokham A, Tsuda H, Shimada Y, Imamura M, Ohki M, Inazawa J (2001) Identification of ciap1 as a candidate target gene within an amplicon at 11q22 in esophageal squamous cell carcinomas. Cancer Res 61:6629–6634

Kallioniemi A, Kallioniemi OP, Sudar D, Rutovitz D, Gray JW, Waldman F, Pinkel D (1992) Comparative genomic hybridization for molecular cytogenetic analysis of solid tumors. Science 258:818–821

Kim M, Gans JD, Nogueira C, Wang A, Paik JH, Feng B, Brennan C, Hahn WC, Cordon-Cardo C, Wagner SN, Flotte TJ, Duncan LM, Granter SR, Chin L (2006) Comparative oncogenomics identifies nedd9 as a melanoma metastasis gene. Cell 125:1269–1281

Liu ML, Von Lintig FC, Liyanage M, Shibata MA, Jorcyk CL, Ried T, Boss GR, Green JE (1998) Amplification of ki-ras and elevation of map kinase activity during mammary tumor progression in c3(1)/sv40 tag transgenic mice. Oncogene 17:2403–2411

Maser RS, Choudhury B, Campbell PJ, Feng B, Wong KK, Protopopov A, O'Neil J, Gutierrez A, Ivanova E, Perna I, Lin E, Mani V, Jiang S, McNamara K, Zaghlul S, Edkins S, Stevens C, Brennan C, Martin ES, Wiedemeyer R, Kabbarah O, Nogueira C, Histen G, Aster J, Mansour M, Duke V, Foroni L, Fielding AK, Goldstone AH, Rowe JM, Wang YA, Look AT, Stratton MR, Chin L, Futreal PA, DePinho RA (2007) Chromosomally unstable mouse tumours have genomic alterations similar to diverse human cancers. Nature 447:966–971

Mitelman F, Johansson B, Mertens FE (2008) Mitelman database of chromosome aberrations in cancer. http://www.cgap.nci.nih.gov/Chromosomes/Mitelman.

Montagna C, Andrechek ER, Padilla-Nash H, Muller WJ, Ried T (2002) Centrosome abnormalities, recurring deletions of chromosome 4, and genomic amplification of her2/neu define mouse mammary gland adenocarcinomas induced by mutant her2/neu. Oncogene 21:890–898

Montagna C, Lyu MS, Hunter K, Lukes L, Lowther W, Reppert T, Hissong B, Weaver Z, Ried T (2003) The septin 9 (msf) gene is amplified and overexpressed in mouse mammary gland adenocarcinomas and human breast cancer cell lines. Cancer Res 63:2179–2187

Parssinen J, Kuukasjarvi T, Karhu R, Kallioniemi A (2007) High-level amplification at 17q23 leads to coordinated overexpression of multiple adjacent genes in breast cancer. Br J Cancer 96:1258–1264

Pinkel D, Segraves R, Sudar D, Clark S, Poole I, Kowbel D, Collins C, Kuo WL, Chen C, Zhai Y, Dairkee SH, Ljung BM, Gray JW, Albertson DG (1998) High resolution analysis of DNA copy number variation using comparative genomic hybridization to microarrays. Nat Genet 20:207–211

Poremba C, Bocker W, Willenbring H, Schafer KL, Otterbach F, Burger H, Diallo R, Dockhorn-Dworniczak B (1998) Telomerase activity in human proliferative breast lesions. Int J Oncol 12:641–648

Rudolph KL, Millard M, Bosenberg MW, DePinho RA (2001) Telomere dysfunction and evolution of intestinal carcinoma in mice and humans. Nat Genet 28:155–159

Ruivenkamp CA, van Wezel T, Zanon C, Stassen AP, Vlcek C, Csikos T, Klous AM, Tripodis N, Perrakis A, Boerrigter L, Groot PC, Lindeman J, Mooi WJ, Meijjer GA, Scholten G, Dauwerse H, Paces V, van Zandwijk N, van Ommen GJ, Demant P (2002) Ptprj is a candidate for the mouse colon-cancer susceptibility locus scc1 and is frequently deleted in human cancers. Nat Genet 31:295–300

Shi YP, Mohapatra G, Miller J, Hanahan D, Lander E, Gold P, Pinkel D, Gray J (1997) Fish probes for mouse chromosome identification. Genomics 45:42–47

Snijders AM, Hermsen MA, Baughman J, Buffart TE, Huey B, Gajduskova P, Roydasgupta R, Tokuyasu T, Meijer GA, Fridlyand J, Albertson DG (2008) Acquired genomic aberrations associated with methotrexate resistance vary with background genomic instability. Genes Chromosomes Cancer 47:71–83

Snijders AM, Schmidt BL, Fridlyand J, Dekker N, Pinkel D, Jordan RC, Albertson DG (2005) Rare amplicons implicate frequent deregulation of cell fate specification pathways in oral squamous cell carcinoma. Oncogene 24:4232–4242

Solinas-Toldo S, Lampel S, Stilgenbauer S, Nickolenko J, Benner A, Dohner H, Cremer T, Lichter P (1997) Matrix-based comparative genomic hybridization: biochips to screen for genomic imbalances. Genes Chromosomes Cancer 20:399–407

Weaver Z, Montagna C, Xu X, Howard T, Gadina M, Brodie SG, Deng CX, Ried T (2002) Mammary tumors in mice conditionally mutant for brca1 exhibit gross genomic instability and centrosome amplification yet display a recurring distribution of genomic imbalances that is similar to human breast cancer. Oncogene 21:5097–5107

Weaver ZA, McCormack SJ, Liyanage M, du Manoir S, Coleman A, Schrock E, Dickson RB, Ried T (1999) A recurring pattern of chromosomal aberrations in mammary gland tumors of mmtv-cmyc transgenic mice. Genes Chromosomes Cancer 25:251–260

Weber RG, Scheer M, Born IA, Joos S, Cobbers JM, Hofele C, Reifenberger G, Zoller JE, Lichter P (1998) Recurrent chromosomal imbalances detected in biopsy material from oral premalignant and malignant lesions by combined tissue microdissection, universal DNA amplification, and comparative genomic hybridization. Am J Pathol 153:295–303

Zender L, Spector MS, Xue W, Flemming P, Cordon-Cardo C, Silke J, Fan ST, Luk JM, Wigler M, Hannon GJ, Mu D, Lucito R, Powers S, Lowe SW (2006) Identification and validation of oncogenes in liver cancer using an integrative oncogenomic approach. Cell 125:1253–1267

Chapter 9
Characterization of Chromosomal Translocations in Mouse Models of Hematological Malignancies Using Spectral Karyotyping, FISH, and Immunocytochemistry

Thomas Ried and Michael J. Difilippantonio

9.1 Historical Overview: The Mouse as a Model for the Study of Chromosomal Aberrations in Cancer

The concept that chromosomal aberrations could be responsible for the development of cancer emerged based on observations by David von Hansemann and on far-reaching previsions formulated by Theodor Boveri. Summarized in his article "Über asymmetrische Zellteilung in Epithelkrebsen und deren biologische Bedeuting" (On asymmetrical cell division in epithelial cancers and its biological relevance), von Hansemann was probably the first to recognize that the number of chromatin segments is constant in different normal tissues, yet distributed irregularly in mitoses in malignant tissues and, hence, present in unequal amounts in tumor cells (Hansemann 1890). Boveri, on the other hand, derived his conclusions from the observation of mitotic events in sea urchins and ascaris eggs. His thorough cytological analyses of these model systems led him to conclude that chromosomes have both continuity throughout the cell cycle and, more importantly in the context here, individuality (Ried 2009).

Once Mendel's laws of heredity had been rediscovered, Boveri realized that the chromosome individuality he had postulated would suffice to explain Mendel's laws if the chromosomes were the carriers of genetic information. This led him to formulate the "Chromosome theory of heredity." Boveri applied these hypotheses to the cancer problem in 1914 in his book "Zur Frage der Entstehung solider Tumoren" [The origin of malignant tumors (Boveri 1914, 1929)]. He surmised that apolar

T. Ried
Genetics Branch, Center for Cancer Research, National Institutes of Health/National Cancer Institute, 50 South Drive Room 1408, Bethesda, MD 20892, USA

M.J. Difilippantonio (✉)
Division of Cancer Treatment and Diagnosis, National Cancer Institute, NIH, 31 Center Drive, Suite 3A44, Bethesda, MD 20892, USA
e-mail: difilipm@mail.nih.gov

J.E. Green and T. Ried (eds.), *Genetically Engineered Mice for Cancer Research: Design, Analysis, Pathways, Validation and Pre-Clinical Testing*,
DOI 10.1007/978-0-387-69805-2_9,

mitoses would result in chromosome distributions that could result in a loss of growth inhibiting chromosomes and in the gain of growth promoting chromosomes [reviewed in (Ried 2009)]. Today we know that genomic imbalances resulting from chromosomal aneuploidies are a defining feature of cancers of epithelial origin (Ried et al. 1999; Albertson et al. 2003). Boveri's hypotheses therefore provided the theoretical framework for the emerging discipline of cancer cytogenetics. In practice, however, the development of cytogenetics was severely hampered by technical difficulties. Initially, chromosomes were counted in mitotic cells of stained tissue sections, arguably a tedious task. One major breakthrough was achieved with the introduction of tissue culture methods, yet the precise enumeration of the normal number of chromosomes, not to speak of the analysis of aberrant chromosomes in cancer samples, had to wait for additional methodological improvements, including the use of colchicine to arrest cells, improved media for tissue culture, better fixation methods, and most notably the introduction of hypotonic treatment, which allowed the visualization of mammalian chromosomes with unprecedented clarity [for a review of the development of mammalian cytogenetics see (Hsu 1979; Harris 1995)]. Another truly important milestone was the development of chromosome banding techniques, which enabled the cytogeneticist to not only study chromosome numbers, but also allowed assessment of their structural integrity (Caspersson et al. 1970). Despite these limitations, however, the correct number of the normal mouse chromosome complement was established to be 40 as early as 1923 in the laboratory of T.S. Painter (Painter 1926). The establishment of this baseline, obviously, was necessary for a comparative analysis of chromosomes in murine tumors. For instance, Otto von Winge reported cytological studies on tar-induced mouse carcinomas (Winge 1930). In the 1950s, Makino (Makino 1951; Makino 1952; Yosida 1952) reported chromosomally aberrant mouse ascites tumors, and George Klein and Hauschka and Levan contributed a systematic analysis of numerous ascites tumors (Klein 1951; Hauschka and Levan 1958). Together with Tjio, Levan also established the number of human chromosomes as 46 in 1956 (Tjio and Levan 1956). Numerous studies followed; for instance, Ford, Hamerton, and Mole published a detailed account of chromosomal abnormalities in murine reticular tumors (Ford et al. 1958), and Levan and Biesele (1958) characterized spontaneously transformed mouse embryonic skin cells. Chromosome analyses being performed on tumors in animal models other than mice (McMichael et al. 1963; Fichdzhian and Pogosiants 1963; Vrba and Donner 1964), and postirradiation studies in both mice and humans (Nowell and Hungerford 1964; Barnes et al. 1959) likewise demonstrated the presence of chromosome aberrations.

As was true for human cytogenetics, the introduction of banding techniques greatly invigorated the field of mouse cytogenetics as well (Miller et al. 1971; Dev et al. 1971; Nesbitt and Francke 1971; Schnedl 1971; Zech et al. 1972; Nesbitt and Francke 1973). For instance, Hashmi reported chromosomal heterogeneity in mouse tumor lines (Hashmi et al. 1974), Russell reported the karyotype of a murine sarcoma (Russell et al. 1974), Shepard demonstrated chromosomal aberrations in a mouse myeloma (Shepard et al. 1974), and Gee and Harris (1979) analyzed SV40-transformed murine fibroblasts (the authors of this chapter acknowledge that this is by far not a complete review of the cytogenetic literature of mouse tumors, which is not the goal of this chapter, but merely a listing of a few representative publications). Despite considerable evidence of widespread chromosomal abnormalities in murine tumors, Boveri's hypothesis was not unanimously accepted. In 1960, Bayreuther published an article in which he reported primarily normal karyotypes in a variety of primary mouse and human tumors, and in tumors in other species as well (Bayreuther 1960). These observations prompted him to conclude that "...the results reported are inconsistent with the chromosome mutation theory of carcinogenesis" and "that the abnormal chromosome patterns which characteristically occur in the majority of well-advanced tumours are of secondary importance in the acquirement of malignant properties." In retrospect, this conclusion is perhaps not too surprising, given the technical limitations at that time and the overrepresentation of hematological malignancies in the dataset. The misinterpretations were probably supported by the findings of Miller, who characterized a number of murine leukemias (Miller 1961). As we know today, balanced chromosomal translocations, such as the Philadelphia chromosome or the t(8;14) translocation in Burkitt lymphoma, can occur as sole abnormalities, or accompanied by only a few additional changes, in otherwise stable genomes (Heim and Mitelman 2009). This is in striking contrast to the number of chromosomal aberrations and crude aneuploidy typically observed in carcinomas (Heim and Mitelman 2009). Not unlike the situation in solid tumors in humans, the comprehensive identification of cytogenetic aberrations in mouse models of epithelial cancers proved more difficult than the description of chromosomal abnormalities in models of leukemia and lymphoma. Cytogenetic analyses, in fact, have played a central role in elucidating the consequences of chromosomal translocations in mouse models of hematological malignancies. The seminal contribution of Michael Potter regarding the development and characterization of mouse plasma cell tumors as a relevant model of human myelomas serves as an example (Potter 2007). Not only did these models and the characterization of antigen binding proteins eventually pave the way for the development of monoclonal antibodies, but they were also equally instrumental for the realization that the molecular basis of cancer is a genetic one, and that specific cytogenetic abnormalities can, in fact, be the visible manifestation of specific cancer causing molecular genetic events.

In fact, early studies by Potter and Wiener revealed nonrandom chromosomal abnormalities in pristane-induced plasma cell tumors (Ohno et al. 1979; Wiener et al. 1980). The recurrent nature of these events suggested a causative role for tumorigenesis. Further analyses of the breakpoints in PCTs suggested involvement of the immunoglobulin genes on chromosomes 6 and 12, which are actively transcribed in these cells. The recurrent involvement of mouse chromosome 15 led to the postulation that these genomic rearrangements, through juxtaposition of a tumor-promoting and an actively-transcribed gene, resulted in the deregulated expression of the former (Klein 1951). It turned out that this hypothesis was indeed correct: both, the T(12:15) translocation in the PCT and the human correlate in Burkitt lymphoma, t(8;14), bring the *Myc* oncogene under the transcriptional control of IgH regulatory elements (Potter 2007). The description of cytogenetic abnormalities, therefore, guided the characterization of cancer-causing genetic events. This was the first murine cancer model in which the translocation-induced activation of oncogenes was demonstrated (Crews et al. 1982). Interestingly, the article immediately preceding this one in the same issue of *Science* reported the chromosomal mapping of the *c-ABL* gene to mouse chromosome 2 (Goff et al. 1982). The human gene mapped to chromosome 9. Curiously, another gene on mouse chromosome 2, adenylate kinase (*AK-1*), was also known to map to human chromosome 9, at the telomeric end of the long arm which was the region known to be involved in human chronic myelogenous leukemia (CML). The authors therefore speculated that if *c-ABL* was also located in this distal portion of chromosome 9, then the 9/22 translocation (called the Philadelphia chromosome) "may alter the expression of the human *c-ABL* gene and in turn influence tumor progression in CML." The utilization of the mouse as a model system thus led to the discovery of a fundamental mechanism of oncogene activation in tumors and identification of the first causal cytogenetic aberration in human cancer (Nowell and Hungerford 1962; Rowley 1973). This, of course, had major implications not only for the diagnosis and prognostication of human hematological malignancies, but also for the eventual development of therapies specifically directed at the protein products of these gross chromosomal rearrangements–Gleevac as the most prominent example (Druker 2008).

9.2 Development of Molecular Cytogenetic Techniques for the Analysis of Murine Chromosomes

9.2.1 Spectral Karyotyping

Despite the presence of banding techniques, the more complex karyotypes associated with solid tumors were extremely difficult to decipher, making it an arduous task to identify common chromosome aberrations associated with specific tumor types. The determination of common gains and losses of genomic regions was made possible largely through the development of Comparative Genomic Hybridization, as discussed in the previous chapter (see Chapter 7). While this technique provides extremely useful global genomic information, particularly in the absence of metaphase preparations, it has two important limitations. The first is that it does not provide information pertaining to the underlying mechanism responsible for causing the identified copy number alterations. Gains might be due to the presence of double minute chromosomes (dmin's), homogenously staining regions (hsr's) that result from local copy number increases, or jumping translocations that reflect the promiscuity of a particular region of the genome to translocate to multiple recipient chromosomes. The other limitation is that balanced rearrangements, which as their name reflects do not alter the copy number, will not be detected using CGH [for review (Dorritie et al. 2004)].

Hence, there was a need for another methodology to overcome these barriers. If one could label the DNA from each normal chromosome with a distinct fluorochrome and hybridize it to metaphase chromosome preparations from a tumor, it would be possible to clearly identify the composition of aberrantly rearranged chromosomes. This is the premise behind the development of both multifluor fluorescence *in situ* hybridization (M-FISH) (Speicher et al. 1996) and spectral karyotyping (SKY) (Schröck et al. 1996; Liyanage et al. 1996). Chromosomes from a karyotypically normal individual are labeled with the DNA intercollating fluorochromes Hoechst 33258, which has a preference for A-T base pairs, and chromomycin A3, which prefers G-C base pairs. Thus, longer chromosomes have a greater fluorescence intensity compared to smaller chromosomes, and more G-C-rich chromosomes have a greater orange fluorescence compared to more A-T-rich chromosomes, which will have more fluorescence in the blue wavelength. The labeled chromosome preparation is passed through a fluorescence-activated cell sorter (FACS) and the chromosomes sorted into individual tubes based on their relative fluorescence intensity in each of the two wavelengths. It is thereby possible to purify DNA from each individual chromosome. While this in principle sounds straight forward, it is in fact quite an art to the extent that there are very few people in the world who are able to do this accurately. For those chromosomes that are of similar size and A-T/G-C content, mouse/human (rat/human) hybridoma cell lines that contain only a single human chromosome must be used. In mice, where the size variation between the largest and smallest chromosome is only approximately 2.5-fold (compared to >4-fold in humans), this is particularly problematic; however, the use of specific mouse strains overcomes these difficulties.

The amount of DNA that can be reasonably sorted in this manner is in the nanogram range. In order to be practical, the amount of DNA for each chromosome must be increased through PCR amplification. The methodology that has been found to work best is called degenerate-oligonucleotide primed (DOP) PCR (Telenius et al. 1992a). In fact, a fraction of the primary amplified product is then used to generate a secondary amplified product, and then a fraction of that is sometimes used to generate a tertiary amplified product. One must be careful, however, as the quality of the resulting product decreases with each successive round of amplification due to selection for those PCR products that are amplified most efficiently and must therefore be determined empirically.

The next step is to mark each individual amplified chromosome preparation with a unique label. This is accomplished through the incorporation of nucleotides directly conjugated with either a fluorescent molecule (e.g., Rhodamine 110, Texas Red, Spectrum Orange) or small nonfluorescent molecules called haptens (e.g., Biotin, digoxigenein), which are then detected with antibodies or avidin conjugated to fluorochromes (e.g., Cy5, Cy5.5). Upon imaging, the fluorochromes must have fluorescent properties (excitation and emission) that can be distinguished from one another under a microscope. This is one of the factors limiting the number of useable fluorochromes. Because the number of chromosomes exceeds the number of available fluorochromes, unique combinations of fluorochromes must be used to label each chromosome in a process known as combinatorial labeling.

The labeled nucleotides are incorporated into the DNA during yet another round of DOP-PCR. A separate reaction must be performed for each individual conjugated nucleotide, thus for a chromosome that is combinatorially labeled with three different fluorochromes, three separate DOP-PCR labeling reactions must be performed. An aliquot from each of the labeled chromosomes is then mixed together with an excess of CotI DNA to generate the hybridization probe. The Cot1 DNA consists of highly repetitive DNA sequences found throughout the genome, such as Alu, LINE, and SINE elements. Upon denaturation of the probe to single-stranded DNA and subsequent incubation at 37°C, the highly abundant CotI DNA will base pair with the labeled repeat sequences from the chromosomes, thereby eliminating their potential to hybridize to similar sequences on other chromosomes throughout the genome. If this were allowed to occur, the desired specificity of the combinatorial labeling scheme would be lost. The metaphases prepared from the tumor cells of interest, and fixed on a microscope slide, are also denatured to single-stranded DNA and quick chilled to prevent reannealing of the chromosomes during the subsequent hybridization step. The preannealed probe is placed atop the denatured slide and a coverglass is sealed in place with rubber cement to prevent evaporation. The unique combinatorially labeled sequences in the probe are then allowed over a period of 2 days to find and anneal to their complementary sequences on the metaphase chromosomes. Subsequent washing, detection, and DNA counterstaining steps are then performed prior to microscopic visualization and imaging.

The difference between M-FISH and SKY lies in the mechanics of how these fluorescent combinations are imaged and decoded. M-FISH utilizes a series of excitation/emission filters specific for each fluorochrome. Hence, each metaphase must be imaged multiple times to acquire the fluorescence associated with each fluorochrome. A computer program to identify the origin of DNA in each chromosome of the metaphase uses a look-up table to assign specific colors. SKY, however, uses a single triple band pass filter that allows for the simultaneous excitation of all five fluorochromes. Additionally, the entire spectrum of the metaphase between 400 and 800 nm can be measured in a single exposure using a device called an interferometer attached to the epifluorescence microscope. A single image containing spectral information for each pixel is acquired, and the fluorescent intensities in the green, red, and near infrared emission range are visualized in a standard red, green, blue (RGB) display. Each pixel is assigned a chromosome-specific pseudocolor based on its spectral signature and a look-up table, thereby allowing for the spectral classification of each chromosome. A separate image of the DNA counterstain (DAPI; 4,6-diamidino-2-phenylindole-dihydrochloride) is also acquired which provides a banding pattern for each chromosome. The chromosomes are subsequently classified and aligned in a karyotype table. The resulting image (see Fig. 9.1) is a colorful metaphase spread in which the chromosomal origin of each fragment of DNA in the interrogated sample is defined, thereby allowing for the identification and interpretation of all aberrations in the karyogram.

As with CGH, SKY/M-FISH also has a few inherent limitations. Very small marker or double minute chromosomes cannot always be unambiguously classified. Intrachromosomal alterations such as inversions, small deletions, or duplications do

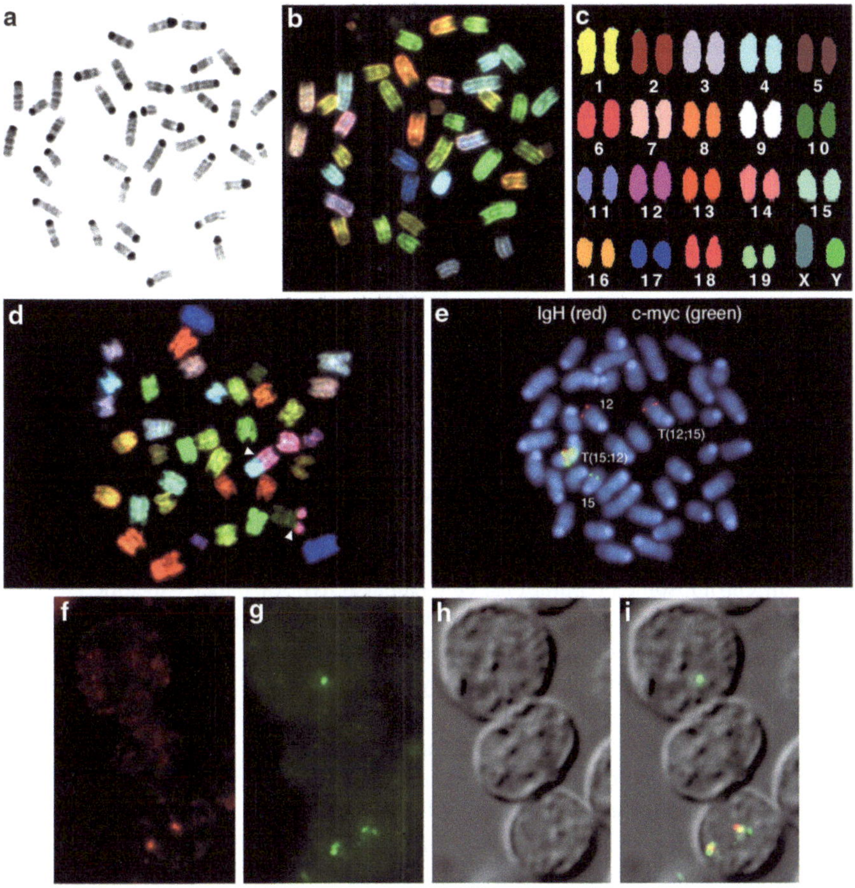

Fig. 9.1 (**a**) Metaphase of normal mouse chromosomes after DAPI-staining. (**b**) SKY analysis of the same metaphase after hybridization with a probe cocktail painting all chromosomes in different colors. (**c**) SKY: Conversion to computer-generated pseudocolors and alignment as a karyotype. (**d**) SKY identifies chromosomal translocations in thymic lymphomas in mice deficient for the ATM tumor suppressor gene. The *arrow* denotes chromosome 14, which carries the genes for the T-cell receptor locus. (**e**) FISH analysis with gene-specific BAC clones shows chromosomal translocations in pro B-cell lymphomas in mice deficient for the DNA repair gene Ku70 and the tumor suppressor gene TP53. The cells carry a translocation that juxtaposes the *c-Myc* oncogene to murine IgH, similar to the situation in human Burkitt lymphoma. (**f**) Immunocytochemistry shows localization of the protein γ-H2X at sites of DNA double strand breaks, followed by FISH with probes for the T-cell receptor locus α (**g**). (**h**) DIC image of the same cells and (**i**) merged image to determine colocalization. Images **f** through **g** represent a single optical section through the cells, hence the absence of two TCR-α signals in all of the cells

not result in macroscopically visible changes in chromosome size or spectral signature and, therefore, escape detection. A combination of molecular cytogenetic methods and banding are therefore required to obtain the most comprehensive analysis of tumor metaphases. However, SKY has been immensely useful for the analysis of chromosomal aberrations in mouse models of human cancer. This is in large part due to the fact that mouse chromosomes are all acrocentric, negating the use of the location of the centromere for identification as is possible in human chromosomes, and are relatively equal in size. Comprehensive karyotype analysis by banding alone, particularly in tumors with complex rearrangements, is therefore extremely challenging.

9.2.2 ICC/FISH

CGH, M-FISH, and SKY are all applications of the technique known as fluorescence *in situ* hybridization, or FISH. Simply defined, it is the hybridization (or annealing) of a labeled nucleic acid probe (DNA or RNA) to a microscope slide containing an unlabeled sample. CGH involves the hybridization of labeled tumor and differently labeled normal DNA to normal chromosomes (described in Chapter 7). Another application is the hybridization of DOP-PCR-labeled flow sorted aberrant tumor chromosomes to normal metaphase spreads to determine the composition of that chromosome. SKY/M-FISH is the hybridization of normal labeled DNA to aberrant chromosome preparations. One can also hybridize locus-specific DNA in the form of a plasmid, phage, BAC or viral genome to metaphase preparations, cytospin preparations, or intact tissue sections from tumor samples or tumor cell lines. These are useful in determining copy number, physical location, or integration site in the genome, if a locus has been deleted/amplified/split at a translocation breakpoint or even sites of active gene transcription.

The detection of proteins within tissues (immunohistochemistry, IHC) or cell preparations (immunocytochemistry, ICC) using antibody reagents is utilized to determine not only the relative abundance of a protein, but also its subcellular localization. This methodology can be combined with FISH in order to determine within the same sample whether co-localization of protein and DNA is occurring. Within the last decade this has proven to be a powerful technique in the areas of telomere maintenance, site-specific recombination in developing lymphocytes, and DNA damage recognition and repair. There are several different protocols that have been successfully followed (Padilla-Nash et al. 2006; Page et al. 1995; Speel et al. 1995; Holland et al. 1996; Mialhe et al. 1996; Brown et al. 1997; Chen et al. 2000; Gaiser et al. 2010). One can either perform FISH and then immunocytochemistry or vice versa. Samples can be fixed in methanol or paraformaldehyde. There are different approaches for cross-linking the antibodies used to detect the proteins or FISH signals, as well as the method used to denature the DNA for hybridization of the nucleic acid probe. These variations are in part dependent upon the protein being detected, the antibody used for detection, and the hybridization efficiency and intensity of the FISH signal. Identifying the protocol and conditions that work best requires a systematic approach.

The different variables for ICC and FISH should each be tested individually before combining them in a single experiment. For instance, hybridization of the nucleic acid probe should be performed on metaphase spreads and interphase nuclei under normal FISH conditions in order to determine the maximum achievable signal.

Likewise, with fixation of the sample and detection of the protein of interest, as some epitope/antibody interactions are preserved in one method and not another. These control experiments are extremely important. Because suboptimal conditions for ICC and FISH are applied when combined in a dual approach, the intensity of each individual signal will also be suboptimal. The next step is to apply these suboptimal methods individually as well to determine which provide the strongest signal intensity. This may include different treatments for antibody cross-linking (e.g., paraformaldehyde or ethylene glycol bis (succinimidyl succinate) – EGS) or DNA denaturation (e.g., formamide, heat, NaOH). Upon combining the ICC and FISH approaches that show the most promise, it is again useful to perform some control experiments. If one is performing FISH followed by ICC, for instance, denaturation of the sample as envisioned for FISH, but rather than hybridization of the FISH probe, proceeding directly to the ICC. This will provide an indication of whether the denaturation will have an adverse effect on the ability to detect the protein of interest without the other confounding variables associated with the FISH hybridization itself and the washes that must be performed. While this stepwise approach may seem time-consuming and unnecessary, it invariably saves time trouble-shooting in the end (for a detailed protocol please visit http://www.riedlab.nci.nih.gov/).

9.3 SKY Analysis of Murine Cancer Models

SKY has been applied to the molecular cytogenetic analysis of numerous mouse models of human cancer. For instance, SKY confirmed the T(12;15) translocation in the above-mentioned mouse plasma cell tumors and detected numerous variant translocations in this particular mouse model and its derivatives (Coleman et al. 1997, 1999a, 1999b, 1999c; Felix et al. 2001; Kovalchuk et al. 2001; Rockwood et al. 2002; Park et al. 2005; McNeil et al. 2005). Homozygous deletion of the mouse *ATM* gene, the homologue of the gene that causes ataxia telangiectasia in humans, results, among many other features that are similar to the ones observed in patients, in the development of thymic lymphomas. SKY analysis immediately resolved the observed cytogenetic changes as translocations involving chromosome 14. Chromosome 14 carries the genes for the T-cell receptor chains α and δ, which are known to be involved in the genesis of lymphomas in patients with ataxia telangiectasia. The SKY analysis, therefore, guided the subsequent use of specific FISH clones, which revealed, with high-resolution, that abnormal rearrangements of the T-cell receptor locus are indeed involved in lymphomagenesis (Barlow et al. 1996; Liyanage et al. 2000; Petiniot et al. 2002), thereby validating this particular model as relevant for the human disease.

SKY was also instrumental in the characterization of cytogenetic abnormalities in a variety of murine cancer models induced by tissue-specific overexpression of oncogenes, including *Myc* (Weaver et al. 1999), *Her2* (Montagna et al. 2002), polyoma middle T antigen (Montagna et al. 2003), E6 and E7 oncoproteins (Schaeffer et al. 2004), and β-catenin (Xu et al. 2008). Likewise, tumors induced by targeted deletion of known or putative tumor suppressor genes, such as *BRCA1* (Weaver et al. 2002), *Ku70/80* (Difilippantonio et al. 2000; Difilippantonio et al. 2002), *H2AX* (Celeste et al. 2002), *Anx7* (Srivastava et al. 2003), and *Men1* (Crabtree et al. 2003), to name just a few, could be successfully analyzed.

These methodologies are not only useful for visualization of the underlying gross chromosomal rearrangements and identification of the genes localized to these regions, but they also provide insight into the underlying mechanism of genomic instability responsible for the creation of these tumor-causing genomic reconfigurations. Genomic amplification of oncogenes in the form of homogeneously staining regions (hsr's) and double minute chromosomes (dmin's) are not uncommon in human solid tumors (Heim and Mitelman 2009), but are relatively rare in hematologic malignancies. In a mouse model of plasmacytoma, utilized SKY and reverse FISH painting of the flow-sorted aberrant chromosomes to identify the presence of the same T(12;15) chromosome rearrangement harboring the *IgH/Myc* fusion localized in two different marker chromosomes within the same tumor (Coleman et al. 1999b). This genomic amplification event, known as a segmental jumping translocation (SJT), is related to jumping translocations which typically occur in areas of highly repetitive DNA such as centromeres, pericentromeric heterochromatin, telomeres, constitutive heterochromatin, etc. (Berger and Bernard 2007).

Pro-B cell lymphomas in $Ku80^{-/-}p53^{-/-}$ or $Ku70^{-/-}p53^{-/-}$ mice similarly contained translocations juxtaposing the *IgH* and *Myc* (Difilippantonio et al. 2002). Because these tumors derive from immature B cells in the bone marrow, compared to the mature peripheral B cells in plasmacytomas, the recombination site within the *IgH* locus is different. Despite this biologic difference in developmental stage, these tumors also contained segmental amplification of the translocation breakpoint. Further interrogation of the chromosome structure by FISH with BAC probes specific for *IgH*, *Myc* and the subtelomeric region of chromosome 15 revealed that the rearrangements were not reciprocal translocations in which the distal regions of chromosomes 12 and 15 were exchanged, but that these nonreciprocal translocations involved duplication of either *IgH* to chromosome 15, or *Myc* to chromomosome 12, resulting in the presence of three chromosomal locations for these genes. In some instances, two normal copies of chromosome 15 were retained while *Myc* was copied to the *IgH* locus on chromosome 12. This pattern of genomic duplication to another chromosome was consistent with a mechanism of DNA repair known as break induced replication (BIR), which was first demonstrated in yeast (Morrow et al. 1997). This event was often followed by an iterative process of localized segmental amplification of the region postulated to occur through a process known as breakage-fusion-bridge (BFB) (Bosco et al. 1998) and subsequent stabilization of the derivative chromosome through the capture of telomeric sequences from yet another chromosome again utilizing BIR (Meltzer et al. 1993).

The variant histone protein H2AX becomes rapidly phosphorylated in response to external DNA damage, and the Mre11/Rad50/NBS1 (MRN) complex forms ionizing radiation-induced foci at sites of DNA double strand breaks. Developing thymocytes undergo a strictly controlled site-specific DNA recombination event, known as V(D)J recombination, responsible for the formation of a T-cell receptor (TCR) gene capable of encoding a protein (Liu et al. 2009). The expression of a functional TCR on the cell surface is necessary for the formation and release of a mature T-cell from the thymus into the periphery. V(D)J recombination was known to involve the formation of a DNA double strand break and repair of this lesion utilizing proteins of the nonhomologous end-joining (NHEJ) repair pathway (Danska et al. 1997). The simultaneous immunocytochemical detection of NBS1 or phosphorylated H2AX and FISH with a TCR-specific probe in developing thymocytes provided the first direct evidence that these proteins, which associate with irradiation-induced DNA double strand breaks, also localize to the sites of endogenous site-specific DNA rearrangement events (Chen et al. 2000). This was subsequently demonstrated to also occur in mature B-cells undergoing class-switch recombination (CSR) of the immunoglobulin (Ig) receptor locus. More importantly, however, the ICC/FISH experiments placed the unknown activity of the putative RNA-editing enzyme activation-induced cytidine deaminase (AID), which is required for CSR, upstream of the initiating DNA double strand break (Petersen et al. 2001). These findings resulted in experiments by numerous groups within the next couple of years defining the enzymatic role of AID (Bransteitter et al. 2003; Chaudhuri et al. 2003; Di Noia and Neuberger 2002; Dickerson et al. 2003; Imai et al. 2003; Petersen-Mahrt et al. 2002; Pham et al. 2003) and more recently the subsequent steps (Rada et al. 2002; Ramiro et al. 2003; Di Noia and Neuberger 2007) of CSR and somatic hypermutation.

The methodologies discussed in this chapter for the analysis of gross chromosomal aberrations and co-localization of proteins and genomic sequences within cells are generally descriptive in nature. However, as in the examples we have discussed, the information they provide serves to guide further scientific investigation. In some instances, this resulted in the identification of a gene whose protein product could be targeted as a therapeutic intervention. Other times they have the capacity to provide insight into an underlying mechanism involved in tumorigenesis. Coupled with the mouse as a model of the human disease, specific hypotheses are readily developed and interrogated.

References

Albertson DG, Collins C, McCormick F, Gray JW (2003) Chromosome aberrations in solid tumors. Nat Genet 34:369–376

Barlow C, Hirotsune S, Paylor R et al (1996) *Atm*-deficient mice: a paradigm of ataxia telangiectasia. Cell 86:159–171

Barnes DW, Ford CE, Gray SM, Loutit JF (1959) Spontaneous and induced changes in cell populations in heavily irradiated mice. In: Progress in Nuclear Biology – Biological Sciences Elsevier, Oxford Series 6:vol 2:1–10

Bayreuther K (1960) Chromosomes in primary neoplastic growth Nature 186:6–9

Berger R, Bernard OA (2007) Jumping translocations. Genes Chromosomes Cancer 46:717–723

Bosco G, Haber JE (1998) Chromosome break-induced DNA replication leads to nonreciprocal translocations and telomere capture. Genetics 150:1037–1047

Boveri T (1914) Zur Frage der Entstehung maligner Tumoren. Gustav Fischer, Jena

Boveri T (1929) The origin of malignant tumors. Williams & Wilkins, Baltimore

Bransteitter R, Pham P, Scharff MD, Goodman MF (2003) Activation-induced cytidine deaminase deaminates deoxycytidine on single-stranded DNA but requires the action of RNase. Proc Natl Acad Sci USA 100:4102–4107

Brown KE, Guest SS, Smale ST, Hahm K, Merkenschlager M, Fisher AG (1997) Association of transcriptionally silent genes with Ikaros complexes at centromeric heterochromatin. Cell 91:845–854

Caspersson T, Zech L, Johansson C, Modest EJ (1970) Identification of human chromosomes by DNA-binding fluorescent agents. Chromosoma 30:215–227

Celeste A, Petersen S, Romanienko PJ et al (2002) Genomic instability in mice lacking histone H2AX. Science 296:922–927

Chaudhuri J, Tian M, Khuong C, Chua K, Pinaud E, Alt FW (2003) Transcription-targeted DNA deamination by the AID antibody diversification enzyme. Nature 422:726–730

Chen HT, Bhandoola A, Difilippantonio MJ et al (2000) Response to RAG-mediated VDJ cleavage by NBS1 and gamma-H2AX. Science 290:1962–1965

Coleman AE, Schrock E, Weaver Z et al (1997) Previously hidden chromosome aberrations in T(12;15)-positive BALB/c plasmacytomas uncovered by multicolor spectral karyotyping. Cancer Res 57:4585–4592

Coleman AE, Forest ST, McNeil N, Kovalchuk AL, Ried T, Janz S (1999a) Cytogenetic analysis of the bipotential murine pre-B cell lymphoma, P388, and its derivative macrophage-like tumor, P388D1, using SKY and CGH. Leukemia 13:1592–1600

Coleman AE, Kovalchuk AL, Janz S, Palini A, Ried T (1999b) Jumping translocation breakpoint regions lead to amplification of rearranged Myc. Blood 93:4442–4444

Coleman AE, Ried T, Janz S (1999c) Recurrent non-reciprocal translocations of chromosome 5 in primary T(12;15)-positive BALB/c plasmacytomas. Curr Top Microbiol Immunol 246:175–180

Collard JG, Philippus E, Tulp A, Lebo RV, Gray JW (1984) Separation and analysis of human chromosomes by combined velocity sedimentation and flow sorting applying single- and dual-laser flow cytometry. Cytometry 5:9–19

Crabtree JS, Scacheri PC, Ward JM et al (2003) Of mice and MEN1: Insulinomas in a conditional mouse knockout. Mol Cell Biol 23:6075–6085

Crews S, Barth R, Hood L, Prehn J, Calame K (1982) Mouse c-myc oncogene is located on chromosome 15 and translocated to chromosome 12 in plasmacytomas. Science 218:1319–1321

Danska JS, Guidos CJ (1997) Essential and perilous: V(D)J recombination and DNA damage checkpoints in lymphocyte precursors. Semin Immunol 9:199–206

Dev VG, Grewal MS, Miller DA, Kouri RE, Hutton JJ, Miller OJ (1971) The quinacrine fluorescence karyotype of Mus musculus and demonstration of strain differences in secondary constrictions. Cytogenetics 10:436–451

Di Noia J, Neuberger MS (2002) Altering the pathway of immunoglobulin hypermutation by inhibiting uracil-DNA glycosylase. Nature 419:43–48

Di Noia JM, Neuberger MS (2007) Molecular mechanisms of antibody somatic hypermutation. Annu Rev Biochem 76:1–22

Dickerson SK, Market E, Besmer E, Papavasiliou FN (2003) AID mediates hypermutation by deaminating single stranded DNA. J Exp Med 197:1291–1296

Difilippantonio MJ, Zhu J, Chen HT et al (2000) DNA repair protein Ku80 suppresses chromosomal aberrations and malignant transformation. Nature 404:510–514

Difilippantonio MJ, Petersen S, Chen HT et al (2002) Evidence for replicative repair of DNA double-strand breaks leading to oncogenic translocation and gene amplification. J Exp Med 196:469–480

Dorritie K, Montagna C, Difilippantonio MJ, Ried T (2004) Advanced molecular cytogenetics in human and mouse. Expert Rev Mol Diagn 4:663–676

Druker BJ (2008) Translation of the Philadelphia chromosome into therapy for CML. Blood 112:4808–4817

Felix K, Kovalchuk AL, Park SS et al (2001) Inducible mutagenesis in TEPC 2372, a mouse plasmacytoma cell line that harbors the transgenic shuttle vector lambdaLIZ. Mutat Res 473:121–136

Fichdzhian BC, Pogosiants EE (1963) Chromosomal characteristics of 3 transplantable leukemias in rats. Vopr Onkol 21:47–51
Ford CE, Hamerton JL, Mole RH (1958) Chromosomal changes in primary and transplanted reticular neoplasms of the mouse. J Cell Physiol Suppl 52:235–262, discussion 62–9
Gaiser T, Berroa-Garcia L, Kemmerling R, Dutta A, Ried T, Heselmeyer-Haddad K (2010) Automated analysis of protein expression and gene amplification within the same cells of paraffin-embedded tumour tissue. Anal Cell Pathol (Amst) 33:105–112
Gee CJ, Harris H (1979) Tumorigenicity of cells transformed by Simian virus 40 and of hybrids between such cells and normal diploid cells. J Cell Sci 36:223–240
Goff SP, D'Eustachio P, Ruddle FH, Baltimore D (1982) Chromosomal assignment of the endogenous proto-oncogene C-abl. Science 218:1317–1319
Hansemann D (1890) Über asymmetrische Zellteilung in Epithelkrebsen und deren biologische Bedeutung. Virchows ArchPathol 119:299–326
Harris H (1995) The cells of the body. A history of somatic cell genetics. Cold Spring Harbor Press, Plainview, NY
Hashmi S, Allderdice PW, Klein G, Miller OJ (1974) Chromosomal heterogeneity in the RAG and MSWBS mouse tumor cell lines. Cancer Res 34:79–88
Hauschka TS, Levan A (1958) Cytologic and functional characterization of single cell clones isolated from the Krebs-2 and Ehrlich ascites tumors. J Natl Cancer Inst 21:77–135
Heim S, Mitelman F (2009) Cancer Cytogenetics. John Wiley & Sons, Hoboken
Holland MS, Mackenzie CD, Bull RW, Silva RF (1996) A comparative study of histological conditions suitable for both immunofluorescence and in situ hybridization in the detection of Herpesvirus and its antigens in chicken tissues. J Histochem Cytochem 44:259–265
Hsu TC (1979) Human and mammalian cytogenetics. An historical perspective. Springer Verlag, New York
Imai K, Slupphaug G, Lee WI et al (2003) Human uracil-DNA glycosylase deficiency associated with profoundly impaired immunoglobulin class-switch recombination. Nat Immunol 4:1023–1028
Klein G (1951) Comparative studies of mouse tumors with respect to their capacity for growth as "ascites tumors" and their average nucleic acid content per cell. Exp Cell Res 2:518–573
Kovalchuk AL, Esa A, Coleman AE et al (2001) Translocation remodeling in the primary BALB/c plasmacytoma TEPC 3610. Genes Chromosomes Cancer 30:283–291
Levan A, Biesele JJ (1958) Role of chromosomes in cancerogenesis, as studied in serial tissue culture of mammalian cells. Ann N Y Acad Sci 71:1022–1053
Liu Y, Zhang L, Desiderio S (2009) Temporal and spatial regulation of V(D)J recombination: interactions of extrinsic factors with the RAG complex. Adv Exp Med Biol 650:157–165
Liyanage M, Coleman A, du Manoir S et al (1996) Nat Genet 14(3):312–315
Liyanage M, Weaver Z, Barlow C et al (2000) Abnormal rearrangement within the alpha/delta T-cell receptor locus in lymphomas from Atm-deficient mice. Blood 96:1940–1946
Makino S (1951) Some observations on the chromosomes in the Yoshida sarcoma cells based on the homoplastic and heteroplastic transplantations; a preliminary report. Gann 42:87–90
Makino S (1952) Cytological studies on cancer. III. The characteristics and individuality of chromosomes in tumor cells of the Yoshida sarcoma which contribute to the growth of the tumor. Gan 43:17–34
McMichael H, Wagner JE, Nowell PC, Hungerford DA (1963) Chromosome Studies of Virus-Induced Rabbit Papillomas and Derived Primary Carcinomas. J Natl Cancer Inst 31:1197–1215
McNeil N, Kim JS, Ried T, Janz S (2005) Extraosseous IL-6 transgenic mouse plasmacytoma sometimes lacks Myc-activating chromosomal translocation. Genes Chromosomes Cancer 43:137–146
Meltzer PS, Guan XY, Trent JM (1993) Telomere capture stabilizes chromosome breakage. Nat Genet 4:252–255
Mialhe A, Cassanelli S, Louis J, Seigneurin D (1996) Methods for simultaneous interphase in situ hybridization and nuclear antigen immunocytochemistry in T47-D cells. J Histochem Cytochem 44:193–197

Miller JF (1961) Etiology and pathogenesis of mouse leukemia. Adv Cancer Res 6:291–368
Miller OJ, Miller DA, Kouri RE et al (1971) Identification of the mouse karyotype by quinacrine fluorescence, and tentative assignment of seven linkage groups. Proc Natl Acad Sci USA 68:1530–1533
Montagna C, Andrechek ER, Padilla-Nash H, Muller WJ, Ried T (2002) Centrosome abnormalities, recurring deletions of chromosome 4, and genomic amplification of HER2/neu define mouse mammary gland adenocarcinomas induced by mutant HER2/neu. Oncogene 21:890–898
Montagna C, Lyu MS, Hunter K et al (2003) The Septin 9 (MSF) gene is amplified and overexpressed in mouse mammary gland adenocarcinomas and human breast cancer cell lines. Cancer Res 63:2179–2187
Morrow DM, Connelly C, Hieter P (1997) "Break copy" duplication: a model for chromosome fragment formation in Saccharomyces cerevisiae. Genetics 147:371–382
Nesbitt M, Francke U (1971) Linkage groups II and XII of the mouse: cytological localization by fluorochrome staining. Science 174:60–62
Nesbitt MN, Francke U (1973) A system of nomenclature for band patterns of mouse chromosomes. Chromosoma 41:145–158
Nowell PC, Hungerford DA (1962) The minute chromosome (Phl) in chronic granulocytic leukemia. Blut 132:65–66
Nowell PC, Hungerford DA (1964) Chromosome changes following irradiation in mammals. Ann N Y Acad Sci 114:252–258
Ohno S, Babonits M, Wiener F, Spira J, Klein G, Potter M (1979) Nonrandom chromosome changes involving the Ig gene-carrying chromosomes 12 and 6 in pristane-induced mouse plasmacytomas. Cell 18:1001–1007
Padilla-Nash HM, Barenboim-Stapleton L, Difilippantonio MJ, Ried T (2006) Spectral karyotyping analysis of human and mouse chromosomes. Nat Protoc 1:3129–3142
Page SL, Earnshaw WC, Choo KH, Shaffer LG (1995) Further evidence that CENP-C is a necessary component of active centromeres: studies of a dic(X; 15) with simultaneous immunofluorescence and FISH. Hum Mol Genet 4:289–294
Painter TS (1926) The chromosomes of rodents. Science 64:336
Park SS, Kim JS, Tessarollo L et al (2005) Insertion of c-Myc into Igh induces B-cell and plasma-cell neoplasms in mice. Cancer Res 65:1306–1315
Petersen S, Casellas R, Reina-San-Martin B et al (2001) AID is required to initiate Nbs1/gamma-H2AX focus formation and mutations at sites of class switching. Nature 414:660–665
Petersen-Mahrt SK, Harris RS, Neuberger MS (2002) AID mutates E. coli suggesting a DNA deamination mechanism for antibody diversification. Nature 418:99–103
Petiniot LK, Weaver Z, Vacchio M et al (2002) RAG-mediated V(D)J recombination is not essential for tumorigenesis in Atm-deficient mice. Mol Cell Biol 22:3174–3177
Pham P, Bransteitter R, Petruska J, Goodman MF (2003) Processive AID-catalysed cytosine deamination on single-stranded DNA simulates somatic hypermutation. Nature 424:103–107
Potter M (2007) The early history of plasma cell tumors in mice, 1954–1976. Adv Cancer Res 98:17–51
Rada C, Williams GT, Nilsen H, Barnes DE, Lindahl T, Neuberger MS (2002) Immunoglobulin isotype switching is inhibited and somatic hypermutation perturbed in UNG-deficient mice. Curr Biol 12:1748–1755
Ramiro AR, Stavropoulos P, Jankovic M, Nussenzweig MC (2003) Transcription enhances AID-mediated cytidine deamination by exposing single-stranded DNA on the nontemplate strand. Nat Immunol 4:452–456
Ried T (2009) Homage to Theodor Boveri (1862–1915): Boveri's theory of cancer as a disease of the chromosomes, and the landscape of genomic imbalances in human carcinomas. Environ Mol Mutagen 50(8):593–601
Ried T, Heselmeyer-Haddad K, Blegen H, Schrock E, Auer G (1999) Genomic changes defining the genesis, progression, and malignancy potential in solid human tumors: a phenotype/genotype correlation. Genes Chromosomes Cancer 25:195–204

Rockwood LD, Torrey TA, Kim JS et al (2002) Genomic instability in mouse Burkitt lymphoma is dominated by illegitimate genetic recombinations, not point mutations. Oncogene 21:7235–7240
Rowley JD (1973) A new consistent chromosomal abnormality in chronic myelogeneous leukemia identified by quinacrine fluorescence and Giemsa staining. Nature 243:290–293
Russell SW, Francke U, Buettner L, Cochrane CG (1974) Modes of growth and spread of a transplantable, virus-producing murine (Moloney) sarcoma: karyotypic analyses. J Natl Cancer Inst 53:801–806
Schaeffer AJ, Nguyen M, Liem A et al (2004) E6 and E7 oncoproteins induce distinct patterns of chromosomal aneuploidy in skin tumors from transgenic mice. Cancer Res 64:538–546
Schnedl W (1971) The karyotype of the mouse Chromosoma 35:111–116
Schröck E, du Manoir S, Veldman T et al (1996) Science 26;273(5274):494–497
Shepard JS, Wurster-Hill DH, Pettengill OS, Sorenson GD (1974) Giemsa-banded chromosomes of mouse myeloma in relationship to oncogenicity. Cytogenet Cell Genet 13:279–309
Speel EJ, Ramaekers FC, Hopman AH (1995) Cytochemical detection systems for in situ hybridization, and the combination with immunocytochemistry, 'who is still afraid of red, green and blue?'. Histochem J 27:833–858
Speicher MR, Gwyn Ballard S, Ward DC (1996) Karyotyping human chromosomes by combinatorial multi-fluor FISH. Nat Genet 12:368–375
Srivastava M, Montagna C, Leighton X et al (2003) Haploinsufficiency of Anx7 tumor suppressor gene and consequent genomic instability promotes tumorigenesis in the Anx7(+/–) mouse. Proc Natl Acad Sci USA 100:14287–14292
Telenius H, Pelear AH, Tunnacliffe A et al (1992a) Cytogenetic analysis by chromosome painting using DOP-PCR amplified flow sorted chromosomes. Genes ChromosomCancer 4:267–3
Telenius H, Carter NP, Bebb CE, Norednskjöld M, Ponder BAJ, Tunnacliffe A (1992b) Degenerate oligonucleotide-primed PCR (DOP-PCR): general amplification of target DNA by a single degenerate primer. Genomics 13:718–725
Tjio JH, Levan A (1956) The chromosome number in man. Hereditas 42:1–6
Toledo F, Buttin G, Debatisse M (1993) The origin of chromosome rearrangements at early stages of AMPD2 gene amplification in Chinese hamster cells. Curr Biol 3:255–264
Vrba M, Donner L (1964) Chromosome numbers and karyotypes of two rat tumours induced by Rous sarcoma virus in vitro. Folia Biol (Praha) 10:373–380
Weaver ZA, McCormack SJ, Liyanage M et al (1999) A recurring pattern of chromosomal aberrations in mammary gland tumors of MMTV-cmyc transgenic mice. Genes Chromosomes Cancer 25:251–260
Weaver Z, Montagna C, Xu X et al (2002) Mammary tumors in mice conditionally mutant for Brca1 exhibit gross genomic instability and centrosome amplification yet display a recurring distribution of genomic imbalances that is similar to human breast cancer. Oncogene 21:5097–5107
Wiener F, Babonits M, Spira J, Klein G, Potter M (1980) Cytogenetic studies on IgA/lambda-producing murine plasmacytomas: regular occurrence of a T(12;15) translocation. Somatic Cell Genet 6:731–738
Winge O (1930) Zytologische Untersuchungen über die Natur maligner Tumoren. II Teerkarzinome bei Mäusen. Z Zellforsch Mikrosk Anat 10:397–423
Xu M, Yu Q, Subrahmanyam R, Difilippantonio MJ, Ried T, Sen JM (2008) Beta-catenin expression results in p53-independent DNA damage and oncogene-induced senescence in prelymphomagenic thymocytes in vivo. Mol Cell Biol 28:1713–1723
Yosida TH (1952) Cytological studies on cancer. V. Heteroplastic transplantations of the Yoshida sarcoma, with special regard to the behaviour of tumor cells. Gan 43:35–43
Zech L, Evans EP, Ford CE, Gropp A (1972) Banding patterns in mitotic chromosomes of tobacco mouse. Exp Cell Res 70:263–268

Chapter 10
Expression Profiling of Mouse Models of Human Cancer: Model Categorization and Guidance for Preclinical Testing

Min Zhu, Aleksandra M. Michalowski, and Jeffrey E. Green

10.1 Overview

Cancer is a very complex disease resulting from an accumulation of multiple genetic aberrations (Vogelstein et al. 1988). The diagnosis and treatment of cancer have primarily relied on the histopathological evaluation of tumor specimens with limited determinations of biomarker expression. Such morphological analyses provide a very descriptive way to characterize and classify subtypes of cancer. However, the application of high-throughput genomic technologies has revolutionized the analysis of cancer and has led to the recognition that cancers can be further subcategorized based upon their genetic complexity as determined by gene expression profiles, miRNA expression, genomic alterations, and proteomic profiles. These technologies have expanded our understandings of the heterogeneous nature of cancer and are providing insights into complex interactions of genes and gene networks. In particular, gene expression profiling provides a very powerful tool to perform large-scale, genome-wide analyses of the cancer transcriptome and is proving to greatly enhance cancer diagnosis, prognosis, and personalized therapy. High-throughput gene expression profiling provides robust molecular information for the more precise classification of tumor subtypes (Golub et al. 1999; Perou et al. 1999), discovery of gene functions (Chu et al. 1998; Hughes et al. 2000), dissection of molecular

M. Zhu • A.M. Michalowski
Transgenic Oncogenesis and Genomics Section, Laboratory of Cancer Biology and Genetics, Center for Cancer Research, National Cancer Institute, Bethesda, MD, USA

J.E. Green (✉)
Transgenic Oncogenesis and Genomics Section, Laboratory of Cancer Biology and Genetics, Center for Cancer Research, National Cancer Institute, Building 37, Room 4054, 37 Convent Drive, Bethesda, MD 20892, USA
e-mail: jegreen@nih.gov

J.E. Green and T. Ried (eds.), *Genetically Engineered Mice for Cancer Research: Design, Analysis, Pathways, Validation and Pre-Clinical Testing*,
DOI 10.1007/978-0-387-69805-2_10, © Springer Science+Business Media, LLC 2012

pathways (Roberts et al. 2000), and evaluation of drug response (Hughes et al. 2000; Gray et al. 1998; Marton et al. 1998).

DNA microarrays contain tens of thousands of discrete DNA sequences in a single slide or chip. The hybridization of sample probes that are generated from tumor RNA onto the microarray provides the means to monitor the expression levels of thousands of genes simultaneously in a single experiment. Analyses of this data can identify "molecular fingerprints" that reflect genotypic and phenotypic differences among large sets of tumors (Granjeaud et al. 1999). Gene expression profiling of human cancer not only classifies tumors into additional subcategories, but also reveals novel biological and molecular markers associated with the clinical course of the disease. Genes found to have important functions in tumor biology may serve as new targets for drug development. Moreover, global gene expression profiling provides a tool to compare the cancer transcriptome to that of normal tissues in order to identify biomarkers that may aid in early detection or evaluating the response to treatment. Importantly, cross-species analyses of cancers have proved valuable in identifying evolutionarily conserved genes and genetic networks that may be especially important targets for eliminating tumor cells. Through the analysis of microarray data from human breast, prostate, lung, and liver carcinomas with their respective counterparts in genetically engineered mouse models, important genes and genetic networks have been identified that have been valuable in understanding and treating human cancers.

Genetically engineered mouse (GEM) models of human cancer have afforded us tremendous opportunities to predictably manipulate the mouse genome in a defined genetic background to experimentally model human cancer. The generation of GEM models has been an invaluable tool to gain insight into the mechanisms underlying tumor initiation and progression. Furthermore, these models are proving to be important systems for performing *in vivo* preclinical studies to assess therapeutic responses to drugs and drug combinations (Van Dyke and Jacks 2002). Gene expression profiling of GEM models of human cancer can identify molecular aberrations associated with the development of tumors that have been initiated by specific oncogenic insults. When these molecular alterations in the GEM models are found to match similar processes in subtypes of human cancers, such models may be particularly appropriate for preclinical testing for that subtype of human cancer. Currently, much effort is devoted to carefully identifying GEM models that are relevant to human cancers and validating them for preclinical studies.

10.2 Gene Expression Profiling Using Microarray Technologies

10.2.1 Microarray Platforms

Although early forms of microarray technology were being developed in the mid-1980s, the advancement and application of this technology did not gain widespread attention until almost a decade later (Ekins and Chu 1999). Over the past 10 years, high-throughput microarray technologies have rapidly advanced and are widely being used in the laboratories around the world. A variety of DNA microarray platforms are commercially available for large-scale gene expression analysis and a complete review of these technologies is beyond the scope of this chapter (Auer et al. 2009; Bilitewski 2009; Heller 2002; Jaluria et al. 2007; Peeters and Van der Spek 2005). Initially, arrays generated by spotting cDNA or expressed sequence tags (ESTs) were commonly employed, but more recently, several array platforms using high-density oligonucleotides are being widely utilized and have improved the quality of these studies.

Oligonucleotide arrays contain a large number of gene-specific oligonucleotides of a predetermined sequence (generally, 25–70 mers), synthesized *in situ* or robotically spotted onto a glass or silicon chip. The high-density GeneChip probe arrays manufactured by Affymetrix (Affymetrix Inc.; http://www.Affymetrix.com) are widely used for gene expression studies. These arrays use 25-mer oligonucleotides that are chemically synthesized on a silicon surface using photolithographic technology. This technology provides the means to produce arrays that contain up to 320,000 features within a $1.28 \times 1.28\ cm^2$ area, with a high sensitivity of signal detection and good inter-array reproducibility (Granjeaud et al. 1999). In addition, oligonucleotide probes offer greater specificity than cDNAs or PCR products, having the capacity to distinguish single-nucleotide polymorphism (Guo et al. 1994) and discern splice variants.

Another commonly used oligonucleotide array platform with excellent sensitivity and specificity has been developed by Agilent (Agilent Inc.; http://www.Agilent.com) which uses an *in situ* ink-jet oligonucleotide synthesis method, thus creating a very flexible microarray platform for gene expression profiling (Hughes et al. 2001). Although short (20–25 base) oligonucleotides theoretically provide the greatest discrimination between related sequences, the Agilent ink-jet array platform uses 60-mer oligonucleotides that also produce signals with high sensitivity and specificity. More recently, arrays using Bead Array Technology by Illumina, Inc., have also become widely used.

Depending on the specific microarray platform applied, the gene expression output of microarrays can be measured using either a dual-channel (two-color) or a single-channel (one-color) system (Jaluria et al. 2007). The spotted cDNA microarrays usually make use of the two-color approach, in which two samples, each labeled with a different fluorophore, are hybridized onto a single slide. As a result, the readout is determined by the relative expression levels from the signal of the two colors expressed as a ratio. In contrast, the oligonucleotide microarrays utilize a single-color approach in which only one sample is hybridized onto a single slide. Thus, the readout generated is the absolute value of the gene expression signals or intensity based. As compared to the two-color system, the one-color system may simplify comparisons of data between arrays and between datasets.

10.2.2 Microarray Data Analysis and Cross-Species Comparison

The statistical analysis of microarray data requires a complex series of preprocessing and quality reviews prior to analyzing the data. Due to multiple processes involved in the production of microarrays, biological sample preparation, hybridization, and image acquisition, the raw microarray signals may acquire a range of random noise and systematic biases which can considerably alter the analysis and interpretation of mRNA transcript levels if not initially corrected. This section discusses how array data can be analyzed. The importance of working with a statistician with expertise in this area cannot be overemphasized, but some fundamental considerations are discussed below.

Data quality can be checked using various diagnostic plots and exploratory visualization tools of both the raw and processed data. Scanned images can be inspected for any serious spatial artifacts, feature intensities out of an accepted range, or grid alignment issues. Boxplots and histograms of intensity distributions extracted from image data are commonly used with the aim of identifying array outliers, determined by differences in the distribution spread and location (Fig. 10.1a). Platform-specific quality metrics and guidelines as to how to assess the validity of data are available. For example, the expected background level of signal intensity, the expression levels of hybridization controls or spiked-in labeling controls using bacterial genes, and the 3′ to 5′ expression ratio of GAPDH and β-actin may all be useful as indicators of RNA quality.

Scatterplots of a probe-wise log ratio against the average intensity (the so-called MA plots) can identify intensity-dependent expression differences between the red and green channels of a dual-channel array or between two single-channel arrays (Fig. 10.1b). These are very useful tools for assessing whether an array is of sufficient quality to be included in the analysis or what particular type of data normalization is required for the given data. If the mass of the distribution in an *MA* plot is not concentrated along the horizontal line but there is no trend in the mean of *M* (the intensities log ratio) as a function of *A* (the average intensity), then a global scaling is appropriate to normalize the expression between arrays or between the red and green channels. A trend identified in the *MA* plot indicates that a nonlinear, intensity-dependent normalization is needed to achieve gene expression comparability between different arrays. Background adjustment and normalization are the main steps in obtaining the final signal intensities from a microarray experiment.

Fig. 10.1 (continued) (**b**) *MA* plots for one dual-channel array (Agilent platform) with raw and normalized data in the *left and right images*, respectively. An average of *red* and *green* channel intensities for each probe is plotted on the *x*-axis and the *red/green* ratio on the *y*-axis. The *blue color* intensity highlights density of data points on the *MA* plot. In addition, a lowess regression line is plotted to the images with *red*. The intensity-dependent bias present in the raw data is removed by the nonlinear normalization. (**c**) Volcano plots of significance against change in expression of a set of 27,338 genes between two experimental conditions using regular student's *t*-test (*left*) and a modified random variance model *t*-test (*right*). On the *y*-axis, negative log10 of *p*-values from a gene-specific *t*-test are plotted and the log2 fold changes in expression on the *x*-axis. Genes with statistically significant differential expression according to a gene-specific *t*-test lie above a horizontal threshold line (the nominal significant level 0.001). Genes with large fold-change values lie outside a pair of vertical threshold lines (twofold change). Significant and downregulated/upregulated genes are colored *green/red*. The increase in statistical power of the RVM *t*-test is visible in shifting the data points higher along the significance dimension. The percentages indicate the fractions of the twofold changes among downregulated and upregulated genes identified as significant with each *t*-test

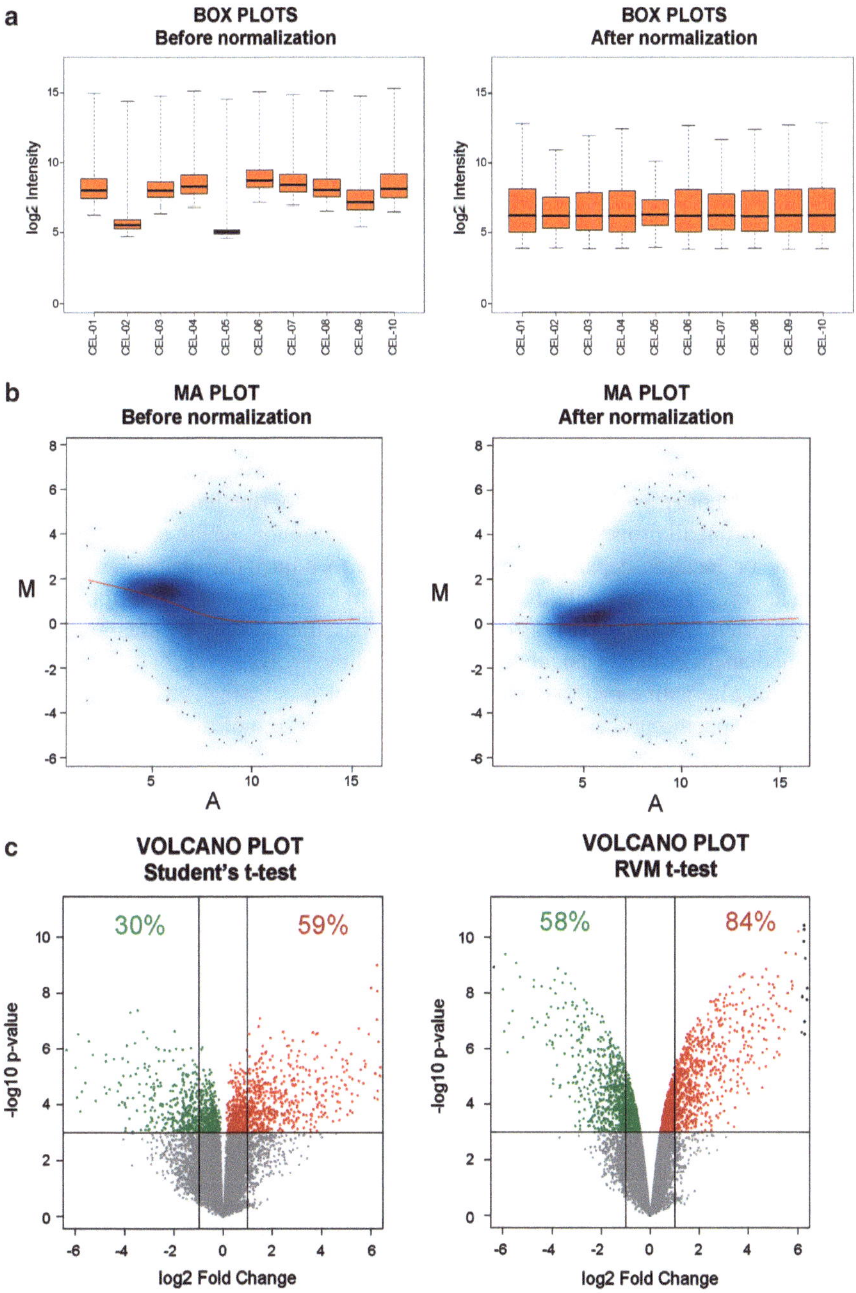

Fig. 10.1 (**a**) Boxplot representation of chip-wise PM log intensity distributions (Affymetrix data). *On the left*, raw data before normalization and *on the right*, the data after RMA normalization are shown. Chips 2 and 5 deviate strongly from the other chips. It is recommended to carefully inspect other quality diagnostics and their impact on further analysis steps for these two chips.

Background correction based on a local or global estimate of the background noise is applied to the observed probe intensities to remove the fluorescence resulting from nonspecific binding and optical noise.

Normalization is a procedure of data transformation aiming to compensate for systematic biases between arrays while retaining variation related to the biology being studied and, in so doing, to standardize array intensities so that the gene expression measurements are comparable across different arrays. Most often, a robust assumption is made that most of the genes on an array do not change in terms of their expression levels between experimental conditions and, therefore, all the genes on an array can be used for estimating normalization factors. The approach usually works well when a large number of randomly selected genes are included on the array. If the assumption is not applicable, a subset of housekeeping genes or exogenous control genes or an invariant set of genes identified within the experiments (using a mathematical algorithm) can be used for normalization.

Commonly used normalization methods for dual-color (dual-channel) arrays are global scaling of array channel ratios using a single factor (median or trimmed mean computed from all probes on an array) and local regression, which derives multiple intensity or location-dependent scaling factors to adjust the channel intensities within an array. Similar methods can be applied to normalization of multiple single-channel arrays when one array is chosen as the baseline array or by creating a "virtual" median array that is calculated across all of the arrays.

Various alternative methods have been used for processing Affymetrix arrays. Due to the specific probe-level and probe set-level design of the Affymetrix platform, the preprocessing of these arrays includes an additional step of summarization of the signal from multiple probes into the final gene-level signal intensity. Many algorithms and robust methods have been introduced for Affymetrix GeneChips in an attempt to correct and diminish the effects of the systematic probe-level signal inconsistencies that may arise from several factors (Gharaibeh et al. 2007; Naef and Magnasco 2003). In the Affymetrix default MAS5 algorithm (Hubbell et al. 2002), probe set-level gene expression is summarized based on local background-corrected probe intensity data using PM and MM probe pair differences and One-step Tukey's Biweight algorithm. The Tukey's algorithm is a method to determine a robust average unaffected by outliers (probe intensities more distant from the median intensity in a probe set contribute less to the average estimate). After the summarization, between-array normalization is carried out for the probe set-level intensities. Numerous model-based methods aim to model chip effects and probe effects. Chip effects are the signals of interest and capture the overall signal of a set of probes. Probe effects/affinities capture the sensitivity (responsiveness) of probes and are assumed to be the same across arrays. The model-based methods require multiple arrays to be used in the calculation of the summarized signal. The most often used model-based method is the Robust Multichip Average (RMA) algorithm that incorporates only perfect-match signals (not mismatch probe signals) to process the probe-level data (Irizarry et al. 2003). RMA corrects probe-level background by fitting a model that is the sum of an exponentially distributed signal and a normally distributed background. The parameters of these two distributions are estimated using a heuristic method, and each PM probe value is adjusted to the expected value of the signal given the PM measurement. RMA, then, normalizes across the arrays in a series by applying quantile normalization at the probe-level and median polish for a robust summarization of the probe set-level intensities.

Common goals in microarray studies are detection of differential gene expression between experimental conditions (class comparisons), identification of new molecular types or patterns (class discovery), and building predictive models for classifying future samples based on gene expression profile (class prediction). Parametric tests for detection of differential gene expression in individual genes are widely used (Student's *t*-test, ANOVA models), sometimes with modifications of variance estimates. High-throughput studies are often carried out with a relatively small number of replicates and, therefore, variance estimates might be inaccurate and the power of statistical tests may be low. Microarray-specific approaches have been developed that include information from many other genes present on the array to improve variance estimates and increase the statistical power of individual tests (Fig. 10.1c). For example, the local-pooled-error (LPE) method estimation is based on pooling within-class variances for genes having similar intensity levels (Jain et al. 2003). The random variance model (RVM) statistic combines gene-specific variance with a common variance estimated from the distribution of the variances across all genes (it assumes that different genes have different variances, but the variances come from the same random distribution, whose parameters can be estimated from the data (Wright and Simon 2003)). In a popular algorithm known as Significance Analysis of Microarrays (SAM), a small constant value ("fudge factor") is determined as the percentile of the gene-wise standard errors that minimizes the coefficient of variation of the modified test statistics (Tusher et al. 2001). The SAM modification is used to overcome bias for genes with small expression differences due to small sample variance at lower expression values.

A crucial issue in microarray analyses is the issue of multiple comparisons and finding a balance between keeping the number of false discoveries low and retaining good sensitivity in detecting truly differentially expressed genes. When differences in gene expression are tested simultaneously for thousands of genes assuming a standard significance level of 0.05 for each performed test, a large number of false discoveries can accumulate (for example, 1,000 false discoveries are expected when 20,000 hypotheses are tested). The traditional procedures for controlling the family-wise error rate (FWER), such as a Bonferroni correction, are too conservative for microarray studies. The Bonferroni correction is defined as the probability of accumulating at least one false positive and for 20,000 hypotheses tested, the required significance level for rejecting a single null hypothesis would be 2.5e-06!

Benjamini and Hochberg (1995) proposed a new measure of the error rate – the false discovery rate (FDR) – which is the expected proportion of false positives among all rejected null hypotheses. This has become one of the most widely used procedures of controlling the number of false discoveries in microarray studies. For example, if 400 genes were selected for the FDR level of 0.05, then on average a 5% rate of false discoveries would be expected among the selected genes. Storey and Tibshirani (2003) extended the FDR concept to a *q*-value estimate, which is the minimal FDR with which a gene can be selected (FDR-corrected *p*-value).

Clustering and heat map visualization are very useful exploratory tools in microarray-based studies when the goal is to find patterns within the data without having prior knowledge about the data (referred to as class discovery or unsupervised learning). An example of class discovery would be the identification of groups of genes that have similar patterns of expression within groups of samples. Biologically, this might result from these genes having similar transcriptional mechanisms or they may be part of a common biological process or pathway. Class discovery may lead to the identification of unanticipated molecular subtypes among previously unclassified samples. Such classifications may lead to new classifications of tumor types with clinical relevance for prognosis or treatment response.

Clustering of array data is a process of grouping samples based upon the expression values of their array features. Sample subsets are related based upon their similarities based upon distance matrices (using Pearson's correlation coefficient or Euclidean distance, for example). Hierarchical clustering calculates all pairwise distances between genes or experimental conditions to merge the most similar pairs of values at each step to generate a tree structure (dendrogram), which displays the relationships among data elements. Hierarchical clustering of both genes and experimental conditions can be visualized as a heat map to enhance the interpretation of the clustering results with the image display of gene expression profiles (Fig. 10.2). Other commonly used clustering algorithms in microarray data analyses are *k*-means clustering and self-organizing maps (Eisen et al. 1998; Tamayo et al. 1999).

Molecular classifiers (class predictors) have been generated in numerous microarray studies with the hope of improving clinical diagnosis, prognosis, and personalizing medical treatment. The main components in developing a multivariate molecular classifier are: (1) the selection of the potentially most relevant genes for differentiating the disease classes; (2) constructing a mathematical model that summarizes the gene signature and assign expression profiles to the given classes (classifier training); and (3) evaluation of the performance of a classifier using an independent test set (split sample validation or cross-validation). Many class prediction algorithms have been applied to microarray data, such as linear predictors (diagonal linear discriminant analysis and compound covariate), support vector machines, artificial neural networks, and random forest or centroid-based methods. Great care is required in the construction and validation of class predictors to avoid overfitting the huge amount of array data from a relatively small number of samples (Simon 2003; Radmacher et al. 2002). Careful validation of a predictor's accuracy is crucial for the successful development of microarray-based diagnostic tools and therapeutic strategies in medicine (Simon 2010; Dupuy and Simon 2007).

GEM models of human cancer provide a unique opportunity to investigate cancer biology, genetics, and therapies, particularly when coupled with high-throughput genomic studies. Even though cross-species hybridizations to the same DNA microarray have been attempted, they are considered a nonstandard approach and are limited to gene expression profiling of species with less well-characterized genomes for which species-specific platforms are not available. The sensitivity and specificity of cross-species hybridizations are largely dependent on the degree of sequence divergence between the respective species, and biologically meaningful microarray

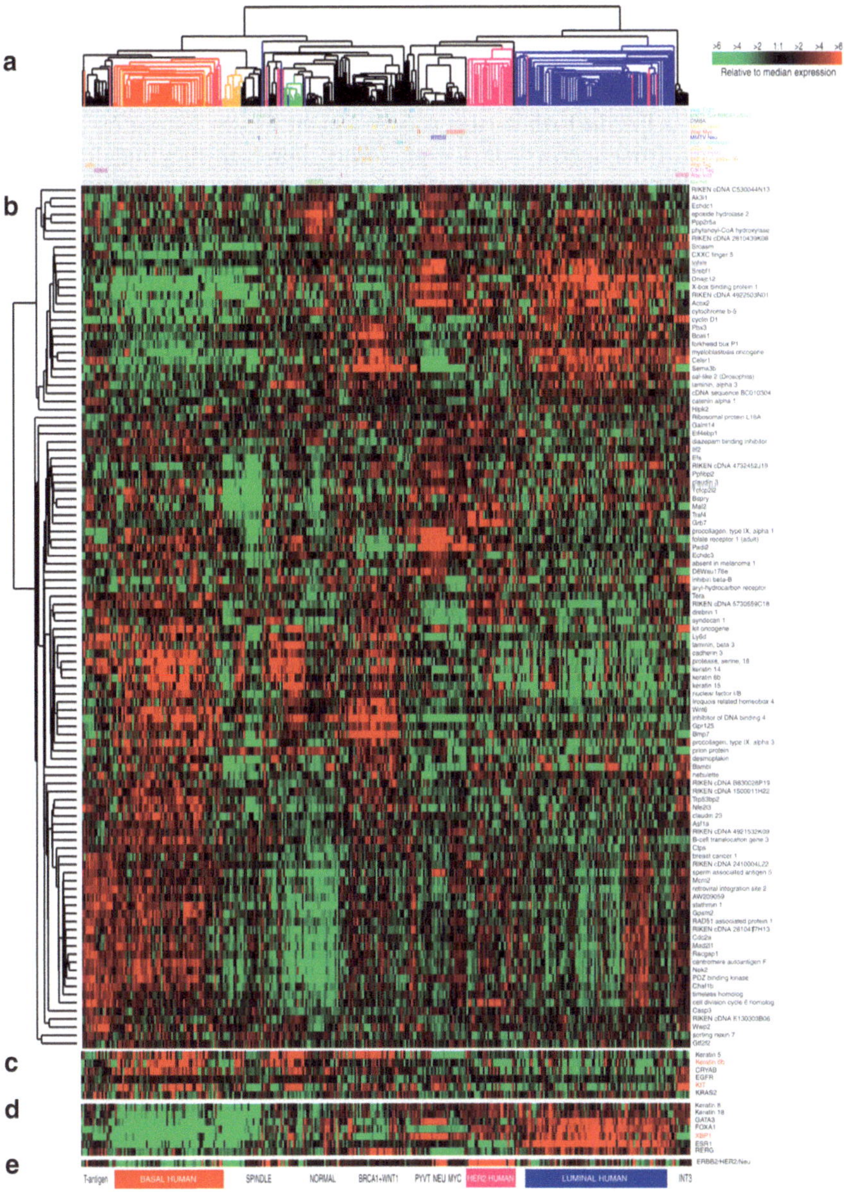

Fig. 10.2 Cluster analysis of mouse and human tumors using the subset of genes common to both species intrinsic lists (106 total genes). (**a**) Experimental sample associated dendrogram color coded according to human tumor subtype and with a matrix below showing murine tumor origins. (**b**) The complete 106 gene cluster diagram. (**c**) Close-up of genes known to be important for human basal-like tumors. (**d**) Close-up of genes known to be important for human luminal tumors, including ER. (**e**) Expression pattern of HER2/ERBB2/NEU. (Figure reproduced from Herschkowitz et al. Genome Biology, (2007) with permission from Biomed Central)

signals are achievable only for a fraction of highly homologous genes (Buckley 2007; Bar-Or et al. 2007). For the traditional genetic model organisms, microarray data are usually collected from species-specific platforms; however, they require meta-analytic approaches to perform comparisons of gene expression between species.

Identifying gene homology between the species and how this is represented on the arrays for each species is a critical first step. Orthologs are commonly of interest as they are genes derived from a single ancestral gene in the last common ancestor of the compared species and are likely to perform equivalent functions across the species. A number of databases provide information and tools enabling automated identification of orthologous genes comparable between microarray platforms, for example HomoloGene, Ensembl, COGs, MGI, and NetAffx.

It has been shown that commonly used normalization procedures, while efficient in removing technical biases within a set of experiments, are not sufficient to remove artifacts related to the use of different microarray platforms or array/experiment batches, even within the same species. Therefore, additional adjustments are necessary to ensure cross-species data comparability when direct comparisons of gene expression profiles are to be carried out. Simple methods of gene-wise adjustments can be applied to remove overall differences between platform-specific estimates of intensity levels or ratios while preserving relative differences between samples in each species (median or mean centering which sets the average expression of a gene in each species to zero or z-score transformation which in addition normalizes gene standard deviation to unity in each species). More complex approaches have been developed for removing batch effects in microarray experiments based on statistical discrimination methods, like singular value decomposition or distance weighted discrimination, applicable to integration of cross-species data (Luo et al. 2010; Benito et al. 2004). Finally, cross-species analyses of gene expression can also be performed separately in each species and derived gene signatures tested for common relevance of the model and human expression patterns, without direct integration of microarray datasets.

10.3 Expression Profiling of Mouse Tumor Models and Comparison to Human Cancers

Significant success has been achieved in using gene expression profiling to identify genetic signatures and networks in mouse models that have relevance to human cancers. Such studies provide important insights into what molecular aberrations result in tumors from specific initiating oncogenic perturbations, identify potentially useful markers of cancer progression and potential drug targets, and provide insights into lineage specification of tumor subtypes. Data from mouse models also serve another critical function – to help serve as a filter to sift through variations in human gene expression data arising from heterogeneous populations. Identification of conserved gene expression and genetic networks in GEM tumors appropriately compared to specific subtypes of human cancers has provided a means to focus on relevant syngeneic relationships related to cancer biology.

10.3.1 Mammary Cancer

Breast cancer is the most frequently diagnosed cancer in women and is the second-leading cause of cancer death in the female population of the USA (American Cancer Society 2007). The phenotypic diversity of human breast tumors suggests a corresponding diversity in gene expression patterns. Global gene expression analyses of human breast cancers have uncovered significant molecular differences between primary tumors and identified at least five major tumor subtypes, thereby providing a distinctive molecular portrait of breast cancer (Perou et al. 2000). Two subtypes, with features of mammary epithelial basal cells and designated basal-type breast cancers, are estrogen receptor (ER) negative with generally poor patient outcomes; one of these two subtypes is defined by the high expression of HER2/ERBB2/NEU ($HER2^+/ER^-$), whereas the other major subtype is ER^-, PR^-, and $HER2^-$. Other major categories of breast cancer based upon their molecular expression patterns share features of luminal mammary epithelial cells and are designated within the "luminal" subtype. This group of tumors has been subdivided into three subtypes with the tumors expressing the highest levels of ER and having a favorable prognosis designated "luminal A" tumors, where "luminal B" tumors have lower ER expression and "luminal C" tumors express little if any ER and have a generally poor prognosis (Sorlie et al. 2001; Sorlie et al. 2003). These analyses suggest that human breast cancers are quite heterogeneous and may arise from distinct types of tumor progenitor cells. Since each subtype harbors distinct molecular signatures, their responses to therapies are also different from one another. To gain a more fundamental understanding of the heterogeneous nature of breast cancer and develop improved therapeutic modalities, relevant experimental animal models are needed to model specific subtypes of breast cancer.

Significant progress in the ability to genetically engineer mice has led to the generation of models that recapitulate many properties of human breast cancers. Relevant oncogenic and tumor-suppressive networks have been manipulated to develop GEM models with molecular mimicry to human mammary carcinoma. Additionally, in contrast to the great genetic variability found in human populations, GEM models provide a system to study the effects of specific oncogenes, tumor-suppressor genes, and signaling networks in a well-defined genetic background, thus offering particular advantages for studying the mechanisms of tumorigenesis initiated by a specific gene or combination of genes.

GEM models for mammary carcinoma can broadly, though simplistically, be classified into two groups. One group is generated using standard transgenic technology to overexpress a variety of oncogenes, such as Myc, Ras, Her2/Neu, Cyclin D1, Notch, and Wnt under the regulatory control of specific promoters, which differ in their temporal expression and cellular compartment where the oncogene is expressed. The commonly used promoters include the long terminal repeat of the mouse mammary tumor virus (MMTV-LTR) and promoters from milk protein genes (WAP, β-lactoglobulin, β-casein). All these control elements target the mammary epithelium, and they are stimulated by lactogenic hormones. The second group of GEM models is constructed using knockout or knock-in

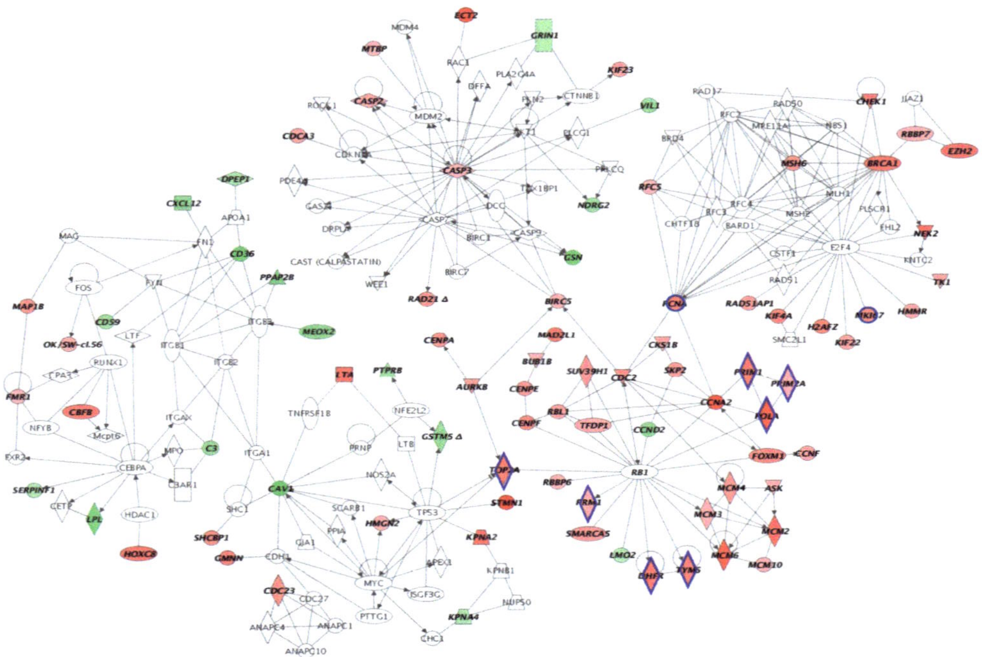

Fig. 10.3 An intrinsic biological network associated with expression of the SV40 T/t-antigen oncoproteins in the GEM models. Of the 120 known genes in the SV40 T/t-antigen gene cluster, 85 of the genes formed biological networks that could be related to the tumor-suppressor genes p53 and pRB nodes. Upregulated (*red*) and downregulated (*green*) genes are shown. Genes highlighted in *blue* are proliferation markers, Ki67 and PCNA, and those that are highlighted in purple are potential chemotherapeutic targets. Reproduced from Deeb et al., *Cancer Research*, 2007, with permission from AACR

technology, in which deletions or mutations of tumor suppressors, such as Brca1, p53, and Rb, have been introduced into the mouse germ line (see Chapter 2; Cardiff et al. 2000). Additionally, numerous models with compound alterations in their genomes have been generated by crossing various GEM models together. The ability to target gene expression in a temporal and tissue-specific manner has been particularly useful to study tumor dependency on oncogene expression (see Chapter 25) (Christophorou et al. 2005; D'Cruz et al. 2001; Ewald et al. 1996; Gunther et al. 2002; Hennighausen et al. 1995; Moody et al. 2002; Ventura et al. 2007; Xue et al. 2007). Such studies have demonstrated that under certain circumstances, tumor lesions completely regress if the oncogenic stimulus is removed.

The application of global gene expression profiling techniques to various GEM models of mammary carcinoma has led to the discovery of both common and gene-specific events associated with tumor initiation and progression. Our laboratory examined the gene expression profiles of various GEM models of mammary tumors initiated by six specific oncogenes: MMTV-c-myc, MMTV-neu, MMTV-Ha-ras, MMTV-Polyoma middle T antigen (PyMT), C3(1)/simian virus 40 (SV40) T/t antigen (T-ag), and WAP-SV40/T-ag (Desai et al. 2002). The analysis of over 8,600 unique genes demonstrated that despite different initiating oncogenic signals, the tumors shared great similarities in their gene expression profiles across all of the models. Selection and hierarchical clustering of the most variant genes, however, resulted in separating the mouse models into three distinct groups: (1) T/t antigen-driven tumors, (2) Myc tumors, and (3) Neu, Ras, and PyMT tumors. The unique gene sets within each model cluster offered new insights into potential mechanisms of oncogenic initiation and tumor progression.

In another approach, our laboratory has defined an expression signature specific for the SV40 viral oncoproteins T- and t-antigens (Deeb et al. 2007). This was determined by identifying genes similarly dysregulated in three different types of T/t-antigen-driven epithelial tumors (mammary, lung, and prostate). Although not involved in the etiology of human breast cancer, the expression of T-antigen functions as a targeted knockout of the tumor-suppressor genes p53 and pRb and is therefore highly relevant mechanistically to human tumors with loss of these suppressor gene functions. This *functional* genetic signature, composed primarily of genes regulating cell replication, proliferation, DNA repair, and apoptosis (Fig. 10.3), is coordinately expressed in a subset of human breast, lung, and prostate cancers that appear to be highly aggressive with poor prognosis (Deeb et al. 2007). GEM tumors in which p53, Rb, or BRCA1 are mutated tend to exhibit expression of this signature, but it is not as well-represented in tumors initiated through the overexpression of myc, ras, her2/neu, or polyoma middle T oncogenes using the MMTV LTR. In addition, this genetic signature identifies potential targets for novel therapies that could be directed against these often lethal forms of cancer. Because these genetic targets have been discovered using GEM models for mammary, prostate, and lung cancers, these models are rationale candidates for use in preclinical testing of therapies focused on these biologically important targets.

With such a wide array of GEM models available to study breast cancer, it becomes imperative to determine to what extent any particular model represents human mammary carcinoma. In 1999, a panel of pathologists at the Annapolis conference reviewed 33 GEM models and assessed their morphologic similarities to human breast cancers using traditional histopathological criteria (see Chapter 7) (Cardiff et al. 2000). The utilization of gene expression profiling, however, provides a more global, quantitative, and molecular means of evaluating GEM models of human breast cancer.

Our laboratory has demonstrated that several commonly studied mouse models can be categorized into three groups according to the basal-luminal distinctions identified for human breast cancer: (1) tumors with a basal phenotype (C3(1)/Tag, $p53^{-/-}$, and $BRCA1^{-/-}$; $p53^{+/-}$); (2) ER^- tumors with luminal characteristics (MMTV-myc, -ras, -PyMT, and -her2/neu); and (3) ER^+ tumors with luminal characteristics (generated in a p53 mammary epithelial conditional knockout model). A recent study by Herschkowitz et al. (2007) profiled 13 different GEM models of mammary carcinoma using Agilent microarrays and further performed an unsupervised clustering analysis using the combined mouse–human datasets. Consistent with our data, it was found that GEM models utilizing promoters, such as MMTV and WAP, develop more homogenous tumors that cluster separately from basal tumors but show significant differences to human luminal tumors. GEM models that inhibit cell cycle regulators, such as p53, Rb, and BRCA1, develop tumors that tend to cluster with human basal tumors (Fig. 10.2). Interestingly, the MMTV-Wnt1 GEM model developed a proportion of tumors with both luminal and basal expression patterns. This finding, coupled with previous reports implicating Wnt signaling in stem cell renewal (Lindvall et al. 2006; Reya et al. 2003), suggests that the MMTV-Wnt1 mice may serve as a model to explore the role of mammary progenitor cells in tumor growth.

A major clinical distinction in breast cancer classification and treatment is dependent upon the expression of estrogen receptor. Although several studies have been published that report gene signatures that distinguish these two types of human breast cancers, we have found that the addition of array data from ER^+ and ER^- mouse mammary tumors provides additional power to the development of gene signatures that distinguish these tumor types (Deeb et al. 2007; Shoushtari et al. 2006). Presumably, the genes represented in the 328-gene ER classifier gene set represent biologically significant genes for ER signaling, but also genes that may distinguish the type of progenitor cells from which the tumors arise. Interestingly, the cross-species ER gene set contains more genes than that identified by only analyzing a human array dataset alone, and appears to be somewhat more accurate in classifying human tumors as ER^+ or ER^-. Subsequent pathway analysis of the combined human–mouse ER classifier also identified a vast number of novel interactions that may be critical for distinguishing the luminal from basal lineage of mammary tumors and the evolution and maintenance of ER signaling in tumors. Thus, a cross-species approach to gene expression profiling may significantly improve the identification of biologically relevant genes compared to analyses that utilize data from only one species, especially if the dataset is relatively small.

In addition to the analyses of breast tumors, gene expression technology has also been widely utilized in studies of other tumor types, including human prostate, lung, liver carcinomas, and their respective murine counterparts, which are briefly discussed in the following sections. The gene signatures among and across different cancer models have provided both biological and clinical insights for cancer prognosis, diagnosis, and therapy.

10.3.2 Prostate Cancer

Prostate cancer (PrCa) is a highly prevalent disease and a leading cause of cancer deaths in men in the Western world (Greenlee et al. 2001). The first transgenic mouse model of prostate cancer developed in our lab expressed the SV40 T/t-antigens under the regulation of the rat C3(1) prostatein promoter (C3(1)/Tag) in which prostate intraepithelial neoplasia (PIN) lesions form at about 2 months of age and the emergence of microscopic invasive carcinomas is observed after about 7 months of age (Maroulakou et al. 1994; Shibata et al. 1996). Cell lines derived from low-grade PIN, high-grade PIN, invasive carcinoma, and a lung metastasis of C3(1)/Tag mice exhibited increasing rates of cell growth, tumorigenicity, invasiveness, and angiogenesis. cDNA microarray analysis of 8,700 features revealed correlations between the tumorigenicity of the C3(1)/Tag-Pr cells and changes in the expression levels of genes regulating cell growth, angiogenesis, and invasion. The gene expression profiling has also identified novel genes that may be involved in mechanisms of prostate oncogenesis or serve as potential biomarkers or therapeutic targets for prostate cancer. Examples include the L1 cell adhesion molecule, metastasis-associated gene (MTA-2), Rab-25, and tumor-associated signal transducer-2 (Trop-2). Moreover, many genes identified in the Pr cell line model have been shown to be altered in human prostate cancer, as well as other types of human cancers. In particular, selenoprotein-P expression was found to be inversely related to tumorigenicity, suggesting a potential mechanism for the role of selenium in prostate cancer prevention. Based upon the findings in the GEM model, we demonstrated that selenoprotein-P expression is significantly reduced in a subset of human PrCa (Calvo et al. 2002).

A highly robust model for aggressive PrCa using the probasin promoter driving the expression of the SV40 T/t-angitens (transgenic adenocarcinoma mouse prostate (TRAMP)) has been the most widely used model for PrCa studies (Gingrich et al. 1996) and progresses through all stages of prostate lesion development with primarily lymph node and lung metastases. This model develops neuroendocrine prostate tumors and phyllodes seminal vesicle tumors that need to be distinguished from adenocarcinomas of the prostate. Interestingly, gene expression profiling analysis detected increased levels of the polycomb gene Bmi-1 mRNA expression in samples derived from late-stage invasive primary tumors and multiple distant metastatic lesions of the TRAMP mice (Glinsky et al. 2005) which correlated to increased levels of Bmi-1 expression found in human prostate cancer cell lines (Glinsky et al. 2003). This suggested that a Bmi-1 oncogene-driven signature, previously shown to be associated with activation of the self-renewal program of stem cells, might promote tumor progression. Furthermore, by applying a cross-species mouse/human comparative genomics analysis, a 11-gene Bmi-1 signature was identified in distant metastatic lesions. Moreover, it was demonstrated that this 11-gene signature in primary tumors is a powerful predictor of a short interval to disease recurrence, distant metastasis, and death after therapy in cancer patients diagnosed with 11 distinct types of cancer, including 5 epithelial malignancies (prostate, breast, lung, ovarian, and bladder cancers) and 5 nonepithelial malignancies (lymphoma, mesothelioma, medulloblastoma, glioma, and acute myeloid leukemia) (Glinsky et al. 2005) (Fig. 10.4).

Ellwood-Yen et al. generated a probasin-Myc GEM model that has provided important insights into the role of Myc in prostatic tumor initiation (Ellwood-Yen et al. 2003). Gene expression profiling before and after the well-defined transition from normal prostate gland to PIN identified a Myc PrCa expression signature, including loss of NKX3.1, which is a putative tumor-suppressor gene in human PrCa (He et al. 1997). After integrating the mouse signature with human prostate tumor databases, it was found that both murine and human tumors with high Myc activity had concurrent upregulation of Pim-1, a serine–threonine kinase previously implicated in Myc-initiated lymphomagenesis (van Lohuizen et al. 1991). More recently, elegant prostate cancer models have been developed through targated deletions of PTEN and NKX3.1 (Shen and Abate - Shen 2010).

10.3.3 Lung Cancer

Lung cancer is the leading cause of cancer death worldwide, accounting for more solid tumor deaths than breast, pancreatic, prostate, and colorectal combined (Landis et al. 1999).

Various GEM models have been developed to represent different types of human lung cancer through targeted overexpression of various oncogenes or deletion of tumor-suppressor genes in mice. The generation of mouse strains, where oncogenes can be conditionally overexpressed or tumor-suppressor genes can be conditionally mutated, enables the somatic induction of these genetic alterations in a locotemporal fashion. Models have been developed, where oncogenesis is induced in a limited number of target cells, thereby more closely mimicking the sporadic tumorigenic process of human lung cancer (Meuwissen and Berns 2005; Wakamatsu et al. 2007).

An elegant GEM model of lung adenocarcinoma (KrasLA) was created in which a latent, mutated Kras2 allele is sporadically activated (Johnson et al. 2001). Since a mutation in KRAS is found in approximately 30% of non-small-cell lung cancer (NSCLC) (Sweet-Cordero et al. 2006), the KrasLA mice provide an excellent model to study KRAS2-mediated transformation in the lung cancer. Affymetrix gene expression profiling of the KrasLA model of lung adenocarcinoma demonstrated a significant distinction between mouse lung tumors and normal lung. Furthermore, gene set enrichment analysis (GSEA) of Kras2-mediated mouse lung cancer identified similar enrichments of both up- and downregulated gene sets in human lung adenocarcinoma, thereby supporting the utility of the mouse model. In addition, by using this approach to integrate both mouse and human data, a gene expression signature of KRAS2 mutation in human lung cancer was uncovered which was not identifiable when analyzing human tumors with known KRAS2 mutation status alone. This lends further credence to the idea that these genes have a role in controlling aspects of KRAS2-mediated transformation. The KrasLA mouse model, together with the identification of KRAS2 activation signature in human tumors, should help in the development of anti-Ras pathway therapeutic strategies (Sweet-Cordero et al. 2005).

To better understand the mechanisms of NSCLC metastasis, gene expression profiling was performed on tumors from a K-ras^{G12D} and p53^{R172H} mouse model of human lung adenocarcinoma that develops metastatic disease. A gene expression signature in distant metastases relative to matched lung tumors in this mouse model was identified in a subset of NSCLC patients who had a poorer prognosis than those who did not express this signature. Thus, the K-ras^{G12D}; p53^{R172H} mice proved useful for studying human lung adenocarcinoma and the propensity for metastasis (Gibbons et al. 2009).

Since the survival of patients with NSCLC decreases markedly with advanced stage at diagnosis, early detection is the major determinant of patient outcome. Gene expression analysis was performed in the MMTV-IGF-II transgenic model of lung cancer to identify markers of early lung tumors. Although the MMTV promoter directs transgene expression primarily to the mammary gland, these transgenic mice expressed IGF-II in the lung, and the tumors arising in these mice histologically resemble human pulmonary adenocarcinomas (Moorehead et al. 2003). Microarray analysis identified nine genes consistently elevated in these murine lung tumors. Protein overexpression of three of these genes – microsomal glutathione-s-transferase 1 (Mgst1), cathepsin H, and syndecan 1 – was confirmed in early stage, node-negative human lung adenocarcinomas and squamous cell carcinomas by immunohistochemistry. These proteins, originally identified through analysis of a GEM model, may potentially be markers for detecting patients with early-stage lung tumors (Linnerth et al. 2005).

10.3.4 Liver Cancer

Cross-species comparison of global expression patterns of orthologous genes in human and murine hepatocellular carcinomas (HCCs) determined that certain mouse HCC models closely reproduce specific subgroups of human HCCs (Lee et al. 2004, 2005). Global gene expression patterns of 68 HCCs from 7 different mouse models and 91 human HCCs from predefined subclasses were used to obtain direct comparison of the molecular features of mouse and human HCCs. Gene expression patterns in HCCs from Myc, E2f1, and Myc; E2f1 transgenic mice were most similar to those of the better survival group of human HCCs, whereas the expression patterns in HCCs from Myc; Tgfa transgenic mice and in diethylnitrosamine-induced mouse HCCs were most similar to those of the poorer survival group of human HCCs. Gene expression patterns in HCCs from Acox1$^{-/-}$ mice and in ciprofibrate-induced HCCs were least similar to those observed in human

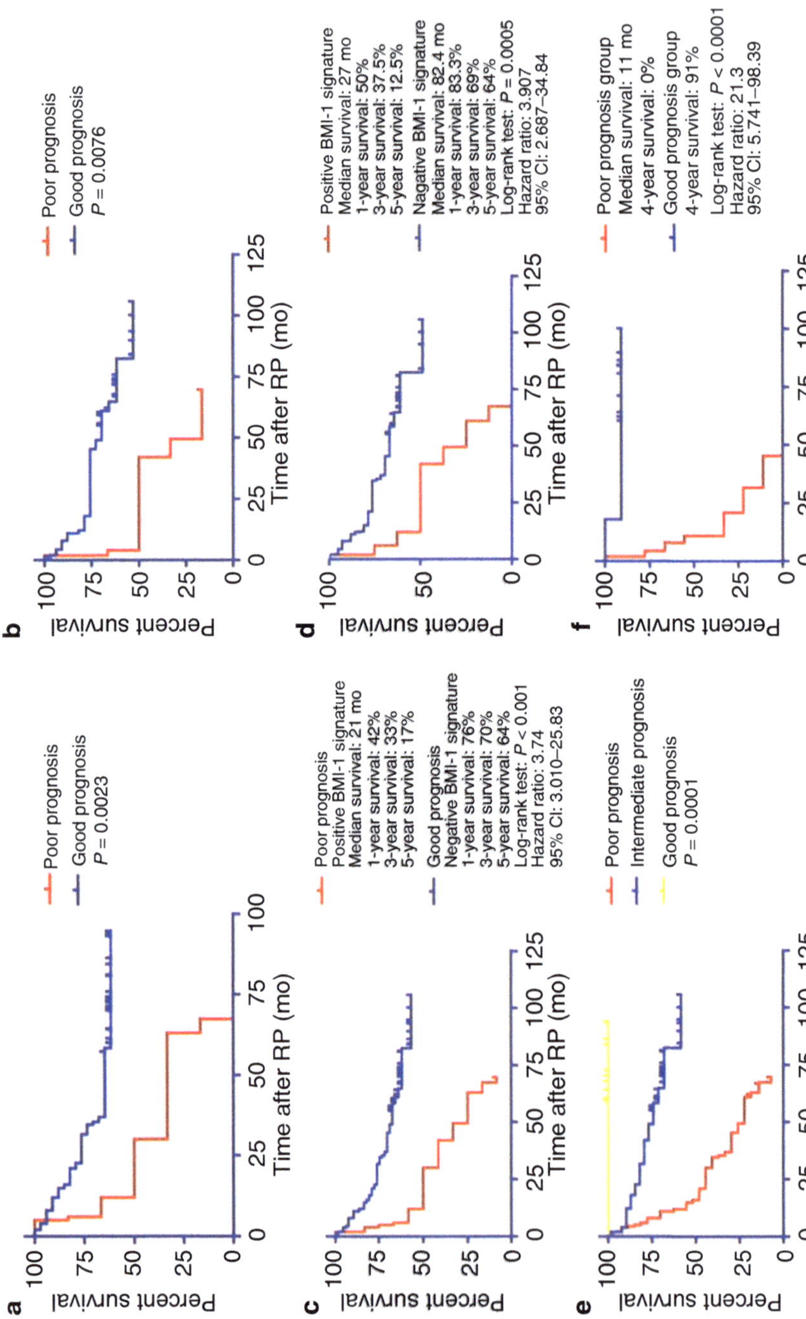
a
Percent survival
Time after RP (mo)
Poor prognosis
Good prognosis
P = 0.0023
b
Percent survival
Time after RP (mo)
Poor prognosis
Good prognosis
P = 0.0076
c
Percent survival
Time after RP (mo)
Poor prognosis
Positive BMI-1 signature
Median survival: 21 mo
1-year survival: 42%
3-year survival: 33%
5-year survival: 17%
Good prognosis
Negative BMI-1 signature
1-year survival: 76%
3-year survival: 70%
5-year survival: 64%
Log-rank test: P < 0.001
Hazard ratio: 3.74
95% CI: 3.010–25.83
d
Percent survival
Time after RP (mo)
Positive BMI-1 signature
Median survival: 27 mo
1-year survival: 50%
3-year survival: 37.5%
5-year survival: 12.5%
Nagative BMI-1 signature
Median survival: 82.4 mo
1-year survival: 83.3%
3-year survival: 69%
5-year survival: 64%
Log-rank test: P = 0.0005
Hazard ratio: 3.907
95% CI: 2.687–34.84
e
Percent survival
Time after RP (mo)
Poor prognosis
Intermediate prognosis
Good prognosis
P = 0.0001
f
Percent survival
Time after RP (mo)
Poor prognosis group
Median survival: 11 mo
4-year survival: 0%
Good prognosis group
4-year survival: 91%
Log-rank test: P < 0.0001
Hazard ratio: 21.3
95% CI: 5.741–98.39

Fig. 10.4 Classification of prostate cancer patients into subgroups with distinct therapy outcome based on expression profile of the 11-gene metastatic TRAMP tumor sample (MTTS)/peripheral nervous system (PNS) signature. (**a–c**) Kaplan–Meier analysis of the probability that patients would remain disease-free among 79 prostate cancer patients constituting clinical outcome set 2, according to whether they had a good-prognosis or a poor-prognosis signature as defined by the expression profiles of the 11-gene MTTS/PNS signature. The patients' stratification cutoff value of 0.4 was defined in the training set of 40 patients (19 poor prognosis and 21 good prognosis); (**a**) validated in a test set of 39 patients (18 poor prognosis and 21 good prognosis); (**b**) and confirmed in an entire cohort of 79 patients (**c**); (**d**) Kaplan–Meier survival curves for distinct subgroups of prostate cancer patients diagnosed with early-stage disease (stages 1c and 2a); (**e**) Kaplan–Meier survival curves for 79 prostate cancer patients stratified into distinct subgroups using a weighted survival predictor score algorithm; (**f**) Kaplan–Meier survival curves for 20 prostate cancer patients stratified into distinct subgroups using Q-RT-PCR assay of the 11-gene signature. Figure reproduced from Glinsky et al., *Journal of Clinical Investigation*, 2005, with permission

HCCs. Moreover, gene expression profiling of c-Myc, E2f1, and c-Myc; E2f1 transgenic mice identified oncogene-specific gene expression signatures at an early dysplastic stage of hepatocarcinogenesis, suggesting that E2f1, c-Myc, and their combination may promote liver tumor development by distinct mechanisms (Coulouarn et al. 2006).

Katzenellenbogen et al. investigated hepatocarcinogenesis in Mdr2-knockout mice, a model of inflammation-associated HCC. Gene expression profiling showed that although these mice differ from other published murine HCC models, they may serve as a model for the β-catenin-negative subgroup of human HCCs characterized by low nuclear cyclin D1 levels in tumor cells and by downregulation of multiple tumor-suppressor genes (Katzenellenbogen et al. 2007). Moreover, gene expression profiling with HCCs derived from $Mdr2^{-/-}$ and IκB-SR $Mdr2^{-/-}$ mice identified S100A8 and S100A9 as novel nuclear factor kappa B target genes during malignant progression of murine and human liver carcinogenesis (Nemeth et al. 2009).

10.3.5 Pancreatic Cancer

Pancreatic cancer (PCa) is one of the most aggressive human solid tumors, with rapid growth and metastatic spread as well as resistance to chemotherapeutic drugs, leading rapidly to virtually incurable disease (Koliopanos et al. 2008). The survival rate of PCa patients is among the lowest in all kinds of cancer patients with a 5-year survival rate of <5% (Faivre et al. 1998). There are several subtypes of PCa: (1) pancreatic adenocarcinoma, which is the most common PCa; (2) pancreatic endocrine tumors, including insulinomas; and (3) other tumors of the exocrine pancreas, including acinar and serous cystic tumors (Koliopanos et al. 2008).

Gene expression profiling of PCa has revealed the molecular changes associated with tumor progression from intraductal proliferation to invasive ductal carcinoma. The accumulation of genetic alterations in cancer-associated genes, such as K-ras activation and HER-2/neu expression (early events), p16 inactivation (intermediate event) and p53, DPC4, and BRCA2 inactivation (late events), correlates with increasing degrees of cytological and architectural atypia in the carcinogenesis process (Koliopanos et al. 2008).

Several GEM models have been created to study the genetic lesions implicated in human PCa. A transgenic mouse model with activated Kras and Ink4a/Arf deficiency displayed phenotypes of metastatic pancreatic ductal adenocarcinoma (Aguirre et al. 2003). An SV40 Tag transgenic mouse model was created with classic features of pancreatic cystic neoplasms owing to the expression of SV40 Tag (Sun et al. 2006). Feng et al. performed gene expression profiling in these mice using high-density oligonucleotide microarray and identified a tumor-associated gene signature involving the IGF, Shh, and Wnt signal pathway (Feng et al. 2007). Fontaniere et al. performed gene expression profiling in insulinomas developed in β-cell-specific multiple endocrine neoplasia type 1 (*Men1*) mutant mice at early and late stages. It was shown that in late-stage insulinomas, 56 genes are upregulated and 194 are downregulated more than fourfold compared with normal pancreatic islets. Clustering analysis reveals the deregulation of *Hox* gene family and the genes involved in cell proliferation and cell cycle control. Thus, their data revealed both early genetic and epigenetic events involved in pancreatic β-cell tumorigenesis related to *MEN*1 inactivation (Fontaniere et al. 2006).

10.3.6 Colorectal Cancer

Colorectal cancer (CRC) is the third most common cancer in men and women and accounts for 11% of all cancer deaths (Reichling et al. 2005). Several mutant and GEM models of intestinal neoplasia have been developed to study human CRC, which includes mutations in the adenomatous polyposis coli (APC) gene and Wnt signaling pathway, alterations in the mismatch repair genes, alterations in the TGF-β, KRAS, and toll-like receptor signaling pathway, immunodeficient mice with colitis, and carcinogen-treated mice (Boivin et al. 2003; Nandan and Yang 2010).

Approximately 85% of human colorectal tumors carry biallelic-inactivating mutations of APC (Oving and Clevers 2002), a tumor suppressor regulating the Wnt signaling pathway in normal intestinal epithelium (Reichling et al. 2005). A mutant Apc mouse model was developed and designated as the $Apc^{min/+}$ model (*Min* stands for "multiple intestinal neoplasia"). The mutant *Min* mice are predisposed to spontaneous intestinal adenomas and carcinomas in an autosomal dominant fashion with full penetrance (Moser et al. 1990). Several studies have compared the gene expression profiles of tumors from *Apc*-mutant mouse models of gastrointestinal tumorigenesis to that of human colorectal adenomas and carcinomas (Reichling et al. 2005; Leclerc et al. 2004; Martinez et al. 2005; Paoni et al. 2003). Paoni et al. performed gene expression profiling in the $Apc^{Min/+}$ mouse to study the transition from normal intestinal epithelia to adenomas and carcinomas. It was found that tumors have altered transcript abundance for members of several pathways that influence cell growth and proliferation. Furthermore, comparison of gene expression between adenomas and carcinomas revealed nine differentially expressed transcripts. Since the changes in gene expression observed in this study are directly associated with a deficiency in APC, the data provide new insights into how loss of this important tumor suppressor translates into benign and malignant tumor growth (Paoni et al. 2003). Another study by Reichling et al. reported that the transcriptional profiles of intestinal tumors in $Apc^{Min/+}$ mouse are unique from those of the embryonic intestine and identified novel gene targets dysregulated in human colorectal tumors. By performing cDNA microarrays, 114 genes were identified with altered levels of expression in Apc^{Min} mouse adenomas from the duodenum, jejunum, and colon when compared with normal adult intestine. Among them, 24 of these genes were differentially expressed in adenomas but not in embryonic or postnatal whole intestine. Six of the genes, Igfbp5, Lcn2, Ly6d,

N4wbp4, S100c, and Sox4, were found differentially expressed in human CRC cell lines and adenocarcinomas (Reichling et al. 2005).

In order to dissect the mechanistic and etiologic differences between proximal and distal colonic cancer, Sodir et al. created $Apc^{Min/+}$Smad3$^{-/-}$ mice by crossing $Apc^{Min/+}$ and Smad3 mutant mice, thus providing an alternative model to $Apc^{Min/+}$ mice to study the familial adenomatous polyposis (FAP) disease and distal sporadic CRC. Transcriptional profiling revealed higher expression of several TGF-β activators in the normal distal mucosa than in proximal mucosa, suggesting a stronger reliance on TGF-β-mediated growth control in the distal than in the proximal colon (Sodir et al. 2006).

10.4 Summary and Future Challenges

High-throughput genomic analyses have revolutionized cancer biology. The application of these technologies to investigate mechanisms of oncogenesis in GEM models has provided important insights into the evolution of the cancer transcriptome based upon specific genetic aberrations introduced into the mouse germ line. The complexities of genetic interactions have also been elucidated through rational crosses of many GEM models to mimic similar abnormalities found in human cancers. Gene expression profiling has been an invaluable tool for deciphering in what ways GEM models recapitulate genomic changes that occur in human cancers. It has become clear that what was previously categorized morphologically as one type of cancer actually includes multiple molecular subtypes of that class of cancer. Therefore, one GEM model does not represent all subtypes of a given human cancer. Genomic approaches, however, allow us to more precisely identify which models share molecular similarities with human cancer counterparts, an essential first step in utilizing GEM models for preclinical testing. More recent advances in genomic technologies that identify patterns of miRNA expression and genome amplifications and deletions using array CGH continue to advance our understanding of how best to utilize GEM models for cancer research.

Acknowledgments This research was supported by the Intramural Research Program of the NIH, Center for Cancer Research, National Cancer Institute.

References

Aguirre AJ et al (2003) Activated Kras and Ink4a/Arf deficiency cooperate to produce metastatic pancreatic ductal adenocarcinoma. Genes Dev 17(24):3112–3126

Auer H, Newsom DL, Kornacker K (2009) Expression profiling using affymetrix GeneChip microarrays. Methods Mol Biol 509:35–46

Bar-Or C, Czosnek H, Koltai H (2007) Cross-species microarray hybridizations: a developing tool for studying species diversity. Trends Genet 23(4):200–207
Benito M et al (2004) Adjustment of systematic microarray data biases. Bioinformatics 20(1):105–114
Benjamini Y, Hochberg Y (1995) Controlling the false discovery rate: a practical and powerful approach to multiple testing. J R Statist Soc B 57(1):289–300
Bilitewski U (2009) DNA microarrays: an introduction to the technology. Methods Mol Biol 509:1–14
Boivin GP et al (2003) Pathology of mouse models of intestinal cancer: consensus report and recommendations. Gastroenterology 124(3):762–777
Buckley BA (2007) Comparative environmental genomics in non-model species: using heterologous hybridization to DNA-based microarrays. J Exp Biol 210(Pt 9):1602–1606
Calvo A et al (2002) Alterations in gene expression profiles during prostate cancer progression: functional correlations to tumorigenicity and down-regulation of selenoprotein-P in mouse and human tumors. Cancer Res 62(18):5325–5335
American Cancer Society (2007) Cancer facts and figures. American Cancer Society, Atlanta
Cardiff RD et al (2000) The mammary pathology of genetically engineered mice: the consensus report and recommendations from the Annapolis meeting. Oncogene 19(8):968–988
Christophorou MA et al (2005) Temporal dissection of p53 function in vitro and in vivo. Nat Genet 37(7):718–726
Chu S et al (1998) The transcriptional program of sporulation in budding yeast. Science 282(5389):699–705
Coulouarn C et al (2006) Oncogene-specific gene expression signatures at preneoplastic stage in mice define distinct mechanisms of hepatocarcinogenesis. Hepatology 44(4):1003–1011
D'Cruz CM et al (2001) c-MYC induces mammary tumorigenesis by means of a preferred pathway involving spontaneous Kras2 mutations. Nat Med 7(2):235–239
Deeb KK et al (2007) Identification of an integrated SV40 T/t-antigen cancer signature in aggressive human breast, prostate, and lung carcinomas with poor prognosis. Cancer Res 67(17):8065–8080
Desai KV et al (2002) Initiating oncogenic event determines gene-expression patterns of human breast cancer models. Proc Natl Acad Sci USA 99(10):6967–6972
Dupuy A, Simon RM (2007) Critical review of published microarray studies for cancer outcome and guidelines on statistical analysis and reporting. J Natl Cancer Inst 99(2):147–157
Eisen MB et al (1998) Cluster analysis and display of genome-wide expression patterns. Proc Natl Acad Sci USA 95(25):14863–14868
Ekins R, Chu FW (1999) Microarrays: their origins and applications. Trends Biotechnol 17(6):217–218
Ellwood-Yen K et al (2003) Myc-driven murine prostate cancer shares molecular features with human prostate tumors. Cancer Cell 4(3):223–238
Ewald D et al (1996) Time-sensitive reversal of hyperplasia in transgenic mice expressing SV40 T antigen. Science 273(5280):1384–1386
Faivre J et al (1998) Survival of patients with primary liver cancer, pancreatic cancer and biliary tract cancer in Europe. Eur J Cancer 34(14):2184–2190
Feng J et al (2007) Gene expression analysis of pancreatic cystic neoplasm in SV40Tag transgenic mice model. World J Gastroenterol 13(15):2218–2222
Fontaniere S et al (2006) Gene expression profiling in insulinomas of Men1 beta-cell mutant mice reveals early genetic and epigenetic events involved in pancreatic beta-cell tumorigenesis. Endocr Relat Cancer 13(4):1223–1236
Gharaibeh RZ, Fodor AA, Gibas CJ (2007) Software note: using probe secondary structure information to enhance Affymetrix GeneChip background estimates. Comput Biol Chem 31(2):92–98
Gibbons DL et al (2009) Expression signatures of metastatic capacity in a genetic mouse model of lung adenocarcinoma. PLoS One 4(4):e5401
Gingrich JR et al (1996) Metastatic prostate cancer in a transgenic mouse. Cancer Res 56(18):4096–4102
Glinsky GV et al (2003) Microarray analysis of xenograft-derived cancer cell lines representing multiple experimental models of human prostate cancer. Mol Carcinog 37(4):209–221

Glinsky GV, Berezovska O, Glinskii AB (2005) Microarray analysis identifies a death-from-cancer signature predicting therapy failure in patients with multiple types of cancer. J Clin Invest 115(6):1503–1521

Golub TR et al (1999) Molecular classification of cancer: class discovery and class prediction by gene expression monitoring. Science 286(5439):531–537

Granjeaud S, Bertucci F, Jordan BR (1999) Expression profiling: DNA arrays in many guises. Bioessays 21(9):781–790

Gray NS et al (1998) Exploiting chemical libraries, structure, and genomics in the search for kinase inhibitors. Science 281(5376):533–538

Greenlee RT et al (2001) Cancer statistics, 2001. CA Cancer J Clin 51(1):15–36

Gunther EJ et al (2002) A novel doxycycline-inducible system for the transgenic analysis of mammary gland biology. FASEB J 16(3):283–292

Guo Z et al (1994) Direct fluorescence analysis of genetic polymorphisms by hybridization with oligonucleotide arrays on glass supports. Nucleic Acids Res 22(24):5456–5465

He WW et al (1997) A novel human prostate-specific, androgen-regulated homeobox gene (NKX3.1) that maps to 8p21, a region frequently deleted in prostate cancer. Genomics 43(1):69–77

Heller MJ (2002) DNA microarray technology: devices, systems, and applications. Annu Rev Biomed Eng 4:129–153

Hennighausen L et al (1995) Conditional gene expression in secretory tissues and skin of transgenic mice using the MMTV-LTR and the tetracycline responsive system. J Cell Biochem 59(4):463–472

Herschkowitz JI et al (2007) Identification of conserved gene expression features between murine mammary carcinoma models and human breast tumors. Genome Biol 8(5):R76

Hubbell E, Liu WM, Mei R (2002) Robust estimators for expression analysis. Bioinformatics 18(12):1585–1592

Hughes TR et al (2000) Functional discovery via a compendium of expression profiles. Cell 102(1):109–126

Hughes TR et al (2001) Expression profiling using microarrays fabricated by an ink-jet oligonucleotide synthesizer. Nat Biotechnol 19(4):342–347

Irizarry RA et al (2003) Exploration, normalization, and summaries of high density oligonucleotide array probe level data. Biostatistics 4(2):249–264

Jain N et al (2003) Local-pooled-error test for identifying differentially expressed genes with a small number of replicated microarrays. Bioinformatics 19(15):1945–1951

Jaluria P et al (2007) A perspective on microarrays: current applications, pitfalls, and potential uses. Microb Cell Fact 6:4

Johnson L et al (2001) Somatic activation of the K-ras oncogene causes early onset lung cancer in mice. Nature 410(6832):1111–1116

Katzenellenbogen M et al (2007) Molecular mechanisms of liver carcinogenesis in the mdr2-knockout mice. Mol Cancer Res 5(11):1159–1170

Koliopanos A et al (2008) Molecular aspects of carcinogenesis in pancreatic cancer. Hepatobiliary Pancreat Dis Int 7(4):345–356

Landis SH et al (1999) Cancer statistics, 1999. CA Cancer J Clin 49(1):8–31, 1

Leclerc D et al (2004) ApcMin/+ mouse model of colon cancer: gene expression profiling in tumors. J Cell Biochem 93(6):1242–1254

Lee JS et al (2004) Application of comparative functional genomics to identify best-fit mouse models to study human cancer. Nat Genet 36(12):1306–1311

Lee JS, Grisham JW, Thorgeirsson SS (2005) Comparative functional genomics for identifying models of human cancer. Carcinogenesis 26(6):1013–1020

Lindvall C et al (2006) The Wnt signaling receptor Lrp5 is required for mammary ductal stem cell activity and Wnt1-induced tumorigenesis. J Biol Chem 281(46):35081–35087

Linnerth NM, Sirbovan K, Moorehead RA (2005) Use of a transgenic mouse model to identify markers of human lung tumors. Int J Cancer 114(6):977–982

Luo J et al (2010) A comparison of batch effect removal methods for enhancement of prediction performance using MAQC-II microarray gene expression data. Pharmacogenomics J 10(4):278–291

Maroulakou IG et al (1994) Prostate and mammary adenocarcinoma in transgenic mice carrying a rat C3(1) simian virus 40 large tumor antigen fusion gene. Proc Natl Acad Sci USA 91(23): 11236–11240

Martinez C et al (2005) Expression profiling of murine intestinal adenomas reveals early deregulation of multiple matrix metalloproteinase (Mmp) genes. J Pathol 206(1):100–110

Marton MJ et al (1998) Drug target validation and identification of secondary drug target effects using DNA microarrays. Nat Med 4(11):1293–1301

Meuwissen R, Berns A (2005) Mouse models for human lung cancer. Genes Dev 19(6):643–664

Moody SE et al (2002) Conditional activation of Neu in the mammary epithelium of transgenic mice results in reversible pulmonary metastasis. Cancer Cell 2(6):451–461

Moorehead RA et al (2003) Transgenic overexpression of IGF-II induces spontaneous lung tumors: a model for human lung adenocarcinoma. Oncogene 22(6):853–857

Moser AR, Pitot HC, Dove WF (1990) A dominant mutation that predisposes to multiple intestinal neoplasia in the mouse. Science 247(4940):322–324

Naef F, Magnasco MO (2003) Solving the riddle of the bright mismatches: labeling and effective binding in oligonucleotide arrays. Phys Rev E Stat Nonlin Soft Matter Phys 68(1 Pt 1):011906

Nandan MO, Yang VW (2010) Genetic and chemical models of colorectal cancer in mice. Curr Colorectal Cancer Rep 6(2):51–59

Nemeth J et al (2009) S100A8 and S100A9 are novel nuclear factor kappa B target genes during malignant progression of murine and human liver carcinogenesis. Hepatology 50(4):1251–1262

Oving IM, Clevers HC (2002) Molecular causes of colon cancer. Eur J Clin Invest 32(6):448–457

Paoni NF et al (2003) Transcriptional profiling of the transition from normal intestinal epithelia to adenomas and carcinomas in the APCMin/+ mouse. Physiol Genomics 15(3):228–235

Peeters JK, Van der Spek PJ (2005) Growing applications and advancements in microarray technology and analysis tools. Cell Biochem Biophys 43(1):149–166

Perou CM et al (1999) Distinctive gene expression patterns in human mammary epithelial cells and breast cancers. Proc Natl Acad Sci USA 96(16):9212–9217

Perou CM et al (2000) Molecular portraits of human breast tumours. Nature 406(6797):747–752

Radmacher MD, McShane LM, Simon R (2002) A paradigm for class prediction using gene expression profiles. J Comput Biol 9(3):505–511

Reichling T et al (2005) Transcriptional profiles of intestinal tumors in Apc(Min) mice are unique from those of embryonic intestine and identify novel gene targets dysregulated in human colorectal tumors. Cancer Res 65(1):166–176

Reya T et al (2003) A role for Wnt signalling in self-renewal of haematopoietic stem cells. Nature 423(6938):409–414

Roberts CJ et al (2000) Signaling and circuitry of multiple MAPK pathways revealed by a matrix of global gene expression profiles. Science 287(5454):873–880

Shen MM and Abate – Shen C (2010) Molecular Genetics for Prostate Cancer: new prospects for old challanges. Genes Dev 24(18):1967–2000

Shibata MA et al (1996) Progression of prostatic intraepithelial neoplasia to invasive carcinoma in C3(1)/SV40 large T antigen transgenic mice: histopathological and molecular biological alterations. Cancer Res 56(21):4894–4903

Shoushtari AN, Michalowska AM, Green JE (2006) Comparing genetically engineered mouse mammary cancer models with human breast cancer by expression profiling. Breast Dis 28:39–51

Simon R (2003) Using DNA microarrays for diagnostic and prognostic prediction. Expert Rev Mol Diagn 3(5):587–595

Simon R (2010) Clinical trials for predictive medicine: new challenges and paradigms. Clin Trials 7(5):516–524

Sodir NM et al (2006) Smad3 deficiency promotes tumorigenesis in the distal colon of ApcMin/+ mice. Cancer Res 66(17):8430–8438

Sorlie T et al (2001) Gene expression patterns of breast carcinomas distinguish tumor subclasses with clinical implications. Proc Natl Acad Sci USA 98(19):10869–10874

Sorlie T et al (2003) Repeated observation of breast tumor subtypes in independent gene expression data sets. Proc Natl Acad Sci USA 100(14):8418–8423

Storey JD, Tibshirani R (2003) Statistical significance for genomewide studies. Proc Natl Acad Sci USA 100(16):9440–9445
Sun Q et al (2006) Generation and characterization of a transgenic mouse model for pancreatic cancer. World J Gastroenterol 12(17):2785–2788
Sweet-Cordero A et al (2005) An oncogenic KRAS2 expression signature identified by cross-species gene-expression analysis. Nat Genet 37(1):48–55
Sweet-Cordero A et al (2006) Comparison of gene expression and DNA copy number changes in a murine model of lung cancer. Genes Chromosomes Cancer 45(4):338–348
Tamayo P et al (1999) Interpreting patterns of gene expression with self-organizing maps: methods and application to hematopoietic differentiation. Proc Natl Acad Sci USA 96(6):2907–2912
Tusher VG, Tibshirani R, Chu G (2001) Significance analysis of microarrays applied to the ionizing radiation response. Proc Natl Acad Sci USA 98(9):5116–5121
Van Dyke T, Jacks T (2002) Cancer modeling in the modern era: progress and challenges. Cell 108(2):135–144
van Lohuizen M et al (1991) Identification of cooperating oncogenes in E mu-myc transgenic mice by provirus tagging. Cell 65(5):737–752
Ventura A et al (2007) Restoration of p53 function leads to tumour regression in vivo. Nature 445(7128):661–665
Vogelstein B et al (1988) Genetic alterations during colorectal-tumor development. N Engl J Med 319(9):525–532
Wakamatsu N et al (2007) Overview of the molecular carcinogenesis of mouse lung tumor models of human lung cancer. Toxicol Pathol 35(1):75–80
Wright GW, Simon RM (2003) A random variance model for detection of differential gene expression in small microarray experiments. Bioinformatics 19(18):2448–2455
Xue W et al (2007) Senescence and tumour clearance is triggered by p53 restoration in murine liver carcinomas. Nature 445(7128):656–660

Chapter 11
Imaging Mouse Models of Human Cancer

Jennifer A. Prescher and Christopher H. Contag

11.1 Introduction

While long-recognized as vital components of cancer research, mouse models largely remained a "black box" until the development of advanced tools for preclinical imaging (Edinger et al. 2002; Van Dyke and Jacks 2002; Hirst and Balmain 2004; Lyons 2005). The dynamics of cancer progression, metastatic spread, and therapeutic response were difficult to study without noninvasive access to real-time information in living animals. Histology and other ex vivo analyses provided some insight into the molecular features of malignancy, but required biopsy and invasive acquisition, or were limited to terminal samples at necropsy. Moreover, such measurements offered only a static snapshot of disease and therapeutic outcome, and did not capture the active nature of malignancy or the dynamic changes associated with treatment. Overcoming these limitations required a set of tools that could probe tumor cells in their native habitat, and track molecular and biochemical changes accompanying tumor growth and regression in real time.

Dedicated small animal imaging equipment and reagents are now available to non-invasively measure a broad range of tumor-relevant parameters in mice, from the early onset of malignancy, through tumor growth and dispersion, and during therapeutic intervention (Weissleder 2002; Massoud and Gambhir 2003; Lyons 2005; Atri 2006; Helms et al. 2006; Weissleder 2006; Contag 2007; Stell et al. 2007; Weissleder

J.A. Prescher
Molecular Imaging Program at Stanford, Stanford School of Medicine, Stanford, CA, USA

C.H. Contag (✉)
Molecular Imaging Program at Stanford, Stanford School of Medicine, Stanford, CA, USA

Department of Pediatrics, Stanford School of Medicine, Stanford, CA 94305, USA

Department of Microbiology and Immunology, Stanford School of Medicine,
Stanford, CA 94305, USA
e-mail: ccontag@stanford.edu

J.E. Green and T. Ried (eds.), *Genetically Engineered Mice for Cancer Research: Design, Analysis, Pathways, Validation and Pre-Clinical Testing*,
DOI 10.1007/978-0-387-69805-2_11,

and Pittet 2008). Macroscopic features such as tumor size and location can be tracked *in vivo*, along with microscopic events such as cell movement and gene expression. Furthermore, these measurements can be acquired serially from the same mouse, enabling dynamic observation without the need for repeated biopsy, surgical manipulation, or euthanasia. The ability to noninvasively visualize a multitude of relevant parameters in mouse models is enabling researchers to unravel the multistep, multicomponent nature of malignancy. In turn, this knowledge is spurring the development of new cancer treatments for human patients that address the molecular basis of cancer and take aim at the root of the disease (Herschman 2003a, b; Weissleder 2006).

This chapter provides both an introduction to the most well-established techniques for *in vivo* imaging, and an overview of where they can be integrated into studies with mouse models of human cancer. Some of the imaging techniques, including computed tomography (CT), magnetic resonance imaging (MRI), positron emission tomography (PET), and single photon computed tomography (SPECT), are based on clinical imaging tools that have been miniaturized for use with laboratory rodents. Other modalities have been specifically designed for preclinical imaging where the ability to screen large numbers of animals with relative ease is emphasized. These modalities are largely based on optical methods with visible or near infrared (NIR) light, and are a main focus of this chapter.

Instead of framing the discussion around specific imaging techniques, we have organized the chapter by experiment type. This should assist researchers in selecting appropriate imaging modalities for their particular studies. We first highlight techniques for measuring anatomical features such as tumor size, and discuss contrast agents that are available to assess physical changes in mouse tumor models. The strengths and weaknesses of the methods are noted, along with examples of their use in mouse models. We next showcase imaging approaches that reveal tumor-relevant physiological parameters, including vascular flow and metabolism. Last, we highlight imaging strategies that can visualize cellular and molecular events *in vivo*, placing a strong emphasis on the use of optical reporters to track tumor cell growth, assess cellular physiology, and monitor therapeutic outcome.

Imaging technologies, combined with sophisticated mouse models of human cancer, are providing researchers with an unprecedented look into tumorigenesis and mechanisms of disease progression, and providing useful measures of therapeutic outcome in models where tumors develop spontaneously and at a variety of tissue sites. The continued development of novel molecular probes and creative combinations of imaging strategies will expand the imaging toolbox, and these advances will substantially increase our understanding of malignancy and reveal new methods to combat cancer.

11.2 Imaging in Living Systems: Basic Principles

All small animal imaging methods use discrete forms of energy to interrogate living tissues and derive visual information. Some imaging modalities, including CT, MRI, and ultrasound (US), extract primarily anatomical information. Tumor sizes and the locations of metastatic foci in mouse models have been routinely screened using these methods (Ritman 2004; Ferrara et al. 2007). Other techniques such as functional MRI (fMRI) capture not only physical metrics, but also tumor-relevant physiological information such as blood flow and metabolism (Sosnovik and Weissleder 2007). Still other techniques derive cellular and molecular information; radionuclide imaging (PET and SPECT) and some MRI approaches fall into this category, as do optical methods. Optical imaging strategies exploit energy in the peri-visible region of the electromagnetic spectrum, including UV, visible, and NIR light, and were largely developed for monitoring changes in cellular and molecular processes *in vitro*. A number of these optical methods can be adapted for imaging in live laboratory animals, and have been used in combination with mouse models of cancer (Contag 2007; Luker and Luker 2008). Imaging technologies differ not only in the types of information that can be acquired from living subjects, but also in their resolution, sensitivity, ease-of-use, and cost. Several of these features are compared in Table 11.1.

Few cellular processes or molecules possess features that permit their direct detection *in vivo*. However, a variety of probes exist to "report" on physiological and molecular processes by providing detectable signals that correlate with the biological event of interest (Massoud and Gambhir 2003; Gross and Piwnica-Worms 2005). Some imaging probes are initially distributed throughout the body, then sequestered in tissues based on intrinsic physical properties such as leaky vasculature or heightened metabolism (Jaffer and Weissleder 2005). These agents are routinely used in clinical practice to determine blood volume, define tumor margins, and assess metabolic activity in humans, and they can be employed for similar studies in mouse models. Other imaging probes are targeted to specific cells or molecules and used to track the expression of the labeled molecular species *in vivo*. Additionally, distinct cell populations, gene products, and biochemical processes can be visualized using genetically encoded reporter proteins such as the green fluorescent protein (GFP) and luciferase (Gross and Piwnica-Worms 2005). These reporters produce optical signatures that correlate with cell number or the amount of protein expressed in a given study, as determined by the design of the genetic construct. Reporter genes have been used to track anatomic, physiological, and cellular/molecular changes accompanying tumor progression in mouse models, and specific examples are provided in later sections. Collectively, these tools for noninvasive imaging have enabled researchers to peer into mouse models and see disease mechanisms and host responses "in action."

11.3 Imaging Anatomic Features in Mouse Models

Advances in genetics have produced sophisticated mouse models that closely mimic human disease (Jonkers and Berns 2002; Van Dyke and Jacks 2002; Frese and Tuveson 2007). In some models, lesions arise at predictable times and in predetermined cell types and organs. In other models, tumors arise spontaneously within a target organ.

Table 11.1 Comparison of small animal imaging techniques

Technique	Radiation utilized	Spatial resolution	Depth limit	Time	Sensitivity (*M*)	Target	Common imaging agents	Cost	Advantages	Disadvantages
Computed tomography (CT)	X-rays	50–200 μm	None	Minutes	Not well characterized	Mainly anatomic	Electron-dense materials (e.g., iodine)	$$	Excellent method for anatomical lung/bone and tumor imaging	Radiation exposure, poor soft-tissue resolution, limited functional/molecular applications
Ultrasound	Soundwaves	50–500 μm	mm to cm	Seconds to minutes	Not well characterized	Anatomic, physiological	Microbubbles, liposomes, perfluorocarbons	$	Vascular and interventional imaging	Limited depth penetration and functional/molecular applications
Magnetic resonance imaging (MRI)	Radiowaves	10–100 μm	None	Minutes to hours	$>10^{-5}$	Anatomic, physiological, cellular/molecular	Paramagnetic chelates, magnetic particles	$$$	Highest spatial resolution, versatile imaging modality	Requires expert training, long processing times, low sensitivity

PET	High-energy γ-rays	1–2 mm	None	Seconds to minutes	10^{-11} to 10^{-12}	Physiological, cellular/molecular	$^{18}F^{-}$, $^{64}Cu^{-}$, or ^{11}C-labeled compounds	$$$	Quantitative, high sensitivity, versatile imaging modality	Radiation exposure, on-site production facilities required, low spatial resolution, cost
SPECT	Lower energy γ-rays	1–2 mm	None	Minutes to hours	10^{-10} to 10^{-11}	Physiological, cellular/molecular	^{99m}Tc- or ^{111}In-labeled compounds	$$	Simultaneous imaging of multiple probes, wide assortment of probes available	Radiation exposure, low spatial resolution
Whole-body fluorescence	Visible/near-infrared light	1–3 mm	<1 cm	Seconds to minutes	$\sim 10^{-12}$	Physiological, cellular/molecular	Fluorophores, fluorescent proteins	$ to $$	Rapid screening of cellular and molecular events in surface tumors, cost-efficient	Surface-weighted, low spatial resolution, limited availability of suitable probes

(continued)

Table 11.1 (continued)

Technique	Radiation utilized	Spatial resolution	Depth limit	Time	Sensitivity (*M*)	Target	Common imaging agents	Cost	Advantages	Disadvantages
Intravital microscopy	Visible/near-infrared light	1 μm	400–800 μm	Seconds to hours	$\sim 10^{-12}$	Anatomical, physiological, cellular/molecular	Fluorophores, fluorescent proteins	$$$	Rapid screening of cellular and molecular events in surface tumors	Limited depth and fields of view, limited availability of suitable probes, cost
Bioluminescence	Visible light	mm	cm	Seconds to minutes	$< 10^{-12}$	Cellular/molecular	Luciferins	$ to $$	High sensitivity, rapid screening of cellular and molecular events	Surface-weighted, substrate injection required

Adapted from Massoud and Gambhir (2003), Helms et al. (2006), and Weissleder and Pittet (2008)

There is some variability in all of these models, though, regarding the onset of disease. Tumors often form in only a subset of animals, and the development of nonsuperficial lesions can be difficult to chart using palpation or caliper measurements. Thus, detecting the initial stages of cancers and monitoring their progression over time can present significant challenges. In many cases, animals must be sacrificed and dissected to determine the location and size of tumors within an experimental cohort (Singh and Johnson 2006; Vignjevic et al. 2007). The number of animals sacrificed can be quite large, depending on the parameter being measured (e.g., tumor volume or angiogenesis). Small animal imaging techniques such as CT, US, and MRI can noninvasively locate and monitor tumor development via changes in anatomic structure, thus guiding biopsy or necropsy analyses and reducing the number of animals required per experiment. CT, US, and MRI have been widely used in clinical practice to detect tumors in humans and, in recent years, have been adapted to imaging small animals (Masuda et al. 2008). The use of these tools in rodent models obviously enriches preclinical studies, but also drives the innovation of potentially useful reagents and methods for clinical application.

CT utilizes a beam of high-energy X-rays and a digital detector to capture a subject's anatomical features. X-rays are differentially absorbed by biological materials including bone, fat, and water, and the distinctive absorption patterns of these structures can be used to generate high-resolution, 3D-reconstructed images of physical structures (Fig. 11.1a). Since X-rays easily penetrate all soft tissues, CT cannot distinguish between most tumor tissue and surrounding organs. However, additional tissue contrast can be achieved with iodinated reagents and other electron-dense materials (Rutten and Prokop 2007). These reagents accumulate nonspecifically in a variety of tissues and partially attenuate X-rays, providing signal enhancement for tumors and other soft tissues. Contrast-enhanced CT is a relatively rapid imaging technique, and therefore useful for screening large groups of animals for lung tumors and other macroscopic abnormalities (Ritman 2004). However, the radiation dose is not negligible and can limit repeated imaging of individual animals. While some CT-specific contrast agents are being developed for monitoring physiological and molecular features, this modality is almost exclusively used for anatomic imaging (Ritman 2002; Perkins and Missailidis 2007).

US imaging and MRI have also been used to probe tumor volumes and morphological features in mouse models. Both of these techniques utilize nonionizing forms of radiation and can detect tumors in tissues that are more refractory to CT imaging. US employs soundwaves to image physical features and is among the most user-friendly and inexpensive techniques for mapping tumor size and location (Hauff et al. 2008). However, soundwaves are not transmitted through bones, lungs, and deep tissues, restricting the use of ultrasound to probing relatively shallow lesions.

MRI is less tissue-restricted than CT or ultrasound, and has been used to image a broad range of tumor types in mouse models (Nieman et al. 2005). MRI is based on the principle that unpaired nuclear spins (such as hydrogen atoms in water, metabolites, or other biomolecules) align themselves when placed in an external magnetic field. An input pulse of energy can alter the alignment of the spins, and their return to baseline emits low-energy radiation that is recorded as a change in electromagnetic flux. Water molecules and protons in different types of tissues emit unique signals that can be measured and used to reconstruct anatomical information.

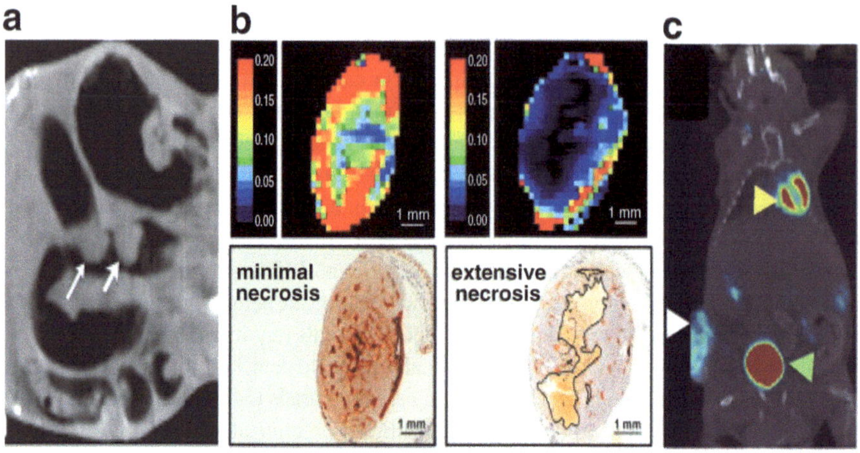

Fig. 11.1 Examples of anatomic and functional imaging in mouse models of cancer. (**a**) Contrast-enhanced CT image (coronal view) of a tumor-bearing mouse. *Arrows* indicate two colonic tumors identified in the intestinal track. (Pickhardt et al. 2005) (**b**) Functional MRI imaging of human melanoma xenografts with necrotic regions. Mice bearing E-13 tumors were injected with a gadolinium contrast agent, and the tumor tissue was evaluated using dynamic contrast-enhanced MRI (DCE-MRI) to report on blood perfusion and tumor oxygenation status. The *dim areas* of the images (*upper panel*) indicate areas of necrosis. The corresponding histological stains of the resected tumors are shown in the *lower panels*, with the necrotic regions outlined in *black*. (**c**) PET/CT scan of an osteosarcoma-bearing mouse injected with radiolabeled FDG (for PET imaging) and Fenestra (for CT contrast). FDG was taken up in the tumor (*white arrow*) and heart (*yellow arrow*) and excreted into the bladder (*green arrow*). The PET image is overlaid on the CT scan for anatomical reference (Arvanitis et al. 2008)

Additionally, these intrinsic signals can be enhanced with exogenous reagents such as paramagnetic metal cations (e.g., chelated gadolinium) or magnetic nanoparticles (e.g., iron oxide). These probes have distinct distribution patterns and must be carefully selected for a given experiment (Frullano and Meade 2007). Contrast-enhanced MRI has been successfully employed in monitoring tumor growth, bulk tumor regression in response to therapies, and metastatic dispersion in mouse models (Sosnovik and Weissleder 2007). MRI probes can also be functionalized to target specific cells and tissues as discussed in later sections. While versatile, MRI has generally been less accessible to basic researchers than CT or US due to instrument size, cost, and required training. However, imaging instruments based on permanent and portable magnets are being developed, and these modifications may serve to make MRI more user friendly and less costly (Inoue et al. 2006).

11.4 Monitoring Physiological Function in Mouse Tumors

In addition to providing high-resolution images of tumors, MRI is capable of probing functional parameters relevant to cancer progression. For example, blood flow and other vascular properties are often altered in malignant tissue, and water mobility can change in response to effective therapies (Fournier et al. 2007; Hamstra et al. 2007).

The timing and duration of MRI pulse sequences and recording events can be varied to produce images that report on these parameters (Fig. 11.1b). Changes in diffusion rates, tissue oxygenation levels, and blood flow have all been measured using such fMRI approaches in mouse and rat tumor models (Barrett et al. 2007). Monitoring the physiological status of tumors with fMRI has direct application to studies of tumor angiogenesis and hypoxia, and is translatable to the clinic.

Surveying tumor metabolism with MRI. MRI equipment can also be tuned to record signals from distinct chemical nuclei, producing unique spectra for different molecules (Raman et al. 2007). These chemical resonances thus provide information on tissue composition and can be used to distinguish malignant and normal tissues. The chemical signatures can also be tracked over time to probe the metabolic fates of tumor-relevant species (Kurhanewicz et al. 2008). This technique, referred to as MR spectroscopy or MR spectroscopic imaging, has been used to monitor levels of citrate, choline, and *N*-acetyl aspartate in mouse brain tissue (Meric et al. 2004; Braakman et al. 2008; Simoes et al. 2008). The relative levels of these metabolites change with the onset of malignancy, and their fluctuations provide a real-time measurement of tumor physiology. Metabolic alterations often precede global diminishment in tumor size in response to cancer treatment, so this imaging method can also provide an early readout of therapeutic efficacy.

MRI is an extraordinarily versatile technique to image biochemical function in mouse models, but this modality is plagued by low molecular sensitivity. Generally, only the most abundant endogenous metabolites can be visualized via MR spectroscopy. The metabolic fates of less abundant species can be tracked using exogenously supplied molecules enriched with ^{13}C, ^{19}F, or other spin-active nuclei. However, the concentrations of probes required for these studies (approx. mM) are often toxic to mice (Gillies and Morse 2005). Strategies to enhance the sensitivity of MRI for metabolite tracking are being developed and include hyperpolarization of tracer molecules. In this approach, molecules are cooled to temperatures near absolute zero and warmed rapidly prior to injection (Day et al. 2007). Although this technique can increase the sensitivity of MRI up to 10,000-fold (for hyperpolarized ^{13}C), the half-life of polarization is very short (~30–40 s maximum) (Kurhanewicz et al. 2008). Thus, hyperpolarization techniques will find utility in several niches, but currently are not broadly applicable to small animal imaging.

Metabolic imaging with radioactive probes. The metabolic activity of tumor tissue can also be assessed using radioactive imaging agents that report on tumor physiology. Most of these reagents are designed to hijack biochemical pathways that are upregulated in tumor tissue. Radioactive isotopes emit high-energy photons that easily penetrate all tissues, and some of these – positron emitters – can be detected using PET, and others – gamma emitters – can be detected using SPECT. Isotopes detected by PET and SPECT are not abundant in living tissue, but can be chemically affixed to sugars, hormones, or other small molecules that are supplied to mice for *in vivo* tracking (Herschman 2003a, b). Such radionuclide imaging approaches are exquisitely sensitive, so only trace amounts of the probes (typically <pM concentrations) are required. Additionally, many isotopes used for radionuclide imaging are surrogates for naturally occurring atoms, so their presence on labeled molecules does not typically interfere with the metabolic pathway being investigated.

PET isotopes (e.g., ^{18}F, ^{11}C, ^{64}Cu) decay by positron emission. The ejected positrons collide with nearby electrons, expelling two gamma rays in opposite directions. The concomitant detection of gamma ray pairs enables localization of the radionuclide source within the tissue. Because PET isotopes are usually short-lived, their use requires on-site radiochemistry facilities for production. SPECT imaging utilizes isotopes that generate a single photon (e.g., ^{111}In, ^{123}I) for direct detection. To localize the source in the body, pinhole collimators are utilized to reject scattered signal. This results in reduced sensitivity relative to PET, but the use of multipinhole collimators has increased the sensitivity of SPECT systems and shortened data acquisition times, and these instruments are readily available for use with laboratory mice.

The most widely used PET tracer for metabolic imaging in humans and mouse models is the ^{18}F-labeled glucose analog fluorodeoxyglucose (FDG) (Gambhir 2002). Systemically administered FDG is accessible to essentially all tissues and is transported into cells via the glucose transporter. FDG is retained in cells only if phosphorylated by cellular kinases. Thus, FDG accumulation reflects the metabolic demand of the cell as determined by increased glucose transport and retention. Increased glucose metabolism is a distinguishing feature of growing tumors, and FDG preferentially accumulates in these tissues, along with other metabolically active sites such as heart and brain tissue. Similar probes for detecting metabolically active tumors have been developed for SPECT imaging, including ^{99}Tc-sestamibi and ^{111}In-oxine (Herschman 2003a, b). These radiolabeled compounds can also be exogenously loaded into cells and used for short-term cell trafficking studies in mouse tumor models (Swirski et al. 2006).

While offering a window of nearly unlimited depth within the body to probe tumor physiology, both PET and SPECT are characterized by poor spatial resolution. Photons emitted from radiotracers can be easily detected, but their assignment to distinct locations within small animals remains challenging and limited to resolutions of 1–2 mm (Gambhir 2002; Weissleder 2002). New designs for SPECT and PET detectors and algorithms for assigning radioactive signals to ever-smaller voxels are being developed. Hybrid PET/CT and SPECT/CT instruments for small animal research have also been designed to allow lower resolution functional information from radionuclide imaging to be overlaid with higher resolution anatomic detail from CT (Cherry 2006; Franc et al. 2008). Additionally, PET-MR scanners are being developed to fuse disease-specific, molecular information with ultrafine anatomic reference images (Cherry 2006; Hofmann et al. 2008; Townsend 2008).

11.5 Visualizing Cells and Molecular Events *In Vivo*

Genomics, proteomics, and other broad screening efforts continue to catalog molecules involved in cancer progression, and the use of these massively parallel bioassays is increasing the number of candidate biomarkers and therapeutic targets that need to be evaluated in the context of living subjects. Dissecting how these molecules act in concert and contribute to distinct stages of tumorigenesis requires the ability to track their expression *in vivo*. Both molecular and cellular targets have been tagged and noninvasively monitored using specific probes and genetic tags in living animals. Such studies are shedding light on how distinct groups of molecules and cells orchestrate tumor behavior, and are highlighting new avenues for therapeutic intervention.

Imaging with targeted injectable probes. Tumor-specific antigens and other potential biomarkers can be directly detected with targeted agents equipped with imaging tracers. For example, antibodies decorated with radiolabels, magnetic particles, and other contrast agents have been used to image a diverse array of protein targets (Gao et al. 2004; Giepmans et al. 2006; Popovtzer et al. 2008). When administered to mice, the antibodies circulate broadly and bind their respective targets, if accessible, and the imaging motif confers visualization of the tagged species. Labeled antibodies of this sort have been used to image cell surface receptors such as Her2 and various adhesion molecules overexpressed on tumor vasculature and cancer cells (Costantini et al. 2007; Olafsen et al. 2007). Antibodies can be generated with specificity for an incredible range of epitopes and, in this regard, they are attractive reagents for *in vivo* imaging. However, antibody architectures have been selected via evolution for long circulation times (to increase the likelihood of target binding), an undesirable feature for imaging purposes as circulating, unbound probe generates background signal. Moreover, their large size hinders access to many antigens inside cells and outside of the vasculature. Smaller versions of these proteins (e.g., mini-bodies and affibodies) retain the desired features of antibodies, but have characteristics that are more well suited for imaging (Sundaresan et al. 2003; Nilsson and Tolmachev 2007). Additionally, small molecules including therapeutic drugs, peptides, and DNA-targeting aptamers can be outfitted with radiolabels or other detectable probes and used to image intracellular and extravascular targets (Sharma et al. 2005; Weiner and Thakur 2005; Benvin et al. 2007; Perkins and Missailidis 2007).

As the "omics" technologies continue to paint a more complete picture of tumorigenesis, small numbers of cells and combinations of gene products will need to be viewed simultaneously *in vivo*. However, multiparametric imaging with radioactive probes or MRI agents remains either infeasible or impractical. All positron-emitting isotopes expel two gamma rays separated by 180°, meaning that the energy emitted from one PET isotope cannot be spectrally resolved from a second isotope. It is possible to detect multiple isotopes that emit different photon energies in SPECT, but the nuclides still require special handling. A recent report suggests that some MRI probes can be spectrally resolved and used together in a single subject (Zabow et al. 2008). However, most MRI probes are inherently insensitive to molecular changes, and usually need to be injected directly into tumor masses or administered at high (often lethal) concentrations for visualization *in vivo*.

Fluorescent dyes represent a convenient alternative to radioactive or MRI probes for multiparametric imaging. Fluorescent dyes, such as radioactive isotopes, can be chemically linked to a multitude of tumor-targeting reagents for optical imaging in small animal models (Sevick-Muraca et al. 2002; Ntziachristos et al. 2003;

Ntziachristos 2006; Rao et al. 2007). The dyes absorb external light capable of penetrating mammalian tissue, and they subsequently emit photons that can be detected with a charge-coupled device (CCD) camera. A wide assortment of small molecule dyes is available for targeted imaging studies *in vivo*, and most of the dyes can be spectrally resolved (Ntziachristos 2006). However, the same fluorescent reagents available to the cell biologist for microscopy and flow cytometry are not equally useful for *in vivo* imaging (Makale 2007). These agents require excitation light and emit light in the blue-green region of the visible spectrum (~450–550 nm). Blue-green light and wavelengths below 600 nm are largely absorbed by hemoglobin, reducing the amount of detectable signal as light transverses mammalian tissue (Benaron et al. 1997; Rice et al. 2001). Signal attenuation and tissue autofluorescence (noise) are minimized in the NIR region of the spectrum (650–1,100 nm) where biomolecules have their lowest absorption coefficients and water absorption is minimal. In this wavelength range, it may be possible to localize a focus of several thousand cells labeled with fluorescent dyes, depending on the depth in the body and the optical properties of the overlying tissues (Sevick-Muraca et al. 2002; Frangioni 2003). Thus, optical imaging can approximate the sensitivity of radionuclide imaging in small animals and exceeds the multiplexing capability of SPECT, while circumventing the difficulties of handling radioactive materials.

Quantum dots also emit light in the NIR region and provide intense, stable signals with potential for imaging cells and tumor antigens *in vivo* (Larson et al. 2003; Gao et al. 2004; Michalet et al. 2005), although many of these particles excite at short wavelengths. The physical properties of quantum dots restrict their use to mostly vascular spaces, but smaller particles with improved tissue accessibility, and more optimal excitation wavelengths are being developed (Howarth et al. 2008). The parallel use of fluorophores and quantum dots *in vivo* has enabled multiplexed cell trafficking studies (Voura et al. 2004), and improvements in detector technologies and imaging probe technology will continue to increase the number of species that can be simultaneously imaged in a living subject (Sevick-Muraca et al. 2002).

Currently, the number of targeted probes available for *in vivo* imaging is quite small (Weissleder and Pittet 2008), as the demands placed on these reagents are high. The molecules must be able to break through biological barriers to reach their intended targets and, upon arrival, react or bind with high specificity. The probes must also be sufficiently stable to persist for a suitable period of time in a mouse, even with cell division, and remain sensitive to subtle changes in target expression levels. Additionally, they must show minimal nonspecific reactivity with off-target tissues, and unbound probe must be cleared quickly to ensure adequate signal-to-noise ratios for detection. Very few probes meet all of these criteria, but ongoing efforts using high-throughput screens of chemical libraries, yeast and phage display, ligand evolution technologies, and nanotechnology screening platforms are expanding the pool of targeted imaging probes (Cai and Chen 2007; Nie et al. 2007; Hsiung et al. 2008)

Some of the limitations associated with conventional targeted imaging agents can be overcome with activatable probes for visualizing tumor-specific targets. These "smart" probes produce minimal or no signal until they are activated by the

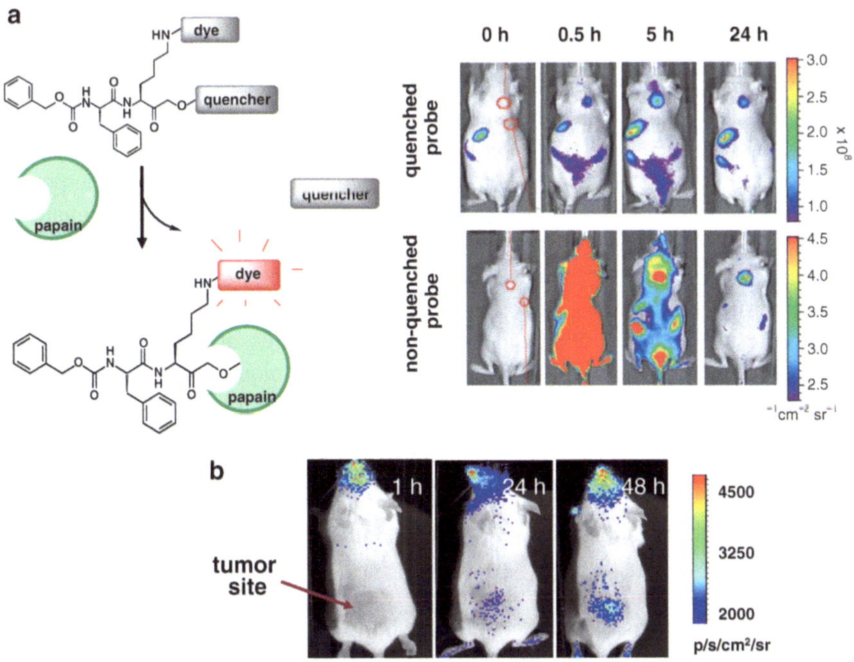

Fig. 11.2 Examples of cellular and molecular imaging in mouse models. (**a**) A "smart" near-infrared probe was designed to covalently target the papain family of cysteine proteases. Upon interaction of the quenched probe with the protease, the quencher is released, and the dye can be detected via fluorescence imaging. The probe was used to image tumor-associated protease activity *in vivo*. Mice bearing grafted breast tumors (*top circle in each panel*) were injected with either the quenched probe or a nonquenched control probe. Whole-body fluorescence images were acquired using a Cy5.5 filter set. Retention of both probes at the tumor site was observed, although higher signal-to-noise ratios were achieved with the "smart" probe, and the maximum signal was achieved more rapidly (Blum et al. 2007). (**b**) Bioluminescence imaging of immune cell trafficking *in vivo*. Fluc-labeled cytotoxic T cells were administered to mice bearing A120 tumor cells. The trafficking patterns of the T cells were monitored over time via luciferin injection and bioluminescence imaging. The cells were observed to accumulate at the tumor site 24–48 h postinjection (Thorne et al., unpublished)

target (often an enzyme), which may be elevated at sites of tumor growth. Such probes have the potential to provide greatly enhanced signal-to-noise ratios for imaging of tumor-associated enzymes that are secreted by malignant cells or associated stroma. In one example, a quenched NIR fluorophore was used to covalently target the papain-family of cysteine proteases, a class of enzymes overproduced by tumor cells (Blum et al. 2007). Upon cleavage by papain in a mouse tumor model, the targeted probe released its quencher, and fluorescence signal was observed at the tumor site (Fig. 11.2a). Similar activatable fluorescence agents have been used to image matrix metalloproteinases (Weissleder et al. 1999), cathepsins (Tung et al. 2000; Bremer et al. 2002; Grimm et al. 2005), and other tumor-relevant protease activities *in vivo* (Jiang et al. 2004). "Smart" probes that increase signal-to-noise ratios in MRI have also been reported (Meade et al. 2003).

Imaging with genetic reporters. A number of experimental strategies require permanently labeled cells or genes for imaging over extended periods of time. "Smart" probes and chemical tags may be diluted with cell division or lost to membrane turnover, and such probes are not easily amplified. An alternative is to incorporate reporter genes into the genomes of cells and animals and use the optical signatures of their encoded proteins to monitor the desired biological process. Such genetic strategies can be used to "mark" cells or proteins for long-term visualization and report on molecular events. Importantly, genetic tags propagate with cell division, providing stable sources of signal for longitudinal studies, and strategies exist for amplification and activation of genetic tags, enabling unique imaging experiments that are not feasible with other approaches (Gross and Piwnica-Worms 2005). Genetic reporters are also modular and can be used to label a variety of targets, eliminating the need to synthesize a new probe for each event to be studied.

The use of genetic tags in small animal imaging builds on the same technologies commonly used by cell biologists for understanding gene function and revealing cell physiology. These tags encode protein products that can be visualized after illumination with an excitation wavelength (i.e., fluorescent proteins), or enzymes that require exogenous substrates to produce optical signals (i.e., luciferases) (Gross and Piwnica-Worms 2005). For cell tracking purposes, the reporter genes are placed under the control of a constitutively active promoter that is ubiquitously expressed in many if not all tissues (e.g., CMV, β-actin, or ubiquitin) to ensure constant expression of the protein. This approach has been extensively employed for visualizing cell trafficking patterns (Mandl et al. 2002), tumor growth (Edinger et al. 2002), and tissue transplantation and regeneration (Cao et al. 2005). However, this approach is particularly challenging for cells that differentiate after transfer into animals as there are few, if any promoters that are truly expressed in a constitutive and ubiquitous manner.

Genetic reporters can also be cloned into regulatory sequences of genes (e.g., promoters/enhancers), and used as *in vivo* reporters of transcriptional activity (Blasberg and Gelovani-Tjuvajev 2002). In these cases, the reporter is designed to mirror the transcription of the native gene for studying cellular responses to various stimuli, including cytokines and thermal stress (O'Connell-Rodwell et al. 2004, 2008). The use of regulated genetic reporters in tumor cells enables visualization of tumor-relevant genes as they turn on and off in response to various stimuli, including oxygen levels (Safran et al. 2006), genotoxic stress, and UV irradiation. These reporters have been elegantly applied to studying a variety of events in mouse tumor models, including oncogenic transformation, tumor suppression, therapeutic efficacy, and cell activation (Gross and Piwnica-Worms 2005). Fluorescent proteins, luciferases, and other imaging tags can also be engineered into direct fusion with genes of interest to monitor discrete protein targets and their roles in signal transduction cascades and protein–protein interactions (Villalobos et al. 2007; Luker et al. 2009).

The gene sequences for both constitutive and regulatable reporters are typically introduced into cell populations ex vivo using standard gene transfer strategies including viral and nonviral gene transfer. The cells are then introduced into mice for subsequent visualization and tracking. Transgenes can be directly delivered to some cells and tissues *in vivo*, as is the case with DNA constructs targeted to liver cells via hydrodynamic injection. Other techniques to label endogenous cells with reporter genes *in vivo* are being developed, but further optimization of these systems are necessary to achieve broad applicability (Yang et al. 2008). Fluorescent and bioluminescent proteins are the most widely used genetic reporters in small animal imaging, and they are showcased in the following sections.

Fluorescent proteins. A collection of fluorescent proteins is available that spans virtually the entire visible spectrum, and these proteins have been used as reporters for imaging multiple cell types *in vivo* (Chudakov et al. 2005; Hoffman 2005; Shaner et al. 2005; Giepmans et al. 2006; Stewart 2006). Like small molecule dyes for targeted imaging, fluorescent proteins require external excitation light in order to emit detectable photons. Fluorescent protein reporters have been used to monitor tumor cell growth and cell locomotion in intact animals using macroscopic imaging and within individual blood vessels and tissues using intravital microscopy. GFP and fluorescent proteins that are excited at shorter wavelengths can be used as genetic reporters, but they are largely limited to monitoring events at more superficial sites (e.g., subcutaneous tumor xenografts) or surgically exposed tissues where the detector can be positioned near the site of interest (Jain et al. 2002; Halin et al. 2005). In one example of the latter approach, intravital fluorescence microscopy was used to catalog the interactions of GFP-labeled cytotoxic T cells (CTLs) with grafted tumor cells (Mrass et al. 2006; Boissonnas et al. 2007). CTLs were observed to migrate from peripheral blood vessels and contact the tumor mass. The CTLs moved throughout the tumor microenvironment in a random fashion and paused to sample surface antigens expressed by the tumor cells. Only antigen-specific CTLs infiltrated deeply into the tumor, indicating that the sustained, directional movement of these cells was dependent on the presence of cognate antigen. Intravital microscopy has also been used to monitor circulating cells within the bone marrow of live animals, along with tumor cell chemotaxis (Alexandrakis et al. 2004; Georgakoudi et al. 2004; Halin et al. 2005). In these studies, a combination of fluorescent proteins and fluorophore-labeled antibodies was used to distinguish among cells and proteins of interest. Additional insight into cellular behavior at a microscopic level is anticipated with the development of improved NIR dyes and further advances in intravital multiphoton microscopy (Cahalan et al. 2002; Swirski et al. 2007; Bullen 2008).

DsRed and other red-fluorescent proteins are more useful than GFP for whole body imaging as their excitation and emission wavelengths are sufficient to penetrate mammalian tissue to reasonable depths (Merzlyak et al. 2007; Shcherbo et al. 2007). However, high levels of fluorescent protein expression are required to achieve signal in deep tissues, and this can lead to cellular toxicity (Strack et al. 2008). The development of infrared-emitting fluorescent proteins would enhance the capabilities of fluorescence imaging in mouse models, and more research efforts in this area are anticipated.

Bioluminescent proteins. Reporter genes for bioluminescence imaging (BLI) comprise the second major class of optical imaging reporters. Similar to fluorescence, BLI utilizes visible light to generate images. However, BLI requires no incident radiation; instead, this imaging modality utilizes internal light-producing enzymes known as luciferases. Luciferases emit light during the catalytic oxidation of small molecule substrates. Mammalian tissues do not naturally emit large amounts of light, so the signal-to-noise ratio in BLI is extremely high. In some cases, as few as three to ten bioluminescent cells have been detected *in vivo* (Rabinovich et al. 2008). Several luciferases and their corresponding substrates have been identified, but only a handful of these enzyme–substrate pairs have been widely used for small animal BLI. These include the luciferase–luciferin pairs from the North American firefly, the marine organisms *Renilla reniformis* and *Gaussia princeps*, and the bacterium *Photorhabdus luminescens* (Contag et al. 1995; Contag and Bachmann 2002; Contag and Ross 2002). Firefly luciferase (Fluc) catalyzes the oxidation of the small molecule luciferin in the presence of oxygen and ATP and produces light near 612 nm in living systems (Zhao et al. 2005). Renilla luciferase (Rluc) and Gaussia luciferase (Gluc) are biochemically distinct from Fluc and use the small molecule coelenterazine to produce light near 481 nm. Rluc and Gluc do not depend on ATP for light production, and can be used to image extracellular events where ATP stores are lacking (Venisnik et al. 2006). The bacterial luciferase (Lux) is useful for labeling and tracking bacteria in living systems, and has been used to evaluate bacteriotherapies for cancer treatment (Min et al. 2008). Interestingly, the substrate for Lux is the only luciferin for which the biosynthetic pathway is known. This pathway can be engineered into bacteria along with Lux to produce strains that glow constitutively (as they synthesize both the enzyme and substrate required for light production). For cells labeled with Fluc, Rluc, or Gluc, the appropriate luciferin substrate must be supplied exogenously to generate luminescence. There is obviously much interest in extending the bacterial bioluminescence system to other cell types, and strategies to optimize the expression of Lux and its substrate in eukaryotic cells are being explored (Patterson et al. 2005).

To date, Fluc has been the most widely utilized luciferase for tracking cells and gene products *in vivo* owing to its facile expression in mammalian cells, bioavailable substrate, and favorable spectral properties. Fluc emits more red-shifted light than Rluc or Gluc, and these wavelengths are better able to penetrate mammalian tissues. The use of Rluc and Gluc *in vivo* is further hampered by the fast turnover and degradation of their substrates (Zhao et al. 2004). A mutant version of Rluc (hRluc) was recently reported to emit more red-shifted light than the parent enzyme, and this reporter will likely be used more extensively in the near future (Loening et al. 2007). It is also important to note that the bioluminescence signals from Fluc and Rluc are not spectrally resolved. However, the two enzymes are specific for their respective substrates and can be used together in a single subject when the substrates are supplied sequentially to the animal. This approach was taken to simultaneously monitor tumor burden (using Fluc-labeled glioma cells) and viral transduction (using Rluc) *in vivo* (Shah et al. 2003). The ability to monitor more than two biological processes in a single animal would enhance the utility of BLI, and efforts to increase the number of luciferase–luciferin pairs suitable for *in vivo* imaging are ongoing (Prescher and Contag, unpublished).

The exquisite sensitivity and user-friendly features of BLI permit rapid, high-throughput imaging, and this modality has been extensively employed to monitor tumorigenesis in a wide range of mouse models. A variety of tumor cell lines stably expressing Fluc have been developed and used for tracking *in vivo*. When grafted into mice, these cells can be visualized through their entire life cycle, from the early onset of disease, to progression and response to therapy (Sweeney et al. 1999; Edinger et al. 2002). Because BLI can track small numbers of cells over an entire mouse, this modality has also facilitated studies of cellular metastases, immune cell trafficking, and minimal residual disease in various tumor models (Luker et al. 2002; Shachaf and Felsher 2005; Fan et al. 2008) (Fig. 11.2b). Bioluminescence has been particularly useful for monitoring molecular mechanisms of oncogenesis in spontaneous tumor models, where highly sensitive imaging techniques are necessary to capture the initial and early events in tumor formation. For example, a conditional transgenic mouse model for retinoblastoma-dependent sporadic cancer was labeled with Fluc as a genetic reporter (Vooijs et al. 2002). BLI enabled the development of initial pituitary tumors to be monitored, along with the full course of disease progression and therapeutic response. More recently, BLI has been applied to unraveling the tumorigenic and metastatic potential of human breast cancer stem cells implanted into immunodeficient mice (Liu et al., unpublished). Coupling bioluminescence with other tumor models will continue to reveal the molecular underpinnings of disease and accelerate the development of more effective anticancer treatments to target these events.

Genetic reporters for other imaging modalities. Reporter genes for MRI and PET/SPECT imaging have been described, although they have not been as widely adopted as optical reporters for imaging mouse models. One reporter gene for MRI encodes a lysine-rich protein that promotes rapid proton exchange with surrounding water molecules. This exchange of spin-active nuclei generates detectable magnetic contrast (Gilad et al. 2007). Transferrin receptor has also been used as a reporter gene for cell tracking studies. Cells engineered to express this protein are able to concentrate exogenously supplied magnetic particles, facilitating their detection by MRI (Bulte et al. 1999). One of the most familiar PET/SPECT reporter genes encodes a thymidine kinase from the herpes simplex virus (HSV-TK). By expressing HSV-TK in a target cell, it is possible to image gene expression patterns using radiolabeled versions of thymidine. The thymidine analogs are phosphorylated by the viral TK and, thus, trapped inside cells expressing the reporter gene. The use HSV-TK has enabled long-term studies of cell trafficking in mice, not possible with FDG and other radioactive probes (Acton and Zhou 2005).

11.5.1 Monitoring Therapeutic Intervention at the Cellular and Molecular Level in Mouse Models

As noted in previous sections, imaging technologies can be used to monitor tumor responses to treatment and provide an early indication of therapeutic efficacy. These strategies can be employed on a morphological scale (using CT, MRI, and US) or a biochemical scale (using fMRI or PET/SPECT imaging), but imaging tumor response at a cellular and molecular level provides added insight into the mechanisms and actions of cancer therapies. For example, gene-specific promoters driving luciferase (e.g., HIF, p53, and NF-kB) permit the screening of drugs that activate or block transcriptional activity of a variety of cancer-related genes, and thus provide an early readout of therapeutic efficacy (Zhang et al. 2007). Similarly, the effects of tumor treatments on estrogen signaling (Kanno et al. 2007), proteasome function (Luker et al. 2003), and metabolite production (Wurdinger et al. 2008) can be studied with luciferase reporters. Complementary approaches in determining the efficacy of cytotoxic drugs involve the real-time imaging of cell death with caspase-3 cleavable reporters (Laxman et al. 2002) and fluorescent ligands targeted to annexin V (Ntziachristos et al. 2004).

In addition to providing information about how tumors respond to therapeutics, mouse models provide a useful platform for evaluating the pharmacological properties of the therapies themselves. Radiolabels have been used for decades in this regard to monitor the metabolism, biodistribution, and excretion profiles of pharmaceuticals in mice. The goal of these studies is to understand the distribution and clearance properties of cancer-targeting agents and improve their pharmacokinetic profiles (Akins and Dubey 2008). In recent years, fluorescence and BLI have been applied to tracking therapeutic proteins, viral particles, nanomaterials, and other cancer drugs (Shah and Weissleder 2005). In an excellent example, fluorescence imaging and BLI were used to assess a combined immuno- and viro-therapy in mouse tumor models. This dual therapy comprised cytotoxic T cells carrying lytic virus particles engineered to replicate preferentially in cancer cells over nontransformed cells (Thorne et al. 2006). To evaluate the tumor-targeting potential of the dual therapy *in vivo*, the immune cells were labeled with Fluc, infected with GFP-tagged virus particles, and infused into tumor-bearing mice. BLI revealed effective tumor localization by the immune cells, and fluorescence imaging confirmed the release of the virus at the tumor site and infection of tumor cells. In parallel experiments, BLI indicated that the combined therapy could effectively reduce tumor burden in mice bearing grafted Fluc-labeled breast cancer cells (Thorne et al. 2006).

BLI has also guided efforts to improve drug delivery to specific cells and tissues *in vivo.* In these studies, luciferin is used as a surrogate drug in combination with luciferase-expressing cells and tissues, and the distribution and duration of the "therapy" can be assayed via light emission. Luciferin-based nanoparticles, peptide conjugates, and other scaffolds have been assayed for drug release and tissue-targeting properties in these models, and promising new materials and formulations for targeted delivery were identified (Jones et al. 2006; Wender et al. 2007; Jacobson et al. 2008).

11.6 Illuminating Cancer Biology with Combination Probes and Multiparameter Imaging

No single imaging method embodies all the desired features for monitoring tumors in mice, but combinations of modalities can be used to image over a variety of levels – from anatomic to molecular – and maximize the information gained in a single study. Merging these techniques with histology and other ex vivo assays will ultimately provide a more holistic picture of cancer progression. To streamline multimodality imaging efforts, a series of constructs have been developed that combine different classes of reporter genes. These multifunction reporters enable *in vivo* measurements to be obtained over a range of scales – from macroscopic to microscopic – and coupled to ex vivo assays. For example, a single construct linking Fluc and GFP can be used to integrate longitudinal, whole-body analyses (using BLI) with *in vivo* microscopy (using fluorescence imaging) and ex vivo analyses such as flow cytometry and fluorescence microscopy. This strategy was elegantly employed by Massague and colleagues to dissect the roles of specific microRNAs in breast cancer metastases (Tavazoie et al. 2008). Fluc/GFP-labeled breast cancer cells were engineered to express various combinations of microRNAs and introduced into immunocompromised mice. The animals were serially monitored by BLI, and once sites of metastases were identified, the relevant tissues were harvested and analyzed for the presence of cancer cell infiltrates using fluorescence microscopy.

Triple fusion transgenes linking luciferase (for BLI), GFP (for fluorescence assays), and HSV-TK (for PET) have also been described (Ray et al. 2007), along with fusions incorporating beta-galactosidase for histological analyses (Wehrman et al. 2006). A number of combinations of reporter genes have been reported providing the cancer biologist with a wide range of tools and imaging approach. Additional reporter gene fusions are likely to be developed as multimodality instrumentation (e.g., PET-MRI and PET-CT) becomes more widely available. Collectively, these probes will provide information about mouse models of cancer unable to be gleaned from a single reporter alone.

11.7 Summary and Future Perspectives

Cancer is a multifaceted disease influenced by diverse cell types and environmental stimuli that often cannot be reproduced even with the most sophisticated culture techniques (Karnoub et al. 2007; Ma et al. 2007; Mani et al. 2008; McAllister et al. 2008). Thus, imaging events as they occur in the living body, where the contextual influences of the tumor microenvironment remain intact, provides insights into malignancy that are not possible in analyses with cultured cells or excised tissues. Imaging approaches are enabling researchers to noninvasively measure a broad range of tumor-relevant parameters in mice, from the early onset of disease, through tumor growth, and during treatment. Moreover, these techniques are able to capture information on a variety of scales ranging from anatomic features to gene expression patterns. The ability to visualize a multitude of events *in vivo* is transforming the ways mouse models are used in oncology research, and providing a window into the mechanistic basis of cancer.

In the future, researchers will benefit from continued efforts to develop new reporter probes and instrumentation for tracking multiple cellular and molecular activities *in vivo*. These efforts include the development of tools and instrumentation that exploit different modes of radiation to generate contrast (including photoacoustic and Raman imaging) or employ creative combinations of existing modalities. We also anticipate the instrumentation to become interfaced with other tools for molecular measurement, including flow cytometry and mass spectrometry. Additionally, the development of tools and general strategies to visualize nonproteinaceous biomolecules including lipids and glycans will provide a more expansive look at tumorigenesis (Bonasio et al. 2007).

Using combinations of imaging strategies to measure dynamic, tumor-specific parameters within living animals has enormous implications for oncology research. With these tools, researchers can begin to deconvolute the intertwined biological networks that give rise to human cancers, and more rapidly design and test directed interventions and combination therapies. The ability to peer inside mouse models and visualize tumor progression and regression in real time will translate into improvements in the way cancers are detected, diagnosed, and treated in human patients.

Acknowledgments J.A.P. is supported by grants from the Susan G. Komen Foundation and the National Cancer Institute. The authors also thank M. Helms and M. Sellmyer for their helpful comments.

References

Acton PD, Zhou R (2005) Imaging reporter genes for cell tracking with PET and SPECT. Q J Nucl Med Mol Imaging 49(4):349–360

Akins EJ, Dubey P (2008) Noninvasive imaging of cell-mediated therapy for treatment of cancer. J Nucl Med 49(Suppl 2):180S–195S

Alexandrakis G, Brown EB et al (2004) Two-photon fluorescence correlation microscopy reveals the two-phase nature of transport in tumors. Nat Med 10(2):203–207

Arvanitis C, Bendapudi PK et al (2008) Cancer Biol Ther 7(12):1947–1951

Atri M (2006) New technologies and directed agents for applications of cancer imaging. J Clin Oncol 24(20):3299–3308

Barrett T, Brechbiel M et al (2007) MRI of tumor angiogenesis. J Magn Reson Imaging 26(2):235–249

Benaron DA, Cheong WF et al (1997) Tissue optics. Science 276(5321):2002–2003

Benvin AL, Creeger Y et al (2007) Fluorescent DNA nanotags: supramolecular fluorescent labels based on intercalating dye arrays assembled on nanostructured DNA templates. J Am Chem Soc 129(7):2025–2034

Blasberg RG, Gelovani-Tjuvajev J (2002) In vivo molecular-genetic imaging. J Cell Biochem Suppl 39:172–183

Blum G, von Degenfeld G et al (2007) Noninvasive optical imaging of cysteine protease activity using fluorescently quenched activity-based probes. Nat Chem Biol 3(10):668–677
Boissonnas A, Fetler L et al (2007) In vivo imaging of cytotoxic T cell infiltration and elimination of a solid tumor. J Exp Med 204(2):345–356
Bonasio R, Carman CV et al (2007) Specific and covalent labeling of a membrane protein with organic fluorochromes and quantum dots. Proc Natl Acad Sci USA 104(37):14753–14758
Braakman N, Oerther T et al (2008) High resolution localized two-dimensional MR spectroscopy in mouse brain in vivo. Magn Reson Med 60(2):449–456
Bremer C, Tung CH et al (2002) Imaging of differential protease expression in breast cancers for detection of aggressive tumor phenotypes. Radiology 222(3):814–818
Bullen A (2008) Microscopic imaging techniques for drug discovery. Nat Rev Drug Discov 7(1):54–67
Bulte JW, Zhang S et al (1999) Neurotransplantation of magnetically labeled oligodendrocyte progenitors: magnetic resonance tracking of cell migration and myelination. Proc Natl Acad Sci USA 96(26):15256–15261
Cahalan MD, Parker I et al (2002) Two-photon tissue imaging: seeing the immune system in a fresh light. Nat Rev Immunol 2(11):872–880
Cai W, Chen X (2007) Nanoplatforms for targeted molecular imaging in living subjects. Small 3(11):1840–1854
Cao YA, Bachmann MH et al (2005) Molecular imaging using labeled donor tissues reveals patterns of engraftment, rejection, and survival in transplantation. Transplantation 80(1):134–139
Cherry SR (2006) Multimodality in vivo imaging systems: twice the power or double the trouble? Annu Rev Biomed Eng 8:35–62
Chudakov DM, Lukyanov S et al (2005) Fluorescent proteins as a toolkit for in vivo imaging. Trends Biotechnol 23(12):605–613
Contag CH (2007) In vivo pathology: seeing with molecular specificity and cellular resolution in the living body. Annu Rev Pathol 2:277–305
Contag CH, Bachmann MH (2002) Advances in in vivo bioluminescence imaging of gene expression. Annu Rev Biomed Eng 4:235–260
Contag CH, Contag PR et al (1995) Photonic detection of bacterial pathogens in living hosts. Mol Microbiol 18(4):593–603
Contag CH, Ross BD (2002) It's not just about anatomy: in vivo bioluminescence imaging as an eyepiece into biology. J Magn Reson Imaging 16(4):378–387
Costantini DL, Chan C et al (2007) (111)In-labeled trastuzumab (Herceptin) modified with nuclear localization sequences (NLS): an Auger electron-emitting radiotherapeutic agent for HER2/neu-amplified breast cancer. J Nucl Med 48(8):1357–1368
Day SE, Kettunen MI et al (2007) Detecting tumor response to treatment using hyperpolarized 13C magnetic resonance imaging and spectroscopy. Nat Med 13(11):1382–1387
Edinger M, Cao YA et al (2002) Advancing animal models of neoplasia through in vivo bioluminescence imaging. Eur J Cancer 38(16):2128–2136
Fan F, Binkowski BF et al (2008) Novel genetically encoded biosensors using firefly luciferase. ACS Chem Biol 3(6):346–351
Ferrara K, Pollard R et al (2007) Ultrasound microbubble contrast agents: fundamentals and application to gene and drug delivery. Annu Rev Biomed Eng 9:415–447
Fournier LS, Cuenod CA et al (2007) Imaging of response to treatment in oncology. J Radiol 88(6):829–843
Franc BL, Acton PD et al (2008) Small-animal SPECT and SPECT/CT: important tools for preclinical investigation. J Nucl Med 49(10):1651–1663
Frangioni JV (2003) In vivo near-infrared fluorescence imaging. Curr Opin Chem Biol 7(5):626–634
Frese KK, Tuveson DA (2007) Maximizing mouse cancer models. Nat Rev Cancer 7(9):645–658
Frullano L, Meade TJ (2007) Multimodal MRI contrast agents. J Biol Inorg Chem 12(7):939–949

Gambhir SS (2002) Molecular imaging of cancer with positron emission tomography. Nat Rev Cancer 2(9):683–693
Gao X, Cui Y et al (2004) In vivo cancer targeting and imaging with semiconductor quantum dots. Nat Biotechnol 22(8):969–976
Georgakoudi I, Solban N et al (2004) In vivo flow cytometry: a new method for enumerating circulating cancer cells. Cancer Res 64(15):5044–5047
Giepmans BN, Adams SR et al (2006) The fluorescent toolbox for assessing protein location and function. Science 312(5771):217–224
Gilad AA, McMahon MT et al (2007) Artificial reporter gene providing MRI contrast based on proton exchange. Nat Biotechnol 25(2):217–219
Gillies RJ, Morse DL (2005) In vivo magnetic resonance spectroscopy in cancer. Annu Rev Biomed Eng 7:287–326
Grimm J, Kirsch DG et al (2005) Use of gene expression profiling to direct in vivo molecular imaging of lung cancer. Proc Natl Acad Sci USA 102(40):14404–14409
Gross S, Piwnica-Worms D (2005) Spying on cancer: molecular imaging in vivo with genetically encoded reporters. Cancer Cell 7(1):5–15
Halin C, Rodrigo Mora J et al (2005) In vivo imaging of lymphocyte trafficking. Annu Rev Cell Dev Biol 21:581–603
Hamstra DA, Rehemtulla A et al (2007) Diffusion magnetic resonance imaging: a biomarker for treatment response in oncology. J Clin Oncol 25(26):4104–4109
Hauff P, Reinhardt M et al (2008) Ultrasound basics. Handb Exp Pharmacol (185 Pt 1):91–107
Helms MW, Brandt BH et al (2006) Options for visualizing metastatic disease in the living body. Contrib Microbiol 13:209–231
Herschman HR (2003a) Micro-PET imaging and small animal models of disease. Curr Opin Immunol 15(4):378–384
Herschman HR (2003b) Molecular imaging: looking at problems, seeing solutions. Science 302(5645):605–608
Hirst GL, Balmain A (2004) Forty years of cancer modelling in the mouse. Eur J Cancer 40(13):1974–1980
Hoffman RM (2005) The multiple uses of fluorescent proteins to visualize cancer in vivo. Nat Rev Cancer 5(10):796–806
Hofmann M, Steinke F et al (2008) MRI-based attenuation correction for PET/MRI: a novel approach combining pattern recognition and atlas registration. J Nucl Med 49(11):1875–1883
Howarth M, Liu W et al (2008) Monovalent, reduced-size quantum dots for imaging receptors on living cells. Nat Methods 5(5):397–399
Hsiung PL, Hardy J et al (2008) Detection of colonic dysplasia in vivo using a targeted heptapeptide and confocal microendoscopy. Nat Med 14(4):454–458
Inoue Y, Nomura Y et al (2006) Imaging living mice using a 1-T compact MRI system. J Magn Reson Imaging 24(4):901–907
Jacobson GB, Shinde R et al (2008) Sustained release of drugs dispersed in polymer nanoparticles. Angew Chem Int Ed Engl 47(41):7880–7882
Jaffer FA, Weissleder R (2005) Molecular imaging in the clinical arena. JAMA 293(7):855–862
Jain RK, Munn LL et al (2002) Dissecting tumour pathophysiology using intravital microscopy. Nat Rev Cancer 2(4):266–276
Jiang T, Olson ES et al (2004) Tumor imaging by means of proteolytic activation of cell-penetrating peptides. Proc Natl Acad Sci USA 101(51):17867–17872
Jones LR, Goun EA et al (2006) Releasable luciferin-transporter conjugates: tools for the real-time analysis of cellular uptake and release. J Am Chem Soc 128(20):6526–6527
Jonkers J, Berns A (2002) Conditional mouse models of sporadic cancer. Nat Rev Cancer 2(4):251–265
Kanno A, Yamanaka Y et al (2007) Cyclic luciferase for real-time sensing of caspase-3 activities in living mammals. Angew Chem Int Ed Engl 46(40):7595–7599
Karnoub AE, Dash AB et al (2007) Mesenchymal stem cells within tumour stroma promote breast cancer metastasis. Nature 449(7162):557–563

Kurhanewicz J, Bok R et al (2008) Current and potential applications of clinical 13C MR spectroscopy. J Nucl Med 49(3):341–344
Larson DR, Zipfel WR et al (2003) Water-soluble quantum dots for multiphoton fluorescence imaging in vivo. Science 300(5624):1434–1436
Laxman B, Hall DE et al (2002) Noninvasive real-time imaging of apoptosis. Proc Natl Acad Sci USA 99(26):16551–16555
Loening AM, Wu AM et al (2007) Red-shifted Renilla reniformis luciferase variants for imaging in living subjects. Nat Methods 4(8):641–643
Luker GD, Luker KE (2008) Optical imaging: current applications and future directions. J Nucl Med 49(1):1–4
Luker GD, Pica CM et al (2003) Imaging 26S proteasome activity and inhibition in living mice. Nat Med 9(7):969–973
Luker GD, Sharma V et al (2002) Noninvasive imaging of protein-protein interactions in living animals. Proc Natl Acad Sci USA 99(10):6961–6966
Luker KE, Gupta M et al (2009) Imaging chemokine receptor dimerization with firefly luciferase complementation. FASEB J 23(3):823–834
Lyons SK (2005) Advances in imaging mouse tumour models in vivo. J Pathol 205(2):194–205
Ma L, Teruya-Feldstein J et al (2007) Tumour invasion and metastasis initiated by microRNA-10b in breast cancer. Nature 449(7163):682–688
Makale M (2007) Intravital imaging and cell invasion. Methods Enzymol 426:375–401
Mandl S, Schimmelpfennig C et al (2002) Understanding immune cell trafficking patterns via in vivo bioluminescence imaging. J Cell Biochem Suppl 39:239–248
Mani SA, Guo W et al (2008) The epithelial-mesenchymal transition generates cells with properties of stem cells. Cell 133(4):704–715
Massoud TF, Gambhir SS (2003) Molecular imaging in living subjects: seeing fundamental biological processes in a new light. Genes Dev 17(5):545–580
Masuda H, Okano HJ et al (2008) In vivo imaging in humanized mice. Curr Top Microbiol Immunol 324:179–196
McAllister SS, Gifford AM et al (2008) Systemic endocrine instigation of indolent tumor growth requires osteopontin. Cell 133(6):994–1005
Meade TJ, Taylor AK et al (2003) New magnetic resonance contrast agents as biochemical reporters. Curr Opin Neurobiol 13(5):597–602
Meric P, Autret G et al (2004) In vivo 2D magnetic resonance spectroscopy of small animals. MAGMA 17(3–6):317–338
Merzlyak EM, Goedhart J et al (2007) Bright monomeric red fluorescent protein with an extended fluorescence lifetime. Nat Methods 4(7):555–557
Michalet X, Pinaud FF et al (2005) Quantum dots for live cells, in vivo imaging, and diagnostics. Science 307(5709):538–544
Min JJ, Nguyen VH et al (2008) Quantitative bioluminescence imaging of tumor-targeting bacteria in living animals. Nat Protoc 3(4):629–636
Mrass P, Takano H et al (2006) Random migration precedes stable target cell interactions of tumor-infiltrating T cells. J Exp Med 203(12):2749–2761
Nie S, Xing Y et al (2007) Nanotechnology applications in cancer. Annu Rev Biomed Eng 9:257–288
Nieman BJ, Bock NA et al (2005) Magnetic resonance imaging for detection and analysis of mouse phenotypes. NMR Biomed 18(7):447–468
Nilsson FY, Tolmachev V (2007) Affibody molecules: new protein domains for molecular imaging and targeted tumor therapy. Curr Opin Drug Discov Devel 10(2):167–175
Ntziachristos V (2006) Fluorescence molecular imaging. Annu Rev Biomed Eng 8:1–33
Ntziachristos V, Bremer C et al (2003) Fluorescence imaging with near-infrared light: new technological advances that enable in vivo molecular imaging. Eur Radiol 13(1):195–208
Ntziachristos V, Schellenberger EA et al (2004) Visualization of antitumor treatment by means of fluorescence molecular tomography with an annexin V-Cy5.5 conjugate. Proc Natl Acad Sci USA 101(33):12294–12299

O'Connell-Rodwell CE, Mackanos MA et al (2008) In vivo analysis of heat-shock-protein-70 induction following pulsed laser irradiation in a transgenic reporter mouse. J Biomed Opt 13(3):030501
O'Connell-Rodwell CE, Shriver D et al (2004) A genetic reporter of thermal stress defines physiologic zones over a defined temperature range. FASEB J 18(2):264–271
Olafsen T, Gu Z et al (2007) Targeting, imaging, and therapy using a humanized antiprostate stem cell antigen (PSCA) antibody. J Immunother 30(4):396–405
Patterson SS, Dionisi HM et al (2005) Codon optimization of bacterial luciferase (lux) for expression in mammalian cells. J Ind Microbiol Biotechnol 32(3):115–123
Perkins AC, Missailidis S (2007) Radiolabelled aptamers for tumour imaging and therapy. Q J Nucl Med Mol Imaging 51(4):292–296
Pickhardt PJ, Halberg RB et al (2005) Proc Natl Acad Sci USA 102(9):3419–3422
Popovtzer R, Agrawal A et al (2008) Targeted gold nanoparticles enable molecular CT imaging of cancer. Nano Lett 8(12):4593–4596
Rabinovich BA, Ye Y et al (2008) Visualizing fewer than 10 mouse T cells with an enhanced firefly luciferase in immunocompetent mouse models of cancer. Proc Natl Acad Sci USA 105(38):14342–14346
Raman V, Pathak AP et al (2007) Magnetic resonance imaging and spectroscopy of transgenic models of cancer. NMR Biomed 20(3):186–199
Rao J, Dragulescu-Andrasi A et al (2007) Fluorescence imaging in vivo: recent advances. Curr Opin Biotechnol 18(1):17–25
Ray P, Tsien R et al (2007) Construction and validation of improved triple fusion reporter gene vectors for molecular imaging of living subjects. Cancer Res 67(7):3085–3093
Rice BW, Cable MD et al (2001) In vivo imaging of light-emitting probes. J Biomed Opt 6(4):432–440
Ritman EL (2002) Molecular imaging in small animals – roles for micro-CT. J Cell Biochem Suppl 39:116–124
Ritman EL (2004) Micro-computed tomography-current status and developments. Annu Rev Biomed Eng 6:185–208
Rutten A, Prokop M (2007) Contrast agents in X-ray computed tomography and its applications in oncology. Anticancer Agents Med Chem 7(3):307–316
Safran M, Kim WY et al (2006) Mouse model for noninvasive imaging of HIF prolyl hydroxylase activity: assessment of an oral agent that stimulates erythropoietin production. Proc Natl Acad Sci USA 103(1):105–110
Sevick-Muraca EM, Houston JP et al (2002) Fluorescence-enhanced, near infrared diagnostic imaging with contrast agents. Curr Opin Chem Biol 6(5):642–650
Shachaf CM, Felsher DW (2005) Tumor dormancy and MYC inactivation: pushing cancer to the brink of normalcy. Cancer Res 65(11):4471–4474
Shah K, Tang Y et al (2003) Real-time imaging of TRAIL-induced apoptosis of glioma tumors in vivo. Oncogene 22(44):6865–6872
Shah K, Weissleder R (2005) Molecular optical imaging: applications leading to the development of present day therapeutics. NeuroRx 2(2):215–225
Shaner NC, Steinbach PA et al (2005) A guide to choosing fluorescent proteins. Nat Methods 2(12):905–909
Sharma V, Prior JL et al (2005) Characterization of a 67Ga/68Ga radiopharmaceutical for SPECT and PET of MDR1 P-glycoprotein transport activity in vivo: validation in multidrug-resistant tumors and at the blood-brain barrier. J Nucl Med 46(2):354–364
Shcherbo D, Merzlyak EM et al (2007) Bright far-red fluorescent protein for whole-body imaging. Nat Methods 4(9):741–746
Simoes RV, Martinez-Aranda A et al (2008) Preliminary characterization of an experimental breast cancer cells brain metastasis mouse model by MRI/MRS. MAGMA 21(4):237–249
Singh M, Johnson L (2006) Using genetically engineered mouse models of cancer to aid drug development: an industry perspective. Clin Cancer Res 12(18):5312–5328

Sosnovik DE, Weissleder R (2007) Emerging concepts in molecular MRI. Curr Opin Biotechnol 18(1):4–10
Stell A, Biserni A et al (2007) Cancer modeling: modern imaging applications in the generation of novel animal model systems to study cancer progression and therapy. Int J Biochem Cell Biol 39(7–8):1288–1296
Stewart CN Jr (2006) Go with the glow: fluorescent proteins to light transgenic organisms. Trends Biotechnol 24(4):155–162
Strack RL, Strongin DE et al (2008) A noncytotoxic DsRed variant for whole-cell labeling. Nat Methods 5(11):955–957
Sundaresan G, Yazaki PJ et al (2003) 124I-labeled engineered anti-CEA minibodies and diabodies allow high-contrast, antigen-specific small-animal PET imaging of xenografts in athymic mice. J Nucl Med 44(12):1962–1969
Sweeney TJ, Mailander V et al (1999) Visualizing the kinetics of tumor-cell clearance in living animals. Proc Natl Acad Sci USA 96(21):12044–12049
Swirski FK, Berger CR et al (2007) A near-infrared cell tracker reagent for multiscopic in vivo imaging and quantification of leukocyte immune responses. PLoS One 2(10):e1075
Swirski FK, Pittet MJ et al (2006) Monocyte accumulation in mouse atherogenesis is progressive and proportional to extent of disease. Proc Natl Acad Sci USA 103(27):10340–10345
Tavazoie SF, Alarcon C et al (2008) Endogenous human microRNAs that suppress breast cancer metastasis. Nature 451(7175):147–152
Thorne SH, Negrin RS et al (2006) Synergistic antitumor effects of immune cell-viral biotherapy. Science 311(5768):1780–1784
Townsend DW (2008) Multimodality imaging of structure and function. Phys Med Biol 53(4):R1–R39
Tung CH, Mahmood U et al (2000) In vivo imaging of proteolytic enzyme activity using a novel molecular reporter. Cancer Res 60(17):4953–4958
Van Dyke T, Jacks T (2002) Cancer modeling in the modern era: progress and challenges. Cell 108(2):135–144
Venisnik KM, Olafsen T et al (2006) Bifunctional antibody-Renilla luciferase fusion protein for in vivo optical detection of tumors. Protein Eng Des Sel 19(10):453–460
Vignjevic D, Fre S et al (2007) Conditional mouse models of cancer. Handb Exp Pharmacol (178):263–287
Villalobos V, Naik S et al (2007) Current state of imaging protein-protein interactions in vivo with genetically encoded reporters. Annu Rev Biomed Eng 9:321–349
Vooijs M, Jonkers J et al (2002) Noninvasive imaging of spontaneous retinoblastoma pathway-dependent tumors in mice. Cancer Res 62(6):1862–1867
Voura EB, Jaiswal JK et al (2004) Tracking metastatic tumor cell extravasation with quantum dot nanocrystals and fluorescence emission-scanning microscopy. Nat Med 10(9):993–998
Wehrman TS, von Degenfeld G et al (2006) Luminescent imaging of beta-galactosidase activity in living subjects using sequential reporter-enzyme luminescence. Nat Methods 3(4):295–301
Weiner RE, Thakur ML (2005) Radiolabeled peptides in oncology: role in diagnosis and treatment. BioDrugs 19(3):145–163
Weissleder R (2002) Scaling down imaging: molecular mapping of cancer in mice. Nat Rev Cancer 2(1):11–18
Weissleder R (2006) Molecular imaging in cancer. Science 312(5777):1168–1171
Weissleder R, Pittet MJ (2008) Imaging in the era of molecular oncology. Nature 452(7187):580–589
Weissleder R, Tung CH et al (1999) In vivo imaging of tumors with protease-activated near-infrared fluorescent probes. Nat Biotechnol 17(4):375–378
Wender PA, Goun EA et al (2007) Real-time analysis of uptake and bioactivatable cleavage of luciferin-transporter conjugates in transgenic reporter mice. Proc Natl Acad Sci USA 104(25):10340–10345
Wurdinger T, Badr C et al (2008) A secreted luciferase for ex vivo monitoring of in vivo processes. Nat Methods 5(2):171–173

Yang L, Yang H et al (2008) Engineered lentivector targeting of dendritic cells for in vivo immunization. Nat Biotechnol 26(3):326–334
Zabow G, Dodd S et al (2008) Micro-engineered local field control for high-sensitivity multispectral MRI. Nature 453(7198):1058–1063
Zhang L, Lee KC et al (2007) Molecular imaging of Akt kinase activity. Nat Med 13(9):1114–1119
Zhao H, Doyle TC et al (2005) Emission spectra of bioluminescent reporters and interaction with mammalian tissue determine the sensitivity of detection in vivo. J Biomed Opt 10(4):41210
Zhao H, Doyle TC et al (2004) Characterization of coelenterazine analogs for measurements of Renilla luciferase activity in live cells and living animals. Mol Imaging 3(1):43–54

Chapter 12
Identifying Mammary Epithelial Stem and Progenitor Cells

Andrew O. Giacomelli, Robin M. Hallett, and John A. Hassell

12.1 Introduction

The adult mouse mammary gland is composed of a branched ductal epithelium, which is embedded in a fatty stroma, termed the fat pad, comprising largely adipocytes but also fibroblasts, macrophages, and endothelial cells (Daniel and Silberstein 1987; Sakakura 1991). The various branches of the mammary tree are connected to a primary duct that terminates at the skin surface through the nipple. The mammary epithelium initially develops during embryogenesis from the ectoderm by cell migration and proliferation thus forming mammary buds, which subsequently sprout a rudimentary ductal tree that is present at birth (Sakakura 1987). Further mammary gland development is completed postnatally in defined stages (puberty, pregnancy, lactation, and involution), which are coupled to the sexual maturation and reproductive status of the animal (Fig. 12.1) (Hennighausen and Robinson 1998).

Adult mammary ducts comprise a continuous bilayer of two differentiated cell types (Sakakura 1987). Contractile myoepithelial cells constitute the outer surface of ducts, whereas luminal epithelial cells make up their inner layer. Alveoli, sack-like structures that develop during pregnancy from the sides and ends of ducts, are also composed of an outer basket-like layer of myoepithelial cells and a contiguous interior surface of secretory luminal epithelial cells that secrete milk proteins during lactation. These differentiated cells together constitute ~90% of all the epithelial cells in the mouse mammary tree (Smalley et al. 1998).

A.O. Giacomelli • R.M. Hallett
Department of Biochemistry, McMaster University, ON, Canada

J.A. Hassell (✉)
Department of Biochemistry, McMaster University, ON, Canada

Centre for Functional Genomics, McMaster University, 1200 Main Street West, Hamilton, ON, L8N 3Z5 Canada
e-mail: hassell@mcmaster.ca

J.E. Green and T. Ried (eds.), *Genetically Engineered Mice for Cancer Research: Design, Analysis, Pathways, Validation and Pre-Clinical Testing*,
DOI 10.1007/978-0-387-69805-2_12, © Springer Science+Business Media, LLC 2012

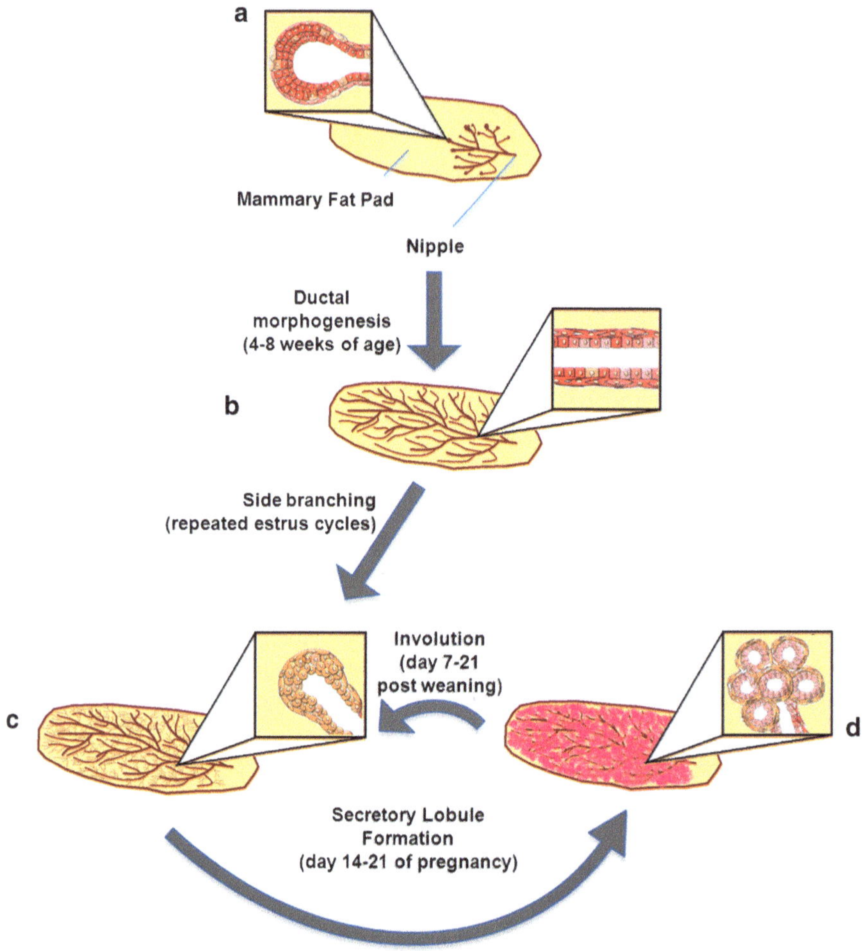

Fig. 12.1 Postnatal development of the mouse mammary epithelium. At the onset of puberty in the mammary glands of female mice, terminal end-buds (TEBs – inset **a**) form at the tips of a rudimentary mammary tree, which was present at birth. TEBs comprise abundant mammary epithelial stem cells, cap cells (ductal myoepithelial progenitors), and body cells (ductal luminal progenitors). At the end of puberty (**b**), the ductal tree extends to the extremities of the fat pad and the majority of ducts comprise a contiguous layer of tightly packed luminal cells separated from the basement membrane by a layer of contractile myoepithelial cells (inset **b**). During successive estrus cycles, the ductal tree undergoes repeated side branching to yield alveolar buds (inset **c**). During pregnancy, secretory lobules (mature alveoli – inset **d**) develop from these buds and secrete milk into the epithelial lumen throughout lactation. When pups cease nursing, milk stasis signals cells of the alveoli to undergo programmed cell death. This process, termed involution, returns the mammary gland to a state, which resembles that of a sexually mature nulliparous female. The process of secretory lobule formation, lactation, and involution occurs at each subsequent pregnancy. Epithelial cell types in inset images are as they appear in Fig. 12.2

Terminal end buds (TEBs), which develop at the onset of puberty, are bulbous structures at the ends of ducts and are composed of a near solid mass of myoepithelial and luminal progenitor cells that are required for ductal elongation and branching (Williams and Daniel 1983; Chepko and Smith 1999). Morphological analyses and immunostaining with antibodies to lineage-restricted mammary epithelial cell markers suggest that cap cells, which are located on the periphery of TEBs, are the progenitors of myoepithelial cells, whereas body cells, situated centrally within the TEBs, are the progenitors of the luminal epithelial cells of the subtending duct (Williams and Daniel 1983). TEBs regress at the time of sexual maturity when the ducts reach the limits of the fat pad.

Mammary epithelial stem cells are thought to be required for organogenesis, and to maintain homeostasis and regenerative capacity of the mammary epithelium during successive reproductive cycles (Smith and Chepko 2001; Smith and Boulanger 2003). The existence of mouse mammary epithelial stem cells was first inferred from the pioneering studies of DeOme and colleagues, who showed that transplant of mammary epithelial tissue fragments into the epithelium-free (cleared) fat pad of syngeneic hosts resulted in outgrowths that recapitulated the morphology of the mammary tree in virgin recipients (DeOme et al. 1959; Daniel 1973; Daniel et al. 1983). These outgrowths further developed during pregnancy to yield increased branching ducts and alveoli secreting milk proteins characteristic of the normal mammary gland during pregnancy. Transplant of such mammary-tissue seeded outgrowths into females that subsequently become pregnant could be repeated between five and eight times before propagation of the outgrowths failed; the serial transplantation limit is thought to be due to the exhaustion of the stem cell pool and suggests that these cells have a high but finite capacity for self-renewal (Smith and Boulanger 2002; Bruno and Smith 2010).

Mammary epithelial tissue fragments give rise to outgrowths in female hosts independent of the region or developmental stage from which they are isolated, suggesting that stem cells are located throughout the mammary tree and persist during all phases of postnatal mammary gland development (Smith and Medina 1988). Interestingly, the frequency of mammary epithelial stem cells in 3-week or 26-month-old virgin females is similar, as is their capacity for self-renewal. Outgrowths with similar capacity for serial transplantation exist in virgin or parous females independent of their age (Bruno and Smith 2010). Collectively, these findings suggest that mammary epithelial stem cells are generally quiescent and can be stimulated to self-renew during postnatal mammary gland development.

Retroviral marking experiments showed that an entire functional mammary tree capable of serial transplantation can originate from an individual stem cell (Kordon and Smith 1998). Limiting dilution cell transplant experiments of dispersed mammary epithelial cells into the epithelium-free fat pads of mice demonstrate that stem cells comprise between 0.01 and 0.04% of all the cells in the mammary epithelium (Smith 1996; Shackleton et al. 2006). Direct demonstration that single mammary stem cells seed the growth of mammary trees was achieved by partial purification of these cells using fluorescence-activated cell sorting (FACS) and injection of visually identified single cells into cleared fat pads of recipient mice (Stingl et al. 2006).

Two functional classes of progenitor cells, duct-limited and lobule-limited progenitors, were identified by transplantation of between 5,000 and 10,000 dispersed mammary epithelial cells into the epithelial-free fat pads of recipient females (Smith 1996; Smith and Boulanger 2002). Both of these functional progenitor classes are less abundant than mammary epithelial stem cells; moreover, duct-limited progenitors are much less prevalent than their lobule-limited counterparts. Both functional progenitor cell types occur in genetically marked clonal outgrowths during their serial transplantation at roughly the same frequency as that found in the mammary gland thus demonstrating that the duct- and lobule-limited progenitor cells originate from mammary epithelial stem cells (Bruno and Smith 2010). Moreover, the capacity of outgrowths seeded by mammary epithelial cells to generate ducts or lobules is lost independently during serial transplantation implying that the two progenitor classes have different proliferative capacities (Bruno and Smith 2010).

Mouse mammary epithelial progenitor cells can also be identified, and their developmental potential and frequency estimated by two-dimensional colony-forming assays performed *in vitro* (Smalley et al. 1998; Kurpios et al. 2009). Four types of progenitor cells have been enumerated by these assays and include two multipotent progenitors, which form morphologically distinct colonies comprising cells that express either luminal or myoepithelial markers. The two other morphologically distinct colonies comprise only luminal epithelial cells or myoepithelial cells and thus originate from different unipotent progenitors. Colony-forming assays performed with limiting dilutions of primary mammary epithelial cells suggest that progenitor cells comprise a minimum of 2–5% of the mammary epithelial cells in sexually mature virgin mice.

Whereas functional *in vivo* assays represent the gold standard for defining mammary epithelial stem and progenitor cell subpopulations, various other methods in addition to those described above including functional *in vitro* clonogenic assays such as the formation of spherical, duct-like, or acinar structures in Matrigel, and sphere-formation in chemically defined medium, as well as analyses of the expression of lineage-restricted markers *in situ* have all contributed to our understanding of the mammary epithelial cell hierarchy. Here, we review the recent literature focusing on methods that have been used to partially purify individual mouse mammary epithelial stem and progenitor cell populations, and assays that have been employed to identify these cells. We refer the reader to several excellent reviews that have recently been published for additional perspectives on mammary epithelial stem and progenitor cells in both the mouse and human mammary glands (Visvader 2009; Bruno and Smith 2010; Petersen and Polyak 2010).

12.2 Preparation and Isolation of Mammary Epithelial Cell Subpopulations

Stem and progenitor cells are generally rare among the cells of the mammary epithelium. Hence, rigorous characterization of their phenotypic and molecular properties requires that they should be separated from the bulk cell population prior to analysis. Typically, the latter is accomplished using FACS, a technique that can fractionate cells based on their expression of particular cell surface antigens, retention of small molecular weight dyes (i.e., Hoechst), or the activity of intracellular enzymes (Hulett et al. 1969; Alexander et al. 2009). The choice of a candidate marker or combination of markers for the sorting protocol is usually informed by some prior knowledge of the population of interest and depends on the availability of suitable fluorescent probes.

The successful application of FACS requires that viable cells be isolated from tissues, and that these survive the sorting protocol; these procedures warrant special consideration when cells from solid tissues are characterized (Sleeman et al. 2006; Alexander et al. 2009). The mammary epithelium comprises tightly linked epithelial cells and is surrounded by a supportive fatty stroma and extracellular matrix proteins, principally comprising collagens and laminins (Sheffield 1988). Because of this complex architecture, a series of mild mechanical and enzymatic dissociation steps must be performed to isolate viable dispersed cells (Smith 1996). Unfortunately, the dissociation of mammary tissue can destroy cell surface proteins that may be required for stem/progenitor cell isolation or engraftment, and may thus complicate downstream analyses. Despite these shortcomings, coupling FACS to limiting-dilution cell transplantation assays *in vivo* and clonogenic assays *in vitro* has allowed researchers to assemble markers that can be used to identify and segregate defined epithelial cell populations within the mammary gland (Fig. 12.2 and Table 12.1) (also reviewed in Visvader 2009).

Quantitative functional assays are required to determine the frequency of stem or particular progenitor cells in the FACS-sorted fractions and provide a measure of the usefulness of marker combinations in purifying these cells. The robustness of these functional assays needs to be determined to provide assurance that they accurately read out stem or progenitor cell frequency. Contingent on the frequency of the cell of interest identified by a marker or combination thereof, an estimate of the relative abundance of these cells may be inferred from the frequency of cells that express the marker. However, few markers currently exist that uniquely identify stem or specific progenitor cells either individually or in combination with others. For example, stem cells represent less than 10% of the total cell population in the most highly purified cell fractions, underscoring the paramount importance of using functional assays rather than biomarkers to define these cells.

12.3 Purification of Mouse Mammary Epithelial Stem Cells

Mouse mammary epithelial cells can be fractionated based on the expression of CD24 to separate myoepithelial lineage and luminal lineage cells (Sleeman et al. 2006); the $CD24^{lo}$ cell population comprises stem cells and myoepithelial-lineage restricted cells, whereas $CD24^{hi}$ cells comprises luminal-lineage cells including progenitor cells that exhibit clonogenic capacity *in vitro* (Sleeman et al. 2007). The $CD24^{+/med}$ fraction can be further subdivided on the basis of CD29 or CD49f expression

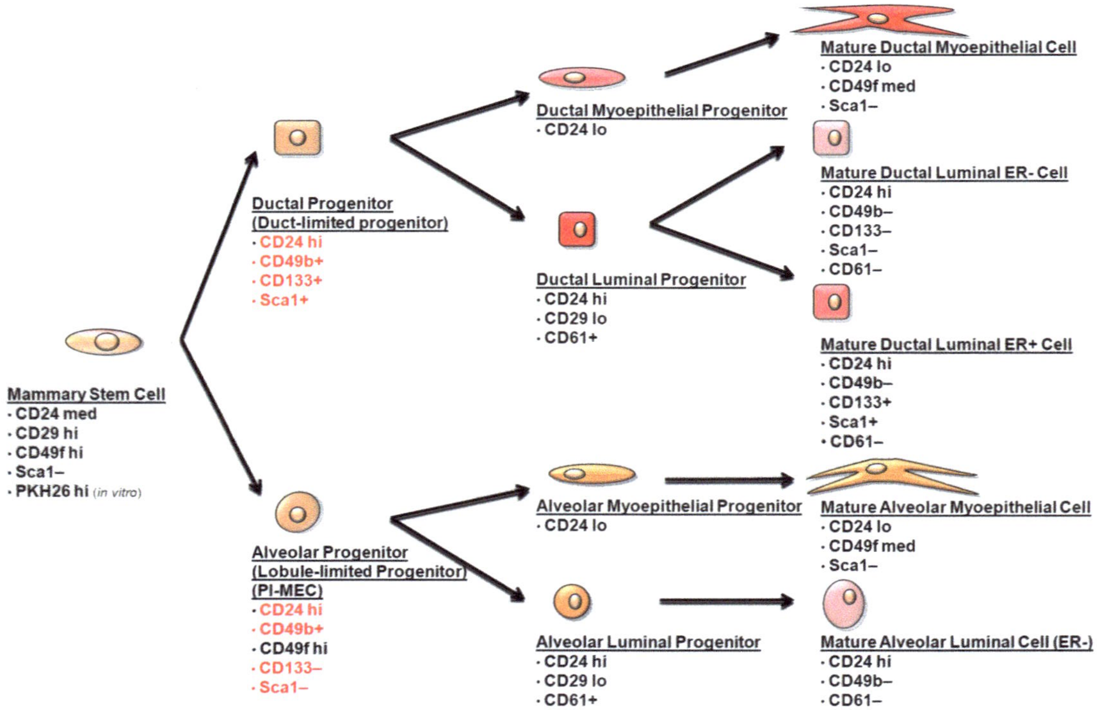

Fig. 12.2 Surface markers to partially purify epithelial cell subpopulations from the mouse mammary gland. Multipotent mammary epithelial stem cells can give rise to all the cell types of the mature mammary gland through various progenitor cell intermediates. Using a combination of *in situ* staining, fluorescence-activated cell sorting, and functional stem and progenitor cell assays, a variety of surface markers have been described which can enrich for these various epithelial cell populations. Markers in *red* are inferred

Table 12.1 Mammary Epithelial Cell Markers.

Cell-surface marker	Pseudonyms	Cell types marked in the mouse mammary gland
CD10	CALLA Neprilysin Membrane metallo endopeptidase	+: Myoepithelial-lineage cells
CD24	Heat stable antigen	Hi: Luminal-lineage cells Lo: Myoepithelial-lineage cells –: Nonepithelial cells
CD29	β1-Integrin Fibronectin receptor β	Hi: Mammary stem cells Lo: Mammary colony-forming cells
CD44	Hyaluronic acid receptor	Hi: Myoepithelial-lineage cells, primitive luminal-lineage cells Lo: Differentiated luminal-lineage cells
CD49b	α2-Integrin	+: Multipotent colony-forming cells
CD49f	α6-Integrin	Hi: Mammary stem cells Lo: Mammary colony-forming cells
CD61	β3-Integrin	+: Luminal-lineage-limited colony-forming cells
CD90	Thy1	+: Myoepithelial progenitor cells
CD133	Prominin-1	+: $ER\alpha^{+}$ multipotent colony-forming cells –: $ER\alpha^{-}$ multipotent colony-forming cells
MUC1	Mucin-1	+: Luminal-lineage cells –: Myoepithelial-lineage cells
MUC4	Mucin-4	+: Luminal-lineage cells –: Myoepithelial-lineage cells
EpCAM	CD326 Epithelial specific antigen Epithelial cell adhesion molecule Tumor-associated calcium signal transducer 1	Hi: Luminal-lineage cells Lo: Basally located cells
Sca1	Stem cell antigen	+: $ER\alpha^{+}$ multipotent colony-forming cells –: $ER\alpha^{-}$ multipotent colony-forming cells
Intracellular marker		
α-SMA	α-Smooth muscle actin	+: Myoepithelial-lineage cells, Fibroblasts –: Luminal-lineage cells
CK5	Keratin, type II cytoskeletal 5	+: Myoepithelial-lineage cells –: Luminal-lineage cells
CK6	Keratin, type II cytoskeletal 6	+: Myoepithelial-lineage cells –: Luminal-lineage cells
CK8	Keratin, type II cytoskeletal 8	+: Luminal-lineage cells –: Myoepithelial-lineage cells
CK14	Keratin, type I cytoskeletal 14 Epidermolysis bullosa simplex	+: Myoepithelial-lineage cells –: Luminal-lineage cells
CK18	Keratin, type I cytoskeletal 18	+: Luminal-lineage cells –: Myoepithelial-lineage cells
CK19	Keratin, type I cytoskeletal 19	+: Luminal-lineage cells –: Myoepithelial-lineage cells

to enrich for mammary epithelial stem cells. Single cells from lineage-depleted (Ter119$^-$CD45$^-$CD31$^-$) CD24$^+$CD29^{hi} (Shackleton et al. 2006) or CD24^{med}CD49f^{hi} Sca1$^-$ (Stingl et al. 2006) cell fractions are capable of producing complete mammary outgrowths that comprise both luminal and myoepithelial cells. In response to hormonal cues during pregnancy, these outgrowths undergo extensive alveolar development, and at parturition produce and secrete milk proteins and lipid droplets into the epithelial lumen – an indication that the transplanted stem cells morphologically and functionally differentiated. Importantly, cells isolated from single cell-generated primary outgrowths yielded multiple complete secondary and tertiary outgrowths, confirming that the cell that formed the primary outgrowth possessed extensive self-renewal capacity (Shackleton et al. 2006). Evidence for the clonality of the outgrowths was also established by transplanting mixtures of 140 Lin$^-$CD29^{hi}CD24$^+$ mammary epithelial cells, 70 from a LacZ-positive donor, and 70 from a LacZ-negative donor, and demonstrating that nearly all the resulting outgrowths comprised either LacZ-positive or LacZ-negative cells, but not both (Shackleton et al. 2006). It should be noted, however, that despite the significant enrichment achieved using these purification strategies, stem cells constitute but a minor fraction of the Lin$^-$CD29^{hi}CD24$^+$ and Lin$^-$CD49f^{hi}CD24^{med}Sca1$^-$ sorted populations (about 1/60 and 1/20, respectively). Hence co-expression of CD49f or CD29 with low to medium levels of CD24 is an attribute of cells other than stem cells and thus expression of these markers is insufficient to uniquely identify stem cells. Furthermore, molecular analyses of these populations might not be expected to identify stem cell-specific attributes. Indeed, no significant difference in the expression of lineage markers was identified in the profile of the Lin$^-$CD49f^{hi}CD24^{med} population compared to the Lin$^-$CD49f^{lo}CD24^{lo} population (Stingl et al. 2006). It is also noteworthy that the stem cell engraftment assay as practiced by these investigators was robust, because a single stem cell yielded complete functional mammary trees capable of serial transplantation following its orthotopic transplantation into cleared fat pads.

Purer populations of mammary epithelial stem cells have been isolated more recently using a technique that exploits their likelihood of undergoing asymmetric self-renewal *in vitro* during the process of mammosphere formation (Cicalese et al. 2009). Cicalese et al. built on earlier findings of Dontu et al., which showed that a minor cell population isolated from human breast reduction mammoplasties forms clonal nonadherent mammospheres when cultured in chemically defined medium (Dontu et al. 2003). Dissociating human mammospheres and replating the dispersed cells at low density on a collagen substratum in differentiation-inducing conditions resulted in the outgrowth of colonies that contain both luminal (EpCAM$^+$, MUC1$^+$, CK18$^+$) and myoepithelial (CALLA$^+$, A-SMA$^+$, CK14$^+$) cells; roughly 10% of the mammosphere-derived cells formed colonies. When overlaid with a layer of Matrigel and exposed to lactogenic hormones, some of the colonies produced alveolar-like, acinar structures that secreted milk proteins into their lumen. Collectively, these results suggested that human mammospheres comprise a population of multipotent progenitor cells. Furthermore, the capacity of a small fraction (1/100) of mammosphere-resident cells to self-renew in culture and yield new spheres following dissociation and reseeding of the dissociated cells suggested that human mammospheres also comprise mammary epithelial stem cells. This method of establishing and propagating mammospheres was subsequently extended to mouse mammary epithelial cells (Youn et al. 2005; Liao et al. 2007).

To determine whether mouse mammospheres comprised functional mammary epithelial stem cells and to learn whether these cells could be fractionated from the other mammosphere-resident cells, Cicalese and colleagues used PKH26, a membrane-binding fluorescent dye that segregates equally to daughter cells upon mitosis (Lanzkron et al. 1999; Cicalese et al. 2009; Pece et al. 2010). When administered in a pulse-chase manner in culture, the dye can act as an indicator of proliferation history at the single-cell level; cells that undergo few divisions during the chase period retain high levels of the dye, whereas the dye is diluted to low or undetectable levels in those cells that proliferated during the chase. When freshly isolated mouse mammary epithelial cells were seeded in mammosphere-forming conditions and labeled with PKH26, greater than 99% of the cells initially stained brightly for the dye (Cicalese et al. 2009). However, after an additional passage, fewer than 60% of cells retained detectable PKH26 levels and the staining intensities were distributed over a broad range, suggesting that there was substantial heterogeneity in the proliferation rates of cells within mammospheres.

Time-lapse fluorescence microscopy confirmed that following dissociation, a minority of PKH26-stained cells survived anoikis and divided to produce daughter cells with distinct fates – one cell subpopulation remained quiescent and thus $PKH26^{hi}$, whereas the other proliferated extensively to generate $PKH26^{lo/-}$ cells. When second generation mammospheres were fractionated by FACS and subjected to limiting-dilution mammary fat pad repopulation assays, those cells that stained most brightly with PKH26 (top 1%) were highly enriched (on the order of 1/3) in functional mammary stem cells. By contrast, no outgrowths were obtained when cells from other PKH26-subsets were transplanted. Collectively, these data support the notion that mammosphere-initiating cells comprise a population of bona fide stem cells, and by inference that sphere-formation provides an *in vitro* surrogate assay for stem cell self-renewal. Moreover, this study demonstrated that PKH26 could be used as a single agent to substantially enrich for mammary epithelial stem cells to levels approximating 30% of the mammosphere-derived, FACS-sorted cell population.

12.4 Prospective Isolation of Human Mammary Stem Cells

Analyses of mammary epithelial stem cells from the human gland have proven challenging, primarily because cross species transplantation of human cells into immune-compromised mice may not be permissive for stem cell engraftment, self-renewal, and/or differentiation. The xenograft transplantation assay may also underestimate human stem cell frequencies for these very same reasons. The adipose cell-rich mouse mammary fat pad, the site of choice for transplantation of mouse mammary epithelial cells, does not readily lend itself to engraftment of their human counterparts, which are adapted to function within the more fibrous tissue of the human breast and may require species-specific growth and/or differentiation factors (Outzen and Custer 1975; Sheffield 1988; Kuperwasser et al. 2004; Proia and Kuperwasser 2006). To overcome this barrier, artificial microenvironments that more accurately mimic the human mammary stroma have been developed, leading to an improvement in the efficiency and extent of engraftment by human mammary epithelial cells. One such technique, referred to as humanizing the mammary fat pad, involves injecting a mixture of irradiated and un-irradiated human fibroblasts into the cleared mammary fat pads of immune-compromised mice 14-days prior to transplantation of human mammary epithelial cells (Kuperwasser et al. 2004). To ensure full development of the xenografted cells, normal fibroblasts must be co-injected with the mammary epithelial cells in this second transplant surgery. Furthermore, the entire procedure must be completed prior to the end of puberty; otherwise the mice require hormone supplementation. This procedure, although labor-intensive, yields epithelial structures that morphologically resemble those of the human mammary gland, comprise cells that express markers of mature human mammary epithelial cells, and secrete milk proteins into the epithelial lumen in response to pregnancy. Because the latter features are measures of developmental potential, this assay can be used to read-out the differentiation capacity of human mammary epithelial stem cells and to estimate their frequency (Kuperwasser et al. 2004).

An analogous strategy has been used to generate mammary outgrowths by transplanting a mixture of human mammary epithelial cells and immortalized or irradiated fibroblasts into collagen gels, and transplanting these under the kidney capsules of hormone-supplemented mice (Eirew et al. 2008). This latter technique yields structures that are comparable to those generated in the humanized fat pad and avoids the need for multiple rounds of surgery. It should be noted, however, that the tissues generated by human mammary cells in either of these two xenograft assays are not nearly as well developed as those generated by mouse mammary cells in mouse mammary fat pad repopulation assays. Indeed orthotopic transplant of mouse mammary epithelial cells into syngeneic mice yields complete mammary trees comprising ducts and alveoli that secrete milk proteins in pregnant hosts; only the nipple fails to form in such outgrowths. By contrast the structures formed by human breast epithelial cells are rudimentary in nature and rarely fill greater than 10% of the humanized fat pad.

The first attempt to purify human mammary stem cells by FACS utilized the Aldefluor assay, which had previously been shown to enrich for hematopoietic (Armstrong et al. 2004) and neural stem cells (Corti et al. 2006). In cells with high aldehyde dehydrogenase (ALDH) activity, the pro-fluorescent Aldefluor reagent is oxidized to a fluorescent derivative, which remains intracellular. When dispersed cells isolated from breast reduction mammoplasties were incubated with the Aldefluor reagent and then separated into $ALDH^+$ and $ALDH^-$ subsets, only the $ALDH^+$ cells were capable of forming mammospheres in suspension culture and generating outgrowths in the humanized fat pads of NOD/SCID mice

(Ginestier et al. 2007). Although the same minimum number of cells was required for engraftment of bulk and ALDH$^+$ populations, the ALDH$^+$ subset yielded approximately ten times as many ducts per gland, suggesting that this population was enriched in mammary repopulating cells. The self-renewal capacity of these outgrowths was not tested by serial transplantation, so it is unclear whether this cell purification strategy permits a true enrichment of human mammary stem cells.

Another means to prospectively isolate human mammary epithelial stem cells utilized the mouse mammary stem cell marker CD49f in combination with the epithelial-specific antigen, EpCAM (Eirew et al. 2008). In kidney capsule transplants, the CD49f$^+$EpCAM$^{lo/-}$ cell fraction produced primary outgrowths more efficiently than either the CD49f$^+$EpCAM^{hi} or the CD49f$^-$ fractions. Furthermore, the CD49f$^+$EpCAM$^{lo/-}$-derived outgrowths exhibited a higher efficiency of generating secondary outgrowths and contained a greater number of colony-forming progenitor cells than the other cell fractions. However, it should be noted that the CD49f$^+$EpCAM^{hi} and CD49f$^-$ fractions also gave rise to serially transplantable outgrowths, illustrating that this marker combination may not uniquely identify mammary stem cells. Interestingly, these two markers had previously been used in combination to identify human bipotent progenitors in two-dimensional colony-forming assays, however, the majority of the cells that formed colonies under these conditions were located in the CD49f$^+$EpCAM^{hi} subpopulation (Stingl et al. 2001).

PKH26 has also been used to partially purify human mammary epithelial stem cells and to identify other cell-surface markers of these cells (Pece et al. 2010). Similar to the murine system, human mammosphere-initiating cells were found to remain quiescent after undergoing a single division in suspension culture. The PKH26^{hi} phenotype thus provided a means to isolate these cells and assay their properties. When replated in serum-free, chemically defined medium, only the PKH26^{hi} cells had the capacity to form secondary mammospheres, doing so with an efficiency of about 25%. Because the total number of human mammospheres in the bulk population was found to consistently decline by ~75% at each passage, whereas sphere diameter remained unchanged, this corresponded to the maximum expected sphere-forming efficiency of any mammosphere-resident population.

Using a panel of well-established markers, it was determined that human PKH26^{hi} cells express markers of both luminal (CD24$^+$EpCAM$^+$) and myoepithelial (CD49f$^+$CK5$^+$TP63$^+$) lineages, but are devoid of mature differentiation markers such as MUC1, E-cadherin, and alpha-smooth muscle actin (α-SMA). When subjected to colony forming assays, the majority of PKH26^{hi} cells gave rise to bi-lineage colonies containing cells that stained for markers of more mature differentiated mammary epithelial cells. Interestingly, cells from these colonies could be picked and replated in defined medium to give rise to mammospheres with high efficiency. By contrast, the PKH26$^{lo/-}$ cells produced primarily luminal cell-containing colonies, and some colonies comprising only myoepithelial cells; cells from these colonies were incapable of forming mammospheres. Collectively, these data support the notion that PKH26^{hi} cells are enriched in stem cells and/or multipotent mammary progenitor cells capable of self-renewal *in vitro*.

Limiting-dilution cell transplantation assays were carried out and confirmed that the PKH26^{hi} subset was enriched in mammary repopulating activity (Pece et al. 2010). When transplanted into the humanized fat pads of NOD/SCID mice, PKH26^{hi} cells engrafted and produced outgrowths when as few as ten cells were transplanted. By contrast, PKH26$^{lo/-}$ cells did not engraft, even when 100,000 cells were transplanted. However, the capacity of the outgrowths seeded by the PKH26^{hi} cell population to be serially transplanted was not tested and hence the nature of the cells that generated the grafts is uncertain.

Nonetheless, by separating PKH26^{hi} from PKH26$^{lo/-}$ cell populations and performing global gene expression profiling, the authors identified a number of genes that were preferentially expressed in the PKH26^{hi} population. From these differentially expressed genes, these investigators identified genes encoding cell surface proteins including the Notch ligands DLL1, Jag-1, and DNER, which they predicted would mark human mammary epithelial stem cells in primary tissue. As expected, antibodies against these proteins often identified single cells within mammospheres and rare cells in normal human mammary gland sections. Notably, sorted primary CD49f^{hi}CD24^{hi}DNER^{hi} mammary epithelial cells were capable of reconstituting mammary epithelial structures in mice, whereas CD49f^{hi}CD24^{hi}DNER^{lo} cells were not, consistent with the notion that DNER expression can be used to distinguish cells with mammary repopulating activity from those lacking this capacity and/or that DNER expression is required for human mammary epithelial stem cell engraftment. Finally, CD49f^{+}DLL1^{hi}DNER^{hi} cells formed mammospheres 530-fold more efficiently than bulk mammary epithelial cells, and greater than twice as efficiently as either CD49f^{hi}CD24^{hi}DLL1^{hi} or CD49f^{hi}CD24^{hi}DNER^{hi} cells, illustrating the advantage of using these markers in combination to enrich for human mammary epithelial stem cells.

12.5 Identifying and Purifying Mouse Mammary Epithelial Progenitor Cell Populations

12.5.1 Progenitors Identified by Cell Transplantation

As recounted in Sect. 12.1, the mouse mammary epithelium comprises at least two distinct classes of functionally defined progenitor cells, which are capable of generating either duct- or lobule-limited outgrowths when transplanted orthotopically into the epithelium-free mammary fat pads of recipient mice (Smith 1996; Kordon and Smith 1998; Booth et al. 2007, 2008). The resulting structures are histologically indistinguishable from those that occur in the normal mammary gland, comprising an inner layer of luminal cells and an outer layer of myoepithelial cells (Smith and Boulanger 2002; Boulanger et al. 2005; Booth et al. 2008). Moreover, the expression of steroid hormone receptors is heterogeneous among cells of the luminal layer, a characteristic of the normal mammary epithelium (Boulanger et al. 2005).

An additional class of lobule-limited progenitors have been identified in the mammary glands of doubly transgenic WAP-Cre/Rosa26-LacZ (Wagner et al. 2002) or WAP-Cre/CAG-GFP (Matulka et al. 2007) mice. In these mice, Cre-recombinase is expressed under the transcriptional control of the whey acidic protein (WAP) promoter and LoxP sites flank a transcription/translation stop sequence situated between either a Rosa26 or chicken actin gene (CAG) promoter and a downstream β-galactosidase (LacZ) or green fluorescent protein (GFP) reporter. Activation of the WAP promoter and thus Cre expression leads to recombination of the LoxP sites, excision of the stop sequence, and constitutive expression of the reporter gene in these cells and their progeny, irrespective of continued WAP promoter activity.

It was originally thought that the WAP promoter would only become active in fully differentiated secretory mammary epithelial cells during pregnancy, and that WAP-expressing cells would be completely eliminated during involution. However, in the aforementioned transgenic mice, rare marked cells are in fact present in the mammary epithelium of sexually mature nulliparous females, representing between 0.8 and 4% of the total epithelial cell population. Whereas lactation leads to a dramatic expansion of the LacZ/GFP marked cell population, involution does not return the gland to its nulliparous state, and in fact greater than 20% of the LacZ/GFP-positive cells remain up to 4 weeks after the pups have been weaned; hence, the designation of these cells as Parity Induced or Parity Identified Mammary Epithelial Cells (PI-MECs). Notably, in the resting gland PI-MECs are localized to terminal ducts and alveolar units, and in subsequent pregnancies expand to reconstitute secretory alveoli. Furthermore, when transplanted along with other supporting mammary epithelial cells, PI-MECs contribute substantial progeny to the chimeric outgrowths, including luminal, but not basal cells of the ducts as well as all alveolar cell types. It is unclear whether parity leads to an expansion in the absolute number of alveolar progenitors, or whether all genetically marked cells possess the capacity to proliferate and generate alveolar progeny in subsequent pregnancies. Indeed, it is entirely possible that following involution these alveolar progenitors comprise only a small subset (i.e., 0.8–4%) of the total epithelial population.

12.5.2 Progenitors Identified by Functional Assays In Vitro

A subset of primary mouse mammary epithelial cells form two-dimensional colonies when plated on a collagen substratum in the presence of serum, insulin, epidermal growth factor, and cholera toxin, which can be distinguished based on their morphology and the expression of lineage-specific markers of their constituent cells (Smalley et al. 1998; Kurpios et al. 2009). Type A colonies comprise dispersed cells that stain for the luminal markers CK8, CK18, and/or CK19. Type B colonies comprise tightly arranged, compact cuboidal cells that stain for luminal markers surrounded by a halo of cells that stain for the myoepithelial markers A-SMA or CK14. Type C colonies comprise cells that exhibit a variety of morphological phenotypes including large ovoid, and compact as well as elongated cuboidal cells.

Cells that express markers of luminal, myoepithelial, or both lineages are readily identifiable in these colonies. Finally, type D colonies consist of large, spread-out cells that stain for markers of myoepithelial cells but not those of luminal epithelial cells. These findings suggest that at least four distinct clonogenic progenitors exist in the mouse mammary epithelium, two of these are unipotent (luminal type A and myoepithelial type D), and two are multipotent (types B and C).

Progenitor cell populations also have been identified based on their capacity to form morphologically diverse three-dimensional structures when embedded in Matrigel (Shackleton et al. 2006; Stingl et al. 2006; Asselin-Labat et al. 2007). Some of the structures are spherical and comprise a solid mass of cells that express markers of luminal and/or myoepithelial cells, others are acinar-like and constitute a single layer of epithelial cells and a hollow lumen – these alveolar-like acini comprise cells that only express luminal markers and secrete milk proteins into the lumen in response to lactogenic hormones, and some are duct-like comprising both luminal and myoepithelial cells (Shackleton et al. 2006). The relationship among the progenitor cells that form duct- and lobule-limited structures after transplant into cleared fat pads and those that form two- or three-dimensional colonies *in vitro* is not clear. We speculate that the duct- and lobule-limited progenitors may be the same as those that form solid spheres and duct-like structures in Matrigel; in light of the fact that these various structures formed *in vivo* and in Matrigel comprise cells of diverse mammary epithelial lineages, it seems reasonable to hypothesize that they are seeded by the multipotent progenitor cells that form type B and C colonies. The acinar structures formed in Matrigel comprise only luminal epithelial cells that are capable of secreting milk proteins. Such acinar structures are not observed after cell transplantation into cleared fat pads; only type A colonies exclusively comprise luminal cells. In consequence it is not clear whether the same unipotent luminal progenitor cell spawns both acinar structures in Matrigel and type A colonies. The relationship among these various progenitor cells may be uncovered by using fractionated mammary epithelial cell populations in all three functional assays described above.

12.6 Identification of Progenitor Cells in the Mouse Mammary Epithelium

Analyses of mammary epithelial cell populations fractionated by cell sorting with various antibodies to cell-surface proteins reveal that distinct progenitor cell subpopulations can be identified as defined by two- and three-dimensional (Matrigel) colony-forming assays defined above. Antibodies to CD24 have been used to separate mammary epithelial cells into $CD24^{lo}$ and $CD24^{hi}$ subpopulations (Sleeman et al. 2006, 2007). The $CD24^{lo}$ subpopulation comprises mammary repopulating cells and myoepithelial-lineage restricted cells, whereas the $CD24^{hi}$ fraction comprises luminal-lineage cells. Mammary epithelial cells are not found in the $CD24^{-}$ cell fraction, which is instead composed of a variety of cells including fibroblasts, adipocytes, and other cells found in the mammary gland. Two-dimensional clonogenic assays demonstrated that the $CD24^{lo}$ population formed type D colonies, which are seeded by myoepithelial progenitor cells, but no other colony subtype (Sleeman et al. 2007). The cellular composition of colonies was defined by staining with antibodies to CK14 (myoepithelial) and CK18 (luminal). The fact that the $CD24^{lo}$ cell subpopulation did not contain cells that seeded type B or C colonies, which comprise both myoepithelial and luminal cells, implies that the mammary repopulating cells or stem cells present in this fraction, which might be predicted to form colonies comprising cells of both lineages, do not form adherent colonies under these conditions. Alternatively colonies formed by the stem cell population may have been infrequent relative to the predominant unipotent myoepithelial progenitors present in this $CD24^{lo}$ cell population.

Further fractionation of the $CD24^{hi}$ subpopulation with antibodies to Prominin (CD133) demonstrated that cells exhibiting a $CD24^{hi}CD133^{+}$ immunophenotype, are also characterized by surface expression of Sca1 and nuclear expression of the estrogen receptor. About 15% of these cells yielded adherent colonies that comprised both myoepithelial and luminal cells, however, the nature of the colonies, whether B and/or C, formed by the $CD24^{lo}CD133^{+}$ subpopulation was not reported. $CD24^{hi}CD133^{-}$ cells, which did not express Sca1 or the estrogen receptor also formed two-dimensional colonies, and all the colonies seeded by this subpopulation were composed of both myoepithelial and luminal epithelial cells. Hence, antibodies to CD133 were capable of subdividing the $CD24^{hi}$ population into two multipotent progenitor cell classes. As mentioned, expression of the estrogen receptor is restricted to the $CD24^{+}CD29^{Lo}$ subpopulation, which is poor in stem and myoepithelial progenitor cells and enriched in luminal progenitor cells (Asselin-Labat et al. 2006). However, functional analyses of the cells comprising the $CD24^{+}CD29^{Lo}$ cell population was not reported in these latter studies, and hence the relationship of the $CD24^{hi}CD133^{+}$ cell population to the $CD24^{hi}CD29^{lo}$ cell fraction is not clear. Nonetheless, taken together these findings illustrate the feasibility of combining cell sorting with various antibodies to cell surface antigens in combination with functional assays to elucidate the hierarchical cellular organization in the mammary epithelium.

As recounted previously antibodies to CD24 and those that bind to the CD29 or CD49f cell surface antigens can be used in combination to enrich for mammary epithelial stem cells ($CD24^{+}CD29^{hi}$ and $CD24^{med}CD49f^{hi}$) (Shackleton et al. 2006; Stingl et al. 2006). Progenitor cell populations were also identified in the aforementioned studies by assessing the capacity of various cell subpopulations to form colonies or structures in two- and three-dimensional assays.

Approximately 30% of the $CD24^{+}CD29^{hi}$ cell subpopulation formed two-dimensional colonies, however, the cellular composition of these colonies was not reported (Shackleton et al. 2006). The latter cell subpopulation formed principally solid spherical structures in three-dimensional Matrigel assays as well as ductal structures and rare acinar structures. The solid spherical and duct-like structures formed in Matrigel comprised both myoepithelial (CK14-positive) and luminal (CK18-positive) cells demonstrating that the $CD24^{+}CD29^{hi}$ cell subpopulation includes multipotent progenitor cells. It is not clear whether the stem cells within the $CD24^{+}CD29^{hi}$ population might also have formed adherent colonies or structures in Matrigel, but it is noteworthy that the stem cells in the $CD24^+CD29^{hi}$ cell subpopulation constitute no more than 5% of the total cell population of this fraction (Shackleton et al. 2006). Interestingly, the $CD24^+CD29^{lo}$ subset comprised ~5% progenitor cells as defined by two-dimensional colony-forming cells and seeded principally acinar structure that contained only luminal-lineage cells (CK18-positive). The morphology and cellular composition of the adherent colonies formed in two dimensions were not reported. Hence, the $CD24^+CD29^{hi}$ cell subpopulation comprises primarily multipotent progenitor cells, whereas the $CD24^+CD29^{lo}$ fraction is enriched in unipotent luminal progenitor cells.

Antibodies to CD49f similarly fractionated the $CD24^+$ cell population into one fraction comprising cells that formed solid structures in Matrigel, some with an irregular lumen, and others with a branched ductal appearance ($CD24^{med}CD49f^+$) (Stingl et al. 2006). By contrast, the $CD24^+CD49f^{lo}$ subpopulation predominantly formed acinar structures comprising only luminal epithelial cells (Stingl et al. 2006). The capacity of the various subpopulations to form colonies in two dimensions was not reported. Hence, the antibodies to CD29 and those to CD49f identified similar if not identical $CD24^+$ subpopulations. These latter results are not surprising because CD49f (alpha-6 integrin) and CD29 (beta-1 integrin) form heterodimers and are likely to be expressed on the surface of the same cells.

The $CD24^+CD29^{lo}$ subpopulation can be further divided by cell sorting using antibodies to CD61 (beta-3 integrin) (Asselin-Labat et al. 2007). The $CD61^+$ subpopulation constitutes about 22% of all the cells present in the $CD29^{lo}$ population; 30% of these $CD61^+$ cells were capable of forming colonies in two dimensions, whereas 18% formed acinar structures in Matrigel that comprised only luminal epithelial cells. By contrast, the $CD61^-$ subfraction constituted >50% of the cells in the $CD29^{lo}$ subpopulation; ~12% of these cells formed colonies in two dimensions but these were smaller than those arising from the $CD61^+$ fraction and very few (~2%) of these cells formed structures in Matrigel. The nature of the cells (luminal or myoepithelial) in the $CD61^-$ subpopulation was not reported. These latter data further illustrate that unipotent luminal progenitor cells are located in the nonstem cell fractions and that these progenitor cells can be enriched using antibodies to CD61.

The mammary epithelial progenitor cells that give rise to duct-limited and lobule-limited structures observed after transplantation of limiting cell dilutions have not been purified based on the expression of cell surface markers. However, an early attempt to isolate mammary epithelial stem cells using the hematopoietic stem cell marker, stem cell antigen 1 (Sca1), may have inadvertently identified duct-limited progenitors (Welm et al. 2002). $Sca1^+$ mammary epithelial cells proved capable of engrafting when 1,000 and up to 50,000 cells were injected, resulting in outgrowths that either partially or completely filled the fat pad in virgin recipients. The formation of secretory lobules in full-pregnant females that had been transplanted with these cells was not shown.

Different mouse mammary epithelial stem cell purification protocols developed some years later showed that single cells capable of mammary repopulating activity were in fact $Sca1^-$ (Shackleton et al. 2006; Stingl et al. 2006), thus calling into question the validity of these earlier findings. However, in these later studies the transplanted Sca1$^+$ cells were purified from the stem cell-enriched, CD49f^{hi} or CD29^{hi}CD24$^+$ subpopulations; whereas cells in the CD49f^{lo} or CD29^{lo}CD24$^+$ fractions were excluded from these analyses. Because Sca1 expression is observed primarily in luminal cells of the mammary epithelium (Welm et al. 2002), it could be argued that the Sca1$^+$ subpopulation exhibiting mammary repopulating activity might have been located in these discarded fractions. Indeed, because Sca1 and CD133 are co-expressed in CD24^{hi} cells, and because 15% of purified CD24^{hi}CD133$^+$ cells form adherent colonies *in vitro* composed of both myoepithelial- and luminal-lineage cells (Sleeman et al. 2007; Li et al. 2009), it is conceivable that the Sca1$^+$ cells capable of giving rise to duct-limited outgrowth *in vivo* might also be CD49f^{lo} or CD29^{lo}CD24$^+$. Indeed it will be interesting to determine whether antibodies to Sca-1 and CD133 can divide the CD24^{hi} fraction into duct-limited or lobule-limited progenitor cell populations.

12.7 Perspectives

The mammary glands of mice and humans are complex and dynamic tissues, composed of a variety of epithelial cell types that can be distinguished from one another based on their morphology, location in the gland, and marker expression. Functional analyses including cell transplantation and colony-forming assays have identified mammary epithelial stem cells as well as a variety of progenitor cells. However, our understanding of the hierarchical relationship among these mammary epithelial progenitor cells remains primitive. The marriage of FACS with functional assays has been instrumental in identifying cell-surface markers of stem and progenitor cells, which have in turn enhanced our ability to purify these cell populations for more detailed molecular analyses. Collectively, these efforts have not only enhanced our understanding of the mammary epithelial stem and progenitor cell compartment but also provided a better understanding of breast cancer etiology.

References

Alexander CM, Puchalski J, Klos KS, Badders N, Ailles L, Kim CF, Dirks P, Smalley MJ (2009) Separating stem cells by flow cytometry: reducing variability for solid tissues. Cell Stem Cell 5:579–583

Armstrong L, Stojkovic M, Dimmick I, Ahmad S, Stojkovic P, Hole N, Lako M (2004) Phenotypic characterization of murine primitive hematopoietic progenitor cells isolated on basis of aldehyde dehydrogenase activity. Stem Cells 22:1142–1151

Asselin-Labat ML, Shackleton M, Stingl J, Vaillant F, Forrest NC, Eaves CJ, Visvader JE, Lindeman GJ (2006) Steroid hormone receptor status of mouse mammary stem cells. J Natl Cancer Inst 98:1011–1014

Asselin-Labat ML, Sutherland KD, Barker H, Thomas R, Shackleton M, Forrest NC, Hartley L, Robb L, Grosveld FG, van der Wees J et al (2007) Gata-3 is an essential regulator of mammary-gland morphogenesis and luminal-cell differentiation. Nat Cell Biol 9:201–209

Booth BW, Boulanger CA, Smith GH (2007) Alveolar progenitor cells develop in mouse mammary glands independent of pregnancy and lactation. J Cell Physiol 212:729–736

Booth BW, Boulanger CA, Smith GH (2008) Stem cells and the mammary microenvironment. Breast Dis 29:57–67

Boulanger CA, Wagner KU, Smith GH (2005) Parity-induced mouse mammary epithelial cells are pluripotent, self-renewing and sensitive to TGF-beta1 expression. Oncogene 24:552–560

Bruno, R.D., and Smith, G.H. 2010. Functional Characterization of Stem Cell Activity in the Mouse Mammary Gland. Stem Cell Rev.

Chepko G, Smith GH (1999) Mammary epithelial stem cells: our current understanding. J Mammary Gland Biol Neoplasia 4:35–52

Cicalese A, Bonizzi G, Pasi CE, Faretta M, Ronzoni S, Giulini B, Brisken C, Minucci S, Di Fiore PP, Pelicci PG (2009) The tumor suppressor p53 regulates polarity of self-renewing divisions in mammary stem cells. Cell 138:1083–1095

Corti S, Locatelli F, Papadimitriou D, Donadoni C, Salani S, Del Bo R, Strazzer S, Bresolin N, Comi GP (2006) Identification of a primitive brain-derived neural stem cell population based on aldehyde dehydrogenase activity. Stem Cells 24:975–985

Daniel CW (1973) Finite growth span of mouse mammary gland serially propagated in vivo. Experientia 29:1422–1424

Daniel CW, Shannon JM, Cunha GR (1983) Transplanted mammary epithelium grows in association with host stroma: aging of serially transplanted mammary gland is intrinsic to epithelial cells. Mech Ageing Dev 23:259–264

Daniel CW, Silberstein GB (1987) Postnatal development of the rodent mammary gland. Plenum, New York

DeOme KB, Faulkin LJ Jr, Bern HA, Blair PB (1959) Development of mammary tumors from hyperplastic alveolar nodules transplanted into gland-free mammary fat pads of female C3H mice. Cancer Res 19:515–520

Dontu G, Abdallah WM, Foley JM, Jackson KW, Clarke MF, Kawamura MJ, Wicha MS (2003) In vitro propagation and transcriptional profiling of human mammary stem/progenitor cells. Genes Dev 17:1253–1270

Eirew P, Stingl J, Raouf A, Turashvili G, Aparicio S, Emerman JT, Eaves CJ (2008) A method for quantifying normal human mammary epithelial stem cells with in vivo regenerative ability. Nat Med 14:1384–1389

Ginestier C, Hur MH, Charafe-Jauffret E, Monville F, Dutcher J, Brown M, Jacquemier J, Viens P, Kleer CG, Liu S et al (2007) ALDH1 is a marker of normal and malignant human mammary stem cells and a predictor of poor clinical outcome. Cell Stem Cell 1:555–567

Hennighausen L, Robinson GW (1998) Think globally, act locally: the making of a mouse mammary gland. Genes Dev 12:449–455

Hulett HR, Bonner WA, Barrett J, Herzenberg LA (1969) Cell sorting: automated separation of mammalian cells as a function of intracellular fluorescence. Science 166:747–749

Kordon EC, Smith GH (1998) An entire functional mammary gland may comprise the progeny from a single cell. Development 125:1921–1930

Kuperwasser C, Chavarria T, Wu M, Magrane G, Gray JW, Carey L, Richardson A, Weinberg RA (2004) Reconstruction of functionally normal and malignant human breast tissues in mice. Proc Natl Acad Sci USA 101:4966–4971

Kurpios NA, MacNeil L, Shepherd TG, Gludish DW, Giacomelli AO, Hassell JA (2009) The Pea3 Ets transcription factor regulates differentiation of multipotent progenitor cells during mammary gland development. Dev Biol 325:106–121

Lanzkron SM, Collector MI, Sharkis SJ (1999) Hematopoietic stem cell tracking in vivo: a comparison of short-term and long-term repopulating cells. Blood 93:1916–1921

Li W, Ferguson BJ, Khaled WT, Tevendale M, Stingl J, Poli V, Rich T, Salomoni P, Watson CJ (2009) PML depletion disrupts normal mammary gland development and skews the composition of the mammary luminal cell progenitor pool. Proc Natl Acad Sci USA 106:4725–4730

Liao MJ, Zhang CC, Zhou B, Zimonjic DB, Mani SA, Kaba M, Gifford A, Reinhardt F, Popescu NC, Guo W et al (2007) Enrichment of a population of mammary gland cells that form mammospheres and have in vivo repopulating activity. Cancer Res 67:8131–8138

Matulka LA, Triplett AA, Wagner KU (2007) Parity-induced mammary epithelial cells are multipotent and express cell surface markers associated with stem cells. Dev Biol 303:29–44
Outzen HC, Custer RP (1975) Growth of human normal and neoplastic mammary tissues in the cleared mammary fat pad of the nude mouse. J Natl Cancer Inst 55:1461–1466
Pece S, Tosoni D, Confalonieri S, Mazzarol G, Vecchi M, Ronzoni S, Bernard L, Viale G, Pelicci PG, Di Fiore PP (2010) Biological and molecular heterogeneity of breast cancers correlates with their cancer stem cell content. Cell 140:62–73
Petersen OW, Polyak K (2010) Stem cells in the human breast. Cold Spring Harb Perspect Biol 2:a003160
Proia DA, Kuperwasser C (2006) Reconstruction of human mammary tissues in a mouse model. Nat Protoc 1:206–214
Sakakura T (1987) Mammary embryogenesis. Plenum Publishing Corporation, New York
Sakakura T (1991) New aspects of stroma-parenchyma relations in mammary gland differentiation. Int Rev Cytol 125:165–202
Shackleton M, Vaillant F, Simpson KJ, Stingl J, Smyth GK, Asselin-Labat ML, Wu L, Lindeman GJ, Visvader JE (2006) Generation of a functional mammary gland from a single stem cell. Nature 439:84–88
Sheffield LG (1988) Organization and growth of mammary epithelia in the mammary gland fat pad. J Dairy Sci 71:2855–2874
Sleeman KE, Kendrick H, Ashworth A, Isacke CM, Smalley MJ (2006) CD24 staining of mouse mammary gland cells defines luminal epithelial, myoepithelial/basal and non-epithelial cells. Breast Cancer Res 8:R7
Sleeman KE, Kendrick H, Robertson D, Isacke CM, Ashworth A, Smalley MJ (2007) Dissociation of estrogen receptor expression and in vivo stem cell activity in the mammary gland. J Cell Biol 176:19–26
Smalley MJ, Titley J, O'Hare MJ (1998) Clonal characterization of mouse mammary luminal epithelial and myoepithelial cells separated by fluorescence-activated cell sorting. In Vitro Cell Dev Biol Anim 34:711–721
Smith GH (1996) Experimental mammary epithelial morphogenesis in an in vivo model: evidence for distinct cellular progenitors of the ductal and lobular phenotype. Breast Cancer Res Treat 39:21–31
Smith GH, Boulanger CA (2002) Mammary stem cell repertoire: new insights in aging epithelial populations. Mech Ageing Dev 123:1505–1519
Smith GH, Boulanger CA (2003) Mammary epithelial stem cells: transplantation and self-renewal analysis. Cell Prolif 36(Suppl 1):3–15
Smith GH, Chepko G (2001) Mammary epithelial stem cells. Microsc Res Tech 52:190–203
Smith GH, Medina D (1988) A morphologically distinct candidate for an epithelial stem cell in mouse mammary gland. J Cell Sci 90(Pt 1):173–183
Stingl J, Eaves CJ, Zandieh I, Emerman JT (2001) Characterization of bipotent mammary epithelial progenitor cells in normal adult human breast tissue. Breast Cancer Res Treat 67:93–109
Stingl J, Eirew P, Ricketson I, Shackleton M, Vaillant F, Choi D, Li HI, Eaves CJ (2006) Purification and unique properties of mammary epithelial stem cells. Nature 439:993–997
Visvader JE (2009) Keeping abreast of the mammary epithelial hierarchy and breast tumorigenesis. Genes Dev 23:2563–2577
Wagner KU, Boulanger CA, Henry MD, Sgagias M, Hennighausen L, Smith GH (2002) An adjunct mammary epithelial cell population in parous females: its role in functional adaptation and tissue renewal. Development 129:1377–1386
Welm BE, Tepera SB, Venezia T, Graubert TA, Rosen JM, Goodell MA (2002) Sca-1(pos) cells in the mouse mammary gland represent an enriched progenitor cell population. Dev Biol 245:42–56
Williams JM, Daniel CW (1983) Mammary ductal elongation: differentiation of myoepithelium and basal lamina during branching morphogenesis. Dev Biol 97:274–290
Youn BS, Sen A, Kallos MS, Behie LA, Girgis-Gabardo A, Kurpios N, Barcelon M, Hassell JA (2005) Large-scale expansion of mammary epithelial stem cell aggregates in suspension bioreactors. Biotechnol Prog 21:984–993

Chapter 13
Differentiation Programs in Development and Cancer

Hosein Kouros-Mehr

13.1 Introduction

The majority of solid tumors, including breast, prostate, colon, and lung cancers, originate from normal epithelium. The differentiation programs of epithelial cells dictate their specialized function, including their cell shape, polarity, arrangement, and architecture. In epithelial malignancies, the differentiation status of a primary tumor strongly predicts its capacity for metastasis formation and resistance to chemotherapeutic agents (Bloom and Richardson 1957; Contesso et al. 1987). Poorly differentiated neoplasias typically harbor higher rates of distant metastasis formation and thus carry poorer prognoses compared to their well-differentiated counterparts. The loss of tumor differentiation is one of the central hallmarks of malignant progression, the process by which a primary tumor acquires the capacity for dissemination and metastasis (Gupta and Massague 2006; Hanahan and Weinberg 2000). Genetic studies in mice and other organisms have uncovered the molecular basis for epithelial differentiation, which is shedding light on the pathogenesis of epithelial malignancies and revealing new strategies for cancer therapeutic development.

Breast cancers originate from normal mammary epithelium, which is composed of two cell types – the luminal and myoepithelial (basal cells). Most breast cancers originate from luminal cells, which are single-layered, columnar epithelial cells that adhere with each other to form a ductal network with a central hollow lumen for milk secretion and passage. Their apical domains line the lumen and are specialized for secretion while their basal domains are anchored to basement membrane and to myoepithelial cells. Myoepithelium lie between luminal cells and basement membrane

H. Kouros-Mehr (✉)
Genentech, 1 DNA Way, MS-60, South San Francisco, CA 94080, USA
e-mail: kourosmehr.hosein@gene.com

J.E. Green and T. Ried (eds.), *Genetically Engineered Mice for Cancer Research: Design, Analysis, Pathways, Validation and Pre-Clinical Testing*,
DOI 10.1007/978-0-387-69805-2_13, © Springer Science+Business Media, LLC 2012

and provide a contractile function for secretion. Both cell types undergo proliferation during stages of the estrus cycle as well as pregnancy, when the mammary ductal tree develops milk-producing alveoli. After lactation, the mammary gland undergoes involution, returning to its native state as a result of widespread cell death and clearance (Sternlicht et al. 2006; Wiseman and Werb 2002). Compared to other organs, the mammary gland has a high rate of turnover and plasticity as it undergoes continual remodeling, which may explain the relatively high incidence of breast cancers and the high preponderance of luminal epithelial neoplasias. Insight into the molecular basis of luminal cell differentiation and reprogramming has shed further light on breast cancer pathogenesis. Here, I explore the molecular basis of cellular differentiation in normal and cancerous mammary glands and highlight the use of genetically engineered mice (GEM) to further explore the regulation of breast cancer differentiation and metastasis.

13.2 Luminal Cell Fate Programming

Luminal cells express a battery of differentiation genes that govern their structure and function. These include intermediate filaments, such as cytokeratins 8, 18, and 19, cell adhesion proteins, such as E-cadherin and Ep-CAM, secretory proteins, such as β-casein, whey acidic protein and lactalbumin, and steroid receptors, such as estrogen receptor (ER) and progesterone receptor (PR). Differentiated luminal cells in a nonpregnant host display low rates of proliferation and produce baseline levels of secretory material. During pregnancy and lactation, luminal cells undergo a cell fate switch termed lobuloalveolar differentiation, during which they undergo rapid increases in cell proliferation and secretion, resulting in an expanded ductal lumen for milk passage. Luminal cells are initially specified in the embryo in the anlagen and also in terminal end buds (TEBs). TEBs develop at the distal ends of primitive ductal epithelium during puberty, and these structures undergo invasion and branching morphogenesis to give rise to the mature ductal tree (Sternlicht et al. 2006). Mammary stem cells, luminal progenitor cells, and myoepithelial progenitor cells are believed to reside in TEBs during development, but the pathways that control their self-renewal and differentiation have not been identified. Further, the mechanisms by which luminal cells activate and maintain the differentiated program have remained elusive until recently.

In other model systems, the expression of differentiation effector genes is governed by the accessibility of their loci to transcriptional machinery, which is regulated in part by transcription factors, histone modifiers, and nuclear scaffolding genes. A key role in cell-fate programming is played by transcription factors, including both transcriptional activators and repressors that form higher order complexes and operate in regulatory networks. These genes act on *cis*-regulatory modules on effector genes to activate and maintain the effector genes of a given cell fate while repressing alternate fates (Levine and Davidson 2005). In embryonic stem (ES) cells, the pluripotent state is regulated by transcription factors, such as Oct4, Sox2, and Nanog (Loh et al. 2006; Wang et al. 2006). These genes, which modulate transcriptional repressors and histone modifiers, maintain ES cells in a quiescent state that is poised for activation (Bernstein et al. 2006; Lee et al. 2006). Differentiation occurs with the upregulation of lineage-specific transcription factor networks and the active suppression of alternate cell fate networks (Nishiyama et al. 2009; Walker et al. 2007). In many cases, the regulatory networks are activated by paracrine and juxtacrine signaling pathways, such as Wnt and Notch, which upregulate the initial nodes of the networks. These genes auto-activate in positive-feedback loops and activate the other nodes in the network, thereby establishing a hierarchical regulatory network governing cell fate. This feed-forward system locks cells into their respective fates and prohibits switching to alternate cell fates during development (Davidson et al. 2002).

In the mammary gland, a transcriptional regulatory network governs the differentiation of the luminal epithelial cell fate. A key node in the network is GATA-3, a member of the GATA family of dual zinc-finger transcription factors that govern the fate of many cell types, including T-cells, erythrocytes, and endodermal tissues (Kouros-Mehr et al. 2008b). An expression profiling screen identified GATA-3 as the most highly expressed transcription factor in the developing mouse mammary epithelium (Kouros-Mehr et al. 2006). GATA-3 expression was restricted to luminal epithelium and was absent in other cell types, such as myoepithelium or stromal cells. In the mammary gland, GATA-3 is necessary for the specification and maintenance of luminal cells from mammary stem/progenitor cells. The conditional deletion of GATA-3 in adult mammary glands leads to a sudden loss of differentiation, resulting in detachment from the basement membrane and cell death in the ductal lumen. GATA-3 activation in mammary stem cells leads to the upregulation of luminal-specific genes, such as milk proteins, ERs, and cell adhesion proteins, such as E-cadherin (Asselin-Labat et al. 2007; Kouros-Mehr et al. 2006). It also induces changes in the cell cycle, such as activation of the CDK inhibitor p18INK4c, to restrain cell proliferation in the differentiated state (Pei et al. 2009). GATA-3 also represses the gene products of alternate cell fates, such as adipocytes, thereby locking in the luminal cell fate during mammary development (Tong et al. 2000).

As in other systems, GATA-3 likely interacts with other lineage-specific transcription factors to guide luminal cell differentiation. These transcription factors operate in complex regulatory networks, either as single factors or in high molecular weight transcriptional complexes with GATA-3 (Hoch et al. 1999). The forkhead box transcription factor FOXA1 is a downstream target of GATA-3 and is recruited to distant enhancer elements to guide lineage-specific transcription (Kouros-Mehr et al. 2006; Lupien et al. 2008). FOXA1 is necessary for the binding of ER to target loci, suggesting a coordinate regulation of GATA-3, FOXA1, and ER in luminal cells (Carroll et al. 2005; Laganiere et al. 2005). A positive cross-regulatory loop may reinforce the expression of these genes as master regulators of the transcriptional network, similar to the functions of GATA and FOX genes in sea urchin development (Davidson et al. 2002; Eeckhoute et al. 2007). Other transcription factors that may operate in the regulatory network include LEF1, MSX2, C/EPβ,

FOXP4, Elf5, EHF, RUNX1, and several members of the Id, Irx, Sox, Stat, and TCFAP-2 families, all of which demonstrate epithelial-specific expression patterns (Kouros-Mehr et al. 2006; Visvader and Lindeman 2003). LEF1 and MSX2 are essential for embryonic mammary development and the initial specification of mammary epithelium. Additional transcription factors, including STAT5a and ELF5, are necessary in postnatal development during the specification of alveolar cells during pregnancy (Visvader and Lindeman 2003). Further work will be necessary to define the other nodes in the luminal regulatory work and to identify binding partners of GATA-3 and FOXA1 in transcriptional complexes.

13.3 Mammary Stem Cells

Differentiated luminal epithelia are derived from multipotential mammary stem/progenitor cells, which have been recently characterized. Various groups have reported the enrichment of mammary stem cells using cell surface markers, such as CD24, CD29, and CD49f. Single cells derived from a $CD29^{hi}CD24^{+}$ subpopulation of mammary cells can self-renew and give rise to a functional mammary gland (Shackleton et al. 2006; Stingl et al. 2006). This is consistent with earlier reports that single retrovirally tagged mammary cells can repopulate the mammary fat pad (Kordon and Smith 1998). Electron micrography and label retention assays suggest that mammary stem cells are small, basally located cells within mature duct epithelium (Chepko and Smith 1997; Welm et al. 2002). Further, they are believed to reside in terminal end buds in the developing mammary gland, though it is unclear if they reside in the body or cap cell layers (or both) (Kenney et al. 2001). Although the existence of mammary stem cells has been proven, the genes that regulate their multipotentiality and self-renewing properties have not been identified. Candidate regulators include the polycomb group protein BMI-1 and members of the Wnt family, including Wnt-5A, which is enriched in terminal end buds (Kouros-Mehr and Werb 2006; Pietersen et al. 2008).

Recent work suggests that the Notch pathway controls the initial specification of luminal epithelium from mammary stem/progenitor cells. In other systems, including T-cells, Notch1 activates the expression of GATA-3 during the establishment of cell fates (Amsen et al. 2007; Fang et al. 2007). In the mammary gland, several Notch ligands are upregulated during commitment of bipotential progenitors to luminal-restricted epithelial cells (Bouras et al. 2008; Raouf et al. 2008). Downstream Notch effectors, including Hey1 and Hey2, are active specifically in luminal progenitors. Downregulation of the Notch effector gene CBF-1 in mammary stem cells leads to increased stem cell activity, while constitutive Notch signaling promotes the expansion of luminal progenitor cells, suggesting that the Notch pathway specifically promotes luminal cell fate commitment from multipotential stem/progenitor cells (Bouras et al. 2008; Welm et al. 2008). The Notch pathway also contributes to the programming of the alveolar cell fate during pregnancy, though detailed functional analysis of this role is lacking (Buono et al. 2006).

Further work is necessary to explore the role of Notch in mammary development and breast cancer and its relationship to the luminal cell transcriptional regulatory network.

13.4 Differentiation in Breast Cancer

Breast cancers are classified according to various tumor markers and histopathologic criteria, including the expression of ER, PR, and ERBB2/Her2, as well as histologic type and tumor grade. This information, together with the involvement of regional lymph nodes, distant lymph nodes, and distant organs establishes the stage of disease, guides clinical decision-making and portends prognosis. Localized breast cancers with negative margins, such as ductal carcinoma *in situ* (DCIS), are typically treated with lumpectomy and radiation therapy, followed by biannual surveillance. Invasive breast cancers with regional lymph node involvement but without metastatic disease (i.e., Stage I–III) are additionally treated with adjuvant therapies and lymph node resection to reduce the likelihood of distant metastasis formation. These adjuvant therapies include chemotherapy, such as doxorubicin, docetaxel, and cyclophosphamide, and endocrine therapies, such as selective ER modulators (e.g., tamoxifen) and aromatase inhibitors. Endocrine therapies are administered to patients whose tumors are ER positive and PR positive. Additionally, patients with Her2-positive tumors receive trastuzumab in the adjuvant setting. Patients with Stage IV breast cancer, who harbor distant metastases, may also be treated with these modalities to control systemic disease and extend life. Additionally, patients with bone metastases receive bisphosphonates and local radiation therapy, and patients with focal brain metastases may undergo surgery for resection. The modalities used to treat Stage IV disease requires an assessment to determine if the benefits of treatment outweigh the risks (NCCN 2009).

Recent molecular analysis indicates that existence of distinct breast cancers subtypes with differing biologies and prognoses. These subtypes include the luminal A, luminal B, ERBB2, basal-like, and apocrine subtypes (Perou et al. 2000; Sorlie et al. 2001). Luminal A tumors are typically low grade, well-differentiated tumors that carry a favorable prognosis. These tumors express markers of differentiated luminal epithelium, include GATA-3, ER, and PR, and possess distinct cytogenetic abnormalities, such as loss of chromosome 16q. In contrast, basal-like breast cancers that lack ER, PR, and Her2 (the so-called triple-negative disease) are high grade cancers that are more resistant to chemotherapy and show an increased capacity for metastasis formation, thereby carrying a relatively poor prognosis. These cancers also display distinct cytogenetic abnormalities and genetic profiles, such as chromosome 6p21-p25 gain and the expression of hypoxia and wound response genetic signatures (Sims et al. 2007). The clustering of "basal-like" tumors into one subtype of breast cancer remains controversial, however, as these tumors comprise a range of histologic subtypes with heterogeneous gene expression programs. It remains unclear whether "basal-like" breast cancers originate from a single mammary cell type, i.e., the basal or myoepithelial cells, or whether they can originate from multiple cell types, including mammary stem/progenitor cells. Further, "basal-like" breast cancer cells may differ substantially from their cell of origin at the time of transformation. These changes occur during malignant progression of the tumor, which is influenced by epithelial–mesenchymal transition, microenvironmental effects, and hypoxia (Finnegan and Carey 2007; Gusterson 2009).

Microarray expression analysis has revealed a number of transcriptional regulators associated with the luminal breast cancer subtype, including GATA-3, FOXA1, TFF3, XBP-1, and ER (Kouros-Mehr et al. 2008b). Of these genes, GATA-3 has emerged as one of the strongest predictors of ER status, tumor grade, and prognosis. Artificial neural network and correlative gene analysis identified GATA-3 as among the best predictors of ER status (Bertucci et al. 2000; Gruvberger et al. 2001). Low GATA-3 expression was strongly predictive of poor clinical outcome, high tumor grade, positive lymph node status, and large tumor size (Mehra et al. 2005). Log-rank analysis of a microarray dataset revealed that GATA-3 expression was the second best predictor of clinical outcome among 8,024 genes, with its low expression indicating a poor prognosis (Jenssen et al. 2002). However, it is unclear if GATA-3 has prognostic significance independent of ER, which itself carries strong prognostic value (Mehra et al. 2005; Voduc et al. 2008). Interestingly, GATA-3 expression is a strong inverse predictor of lung metastasis development in patients, as tumors with high GATA-3 show a remarkably low metastasis rate. This suggests that GATA-3 may prevent tumor cells from dissemination and metastasis, and that loss of GATA-3 may be causally involved in metastasis formation. Indeed, the restoration of GATA-3 in breast cancer cell lines significantly reduced their ability to metastasize when these cells were injected into mice (Dydensborg et al. 2009; Chu et al. 2011). Taken together, these studies implicate GATA-3 as a putative regulator of tumor differentiation and metastasis in breast cancer.

13.5 GEM Models for the Study of Breast Cancer Differentiation and Metastasis

Studies of breast cancer patient samples have established correlative links between gene expression levels and measures of tumor activity, such as the correlation between GATA-3 and tumor differentiation. These clinical studies are limited in that they do not establish causative relationships between genes and tumor processes. For example, clinical studies suggest that tumors with high GATA-3 or ER levels correlate with low metastasis rate, but are these relationships functional or simply correlational? Experiments with cell lines injected into immunocompromised mice indicate that tumors overexpressing GATA-3 can switch from being highly metastatic to low metastasis, suggesting a functional relationship between GATA-3 and metastasis (Dydensborg et al. 2009; Chu et al. 2011). However, these studies are also limited, as cell line metastasis assays do not model all the steps of the metastatic process, including tumor invasion, angiogenesis, tumor dissemination, tumor dormancy, extravasation to distant sites, and growth of metastases. In order to study the genes that functionally regulate metastasis, it is critical to utilize a model in which all steps of metastasis are recapitulated. In this way, it is possible to address which steps of metastasis are rate-limiting and which steps can be targeted to block metastatic outgrowth.

Genetically engineered mouse (GEM) models have emerged as robust models of breast cancer for the study of tumor progression and metastasis formation. In transgenic models of breast cancer, oncogenes, such as Her2/Neu and Ras, are driven by mammary-specific promoters, such as MMTV and WAP, thereby directing oncogenic expression to the mammary gland. A number of GEM models of breast cancer exist, and they have been grouped into distinct clusters based on gene expression patterns. GEM models of luminal breast cancers include MMTV-Neu, WAP-Myc, and MMTV-PyMT while models of basal and mixed lineage tumors include MMTV-Cre, $p53^{+/-}$, and MMTV-Wnt, respectively (Desai et al. 2002; Herschkowitz et al. 2007). These oncogenes are frequently detected in patient samples; for example, amplification of her-2/neu and c-myc are detected in 30 and 17% of breast cancer samples, respectively. The relative tumor incidence and growth can vary among the models; for instance, MMTV-Neu mice develop a small number of tumor foci with a time to palpable tumor of 7–8 months. MMTV-PyMT, however, develop hundreds of tumor foci in their mammary glands, with a time to palpable tumor of 7–8 weeks (Guy et al. 1992a, b).

The GEM models of breast cancer offer significant advantages for the study of metastasis and differentiation. The mammary glands of these mice undergo neoplastic transformation in normal mammary tissue, similar to de novo arising breast cancers in patients. In most models, including MMTV-PyMT and MMTV-Neu, these early neoplastic lesions undergo stages of tumor progression, including tumor invasion and loss of basement membrane. These changes are followed by the later stages of malignant progression, such as angiogenesis, tumor dissemination, and metastasis. Further, these events occur in an orthotopic site (the mammary gland) in an immune competent mouse. In this way, all of the steps in metastasis formation can be recapitulated. Thus, the genes and pathways that control the individual steps of metastasis can be identified, and the rate-limiting steps of metastasis formation can be assessed and targeted (Guy et al. 1992a, b). This is accomplished by treating GEM mice with drugs that target specific pathways or by crossing GEM mice with mice that carry gain-of-function or loss-of-function mutations in specific genes.

A disadvantage of GEM models is multifocal tumor development, making it difficult to study the progression of individual tumor foci over time. This can be circumvented by orthotopically transplanting single hyperplastic foci from GEM mice into the mammary fat pads of wild-type mice (Fig. 13.1a). For instance, during puberty MMTV-PyMT × β-actin-green fluorescent protein (GFP) mice develop focal hyperplasias, which can be microdissected and transplanted into the mammary glands of normal hosts (Kouros-Mehr et al. 2008a). The GFP-positive hyperplasias undergo stereotyped progression to adenoma, early carcinoma, and advanced carcinoma at 1, 2, and 4 months, respectively, with a 2-month median time to palpable tumor (Fig. 13.1b). The adenomas are noninvasive tumors that express markers of differentiation (β-casein, GATA-3) and contain basement membrane.

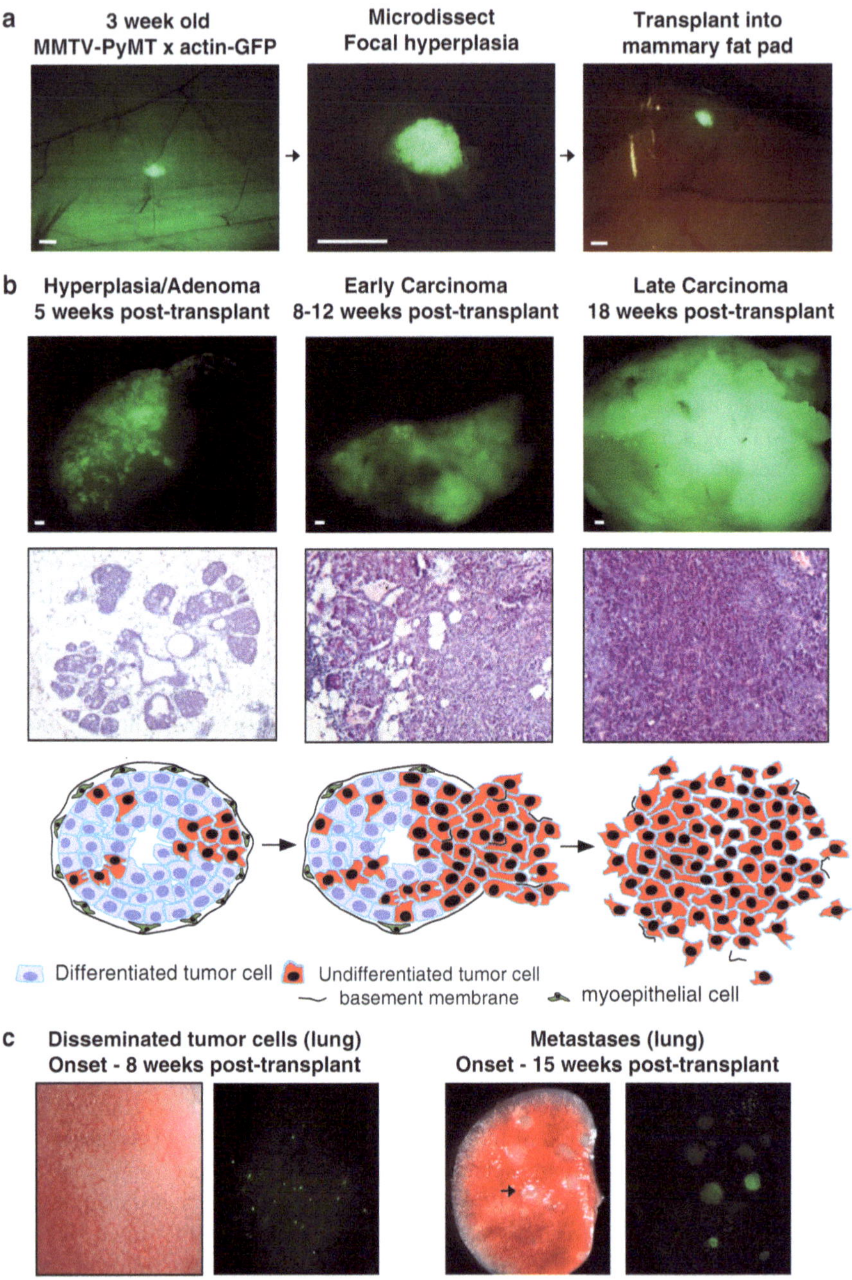

Fig. 13.1 (**a**) Focal hyperplasia isolated from a 3-week-old MMTV-PyMT; β-actin GFP mouse and transplanted into FVB/n wild-type mammary fat pad. (**b**) Representative GFP whole mount, H&E staining, and cartoon model of 5-, 8-, and 18-week tumor outgrowths. (**c**) Representative brightfield and GFP whole mount images of disseminated tumor cells and metastases in lungs of tumor bearing mice

In contrast, early carcinomas display the loss of basement membrane and the onset of tumor dissemination to distant organs, including lung, brain, and liver. Tumor dissemination coincides with the loss of differentiation markers in the primary tumor. By 4 months posttransplant, several thousand disseminated cells are lodged in the lung microvasculature. Interestingly, mastectomy of the primary tumor and subsequent chasing reveals that only a small percentage ($< 0.001\%$) of disseminated cells develops into metastases (Fig. 13.1c). This model enables the stereotyped analysis of the steps underlying tumor dissemination and metastasis formation (Kouros-Mehr et al. 2008a).

Using the orthotopic hyperplasia transplant model, it was shown that GATA-3 was involved in early steps of breast cancer progression while additional events were necessary for metastatic outgrowth (Kouros-Mehr et al. 2008a). The loss of GATA-3 expression in tumors cells occurred in the transition from adenoma to early carcinoma. Disseminated tumor cells lacked GATA-3 expression, suggesting that loss of GATA-3 occurred early in malignant progression. The restoration of GATA-3 in late carcinomas prevented the formation of metastasis, similar to observation in the MDA-MB-231 cell line model (Dydensborg et al. 2009; Chu et al. 2011; Kouros-Mehr et al. 2008a). The targeted ablation of GATA-3 in the early adenomatous tumors led to loss of tumor differentiation and subsequent cell death due to loss of cell survival signals, indicating that loss of GATA-3 itself was not sufficient to promote malignant progression. Instead, it appeared that a GATA-3 negative population that persisted in early tumors expanded during malignant progression and the growth of this population led to the onset of tumor dissemination. Additional steps were required for the GATA-3 negative disseminated tumor cells to form into metastases at distant sites. Such steps may require disseminated cells to evade the immune system, form a premetastatic niche in distant sites, activate cancer-initiating/self-renewal properties, etc. In the hyperplasia transplant model, the disseminated tumor cells and metastases are GFP-positive cells in a wild-type background, which allows them to be collected and studied to elucidate the mechanisms that regulate metastatic outgrowth. A similar approach can also be applied to other GEM models of breast cancer, such as MMTV-Neu and MMTV-Wnt, for further insight into these processes (Kouros-Mehr et al. 2008a).

13.6 Conclusions

The importance of breast cancer differentiation as a prognostic indicator and predictor of distant metastasis formation has been appreciated for decades. Classically, tumor differentiation has been assessed by histopathology, including the histologic appearance of epithelial architecture, duct formation and the presence of secretory material. Recently, the molecular mechanisms that regulate epithelial differentiation have been uncovered. Epithelial cells are specified and maintained in their differentiated state by the activation of transcription factor networks that control the effector genes of a given cell fate. In luminal epithelium, these networks comprise the transcriptional regulators GATA-3, FOXA1, ER, and PR, which likely operate with other transcriptional regulators in a hierarchical fashion to activate the luminal cell fate program while repressing alternate cell fates. In breast cancer, these genes are strong predictors of tumor differentiation and prognosis, as the downregulation of these genes predicts an unfavorable outcome and resistance to therapy. In GEM models of breast cancer, the loss of these regulators, in particular GATA-3, occurs early in tumor development as differentiated tumor cells undergo progression to carcinoma cells with the capacity for tumor dissemination. The precise roles of these genes in the metastastic process can be evaluated in GEM models, which are robust models of the metastatic process. GEM models are useful as hypothesis-generating tools. Ultimately, the insight gleamed from GEM models can be used to interpret clinical data and to guide clinical decision-making. The GEM models also serve as robust preclinical models to evaluate novel therapeutic agents in the adjuvant setting.

References

Amsen D, Antov A, Jankovic D, Sher A, Radtke F, Souabni A, Busslinger M, McCright B, Gridley T, Flavell RA (2007) Direct regulation of Gata3 expression determines the T helper differentiation potential of Notch. Immunity 27:89–99

Asselin-Labat ML, Sutherland KD, Barker H, Thomas R, Shackleton M, Forrest NC, Hartley L, Robb L, Grosveld FG, van der Wees J et al (2007) Gata-3 is an essential regulator of mammary-gland morphogenesis and luminal-cell differentiation. Nat Cell Biol 9:201–209

Bernstein BE, Mikkelsen TS, Xie X, Kamal M, Huebert DJ, Cuff J, Fry B, Meissner A, Wernig M, Plath K et al (2006) A bivalent chromatin structure marks key developmental genes in embryonic stem cells. Cell 125:315–326

Bertucci F, Houlgatte R, Benziane A, Granjeaud S, Adelaide J, Tagett R, Loriod B, Jacquemier J, Viens P, Jordan B et al (2000) Gene expression profiling of primary breast carcinomas using arrays of candidate genes. Hum Mol Genet 9:2981–2991

Bloom HJ, Richardson WW (1957) Histological grading and prognosis in breast cancer; a study of 1409 cases of which 359 have been followed for 15 years. Br J Cancer 11:359–377

Bouras T, Pal B, Vaillant F, Harburg G, Asselin-Labat ML, Oakes SR, Lindeman GJ, Visvader JE (2008) Notch signaling regulates mammary stem cell function and luminal cell-fate commitment. Cell Stem Cell 3:429–441

Buono KD, Robinson GW, Martin C, Shi S, Stanley P, Tanigaki K, Honjo T, Hennighausen L (2006) The canonical Notch/RBP-J signaling pathway controls the balance of cell lineages in mammary epithelium during pregnancy. Dev Biol 293:565–580

Carroll JS, Liu XS, Brodsky AS, Li W, Meyer CA, Szary AJ, Eeckhoute J, Shao W, Hestermann EV, Geistlinger TR et al (2005) Chromosome-wide mapping of estrogen receptor binding reveals long-range regulation requiring the forkhead protein FoxA1. Cell 122:33–43

Chepko G, Smith GH (1997) Three division-competent, structurally-distinct cell populations contribute to murine mammary epithelial renewal. Tissue Cell 29:239–253

Chu IM et al (2011) GATA3 inhibits lysyl oxidase-mediated metastases of human basal triple-negative breast cancer cells. Oncogene doi: 10.1038/onc.2011.382

Contesso G, Mouriesse H, Friedman S, Genin J, Sarrazin D, Rouesse J (1987) The importance of histologic grade in long-term prognosis of breast cancer: a study of 1,010 patients, uniformly treated at the Institut Gustave-Roussy. J Clin Oncol 5:1378–1386

Davidson EH, Rast JP, Oliveri P, Ransick A, Calestani C, Yuh CH, Minokawa T, Amore G, Hinman V, Arenas-Mena C et al (2002) A genomic regulatory network for development. Science 295:1669–1678

Desai KV, Xiao N, Wang W, Gangi L, Greene J, Powell JI, Dickson R, Furth P, Hunter K, Kucherlapati R et al (2002) Initiating oncogenic event determines gene-expression patterns of human breast cancer models. Proc Natl Acad Sci USA 99:6967–6972
Dydensborg AB, Rose AA, Wilson BJ, Grote D, Paquet M, Giguere V, Siegel PM, Bouchard M (2009) GATA3 inhibits breast cancer growth and pulmonary breast cancer metastasis. Oncogene 28:2634–2642
Eeckhoute J, Keeton EK, Lupien M, Krum SA, Carroll JS, Brown M (2007) Positive cross-regulatory loop ties GATA-3 to estrogen receptor alpha expression in breast cancer. Cancer Res 67:6477–6483
Fang TC, Yashiro-Ohtani Y, Del Bianco C, Knoblock DM, Blacklow SC, Pear WS (2007) Notch directly regulates Gata3 expression during T helper 2 cell differentiation. Immunity 27:100–110
Finnegan TJ, Carey LA (2007) Gene-expression analysis and the basal-like breast cancer subtype. Future Oncol 3:55–63
Gruvberger S, Ringner M, Chen Y, Panavally S, Saal LH, Borg A, Ferno M, Peterson C, Meltzer PS (2001) Estrogen receptor status in breast cancer is associated with remarkably distinct gene expression patterns. Cancer Res 61:5979–5984
Gupta GP, Massague J (2006) Cancer metastasis: building a framework. Cell 127:679–695
Gusterson B (2009) Do 'basal-like' breast cancers really exist? Nature reviews 9:128–134
Guy CT, Cardiff RD, Muller WJ (1992a) Induction of mammary tumors by expression of polyomavirus middle T oncogene: a transgenic mouse model for metastatic disease. Mol Cell Biol 12:954–961
Guy CT, Webster MA, Schaller M, Parsons TJ, Cardiff RD, Muller WJ (1992b) Expression of the neu protooncogene in the mammary epithelium of transgenic mice induces metastatic disease. Proc Natl Acad Sci USA 89:10578–10582
Hanahan D, Weinberg RA (2000) The hallmarks of cancer. Cell 100:57–70
Herschkowitz JI, Simin K, Weigman VJ, Mikaelian I, Usary J, Hu Z, Rasmussen KE, Jones LP, Assefnia S, Chandrasekharan S et al (2007) Identification of conserved gene expression features between murine mammary carcinoma models and human breast tumors. Genome Biol 8:R76
Hoch RV, Thompson DA, Baker RJ, Weigel RJ (1999) GATA-3 is expressed in association with estrogen receptor in breast cancer. Int J Cancer 84:122–128
Jenssen TK, Kuo WP, Stokke T, Hovig E (2002) Associations between gene expressions in breast cancer and patient survival. Hum Genet 111:411–420
Kenney NJ, Smith GH, Lawrence E, Barrett JC, Salomon DS (2001) Identification of stem cell units in the terminal end bud and duct of the mouse mammary gland. J Biomed Biotechnol 1:133–143
Kordon EC, Smith GH (1998) An entire functional mammary gland may comprise the progeny from a single cell. Development 125:1921–1930
Kouros-Mehr H, Bechis SK, Slorach EM, Littlepage LE, Egeblad M, Ewald AJ, Pai SY, Ho IC, Werb Z (2008a) GATA-3 links tumor differentiation and dissemination in a luminal breast cancer model. Cancer Cell 13:141–152
Kouros-Mehr H, Kim JW, Bechis SK, Werb Z (2008b) GATA-3 and the regulation of the mammary luminal cell fate. Curr Opin Cell Biol 20:164–170
Kouros-Mehr H, Slorach EM, Sternlicht MD, Werb Z (2006) GATA-3 maintains the differentiation of the luminal cell fate in the mammary gland. Cell 127:1041–1055
Kouros-Mehr H, Werb Z (2006) Candidate regulators of mammary branching morphogenesis identified by genome-wide transcript analysis. Dev Dyn 235:3404–3412
Laganiere J, Deblois G, Lefebvre C, Bataille AR, Robert F, Giguere V (2005) From the cover: location analysis of estrogen receptor alpha target promoters reveals that FOXA1 defines a domain of the estrogen response. Proc Natl Acad Sci USA 102:11651–11656
Lee TI, Jenner RG, Boyer LA, Guenther MG, Levine SS, Kumar RM, Chevalier B, Johnstone SE, Cole MF, Isono K et al (2006) Control of developmental regulators by Polycomb in human embryonic stem cells. Cell 125:301–313
Levine M, Davidson EH (2005) Gene regulatory networks for development. Proc Natl Acad Sci USA 102:4936–4942
Loh YH, Wu Q, Chew JL, Vega VB, Zhang W, Chen X, Bourque G, George J, Leong B, Liu J et al (2006) The Oct4 and Nanog transcription network regulates pluripotency in mouse embryonic stem cells. Nat Genet 38:431–440

Lupien M, Eeckhoute J, Meyer CA, Wang Q, Zhang Y, Li W, Carroll JS, Liu XS, Brown M (2008) FoxA1 translates epigenetic signatures into enhancer-driven lineage-specific transcription. Cell 132:958–970
Mehra R, Varambally S, Ding L, Shen R, Sabel MS, Ghosh D, Chinnaiyan AM, Kleer CG (2005) Identification of GATA3 as a breast cancer prognostic marker by global gene expression meta-analysis. Cancer Res 65:11259–11264
NCCN. (2009). National Comprehensive Cancer Network. Clinlcal practice guidelines in oncology. Breast Cancer 1
Nishiyama A, Xin L, Sharov AA, Thomas M, Mowrer G, Meyers E, Piao Y, Mehta S, Yee S, Nakatake Y et al (2009) Uncovering early response of gene regulatory networks in ESCs by systematic induction of transcription factors. Cell Stem Cell 5:420–433
Pei XH, Bai F, Smith MD, Usary J, Fan C, Pai SY, Ho IC, Perou CM, Xiong Y (2009) CDK inhibitor p18(INK4c) is a downstream target of GATA3 and restrains mammary luminal progenitor cell proliferation and tumorigenesis. Cancer Cell 15:389–401
Perou CM, Sorlie T, Eisen MB, van de Rijn M, Jeffrey SS, Rees CA, Pollack JR, Ross DT, Johnsen H, Akslen LA et al (2000) Molecular portraits of human breast tumours. Nature 406:747–752
Pietersen AM, Evers B, Prasad AA, Tanger E, Cornelissen-Steijger P, Jonkers J, van Lohuizen M (2008) Bmi1 regulates stem cells and proliferation and differentiation of committed cells in mammary epithelium. Curr Biol 18:1094–1099
Raouf A, Zhao Y, To K, Stingl J, Delaney A, Barbara M, Iscove N, Jones S, McKinney S, Emerman J et al (2008) Transcriptome analysis of the normal human mammary cell commitment and differentiation process. Cell Stem Cell 3:109–118
Shackleton M, Vaillant F, Simpson KJ, Stingl J, Smyth GK, Asselin-Labat ML, Wu L, Lindeman GJ, Visvader JE (2006) Generation of a functional mammary gland from a single stem cell. Nature 439:84–88
Sims AH, Howell A, Howell SJ, Clarke RB (2007) Origins of breast cancer subtypes and therapeutic implications. Nature clinical practice 4:516–525
Sorlie T, Perou CM, Tibshirani R, Aas T, Geisler S, Johnsen H, Hastie T, Eisen MB, van de Rijn M, Jeffrey SS et al (2001) Gene expression patterns of breast carcinomas distinguish tumor subclasses with clinical implications. Proc Natl Acad Sci USA 98:10869–10874
Sternlicht MD, Kouros-Mehr H, Lu P, Werb Z (2006) Hormonal and local control of mammary branching morphogenesis. Differentiation 74:365–381
Stingl J, Eirew P, Ricketson I, Shackleton M, Vaillant F, Choi D, Li HI, Eaves CJ (2006) Purification and unique properties of mammary epithelial stem cells. Nature 439:993–997
Tong Q, Dalgin G, Xu H, Ting CN, Leiden JM, Hotamisligil GS (2000) Function of GATA transcription factors in preadipocyte-adipocyte transition. Science 290:134–138
Visvader JE, Lindeman GJ (2003) Transcriptional regulators in mammary gland development and cancer. Int J Biochem Cell Biol 35:1034–1051
Voduc D, Cheang M, Nielsen T (2008) GATA-3 expression in breast cancer has a strong association with estrogen receptor but lacks independent prognostic value. Cancer Epidemiol Biomarkers Prev 17:365–373
Walker E, Ohishi M, Davey RE, Zhang W, Cassar PA, Tanaka TS, Der SD, Morris Q, Hughes TR, Zandstra PW et al (2007) Prediction and testing of novel transcriptional networks regulating embryonic stem cell self-renewal and commitment. Cell Stem Cell 1:71–86
Wang J, Rao S, Chu J, Shen X, Levasseur DN, Theunissen TW, Orkin SH (2006) A protein interaction network for pluripotency of embryonic stem cells. Nature 444:364–368
Welm BE, Dijkgraaf GJ, Bledau AS, Welm AL, Werb Z (2008) Lentiviral transduction of mammary stem cells for analysis of gene function during development and cancer. Cell Stem Cell 2:90–102
Welm BE, Tepera SB, Venezia T, Graubert TA, Rosen JM, Goodell MA (2002) Sca-1(pos) cells in the mouse mammary gland represent an enriched progenitor cell population. Dev Biol 245:42–56
Wiseman BS, Werb Z (2002) Stromal effects on mammary gland development and breast cancer. Science 296:1046–1049

Chapter 14
Roles of p53 and pRB Tumor Suppressor Networks in Human Cancer: Insight from Studies in the Engineered Mouse

Julien Sage, Laura Attardi, and Terry Van Dyke

Abbreviations

GEM	Genetically engineered mice
GOF	Gain-of-function
LOF	Loss-of-function
MEFs	Mouse embryo fibroblasts
RB	Retinoblastoma
SCLC	Small cell lung carcinoma

14.1 Introduction

The retinoblastoma (RB) and p53 tumor suppressor genes were identified more than 20 years ago and have since been implicated in multiple cellular processes, including the control of cell cycle progression, cell death, and cellular differentiation. A number of recent reviews explain in detail how RB and p53 normally prevent the development of cancer (Burkhart and Sage 2008; van den Heuvel and Dyson 2008; Sherr 2004; Vousden and Lu 2002; Levine et al. 2006). Our goal here is not to provide an exhaustive description of how these two factors function in mammalian cells, but

J. Sage
Department of Pediatrics and Genetics, Stanford University, Stanford, CA 94305, USA

L. Attardi
Department of Radiation and Cancer Biology, Stanford University School of Medicine, Stanford, CA 94305, USA

T. Van Dyke (✉)
Mouse Cancer Genetics Program, National Cancer Institute at Frederick, 1050 Boyles Street, Building 560, Room 32-32, Frederick, MD 21702, USA
e-mail: vandyket@mail.nih.gov

J.E. Green and T. Ried (eds.), *Genetically Engineered Mice for Cancer Research: Design, Analysis, Pathways, Validation and Pre-Clinical Testing*,
DOI 10.1007/978-0-387-69805-2_14, © Springer Science+Business Media, LLC 2012

rather to summarize some key aspects of their biology and to discuss areas in the field that are thought-provoking. In particular, we focus on how mouse models associated with perturbed of RB and p53 function have influenced our vision of the mechanisms of tumorigenesis.

p53 was cloned in 1979. Although originally believed to be an oncogene because of studies analyzing a dominant negative mutant version of p53, its character as a tumor suppressor was revealed in 1989 when the wild-type gene was found to suppress transformation (Levine 1989, 2009). *p53* is one of the most commonly mutated genes in human cancers, with mutations observed in over half of all cancers, of a broad range of types (Soussi and Lozano 2005). In addition, individuals with Li-Fraumeni syndrome, who inherit a mutant *p53* allele, are predisposed to cancer (Malkin et al. 1990). Although these two observations clearly linked p53 mutation to cancer, an unequivocal demonstration of its importance in blocking tumorigenesis came from the generation of p53 null mice. *p53* null mice develop cancer (T-cell lymphomas and/or sarcomas) at 100% frequency within a year after birth (Jacks et al. 1994; Harvey et al. 1993; Donehower et al. 1992). While this tumor spectrum does not include many types of human cancers associated with p53 mutation, subsequent work in engineered mice has uncovered several mechanisms by which p53 mutation contributes to a full spectrum of cancers (see below). In addition to the dramatic phenotype of p53-deficient mice, compound mutant mice derived from crossing transgenic mice expressing a particular oncogene or mice expressing reduced levels of a tumor suppressor gene with p53-deficient mice nearly always display an exacerbated phenotype, with a decrease in tumor latency or the appearance of new tumor types (Attardi and Jacks 1999). Notably, most human cancers harbor missense mutations in the p53 gene, and a long-standing debate has been the functional consequences of these mutations, whether they are loss-of-function (LOF), gain-of-function (GOF), or both; this issue has recently been addressed in genetically engineered mice (GEM) (see below). A large body of work has resulted in the identification of p53 interacting proteins and target genes of its transcriptional activity, leading to the elaboration of p53 pathways in mammalian cells. This network, summarized in Fig. 14.1, plays a critical role in the response of cells to a variety of stresses, including DNA damage and oncogenic activity (Sherr 2004; Levine et al. 2006; Van Dyke 2007). When p53 itself is not mutated in human tumor cells, the function of other components of this pathway can be altered, further validating the functional interactions between members of this pathway.

RB was cloned in 1986 from families in which children developed retinoblastoma by the groups of Weinberg, Dryja, Lee, and Gallie as the gene responsible for familial retinoblastoma (Friend et al. 1986; Cavenee et al. 1985; Lee et al. 1987). These RB patients have been found to be at increased risk for additional malignancies, including osteosarcoma (Deshpande and Hinds 2006). In addition, the RB gene is inactivated in a broad range of human sporadic cancers, including lung, breast, prostate, bladder, and liver cancers (Burkhart and Sage 2008). A combination of biochemical, molecular, and genetic approaches has also defined the "RB pathway" in mammalian cells (Fig. 14.2). This pathway plays a central role in the control of cell cycle progression in response to cellular and extracellular signals.

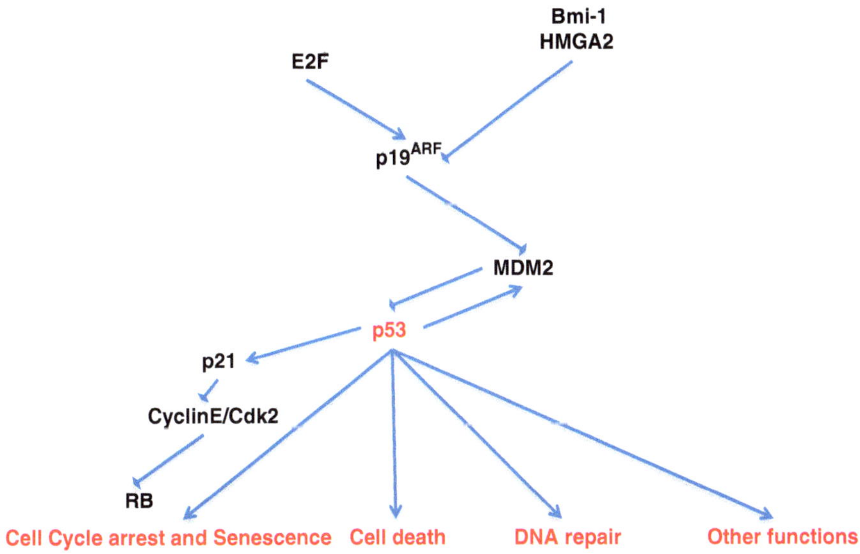

Fig. 14.1 p53 network

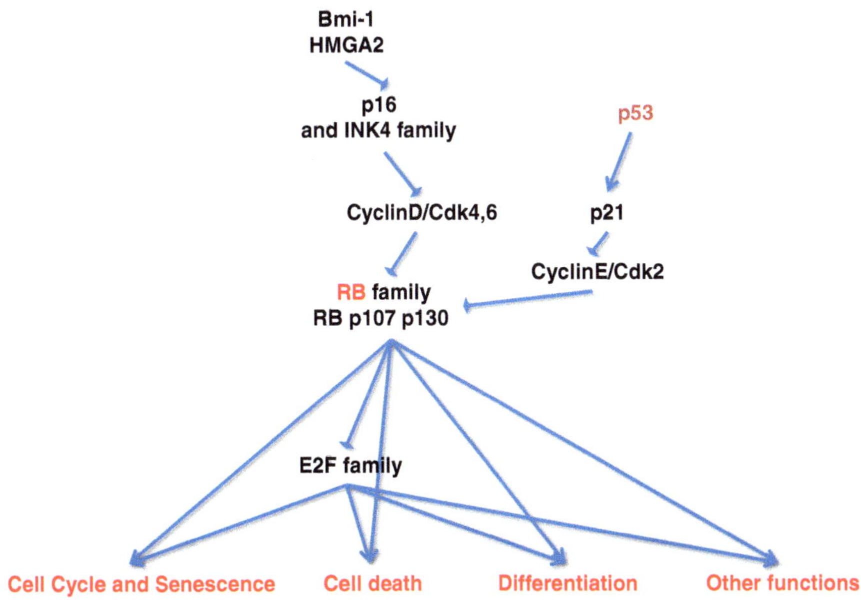

Fig. 14.2 RB network

As with p53, when RB itself is not inactivated in human cancer cells, another member of the pathway is commonly altered.

Because p53 and RB functions are mutated in such a broad range of human cancers, a major effort in the cancer field in the past 25 years has been to explore the mechanisms of action of these two tumor suppressor genes. In particular, GEM have been generated to model human cancers associated with aberrant RB and/or p53 function and to dissect their cellular functions (Johnson and Attardi 2006; Wikenheiser-Brokamp 2006). These mutant mice have provided a wealth of information regarding the mechanisms of tumorigenesis; yet, a number of questions remain to be addressed.

14.2 Modeling Cancers with Mutations in p53

14.2.1 p53 Suppresses Cancer in Many Cell Types

While p53 plays a paramount role in suppressing tumor development in both humans and mice, the majority of the cancers developing in p53-deficient mice are T-cell lymphomas, with sarcomas of a variety of types being the next most common lesion (Jacks et al. 1994). It is intriguing that the *p53* null mice develop lymphomas and sarcomas rather than epithelial cancers. Evidence exists that this is not fully due to a resistance of other tissue types to p53 loss but rather that mice succumb to lymphomas earliest, masking any potential tumor suppressor roles in other tissues. This point is underscored by studies using conditional p53 knockout mice, which have demonstrated that p53 ablation specifically in epithelial tissues, such as the mammary gland, does in fact contribute to epithelial cancer (Lin et al. 2004). The use of conditional p53 knockout mice in the backdrop of other oncogenic lesions has also revealed the development of new tumor types, further illustrating a widespread role for p53 tumor suppressor activity in different tissues. In these cases, mechanistic studies show that "oncogenic stress" generated by a cancer-initiating event triggers a p53 tumor suppressor response such that only under this condition does p53 disruption have an impact. Indeed, p53 mutation in human cancers appears to occur during progression of the disease. For example, activation of K-ras in the lung epithelium combined with p53 loss results in nonsmall cell lung cancer. These aggressive tumors mimic the human lung cancers in which both of these mutations are commonly observed, and tumors display aggressive behavior (Jackson et al. 2005). Thus, these mice highlight the importance of p53 as a tumor suppressor in these tissues, and moreover, provide good models for the cognate human cancers. Alternatively, if a source of genetic instability is combined with p53 deficiency, such as through telomerase deficiency and consequent telomere attrition, then p53 loss promotes epithelial carcinogenesis, in the mammary gland, skin, and small intestine (Artandi et al. 2000). Together, these findings emphasize that p53 is poised to prevent cancer in myriad tissues and highlight the mechanistic understanding that the p53 tumor suppressor must be "activated," thus producing selective pressure for its loss or mutation.

14.2.2 Missense p53 Mutations and p53 Function

An unusual feature of p53 mutations in human cancer is that they are typically missense mutations rather than nonsense mutations (Soussi and Lozano 2005). This observation indicates that there is some selective advantage of p53 point mutants. While originally believed to be simply dominant negative versions of p53 that could inactivate the remaining wild-type protein through tetramerization, it has become clear that mutants may also have GOF activities. Indeed, most human cancers with p53 missense mutations have also deleted the wild-type allele (Soussi and Lozano 2005). Evidence from mouse models has been key in confirming the GOF hypothesis. Knock-in mice expressing p53 point mutants, based on human tumor-derived mutants, with alterations at residue 172 (human 175) and at 270 (human 273), were generated (Olive et al. 2004; Lang et al. 2004). Analysis of these mice, either in heterozygous or homozygous form, showed similar survival profiles to mice heterozygous or homozygous for a p53 null allele. Interestingly, however, there was a difference in spectrum of the tumors, with new tumor types developing in the presence of the mutant allele, some of which metastasized. For example, p53R172H/+ mice on a C57Bl/6 background developed carcinomas at an equivalent frequency to $p53^{+/-}$ mice, but the incidence of metastasis was dramatically increased (Lang et al. 2004). In another study analyzing both p53R172H and p53R270H knock-in mice, survival curves with heterozygous and homozygous mutant mice mirrored those observed with $p53^{+/-}$ and $p53^{-/-}$ mice, respectively, but new tumor types and metastases were observed in the presence of these mutant alleles (Olive et al. 2004). Recently, it was shown that mutant p53 is unstable in normal tissues but becomes stabilized in cancer, adding a new twist on missense p53 (Terzian et al. 2008). Together, these studies emphasize the idea that p53 missense mutants have GOF activities, leading to phenotypes more aggressive than produced by p53 deficiency. The GOF capacity of mutant p53 has been proposed to occur through various mechanisms, such as inappropriate induction of expression of specific genes, or interaction with other proteins (Olive et al. 2004; Lang et al. 2004; Terzian et al. 2008). In the aforementioned p53 knock-in studies, p53 mutant GOF activity was proposed to be due to binding and inactivation of the p53 family members, p63 and p73. Using mouse embryo fibroblasts (MEFs), the authors showed that R172H/R172H MEFs proliferate better and are more readily transformed than p53 null cells, and that this is attributable to inhibition of p63 and p73 (Olive et al. 2004; Lang et al. 2004). Interestingly, these GOF activities are very cell-type specific. In conjunction with K-ras activation, p53R172H and p53R270H were not significantly different from p53 deficiency in promoting lung cancer, but in contrast, both point mutants could dramatically increase the incidence of K-ras-driven sinonasal tumors, suggesting a tissue-specific GOF effect (Jackson et al. 2005). Understanding the mechanisms of p53 GOF activities in different contexts will continue to be an area of great interest with potential for therapeutic interventions.

14.2.3 p53 Family Members in Cancer

As just described, the inhibition of p63 and p73 by p53 missense mutants has been proposed to promote tumorigenesis. In addition to these studies, direct experiments to address the importance of p63 and p73 in cancer have been performed. p63 and p73 exist in a number of different isoforms, generated both by alternative promoter usage and alternate splicing (Bourdon 2007; Deyoung and Ellisen 2007). Most studies have focused on mouse strains designed to ablate all isoforms. Aging $p73^{+/-}$ mice display premature mortality accompanied by a predisposition to various types of cancer, and these phenotypes are enhanced upon crossing onto $p63^{+/-}$ or $p53^{+/-}$ mice (Flores et al. 2005). In addition, deletion of one p73 isoform, TAp73, leads to an enhanced spontaneous predisposition to cancer, particularly lung adenocarcinomas (Tomasini et al. 2008). However, $p73^{+/-}$ and $p73^{+/-}$;$p53^{+/-}$ mice do not exhibit an increased predisposition to radiation-induced lymphomas compared to $p73^{+/+}$ controls (Perez-Losada et al. 2005). The role of p63 in cancer has been more controversial, largely stemming from the analysis of two different p63 knockout lines. In one model, $p63^{+/-}$ mice exhibited decreased survival and an increased risk of spontaneous cancer relative to wild-type controls, and the severity of the phenotypes was increased in the backdrop of p53 heterozygosity, with higher tumor burden and more metastases observed (Flores et al. 2005). In the other model, no spontaneous cancer predisposition was observed upon aging $p63^{+/-}$ mice and, in fact, compound $p63^{+/-}$;$p53^{+/-}$ mice had fewer tumors than $p53^{+/-}$ mice (Keyes et al. 2006). Moreover, in a DMBA/TPA model for skin carcinogenesis, these $p63^{+/-}$ mice did not show enhanced tumor development (Keyes et al. 2006). Additionally, as with the p73-deficient mice, $p63^{+/-}$ and $p63^{+/-}$;$p53^{+/-}$ mice did not show an enhanced susceptibility to radiation-induced lymphoma development relative to $p63^{+/+}$ controls (Perez-Losada et al. 2005). These differences in findings with p63-deficient mice could relate to genetic background or allelic differences in the p63 knockout strains. Clearly defining the roles of p63 and p73 in cancer will help to better elaborate how p53 and the interplay among the family members contribute to tumorigenesis.

14.2.4 Cellular Functions of p53 in Tumor Suppression

p53 functions to suppress cancer by inhibiting cell cycle progression, either transiently or long term (senescence, differentiation) and by promoting apoptosis. (Vousden and Lu 2002; Van Dyke 2007; Johnson and Attardi 2006; Stiewe 2007). Various studies have highlighted the fact that these different functions of p53 may be context-dependent. In particular, analyses of a p53 knock-in strain expressing the mouse version of a human tumor mutant, p53R172P, which can induce growth arrest and maintain genetic stability but lacks the ability to trigger apoptosis, have been revealing (Liu et al. 2004). T-cell lymphoma development is greatly inhibited by this mutant, highlighting the importance of cell cycle restraint and the maintenance of genomic stability for p53 function. However, tumors readily develop in other tissues, suggesting that loss of the apoptotic function of p53 is important in those cases. The importance of p53 triggering apoptosis for p53 tumor suppressor function has been underscored by the analysis of p53 and components of the apoptotic machinery in choroid plexus tumors and mammary carcinomas driven by expression of the RB/p107/p130-inactivating domain of SV40 Large-T Antigen (Lu et al. 2008; Yin et al. 1997; Symonds et al. 1994a, b; Saenz Robles et al. 1994; McCarthy et al. 1994; Chen et al. 1992, see below) and in B-cell lymphomas driven by the Eμ-myc transgene (Schmitt et al. 2002). The notion that the p53 activity relevant for tumor suppression varies according to context is further supported by studies in which p53 was reactivated in p53-deficient tumors. In lymphoma models, p53 reactivation results in apoptosis, while in sarcomas or liver tumors, p53 reactivation causes senescence (Ventura et al. 2007; Xue et al. 2007; Martins et al. 2006). Why different cellular milieus influence the program that p53 engages is an area of great interest.

14.2.5 Signals Triggering p53 Activation in Neoplastic Lesions

A variety of different cellular stresses, including DNA damage, oncogene expression, and hypoxia, can induce p53 function with the consequence of selecting for p53 mutation. In one study, it was hypothesized that double-strand breaks generated during VDJ recombination were the trigger for p53 tumor suppression in thymocytes, thus resulting in lymphoma in a p53-deficient state. However, inhibition of VDJ recombination in a *Rag1*-deficient background did not alter the tumor frequency or latency (Nacht and Jacks 1998; Liao et al. 1998). Another set of experiments examining the importance of acute DNA damage signals for activating p53 utilized mice in which endogenous p53 coding sequences were fused with a tamoxifen-responsive estrogen receptor domain (p53-ER), rendering p53 function regulatable *in vivo* (Christophorou et al. 2005, 2006). In the absence of tamoxifen, p53 is not active, but can be rendered functional by the treatment of mice with tamoxifen. To determine whether acute DNA damage was the critical inducer of p53 tumor suppression in thymocytes, p53 was activated immediately after irradiation or only subsequent to tissue recovery. Given the long-standing view that loss of the p53 acute DNA damage response would drive tumors due to increasing mutation frequency, it was surprising that maintenance of p53 function throughout tissue recovery did not rescue the frequency or timing of thymic lymphoma. Rather, effective tumor suppression required p53 function subsequent to tissue recovery. A complementary study examining the stage at which p53 is required for suppressing irradiation-induced lymphomas, conducted by inactivating p53 either prior or subsequent to irradiation, similarly concluded that p53 is not needed during acute DNA damage treatment to suppress tumorigenesis (Hinkal et al. 2009). The p53ER study, along with studies of fibrosarcoma suppression by p53, showed that p53 tumor suppression was abrogated in a p19ARF-deficient background, indicating that "oncogenic stress" mediated by ARF, and not acute DNA damage, was the trigger for p53 tumor suppression (Van Dyke 2007; Christophorou et al. 2006; Efeyan et al. 2006). Collectively, these studies suggest that oncogenic signals may be more relevant than acute DNA damage for activating p53 in nascent tumors, at least in these specific models. The role of other stresses, including hypoxia and chronic low level DNA damage, remains to be addressed.

14.3 Modeling Cancers with Mutations in RB

14.3.1 Absence of Retinoblastoma Development in RB Mutant Mice

The RB gene was identified in families in which children developed retinoblastoma: these patients inherit one mutant allele of RB, and it is thought that loss of heterozygosity during retinal development initiates retinoblastoma in these children. In an effort to model retinoblastoma in mice, $RB^{+/-}$ mutant mice were generated. Strikingly, however, these mutant mice never developed retinoblastoma (Jacks et al. 1992; Lee et al. 1992; Clarke et al. 1992). The same observation was made in chimeric mice born with $RB^{-/-}$ cells in their retina and mice with conditional inactivation of RB in the eye using the Cre/lox system (Maandag et al. 1994; Williams et al. 1994; Chen et al. 2004; Zhang et al. 2004; MacPherson et al. 2004). Thus, the absence of retinoblastoma development in mice is not due to a reduced rate of loss of heterozygosity in mouse retinal progenitors compared to human retinal cells. Importantly, loss of RB function in mice leads to the development of pituitary, thyroid, and adrenal gland tumors, and tumors developing in $RB^{+/-}$ mice display loss of heterozygosity (reviewed in Wikenheiser-Brokamp 2006; Dannenberg and te Riele 2006). These experiments directly demonstrate that Rb is a tumor suppressor but also raise numerous questions regarding the importance of the cellular context in which Rb acts to normally suppress tumorigenesis in humans.

14.3.2 Functional Overlap Within the RB Gene Family

Why do mice fail to develop retinoblastoma upon loss of RB function in their eye? A partial answer to this question has come from the identification of two proteins structurally related to RB in mammalian cells. p107 and p130 are controlled by the same kinase activities that control RB function, the three family members can be bound and inactivated by the same viral oncoproteins, such as adenovirus E1A, SV40 Large T, or HPV E7, and they share a number of binding partners, including some members of the E2F family of transcription factors (Fig. 14.1) (reviewed in Wikenheiser-Brokamp 2006; Dannenberg and te Riele 2006; Cobrinik 2005; Claudio et al. 2002; Classon and Dyson 2001). These observations raised the possibility that p107 and/or p130 may functionally compensate for loss of RB function in mouse retinal cells. Indeed, combined inactivation of RB and p107 or RB and p130 in the retina of mice resulted in retinoblastoma development, which provided long-sought models for this tumor type (Chen et al. 2004; MacPherson et al. 2004, 2007; Ajioka et al. 2007). These results have been expanded to other tissues and organs: compound inactivation of RB family genes in mice promotes tumor development and can provide mouse models for human cancers associated with loss of RB function (reviewed in Wikenheiser-Brokamp 2006; Dannenberg and te Riele 2006). Similarly, transgenic mice expressing a fragment of the SV40 Large T oncoprotein (T121) that specifically inactivates the three members of the RB family develop tumors that are associated with loss of RB function in humans (Symonds et al. 1994a; Saenz Robles et al. 1994; Pan et al. 1998; Xiao et al. 2002; Simin et al. 2004).

14.3.3 Molecular Basis of the Functional Overlap Between RB Family Members

Several mechanisms could explain the extent of the functional overlap between RB, p107, and p130. First, there can be redundancy such that loss of one protein results in functional replacement by a family member that is present in the same cells and normally performs an identical function. A nonmutually exclusive possibility is that loss of one family member results in the induction of a compensatory activity, either because the expression of the other family member is changed or because this other family member now performs a function that it does not normally perform.

Thus, one possible interpretation of these experiments underscoring the functional overlap between RB family members is that the pattern of expression of the three family genes is cell type specific. One could envision that cells that do not express p107 and p130 may be more likely to initiate cancer upon loss of RB function because compensation is not possible. To a first approximation, although their levels of expression vary, the three family members are ubiquitously expressed *in vivo* and, therefore, this model is probably not correct (Garriga et al. 1998; Jiang and Zacksenhaus 2002; Smith et al. 1998; Jiang et al. 1997 and references therein). However, loss of RB in many cell types leads to the direct up-regulation of p107 transcription; thus, p107 may be able to compensate for loss of RB even in cells with normally low levels of p107 (Burkhart et al. 2008; Sage et al. 2003). Furthermore, an intriguing observation in human fetal retinal explants suggests that in specific contexts, this feedback loop may not be functional: in these human cells, which may be closely related to the cell of origin for human retinoblastoma, the p107 gene is silent and p107 levels do not increase following knockdown of RB (Donovan et al. 2006). There is no evidence that p130 levels change in response to RB inactivation, and the mechanisms by which p130 may compensate for loss of RB are still unclear; one possibility is that p130 levels are always high enough that they may buffer the effects of loss of RB function in cells.

Thus, a slightly different view of these observations is that the response of cells to loss of RB function is dictated by normal and induced levels of p107 and p130. Under conditions when a low threshold is reached, loss of RB may affect the proliferation of the mutant cells. It is possible that this threshold is different for different cell types, depending on other regulatory networks. It is also possible that different effects of RB (control of cell death, cellular differentiation, cell cycle progression) are set at different thresholds.

14.3.4 Specific Tumor Suppressor Functions of RB, p107, and p130

While these experiments underscore the functional overlap within the RB family, a key goal in the field is to decipher the specific cellular functions of RB, p107, and p130. Strikingly, only RB is mutated in large numbers of human tumors. To our knowledge, there are only two reports of mutations in the p107 gene, both observed in cell lines in culture (Ichimura et al. 2000; Takimoto et al. 1998). Increasing evidence suggests that p130 levels are decreased in human tumors, and p130 mutations have been found in human retinoblastomas and other tumors, including lung cancer, but p130 is not a major tumor suppressor in humans (Masciullo et al. 2008; D'Andrilli et al. 2004; Sanseverino et al. 2003; Paggi and Giordano 2001). p107 and p130 knockout mice have limited phenotypes and are not cancer-prone (LeCouter et al. 1998a, b; Lee et al. 1996; Cobrinik et al. 1996). Loss of p130 has been shown to mildly promote the development of lung cancer in a mouse model of lung adenocarcinoma, although the molecular and cellular basis of this effect is still unknown (Ho et al. 2009). p107 specifically restricts the proliferation of neural progenitors in mice (Vanderluit et al. 2004, 2007), and it would be interesting to know if loss of p107 function could contribute to the development of brain cancer.

While the molecular basis of RB's potent tumor suppressor function compared to p107 and p130 is not fully understood, it may lie in specific binding partners (including E2F transcription factors) and specific target genes, an area of intense investigation (van den Heuvel and Dyson 2008; Morris and Dyson 2001; Iaquinta and Lees 2007; Ferrari et al. 2008). Another area of great interest is the identity of the cells in which RB normally prevents cancer initiation, as this may provide some insights into the mechanisms of tumorigenesis in RB patients. Experiments in mutant mice suggest that retinoblastoma may arise from retinal progenitors (Macpherson 2008) and recent data from mice with mutations in RB family members in the blood compartment also suggest that mutant hematopoietic stem cells may initiate blood cancer (Viatour et al. 2008), but whether RB always suppresses cancer development in stem/progenitor cell population is unclear (Ajioka et al. 2007).

14.3.5 Are Mice and Humans that Different?

These observations raise a second question: are humans and mice so different that the mechanisms of cancer development in the same organs between the two species rely on different mechanisms? For instance, there is no indication that *RB* patients are at any risk of developing pituitary cancer while 100% of *RB* mutant mice develop pituitary adenoma. Interestingly, however, recent observations suggest that the RB pathway is inactivated in human pituitary tumors. The *RB* promoter has been found to be methylated in some cases of human pituitary adenomas (Yoshino et al. 2007; Honda et al. 2003). There is also emerging evidence that HMG2A mutations, which are involved in human pituitary cancer development, result in deregulation of E2F activity, similarly to loss of RB function (Fedele et al. 2006). Similarly, recent reports indicate that p130 is often deleted in human retinoblastoma cells (Priya et al. 2009; Tosi et al. 2005), suggesting that the mouse models relying on combinations of mutations in RB family genes actually mimic the situation observed in humans.

14.4 Modeling Cancers with Mutations in RB and p53

Not surprisingly given the high frequency of mutations in both RB and p53 in human cancers, the two genes are often simultaneously mutated in cancer cells and alterations in both pathways are nearly always present. One striking example is small cell lung carcinoma (SCLC), in which mutations in the *p53* and *RB* genes occur in more than 90% of the cases. This cooperativity can also result from mutations in other components of the RB or p53 pathways. For instance, although p53 is not mutated in human retinoblastomas, the MDM4 oncogene, a negative regulator of p53, is amplified in a significant subset of these RB-deficient tumors (Laurie et al. 2006). Cooperative effects of deficiency for the RB and p53 pathways have also been clearly demonstrated by experiments in mouse models. For example, RB/p53 double mutant mice develop cancers not observed in the single mutant mice (Meuwissen et al. 2003; Zhou et al. 2006; Flesken-Nikitin et al. 2003). Furthermore, conditional deletion of the *RB* and *p53* genes specifically in the lung epithelium of adult mice using intra-tracheal delivery of adenoviral particles expressing the Cre recombinase, results in the development of neuroendocrine SCLCs, with many characteristics of human SCLC. Thus, RB and p53 deficiency often cooperate to promote tumorigenesis, although there are some exceptions (Xiao et al. 2002).

What is the basis for the cooperativity between RB and p53 deficiencies? Because RB and p53 have so many overlapping functions, it is perhaps initially unclear why the two pathways are so often mutated together. However, one simple explanation based on experiments in mouse models is that loss of RB function provokes p53-dependent cell death, and that inactivation of p53 diminishes this cell death. For example, in mouse embryos, loss of RB leads to cell death in several cell types, including the lens and the central nervous system, and concomitant loss of p53 abrogates this death (Macleod et al. 1996). Moreover, in brain tumors induced by expression of T121, p53 is activated and stimulates cell death, but p53 loss results in decreased cell death and tumor progression (Symonds et al. 1994a). Another potential reason for the cooperativity may be because of nonoverlapping functions for RB and p53, which would provide additional selection for combined loss of RB and p53.

14.5 Conclusions

Studies in mouse models have greatly advanced our understanding of both the RB and p53 pathways. Future analyses in ever more sophisticated settings will continue to reveal new facets of these critical tumor suppressors. These studies will include spatially and temporally controlled perturbations in the RB and p53 pathways to better model and study the early stages of cancer in patients. Importantly, mechanistic knowledge of these two tumor suppressor pathways will translate into better therapeutic strategies for cancer patients.

References

Ajioka I, Martins RA, Bayazitov IT, Donovan S, Johnson DA, Frase S et al (2007) Differentiated horizontal interneurons clonally expand to form metastatic retinoblastoma in mice. Cell 131(2):378–390

Artandi SE, Chang S, Lee SL, Alson S, Gottlieb GJ, Chin L et al (2000) Telomere dysfunction promotes non-reciprocal translocations and epithelial cancers in mice. Nature 406(6796):641–645

Attardi LD, Jacks T (1999) The role of p53 in tumour suppression: lessons from mouse models. Cell Mol Life Sci 55(1):48–63

Bourdon JC (2007) p53 Family isoforms. Curr Pharm Biotechnol 8(6):332–336

Burkhart DL, Sage J (2008) Cellular mechanisms of tumour suppression by the retinoblastoma gene. Nat Rev Cancer 8(9):671–682

Burkhart DL, Viatour P, Ho VM, Sage J (2008) GFP reporter mice for the retinoblastoma-related cell cycle regulator p107. Cell Cycle 7(16):2544–2552

Cavenee WK, Hansen MF, Nordenskjold M, Kock E, Maumenee I, Squire JA et al (1985) Genetic origin of mutations predisposing to retinoblastoma. Science 228(4698):501–503

Chen J, Tobin GJ, Pipas JM, Van Dyke T (1992) T-antigen mutant activities in vivo: roles of p53 and pRB binding in tumorigenesis of the choroid plexus. Oncogene 7(6):1167–1175

Chen D, Livne-Bar I, Vanderluit JL, Slack RS, Agochiya M, Bremner R (2004) Cell-specific effects of RB or RB/p107 loss on retinal development implicate an intrinsically death-resistant cell-of-origin in retinoblastoma. Cancer Cell 5(6):539–551

Christophorou MA, Martin-Zanca D, Soucek L, Lawlor ER, Brown-Swigart L, Verschuren EW et al (2005) Temporal dissection of p53 function in vitro and in vivo. Nat Genet 37(7):718–726

Christophorou MA, Ringshausen I, Finch AJ, Swigart LB, Evan GI (2006) The pathological response to DNA damage does not contribute to p53-mediated tumour suppression. Nature 443(7108):214–217

Clarke AR, Maandag ER, van Roon M, van der Lugt NM, van der Valk M, Hooper ML et al (1992) Requirement for a functional Rb-1 gene in murine development. Nature 359(6393):328–330

Classon M, Dyson N (2001) p107 and p130: versatile proteins with interesting pockets. Exp Cell Res 264(1):135–147
Claudio PP, Tonini T, Giordano A (2002) The retinoblastoma family: twins or distant cousins? Genome Biol 3(9):reviews3012
Cobrinik D (2005) Pocket proteins and cell cycle control. Oncogene 24(17):2796–2809
Cobrinik D, Lee MH, Hannon G, Mulligan G, Bronson RT, Dyson N et al (1996) Shared role of the pRB-related p130 and p107 proteins in limb development. Genes Dev 10(13):1633–1644
D'Andrilli G, Masciullo V, Bagella L, Tonini T, Minimo C, Zannoni GF et al (2004) Frequent loss of pRb2/p130 in human ovarian carcinoma. Clin Cancer Res 10(9):3098–3103
Dannenberg JH, te Riele HP (2006) The retinoblastoma gene family in cell cycle regulation and suppression of tumorigenesis. Results Probl Cell Differ 42:183–225
Deshpande A, Hinds PW (2006) The retinoblastoma protein in osteoblast differentiation and osteosarcoma. Curr Mol Med 6(7):809–817
Deyoung MP, Ellisen LW (2007) p63 and p73 in human cancer: defining the network. Oncogene 26(36):5169–5183
Donehower LA, Harvey M, Slagle BL, McArthur MJ, Montgomery CA Jr, Butel JS et al (1992) Mice deficient for p53 are developmentally normal but susceptible to spontaneous tumours. Nature 356(6366):215–221
Donovan SL, Schweers B, Martins R, Johnson D, Dyer MA (2006) Compensation by tumor suppressor genes during retinal development in mice and humans. BMC Biol 4:14
Efeyan A, Garcia-Cao I, Herranz D, Velasco-Miguel S, Serrano M (2006) Tumour biology: policing of oncogene activity by p53. Nature 443(7108):159
Fedele M, Visone R, De Martino I, Troncone G, Palmieri D, Battista S et al (2006) HMGA2 induces pituitary tumorigenesis by enhancing E2F1 activity. Cancer Cell 9(6):459–471
Ferrari R, Pellegrini M, Horwitz GA, Xie W, Berk AJ, Kurdistani SK (2008) Epigenetic reprogramming by adenovirus e1a. Science 321(5892):1086–1088
Flesken-Nikitin A, Choi KC, Eng JP, Shmidt EN, Nikitin AY (2003) Induction of carcinogenesis by concurrent inactivation of p53 and Rb1 in the mouse ovarian surface epithelium. Cancer Res 63(13):3459–3463
Flores ER, Sengupta S, Miller JB, Newman JJ, Bronson R, Crowley D et al (2005) Tumor predisposition in mice mutant for p63 and p73: evidence for broader tumor suppressor functions for the p53 family. Cancer Cell 7(4):363–373
Friend SH, Bernards R, Rogelj S, Weinberg RA, Rapaport JM, Albert DM et al (1986) A human DNA segment with properties of the gene that predisposes to retinoblastoma and osteosarcoma. Nature 323(6089):643–646
Garriga J, Limon A, Mayol X, Rane SG, Albrecht JH, Reddy EP et al (1998) Differential regulation of the retinoblastoma family of proteins during cell proliferation and differentiation. Biochem J 333(Pt 3):645–654
Harvey M, McArthur MJ, Montgomery CA Jr, Bradley A, Donehower LA (1993) Genetic background alters the spectrum of tumors that develop in p53-deficient mice. FASEB J 7(10):938–943
Hinkal G, Parikh N, Donehower LA (2009) Timed somatic deletion of p53 in mice reveals age-associated differences in tumor progression. PLoS One 4(8):e6654, PMCID: 2721630
Ho VM, Schaffer BE, Karnezis AN, Park KS, Sage J (2009) The retinoblastoma gene Rb and its family member p130 suppress lung adenocarcinoma induced by oncogenic K-Ras. Oncogene 28(10):1393–1399
Honda S, Tanaka-Kosugi C, Yamada S, Sano T, Matsumoto T, Itakura M et al (2003) Human pituitary adenomas infrequently contain inactivation of retinoblastoma 1 gene and activation of cyclin dependent kinase 4 gene. Endocr J 50(3):309–318
Iaquinta PJ, Lees JA (2007) Life and death decisions by the E2F transcription factors. Curr Opin Cell Biol 19(6):649–657
Ichimura K, Hanafusa H, Takimoto H, Ohgama Y, Akagi T, Shimizu K (2000) Structure of the human retinoblastoma-related p107 gene and its intragenic deletion in a B-cell lymphoma cell line. Gene 251(1):37–43

Jacks T, Fazeli A, Schmitt EM, Bronson RT, Goodell MA, Weinberg RA (1992) Effects of an Rb mutation in the mouse. Nature 359(6393):295–300
Jacks T, Remington L, Williams BO, Schmitt EM, Halachmi S, Bronson RT et al (1994) Tumor spectrum analysis in p53-mutant mice. Curr Biol 4(1):1–7
Jackson EL, Olive KP, Tuveson DA, Bronson R, Crowley D, Brown M et al (2005) The differential effects of mutant p53 alleles on advanced murine lung cancer. Cancer Res 65(22):10280–10288
Jiang Z, Zacksenhaus E (2002) Coordinated expression of Rb gene family in the mammary gland. Mech Dev 119(Suppl 1):S39–S42
Jiang Z, Zacksenhaus E, Gallie BL, Phillips RA (1997) The retinoblastoma gene family is differentially expressed during embryogenesis. Oncogene 14(15):1789–1797
Johnson TM, Attardi LD (2006) Dissecting p53 tumor suppressor function in vivo through the analysis of genetically modified mice. Cell Death Differ 13(6):902–908
Keyes WM, Vogel H, Koster MI, Guo X, Qi Y, Petherbridge KM et al (2006) p63 heterozygous mutant mice are not prone to spontaneous or chemically induced tumors. Proc Natl Acad Sci USA 103(22):8435–8440, PMCID: 1482510
Lang GA, Iwakuma T, Suh YA, Liu G, Rao VA, Parant JM et al (2004) Gain of function of a p53 hot spot mutation in a mouse model of Li–Fraumeni syndrome. Cell 119(6):861–872
Laurie NA, Donovan SL, Shih CS, Zhang J, Mills N, Fuller C et al (2006) Inactivation of the p53 pathway in retinoblastoma. Nature 444(7115):61–66
LeCouter JE, Kablar B, Whyte PF, Ying C, Rudnicki MA (1998a) Strain-dependent embryonic lethality in mice lacking the retinoblastoma-related p130 gene. Development 125(23):4669–4679
LeCouter JE, Kablar B, Hardy WR, Ying C, Megeney LA, May LL et al (1998b) Strain-dependent myeloid hyperplasia, growth deficiency, and accelerated cell cycle in mice lacking the Rb-related p107 gene. Mol Cell Biol 18(12):7455–7465
Lee WH, Bookstein R, Hong F, Young LJ, Shew JY, Lee EY (1987) Human retinoblastoma susceptibility gene: cloning, identification, and sequence. Science 235(4794):1394–1399
Lee EY, Chang CY, Hu N, Wang YC, Lai CC, Herrup K et al (1992) Mice deficient for Rb are nonviable and show defects in neurogenesis and haematopoiesis. Nature 359(6393):288–294
Lee MH, Williams BO, Mulligan G, Mukai S, Bronson RT, Dyson N et al (1996) Targeted disruption of p107: functional overlap between p107 and Rb. Genes Dev 10(13):1621–1632
Levine AJ (1989) The p53 tumor suppressor gene and gene product. Princess Takamatsu Symp 20:221–230
Levine AJ (2009) The common mechanisms of transformation by the small DNA tumor viruses: the inactivation of tumor suppressor gene products: p53. Virology 384(2):285–293
Levine AJ, Hu W, Feng Z (2006) The P53 pathway: what questions remain to be explored? Cell Death Differ 13(6):1027–1036
Liao MJ, Zhang XX, Hill R, Gao J, Qumsiyeh MB, Nichols W et al (1998) No requirement for V(D)J recombination in p53-deficient thymic lymphoma. Mol Cell Biol 18(6):3495–3501, PMCID: 108930
Lin SC, Lee KF, Nikitin AY, Hilsenbeck SG, Cardiff RD, Li A et al (2004) Somatic mutation of p53 leads to estrogen receptor alpha-positive and -negative mouse mammary tumors with high frequency of metastasis. Cancer Res 64(10):3525–3532
Liu G, Parant JM, Lang G, Chau P, Chavez-Reyes A, El-Naggar AK et al (2004) Chromosome stability, in the absence of apoptosis, is critical for suppression of tumorigenesis in Trp53 mutant mice. Nat Genet 36(1):63–68
Lu X, Yang C, Hill R, Yin C, Hollander MC, Fornace AJ Jr et al (2008) Inactivation of gadd45a sensitizes epithelial cancer cells to ionizing radiation in vivo resulting in prolonged survival. Cancer Res 68(10):3579–3583
Maandag EC, van der Valk M, Vlaar M, Feltkamp C, O'Brien J, van Roon M et al (1994) Developmental rescue of an embryonic-lethal mutation in the retinoblastoma gene in chimeric mice. EMBO J 13(18):4260–4268
Macleod KF, Hu Y, Jacks T (1996) Loss of Rb activates both p53-dependent and independent cell death pathways in the developing mouse nervous system. EMBO J 15(22):6178–6188

Macpherson D (2008) Insights from mouse models into human retinoblastoma. Cell Div 3(1):9

MacPherson D, Sage J, Kim T, Ho D, McLaughlin ME, Jacks T (2004) Cell type-specific effects of Rb deletion in the murine retina. Genes Dev 18(14):1681–1694

MacPherson D, Conkrite K, Tam M, Mukai S, Mu D, Jacks T (2007) Murine bilateral retinoblastoma exhibiting rapid-onset, metastatic progression and N-myc gene amplification. EMBO J 26(3):784–794

Malkin D, Li FP, Strong LC, Fraumeni JF Jr, Nelson CE, Kim DH et al (1990) Germ line p53 mutations in a familial syndrome of breast cancer, sarcomas, and other neoplasms. Science 250(4985):1233–1238

Martins CP, Brown-Swigart L, Evan GI (2006) Modeling the therapeutic efficacy of p53 restoration in tumors. Cell 127(7):1323–1334

Masciullo V, Berardengo E, Boglione A, Sgambato A, Bernardi A, Forni M et al (2008) The retinoblastoma family member pRb2/p130 is an independent predictor of survival in human soft tissue sarcomas. Clin Cancer Res 14(15):4775–4779

McCarthy SA, Symonds HS, Van Dyke T (1994) Regulation of apoptosis in transgenic mice by simian virus 40T antigen-mediated inactivation of p53. Proc Natl Acad Sci USA 91(9):3979–3983

Meuwissen R, Linn SC, Linnoila RI, Zevenhoven J, Mooi WJ, Berns A (2003) Induction of small cell lung cancer by somatic inactivation of both Trp53 and Rb1 in a conditional mouse model. Cancer Cell 4(3):181–189

Morris EJ, Dyson NJ (2001) Retinoblastoma protein partners. Adv Cancer Res 82:1–54

Nacht M, Jacks T (1998) V(D)J recombination is not required for the development of lymphoma in p53-deficient mice. Cell Growth Differ 9(2):131–138

Olive KP, Tuveson DA, Ruhe ZC, Yin B, Willis NA, Bronson RT et al (2004) Mutant p53 gain of function in two mouse models of Li–Fraumeni syndrome. Cell 119(6):847–860

Paggi MG, Giordano A (2001) Who is the boss in the retinoblastoma family? The point of view of Rb2/p130, the little brother. Cancer Res 61(12):4651–4654

Pan H, Yin C, Dyson NJ, Harlow E, Yamasaki L, Van Dyke T (1998) Key roles for E2F1 in signaling p53-dependent apoptosis and in cell division within developing tumors. Mol Cell 2(3):283–292

Perez-Losada J, Wu D, DelRosario R, Balmain A, Mao JH (2005) p63 and p73 do not contribute to p53-mediated lymphoma suppressor activity in vivo. Oncogene 24(35):5521–5524

Priya K, Jada SR, Quah BL, Quah TC, Lai PS (2009) High incidence of allelic loss at 16q12.2 region spanning RBL2/p130 gene in retinoblastoma. Cancer Biol Ther 8(8):714–717

Saenz Robles MT, Symonds H, Chen J, Van Dyke T (1994) Induction versus progression of brain tumor development: differential functions for the pRB- and p53-targeting domains of simian virus 40T antigen. Mol Cell Biol 14(4):2686–2698

Sage J, Miller AL, Perez-Mancera PA, Wysocki JM, Jacks T (2003) Acute mutation of retinoblastoma gene function is sufficient for cell cycle re-entry. Nature 424(6945):223–228

Sanseverino F, Torricelli M, Petraglia F, Giordano A (2003) Role of the retinoblastoma family in gynecological cancer. Cancer Biol Ther 2(6):636–641

Schmitt CA, Fridman JS, Yang M, Baranov E, Hoffman RM, Lowe SW (2002) Dissecting p53 tumor suppressor functions in vivo. Cancer Cell 1(3):289–298

Sherr CJ (2004) Principles of tumor suppression. Cell 116(2):235–246

Simin K, Wu H, Lu L, Pinkel D, Albertson D, Cardiff RD et al (2004) pRb inactivation in mammary cells reveals common mechanisms for tumor initiation and progression in divergent epithelia. PLoS Biol 2(2):E22, PMCID: 340938

Smith EJ, Leone G, Nevins JR (1998) Distinct mechanisms control the accumulation of the Rb-related p107 and p130 proteins during cell growth. Cell Growth Differ 9(4):297–303

Soussi T, Lozano G (2005) p53 mutation heterogeneity in cancer. Biochem Biophys Res Commun 331(3):834–842

Stiewe T (2007) The p53 family in differentiation and tumorigenesis. Nat Rev Cancer 7(3):165–168

Symonds H, Krall L, Remington L, Saenz-Robles M, Lowe S, Jacks T et al (1994a) p53-dependent apoptosis suppresses tumor growth and progression in vivo. Cell 78(4):703–711

Symonds H, Krall L, Remington L, Saenz Robles M, Jacks T, Van Dyke T (1994b) p53-dependent apoptosis in vivo: impact of p53 inactivation on tumorigenesis. Cold Spring Harb Symp Quant Biol 59:247–257

Takimoto H, Tsukuda K, Ichimura K, Hanafusa H, Nakamura A, Oda M et al (1998) Genetic alterations in the retinoblastoma protein-related p107 gene in human hematologic malignancies. Biochem Biophys Res Commun 251(1):264–268

Terzian T, Suh YA, Iwakuma T, Post SM, Neumann M, Lang GA et al (2008) The inherent instability of mutant p53 is alleviated by Mdm2 or p16INK4a loss. Genes Dev 22(10):1337–1344, PMCID: 2377188

Tomasini R, Mak TW, Melino G (2008) The impact of p53 and p73 on aneuploidy and cancer. Trends Cell Biol 18(5):244–252

Tosi GM, Trimarchi C, Macaluso M, La Sala D, Ciccodicola A, Lazzi S et al (2005) Genetic and epigenetic alterations of RB2/p130 tumor suppressor gene in human sporadic retinoblastoma: implications for pathogenesis and therapeutic approach. Oncogene 24(38):5827–5836

van den Heuvel S, Dyson NJ (2008) Conserved functions of the pRB and E2F families. Nat Rev Mol Cell Biol 9(9):713–724

Van Dyke T (2007) p53 and tumor suppression. N Engl J Med 356(1):79–81

Vanderluit JL, Ferguson KL, Nikoletopoulou V, Parker M, Ruzhynsky V, Alexson T et al (2004) p107 regulates neural precursor cells in the mammalian brain. J Cell Biol 166(6):853–863

Vanderluit JL, Wylie CA, McClellan KA, Ghanem N, Fortin A, Callaghan S et al (2007) The Retinoblastoma family member p107 regulates the rate of progenitor commitment to a neuronal fate. J Cell Biol 178(1):129–139

Ventura A, Kirsch DG, McLaughlin ME, Tuveson DA, Grimm J, Lintault L et al (2007) Restoration of p53 function leads to tumour regression in vivo. Nature 445(7128):661–665

Viatour P, Somervaille TC, Venkatasubrahmanyam S, Kogan S, McLaughlin ME, Weissman IL et al (2008) Hematopoietic stem cell quiescence is maintained by compound contributions of the retinoblastoma gene family. Cell Stem Cell 3(4):416–428

Vousden KH, Lu X (2002) Live or let die: the cell's response to p53. Nat Rev Cancer 2(8):594–604

Wikenheiser-Brokamp KA (2006) Retinoblastoma family proteins: insights gained through genetic manipulation of mice. Cell Mol Life Sci 63(7–8):767–780

Williams BO, Schmitt EM, Remington L, Bronson RT, Albert DM, Weinberg RA et al (1994) Extensive contribution of Rb-deficient cells to adult chimeric mice with limited histopathological consequences. EMBO J 13(18):4251–4259

Xiao A, Wu H, Pandolfi PP, Louis DN, Van Dyke T (2002) Astrocyte inactivation of the pRb pathway predisposes mice to malignant astrocytoma development that is accelerated by PTEN mutation. Cancer Cell 1(2):157–168

Xue W, Zender L, Miething C, Dickins RA, Hernando E, Krizhanovsky V et al (2007) Senescence and tumour clearance is triggered by p53 restoration in murine liver carcinomas. Nature 445(7128):656–660

Yin C, Knudson CM, Korsmeyer SJ, Van Dyke T (1997) Bax suppresses tumorigenesis and stimulates apoptosis in vivo. Nature 385(6617):637–640

Yoshino A, Katayama Y, Ogino A, Watanabe T, Yachi K, Ohta T et al (2007) Promoter hypermethylation profile of cell cycle regulator genes in pituitary adenomas. J Neurooncol 83(2):153–162

Zhang J, Gray J, Wu L, Leone G, Rowan S, Cepko CL et al (2004) Rb regulates proliferation and rod photoreceptor development in the mouse retina. Nat Genet 36(4):351–360

Zhou Z, Flesken-Nikitin A, Corney DC, Wang W, Goodrich DW, Roy-Burman P et al (2006) Synergy of p53 and Rb deficiency in a conditional mouse model for metastatic prostate cancer. Cancer Res 66(16):7889–7898

Chapter 15
Mouse Models for Colorectal Cancer

Melanie Kucherlapati, Ken Hung, Mari Kuraguchi, and Raju Kucherlapati

15.1 Introduction

Colorectal cancer (CRC) is one of the most common cancers in the Western world and is one of the major causes of cancer mortality and morbidity. Approximately 150,000 new cases of CRC are diagnosed in the USA alone, and more than 50,000 individuals die from this cancer every year. The lifetime risk in the USA is 5%. Nearly 85% of CRCs are sporadic in nature, and the rest of them are the result of hereditary predispositions.

Several of the genes involved in the initiation and progression of CRC have been identified, and many additional genes implicated. It is now well established that disruption of the *Wnt* signaling pathway plays a key role in these tumors, whether they are found in individuals with sporadic cancer or a hereditary predisposition. Although examination of tumors and tumor-derived cell lines has provided important information about the genetic basis of CRC, experimental manipulation of the genes *in vivo* was necessary to gain a better understanding of their role and interrelationship in the disease process. In this chapter, we describe some of the genetically engineered mouse models for CRC, and how they have shaped our understanding of the molecular basis of this class of cancers.

The development of mouse models that faithfully recapitulate human CRC has followed the development of several different technologies. Initially, as genes were

M. Kucherlapati (✉) • R. Kucherlapati
Division of Genetics, Department of Medicine, Brigham and Women's Hospital, Harvard Medical School, Boston, MA 02115, USA
e-mail: rkucherlapati@partners.org

K. Hung
Department of Gastroenterology, Tufts Medical Center, Boston, MA 02111, USA

M. Kuraguchi
Department of Medical Oncology, Belfer Institute for Applied Cancer Science, Dana Farber Cancer Institute, Boston, MA 02115, USA

J.E. Green and T. Ried (eds.), *Genetically Engineered Mice for Cancer Research: Design, Analysis, Pathways, Validation and Pre-Clinical Testing*, DOI 10.1007/978-0-387-69805-2_15, © Springer Science+Business Media, LLC 2012 identified that played a role in the disease process, either by linkage analysis or by mutational signature, they were used to generate null knockout mouse lines in the hope of easily recreating the disease. Two familial syndromes that permitted identification of CRC genes were Familial adenomatous polyposis coli (FAP) and Hereditary nonpolyposis colorectal cancer (HNPCC), also known as Lynch Syndrome. Mouse lines developed from the modification of the human orthologs included the Adenomatous polyposis coli gene (*Apc*) (identified in FAP families), and Mismatch repair (MMR) gene models (e.g. *Mlh1*, *Msh2*, *Msh3*, *Msh6*) identified in HNPCC families. While these null mouse strains had features that were similar to human CRC, all had serious differences that limited their use as preclinical models. Specifically, all mouse models developed gastrointestinal tumors in the small intestine rather than the large intestine where they are predominantly seen in humans. Furthermore, although CRC and skin tumors were found in a proportion of many MMR mouse models, they frequently died of other cancer predisposition phenotypes such as lymphoma. However, many *Apc* and MMR mouse lines have been established using these initial knockouts to study tumor initiation, multiplicity, and incidence of intestinal cancers and cancers of other tissues. They have also been used to study the cooperative and synergistic effects of various other genes on cancer progression. To circumvent the problems of the null knockouts, conditional knockout technologies have been developed that take advantage of the *LoxP-Cre* recombinase system permitting gene removal in specific tissues (see Chapter 2). These mouse models are valuable tools to evaluate the efficacy of pharmaceutical and biological agents and are likely to provide further insights for prevention and treatment of colorectal and other cancers.

CRC is the second most common cause of cancer deaths. Despite this, the prognosis for the disease can be quite good if identified at an early stage. The 5-year survival rate for patients with Stage I tumors is over 90% whereas in Stage IV tumors (distant metastasis) the survival rate is only 5%. Early detection has the potential to alter the types of the interventions that in turn enhance long-term survival. At present, screening is limited to detecting occult blood in fecal samples, and colonoscopy (also termed "endoscopy" in this review). Fecal DNA analysis and virtual colonoscopy are under evaluation, as is the idea of developing a blood test, able to distinguish early tumorigenesis by proteomics. The genetically modified mice are being used to examine serum proteins from the tumor bearing mutant mice and comparing them with their wild type siblings. These comparisons will provide information about circulating tumor biomarkers that might be useful in detecting early-stage human cancer.

The development of conditional knockout models concomitant with the development of Adenoviral delivery of *Cre* recombinase (*Adenoviral-Cre*) surgically, and endoscopy technologies to monitor tumor growth have recently led to the ability to generate murine CRC tumors with features clearly more analogous to the human disease. Below, we discuss and compare both null and conditional mouse models that have been developed for *Apc* and MMR genes, as well as some constitutive and inducible *Cre*-recombinases used in conditional knockout technology. Additionally, we describe some imaging techniques that have been developed to examine gastrointestinal tumorigenesis in mice.

15.2 APC and Colorectal Cancer

FAP is a well defined inherited CRC syndrome. Patients develop hundreds of adenomatous polyps in the colon, with an inevitable progression to cancer in the third and fourth decades of life if left untreated (Groden et al. 1991; Kinzler and Vogelstein 1996). Additionally, patients often develop extracolonic manifestations, including upper gastrointestinal tract polyps, congenital hypertrophy of the retinal pigment epithelium, desmoid tumors of the skin, thyroid cancers, hepatoblastomas, brain tumors, disorders of the maxillary and skeletal bones, and dental abnormalities (Galiatsatos and Foulkes 2006). FAP is caused by germ-line mutations in the adenomatous polyposis coli (APC) gene, located on human chromosome 5q21. The gene is also mutated in the majority of sporadic CRCs regardless of histological staging, placing it at the initiation point of the adenoma-carcinoma sequence in man (Polakis 2000; Powell et al. 1992). APC is a 2843-aa protein with multiple functional domains that bind a multitude of cellular proteins. It has N-terminal oligomerization and armadillo repeat domains, the latter for binding of APC-stimulated guanine nucleotide exchange factor (ASEF) and a cytoskeletal regulator IQGAP1, 15- and 20-amino acid repeats for β-catenin binding, SAMP repeats for axin binding, a basic domain for microtubule binding and C-terminal domains that bind to EB1 and mammalian homolog of *Drosophila* Discs large (*DLG1*) (Aoki and Taketo 2007). The C-terminal half of APC is required for binding of the nuclear transcriptional repressor, C-terminal binding protein (CtBP) (Hamada and Bienz 2004; Sierra et al. 2006). These multiple binding domains suggest that APC can regulate many cellular functions ranging from the control of the *Wnt* signaling pathway to cell adhesion, migration, apoptosis, chromosomal segregation, and mitosis (Aoki and Taketo 2007; Fodde 2003; Nathke 2004; Senda et al. 2007), all of which may be important in colon cancer.

Although mutations in different parts of the APC gene have been detected in human patients and tumors, many of the mutations occur in a central region of the protein and lead to the formation of a truncation protein where the C-terminal half of the protein is missing (Miyoshi et al. 1992). This loss includes β-catenin and microtubule binding domains, and the transcriptional repressor CtBP. The interaction of APC with β-catenin or microtubules is essential for its tumor suppressor activity. APC normally downregulates the canonical *Wnt* signaling pathway. Its loss results in nuclear β-catenin accumulation, permitting interaction with the T-cell factor family of transcription factors (TCF) that coactivate transcription of *Wnt* target genes. Multiple mechanisms are thought to permit APC inhibition of *Wnt* signaling. APC can complex with axin helping glycogen synthase kinase (GSK) 3β to phosphorylate N-terminal serine/threonine residues of β-catenin, thereby accelerating its degradation through ubiquitylation (Polakis 1999), APC promotes export of β-catenin from the nucleus reducing the amount of nuclear β-catenin (Henderson 2000; Neufeld et al. 2000; Rosin-Arbesfeld et al. 2003), APC binds β-catenin blocking TCF interaction (Neufeld et al. 2000; Rosin-Arbesfeld et al. 2003) and can inhibit β-catenin/TCF-dependent transcription through direct interaction with a repressor complex that includes βTrCP and transcriptional repressor CtBP, by forming a stable complex with additional corepressors

(Hamada and Bienz 2004; Sierra et al. 2006). APC mutations leading to loss of these domains, thus, activate *Wnt* signaling by both increasing levels of β-catenin/TCF complexes in the nucleus and decreasing CtBP-mediated inhibition of the complex followed by activation of other target genes. *Wnt* signaling pathway activation by alterations in β-catenin regulation is very common in human tumors (Moon et al. 2004).

15.3 *Apc* Mouse Models

One of the first *Apc* mouse models, the Apc^{Min} mouse, was generated by random mutagenesis with ethylnitrosourea. This mouse line carries a nonsense mutation in codon 850, leading to a truncated *Apc* polypeptide of approximately 95 kDa. Heterozygous $Apc^{Min/+}$ (Min) animals (C57Bl/6 background) are born normally and subsequently develop over 100 small intestinal tumors (Miyoshi et al. 1992). Mutational analysis of the tumors shows that the wild-type *Apc* allele is lost as the result of either entire chromosome 18 loss or homozygosity based on other mechanisms. The mice have a reduced lifespan of ~150 days and die of intestinal bleeding and severe anemia. They have a low-penetrance phenotype of mammary adenosquamous carcinoma and desmoids (Moser et al. 1993; Shoemaker et al. 1997). Homozygosity for the Apc^{Min} mutation is embryonic lethal at early stages of gestation.

A second mouse model called $Apc^{\Delta 716}$ was generated by introducing a neomycin cassette into *Apc* codon 716. The resulting mutation led to a truncation protein of approximately 80 kDa (Oshima et al. 1995). Similar to Apc^{Min} phenotypically, $Apc^{\Delta 716}$ heterozygotes have many upper GI adenomas associated with allelic loss of the wild-type *Apc* in tumors. Homozygotes are embryonic lethal, however. However, unlike the Apc^{Min} mouse, these animals do not develop extraintestinal manifestations. Both Apc^{Min} and $Apc^{\Delta 716}$ mutations result in expression of stable truncated *Apc* proteins retaining the N-terminal oligomerization and armadillo repeat domains, but lack all domains required for β-catenin binding and downregulation.

The Apc^{1638N} mouse was generated by introducing the neomycin gene in codon 1638, in the transcriptional orientation opposite to that of *Apc* (Fodde et al. 1994). For unknown reasons, homozygous Apc^{1638N} cell lines contain very low amounts (1–2%) of the expected 182 kDa truncated protein, and hence, the Apc^{1638N} allele may be a pseudo-null or severe hypormorph (Smits et al. 1999). $Apc^{1638N/+}$ mice develop fewer intestinal tumors compared to either Apc^{Min} or $Apc^{\Delta 716}$ mice, having on average five to six tumors per intestine on the inbred C57B16 genetic background. Homozygosity is embryonic lethal (Fodde et al. 1994). $Apc^{1638N/+}$ mice show extraintestinal manifestations, including multifocal desmoids, epidermal cysts, mammary tumors, retinal pigment epithelial abnormalities, gastric tumors, and osteomas (Fodde et al. 1994; Marcus et al. 1997; van der Houven van Oordt et al. 1997; Smits et al. 1998). Approximately 70% of the intestinal tumors from this mouse line have undergone allelic loss of the remaining wild-type copy of *Apc* by loss of the entire chromosome 18 (Smits et al. 1997), and the rest have truncation mutations at the wild-type *Apc* allele (Kuraguchi et al. 2000, 2001).

Because $Apc^{1638N/+}$ mice were thought to be hypomorphs a second mouse line was created at codon 1638 called Apc^{1638T}. In this model, the same transcriptional cassette as in Apc^{1638N} was employed for targeted disruption, but the cassette was placed in the same transcriptional orientation as *Apc* itself (Smits et al. 1999). This resulted in a mouse line with stable expression of a truncated *Apc* polypeptide (182 kDa) in 1:1 ratio with the wild-type *Apc* polypeptide (312 kDa). The truncated protein lacks the binding region for tubulin, EB1-like proteins and DLG, which play an important role in chromosomal stability (Fodde et al. 2001; Kaplan et al. 2001). Unlike other mutated *Apc* proteins, Apc^{1638T} retains three of seven 20-amino-acid repeats and one SAMP motif. Heterozygous $Apc^{1638T/+}$ mice remained tumor-free. Moreover, unlike all the models discussed above, despite some growth retardation and developmental defects, homozygous $Apc^{1638T/1638T}$ mice are viable (Smits et al. 1999). These observations suggest that the Apc^{1638T} model retains the domains of APC required to ensure its tumor suppressing and developmental functions. A high degree of chromosomal instability is observed in $Apc^{1638T/1638T}$ ES cells, but this chromosomal instability is apparently not sufficient to trigger tumorigenesis (Fodde et al. 2001). Although some of the functional motifs related to the regulation of the *Wnt* signaling pathway are lost, biochemical assays have shown that the Apc^{1638T} truncated protein is still proficient in the regulation of β-catenin levels (Smits et al. 1999). Two hypomorphic *Apc* alleles, Apc^{neoF} (target cassette in intron 13, forward orientation) and Apc^{neoR} (target cassette in intron 13, reverse orientation) have also been described, whose expression levels were reduced to 10 and 20% of the wild type, respectively (Andreu et al. 2005). The intestinal tumor multiplicities of these heterozygous hypormorphic mice were found even lower than that of $Apc^{1638N/+}$ mice, 1 and 0.26 tumor per 15-month-old animal in Apc^{neoF} and Apc^{neoR}, respectively. Reporter assays carrying the Apc^{Min}, Apc^{1638N}, Apc^{1638T}, and Apc^{1572T} targeted mutations in various genetic combinations and dosages measured by TOPFLASH in ES cells correlate strongly with the phenotypic differences among *Apc* mutant mice (Smits et al. 1999; Kielman et al. 2002). Studies using $Apc^{\Delta 716}$ and two hypomorphic *Apc* alleles, Apc^{neoR} and Apc^{neoF}, have similarly shown that both the β-catenin accumulation and β-catenin/Tcf transcriptional activity were inversely correlated with the amount of *Apc* protein, which in turn regulated intestinal polyp multiplicity (Andreu et al. 2005). These results support the idea that β-catenin regulation represents the main tumor-suppressing function of *Apc*. The differences in the severity of the tumor susceptibility phenotypes and the differences in the spectrum of extraintestinal phenotypes in these models reflects the heterogeneity seen in human FAP families.

In an attempt to further understand tumor heterogeneity susceptibility, *Apc* constitutive knockout models have also been examined with numerous secondary mutations (Friedberg and Meira 2006), including all of the known Mismatch Repair genes. Many modifiers exist that have been shown to make a difference in tumor incidence, multiplicity, or progression, indicating that genetic background plays an important role in disease outcome after tumor initiation by somatic mutation to the remaining *Apc* allele. For example, haploinsufficiency of Flap endonuclease 1 (*Fen1*), a protein that functions in both replication and repair processes, in the intestines of *Apc*$^{1638N/+}$ mice reduced survival and significantly increased tumor progression to adenocarcinomas (Kucherlapati et al. 2002), whereas in *Apc* knockout mouse models the tumors did not progress much beyond adenomas. Exonuclease 1 (*Exo1*), an MMR protein with homology to *Fen1*, acted as a modest tumor suppressor in double *Exo1 Apc*^{1638N} mutants; however, it primarily contributed to the decrease in median survival by impairing the immune response through lymphocyte depletion (Kucherlapati et al. 2007). Intestinal microflora, diet, and inflammation have also been considered possible reasons for the change in observed outcomes.

As mentioned previously, one of the criticisms about the *Apc* knockout mouse models is based on the fact that tumors are mostly restricted to the small intestine rather than the colon. Genetic methods at present have not been entirely successful in changing that outcome. While examining Retinoblastoma (*Rb1*) inactivation in *Apc*^{1638N} mutant mice a reduction in median survival was found due to the significant increase in tumor incidence and multiplicity in the cecum and the proximal colon. Polyps of the papilla of vater were also observed in addition to small intestinal tumors, indicating that loss of the *Rb1* tumor suppressor widened the compartment in which *Apc*^{1638N} tumors appeared. Relative expression profiles of cecal tumors versus duodenal tumors showed distinct gene subsets over and under expressed with substantial expression patterns comparable to human CRC, including recapitulation of embryonic genes. These results indicated that *Rb1* had significant influence over tumor location in the gastrointestinal tract (Kucherlapati et al. 2008). *Rb1* deficiency by itself did not initiate intestinal tumors (Kucherlapati et al. 2006).

15.4 Conditional *Apc* Mutant Model and Roles of *Apc* in Tissues Other than the GI Tract

The mouse models mentioned above harbor the same *Apc* mutation in every cell of the body, unlike sporadic forms of cancers, which account for the majority of CRC cases. To better model sporadic forms of the cancer, our laboratory and others have generated mouse models that carry conditionally mutant or "floxed allele" of *Apc* (*Apc*^{CKO}) where exon 14 of the *Apc* gene is flanked by *LoxP* sites (Shibata et al. 1997; Colnot et al. 2004; Kuraguchi et al. 2006). Upon *Cre* mediated excision of exon 14, a truncated protein (*Apc*$^{\Delta 580}$) can be formed. This mutation was tested for its effect in the germ line. In the heterozygous state, it leads to large numbers of adenomas in the small intestine (Kuraguchi et al. 2006). In the homozygous state, it results in embryonic lethality. Embryonic lethality is circumvented in homozygous conditional mice, which are born and develop normally. The use of Adenoviral-Cre in the *Apc*^{CKO} mouse gives the ability to develop adenomas in the colorectal region alone (Shibata et al. 1997), making the model closer to the human disease and demonstrating clearly that inactivation of *Apc* in the colon can lead to colonic tumors.

Inducible intestinal-specific systems such as tamoxifen-inducible villin promoter, *Vil-CreER*^{T2} (el Marjou et al. 2004) and β-napthoflavone-inducible cytochrome P450 promoter *AhCre* (Ireland et al. 2004) enable conditional deletion of the gene in the adult intestinal epithelium. These studies showed that somatic loss of *Apc* in the adult mice causes perturbed differentiation, migration, proliferation, and apoptosis in the intestinal epithelium (Andreu et al. 2005; Ireland et al. 2004). Recently, Hinoi et al. has shown that the *CDX2* promoter elements confer colonic epithelium-specific transgene expression in the adult mouse. Mice carrying *CDX2P-NLS Cre* recombinase transgene and a floxed *Apc* allele developed colorectal adenomas and carcinomas in ~15–20% of mice (Hinoi et al. 2007). Morphologic and molecular studies of the mouse tumors revealed their similarity to human colorectal tumors, suggesting that such mice may be valuable for studies in CRC prevention, diagnosis, and therapy.

The conditional deletion of the *Apc* gene disrupts homeostasis not only in the intestines, but also in many other tissues. We have previously shown that loss of *Apc* gene in *K14*-expressing embryonic cells causes aberrant morphogenesis in various skin appendages including hair follicles and teeth, and abnormal thymus organogenesis (Kuraguchi et al. 2006). The expression of *K14* promoter is known to initiate in embryonic ectoderm at embryonic day (E) 9.5 and remains active in basal cells of epidermis and other stratified epithelia in adults (Byrne et al. 1994; Wang et al. 1997). The importance of *Apc* function in thymic development has also been demonstrated by thymocyte-specific loss of *Apc* by crossing a different strain of *Apc* conditional mice and *LckCre* transgenic mice (Gounari et al. 2005). Given the importance of regulation of *Wnt* signaling in embryonic pattern formation and morphogenesis of many organs, as well as extracolonic manifestations observed in FAP patients, mechanistic understanding of APC in development and in extracolonic tissues becomes critical to better assess potential adverse events in humans. The availability of conditional *Apc* mice and the use of tissue-specific conditional knockout approach based on *Cre-LoxP* system will be a valuable tool for such studies.

15.5 Mismatch Repair Genes and Cancer

Like FAP, HNPCC is also a well-defined inherited CRC predisposition syndrome. The genes involved in this disease were discovered initially through the recognition of a high mutation rate in the tumor genome. A search was conducted for "mutator" genes that could be directly responsible for progressive changes observed in such cancers. The search centered initially on the human homolog of *Escherichia coli* MutS (Cox et al. 1972) designated MSH2 (Fishel et al. 1993), then on other postreplication MMR genes such as MLH1 (Prolla et al. 1994; Bronner et al. 1994), PMS1 (Prolla et al. 1994), PMS2 (Nicolaides et al. 1994), MSH3 (Acakpo-Satchivi et al. 1997), MSH6 (Acakpo-Satchivi et al. 1997), MLH3 (Flores-Rozas and Kolodner 1998), and Exo1 (Schmutte et al. 1998). The unique mutational signature was termed "RER" (replication error positive) or microsatellite instability (MSI). It was similar to that seen in MSH2-deficient yeast and was identified in a proportion of colorectal tumors from families with HNPCC (Lothe et al. 1993) and Muir–Torre syndrome (Honchel et al. 1994). Muir–Torre patients develop cancer of the proximal colon at an early age as well as cancer of the sebaceous glands and keratoacanthomas of the skin. A small fraction of them also develop lymphomas. Heterozygous mutations in one of the MMR genes were found in HNPCC kindreds, segregating with affected individuals. MSI positive tumors can be sub-classified as MSI-H (high) or MSI-L (low) on the basis of incidence using a panel of five microsatellite markers (Dietmaier et al. 1997; Boland et al. 1998), this likely reflects the importance of the individual MMR protein mutated in the MMR complex. Presence of MSI in tumors is considered a hallmark for mutation in MMR genes. A modified Amsterdam criterion is used to diagnose HNPCC (Lin et al. 2004). Patients are classified as Lynch syndrome I if only CRC is present, and Lynch syndrome II if CRC is accompanied by extracolonic tumors. (In the past the terms "HNPCC" and "Lynch Syndrome" were used interchangeably, more recently Lynch Syndrome is being considered descriptive of MMR gene involvement). Most germ-line alterations are found in MLH1 (50%), MSH2 (40%), and MSH6 (10%); however, other MMR genes may contribute to the process through their ability to complex and interact with these major MMR proteins. Epigenetic silencing of MLH1 has been found to underlie most MMR sporadic cancers (Cunningham et al. 1998; Herman et al. 1998; Kane et al. 1997; Miyakura et al. 2001, 2003). As previously mentioned, the familial syndrome FAP led to the identification of *APC* as an early gene involved in the formation of benign polyps. MMR genes are shown to initiate CRC by causing mutations in both alleles of the *APC* tumor suppressor gene (Reitmair et al. 1996), and thought to contribute to tumor progression by causing mutations to other relevant cancer genes such as k-ras, the AKT/PI3K pathway, TGF-β signaling, and the p53 family of genes.

Since colonic tumors are not the only tumors to have MSI, mutations in MMR genes could function in other cancers through tissue-specific initiator genes similar to *APC* in CRC. The high frequency of insertion and deletion mutations in simple repeat sequences that comprise the MSI mutational signature are also found in tumors of the endometrium, ovary, cervix, stomach, breast, skin, lung, prostate, and bladder, as well as in gliomas, leukemias, and lymphomas (Boland et al. 1998). MSI cell lines from sporadic leukemia, endometrial, ovarian, prostate, and bladder cancers have also been shown defective in strand-specific MMR (Gu et al. 2002; Drummond et al. 1995).

The proteins responsible for prokaryotic MMR are MutS, MutL, MutH, DNA helicase II (MutU/UvrD), four exonucleases (Exo1, ExoVII, Exo1X, and Rec J), single stranded DNA binding protein (SSB), DNA polymerase III holoenzyme, and DNA ligase (Lahue et al. 1989; Burdett et al. 2001). The human equivalent of MutS is comprised of two heterodimeric complexes, hMutSα and hMutSβ. Complex hMutSα is made up of MSH2 & MSH6, and hMutSβ of MSH2 & MSH3 (Fishel and Wilson 1997). MutL human equivalents are made up of three complexes of proteins, hMutLα, hMutLβ, hMutLγ. hMutLα comprises MLH1 and PMS2, hMutLβ of MLH1 and PMS1, and hMutLγ of MLH1 and MLH3 (Kolodner and Marsischky 1999). There is no equivalent of MutH or UvrD in mouse or human MMR, ExoI is the only known mammalian exonuclease, and Pol δ is involved in mammalian post excision DNA resynthesis. Other proteins that function in human MMR are proliferating cell nuclear antigen (PCNA), RPA, HMGB1, RFC, and DNA ligase.

Both hMutSα and hMutSβ complexes have crucial ATPase activities and roles in mismatch recognition and repair initiation (Kunkel and Erie 2005). hMutSα is involved in recognition of base-base mismatches and insertion/deletion mutations of one or two nucleotides with high affinity and the single base pair G/T mismatch with low affinity, while hMutSβ complexes, in addition to recognizing single base insertion/deletion mispairs (IDLs) also recognize larger IDL mispairs. The hMutLα ATPase activity essential for MMR regulates termination of mismatched provoked excision, in addition to playing a role in meiosis. Downstream events in MMR after mismatch recognition have been reviewed extensively (Kunkel and Erie 2005; Modrich and Lahue 1996; Buermeyer et al. 1999; Jiricny 2006; Schofield and Hsieh 2003; Li 2008).

15.6 Mouse Models in MMR Genes

15.6.1 Msh2

Three knockout mouse strains for *Msh2* have been made by insertional mutagenesis. Two were mutated at exons 11 and 12 (de Wind et al. 1995; Reitmair et al. 1995). The selection cassettes functionally disrupted the *Msh2* ATP binding site and abolished ATPase activity. Both strains were viable in the homozygous recessive state with reduced median survival due to the development of lymphomas. The lymphomas were MSI positive. Cell lines established from mouse embronic fibroblasts MEFs showed reduced MMR activity, were MSI positive, and had other characteristics of *Msh2* deficiency including tolerance to methylating agents and loss of suppression of homologous recombination. Neither strain was reported to have colorectal, gastric, or endometrial tumors. A third null knockout model (Smits et al. 2000) targeting *Msh2* exon 7 had complete absence of MSH2 by Western blot analysis. Homozygous recessive animals (72%) developed lymphoma but also additionally 72% developed MSI positive intestinal tumors. A proportion of $Msh2^{\Delta7/\Delta7}$ mice also developed skin tumors. The results of the insertional mutagenesis models appeared to imply that *Msh2* knockout mice might be dying of lymphoma in a temporal manner prior to the full development of gastrointestinal cancer, and that the rate of tissue turnover played a role in the tissue specificity of the tumors. When compound $Msh2^{-/-}/Apc^{+/-}$ animals were generated to examine the effect of MMR loss on the remaining *Apc* allele, it was observed that aberrant crypt foci and adenomas of the GI tract were greatly accelerated and median survival significantly reduced (Reitmair et al. 1996). The molecular mechanism did not appear to include *Apc* loss of heterozygosity (LOH) suggesting that loss of *Apc* was by intragenic mutational inactivation (de Wind et al. 1998). Truncation mutations were found in 82% of adenomas studied in $Msh2^{-/-}/Apc^{Min}$ animals (Sohn et al. 2003).

15.7 Conditional *Msh2* Mutant Model

In order to better model sporadic forms of intestinal cancer that did not contain *Msh2* mutations in all murine tissues, we made a conditional knockout for *Msh2* ($Msh2^{LoxP/LoxP}$) by floxing exon 12 (Kucherlapati et al. 2010). As in the $Msh2^{null}$ knockout strains, this mutation disrupted the essential ATPase domain. We tested the mutation for its effect in the germ line using *EIIa-Cre* and found homozygous animals to be viable through embryonic development, and similar in disease phenotype to the earlier null knockout mice. MMR was measured directly on MEF cell lines from these animals and found to be defective. Further characterization supported the authenticity of the mutation. Using the $Msh2^{LoxP}$ allele in combination with a constitutively active *Villin-Cre* transgene ($VCMsh2^{LoxP/LoxP}$) we were able to avoid the lymphoma phenotype that previously limited earlier models for preclinical use, and produce tumors that were restricted to the intestinal tract. The $VCMsh2^{LoxP/LoxP}$ mice were then mated to mice bearing either an $Msh2^{\Delta 7null}$ or $Msh2^{G674D}$ point mutation to create allelic phase mutants. The $Msh2^{G674D}$ mutation located within the ATPase domain, like the previously published $Msh2^{G674A}$ mutation (Lin et al. 2004), uncouples apoptosis from the MMR dependent DNA damage response. Comparison of the tumors from the allelic phase mutants revealed a remarkable significant difference in multiplicity and size. While $VCMsh2^{LoxP/null}$ mice developed 1.40 ± 0.11 intestinal tumors the $VCMsh2^{LoxP/G674D}$ mice developed 3.43 ± 0.42 tumor ($P < 0.0001$). Tumors from $VCMsh2^{LoxP/null}$ mice had an average diameter of 6.48 ± 0.58 mm while $VCMsh2^{LoxP/G674D}$ tumors averaged 3.25 ± 0.49 mm ($P < 0.003$). Eighty-two percent of the tumors in the null allelic phase mutants had progressed to adenocarcinomas, as compared to 63% of tumors from the $VCMsh2^{LoxP/G674D}$ mice. Both strains of mice were subjected to treatment regimens with cisplatin, or 5-fluorouracil/leucovorin and oxaliplatin (FOLFOX). Tumors from mice with the point mutation responded to both drugs, whereas tumors from the null *Msh2* mouse did not (Fig. 15.1). These data indicated that *Msh2* missense mutations can have distinct effects on intestinal tumorigenesis. They also indicated that the DNA damage response function of MMR is important in early steps of intestinal tumorigenesis, and that loss of apoptosis correlated with increased drug resistance. These studies demonstrated the efficacy of the model in correlating mutation type with

Fig. 15.1 (continued) relative size in cisplatin or FOLFOX treated mice. (**b**) Response of intestinal tumors in $VCMsh2^{LoxP/null}$ and $VCMsh2Lox^{P/G674D}$ allelic phase mice to chemotherapy. Tumors are measured in terms of relativity of magnetic resonance imaging measurements. The tumor size at day 0 is 1, and the relative tumor growth or retardation is scored on the basis of percentage. *Red lines* indicate growth; *green lines* indicate retardation. The number of tumors for each treatment regimen is given as follows: *VCMsh2LoxP/null*: PBS, 7; cisplatin, 8; FOLFOX, 11. *VCMsh2LoxP/G674D*: PBS, 7; cisplatin, 9; FOLFOX, 11 (Gastroenterology 2010;138:993–1002)

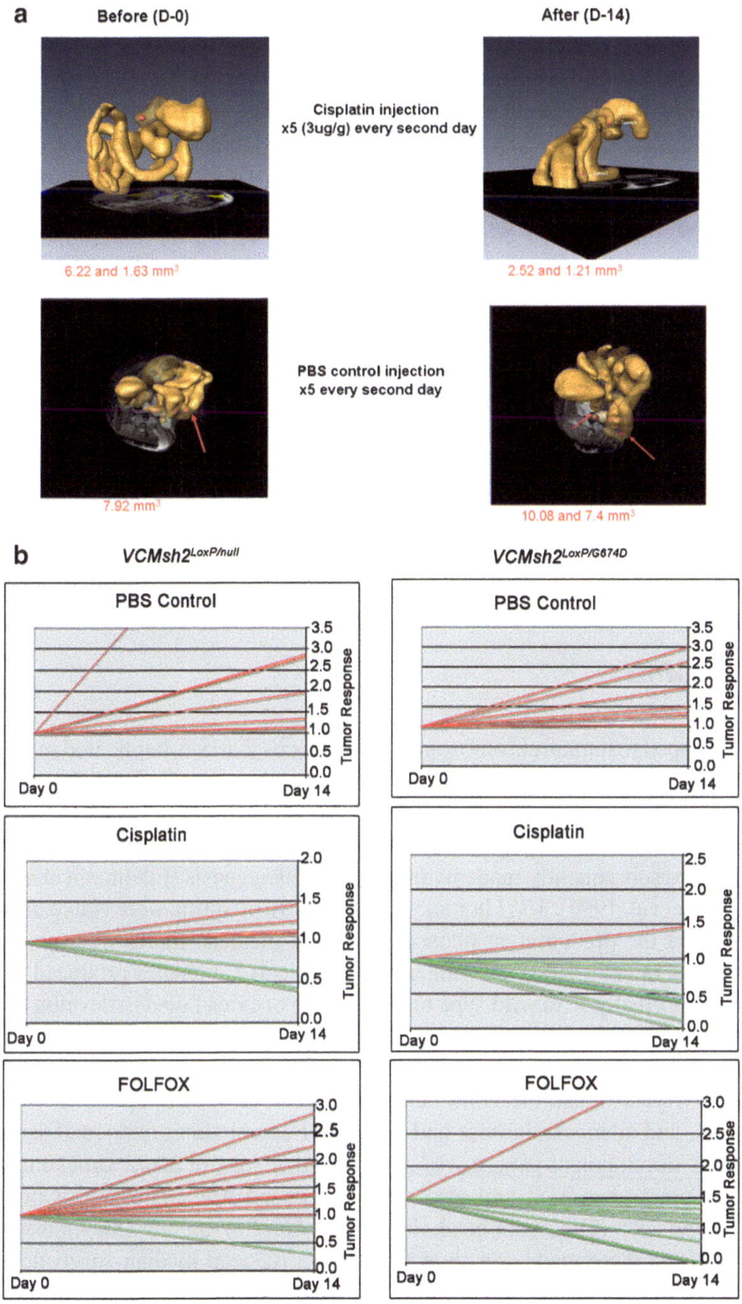

Fig. 15.1 (**a**) Intestinal tumors for chemotherapy are visualized and measured by magnetic resonance imaging (MRI). Two 6-month-old *VCMsh2*$^{LoxP/LoxP}$ mice testing positive for occult blood were subjected to MRI, successfully revealing tumors in each mouse. Two weeks later, during a second MRI the same tumors had grown at the original location in untreated mice, and shrunk in

drug therapy, an approach in synchronization with the development of personalized medicine.

In an attempt to relocate the tumors to the intestine we mated these animals with the Cdx2P-NLS-Cre transgene. A shift of tumors to the large intestine was observed; however, like *Msh2*^{null} mice these animals also developed lymphomas (Kucherlapati et al. 2010). We are now in the process of employing surgical methods of adenoviral-Cre infusion in the attempt to develop colonic tumors exclusively through *Msh2* mutation.

To study the effect of conditional *Msh2* inactivation on skin tumorigenesis *Msh2*^{LoxP} animals were crossed with *K14-Cre* mice that express *Cre*-recombinase in the epidermis, tongue, and thymic epithelium. The resulting tumors somewhat resembled those of Lynch Syndrome II patients (Muir–Torre Syndrome). In *K14CMsh2*$^{LoxP/loxP}$ mice epidermal tumors arose with a frequency of 74% (14/19) by 14 months of age. Approximately half of the animals developed squamous cell carcinoma, a third developed basal cell carcinoma, and one quarter developed papillomas. Five animals had multiple lesions with more than one tumor stage. Other tumors that developed included lymphoma (10.5%, $n=2$) and histiocytic sarcoma (15.8%, $n=3$). Two mice (10.5%) developed intestinal polyps. Median tumor free survival was reduced to 12–14 months.

15.7.1 *Mlh1*

MSH2 mapped to human chromosome 2 near a locus that was implicated in HNPCC (Fishel et al. 1993). Another locus on human chromosome 3 (3p21–23) was also found to be implicated in the disease (Bronner et al. 1994; Papadopoulos et al. 1994) and MLH1, a second MMR gene, was mapped to this region. Two knockout mouse models were subsequently made by insertional mutagenesis (Edelmann et al. 1996, 1999; Baker et al. 1996). *Mlh1* homozygous recessive offspring were viable; however, unlike *Msh2* the predominant phenotype was in meiosis. Males did not produce spermatozoa or spermatids showing arrest in meiosis I. Females produced oocytes and mated normally with wild-type males, but the oocytes failed to develop because of meiotic failure resulting in infertility. Later, MMR proteins MLH3, MSH4, MSH5 were also implicated in meiosis. (For a full review see Svetlanov and Cohen 2004). By 1 year of age the *Mlh1*-deficient mice developed lymphomas, intestinal adenomas, and adenocarcinomas and to a lesser extent skin tumors and sarcomas. The tumor development pattern was similar to that seen in *Msh2* knockout mouse models. Tumors had MSI and also showed reduced *Apc* expression. Compound *Mlh1*$^{-/-}$/*Apc*$^{1638N/+}$ (Edelmann et al. 1999) and *Mlh1*$^{-/-}$/*Apc*$^{Min/+}$ (Shoemaker et al. 2000) animals were made and shown to have a reduced median survival, and an increased GI tumor incidence of 40- to 100-fold. Tumor initiation was increased in this model, but tumor progression was not. The wild type *Apc* allele in these tumors had truncation mutations, allelic loss of *Apc* was not observed. When a large number of mutations in *Apc* were examined a mixture of base substitutions (27%) and frameshifts (73%) were detected, most commonly within dinucleotide repeats (Kuraguchi et al. 2000). The results revealed a characteristic mutational signature in *Apc* that was attributable to *Mlh1* deficiency. An *Mlh1* conditional knockout mouse has been recently made and placed under the direction of *K14-cre*. Like *K14Msh2CKO* mice, these animals developed epidermal basal and squamous cell carcinomas.

15.7.2 Msh6 and Msh3

Numerous HNPCC-like families that did not completely fulfill the original Amsterdam diagnostic criterion were suspected of harboring mutations in other MMR genes. A search conducted for *Msh6* germ-line mutations led to the identification of two families (Akiyama et al. 1997; Miyaki et al. 1997). Later, three common polymorphisms were found in the promoter region that affected promoter activity and increased sensitivity to DNA methylation (Gazzoli and Kolodner 2003). A mouse model for *Msh6* created by insertional mutagenesis developed tumors in the GI tract as well as lymphomas (Acakpo-Satchivi et al. 1997). Cell lines created from the *Msh6* knockout mice were found partially defective for DNA MMR. Repair of mismatches was affected; however, repair of 1, 2, and 4 nucleotide insertion/deletion mutations was unaffected. Tumors did not show MSI. The study concluded that mutations in *Msh6* led to cancer susceptibility where resulting tumors were MSI negative. Perhaps, more importantly, the analysis of these mice led to the first description of *MSH6* as a late-onset cancer susceptibility gene (Bardwell et al. 2004). A second mouse model for *Msh6* was created and examined along with loss of another MMR gene, *Msh3* (Edelmann et al. 2000; de Wind et al. 1999). *Msh3*$^{-/-}$ mice did not have a cancer predisposition on its own, but in combination with *Msh6*$^{-/-}$ loss of *Msh3* was capable of accelerating intestinal tumorigenesis such that the phenotype was similar to *Msh2*$^{-/-}$ and *Mlh1*$^{-/-}$ models. In *Apc*^{1638N} background, *Msh6* deficiency leads to an increase in tumor number, caused by the accumulation of somatic mutations within the wild-type *Apc* allele. The *Apc* mutations found in these tumors are predominantly truncation mutations caused by base substitutions, in the region that affect β-catenin binding or downregulation which are essential for tumor suppression (Kuraguchi et al. 2001). PCNA, a replication protein required for MMR, binds specifically to *Msh6* (Flores-Rozas et al. 2000). Both *Msh6* and *Msh3* proteins contain "PIP" (*P*CNA *i*nteracting *p*rotein) boxes comprised of evolutionarily conserved regions of about 100–600 amino acids in the N terminal region that interact with the sliding clamp that participates in repair (Clark et al. 2007). Another function of the *Msh2–Msh6* complex in addition to MMR has been found in somatic mutation of IgV region genes. The complex is thought necessary to generate antibody diversity and the production of high-affinity neutralizing antibodies (Wiesendanger et al. 2000; Stella et al. 2004).

15.7.3 *Pms1, Pms2, Mlh3*

Mice deficient in *Pms1* are viable and develop normally, with no tumor predisposition or MSI in intestinal cells or MEF cell lines (Prolla et al. 1998; Yao et al. 1999; Narayanan et al. 1997). Animals deficient in *Pms2* develop lymphomas and sarcomas at a low frequency, but do not develop intestinal adenomas or adenocarcinomas as seen in *Mlh1*$^{-/-}$ mice (Prolla et al. 1998; Yao et al. 1998; Narayanan et al. 1998). These findings correlate with the fact that *Pms2* germ-line mutations are infrequently found in HNPCC patients. Although an increase in mutation in mononucleotide tracts occurs, the mutation frequency of *Pms2*$^{-/-}$ tissues is threefold lower than that found in *Mlh1*$^{-/-}$ tissues. Male mice defective in the DNA mismatch repair gene *Pms2* exhibit abnormal chromosome synapsis in meiosis (Baker et al. 1995). The inactivation of *Mlh3* results in mice that are viable but sterile. Spermatocytes reach metaphase then undergo apoptosis, and oocytes fail to complete meiosis I. Mlh3$^{-/-}$ mice show a late-onset cancer phenotype of the gastrointestinal tract and a range of extragastrointestinal tumors including lymphomas and basal cell carcinoma of the skin (Chen et al. 2005; Peng-Chieh et al. 2008).

15.7.4 *Exo1*

Exo1 interacts with MSH2 and MLH1 participating in the excision of mismatched nucleotides. Inactivation of the gene impairs the repair of singe base mismatches and single insertions, but not of dinucleotide insertions. *Exo1*-deficient mice have MSI in mononucleotide repeat sequences only. Animals have a mild cancer predisposition developing lymphomas later in life (Wei et al. 2003). As mentioned earlier *Exo1* deficiency decreases the median survival of *Apc*$^{1638N/+}$ mice; however, the decrease in survival has been found due to impaired immune response as opposed to increased incidence or multiplicity of intestinal tumors. Like *Msh2–Msh6* complexes, *Exo1* has a role in somatic hypermutation (Bardwell et al. 2004), the immune response, and is involved in the development of CRC. Loss of *Exo1* may result in a low-penetrance phenotype in CRC; however, its exact role is undefined.

15.8 A Sporadic Colon Cancer Mouse Model for Translational Research

Recently, *Adenoviral-Cre* and conditional knockouts have been used to create site-specific tumors in the distal colon (Kenneth et al. 2010). These tumors have histological features and gene-expression profiles similar to human CRC (Fig. 15.2). They have been used to test the efficacy of chemotherapeutic reagents, and can be used to identify possible points of intervention in several different key pathways involved in tumorigenesis.

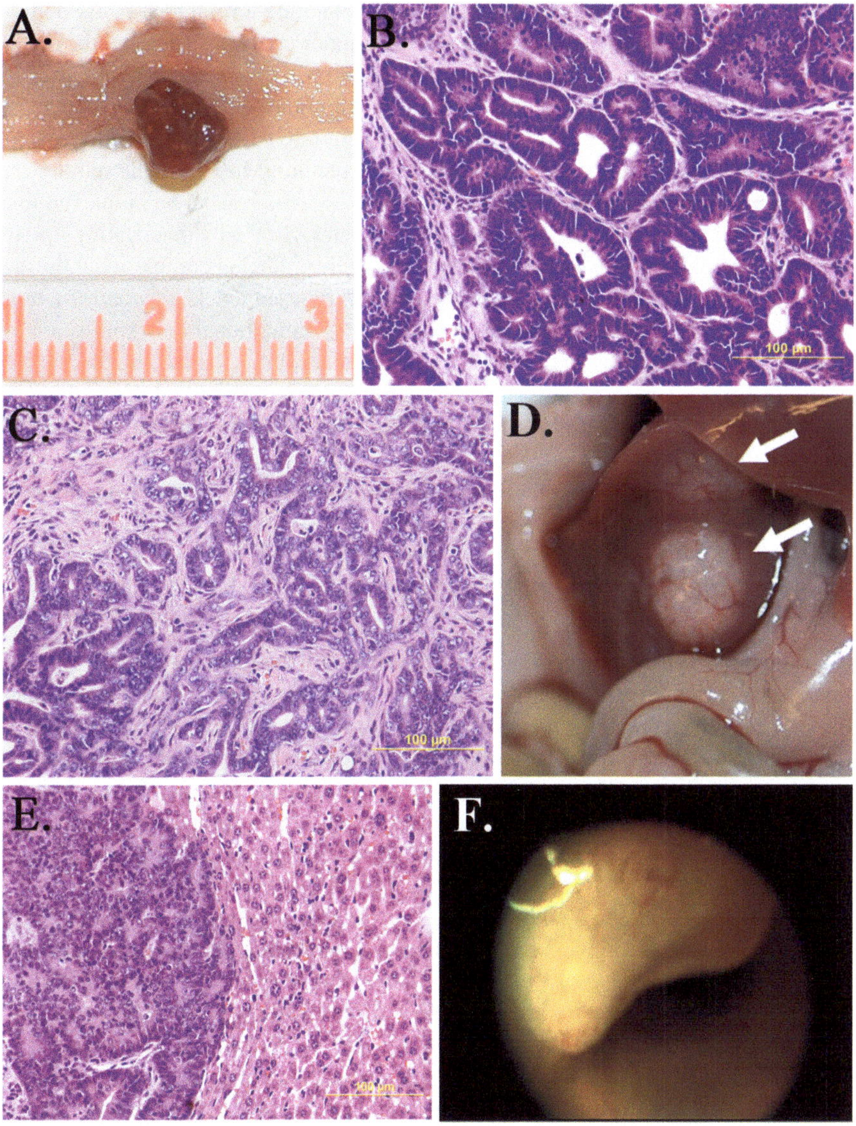

Fig. 15.2 The mouse sporadic colon cancer model recapitulates the entire adenoma–carcinoma–metastasis axis. (**a**) Injection of adenovirus expressing *Cre* recombinase into Apc^{CKO} mice results in discrete tumors restricted to the distal colon. Histological examination of primary tumors demonstrates (**b**) adenomas and (**c**) adenocarcinomas. (**d**) Macroscopic liver metastases are seen that are (**e**) histologically adenocarcinomas. (**f**) Endoscopic image showing natural history of primary tumor growth

The normal features of human sporadic tumors include several factors. By definition these individuals have no family history for tumor predisposition, and the mutation is a stochastic event in the colonic epithelium. Because it is a somatically acquired mutation, the tumors are embedded within normal tissue and result from one or a few polyps that can progressively develop into adenomas and adenocarcinomas. The tumors are localized to the colon or rectum. Most sporadic tumors are the result of mutations in the *APC* gene. All of these features can be mimicked in a mouse model through the use of *Adenoviral-Cre* and *Apc*^{CKO}, by inactivating *Apc* in one or a few random cells in the appropriate colonic epithelial compartment during the adult life of the mouse. To generate a sporadic mouse model, *Adenoviral-Cre* is infused into the colon of the *Apc*^{CKO} by surgery. The conditional knockout is a normal mouse without *Cre* recombinase. This method reliably results in the production of one or two tumors at the site of injection. We have developed and adapted methods for the imaging of these tumors and their development without the need to sacrifice the mouse.

15.8.1 Molecular Imaging in the Mouse

Over the last 10 years there has been much progress in the imaging of small animals that parallels tomographic anatomic imaging on a larger scale. Furthermore imaging processes have gone from anatomic approaches toward imaging at the molecular level (Weissleder and Mahmood 2001). Frequently based on alterations of enzymes, receptors, and metabolic pathways that precede changes in tumor size, such imaging can potentially improve detection and earlier evaluation of therapeutic response. We have taken advantage of these advancements to evaluate mouse colonic tumor development.

Mice that received the *Adenoviral-Cre* into their colons developed tumors solely in the distal colon, and unlike xenograft models in which human tumors are admixed with mouse stroma, these are true spontaneous tumor models, composed of both mouse tumor and supporting mouse stroma. After cleansing with PBS, air is carefully insufflated to distend the colon, while avoiding perforation. Examination of over 200 animals has shown that the custom made colonoscopy imaging system we have can be used to document the progressive growth of individual colonic tumors over time or tumor persistence when using chemotherapeutic reagents. Tumors made in *Apc*^{CKO} are reliably detected as early as 3 weeks after *Adenoviral-Cre* injection and progressively become larger. By 9–12 weeks after injection the tumors begin to completely occlude the intestinal lumen. Because individual mice can be serially followed, this procedure allows the examination of the kinetics of individual tumor growth.

Tumors can also be detected by imaging tumor-specific markers. One such marker for detection is Prosense 680, a protease activated molecular beacon. Prosense 680 is a long circulating graft copolymer that increases the near infared fluorescence intensity after selective enzymatic cleavage of lysine–lysine bonds by cathepsin B *in vivo* (Funovics et al. 2006; Upadhyay et al. 2007). Immunohistochemical analysis indicates that cathepsin B is preferentially expressed in the tumor, as compared to normal mucosa. Prosense 680 can be intravenously injected into colonic-tumor-bearing animals and imaged with near infrared colonoscopy.

Adenovirus expressing β-galactosidase (Adeno-LacZ) was also used to infect colonic crypt cells. Forty-eight hours after the incubation period for the adenovirus, the mice are euthanized and the distal colon removed, longitudinally opened, and stained with X-gal. These studies show that the dominant site of adenoviral infection is the superficial intestinal epithelium. Rare positive staining in the lower colonic crypts is observed, indicating that it might be possible to target the appropriate cells leading to tumor formation.

As additional colon tumor-specific markers are identified, it is possible to test their efficacy in these models and if found to be robust, translated to use in humans.

15.9 Summary and Prospects

The availability of gene targeting methods and the identification of specific genes and pathways involved in CRC permitted and guided the development of a large number of mouse models for the disease. The initial set of models involved introduction of insertional and later point mutations into the germ line. Most of these mouse models developed a cancer predisposition phenotype. These experiments also revealed that different mutations in the *Apc* gene resulted in phenotypes that differed in their degree of severity. These observations provided significant new biological information about genotype-phenotype correlations. The availability of floxed alleles of the *Apc* gene has allowed the generation of models for sporadic CRC. *Wnt* signaling has been shown to be crucial in normal development and in several different malignancies besides colonic tumors. The conditional allele should prove useful in understanding the role of not only *Wnt* signaling but also changes in other pathways as well in many biological processes in many different tissues.

The generation of initial knockout mice with insertion mutations, and later point mutations in each of many MMR genes also proved to be very fruitful, despite lymphoma being the predominant phenotype. Gastrointestinal tumors did occur in these strains as did skin cancers, and the models gave insight into how MMR deficiency might lead to CRC through mutation to the *Apc* gene. The subsequent development of conditional allele for *Msh2* has permitted the development of a suitable model for pre-clinical use for intestinal cancer, and the testing of anticancer agents by the avoidance of the lymphoma phenotype. It has also given insight into tumor response to different drugs in the presence of specific MMR mutations through the creation of allelic phase mutants, strains of mice that carry one conditional allele and a second mutation. Other mutations to both *Apc* and MMR genes that occur in tumor databases can be examined in the future by the same methodology using these and other conditional models. Since MMR-deficient tumors constitute a proportion of all solid tumors, and the molecular basis for tumor initiation in many of these cases is unknown, these models provide an opportunity to explore the molecular basis of MMR-deficient tumors outside the GI tract.

Genetic methods to remove *Apc* and MMR genes from the colon continue to be explored. The conditional knockout mouse strains are now also being placed under the direction of Adenoviral-cre by surgical methods. Mice homozygous for the conditional *Apc* allele are being made that truly develop CRC. This model is expected to be useful in testing anticancer chemotherapeutic reagents, as well as in the identification of novel points for therapeutic intervention. The progressive growth of the tumors is being monitored by colonoscopy imaging system that permits mice to be followed serially with and without tumor-specific markers.

References

Acakpo-Satchivi LJ, Edelmann W, Sartorius C, Lu BD, Wahr PA, Watkins SC, Metzger JM, Leinwand L, Kucherlapati R (1997) J Cell Biol 139:1219–1229
Akiyama Y, Sato H, Yamada T, Nagasaki H, Tsuchiya A, Abe R, Yuasa Y (1997) Cancer Res 57:3920–3923
Andreu P, Colnot S, Godard C, Gad S, Chafey P, Niwa-Kawakita M, Laurent-Puig P, Kahn A, Robine S, Perret C et al (2005) Development 132:1443–1451
Aoki K, Taketo MM (2007) J Cell Sci 120:3327–3335
Baker SM, Bronner CE, Zhang L, Plug AW, Robatzek M, Warren G, Elliott EA, Yu J, Ashley T, Arnheim N et al (1995) Cell 82:309–319
Baker SM, Plug AW, Prolla TA, Bronner CE, Harris AC, Yao X, Christie DM, Monell C, Arnheim N, Bradley A et al (1996) Nat Genet 13:336–342
Bardwell PD, Woo CJ, Wei K, Li Z, Martin A, Sack SZ, Parris T, Edelmann W, Scharff MD (2004) Nat Immunol 5:224–229
Boland CR, Thibodeau SN, Hamilton SR, Sidransky D, Eshleman JR, Burt RW, Meltzer SJ, Rodriguez-Bigas MA, Fodde R, Ranzani GN et al (1998) Cancer Res 58:5248–5257
Bronner CE, Baker SM, Morrison PT, Warren G, Smith LG, Lescoe MK, Kane M, Earabino C, Lipford J, Lindblom A et al (1994) Nature 368:258–261
Buermeyer AB, Deschenes SM, Baker SM, Liskay RM (1999) Annu Rev Genet 33:533–564
Burdett V, Baitinger C, Viswanathan M, Lovett ST, Modrich P (2001) Proc Natl Acad Sci USA 98:6765–6770
Byrne C, Tainsky M, Fuchs E (1994) Development 120:2369–2383
Chen PC, Dudley S, Hagen W, Dizon D, Paxton L, Reichow D, Yoon SR, Yang K, Arnheim N, Liskay RM et al (2005) Cancer Res 65:8662–8670
Clark AB, Deterding L, Tomer KB, Kunkel TA (2007) Nucleic Acids Res 35:4114–4123
Colnot S, Niwa-Kawakita M, Hamard G, Godard C, Le Plenier S, Houbron C, Romagnolo B, Berrebi D, Giovannini M, Perret C (2004) Lab Invest 84:1619–1630
Cox EC, Degnen GE, Scheppe ML (1972) Genetics 72:551–567
Cunningham JM, Christensen ER, Tester DJ, Kim CY, Roche PC, Burgart LJ, Thibodeau SN (1998) Cancer Res 58:3455–3460
de Wind N, Dekker M, Berns A, Radman M, te Riele H (1995) Cell 82:321–330
de Wind N, Dekker M, van Rossum A, van der Valk M, te Riele H (1998) Cancer Res 58:248–255
de Wind N, Dekker M, Claij N, Jansen L, van van Klink Y, Radman M, Riggins G, van der Valk M, van't Wout K, te Riele H (1999) Nat Genet 23:359–362
Dietmaier W, Wallinger S, Bocker T, Kullmann F, Fishel R, Ruschoff J (1997) Cancer Res 57:4749–4756

Drummond JT, Li GM, Longley MJ, Modrich P (1995) Science 268:1909–1912
Edelmann W, Cohen PE, Kane M, Lau K, Morrow B, Bennett S, Umar A, Kunkel T, Cattoretti G, Chaganti R et al (1996) Cell 85:1125–1134
Edelmann W, Yang K, Kuraguchi M, Heyer J, Lia M, Kneitz B, Fan K, Brown AM, Lipkin M, Kucherlapati R (1999) Cancer Res 59:1301–1307
Edelmann W, Umar A, Yang K, Heyer J, Kucherlapati M, Lia M, Kneitz B, Avdievich E, Fan K, Wong E et al (2000) Cancer Res 60:803–807
el Marjou F, Janssen KP, Chang BH, Li M, Hindie V, Chan L, Louvard D, Chambon P, Metzger D, Robine S (2004) Genesis 39:186–193
Fishel R, Wilson T (1997) Curr Opin Genet Dev 7:105–113
Fishel R, Lescoe MK, Rao MR, Copeland NG, Jenkins NA, Garber J, Kane M, Kolodner R (1993) Cell 75:1027–1038
Flores-Rozas H, Kolodner RD (1998) Proc Natl Acad Sci USA 95:12404–12409
Flores-Rozas H, Clark D, Kolodner RD (2000) Nat Genet 26:375–378
Fodde R (2003) Nat Cell Biol 5:190–192
Fodde R, Edelmann W, Yang K, van Leeuwen C, Carlson C, Renault B, Breukel C, Alt E, Lipkin M, Khan PM et al (1994) Proc Natl Acad Sci USA 91:8969–8973
Fodde R, Smits R, Clevers H (2001) NatRev 1:55–67
Friedberg EC, Meira LB (2006) DNA Repair 5:189–209
Funovics MA, Alencar H, Montet X, Weissleder R, Mahmood U (2006) Gastrointest Endosc 64:589–597
Galiatsatos P, Foulkes WD (2006) Am J Gastroenterol 101:385–398
Gazzoli I, Kolodner RD (2003) Mol Cell Biol 23:7992–8007
Gounari F, Chang R, Cowan J, Guo Z, Dose M, Gounaris E, Khazaie K (2005) Nat Immunol 6:800–809
Groden J, Thliveris A, Samowitz W, Carlson M, Gelbert L, Albertsen H, Joslyn G, Stevens J, Spirio L, Robertson M et al (1991) Cell 66:589–600
Gu L, Cline-Brown B, Zhang F, Qiu L, Li GM (2002) Oncogene 21:5758–5764
Hamada F, Bienz M (2004) Dev Cell 7:677–685
Henderson BR (2000) Nat Cell Biol 2:653–660
Herman JG, Umar A, Polyak K, Graff JR, Ahuja N, Issa JP, Markowitz S, Willson JK, Hamilton SR, Kinzler KW et al (1998) Proc Natl Acad Sci USA 95:6870–6875
Hinoi T, Akyol A, Theisen BK, Ferguson DO, Greenson JK, Williams BO, Cho KR, Fearon ER (2007) Cancer Res 67:9721–9730
Honchel R, Halling KC, Schaid DJ, Pittelkow M, Thibodeau SN (1994) Cancer Res 54:1159–1163
Ireland H, Kemp R, Houghton C, Howard L, Clarke AR, Sansom OJ, Winton DJ (2004) Gastroenterology 126:1236–1246
Jiricny J (2006) Nat Rev Mol Cell Biol 7:335–346
Kane MF, Loda M, Gaida GM, Lipman J, Mishra R, Goldman H, Jessup JM, Kolodner R (1997) Cancer Res 57:808–811
Kaplan KB, Burds AA, Swedlow JR, Bekir SS, Sorger PK, Nathke IS (2001) Nat Cell Biol 3:429–432
Kenneth EH, Marco AM, Larissa GR, Wei YC, Michael PR, Alexandra K, Roderick TB, Umar M, Raju K (2010) "/pmc/articles/PMC2824379/?tool=pmcentrez" Development of a mouse model for sporadic and metastatic colon tumors and its use in assessing drug treatment. Proc Natl Acad Sci USA. 26;107(4):1565–1570. Published online 2010 January 4. doi: 10.1073/pnas.0908682107
Kielman MF, Rindapaa M, Gaspar C, van Poppel N, Breukel C, van Leeuwen S, Taketo MM, Roberts S, Smits R, Fodde R (2002) Nat Genet 32:594–605
Kinzler KW, Vogelstein B (1996) Cell 87:159–170
Kolodner RD, Marsischky GT (1999) Curr Opin Genet Dev 9:89–96
Kucherlapati M, Yang K, Kuraguchi M, Zhao J, Lia M, Heyer J, Kane MF, Fan K, Russell R, Brown AM et al (2002) Proc Natl Acad Sci USA 99:9924–9929
Kucherlapati MH, Nguyen AA, Bronson RT, Kucherlapati RS (2006) Cancer Res 66:3576–3583

Kucherlapati M, Nguyen A, Kuraguchi M, Yang K, Fan K, Bronson R, Wei K, Lipkin M, Edelmann W, Kucherlapati R (2007) Oncogene 26:6297–6306
Kucherlapati MH, Yang K, Fan K, Kuraguchi M, Sonkin D, Rosulek A, Lipkin M, Bronson RT, Aronow BJ, Kucherlapati R (2008) Proc Natl Acad Sci USA 105:15493–15498
Kucherlapati MH, Lee K, Nguyen AA, Clark AB, Hou H Jr, Rosulek A, Li H, Yang K, Fan K, Lipkin M et al (2010) Gastroenterology 138:993–1002
Kunkel TA, Erie DA (2005) Annu Rev Biochem 74:681–710
Kuraguchi M, Edelmann W, Yang K, Lipkin M, Kucherlapati R, Brown AM (2000) Oncogene 19:5755–5763
Kuraguchi M, Yang K, Wong E, Avdievich E, Fan K, Kolodner RD, Lipkin M, Brown AM, Kucherlapati R, Edelmann W (2001) Cancer Res 61:7934–7942
Kuraguchi M, Wang XP, Bronson RT, Rothenberg R, Ohene-Baah NY, Lund JJ, Kucherlapati M, Maas RL, Kucherlapati R (2006) PLoS Genet 2:e146
Lahue RS, Au KG, Modrich P (1989) Science 245:160–164
Li GM (2008) Cell Res 18:85–98
Lin DP, Wang Y, Scherer SJ, Clark AB, Yang K, Avdievich E, Jin B, Werling U, Parris T, Kurihara N et al (2004) Cancer Res 64:517–522
Lothe RA, Peltomaki P, Meling GI, Aaltonen LA, Nystrom-Lahti M, Pylkkanen L, Heimdal K, Andersen TI, Moller P, Rognum TO et al (1993) Cancer Res 53:5849–5852
Marcus DM, Rustgi AK, Defoe D, Brooks SE, McCormick RS, Thompson TP, Edelmann W, Kucherlapati R, Smith S (1997) Arch Ophthalmol 115:645–650
Miyaki M, Konishi M, Tanaka K, Kikuchi-Yanoshita R, Muraoka M, Yasuno M, Igari T, Koike M, Chiba M, Mori T (1997) Nat Genet 17:271–272
Miyakura Y, Sugano K, Konishi F, Ichikawa A, Maekawa M, Shitoh K, Igarashi S, Kotake K, Koyama Y, Nagai H (2001) Gastroenterology 121:1300–1309
Miyakura Y, Sugano K, Konishi F, Fukayama N, Igarashi S, Kotake K, Matsui T, Koyama Y, Maekawa M, Nagai H (2003) Genes Chromosomes Cancer 36:17–25
Miyoshi Y, Nagase H, Ando H, Horii A, Ichii S, Nakatsuru S, Aoki T, Miki Y, Mori T, Nakamura Y (1992) Hum Mol Genet 1:229–233
Modrich P, Lahue R (1996) Annu Rev Biochem 65:101–133
Moon RT, Kohn AD, De Ferrari GV, Kaykas A (2004) Nat Rev Genet 5:691–701
Moser AR, Mattes EM, Dove WF, Lindstrom MJ, Haag JD, Gould MN (1993) Proc Natl Acad Sci USA 90:8977–8981
Narayanan L, Fritzell JA, Baker SM, Liskay RM, Glazer PM (1997) Elevated levels of mutation in multiple tissues of mice deficient in the DNA mismatch repair gene Pms2. Proc. Natl Acad. Sci. USA. 94:3122–3127.
Nathke IS (2004) Annu Rev Cell Dev Biol 20:337–366
Neufeld KL, Zhang F, Cullen BR, White RL (2000) EMBO Rep 1:519–523
Nicolaides NC, Papadopoulos N, Liu B, Wei YF, Carter KC, Ruben SM, Rosen CA, Haseltine WA, Fleischmann RD, Fraser CM et al (1994) Nature 371:75–80
Oshima M, Oshima H, Kitagawa K, Kobayashi M, Itakura C, Taketo M (1995) Proc Natl Acad Sci USA 92:4482–4486
Papadopoulos N, Nicolaides NC, Wei YF, Ruben SM, Carter KC, Rosen CA, Haseltine WA, Fleischmann RD, Fraser CM, Adams MD et al (1994) Sciecnce 263:1625–1629
Peng-Chieh C, Mari K, John V, Yuxun W, Kan Y, Robert E, Dan G, Winfried E, Raju K, Steven ML (2008) Novel Roles for MLH3 Deficiency and TLE6-Like Amplification in DNA Mismatch Repair-Deficient Gastrointestinal Tumorigenesis and Progression. PLoS Genet. June; 4(6): e1000092. Published online 2008 June 13. doi: 10.1371/journal.pgen.1000092
Polakis P (1999) Curr Opin Genet Dev 9:15–21
Polakis P (2000) Genes Dev 14:1837–1851
Powell SM, Zilz N, Beazer-Barclay Y, Bryan TM, Hamilton SR, Thibodeau SN, Vogelstein B, Kinzler KW (1992) Nature 359:235–237
Prolla TA, Christie DM, Liskay RM (1994) Mol Cell Biol 14:407–415

Prolla TA, Baker SM, Harris AC, Tsao JL, Yao X, Bronner CE, Zheng B, Gordon M; Reneker J, Arnheim N et al (1998) Nat Genet 18:276–279
Reitmair AH, Schmits R, Ewel A, Bapat B, Redston M, Mitri A, Waterhouse P, Mittrucker HW, Wakeham A, Liu B et al (1995) Nat Genet 11:64–70
Reitmair AH, Cai JC, Bjerknes M, Redston M, Cheng H, Pind MT, Hay K, Mitri A, Bapat BV, Mak TW et al (1996) Cancer Res 56:2922–2926
Rosin-Arbesfeld R, Cliffe A, Brabletz T, Bienz M (2003) EMBO J 22:1101–1113
Schmutte C, Marinescu RC, Sadoff MM, Guerrette S, Overhauser J, Fishel R (1998) Cancer Res 58:4537–4542
Schofield MJ, Hsieh P (2003) Annu Rev Microbiol 57:579–608
Senda T, Iizuka-Kogo A, Onouchi T, Shimomura A (2007) Med Mol Morphol 40:68–81
Shibata H, Toyama K, Shioya H, Ito M, Hirota M, Hasegawa S, Matsumoto H, Takano H, Akiyama T, Toyoshima K et al (1997) Science 278:120–123
Shoemaker AR, Gould KA, Luongo C, Moser AR, Dove WF (1997) Biochim Biophys Acta 1332:F25–F48
Shoemaker AR, Haigis KM, Baker SM, Dudley S, Liskay RM, Dove WF (2000) Oncogene 19:2774–2779
Sierra J, Yoshida T, Joazeiro CA, Jones KA (2006) Genes Dev 20:586–600
Smits R, Kartheuser A, Jagmohan-Changur S, Leblanc V, Breukel C, de Vries A, van Kranen H, van Krieken JH, Williamson S, Edelmann W et al (1997) Carcinogenesis 18:321–327
Smits R, van der Houven van Oordt W, Luz A, Zurcher C, Jagmohan-Changur S, Breukel C, Khan PM, Fodde R (1998) Gastroenterology 114:275–283
Smits R, Kielman MF, Breukel C, Zurcher C, Neufeld K, Jagmohan-Changur S, Hofland N, van Dijk J, White R, Edelmann W et al (1999) Genes Dev 13:1309–1321
Smits R, Hofland N, Edelmann W, Geugien M, Jagmohan-Changur S, Albuquerque C, Breukel C, Kucherlapati R, Kielman MF, Fodde R (2000) Genes Chromosomes Cancer 29:229–239
Stella AM, William WY, Patricia JG (2004) A Role Msh6 But Not Msh3 in Somatic Hypermutation and Class Swith Recombination J Exp Med 5;200(1):61–68
Sohn KJ, Choi M, Song J, Chan S, Medline A, Gallinger S, Kim YI (2003) Carcinogenesis 24:217–224
Svetlanov A, Cohen PE (2004) Exp Cell Res 296:71–79
Upadhyay R, Sheth RA, Weissleder R, Mahmood U (2007) Radiology 245:523–531
van der Houven van Oordt CW, Smits R, Williamson SL, Luz A, Khan PM, Fodde R, van der Eb AJ, Breuer ML (1997) Carcinogenesis 18:2197–2203
Wang X, Zinkel S, Polonsky K, Fuchs E (1997) Proc Natl Acad Sci USA 94:219–226
Wei K, Clark AB, Wong E, Kane MF, Mazur DJ, Parris T, Kolas NK, Russell R, Hou H Jr, Kneitz B et al (2003) Genes Dev 17:603–614
Weissleder R, Mahmood U (2001) Radiology 219:316–333
Wiesendanger M, Kneitz B, Edelmann W, Scharff MD (2000) J Exp Med 191:579–584
Yao X, Buermeyer AB, Narayanan L, Tran D, Baker SM, Prolla TA, Glazer PM, Liskay RM, Arnheim N. (1999) Different mutator phenotypes in Mlh1- versus Pms2-deficient mice. Proc. Natl Acad. Sci. USA. 96:6850–6855

Chapter 16
Src Family Tyrosine Kinases: Implications for Mammary Tumor Progression

Richard Marcotte and William J. Muller

16.1 Introduction

The *S*rc-related *f*amily of nonreceptor tyrosine *k*inases (SFK) includes 11 members (Fyn, Yes, Fgr, Lyn, Hck, Lck, Blk, Brk, Srm, Yrk, and c-Src itself). Most members are expressed in specific tissues, such as hematopoietic cells, while Src, Yes, and Fyn are ubiquitously expressed, although some tissues like brain, bone, and platelets express higher levels. Historically, the v-Src oncogene was identified as the principle transforming gene of Rous Sarcoma virus (Jay et al. 1978; Purchio et al. 1978). Subsequently, it was found that cellular homolog of v-src termed c-Src was present within the genome of cells (Spector et al. 1978; Oppermann et al. 1979; Sefton et al. 1980a, b). Following this, c-Src was determined to be a protein tyrosine kinase that was itself tyrosine-phosphorylated (Hunter and Sefton 1980; Radke et al. 1980; Sefton et al. 1980a, b). These analyses further revealed that v-Src possessed a constitutive active tyrosine kinase, whereas c-Src activity is tightly regulated in normal cells. The constitutive activation of v-Src tyrosine kinase activity is due to a truncation of the C-terminus regulatory domain of the protein. Comparable activating mutations in c-Src have been rarely observed (Daigo et al. 1999; Nilbert and Fernebro 2000), although a few have been reported in colon and endometrial cancers (Irby et al. 1999; Sugimura et al. 2000). Rather the levels of c-Src protein appear to play a critical role in the activation of c-Src. For example, elevated levels of c-Src are seen in colon carcinoma as c-Src expression is low in premalignant tissues and steadily increases to reach maximum expression in metastases (Talamonti et al. 1993).

R. Marcotte
Ontario Cancer Institute, University of Toronto, Toronto, ON M5G 1L7, Canada

W.J. Muller (✉)
Goodman Cancer Center, 1160 Pine Ave., Montreal, QC H3A 1A3, Canada
e-mail: muller@molonc.mcgill.ca

J.E. Green and T. Ried (eds.), *Genetically Engineered Mice for Cancer Research: Design, Analysis, Pathways, Validation and Pre-Clinical Testing*,
DOI 10.1007/978-0-387-69805-2_16, © Springer Science+Business Media, LLC 2012

Although c-Src has been discovered 30 years ago, the mechanisms and regulation of c-Src activity required for transformation remain elusive and whether misregulation of c-Src activity truly promotes a malignant phenotype since the seminal transforming proprieties attributed to c-Src are in fact derived from studies done with v-Src, an activated version of c-Src, or the mutated activated cellular equivalent, c-Src^{Y527F} (see below for more details). In colon cancer cells, c-Src promotes cell growth, but not metastasis (Irby et al. 1997). In a pancreatic cancer model, siRNA targeted toward c-Src did not decrease tumor incidence, but tumor size and metastasis were decreased (Trevino et al. 2006). In breast cancer cells, c-Src is required for both proliferation and migration (Gonzalez et al. 2006), suggesting that the role of c-Src is likely cell type specific. Nonetheless, increasing evidence suggests that c-Src is indeed required for metastasis.

Metastasis is a complex process that requires several steps to be successful in achieving distant growth in a different organ. To reach that distant organ, a tumor cell must degrade a basement membrane, invade the surrounding matrix, intravasate in the blood stream, survive in the blood, extravasate in the distant organ, and finally grow in a foreign site. In cell culture experiments and in xenograft models, c-Src regulates several of these mechanisms, including motility, extracellular matrix degradation, survival, endothelial cells permeability, and angiogenesis. In this chapter, we survey the different role it plays in these processes, which are directly implicated in promoting metastasis with an emphasis on the few results from mouse models and clinical data from human studies.

16.2 Mechanism of Activation of Src Family Tyrosine Kinases

16.2.1 Structure

Src family members share a common structural organization. The N-terminal domain oftentimes referred to as the SH4 domain includes the myristoylation sequence as well as a stretch of 50–70 residues which is not shared among different family members (Fig. 16.1a) (Boggon and Eck 2004). This unique domain is followed by a protein–protein interacting SH3 domain (~60 amino acid residues) that recognizes proline rich sequences bearing the PxxP motif. In the case of c-Src, the SH3 domain binds to class I ligand (RXLPPLP) (Rickles et al. 1994, 1995). The SH3 domain is followed by a phosphorylated tyrosine SH2 binding domain (~100 amino acid residues), where pYEEI represents the optimal phosphopeptide for c-Src (Songyang et al. 1993), although c-Src SH2 domain also interacts with nonoptimal sequences as other parts of c-Src protein contribute to binding affinity. The recognition pocket of the c-Src SH2 domain is highly conserved among SH2 domains and contains an arginine residue (Arg 175 in c-Src) important for electrostatic interaction with the phosphorylated tyrosine. Specificity of SH2 binding is achieved by the C-terminal divergent sequence of the SH2 domain. An example of this includes pY397 from focal adhesion kinase (FAK) (YAEI). These two protein interaction domains are followed by a tyrosine kinase domain, sometimes referred to as the

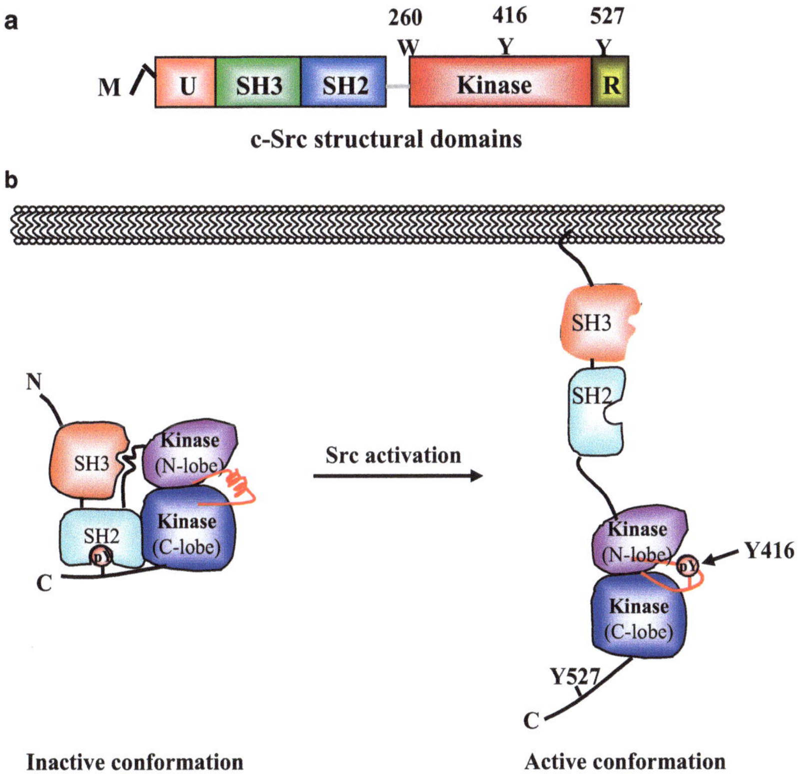

Fig. 16.1 Schematic representation of c-Src. (**a**) Domain structure of c-Src. (**b**) Conformational change in c-Src upon activation

SH1 domain. This domain contains Y416 in the activation loop, which is necessary for kinase activity (Kmiecik and Shalloway 1987). Finally, the C-terminal tail harbors the autoinhibitory Y527, which when phosphorylated by the C-terminal Src kinase (Csk) or its homolog, Csk homologous kinase (Chk), interacts with Arg 175 in the Src SH2 domain, promoting a low-affinity interaction since the sequence surrounding Y527 deviates from the consensus sequence for c-Src SH2 binding (Imamoto and Soriano 1993; Liu et al. 1993; Nada et al. 1993; Davidson et al. 1997). This association promotes an inactive conformation (Fig. 16.1b). This "closed" conformation is also dependent on the association of the SH3 domain with a proline region just upstream of the kinase domain. Upon dephosphorylation of Y527 by tyrosine phosphatase, mutation to phenylalanine, or truncation of the C-terminal tail like v-Src, c-Src autophosphorylates on residue Y416, and adopts an active conformation that has kinase activity (Fig. 16.1b). Phosphorylation of this residue triggers a conformational change in the protein structure such that Glu 310 on helix C can now form a requisite interaction with Lys 295, an interaction that removes Glu 310 from the active site cleft (Boggon and Eck 2004). Mutation of Lys 295 to arginine as well as Tyr 527 to phenylalanine produces a dominant-negative Src protein with an open conformation devoid of enzymatic activity, but still able to interact with other proteins through the SH2 and SH3 domains. Structure analysis suggests a second mechanism, where c-Src could be activated even in the presence of a phosphorylated Y527 residue by binding to a partner through its SH2 or SH3 domain, disrupting the SH2-pY527 interaction. Importantly, in normal basal condition, 90–95% of Src is found in the inactive "closed" conformation. Although Y416 and Y527 are the two main residues subjected to phosphorylation, other residues are also phosphorylated, but their overall contribution to c-Src activity and/or functions are not well understood (Stover et al. 1996; Broome and Hunter 1997; Vadlamudi et al. 2003).

16.2.2 Mode of Activation

To date, only two groups have reported activating mutation of c-Src in human cancers, namely, colon and endometrial cancer (Irby et al. 1999; Sugimura et al. 2000). This mutation creates a truncation at tyrosine 527, therefore abolishing the potential negative regulation of c-Src by Csk (see below). However, the presence of activating mutations in human cancer remains controversial since several other groups were unable to detect any mutations in subsequent studies (Daigo et al. 1999; Nilbert and Fernebro 2000; Laghi et al. 2001). Although mutations are uncommon, c-Src expression and/or activity is increased in several tumor types, including colorectal (Cartwright et al. 1989; Talamonti et al. 1993), breast (Ottenhoff-Kalff et al. 1992), pancreatic (Lutz et al. 1998), ovarian (Wiener et al. 2003), lung (Mazurenko et al. 1992; Masaki et al. 2003), hepatocellular (Masaki et al. 1998), and prostate carcinomas (Lee et al. 2001; Slack et al. 2001) and is likely the preferred mode of activation and transformation. c-Src can also potentiate transformation by integrating signals from different stimuli and facilitating signal transduction of other dominant transforming agent, such as ErbB2 and Ras. Accordingly, c-Src is overexpressed, representing up to 70% of all kinase activity, in the majority of breast cancers also overexpressing a member of the EGFR family (Ottenhoff-Kalff et al. 1992).

Myristoylation of the N-terminus is essential for transformation, suggesting that localization to the plasma membrane, in caveolae and lipid-raft (see below), is required for proper substrate activation and functions. Normally, under resting conditions, c-Src is localized on endosomes in the perinuclear region (Kaplan et al. 1992), a localization dependent on microtubules. Following growth factor stimulation, c-Src relocalizes to the plasma membrane, mostly at focal adhesions (see below); an event controlled by the RhoB family, PI3K, and the actin cytoskeleton (Fincham et al. 1996, 2000a, b; Sandilands et al. 2004) and dependent on c-Src's SH3 domain. Moreover, c-Src co-localizes with RhoB, mDia2, and Wave1/Scar1 on endosomes (reviewed in Frame 2004). Growth stimulation leads to the targeting of only a fraction of these c-Src containing endosomes to the cell periphery. Point mutation in the SH3 domain prevents localization to the plasma membrane likely because this mutation also prevents c-Src from binding to polymerized actin (Fincham et al. 2000a).

Moreover, this peripheral localization is dependent on the integrin-mediated intracellular signaling since cells that are not plated on matrixes supporting integrin engagement do not demonstrate c-Src trafficking. Re-localization of c-Src from the perinuclear region to the plasma membrane is not dependent on c-Src kinase activity, but correlates with the phosphorylation status of Y527 (Kaplan et al. 1994, 1995).

Little is known on how specificity in signaling is attained between the members of Src family. c-Src is not palmitoylated while other Src family members are. This lack of palmitoylation seems responsible for the ability of c-Src to shuttle through the plasma membrane and the endosome/lysosome compartment. A Lyn mutant, engineered to lack palmitoylation, behaves exactly like c-Src in terms of localization (Kasahara et al. 2007). Fyn and Yes are expected to behave like Lyn because they also are palmitoylated. In fact, while Fyn and Yes require actin filaments to localize to the plasma membrane like c-Src, these two proteins are to some extent constitutively located at the plasma membrane. Moreover, Fyn associates with RhoD-positive endosomes, but not with RhoB-positive endosomes while Yes does not localize with either RhoB- or RhoD-containing endosomes. The specificity of endosome localization is dependent on Fyn palmitoylation versus no palmitoylation for c-Src (Sandilands et al. 2007). This specific posttranslational modification associated with different compartmentalization likely explains the nonredundant role seen between these highly homologous family members.

16.2.3 Mechanisms of Activation of c-Src

Activation of c-Src in cancer cells occurs mainly through three different mechanisms: (1) stimulation of a growth factor receptor, (2) removal of the inhibitory phosphotyrosine residue in the C-terminus tail of c-Src by a phosphatase, and (3) inhibition of the activity of the kinases responsible for the phosphorylation of the same tyrosine residue. These three mechanisms are briefly reviewed in the following section.

Historically, c-Src was initially found modified on several residues and activated following simulation of the PDGF receptor (Ralston and Bishop 1985). Since then, several growth factor receptors have been demonstrated to activate c-Src, including the epidermal growth factor receptor family (EGFR and ErbB2), fibroblast growth factor receptor (FGFG), the colony-stimulating factor-1 receptor (CSF-1R), hepatocyte growth factor receptor (MET), and insulin-like growth factor-1 receptor (IGF-1R). Activation of c-Src by these receptors results in cellular responses, such as mitogenesis and survival (reviewed in Bromann et al. 2004).

Activation of c-Src downstream of RTK happens through several mechanisms most of which are not well understood. In the case of the PDGF receptor, c-Src directly interacts with the receptor (Ralston and Bishop 1985; Gould and Hunter 1988; Kypta et al. 1990).This interaction is mediated through its SH2 domain, which interacts with Y579 in the PDGF. This in turn opens up c-Src conformation, leading to its activation. The signal transduction pathway induced by c-Src is largely independent of the canonic MAP kinase pathway. Following PDGF stimulation, c-Src increases mitogenesis through a pathway implicating Stat3, c-Abl, and Myc (Bowman et al. 2001). This signaling pathway requires c-Src located in lipid raft. Interestingly, c-Src activation following PDGF stimulation also induces dorsal ruffles through a different "pool" of c-Src and cross-regulation of G-coupled receptor (Veracini et al. 2006). A number of independent studies have demonstrated that transformation by Src is dependent on Stat3 activation (Schlessinger and Levy 2005), which is in part responsible for the growth-promoting and cell survival function of c-Src (Garcia et al. 2001). In Src-transformed cells, Stat3 is constitutively activated and c-Src is found associated with Stat3 (Cao et al. 1996). Furthermore, c-Src transformation is also dependent on c-Abl, and this can be independent of Stat3 (Sirvent et al. 2007).

c-Src also interacts with two members of the EGFR family, EGFR and ErbB2, although at this point, it is still not clear whether c-Src is activated by binding or is already activated before binding to the receptor. Activity of c-Src is required to phosphorylate Y877 located in the activation loop of ErbB2 (Xu et al. 2007; Marcotte et al. 2009) and this residue is necessary for full activation of the receptor (O'Rourke et al. 1998). ErbB2-induced tumors possess high c-Src kinase activity (Muthuswamy and Muller 1994, 1995; Belsches et al. 1997). Moreover, c-Src immunoprecipitates phosphorylated ErbB2 from these tumors and several established breast cancer cell lines. In these, transformation potential directly correlates with the detection of an ErbB2/c-Src complex (Belsches-Jablonski et al. 2001). Another mechanism through which ErbB2 regulates c-Src is through an increase in c-Src protein synthesis as well as an inhibition of c-Src protein degradation (Tan et al. 2005), hence creating a positive feedback loop and explaining the high c-Src activity and expression levels generally associated with ErbB2-expressing tumors.

Although direct binding of c-Src to EGFR remains controversial, c-Src and EGFR act synergistically to promote proliferation and transformation (Biscardi et al. 1999; Dimri et al. 2007). More specifically, EGFR and c-Src act synergistically to transform fibroblasts and to disrupt lumen formation in 3D cell culture assays of normal mammary cells. While EGFR or c-Src alone leads to the formation of well-rounded acini, combination of these two kinases disrupts proper acini development and promotes an invasive phenotype. Noteworthy, ErbB2 alone induces similar changes. This effect is likely a reflection of the ability of c-Src to phosphorylate EGFR on several tyrosine residues, Y891, Y920 (Stover et al. 1995) and Y845 (Biscardi et al. 1999) located in the kinase activation loop of the receptor and responsible for catalytic activity of the receptor. A likely explanation for c-Src synergism with the EGFR family is the ability of c-Src to regulate the degradation of EGFR and ErbB2 by phosphorylating several proteins implicated in the internalization of activated receptors, such as clathrin and dynamin, but also proteins responsible for ubiquitin-mediated degradation, such as c-Cbl (Wilde et al. 1999; Bao et al. 2003). Therefore, activation of c-Src by ligand-mediated receptor stimulation generates sustained signaling by preventing degradation of the activated receptor. This positive loop likely explains why tumors usually co-overexpress both EGFR and c-Src. At this point, it is still unclear how EGFR or ErbB2 can activate c-Src. Some results suggest that activation of c-Src following EGFR stimulation requires the small GTPases Ras and Ral (Goi et al. 2000) as well as the activation of some phosphatases.

As mentioned above, c-Src kinase activity is regulated by tyrosine phosphorylation, mainly at Y416 and Y527. Accordingly, several protein tyrosine phosphatases, such as RPTPα, PTP1B, PTPε, SHP-1, SHP-2, and others are reported to dephosphorylate tyrosine-527 both *in vitro* and *in vivo* and therefore activating c-Src (reviewed in Roskoski 2005). Of these, RPTPα and PTP1B are clear favorite to regulate tyrosine-527 *in vivo,* since overexpression of these phosphatases in cell line dephosphorylates tyrosine-527 and promotes cell transformation (Zheng et al. 1992). Moreover, PTP1B was isolated from breast cancer cells as a phosphatase able to dephosphorylate Y527 (Bjorge et al. 2000). Cells lacking RPTPα are defective in integrin signaling, in cell spreading and have no c-Src kinase activity (Su et al. 1999). Mice deficient in RPTPα demonstrate a decrease in Src activity, which is caused by an increase in Y527 phosphorylation (Ponniah et al. 1999). Interestingly, RPTPα likely activates c-Src by a displacement mechanism, where phosphorylation of Y789 in RPTPα interacts with c-Src SH2 domain. This liberates Y527 in the C-terminus tail of c-Src and allows dephosphorylation and activation of c-Src.

Tumors derived from transgenic mice overexpressing activated ErbB2 in the mammary glands express high level of both c-Src and RPTPε. When RPTPε is genetically removed from these tumors, c-Src activity is greatly reduced. Accordingly, cells derived from these tumors are morphologically less transformed and proliferate slower than cells expressing RPTPε (Gil-Henn and Elson 2003). This phenotype is rescued by expressing active c-Src in these cells or RPTPε. Surprisingly, while c-Fyn and c-Yes activity are also greatly reduced, reexpression of c-Fyn or c-Yes does not rescue the morphological phenotypes seen in the RPTP$\varepsilon^{-/-}$ cells, suggesting nonoverlapping function between c-Src and these two related family members (Granot-Attas and Elson 2004). This is reminiscent of the mammary tumor virus promoter (MMTV) – polyoma middle T antigen (PyMT) transgenic mice crossed with knock-out c-Src or c-Yes mice. While loss of c-Src almost completely abrogated tumor induction in PyMT transgenic mice, no difference in tumor kinetic is seen when tumor induction is done in a c-Yes$^{-/-}$ background (Guy et al. 1994). Interestingly, ErbB2 activates RPTPε by phosphorylating Y695, which specifically targets the activity of the phosphatase toward c-Src (Berman-Golan and Elson 2007). Moreover, c-Src is "trapped" by a D302A RPTPε mutant, strongly suggesting that c-Src is a substrate for this particular phosphatase.

Other phosphatases potentially implicated in c-Src activity include SHP2, PRL-3, and PP2A. SHP2 and PRL-3 regulate c-Src activity by acting on Csk. Cells devoid of SHP2 are defective in c-Src activation and demonstrate an increased level of Y527 phosphorylation (Zhang et al. 2004). SHP2 regulates Csk activity by dephosphorylating PAG, which prevents Csk re-localization to the plasma membrane. SHP2 is upregulated in breast cancer cells as well as in infiltrating ductal carcinoma suggesting that it might be in part responsible for the high c-Src activity seen in these tumors (Zhou et al. 2008). Phosphatase of regenerating liver 3 (PRL-3) induces cell migration and invasion, through inactivation of Csk. Accordingly, PRL-3 is overexpressed in tumors that also possess increased c-Src activity (Liang et al. 2007). More importantly, PRL-3 is massively overexpressed in liver metastasis of colorectal cancer while its expression is undetectable in normal colon epithelium (Saha et al. 2001). Recently, an elegant RNAi screen against regulatory subunits of protein phosphatase 2A to isolate the regulator of the JNK signaling pathway following UV irradiation identified PR55γ. This subunit regulates the JNK pathway by specifically associating with c-Src, but not with Fyn or Lyn. This binding is sufficient to recruit the catalytic PP2A holoenzyme to c-Src. Overexpression of PR55γ inhibits c-Src activity by dephosphorylating Ser12, and this residue is necessary to induce JNK following UV irradiation (Eichhorn et al. 2007), suggesting that phosphorylation and dephosphorylation of residues other than tyrosine can also regulate c-Src activity. Interestingly, this regulation seems important following UV irradiation and other stresses, but not EGF stimulation.

c-Src inhibitory tyrosine residue Y527 is phosphorylated by two nonreceptor tyrosine kinases, Csk and Chk (reviewed in Chong et al. 2005). Csk is expressed ubiquitously while Chk expression is limited to breast, hematopoietic cells, neurons, and testes (Chong et al. 2005). Given that productive phosphorylation of this regulatory tyrosine residue would result in the shift of c-Src into its inactive conformation, expression of these two closely related tyrosine kinases would act as tumor suppressors. Indeed, there is evidence that Csk expression is decreased in some cancer type, such as hepatocellular carcinoma (Masaki et al. 1999). Consistent with its negative regulatory role elevated expression of Csk in colon carcinoma cells suppresses metastasis *in vivo* (Nagle et al. 2004), and this is associated with a decrease in MMP-2 expression without any effect on cell growth. Moreover, other proteins, such as Csk-binding protein (Cbp) regulating the activity and/or levels of Csk also impacts on the transforming ability of c-Src. Overexpression of Cbp negatively regulates c-Src activity by regulating Csk activity (Jiang et al. 2006). Interestingly, Cbp is tyrosine phosphorylated following growth factor stimulation by c-Src itself. This in turn recruits Csk through its SH2 domain, bringing Csk in close proximity to c-Src, where it phosphorylates Y527. This acts as a negative feedback loop to control c-Src kinase activity. Dephosphorylating Cbp disrupts this negative loop; an event mediated by the tyrosine phosphatase Shp2 (Zhang et al. 2004). On the other hand, Chk is recruited to the plasma membrane by binding directly to growth factor receptor-like ErbB2 (Zrihan-Licht et al. 1997, 1998). Recently, a protein referred to as p140Cap was shown to associate with c-Src through its SH3 domain and inhibit c-Src activity. Decreased expression of p140Cap activates c-Src through downregulation of Csk (Di Stefano et al. 2007). Moreover, overexpression of p140Cap in breast cancer cells acts as a tumor suppressor by inhibiting cell proliferation, cell motility, and invasion, as well as tumor growth in nude mice, suggesting again that c-Src regulates these functions.

The idea that c-Src activity is required downstream of oncogenic receptors has recently been challenged by overexpressing Chk in MMTV-ErbB2-induced mammary gland tumors. In that study, overexpression of Chk did not lead to a delay in tumor latency as would be expected if c-Src activity is required for ErbB2-mediated tumorigenesis (Kaminski et al. 2006). However, given that there was a low level of c-Src activity still detected within these tumors might account for the observed lack of effect on tumor progression. On the other hand, expression analysis of Chk in colon cancer lines and human colon cancer tissues reveals that Chk protein levels are decreased compared to levels in a normal colon cell line (Zhu et al. 2008). This decrease is associated with an increase in c-Src kinase activity. At this point, these conflicting results might reflect a difference in the need for certain tissues for an increase in c-Src activity. The Chk overexpression results contrast sharply with results from the conditional deletion of Csk in the skin (Honda et al. 2007; Yagi et al. 2007). Cells devoid of Csk were altered in their cytoskeleton organization, likely a consequence of an increase in c-Src activity. There was an increase in the formation of podosomes, dorsal ruffles, and lamellipodia, events associated with an increase in invasion potential. Mice lacking Csk expression in the skin ultimately developed hyperplasia, although these never reached the carcinoma stage as reported for mice expressing activated c-Src in the skin (Matsumoto et al. 2003, 2004). Experiments done in cell culture are also consistent with Csk inhibiting c-Src activity and functions activated by c-Src, such as cell scattering, the integrity of cell contacts, and the strength of focal adhesions (Rengifo-Cam et al. 2004).

16.3 c-Src Regulates Several Steps in the Metastasis Process

The most potent oncogenes, such as Ras, Myc, and v-Src, have the ability to simultaneously transduce signaling from several pathways promoting all the steps required for successful colonization of a distant organ. In the next section, the role of c-Src in each step of the metastasis cascade is reviewed with an emphasis on the unique pathway regulating any particular steps.

The most well-characterized step in the metastasis cascade is the motility/invasion step, likely because several *in vitro* assays seem to accurately reflect the intrinsic ability to invade *in vivo*. Historically, the ability of v-Src to promote invasion was acknowledged fairly early after its identification as a transforming agent (Chen et al. 1985). Several cell types transformed with v-Src demonstrate an increase in invasion. Nevertheless, it took several years before the same could be said about the mammalian counterpart. The first formal indication that c-Src (and possibly Fyn and Yes) was required to promote motility and invasion came from studies done with fibroblasts derived from the triple c-Src, Yes, and Fyn knock-out mice. These cells, named SYF (for *S*rc, *Y*es, and *F*yn), which do not express these three SFK, have a greatly reduced motility (Klinghoffer et al. 1999) because of a defect in activating integrin signaling. Integrins are transmembrane heterodimeric receptors made of an α and β-subunit. Different subunit combinations act as receptors for the different extracellular matrix proteins present in the basement membrane, such as laminin and fibronectin. Integrins do not possess intrinsic kinase activity, but rather recruit several kinases to their cytoplasmic tail upon engagement and clustering. This clustering of integrin receptor is referred to as focal adhesions or focal contacts, which act as bridging factors between extracellular matrices and the actin cytoskeleton. A major component of focal adhesions is FAK, which autophosphorylates and is tyrosine-phosphorylated upon integrin engagement. Phosphorylation is partly mediated by c-Src, and this event is necessary for focal adhesion disassembly. Upon engagement of integrins, FAK autophosphorylates on Y397, which acts as a high affinity binding site for the SH2 domain of c-Src. This binding leads to a conformational activation of c-Src, promoting the phosphorylation of several other tyrosine residues in FAK, including Y407, Y576/7, Y861, and Y925, of which only Y925 is truly dependent on c-Src kinase activity (Brunton et al. 2005). Once activated, Src phosphorylates several focal adhesions components, such as talin, cortactin, tensin, paxillin, and p130^{CAS}, which promote the recruitment of several effector proteins, such as CrkL (Li et al. 2003), leading to the activation of small GTPase Rac, Rho, and CDC42. CrkL$^{-/-}$ cells phenocopy the haptotaxis defect seen in SYF cells, which is rescued by reexpressing CrkL in both CrkL$^{-/-}$ or SYF cells (Li et al. 2003). Phosphorylation of Y925 on FAK by Src engages the MAP kinase pathway by recruiting Grb2 to Fak (Schlaepfer and Hunter 1996). Integrin clustering as well as Src activation translocate activated Erk to focal adhesions, which is required for cell spreading, but not attachment per se, and is dependent on the activation of MLCK (Fincham et al. 2000b). Alternatively, p130^{Cas} also modulates c-Src since its activity is increased both in MMTV-p130^{Cas} transgenic mice and in a cross between these mice and an activated ErbB2 (Cabodi et al. 2006).

Other events required for adhesion turnover include recruitment of the protease calpain2 by FAK following c-Src activation. Calpain2 cleaves several proteins in focal adhesion contact, such as FAK itself, and talin, disrupting the link between focal adhesion and the actin network. Moreover, Src increases synthesis of calpain II, which in turn inhibits calpastatin, an endogenous inhibitor of calpain (Carragher et al. 2002). FAK is required to promote the establishment of a complex containing both calpain 2 and p42Erk. This complex is required for the processing of FAK following engagement of integrins, leading to transformation and cell migration (Carragher et al. 2003). This association is not dependent on Fak kinase activity, but on p42Erk activity.

Given all the aforementioned results, a likely scenario is that as cells attach to a cell matrix, c-Src is recruited to the plasma membrane at focal adhesion in an actin- and PI3K-dependent manner. Once at focal adhesion, c-Src acts as a scaffold protein to strengthen focal adhesion by recruiting structural and effector proteins, such as paxillin, p130Cas, cortactin, and Crk. Once the cell body has migrated over the strong adhesion complexes, calpain is activated in an ERK/MAP kinase-dependent manner and disrupts these adhesion structures now at the rear-end by cleaving several of the proteins involved in promoting adhesion, such as Fak and c-Src. Overall, c-Src promotes a cycle of phosphorylation and dephosphorylation that would contribute to a cycle of assembly at the leading edge and disassembly of focal adhesion at the retractory end of cells, allowing cell migration. Overexpression of any of the focal adhesion complex components, including c-Src and FAK (Cary et al. 1996), likely limits the maturation of focal adhesion, promoting motility and invasion of cancer cells. While c-Src overexpression alone was associated with recurrence in colorectal cancer, overexpression level of both FAK and c-Src was associated with faster recurrence and metastasis (de Heer et al. 2008). Moreover, this high expression was conserved in liver metastasis, suggesting that the FAK/c-Src unit and assembly of focal adhesions likely play a significant role in metastasis.

Epithelial cells are characterized by two main features *in vivo*: (1) they form layers of stratified polygonal cells that are joined by cell–cell contacts referred to as tight junctions and (2) these cells are polarized in an apical–basolateral fashion, in part not only due to these tight junctions, but also due to adherens junction. On the other hand, mesenchymal cells are usually motile fusiform single cells attached at focal point. These cells do not express E-Cadherin, a protein implicated in adherens junction formation, and is oftentimes used as a marker for the epithelial to mesenchymal transition (EMT) seen in highly aggressive basal cancer (Hazan et al. 2004). Epithelial cells needs to disrupt these cell–cell contacts to invade surrounding tissue and eventually metastasize to a foreign location. c-Src regulates several proteins located in these adherens junctions.

c-Src is required to induce cell scattering in MDCK following the treatment with either HGF or EGF. This is mainly caused by phosphorylation of p120Ctn, β-catenin, E-cadherin, and p190rhoGAP, all of which are components of adherens junctions (Behrens et al. 1993). These proteins are all heavily phosphorylated in Src-transformed cells. c-Src interacts with p120Ctn and directly phosphorylates Y217 and Y228, which promotes a greater affinity binding site for RhoA. Interestingly, Fyn phosphorylates a different tyrosine residue Y112 which inhibits binding of RhoA to p120Ctn (Castano et al. 2007). To date, this is one of the only differences seen in substrate specificity between c-Src and Fyn. c-Src also phosphorylates E-Cadherin, which disrupts cell–cell adhesion and leads to E-cadherin degradation. This is mediated by Hakai, an E3-ubiquitin ligase which is activated by Src and which promotes ubiquitination of both E-cadherin and β-catenin (Fujita et al. 2002). Moreover, high expression of activated Src in colon carcinoma cells prevents re-localization of E-cadherin to cell–cell contact. This effect is dependent on its kinase activity since overexpression of DN251, while able to promote focal adhesion assembly, is unable to prevent the re-localization (Avizienyte et al. 2002). Blocking antibodies to α_v or β1 also disrupted re-localization of E-cadherin, suggesting that regulation of focal adhesion and cell–cell junction is interdependent, an effect likely mediated by c-Src and dependent on FAK. Interestingly, ablation of β1-integrin in a PyMT-induced breast cancer model had no effect on c-Src activity or localization, while decreasing FAK activity did, suggesting that these effects are likely dependent on cell type and the actual oncogenic pathway responsible for transformation (White et al. 2004; Lahlou et al. 2007). At low level, c-Src strengthens cell–cell adhesion mediated by E-cadherin. At high expression level, c-Src disrupts cell–cell adhesion (McLachlan et al. 2007). Interestingly, when PTEN is overexpressed in cells expressing v-Src, adherens junctions were restored, preventing cell invasion. This result suggests that Src ability to disrupt these structures is dependent on the PI3K pathway (Kotelevets et al. 2001).

More recently, the importance of β1 integrin/c-Src signaling axis in the induction of metastasis was demonstrated through the mammary-specific ablation of β1 integrin in the MMTV/ErbB2 transgenic strain (Huck et al. 2010). Examination of the metastatic potential of these tumors also revealed that β1-integrin-deficient tumors did not metastasize to the lung (Pontier et al. 2010). Like the ILK-deficient tumors, the differences in metastatic potential in the lung were further correlated with differences in the levels of tyrosine phosphorylation of p130Cas, c-Src, FAK, and paxillin (Pontier et al. 2010). In the case of β1-integrin-deficient ErbB2 tumor cells, tyrosine phosphorylation of these key-signaling molecules was completely abrogated (Huck et al. 2010; Pontier et al. 2010). Remarkably, in addition to loss of key adhesion signaling network, the β1-integrin-deficient tumors were incapable of transphosphorylation of EGFR suggesting that β1 integrin is critical for modulating cross talk between ErbB2 and EGFR and c-Src signaling network is particularly intriguing in that we have recently demonstrated that ErbB2 can directly associate with c-Src through a catalytic–catalytic interface (Marcotte et al. 2009). However, the precise molecular mechanism by which this occurs remains to be elucidated.

Consistent with the importance of the c-Src pathway in metastasis, some immediate targets of c-Src-mediated tyrosine phosphorylation cascade also can modulate the metastatic phase of erbB2 mammary tumor progression. Recently it was shown that certain known ErbB2 coupled signaling pathways are likely to impact on the metastatic phenotype. For example, transcription factor Stat3 has been implicated as an important signaling molecule in ErbB2-induced tumor progression (Tan et al. 1997; Guo et al. 2006). Consistent with these functional studies, elevated Stat3 levels have been associated with higher risk of breast cancer recurrence (Diaz et al. 2006). To directly assess the importance of Stat3 in ErbB2 mammary tumor progression, we have recently removed Stat3 from ErbB2-induced tumors using Cre-LoxP1 recombination system. To accomplish this, we have generated transgenic mice in which translation of activated ErbB2 and the Cre recombinase are directed from the same polycistronic mRNA due to the presence of an internal ribosome entry site (IRES) between the two cDNA sequences (NIC strain) (Ursini-Siegel et al. 2008). In this manner, epithelial cells that are targeted for transformation by activated ErbB2 also express Cre and therefore are unable to escape Cre-mediated recombination. Using this unique transgenic approach, we have demonstrated that mammary-specific deletion of Stat3 can have a dramatic and selective impact on ErbB2 mammary tumor metastasis (Ranger et al. 2009). Although tumor onset was not impacted by mammary-specific disruption of Stat3 (Ranger et al. 2009), the Stat3-deficient tumors exhibited a nearly sixfold reduction in tumor metastasis (Ranger et al. 2009). Significantly, the few Stat3-deficient metastatic lesions that were detected were actually intravascular and had not invaded the lung parenchyma (Ranger et al. 2009). Examination of the gene expression profile of these Atat3-deficient ErbB2 tumors demonstrated that Stat3-deficient signature had features of depressed inflammatory response. Consistent with a role for Stat3 in regulation of inflammatory response, disruption of Stat3 in the mammary epithelial compartment can lead to a dramatic impairment of mammary gland involution that is associated with decreases in levels of a number of inflammatory cytokines (Chapman et al. 1999; Stein et al. 2004; Watson 2009). Taken together, these observations suggest that Stat3 function may be required to induce acute inflammation associated with mammary gland involution (Fig. 2). Consistent with this view, it has recently been demonstrated that the involuted mammary gland possesses an inflammatory tumor microenvironment that promotes invasive mammary tumors (Schedin 2006; Schedin et al. 2007).

16.4 c-Src Modulates Both Tumor Cell Intravasation and Extravasation

Another potential mechanism by which c-Src promotes metastasis is through the regulation of angiogenesis and vascular permeability. Once tumors reach a certain size, they require the generation of new blood vessels to provide oxygen and nutrients for further growth, a process call angiogenesis. Vascular endothelial growth factor (VEGF) is one of the major proteins implicated in that process as it acts both on tumor and endothelial cells. For tumor cells to be able to enter a blood vessel, endothelial cells need to be leaky, a process referred to as vascular permeability. c-Src is a key player of VEGF action. VEGF is unable to induce vascular permeability in $src^{-/-}$ mice, even though the development of new blood vessels is relatively normal (Eliceiri et al. 1999). In that particular study, c-Src was required for vascular permeability, but not for tumor growth or angiogenesis (Criscuoli et al. 2005). Interestingly, c-Src is also required in the recipient mouse for extravasation into the lung following injection of tumor cells in an experimental metastasis assay. This effect was specific to c-Src as tumor cells injected in $fyn^{-/-}$ mice display the same extent of lung colonization as wild-type mice. VEGF stimulates Src-mediated phosphorylation of Fak, which is required for endothelial vascular permeability. c-Src also induces VEGF expression at the transcriptional level through Stat3 activation (Niu et al. 2002). c-Src and VEGF are likely implicated in a positive feedback loop since expression of c-SrcDN in established tumors reduces the level of VEGF expression, ultimately inducing apoptosis (Gonzalez et al. 2006). Moreover, c-Src inhibition using specific inhibititory drugs led to a decrease in VEGF and interleukin-8 (a pro-angiogenic factor) secretion in an ovarian cancer model (Han et al. 2006), which greatly affected microvessel density. Effects of VEGF are not limited to endothelial cells. VEGF-C binding to its receptor, flt-4, on lung adenocarcinoma cells induces migration and invasion as well as metastasis *in vivo* in an experimental metastasis assay. This increased invasion was mediated by upregulation of contactin-1 through an Src-p38-MAPK-C/EBP signaling axis (Su et al. 2006). VEGF is not the only secreted factor promoting permeability and extravasation. Recently, βig-h3/TGFB1 (transforming growth factor, β-induced) was shown to promote metastasis by inducing extravasation of colon cancer cells, an effect caused by the dissociation of VE-cadherin on endothelial cells through an integrin-c-Src pathway (Ma et al. 2008). In endothelial cell, c-Src and Csk are complexed with VE-cadherin. Upon VEGF stimulation, Csk is released following the recruitment of Shp2 which induces c-Src activation (Ha et al. 2008). Interestingly, angiopetin-1, also a pro-angiogenic factor, which does not increase vascular permeability, prevents the activation of c-Src by VEGF (Gavard et al. 2008) and therefore protects blood vessels from VEGF-induced leakage. VEGF induces leakage by promoting VE-cadherin endocytosis and degradation, which disrupts adhesion junction between endothelial cells (Gavard and Gutkind 2006).

16.5 c-Src Impacts on Tumor Cell Survival

Once a cell has found its way into the blood vessels, it is no longer attached to an extracellular matrix. At such, this cell needs to overcome anoikis, a term coined for the apoptosis program induced in anchorage-dependent cells following the loss of integrin signaling. Anoikis is induced following loss of integrin engagement through binding to extracellular matrix component. This program is dependent on Bax, a member of the Bcl-2 family which is a major regulator of the apoptotic cascade. v-Src was the first described oncogene to promote resistance to this programmed cell death program (Frisch and Francis 1994), and this was extended to the mammalian equivalent at least in colon tumor cell lines (Windham et al. 2002). In these cell lines, resistance to anoikis was mediated through Akt activation. Src regulation of anoikis is complex and is likely cell type-dependent as Src was shown to both activate anti-apoptotic proteins and inhibit the pro-apoptotic protein activation. Src can inhibit anoikis by upregulating Bcl-xL, an anti-apoptotic effector, through a MEK/MAPK pathway or by preventing the activation of Bax, a pro-apoptotic protein (Gilmore et al. 2000; Coll et al. 2002). Moreover, c-Src also prevents Mcl-1 degradation and Bim induction, again both members of the Bcl-2 family. This event was dependent on the PI3K/Akt pathway (Woods et al. 2007). Recently, c-Src was found to directly phosphorylate Bif, a Bax interacting protein. This event prevented activation of Bax and induction of anoikis (Yamaguchi et al. 2008). Several report using siRNA targeted specifically at c-Src or inhibitors of Src family kinase demonstrate that Src ability to inhibit apoptosis is essential for its ability to promote metastasis (Gonzalez et al. 2006; Nowak et al. 2007; Shor et al. 2007; Zheng et al. 2008). c-Src ability to prevent apoptosis also extends to the regulation of caspase activity. c-Src phosphorylates caspase-8 and inhibits its activity (Cursi et al. 2006). Although unlikely to be involved in anoikis, this event would prevent apoptosis downstream of death receptors, such as Trail and Fas, which are involved in cell death following an immune response.

16.6 Transgenic Mouse Models of Human Breast Cancer Implicate c-Src as a Critical Factor in Mammary Tumor Progression

There are only a few mouse model studies that have directly assessed the role of c-Src in promoting metastasis, likely because of the difficulties in using c-Src knock-out mice. These mice are viable in a mixed background, although only a few mice survived for more than several weeks after birth (Soriano et al. 1991). The surviving mice display osteopetrosis (an increase in bone density) caused by a defect in osteoclasts. When these c-Src knock-out mice are backcrossed into an FVB background, they are embryonic lethal (Marcotte and Muller, unpublished observations), making this mouse model impossible to use for cancer and metastasis studies. On the other hand, double knock-out mice, Src/Fyn and Src/Yes, die perinatally, while Fyn/Yes double knock-out mice are viable, but display severe renal problems (Stein et al. 1994). Nevertheless, mouse embryonic fibroblasts derived from these mice as well as from triple knock-out (TKO) of c-Src, Yes, and Fyn (referred as SYF cells), which die embryonically around day 10 with a severe growth defect, were useful in determining the potential role that c-Src and other family members play in cell adhesion and migration. Other defects in the TKO included their inability to turn the orientation of the germ layers, a wavy neural tube, blood-filled blebs beneath the epidermis, and a tuberous allointis not attached to the chorion. Mice engineered to lack the negative c-Src regulator, Csk, are embryonic lethal, suggesting either that too much c-Src activity is lethal or that Csk possesses other functions unrelated to c-Src regulation (Imamoto and Soriano 1993).

Despite the difficulties in working with the original c-Src knock-out mice, a loss-of-function study was done in the MMTV–PyMT on a mixed genetic background and provided important insight into the role of c-Src in mammary tumor progression. The results revealed that loss of c-Src in that model almost completely abrogated tumor induction (Guy et al. 1994). By contrast, comparable crosses with c-Yes-deficient strains had little impact on tumor progression in this model. These observations suggest that signaling pathways specifically activated by c-Src are critical for induction of mammary gland tumors.

Conversely, ectopic expression of activated c-Src under the strong MMTV is sufficient to induce mammary gland tumors, although tumors were focal in origin and appeared after a long latency period (15 months), suggesting that other genetic events are required to promote tumorigenesis (Webster et al. 1995). In the skin, using keratin promoters, expression of activated c-Src (Y527F) results in severe epidermal hyperplasia and hyperkeratosis, such that mice did not survive past 3 weeks of age (Matsumoto et al. 2003). Expression of wild-type c-Src generated several phenotypes depending on the founder lines. One founder gave rise to a similar phenotype than the activated c-Src transgenic, one founder developed spontaneous squamous cell carcinoma of the skin, and the last founder lines had no phenotype, despite high expression of the transgene. This founder line was more susceptible to TPA-induced skin carcinogenesis with a higher propensity for metastasis to lymph nodes. Overall, these results suggest that the loss of negative regulation of c-Src is likely required to induce tumorigenesis.

16.7 Future Perspectives

Given that c-Src was the first characterized oncogene and the pivotal role it seems to play in tumorigenesis, the lack of potent clinical inhibitor targeted toward c-Src is surprising. This lack of effort likely stems from the fact that c-Src is expressed ubiquitously and that the c-Src knock-out mice only demonstrate a mild phenotype. Lately, a resurgence for a specific c-Src inhibitor has resurfaced, probably with the realization that c-Src is activated virtually in all cancers, and also because inhibition of c-Src using PP2 can resensitize pancreatic adenocarcinoma (Duxbury et al. 2004) and ovarian cells (Pengetnze et al. 2003; Chen et al. 2005) to other chemotherapeutic drug. In some cases, metastasis was completely abrogated (Duxbury et al. 2004; Yezhelyev et al. 2004). In other cases, c-Src inhibitors were able to induce apoptosis in cancer cell lines (Belsches-Jablonski et al. 2001). It is unlikely that c-Src inhibitors will be used as single agent since overexpression of c-Src alone is not tumorigenic, but it does have great potential in combination therapy. Given its role on regulating several synergistic processes responsible for tumorigenesis and metastasis downstream of several growth factor receptors oftentimes mutated or deregulated, a c-Src-specific inhibitor might become the requisite drug in combination with many other therapeutic agents.

References

Avizienyte E, Wyke AW et al (2002) Src-induced de-regulation of E-cadherin in colon cancer cells requires integrin signalling. Nat Cell Biol 4(8):632–638

Bao J, Gur G et al (2003) Src promotes destruction of c-Cbl: implications for oncogenic synergy between Src and growth factor receptors. Proc Natl Acad Sci USA 100(5):2438–2443

Behrens J, Vakaet L et al (1993) Loss of epithelial differentiation and gain of invasiveness correlates with tyrosine phosphorylation of the E-cadherin/beta-catenin complex in cells transformed with a temperature-sensitive v-SRC gene. J Cell Biol 120(3):757–766

Belsches AP, Haskell MD et al (1997) Role of c-Src tyrosine kinase in EGF-induced mitogenesis. Front Biosci 2:d501–d518

Belsches-Jablonski AP, Biscardi JS et al (2001) Src family kinases and HER2 interactions in human breast cancer cell growth and survival. Oncogene 20(12):1465–1475

Berman-Golan D, Elson A (2007) Neu-mediated phosphorylation of protein tyrosine phosphatase epsilon is critical for activation of Src in mammary tumor cells. Oncogene 26(49): 7028–7037

Biscardi JS, Maa MC et al (1999) c-Src-mediated phosphorylation of the epidermal growth factor receptor on Tyr845 and Tyr1101 is associated with modulation of receptor function. J Biol Chem 274(12):8335–8343

Bjorge JD, Pang A et al (2000) Identification of protein-tyrosine phosphatase 1B as the major tyrosine phosphatase activity capable of dephosphorylating and activating c-Src in several human breast cancer cell lines. J Biol Chem 275(52):41439–41446

Boggon TJ, Eck MJ (2004) Structure and regulation of Src family kinases. Oncogene 23(48): 7918–7927

Bowman T, Broome MA et al (2001) Stat3-mediated Myc expression is required for Src transformation and PDGF-induced mitogenesis. Proc Natl Acad Sci USA 98(13):7319–7324

Bromann PA, Korkaya H et al (2004) The interplay between Src family kinases and receptor tyrosine kinases. Oncogene 23(48):7957–7968

Broome MA, Hunter T (1997) The PDGF receptor phosphorylates Tyr 138 in the c-Src SH3 domain in vivo reducing peptide ligand binding. Oncogene 14(1):17–34

Brunton VG, Avizienyte E et al (2005) Identification of Src-specific phosphorylation site on focal adhesion kinase: dissection of the role of Src SH2 and catalytic functions and their consequences for tumor cell behavior. Cancer Res 65(4):1335–1342

Cabodi S, Tinnirello A et al (2006) p130Cas as a new regulator of mammary epithelial cell proliferation, survival, and HER2-neu oncogene-dependent breast tumorigenesis. Cancer Res 66(9):4672–4680
Cao X, Tay A et al (1996) Activation and association of Stat3 with Src in v-Src-transformed cell lines. Mol Cell Biol 16(4):1595–1603
Carragher NO, Westhoff MA et al (2003) A novel role for FAK as a protease-targeting adaptor protein: regulation by p42 ERK and Src. Curr Biol 13(16):1442–1450
Carragher NO, Westhoff MA et al (2002) v-Src-induced modulation of the calpain-calpastatin proteolytic system regulates transformation. Mol Cell Biol 22(1):257–269
Cartwright CA, Kamps MP et al (1989) pp 60c-src activation in human colon carcinoma. J Clin Invest 83(6):2025–2033
Cary LA, Chang JF et al (1996) Stimulation of cell migration by overexpression of focal adhesion kinase and its association with Src and Fyn. J Cell Sci 109(Pt 7):1787–1794
Castano J, Solanas G et al (2007) Specific phosphorylation of p120-catenin regulatory domain differently modulates its binding to RhoA. Mol Cell Biol 27(5):1745–1757
Chapman RS, Lourenco PC et al (1999) Suppression of epithelial apoptosis and delayed mammary gland involution in mice with a conditional knockout of Stat3. Genes Dev 13(19):2604–2616
Chen T, Pengetnze Y et al (2005) Src inhibition enhances paclitaxel cytotoxicity in ovarian cancer cells by caspase-9-independent activation of caspase-3. Mol Cancer Ther 4(2):217–224
Chen WT, Chen JM et al (1985) Local degradation of fibronectin at sites of expression of the transforming gene product pp 60src. Nature 316(6024):156–158
Chong YP, Mulhern TD et al (2005) C-terminal Src kinase (CSK) and CSK-homologous kinase (CHK) – endogenous negative regulators of Src-family protein kinases. Growth Factors 23(3):233–244
Coll ML, Rosen K et al (2002) Increased Bcl-xL expression mediates v-Src-induced resistance to anoikis in intestinal epithelial cells. Oncogene 21(18):2908–2913
Criscuoli ML, Nguyen M et al (2005) Tumor metastasis but not tumor growth is dependent on Src-mediated vascular permeability. Blood 105(4):1508–1514
Cursi S, Rufini A et al (2006) Src kinase phosphorylates Caspase-8 on Tyr380: a novel mechanism of apoptosis suppression. EMBO J 25(9):1895–1905
Daigo Y, Furukawa Y et al (1999) Absence of genetic alteration at codon 531 of the human c-src gene in 479 advanced colorectal cancers from Japanese and Caucasian patients. Cancer Res 59(17):4222–4224
Davidson D, Chow LM et al (1997) Chk, a Csk family tyrosine protein kinase, exhibits Csk-like activity in fibroblasts, but not in an antigen-specific T-cell line. J Biol Chem 272(2):1355–1362
de Heer P, Koudijs MM et al (2008) Combined expression of the non-receptor protein tyrosine kinases FAK and Src in primary colorectal cancer is associated with tumor recurrence and metastasis formation. Eur J Surg Oncol 34(11):1253–1261
Di Stefano P, Damiano L et al (2007) p140Cap protein suppresses tumour cell properties, regulating Csk and Src kinase activity. EMBO J 26(12):2843–2855
Diaz N, Minton S et al (2006) Activation of stat3 in primary tumors from high-risk breast cancer patients is associated with elevated levels of activated SRC and survivin expression. Clin Cancer Res 12(1):20–28
Dimri M, Naramura M et al (2007) Modeling breast cancer-associated c-Src and EGFR overexpression in human MECs: c-Src and EGFR cooperatively promote aberrant three-dimensional acinar structure and invasive behavior. Cancer Res 67(9):4164–4172
Duxbury MS, Ito H et al (2004) Inhibition of SRC tyrosine kinase impairs inherent and acquired gemcitabine resistance in human pancreatic adenocarcinoma cells. Clin Cancer Res 10(7):2307–2318
Eichhorn PJ, Creyghton MP et al (2007) A RNA interference screen identifies the protein phosphatase 2A subunit PR55gamma as a stress-sensitive inhibitor of c-SRC. PLoS Genet 3(12):e218
Eliceiri BP, Paul R et al (1999) Selective requirement for Src kinases during VEGF-induced angiogenesis and vascular permeability. Mol Cell 4(6):915–924
Fincham VJ, Brunton VG et al (2000a) The SH3 domain directs acto-myosin-dependent targeting of v-Src to focal adhesions via phosphatidylinositol 3-kinase. Mol Cell Biol 20(17):6518–6536

Fincham VJ, James M et al (2000b) Active ERK/MAP kinase is targeted to newly forming cell-matrix adhesions by integrin engagement and v-Src. EMBO J 19(12):2911–2923
Fincham VJ, Unlu M et al (1996) Translocation of Src kinase to the cell periphery is mediated by the actin cytoskeleton under the control of the Rho family of small G proteins. J Cell Biol 135(6 Pt 1):1551–1564
Frame MC (2004) Newest findings on the oldest oncogene; how activated src does it. J Cell Sci 117(Pt 7):989–998
Frisch SM, Francis H (1994) Disruption of epithelial cell-matrix interactions induces apoptosis. J Cell Biol 124(4):619–626
Fujita Y, Krause G et al (2002) Hakai, a c-Cbl-like protein, ubiquitinates and induces endocytosis of the E-cadherin complex. Nat Cell Biol 4(3):222–231
Garcia R, Bowman TL et al (2001) Constitutive activation of Stat3 by the Src and JAK tyrosine kinases participates in growth regulation of human breast carcinoma cells. Oncogene 20(20): 2499–2513
Gavard J, Gutkind JS (2006) VEGF controls endothelial-cell permeability by promoting the beta-arrestin-dependent endocytosis of VE-cadherin. Nat Cell Biol 8(11):1223–1234
Gavard J, Patel V et al (2008) Angiopoietin-1 prevents VEGF-induced endothelial permeability by sequestering Src through mDia. Dev Cell 14(1):25–36
Gil-Henn H, Elson A (2003) Tyrosine phosphatase-epsilon activates Src and supports the transformed phenotype of Neu-induced mammary tumor cells. J Biol Chem 278(18):15579–15586
Gilmore AP, Metcalfe AD et al (2000) Integrin-mediated survival signals regulate the apoptotic function of Bax through its conformation and subcellular localization. J Cell Biol 149(2):431–446
Goi T, Shipitsin M et al (2000) An EGF receptor/Ral-GTPase signaling cascade regulates c-Src activity and substrate specificity. EMBO J 19(4):623–630
Gonzalez L, Agullo-Ortuno MT et al (2006) Role of c-Src in human MCF7 breast cancer cell tumorigenesis. J Biol Chem 281(30):20851–20864
Gould KL, Hunter T (1988) Platelet-derived growth factor induces multisite phosphorylation of pp 60c-src and increases its protein-tyrosine kinase activity. Mol Cell Biol 8(8):3345–3356
Granot-Attas S, Elson A (2004) Protein tyrosine phosphatase epsilon activates Yes and Fyn in Neu-induced mammary tumor cells. Exp Cell Res 294(1):236–243
Guo W, Pylayeva Y et al (2006) Beta 4 integrin amplifies ErbB2 signaling to promote mammary tumorigenesis. Cell 126(3):489–502
Guy CT, Muthuswamy SK et al (1994) Activation of the c-Src tyrosine kinase is required for the induction of mammary tumors in transgenic mice. Genes Dev 8(1):23–32
Ha CH, Bennett AM et al (2008) A novel role of vascular endothelial cadherin in modulating c-Src activation and downstream signaling of vascular endothelial growth factor. J Biol Chem 283(11):7261–7270
Han LY, Landen CN et al (2006) Antiangiogenic and antitumor effects of SRC inhibition in ovarian carcinoma. Cancer Res 66(17):8633–8639
Hazan RB, Qiao R et al (2004) Cadherin switch in tumor progression. Ann N Y Acad Sci 1014:155–163
Honda K, Sakaguchi T et al (2007) Epidermal hyperplasia and papillomatosis in mice with a keratinocyte-restricted deletion of csk. Carcinogenesis 28(10):2074–2081
Huck L, Pontier SM et al (2010) beta1-integrin is dispensable for the induction of ErbB2 mammary tumors but plays a critical role in the metastatic phase of tumor progression. Proc Natl Acad Sci USA 107(35):15559–15564
Hunter T, Sefton BM (1980) Transforming gene product of Rous sarcoma virus phosphorylates tyrosine. Proc Natl Acad Sci USA 77(3):1311–1315
Imamoto A, Soriano P (1993) Disruption of the csk gene, encoding a negative regulator of Src family tyrosine kinases, leads to neural tube defects and embryonic lethality in mice. Cell 73(6):1117–1124
Irby R, Mao W et al (1997) Overexpression of normal c-Src in poorly metastatic human colon cancer cells enhances primary tumor growth but not metastatic potential. Cell Growth Differ 8(12):1287–1295

Irby RB, Mao W et al (1999) Activating SRC mutation in a subset of advanced human colon cancers. Nat Genet 21(2):187–190
Jay G, Shiu RP et al (1978) Identification of a transformation-specific protein induced by a Rous sarcoma virus. Cell 13(3):527–534
Jiang LQ, Feng X et al (2006) Csk-binding protein (Cbp) negatively regulates epidermal growth factor-induced cell transformation by controlling Src activation. Oncogene 25(40):5495–5506
Kaminski R, Zagozdzon R et al (2006) Role of SRC kinases in Neu-induced tumorigenesis: challenging the paradigm using Csk homologous kinase transgenic mice. Cancer Res 66(11):5757–5762
Kaplan KB, Bibbins KB et al (1994) Association of the amino-terminal half of c-Src with focal adhesions alters their properties and is regulated by phosphorylation of tyrosine 527. EMBO J 13(20):4745–4756
Kaplan KB, Swedlow JR et al (1995) c-Src enhances the spreading of src–/– fibroblasts on fibronectin by a kinase-independent mechanism. Genes Dev 9(12):1505–1517
Kaplan KB, Swedlow JR et al (1992) Association of p60c-src with endosomal membranes in mammalian fibroblasts. J Cell Biol 118(2):321–333
Kasahara K, Nakayama Y et al (2007) Rapid trafficking of c-Src, a non-palmitoylated Src-family kinase, between the plasma membrane and late endosomes/lysosomes. Exp Cell Res 313(12):2651–2666
Klinghoffer RA, Sachsenmaier C et al (1999) Src family kinases are required for integrin but not PDGFR signal transduction. EMBO J 18(9):2459–2471
Kmiecik TE, Shalloway D (1987) Activation and suppression of pp 60c-src transforming ability by mutation of its primary sites of tyrosine phosphorylation. Cell 49(1):65–73
Kotelevets L, van Hengel J et al (2001) The lipid phosphatase activity of PTEN is critical for stabilizing intercellular junctions and reverting invasiveness. J Cell Biol 155(7):1129–1135
Kypta RM, Goldberg Y et al (1990) Association between the PDGF receptor and members of the src family of tyrosine kinases. Cell 62(3):481–492
Laghi L, Bianchi P et al (2001) Lack of mutation at codon 531 of SRC in advanced colorectal cancers from Italian patients. Br J Cancer 84(2):196–198
Lahlou H, Sanguin-Gendreau V et al (2007) Mammary epithelial-specific disruption of the focal adhesion kinase blocks mammary tumor progression. Proc Natl Acad Sci USA 104(51): 20302–20307
Lee LF, Guan J et al (2001) Neuropeptide-induced androgen independence in prostate cancer cells: roles of nonreceptor tyrosine kinases Etk/Bmx, Src, and focal adhesion kinase. Mol Cell Biol 21(24):8385–8397
Li L, Guris DL et al (2003) Translocation of CrkL to focal adhesions mediates integrin-induced migration downstream of Src family kinases. Mol Cell Biol 23(8):2883–2892
Liang F, Liang J et al (2007) PRL3 promotes cell invasion and proliferation by down-regulation of Csk leading to Src activation. J Biol Chem 282(8):5413–5419
Liu X, Brodeur SR et al (1993) Regulation of c-Src tyrosine kinase activity by the Src SH2 domain. Oncogene 8(5):1119–1126
Lutz MP, Esser IB et al (1998) Overexpression and activation of the tyrosine kinase Src in human pancreatic carcinoma. Biochem Biophys Res Commun 243(2):503–508
Ma C, Rong Y et al (2008) Extracellular matrix protein betaig-h3/TGFBI promotes metastasis of colon cancer by enhancing cell extravasation. Genes Dev 22(3):308–321
Marcotte R, Zhou L et al (2009) c-Src associates with ErbB2 through an interaction between catalytic domains and confers enhanced transforming potential. Mol Cell Biol 29(21):5858–5871
Masaki T, Igarashi K et al (2003) pp 60c-src activation in lung adenocarcinoma. Eur J Cancer 39(10):1447–1455
Masaki T, Okada M et al (1998) pp 60c-src activation in hepatocellular carcinoma of humans and LEC rats. Hepatology 27(5):1257–1264
Masaki T, Okada M et al (1999) Reduced C-terminal Src kinase (Csk) activities in hepatocellular carcinoma. Hepatology 29(2):379–384
Matsumoto T, Jiang J et al (2003) Targeted expression of c-Src in epidermal basal cells leads to enhanced skin tumor promotion, malignant progression, and metastasis. Cancer Res 63(16):4819–4828

Matsumoto T, Kiguchi K et al (2004) Development of transgenic mice that inducibly express an active form of c-Src in the epidermis. Mol Carcinog 40(4):189–200
Mazurenko NN, Kogan EA et al (1992) Expression of pp 60c-src in human small cell and non-small cell lung carcinomas. Eur J Cancer 28(2–3):372–377
McLachlan RW, Kraemer A et al (2007) E-cadherin adhesion activates c-Src signaling at cell-cell contacts. Mol Biol Cell 18(8):3214–3223
Muthuswamy SK, Muller WJ (1994) Activation of the Src family of tyrosine kinases in mammary tumorigenesis. Adv Cancer Res 64:111–123
Muthuswamy SK, Muller WJ (1995) Activation of Src family kinases in Neu-induced mammary tumors correlates with their association with distinct sets of tyrosine phosphorylated proteins in vivo. Oncogene 11(9):1801–1810
Nada S, Yagi T et al (1993) Constitutive activation of Src family kinases in mouse embryos that lack Csk. Cell 73(6):1125–1135
Nagle JA, Ma Z et al (2004) Involvement of insulin receptor substrate 2 in mammary tumor metastasis. Mol Cell Biol 24(22):9726–9735
Nilbert M, Fernebro E (2000) Lack of activating c-SRC mutations at codon 531 in rectal cancer. Cancer Genet Cytogenet 121(1):94–95
Niu G, Wright KL et al (2002) Constitutive Stat3 activity up-regulates VEGF expression and tumor angiogenesis. Oncogene 21(13):2000–2008
Nowak D, Boehrer S et al (2007) Src kinase inhibitors induce apoptosis and mediate cell cycle arrest in lymphoma cells. Anticancer Drugs 18(9):981–995
O'Rourke DM, Nute EJ et al (1998) Inhibition of a naturally occurring EGFR oncoprotein by the p185neu ectodomain: implications for subdomain contributions to receptor assembly. Oncogene 16(9):1197–1207
Oppermann H, Levinson AD et al (1979) Uninfected vertebrate cells contain a protein that is closely related to the product of the avian sarcoma virus transforming gene (src). Proc Natl Acad Sci USA 76(4):1804–1808
Ottenhoff-Kalff AE, Rijksen G et al (1992) Characterization of protein tyrosine kinases from human breast cancer: involvement of the c-src oncogene product. Cancer Res 52(17):4773–4778
Pengetnze Y, Steed M et al (2003) Src tyrosine kinase promotes survival and resistance to chemotherapeutics in a mouse ovarian cancer cell line. Biochem Biophys Res Commun 309(2):377–383
Ponniah S, Wang DZ et al (1999) Targeted disruption of the tyrosine phosphatase PTPalpha leads to constitutive downregulation of the kinases Src and Fyn. Curr Biol 9(10):535–538
Pontier SM, Huck L et al (2010) Integrin-linked kinase has a critical role in ErbB2 mammary tumor progression: implications for human breast cancer. Oncogene 29(23):3374–3385
Purchio AF, Erikson E et al (1978) Identification of a polypeptide encoded by the avian sarcoma virus src gene. Proc Natl Acad Sci USA 75(3):1567–1571
Radke K, Gilmore T et al (1980) Transformation by Rous sarcoma virus: a cellular substrate for transformation-specific protein phosphorylation contains phosphotyrosine. Cell 21(3):821–828
Ralston R, Bishop JM (1985) The product of the protooncogene c-src is modified during the cellular response to platelet-derived growth factor. Proc Natl Acad Sci USA 82(23):7845–7849
Ranger JJ, Levy DE et al (2009) Identification of a Stat3-dependent Transcription regulatory Network involved in metastatic progression. Cancer Res 69(17):6823
Rengifo-Cam W, Konishi A et al (2004) Csk defines the ability of integrin-mediated cell adhesion and migration in human colon cancer cells: implication for a potential role in cancer metastasis. Oncogene 23(1):289–297
Rickles RJ, Botfield MC et al (1994) Identification of Src, Fyn, Lyn, PI3K and Abl SH3 domain ligands using phage display libraries. EMBO J 13(23):5598–5604
Rickles RJ, Botfield MC et al (1995) Phage display selection of ligand residues important for Src homology 3 domain binding specificity. Proc Natl Acad Sci USA 92(24):10909–10913
Roskoski R Jr (2005) Src kinase regulation by phosphorylation and dephosphorylation. Biochem Biophys Res Commun 331(1):1–14
Saha S, Bardelli A et al (2001) A phosphatase associated with metastasis of colorectal cancer. Science 294(5545):1343–1346

Sandilands E, Akbarzadeh S et al (2007) Src kinase modulates the activation, transport and signalling dynamics of fibroblast growth factor receptors. EMBO Rep 8(12):1162–1169

Sandilands E, Cans C et al (2004) RhoB and actin polymerization coordinate Src activation with endosome-mediated delivery to the membrane. Dev Cell 7(6):855–869

Schedin P (2006) Pregnancy-associated breast cancer and metastasis. Nat Rev Cancer 6(4):281–291

Schedin P, O'Brien J et al (2007) Microenvironment of the involuting mammary gland mediates mammary cancer progression. J Mammary Gland Biol Neoplasia 12(1):71–82

Schlaepfer DD, Hunter T (1996) Evidence for in vivo phosphorylation of the Grb2 SH2-domain binding site on focal adhesion kinase by Src-family protein-tyrosine kinases. Mol Cell Biol 16(10):5623–5633

Schlessinger K, Levy DE (2005) Malignant transformation but not normal cell growth depends on signal transducer and activator of transcription 3. Cancer Res 65(13):5828–5834

Sefton BM, Hunter T et al (1980a) Relationship of polypeptide products of the transforming gene of Rous sarcoma virus and the homologous gene of vertebrates. Proc Natl Acad Sci USA 77(4):2059–2063

Sefton BM, Hunter T et al (1980b) Evidence that the phosphorylation of tyrosine is essential for cellular transformation by Rous sarcoma virus. Cell 20(3):807–816

Shor AC, Keschman EA et al (2007) Dasatinib inhibits migration and invasion in diverse human sarcoma cell lines and induces apoptosis in bone sarcoma cells dependent on SRC kinase for survival. Cancer Res 67(6):2800–2808

Sirvent A, Boureux A et al (2007) The tyrosine kinase Abl is required for Src-transforming activity in mouse fibroblasts and human breast cancer cells. Oncogene 26(52):7313–7323

Slack JK, Adams RB et al (2001) Alterations in the focal adhesion kinase/Src signal transduction pathway correlate with increased migratory capacity of prostate carcinoma cells. Oncogene 20(10):1152–1163

Songyang Z, Shoelson SE et al (1993) SH2 domains recognize specific phosphopeptide sequences. Cell 72(5):767–778

Soriano P, Montgomery C et al (1991) Targeted disruption of the c-src proto-oncogene leads to osteopetrosis in mice. Cell 64(4):693–702

Spector DH, Varmus HE et al (1978) Nucleotide sequences related to the transforming gene of avian sarcoma virus are present in DNA of uninfected vertebrates. Proc Natl Acad Sci USA 75(9):4102–4106

Stein PL, Vogel H et al (1994) Combined deficiencies of Src, Fyn, and Yes tyrosine kinases in mutant mice. Genes Dev 8(17):1999–2007

Stein T, Morris JS et al (2004) Involution of the mouse mammary gland is associated with an immune cascade and an acute-phase response, involving LBP, CD14 and STAT3. Breast Cancer Res 6(2):R75–R91

Stover DR, Becker M et al (1995) Src phosphorylation of the epidermal growth factor receptor at novel sites mediates receptor interaction with Src and P85 alpha. J Biol Chem 270(26):15591–15597

Stover DR, Furet P et al (1996) Modulation of the SH2 binding specificity and kinase activity of Src by tyrosine phosphorylation within its SH2 domain. J Biol Chem 271(21):12481–12487

Su J, Muranjan M et al (1999) Receptor protein tyrosine phosphatase alpha activates Src-family kinases and controls integrin-mediated responses in fibroblasts. Curr Biol 9(10):505–511

Su JL, Yang PC et al (2006) The VEGF-C/Flt-4 axis promotes invasion and metastasis of cancer cells. Cancer Cell 9(3):209–223

Sugimura M, Kobayashi K et al (2000) Mutation of the SRC gene in endometrial carcinoma. Jpn J Cancer Res 91(4):395–398

Talamonti MS, Roh MS et al (1993) Increase in activity and level of pp 60c-src in progressive stages of human colorectal cancer. J Clin Invest 91(1):53–60

Tan M, Li P et al (2005) ErbB2 promotes Src synthesis and stability: novel mechanisms of Src activation that confer breast cancer metastasis. Cancer Res 65(5):1858–1867

Tan M, Yao J et al (1997) Overexpression of the c-erbB-2 gene enhanced intrinsic metastasis potential in human breast cancer cells without increasing their transformation abilities. Cancer Res 57(6):1199–1205

Trevino JG, Summy JM et al (2006) Inhibition of SRC expression and activity inhibits tumor progression and metastasis of human pancreatic adenocarcinoma cells in an orthotopic nude mouse model. Am J Pathol 168(3):962–972

Ursini-Siegel J, Hardy WR et al (2008) ShcA signalling is essential for tumour progression in mouse models of human breast cancer. EMBO J 27(6):910–920

Vadlamudi RK, Sahin AA et al (2003) Heregulin and HER2 signaling selectively activates c-Src phosphorylation at tyrosine 215. FEBS Lett 543(1–3):76–80

Veracini L, Franco M et al (2006) Two distinct pools of Src family tyrosine kinases regulate PDGF-induced DNA synthesis and actin dorsal ruffles. J Cell Sci 119(Pt 14):2921–2934

Watson CJ (2009) Immune cell regulators in mouse mammary development and involution. J Anim Sci 87(13 Suppl):35–42

Webster MA, Cardiff RD et al (1995) Induction of mammary epithelial hyperplasias and mammary tumors in transgenic mice expressing a murine mammary tumor virus/activated c-src fusion gene. Proc Natl Acad Sci USA 92(17):7849–7853

White DE, Kurpios NA et al (2004) Targeted disruption of beta1-integrin in a transgenic mouse model of human breast cancer reveals an essential role in mammary tumor induction. Cancer Cell 6(2):159–170

Wiener JR, Windham TC et al (2003) Activated SRC protein tyrosine kinase is overexpressed in late-stage human ovarian cancers. Gynecol Oncol 88(1):73–79

Wilde A, Beattie EC et al (1999) EGF receptor signaling stimulates SRC kinase phosphorylation of clathrin, influencing clathrin redistribution and EGF uptake. Cell 96(5):677–687

Windham TC, Parikh NU et al (2002) Src activation regulates anoikis in human colon tumor cell lines. Oncogene 21(51):7797–7807

Woods NT, Yamaguchi H et al (2007) Anoikis, initiated by Mcl-1 degradation and Bim induction, is deregulated during oncogenesis. Cancer Res 67(22):10744–10752

Xu W, Yuan X et al (2007) Loss of Hsp90 association up-regulates Src-dependent ErbB2 activity. Mol Cell Biol 27(1):220–228

Yagi R, Waguri S et al (2007) C-terminal Src kinase controls development and maintenance of mouse squamous epithelia. EMBO J 26(5):1234–1244

Yamaguchi H, Woods NT et al (2008) SRC directly phosphorylates Bif-1 and prevents its interaction with Bax and the initiation of anoikis. J Biol Chem 283(27):19112–19118

Yezhelyev MV, Koehl G et al (2004) Inhibition of SRC tyrosine kinase as treatment for human pancreatic cancer growing orthotopically in nude mice. Clin Cancer Res 10(23):8028–8036

Zhang SQ, Yang W et al (2004) Shp2 regulates SRC family kinase activity and Ras/Erk activation by controlling Csk recruitment. Mol Cell 13(3):341–355

Zheng X, Resnick RJ et al (2008) Apoptosis of estrogen-receptor negative breast cancer and colon cancer cell lines by PTP alpha and src RNAi. Int J Cancer 122(9):1999–2007

Zheng XM, Wang Y et al (1992) Cell transformation and activation of pp 60c-src by overexpression of a protein tyrosine phosphatase. Nature 359(6393):336–339

Zhou X, Coad J et al (2008) SHP2 is up-regulated in breast cancer cells and in infiltrating ductal carcinoma of the breast, implying its involvement in breast oncogenesis. Histopathology 53(4):389–402

Zhu S, Bjorge JD et al (2008) Decreased CHK protein levels are associated with Src activation in colon cancer cells. Oncogene 27(14):2027–2034

Zrihan-Licht S, Deng B et al (1998) Csk homologous kinase, a novel signaling molecule, directly associates with the activated ErbB-2 receptor in breast cancer cells and inhibits their proliferation. J Biol Chem 273(7):4065–4072

Zrihan-Licht S, Lim J et al (1997) Association of csk-homologous kinase (CHK) (formerly MATK) with HER-2/ErbB-2 in breast cancer cells. J Biol Chem 272(3):1856–1863

Chapter 17
Maspin and Suppression of Tumor Metastasis

Lauren Reinke and Ming Zhang

17.1 Introduction

Maspin is a member of the family of *Ser*ine *P*rotease *In*hibitors or serpins. It was first reported in 1994 as a tumor-suppressing serpin, and was named the *Ma*mmary *Serpin* after its source of identification, the mammary epithelia (Zou et al. 1994). Subsequent research resulted in maspin's characterization as a potent tumor suppressor of both breast and prostate cancers due to its ability to inhibit tumor cell invasion, promote apoptosis, and prevent angiogenesis (Jiang et al. 2002; Latha et al. 2005; Sheng et al. 1996; Zhang et al. 2000b). Maspin is highly expressed in normal mammary and prostate epithelial cells, but is reduced or absent in breast and prostate cancer tissue samples and cell lines (Hall et al. 2008; Sheng et al. 1996; Zou et al. 1994). Current research strives to elucidate the mechanisms underlying maspin's tumor suppressive properties. Since tumor progression, particularly metastasis, is best studied in animal models, several mouse models have been generated and utilized to produce a wealth of information regarding the complex roles that maspin plays in both tumor progression and normal development. By continuing to take advantage of animal models as powerful research tools, investigators will be able to further enhance our understanding of maspin's roles in these crucial processes. Additionally, such studies can be utilized to develop maspin's potential as both a diagnostic marker for cancer progression and an attractive target for gene therapy in the fight against breast and prostate cancers.

L. Reinke • M. Zhang (✉)
Department of Molecular Pharmacology and Biological Chemistry,
Northwestern University Feinberg School of Medicine, Robert H. Lurie
Comprehensive Cancer Center, 303 E. Superior Street, Chicago, IL 60611, USA
e-mail: m-zhang@northwestern.edu

J.E. Green and T. Ried (eds.), *Genetically Engineered Mice for Cancer Research: Design, Analysis, Pathways, Validation and Pre-Clinical Testing*,
DOI 10.1007/978-0-387-69805-2_17, © Springer Science+Business Media, LLC 2012

17.2 Classification and Structure

Serpins comprise a diverse protein family with numerous biological functions. Within this family, maspin has been grouped with the ov-serpin subfamily, as it exhibits significant sequence similarity to chicken ovalbumin (Zou et al. 1994). Maspin is unique among human ov-serpins in that it is classified as a noninhibitory serpin, as there are currently no known target proteases for maspin. Maspin's unique structure further supports its classification as a noninhibitory serpin. Some of these key structural differences are described below.

The crystal structure of maspin reveals that it adopts the typical native serpin fold, as it comprises three antiparallel β sheets (A, B, and C) surrounded by nine α helices (hA–hI) (Law et al. 2005). However, its reactive site loop (RSL) is different than that of most serpins in terms of length, composition, and placement. All true inhibitory serpins accomplish protease inhibition via a conserved RSL. In these instances, the target protease binds to and cleaves the serpin's exposed RSL, thus becoming incorporated into a serpin-protease complex after a large-scale conformational transformation (Domann et al. 2000). This action is facilitated by a partial opening "breach" between strands 3 and 5 of β-sheet A into which the RSL may be partially inserted in the native state. Protease inhibition is thought to arise from the ability of the RSL to fully insert between strands 3 and 5 of β-sheet A, dragging the target protease with it and conformationally "crushing" it against the body of the molecule (Huntington et al. 2000). This dramatic conformational change is known as the stressed (S) to relaxed (R) transition and mutations within a typical serpin's RSL that interfere with this transition abolish or seriously compromise inhibitory activity (Law et al. 2005). The solved crystal structures of maspin indicate that the "breach" between the third and fifth strands of β-sheet A is closed and the RSL is expelled fully from the sheet, possibly denying access to the RSL following interaction with any potential target protease. Consequently, maspin does not undergo the S to R transition and is therefore unable to inactivate proteases in a typical serpin-like manner. Rather, a high resolution crystal structure revealed that maspin is capable of undergoing a novel major conformational change in and around the G α-helix, enabling the molecule to switch between an open and closed conformation. This rearrangement exposes two negatively charged glutamate residues that contribute to a newly formed elongated, negatively charged patch. Pigment epithelium-derived factor (PEDF), a related ov-serpin known to inhibit angiogenesis, possesses a comparable region that has been implicated in collagen binding. These findings support previous reports that maspin binds to types I and III collagen (Blacque and Worrall 2002) and further substantiate its role in preventing angiogenesis.

Studies demonstrate that maspin is able to self-associate in a manner indicative of polymerization, though the relevance of this is currently unknown. Crystal structure indicates that such polymerization may occur by a unique mechanism involving dimers of tetrameric maspin formed by intermolecular contacts predominantly via the RSLs (Al-Ayyoubi et al. 2004). Polymerization is dependent upon the presence of the RSL, but not its integrity (Al-Ayyoubi et al. 2004; Pemberton et al. 1995).

Interestingly, research has demonstrated that the maspin RSL is necessary for some of its tumor suppressive functions, including the inhibition of tumor cell migration and invasion. The maspin RSL also plays a key role in maspin interactions at the cell surface. In addition, recent studies have demonstrated that the RSL is required for the role of secreted maspin in the inhibition of metastasis (Khalkhali-Ellis and Hendrix 2007).

Based on the fact that maspin is *downregulated* during the progression from a normal cell phenotype to a tumorigenic, invasive, and metastatic phenotype, it has been classified as a type II tumor suppressor molecule (Sager et al. 1994). This is in contrast to the genes classified as type I tumor suppressors due to the fact that they are typically *inactivated* in the development of cancer. The majority of tumor suppressor genes described to date are type I tumor suppressors that are typically inactivated due to "two hit" events, namely, the global loss of the region containing one allele (detected by loss of heterozygosity) and subtle mutation in the other. Conversely, type II tumor suppressors such as maspin are not expected to harbor any mutations within their coding regions. Instead, various regulatory events, including epigenetic modifications are responsible for the downregulation of maspin transcription that occurs in cancer development and progression. Importantly, this implies that it is highly unlikely that the protein contains disruptions in its structure or function that would contribute to the complexity of mutational phenotypes that exist for other diseases. Consequently, maspin is an attractive target for gene therapy in the treatment of cancer, as reactivation of the gene and replacement of the protein in the target cell population are both viable options.

17.3 Biological Functions

Maspin expression is limited to certain types of epithelial cells, including those of the breast, prostate, epidermis, and lung, as well as stromal cells of the cornea (Bailey et al. 2006). While maspin is predominantly cytoplasmic, it also localizes to other cellular compartments and is secreted (Khalkhali-Ellis and Hendrix 2007). Due to its broad localization pattern, maspin has been implicated in numerous pathways and processes involved in both tumor progression and normal development.

For example, maspin is able to inhibit cell invasion and metastasis in breast and prostate cancer cells (Bailey et al. 2006). Metastasis is a complex process involving cell migration, local invasion, intravasation and extravasation of the capillary vasculature, growth at secondary sites, and angiogenesis (Li et al. 2008). Metastasis is also the cause of the majority of breast cancer-related deaths and has thus been implicated as a key process in cancer progression and a critical target for cancer treatment and prevention. Therefore, a great deal of research has focused on elucidating the various mechanisms utilized by maspin to exert its antimetastatic effects. For example, studies have shown that maspin is able to alter the integrin profile of the cell, leading to increased adherence to fibronectin and a subsequent decreased ability to move through fibronectin matrices (Bailey et al. 2006). In addition, Odero and colleagues utilized highly invasive MDA-MB-231 breast cancer cells to demonstrate that signaling pathways involved in motility and invasion are modified upon treatment with recombinant maspin (Odero-Marah et al. 2003). Specifically, they showed that recombinant maspin stimulates a decrease in Rac1 activity accompanied by an increase in PI3K and ERK1/2 activities in these cells, which do not express endogenous maspin. Consequently, the maspin-treated cells exhibited an increase in cell adhesion and a decrease in cell motility (Odero-Marah et al. 2003; Shi et al. 2007). Additionally, the cells displayed morphological changes indicative of a reversion to a more epithelial-like phenotype, including increased focal adhesions and stress fibers. Other studies by Shi et al. further supported these findings by demonstrating that maspin controls cell migration by regulating Rho GTPase signaling pathways (Shi et al. 2007).

Another anti-invasive mechanism employed by maspin involves regulation of the urokinase plasminogen activator (uPA)/urokinase plasminogen activator receptor (uPAR) complex located on the cell surface. The uPA/uPAR complex plays a key role in the conversion of plasminogen to the active protease plasmin. Plasmin displays broad specificity and is able to cleave several extracellular matrix proteins, including fibronectin, fibrin, and laminin (Bailey et al. 2006; Domann et al. 2000). While maspin does not directly inhibit uPA activity, studies have demonstrated maspin's ability to reduce cell surface-associated uPA/uPAR by inducing its internalization in both prostate (Bailey et al. 2006; Biliran and Sheng 2001; Domann et al. 2000; Yin et al. 2006) and breast cancer cells (Amir et al. 2005; Bailey et al. 2006). Recent research by Yin et al. showed that maspin regulates uPA/uPAR activity through its binding to the pro-uPA zymogen, thereby inhibiting its activation (Bailey et al. 2006; Yin et al. 2006). Interestingly, this interaction is dependent upon an intact maspin RSL. These results imply that maspin plays a critical role in the regulation of extracellular matrix proteolysis, a key step in the metastatic cascade.

Maspin also functions as an effective inhibitor of angiogenesis. *In vitro*, maspin acts directly on cultured endothelial cells to impede their migration toward basic fibroblastic growth factor (bFGF) and vascular endothelial growth factor (VEGF), which serve as critical chemoattractants during angiogenesis (Zhang et al. 2000b). Additionally, maspin prevents endothelial cells from forming tubes in a matrigel assay. *In vivo*, purified maspin effectively inhibits neovascularization and reduces the density of tumor-associated microvessels (Zhang et al. 2000b). Importantly, maspin's anti-angiogenic properties are not dependent upon an intact RSL: recombinant maspin proteins with mutated RSL domains that are unable to block the migration of other cells still maintain the ability to inhibit endothelial cell migration and mitogenesis. Additionally, maspin mutants with defective RSLs also retain the ability to inhibit neovascularization *in vivo* (Zhang et al. 2000b). These findings suggest that maspin employs a different mechanism to exert its antimigratory effect on endothelial cells than it does to impede the invasion of breast cancer cells, a process that requires an intact RSL. In support of this model, results from a clinical study indicate that both the expression and localization of maspin play a role in ovarian cancer angiogenesis and progression (Solomon et al. 2006). Specifically, the study revealed that in patients with advanced ovarian serous carcinoma, high nuclear expression of maspin was associated with prolonged survival and reduced angiogenesis markers, such as VEGF. Thus intracellular maspin, specifically maspin localized to the nucleus, may function to inhibit angiogenesis independent of the antimetastatic effects exerted by cell surface-associated maspin.

In addition to its antimetastatic and anti-angiogenic properties, maspin also promotes apoptosis, or programmed cell death. This was first demonstrated in studies that utilized a transgenic mouse model in which maspin is targeted to mammary epithelial cells by the whey acidic protein (WAP) promoter for overexpression (Zhang et al. 1999). These WAP-maspin transgenic mice were crossed with a strain of oncogenic WAP-Simian Virus 40 large T-antigen (TAg) mice to investigate whether increased levels of maspin protect against tumor progression *in vivo*. Maspin overexpression reduced tumor growth and was accompanied by an increased rate of apoptosis of both preneoplastic and carcinomatous mammary epithelial cells. Subsequent studies have shown that maspin sensitizes breast cancer cells to staurosporine and serum-deprivation induced apoptosis (Jiang et al. 2002; Latha et al. 2005). It is likely that this effect is due to maspin's activation of caspase pathways, as maspin-expressing tumor cells exhibit a reduced level of anti-apoptotic Bcl-2 protein and an increased level of pro-apoptotic Bax protein (Zhang et al. 2005). Specifically, maspin alters the delicate balance of these proteins via selective control of Bcl-2 and Bax stability. It is important to note that intracellular maspin, but not secreted maspin, is necessary to sensitize breast cancer cells to apoptosis (Bailey et al. 2006; Jiang et al. 2002; Latha et al. 2005), again highlighting the fact that maspin's tumor-suppressive properties are intimately linked to its localization. Maspin is also capable of inducing apoptosis in other cell types. For example, Li and colleagues demonstrated that when maspin overexpression is targeted to endothelial cells *in vivo*, it actively induces endothelial cell apoptosis (Li et al. 2005). In this particular study, the intravascular administration of adenovirus-maspin to mice bearing mammary tumors disrupted tumor-induced angiogenesis by promoting endothelial cell apoptosis. These findings present an elegant example of the ways in which maspin's multiple tumor-suppressive properties can function in concert to inhibit cancer progression.

Until recently, the fate and function of secreted maspin was largely unknown. The secretion of the protein appears dependent upon an uncleaved facultative secretion signal encoded by the first 50 amino acids. The Zhang Lab showed that secreted maspin acts on the surface of mammary epithelial cells to promote cell adhesion and that this effect is dependent on integrin (Cella et al. 2006). Recent studies by Hendrix and colleagues indicate that normal mammary epithelial cells secrete maspin that is incorporated into the extracellular matrix (Khalkhali-Ellis and Hendrix 2007). These findings imply that maspin may potentially regulate mammary matrix remodeling occurring under both normal and pathologic conditions. Significantly, one of the hallmarks of cancer progression is the altered deposition and degradation of the extracellular matrix. Presumably, cancer cells and the surrounding stromal environment secrete proteolytic enzymes that are believed to degrade the ECM, thus facilitating cancer cell invasion and metastasis (Liotta et al. 1983). One of the enzymes implicated in this process is the lysosomal aspartyl protease cathepsin D

(Capony et al. 1989; Ren and Sloane 1996). In pathologic conditions such as cancer, cathepsin D is excessively produced and aberrantly secreted (Capony et al. 1989; Liaudet-Coopman et al. 2006), and its plasma concentration is elevated in patients with metastatic breast cancer (Brouillet et al. 1997; Liaudet-Coopman et al. 2006). Hendrix and colleagues demonstrated that the presence of maspin in a collagen matrix prevents matrix incorporation of cathepsin D, thus reducing matrix degradation (Khalkhali-Ellis and Hendrix 2007). Additionally, their studies showed that an intact maspin RSL is necessary to reduce cathepsin D-mediated matrix degradation, as mutations in the RSL diminish this inhibitory effect. This finding further highlights the important role that an intact RSL plays in maspin's antimetastatic properties, including the regulation of adhesion, migration, and motility.

To further investigate maspin's multiple tumor-suppressive functions at the molecular level, several laboratories have utilized a maspin-baited yeast two-hybrid system in an effort to identify its intracellular binding partners (Bailey et al. 2005, 2006; Yin et al. 2005). Although various candidate proteins that play diverse roles in numerous cellular processes have been identified, several common themes have emerged. For example, two enzymes of the glutathione (GSH) redox system have been pinpointed as putative maspin-binding proteins, suggesting that maspin may play an active role in maintaining cellular homeostasis and responding to cellular stress. These proteins, glutathione S-transferase (GST) (Bailey et al. 2006; Yin et al. 2005) and glutathione peroxidase (Gpx) (Bailey et al. 2006) (Bailey, unpublished observations), are key components of the GSH redox system, one of the cell's primary defense mechanisms against antioxidants. Importantly, alterations to the activity of enzymes involved in this system may be indicative of reduced cellular defense that can lead to the development of numerous diseases, including cancer (Bailey et al. 2006; Rahman et al. 1999). Yin and colleagues demonstrated that through a direct protein–protein interaction maspin is able to enhance GST activity, thus protecting the cell against oxidation-induced damage. Furthermore, by augmenting GST activity, maspin may play a role in reducing the amount of reactive oxygen species (ROS) generated in response to oxidative stress (Bailey et al. 2006). Due to the fact that the generation of ROS is linked to hypoxia-induced angiogenesis, this may represent another mechanism by which maspin inhibits angiogenesis.

Yeast two-hybrid studies also identified various transcription factors as putative maspin-interacting partners, implying that maspin may exert a wide range of transcriptional control in the cell. Interestingly, several of these transcription factors have been linked to stress response, indicating that maspin may also utilize transcriptional regulation as a means of mediating responses to cellular stress (Bailey et al. 2006). For example, maspin was shown to interact with GC-binding factor 2 (GCF2), a transcriptional repressor whose expression is induced following tissue injury (Bailey et al. 2006). In addition to transcriptional regulators involved in stress response, the broad-range transcription factor TFIID was also identified as a putative maspin-interacting protein. Taken together, these binding partners suggest a wide-ranging role for maspin in transcriptional regulation, particularly in transcriptional repression.

Consistent with these observations, several lines of evidence point to a role for maspin in the regulation of chromatin remodeling. For example, Bailey et al. utilized microarray analysis of MDA-MB-231 cells to demonstrate that maspin reexpression results in the upregulation of SMARCA2, a SWI/SNF-related gene involved in chromatin remodeling (Bailey et al. 2006). Additionally, the aforementioned yeast two-hybrid study conducted by Yin et al. suggested that maspin may interact with the major histone deacetylase, HDAC1. Subsequently, Sheng and colleagues confirmed this interaction and provided the first evidence that maspin exerts an inhibitory effect on HDAC1 (Li et al. 2006). HDAC1 plays a critical role in transcriptional repression by facilitating histone deacetylation that leads to chromatin condensation, thus preventing the access of transcription factors to DNA. Notably, HDAC1-mediated epigenetics play an integral role in both development (Pillai et al. 2004) and tumor progression (Kelly and Marks 2005). In addition, Sheng and his colleagues also demonstrated that maspin's inhibitory effect on HDAC1 seems to depend on GST, providing an additional link between maspin and the cellular stress response. GST utilizes glutathione to catalyze the hydrogenation of sulfide groups or disulfide bridges. Previous studies indicate that the active HDAC1 conformation is induced by a divalent cation, preferably Zn^{2+}, that binds to the sulfide group of oxidized cysteine residues present at the HDAC1 catalytic site (Sherman et al. 1999). Thus, GST may be able to inactivate HDAC1 by mediating the sulfide reduction required to displace the metal ion. Given maspin's ability to interact with both of these proteins, it seems plausible that maspin may exert its inhibitory effects on HDAC1 by enhancing GST activity. Thus, loss of maspin expression could result in both an impairment of the cell's response to stress such as oxidative damage, as well as aberrant transcriptional silencing due to misregulated chromatin conformation. Indeed, Sheng et al. demonstrated that HDAC1 is upregulated as maspin is downregulated in prostate tumor tissues (Li et al. 2006). Based on their data, the authors suggested that this inverse correlation in cancerous cells might result in epigenetic changes that favor dedifferentiation and apoptosis-resistance, thus promoting tumor progression and metastasis.

In further support of maspin's role as a transcriptional regulator, yeast two-hybrid studies by Bailey and colleagues also identified Interferon Regulatory Factor 6 (IRF6) as a maspin-interacting protein (Bailey et al. 2005). IRF6 belongs to the IRF family of transcription factors, a group of proteins that has been studied primarily in the context of innate immunity. These analyses have revealed that IRFs function through toll-like receptor signaling to control the activation of type I interferons, as well as to help facilitate NF-κB activation and signaling in response to pathogenic stimuli (Tsujimura et al. 2004). Significantly, aside from their role in host defense, IRFs have been implicated in apoptosis, tumor suppression, and embryonic development (Bailey et al. 2006; Kondo et al. 2002; Tanaka et al. 1994). Bailey and colleagues reported that the maspin-IRF6 interaction occurs within the cytoplasmic compartment in normal mammary epithelial cells (Bailey et al. 2006). Furthermore, they demonstrated that IRF6 levels decrease during breast cancer progression in a manner similar to that observed for maspin, implying that both genes may be under the control of similar regulatory pathways. The authors also showed that reexpressing IRF6 in MDA-MB-231 cells results in morphological changes associated with alterations in the expression and localization of proteins involved in cellular architecture and phenotype, such as vimentin and N-cadherin. However, when maspin was co-expressed with IRF6 in these cells, such phenotypic changes did not occur, suggesting that maspin is able to effectively inhibit IRF6 function. The authors have therefore proposed a model in which the maspin-IRF6 interaction sequesters IRF6 in the cytoplasm, thus preventing its translocation to the nucleus. In support of this model, the serpin plasminogen activator inhibitor 2 (PAI-2) has been shown to negatively regulate IRF3 activity in a similar manner. As Bailey et al. report, it is significant to note that maspin interacts with IRF6 via the conserved IRF association domain (IAD), indicating that maspin may be able to exert regulatory effects on additional transcription factors belonging to the nine-member IRF family. As these proteins are intimately involved in mediating the cell's response to pathogenic insult, such interactions would represent another mechanism through which maspin plays a critical role in the cellular stress response.

Clearly, maspin is a multifaceted protein that is capable of playing several key roles in regulating metastasis (see Fig. 17.1). However, it is important to note that although maspin acts as a potent inhibitor of metastasis in both breast and prostate cancer development, its expression is maintained during ovarian, lung, and pancreatic carcinogenesis (Affara and Coussens 2007). These seemingly conflicting data imply that maspin may act in a tissue-specific manner to regulate metastatic potential. Investigations that aim to better understand maspin's tissue-specific activity are currently underway.

17.4 Regulation of Maspin Gene Expression

The proper regulation of maspin is critical for its role in normal development as well as its numerous tumor-suppressive functions in adult tissues. Several methods of maspin regulation have been identified, including DNA methylation and chromatin organization, both positive and negative transcriptional regulation, and regulation by various compounds, such as Tamoxifen and nitric oxide. These findings indicate that a complex array of processes is at work to ensure appropriate maspin expression. Misregulation can lead to decreased maspin expression, thus triggering or potentiating cancer progression. Therefore, these abundant pathways represent attractive therapeutic targets for the treatment and prevention of breast and pancreatic cancer.

Initial studies indicated that the differential maspin expression in normal versus carcinoma-derived mammary epithelial cells was regulated at the transcriptional level (Domann et al. 2000). Specifically, these observations showed that Ets and Ap1 sites in the maspin promoter that are active in regulating its expression in normal mammary epithelial cells are inactive in tumor cells (Zhang et al. 1997a). Consequently, the enhancing function by Ets and AP1 is decreased in tumor cells and abolished in invasive tumor cells. While these findings demonstrated that the loss of maspin expression during tumor progression results at least in part from the absence of transactivation through Ets and Ap1 sites, the exact mechanism behind

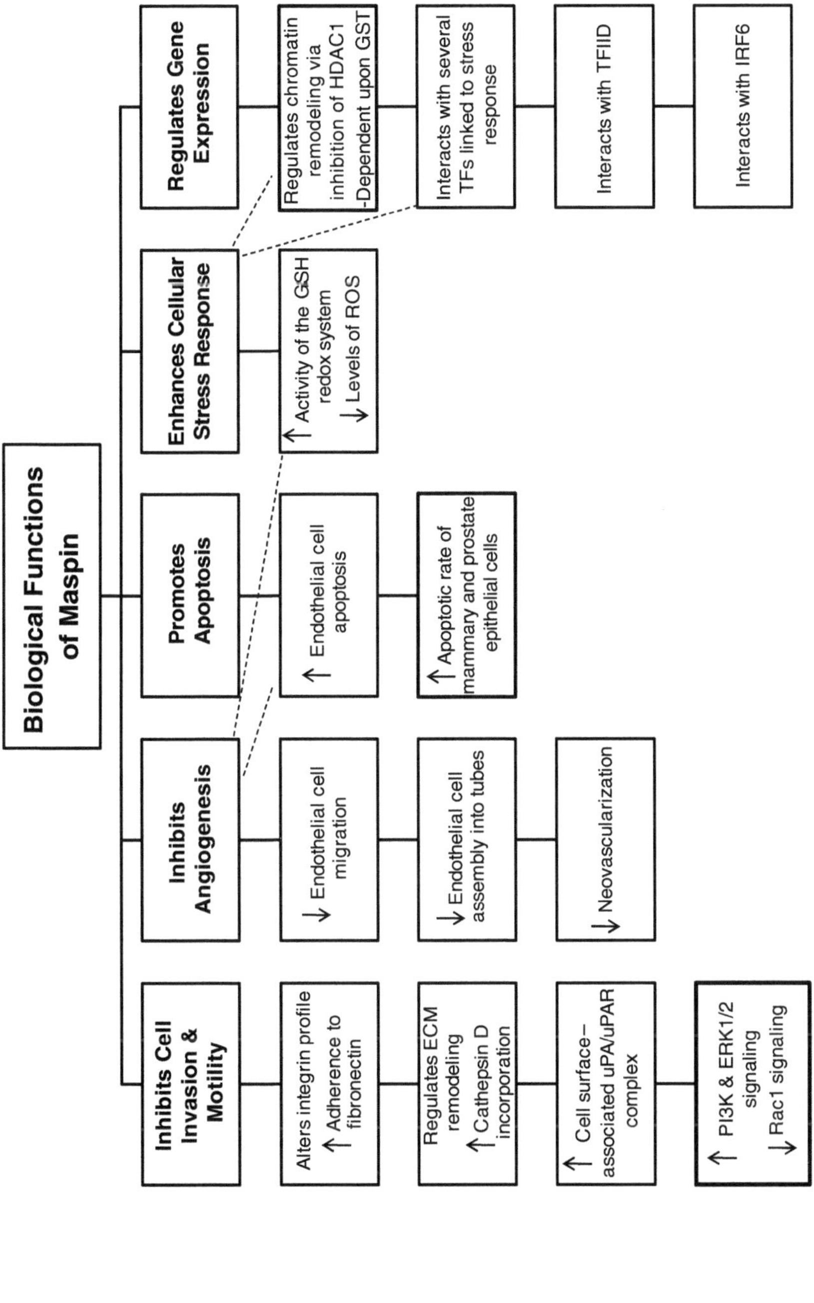

Fig. 17.1 Summary of the known biological functions of maspin

this process was unclear. In addition, previous studies by Zou and colleagues had identified p53 signaling as a regulatory mechanism for maspin expression (Zou et al. 2000). p53 is well-established as a critical tumor suppressor with myriad roles, including cell cycle regulation, apoptosis, and angiogenesis. Zou et al. were able to demonstrate that adenoviral delivery of wild-type p53 to breast and prostate cancer cell lines could induce maspin expression. Additionally, they identified a putative p53 consensus-binding site in the maspin promoter by using a reporter construct driven by varying regions of the promoter. By pinpointing maspin as a downstream target of p53, the authors provided evidence that the role of p53 in cancer cell invasion and metastasis may be due to its ability to regulate maspin expression. They subsequently hypothesized that cancer cells expressing mutant p53 may be more likely to metastasize, partially due to the inability of these cells to upregulate maspin expression.

A groundbreaking study by Futscher and his colleagues shed some light on this situation by providing evidence that cytosine methylation of the maspin promoter plays a crucial role in the establishment and maintenance of normal cell type-specific maspin expression (Futscher et al. 2002). This is evidenced by the fact that in normal cells expressing maspin, the promoter is unmethylated and the promoter region has acetylated histones and an accessible chromatin structure. Conversely, normal cells that do not express maspin possess a completely methylated promoter that is occupied by hypoacetylated histones, resulting in an inaccessible chromatin structure and the repression of transcription (see Chapter 18). Notably, this repression can be relieved by inhibiting DNA methylation. DNA methylation also plays a critical role in controlling maspin expression during cancer progression. This is highlighted by the fact that aberrant methylation of the maspin promoter is associated with the silencing of maspin gene expression in breast cancer cells (Domann et al. 2000). Futscher and colleagues thus present a model for maspin regulation that beautifully integrates the methylation state of the maspin promoter region with the previously observed transcriptional activation of maspin expression by various transcription factors, including AP1 and p53. According to Futcher's model, cells that express maspin possess an unmethylated promoter that is occupied by AP1 and p53. Additionally, the histones are acetylated, providing the open chromatin structure required for maspin expression. Conversely, the maspin promoter in cells that do not express the protein is completely methylated, facilitating the binding of methyl CpG-binding proteins (MeCPs) that can recruit histone deacetylases (HDACs) and chromatin remodeling complexes. Consequently, the chromatin adopts a transcriptionally inactive conformation and maspin is not expressed in these cells. Thus, this model stipulates that methylation is a primary impediment to maspin expression and consequently determines its cell-type specificity (Costello and Vertino 2002).

Recent studies by Maekawa and colleagues indicate that the activating transcription factor-2 (ATF-2) also plays a critical role in maspin regulation (Maekawa et al. 2008). ATF-2 is able to form homodimers or heterodimers with c-Jun and can bind to the cyclic AMP response element (CRE) to activate several different genes. Significantly, prior studies have established that heterozygous Atf-2 mutant mice (Atf-$2^{+/-}$) are highly prone to mammary tumors. In addition, Maekawa and colleagues have previously demonstrated that maspin is downregulated in mammary tumors arisen in Atf-$2^{+/-}$ mice. Their recent report demonstrates that ATF-2 directly binds to a CRE-like element present in the maspin promoter, thus activating its transcription independent of other transcriptional activators, such as p53. Consequently, it is possible that disrupting both the ATF-2 and p53 pathways may entirely impair maspin expression. While both ATF-2 and p53 are activated by stress, different stressors may preferentially activate one or the other. The authors thus speculate that ATF-2, like p53, may be involved in the development and progression of numerous types of cancers.

Additionally, early studies indicated that maspin expression is regulated by a negative hormonal responsive element (HRE) located within the maspin promoter. This negative element is active in both normal and carcinoma-derived prostate epithelial cells (Zhang et al. 1997b). In addition, electrophoresis mobility shift assay (EMSA) experiments confirmed the binding of androgen receptor to the HRE site in both of these cell types. Thus, loss of maspin expression during prostate cancer progression apparently results from both the absence of transactivation through Ets, Ap1, and p53 elements as well as the presence of transcriptional repression through the negative HRE element recognized by androgen receptor. Recent studies by Khalkhali-Ellis and colleagues have also established a role for the HRE in the regulation of maspin expression in mammary epithelial cells (Khalkhali-Ellis et al. 2004). Specifically, these studies demonstrate that the common breast cancer drug tamoxifen is able to induce maspin expression both *in vitro* and *in situ*. Tamoxifen, frequently used in the treatment and prevention of breast cancer, is thought to function by competing with estrogen for binding to the estrogen receptor. Khalkhali-Ellis et al. show that tamoxifen-mediated induction of maspin expression occurs chiefly through the maspin HRE, implying that hormones also play a critical role in the regulation of maspin expression in mammary epithelial cells. The authors further support this theory by demonstrating that 17β-estradiol exerts an inhibitory effect on maspin expression in normal primary mammary epithelial cells, promoting the opposite effect of tamoxifen treatment. These findings also provide further support for the role of tamoxifen as an estrogen antagonist (Bailey et al. 2006).

In a different study, Khalkhali-Ellis and colleagues demonstrated that nitric oxide (NO) plays a role in the regulation of maspin expression in both normal mammary epithelial and breast cancer cells (Khalkhali-Ellis and Hendrix 2003). Specifically, they found that NO induces maspin expression in the breast cancer MCF-7 cell line, which normally does not express maspin. Maspin expression is detected at both the mRNA and protein levels and is highest following treatment with the nitric oxide donor NOC-12. This induced expression results in reduced cell motility and invasion concurrent with an increase in apoptosis. Interestingly, it appears as though the anti-invasive and anti-apoptotic effects of NO are specific for cancer cells, as the treatment of normal epithelial cells with NOC-12 produces the opposite effect. In these cells, the treatment with NOC-12 results in a slight decrease in maspin expression. Additionally, NOC-12 treatment results in the generation of a 38-kDa maspin fragment, implying that NO may cause the removal of the maspin RSL and may direct its lysosomal degradation. This study suggests an intriguing role for NO in the differential regulation of maspin in normal versus cancerous cells.

In studies that provided the crucial – though previously unknown – mechanistic link between inflammation and metastasis, Luo and colleagues demonstrated that IκB kinase α (IKK α) functions to repress maspin, thus controlling prostate cancer metastasis (Luo et al. 2007). Although epidemiological studies have long provided clinical evidence that there is a causative link between chronic inflammation and cancer (Affara and Coussens 2007), the precise nature of this relationship at the molecular level remained a mystery. In recent years, the use of immune-competent mouse models of multistage carcinogenesis has facilitated the elucidation of the molecular mechanisms linking inflammation and the development of epithelial cancer (Affara and Coussens 2007; de Visser et al. 2006; Karin 2006). To this end, Luo et al. utilized a mouse model harboring a mutation that prevents IKKα activation. In this $Ikk\alpha^{AA/AA}$ mouse line, the serine residues whose phosphorylation is required for IKKα activation have been replaced with alanines. To examine the role that IKKα plays in cancer progression, these mice were crossed with transgenic adenocarcinoma mouse prostate (TRAMP) mice, in which the SV40 large T antigen is selectively expressed in the prostate epithelium. TRAMP mice are known to develop metastatic prostate cancer with a high frequency and thus serve as ideal models for evaluating cancer progression. Luo and colleagues demonstrated that while single mutant (WT/TRAMP) mice develop prostate cancer fairly early and begin dying at approximately 22 weeks of age, homozygosity for the $Ikk\alpha^{AA/AA}$ allele prolongs tumor onset and also delays mortality. Interestingly, while the $Ikk\alpha^{AA}$ mutation also affects the growth rate of primary cancer, its most profound effect is on metastogenesis.

Using qRT-PCR to evaluate differences in gene expression levels between these genotypes, the authors found that the only gene exhibiting striking and consistent differences was maspin. Having established that IKKα exerts its pro-metastatic effects via transcriptional repression of the maspin gene, they sought to elucidate the mechanism behind this intriguing process. Their further investigations led them to propose a model that integrates the previous observation that the infiltration of macrophages can promote tumorigenesis. This model highlights the key changes that cause early prostate tumor cells, which express high levels of maspin, to lose maspin expression and become irreversibly committed to a metastatic fate. This critical process is initiated by the infiltration of T lymphocytes and macrophages that secrete the pro-inflammatory cytokine *r*eceptor *a*ctivator of *N*FκB *l*igand (RANKL). RANKL binds to its receptor, RANK, triggering phosphorylation of IKKα. This phosphorylation event induces the nuclear localization of IKKα, where it is able to interact with the maspin promoter and cause transient repression of maspin expression. Subsequently, the maspin gene is permanently silenced by methylation, thus committing the prostate cancer cells to a metastatic fate (Affara and Coussens 2007; Luo et al. 2007). This novel mode of maspin regulation provides an insightful link between inflammation and cancer development at the molecular level.

Clearly, maspin expression is tightly regulated at numerous levels (see Fig. 17.2). While there is a great deal of crosstalk between several of these regulatory mechanisms, each pathway is also distinct in its own right. Thus, each merits careful investigation and represents a potential avenue to be exploited as a therapeutic target for cancer treatment and prevention.

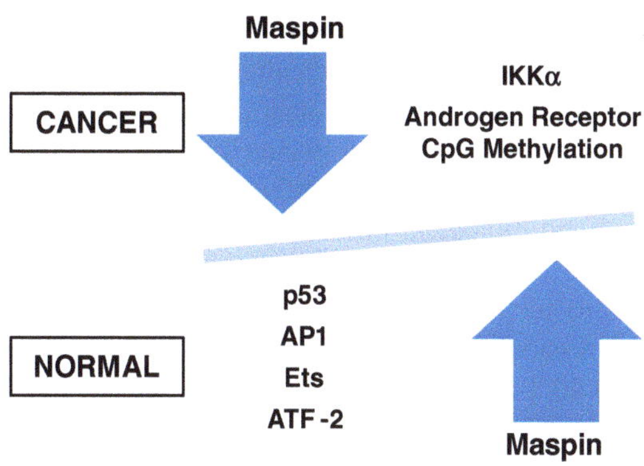

Fig. 17.2 Regulation of maspin gene expression

17.5 Modeling Human Breast Cancer Metastasis in Mice: Maspin as a Paradigm

The majority of cancer deaths result not from the primary tumor, but from metastasis. For example, studies indicate that approximately 60–70% of patients who have died or are dying due to breast cancer exhibit bone metastases (Koblinski et al. 2005; Solomayer et al. 2000). This striking statistic highlights the vital role that metastasis plays in cancer progression and mortality. Thus, it is of critical importance to elucidate the mechanism behind this complicated process both to enhance our understanding of basic biological processes and to facilitate the therapeutic intervention required to combat cancer. Due to the fact that metastasis is an extremely complicated, multistep process, it is best studied in an animal model. In order to facilitate these studies, it is critical to establish an appropriate animal model in which the tumor cells are invasive and possess the ability to metastasize to other organs (Shi et al. 2003). Additionally, the mice utilized should have an intact immune system so that tumor invasion and metastasis occur in a manner analogous to cancer progression in human patients. Researchers have employed several such models to characterize the role of maspin in breast cancer progression. These various mammary tumor metastasis models have provided a wealth of information regarding the key roles that maspin plays in the prevention of tumor growth, invasion, and metastasis. These findings offer promising insight into the treatment and prevention of cancer, exemplifying the benefits of utilizing mouse models to study human disease.

To characterize maspin's *in vivo* activity during mammary development and tumor progression, the Zhang Lab generated transgenic mice exhibiting targeted overexpression of maspin in mammary epithelial cells. This was accomplished by placing the maspin gene under the control of a mammary-specific WAP promoter (Zhang et al. 1999). Subsequently, these WAP-maspin transgenic mice were used to demonstrate that the overexpression of maspin in mammary epithelial cells during pregnancy inhibits mammary gland development and induces apoptosis. These mice exhibit impaired differentiation as well as a reduced number of lobular alveolar structures during late pregnancy. In addition, at midpregnancy alveolar cells from the transgenic mammary glands display an increased rate of apoptosis without any changes in the rate of proliferation. These experiments reveal that maspin has a profound biological influence on the growth and function of mammary epithelial cells *in vivo*. Since the primary goal of generating WAP-maspin transgenic mice was to test the protective role of maspin overexpression on mammary tumor progression, these mice were crossed with a strain of oncogenic WAP-Simian Virus (SV) 40 T-antigen mice (Zhang et al. 2000a). WAP-TAg transgenic mice develop mammary tumors with 100% frequency and thus represent an incredibly useful tool for examining specific mechanisms of tumor progression at both early and late time points (Li et al. 1996; Tzeng et al. 1993). The SV40 TAg initiates tumorigenesis through the inactivation of both p53 and the pRb related family of proteins (Dyson et al. 1989; Li et al. 2000; Mietz et al. 1992). Thus, the WAP-TAg transgenic mice possess specific features of some human breast cancers.

The results from this study show that overexpression of maspin through a transgenic approach can partially block tumor progression. For example, bitransgenic maspin/TAg mammary tumors grow considerably slower than their TAg controls. Growth inhibition is accompanied by an increase in apoptosis as well as a reduction in microvessel density. Significantly, maspin overexpression increases apoptosis of both normal mammary epithelial cells during pregnancy and malignant breast cancer cells. This study also indicates that the overexpression of maspin dramatically reduces tumor metastasis to the lung. Taken together, these data demonstrate that the targeted overexpression of maspin can inhibit tumor progression *in vivo*, likely though a combination of increased apoptosis, decreased angiogenesis, and inhibition of tumor cell migration (Zhang et al. 2000a).

While the SV40 TAg model displays numerous strengths as a tool for studying cancer progression, there are limitations to this model as well (Shi et al. 2003). For example, maspin expression is dependent upon the WAP promoter, which is only strongly active during pregnancy. Thus, both SV40 TAg and bitransgenic mice require mating with male mice throughout the study to continually activate transgene expression. In addition, once mammary cells become tumorigenic, TAg expression may become independent of the WAP promoter (Li et al. 1996; Tzeng et al. 1993). Such a change would shift the balance more toward tumorigenesis and could potentially influence experimental results. Furthermore, because the expression of TAg inactivates p53 and endogenous maspin expression is regulated by p53 (Zou et al. 2000), the activation of TAg results in a decrease in the expression of endogenous maspin. These compounding effects can also alter the balance between positive (oncogenic) and negative (tumor-suppressive) factors. In order to counteract such a potent oncogenic effect, the level of maspin expression must be increased in tumors by either systemic delivery or by placing the maspin transgene under the control of a constitutive promoter. Consequently, a new breast tumor mouse model was generated to accomplish this goal.

To generate this new syngeneic mammary tumor model, the Zhang Lab took advantage of the fact that the mammary gland is a natural site for the implantation of both normal and neoplastic cells (Medina 1996; Medina and Daniel 1996). In an attempt to establish a highly invasive implantation tumor model, TM40D mammary tumor cells orthotopically implanted into the mammary glands of BALB/c mice (Shi et al. 2001). Tumors subsequently grew within the gland and became highly invasive and metastatic to other organs. Results demonstrate that this model produces a 75% rate of invasion and metastasis, making it a very attractive system in which to study breast cancer progression. To utilize this model to investigate whether maspin is able to block tumor growth *in vivo*, maspin transfectants were established using the TM40D cell line. Two experiments were performed: the first used maspin stable clones carrying an expression plasmid, while the second employed retrovirus-transfected clones in which the maspin gene was stably integrated into the genome. The retrovirus maspin transfectants were utilized because of the possibility that some maspin plasmid transfectants might lose the plasmid without antibiotic selection *in vivo*. In the first experiment, growth rate was examined when tumors were relatively small and still in the exponential growth phase so as to study the effect of maspin on early tumor growth and local invasion. Whereas the control tumors grew very rapidly, the maspin transfectants displayed a slow, flat growth curve. Tumor growth patterns were also examined for both control and maspin transfectants. In particular, studies focused on the analysis of tumor encapsulation and necrosis, two histologic parameters widely used by tumor biologists. Aggressive tumor cells are highly invasive and thus more likely to break the surrounding capsule and invade into the mammary fat pad (Ng et al. 1992). Additionally, aggressive tumors are associated with the presence of excessive necrosis. Results demonstrated that control tumors indeed exhibit less encapsulation and more necrosis than maspin transfectants, indicating that an increased level of maspin is associated with better prognosis. This finding supports previous data indicating that a higher maspin level is correlated with low tumor cell invasion and a better prognosis (Ferrucci et al. 2007; Xia et al. 2000).

The second experiment utilized clones from retrovirus-infected maspin or vector control, and tumors were allowed to grow for greater than 5 weeks so that tumor invasion and distant metastases could be monitored. These maspin transfectants were even more effective at inhibiting tumor development than the plasmid maspin clones. Furthermore, the maspin retroviral transfectants did not display any signs of local invasion or metastasis, whereas the control mice exhibited both local invasion and numerous lung metastases. By taking advantage of the fact that TM40D cells in this syngeneic implantation model are highly invasive, these studies were able to demonstrate that maspin alone, when introduced into tumor cells, has the ability to inhibit primary tumor growth, tumor invasion, and metastasis. Significantly, these studies imply that maspin may serve as an important antitumor and antimetastasis reagent in cancer therapy. Consequently, researchers continue to focus on targeting maspin for effective cancer therapy.

Cancer gene therapy requires both a good animal model and an effective delivery system that exhibits low toxicity (McCormick 2001). Since there are very few breast cancer metastasis models available for gene therapy, the Zhang Lab established such a model to facilitate studies investigating the therapeutic value of the maspin gene in breast cancer therapy (Shi et al. 2002). Similar to the TM40D model, these mice were generated by orthotopically implanting tumor cells into syngeneic mice. This model, however, utilizes mammary tumor cells that were initially isolated from MMTV-polyoma virus middle T transgenic mice and were selected *in vitro* for high invasiveness. In these mice, tumors develop in 100% of the transplanted sites and display aggressive growth and high invasiveness. Importantly, they are highly metastatic: following the relatively short period of 1 month, all tumors in the control group metastasized to the lung. This model was used to demonstrate that systemic delivery of a maspin DNA:liposome complex inhibits the growth and metastasis of breast tumors. Specifically, both the tumor size and overall tumor growth rate were significantly decreased for maspin-treated tumors relative to the controls. In addition, a significant decrease in lung metastasis was observed in the maspin-treated mice, as evidenced by a decrease in both the size and number of lung tumor foci. TUNEL assays revealed that this significant inhibition of both primary tumor growth and metastasis is mediated by increased apoptosis in the maspin-treated tumors.

Due to the fact that low delivery efficiency and toxic side effects have been two major problems that hinder the process of gene therapy (Hellgren et al. 2000; O'Brien et al. 1999; Rochlitz 2001; Roth and Cristiano 1997), this system was tested to ensure that these potential obstacles would not present any problems. Indeed, these DNA:liposome complexes exhibit a high delivery efficiency, particularly to the lung and mammary gland. Additionally, maspin:liposome therapy does not produce any toxic side effects, as evidenced by the fact that treated FVB mice that did not undergo tumor cell implantation maintain normal reproductive function and do not exhibit any gross abnormality in tissue organs upon biopsy. Given that all existing evidence associates maspin with "protective" roles *in vivo*, the authors suspect that the delivery of exogenous maspin is unlikely to cause adverse effects in patients. Thus, maspin:liposome treatment offers an effective and nontoxic therapy for the treatment of breast cancer.

In order to further investigate the mechanism by which systemically delivered maspin inhibits tumor growth and metastasis, the Zhang Lab again utilized the TM40D mouse model. To test the hypothesis that maspin exerts its effect through vascular endothelial cells, an adenovirus maspin construct (Ad-Mp) was used to target maspin overexpression to both proliferating neovessels and nonproliferating mature vessels (Li et al. 2005). The results demonstrate that this targeted maspin overexpression actively induces endothelial cell apoptosis. Intravascular administration of Ad-Mp to mice bearing mammary tumors disrupts tumor-induced angiogenesis. However, maspin acts selectively on neovascular endothelial cells and not on mature vessels. This selectivity is especially significant because in the well-developed vascular systems of normal adult mice, there is no neovascular formation unless induced by pathological conditions, such as wound healing or tumorigenesis. Conversely, in mice bearing mammary tumors, active angiogenesis occurs in order to provide the tumor cells with an adequate supply of nutrients. Thus, maspin's selective action against neovascular cells is of benefit for cancer therapy since delivery of maspin to other organs does not exert a toxic effect. This study further supports previous reports that maspin is unlikely to cause adverse effect *in vivo*. Additionally, it implies that targeting tumor neovascular endothelial cells with maspin may offer another effective option for cancer gene therapy.

17.6 Conclusion

Since its discovery in 1994 as a gene that is downregulated in malignant breast tissue, several hundred articles have emerged that link the reduction or loss of maspin expression with other solid tumors and detail the molecular mechanisms of maspin-mediated inhibition of cell invasion, promotion of apoptosis, and blockade of angiogenesis. Clinically relevant findings demonstrate that decreased maspin expression correlates with an increase in the severity of both breast and prostate cancer, resulting in a poor prognosis (Maass et al. 2000; Pierson et al. 2002). Clearly, it is of critical importance to elucidate the precise molecular mechanisms underlying maspin's involvement in these devastating diseases. Thus, considerable effort has focused on utilizing *in vivo* models in an attempt to obtain a better understanding of both maspin expression and function as they pertain to cancer pathogenesis. In particular, mouse models have emerged as critical tools that facilitate investigations into the numerous roles that maspin plays in the complex processes surrounding cancer development, progression, and metastasis. Future studies will serve to further elucidate the mechanisms underlying these processes. In addition, research will focus on ways in which this crucial tumor suppressor can be utilized in cancer prevention, diagnosis, and treatment.

References

Affara NI, Coussens LM (2007) IKKalpha at the crossroads of inflammation and metastasis. Cell 129:25–26

Al-Ayyoubi M, Gettins PG, Volz K (2004) Crystal structure of human maspin, a serpin with antitumor properties: reactive center loop of maspin is exposed but constrained. J Biol Chem 279:55540–55544

Amir S, Margaryan NV, Odero-Marah V, Khalkhali-Ellis Z, Hendrix MJ (2005) Maspin regulates hypoxia-mediated stimulation of uPA/uPAR complex in invasive breast cancer cells. Cancer Biol Ther 4:400–406

Bailey CM, Khalkhali-Ellis Z, Kondo S, Margaryan N, Seftor RE, Wheaton WW, Amir S, Pins MR, Schutte BC, Hendrix MJ (2005) Maspin binds directly to interferon regulatory factor 6: identification of a novel serpin partnership. J Biol Chem 280(40):34210–34217

Bailey CM, Khalkhali-Ellis Z, Seftor EA, Hendrix MJ (2006) Biological functions of maspin. J Cell Physiol 209:617–624

Biliran H Jr, Sheng S (2001) Pleiotrophic inhibition of pericellular urokinase-type plasminogen activator system by endogenous tumor suppressive maspin. Cancer Res 61:8676–8682

Blacque OE, Worrall DM (2002) Evidence for a direct interaction between the tumour suppressor serpin maspin, and types I and III collagen. J Biol Chem 277(13):10783–10788
Brouillet JP, Dufour F, Lemamy G, Garcia M, Schlup N, Grenier J, Mani JC, Rochefort H (1997) Increased cathepsin D level in the serum of patients with metastatic breast carcinoma detected with a specific pro-cathepsin D immunoassay. Cancer 79:2132–2136
Capony F, Rougeot C, Montcourrier P, Cavailles V, Salazar G, Rochefort H (1989) Increased secretion, altered processing, and glycosylation of pro-cathepsin D in human mammary cancer cells. Cancer Res 49:3904–3909
Cella N, Contreras A, Latha K, Rosen JM, Zhang M (2006) Maspin is physically associated with [beta]1 integrin regulating cell adhesion in mammary epithelial cells. FASEB J 20:1510–1512
Costello JF, Vertino PM (2002) Methylation matters: a new spin on maspin. Nat Genet 31:123–124
de Visser KE, Eichten A, Coussens LM (2006) Paradoxical roles of the immune system during cancer development. Nat Rev Cancer 6:24–37
Domann FE, Rice JC, Hendrix MJ, Futscher BW (2000) Epigenetic silencing of maspin gene expression in human breast cancers. Int J Cancer 85:805–810
Dyson N, Buchkovich K, Whyte P, Harlow E (1989) The cellular 107 K protein that binds to adenovirus E1A also associates with the large T antigens of SV40 and JC virus. Cell 58:249–255
Ferrucci PF, Rabascio C, Gigli F, Corsini C, Giordano G, Bertolini F, Martinelli G (2007) A new comprehensive gene expression panel to study tumor micrometastasis in patients with high-risk breast cancer. Int J Oncol 30:955–962
Futscher BW, Oshiro MM, Wozniak RJ, Holtan N, Hanigan CL, Duan H, Domann FE (2002) Role for DNA methylation in the control of cell type specific maspin expression. Nat Genet 31:175–179
Hall DC, Johnson-Pais TL, Grubbs B, Bernal R, Leach RJ, Padalecki SS (2008) Maspin reduces prostate cancer metastasis to bone. Urol Oncol 26(6):652–658
Hellgren I, Drvota V, Pieper R, Enoksson S, Blomberg P, Islam KB, Sylven C (2000) Highly efficient cell-mediated gene transfer using non-viral vectors and FuGene6: in vitro and in vivo studies. Cell Mol Life Sci 57:1326–1333
Huntington JA, Read RJ, Carrell RW (2000) Structure of a serpin-protease complex shows inhibition by deformation. Nature 407:923–926
Jiang N, Meng Y, Zhang S, Mensah-Osman E, Sheng S (2002) Maspin sensitizes breast carcinoma cells to induced apoptosis. Oncogene 21:4089–4098
Karin M (2006) Nuclear factor-kappaB in cancer development and progression. Nature 441:431–436
Kelly WK, Marks PA (2005) Drug insight: histone deacetylase inhibitors – development of the new targeted anticancer agent suberoylanilide hydroxamic acid. Nat Clin Pract Oncol 2:150–157
Khalkhali-Ellis Z, Christian AL, Kirschmann DA, Edwards EM, Rezaie-Thompson M, Vasef MA, Gruman LM, Seftor RE, Norwood LE, Hendrix MJ (2004) Regulating the tumor suppressor gene maspin in breast cancer cells: a potential mechanism for the anticancer properties of tamoxifen. Clin Cancer Res 10:449–454
Khalkhali-Ellis Z, Hendrix MJ (2003) Nitric oxide regulation of maspin expression in normal mammary epithelial and breast cancer cells. Am J Pathol 162:1411–1417
Khalkhali-Ellis Z, Hendrix MJ (2007) Elucidating the function of secreted maspin: inhibiting cathepsin D-mediated matrix degradation. Cancer Res 67:3535–3539
Koblinski JE, Kaplan-Singer BR, VanOsdol SJ, Wu M, Engbring JA, Wang S, Goldsmith CM, Piper JT, Vostal JG, Harms JF et al (2005) Endogenous osteonectin/SPARC/BM-40 expression inhibits MDA-MB-231 breast cancer cell metastasis. Cancer Res 65:7370–7377
Kondo S, Schutte BC, Richardson RJ, Bjork BC, Knight AS, Watanabe Y, Howard E, de Lima RL, Daack-Hirsch S, Sander A et al (2002) Mutations in IRF6 cause Van der Woude and popliteal pterygium syndromes. Nat Genet 32:285–289
Latha K, Zhang W, Cella N, Shi HY, Zhang M (2005) Maspin mediates increased tumor cell apoptosis upon induction of the mitochondrial permeability transition. Mol Cell Biol 25:1737–1748

Law RH, Irving JA, Buckle AM, Ruzyla K, Buzza M, Bashtannyk-Puhalovich TA, Beddoe TC, Nguyen K, Margaret Worrall D, Bottomley SP et al (2005) The high resolution crystal structure of the human tumour suppressor maspin reveals a novel conformational switch in the G-helix. J Biol Chem 280(23):22356–22364
Li M, Hu J, Heermeier K, Hennighausen L, Furth PA (1996) Expression of a viral oncoprotein during mammary gland development alters cell fate and function: induction of p53-independent apoptosis is followed by impaired milk protein production in surviving cells. Cell Growth Differ 7:3–11
Li M, Lewis B, Capuco AV, Laucirica R, Furth PA (2000) WAP-TAg transgenic mice and the study of dysregulated cell survival, proliferation, and mutation during breast carcinogenesis. Oncogene 19:1010–1019
Li X, Yin S, Meng Y, Sakr W, Sheng S (2006) Endogenous inhibition of histone deacetylase 1 by tumor-suppressive maspin. Cancer Res 66:9323–9329
Li Z, Schem C, Shi YH, Medina D, Zhang M (2008) Increased COX2 expression enhances tumor-induced osteoclastic lesions in breast cancer bone metastasis. Clin Exp Metastasis 25(4):389–400
Li Z, Shi HY, Zhang M (2005) Targeted expression of maspin in tumor vasculatures induces endothelial cell apoptosis. Oncogene 24(12):2008–2019
Liaudet-Coopman E, Beaujouin M, Derocq D, Garcia M, Glondu-Lassis M, Laurent-Matha V, Prebois C, Rochefort H, Vignon F (2006) Cathepsin D: newly discovered functions of a long-standing aspartic protease in cancer and apoptosis. Cancer Lett 237:167–179
Liotta LA, Rao CN, Barsky SH (1983) Tumor invasion and the extracellular matrix. Lab Invest 49:636–649
Luo JL, Tan W, Ricono JM, Korchynskyi O, Zhang M, Gonias SL, Cheresh DA, Karin M (2007) Nuclear cytokine-activated IKKalpha controls prostate cancer metastasis by repressing Maspin. Nature 446:690–694
Maass N, Hojo T, Zhang M, Sager R, Jonat W, Nagasaki K (2000) Maspin – a novel protease inhibitor with tumor-suppressing activity in breast cancer. Acta Oncol 39:931–934
Maekawa T, Sano Y, Shinagawa T, Rahman Z, Sakuma T, Nomura S, Licht JD, Ishii S (2008) ATF-2 controls transcription of Maspin and GADD45 alpha genes independently from p53 to suppress mammary tumors. Oncogene 27:1045–1054
McCormick F (2001) Cancer gene therapy: fringe or cutting edge? Nat Rev Cancer 1:130–141
Medina D (1996) The mammary gland: a unique organ for the study of development and tumorigenesis. J Mammary Gland Biol Neoplasia 1:5–19
Medina D, Daniel C (1996) Experimental models of development, function, and neoplasia. J Mammary Gland Biol Neoplasia 1:3–4
Mietz JA, Unger T, Huibregtse JM, Howley PM (1992) The transcriptional transactivation function of wild-type p53 is inhibited by SV40 large T-antigen and by HPV-16 E6 oncoprotein. EMBO J 11:5013–5020
Ng IO, Lai EC, Ng MM, Fan ST (1992) Tumor encapsulation in hepatocellular carcinoma. A pathologic study of 189 cases. Cancer 70:45–49
O'Brien T, Karlsen AE, Andersen HU, Mandrup-Poulsen T, Nerup J (1999) Absence of toxicity associated with adenoviral-mediated transfer of the beta-galactosidase reporter gene to neonatal rat islets in vitro. Diabetes Res Clin Pract 44:157–163
Odero-Marah VA, Khalkhali-Ellis Z, Chunthapong J, Amir S, Seftor RE, Seftor EA, Hendrix MJ (2003) Maspin regulates different signaling pathways for motility and adhesion in aggressive breast cancer cells. Cancer Biol Ther 2:398–403
Pemberton PA, Wong DT, Gibson HL, Kiefer MC, Fitzpatrick PA, Sager R, Barr PJ (1995) The tumor suppressor maspin does not undergo the stressed to relaxed transition or inhibit trypsin-like serine proteases. Evidence that maspin is not a protease inhibitory serpin. J Biol Chem 270:15832–15837
Pierson CR, McGowen R, Grignon D, Sakr W, Dey J, Sheng S (2002) Maspin is up-regulated in premalignant prostate epithelia. Prostate 53:255–262
Pillai R, Coverdale LE, Dubey G, Martin CC (2004) Histone deacetylase 1 (HDAC-1) required for the normal formation of craniofacial cartilage and pectoral fins of the zebrafish. Dev Dyn 231:647–654

Rahman Q, Abidi P, Afaq F, Schiffmann D, Mossman BT, Kamp DW, Athar M (1999) Glutathione redox system in oxidative lung injury. Crit Rev Toxicol 29:543–568
Ren WP, Sloane BF (1996) Cathepsins D and B in breast cancer. Cancer Treat Res 83:325–352
Rochlitz CF (2001) Gene therapy of cancer. Swiss Med Wkly 131:4–9
Roth JA, Cristiano RJ (1997) Gene therapy for cancer: what have we done and where are we going? J Natl Cancer Inst 89:21–39
Sager R, Sheng S, Anisowicz A, Sotiropoulou G, Zou Z, Stenman G, Swisshelm K, Chen Z, Hendrix MJ, Pemberton P et al (1994) RNA genetics of breast cancer: maspin as paradigm. Cold Spring Harb Symp Quant Biol 59:537–546
Sheng S, Carey J, Seftor EA, Dias L, Hendrix MJ, Sager R (1996) Maspin acts at the cell membrane to inhibit invasion and motility of mammary and prostatic cancer cells. Proc Natl Acad Sci USA 93:11669–11674
Sherman JM, Stone EM, Freeman-Cook LL, Brachmann CB, Boeke JD, Pillus L (1999) The conserved core of a human SIR2 homologue functions in yeast silencing. Mol Biol Cell 10:3045–3059
Shi HY, Liang R, Templeton NS, Zhang M (2002) Inhibition of breast tumor progression by systemic delivery of the maspin gene in a syngeneic tumor model. Mol Ther 5:755–761
Shi HY, Stafford LJ, Liu Z, Liu M, Zhang M (2007) Maspin controls mammary tumor cell migration through inhibiting Rac1 and Cdc42, but not the RhoA GTPase. Cell Motil Cytoskeleton 64:338–346
Shi HY, Zhang W, Liang R, Abraham S, Kittrell FS, Medina D, Zhang M (2001) Blocking tumor growth, invasion, and metastasis by maspin in a syngeneic breast cancer model. Cancer Res 61:6945–6951
Shi HY, Zhang W, Liang R, Kittrell F, Templeton NS, Medina D, Zhang M (2003) Modeling human breast cancer metastasis in mice: maspin as a paradigm. Histol Histopathol 18:201–206
Solomayer EF, Diel IJ, Meyberg GC, Gollan C, Bastert G (2000) Metastatic breast cancer: clinical course, prognosis and therapy related to the first site of metastasis. Breast Cancer Res Treat 59:271–278
Solomon LA, Munkarah AR, Schimp VL, Arabi MH, Morris RT, Nassar H, Ali-Fehmi R (2006) Maspin expression and localization impact on angiogenesis and prognosis in ovarian cancer. Gynecol Oncol 101:385–389
Tanaka N, Ishihara M, Kitagawa M, Harada H, Kimura T, Matsuyama T, Lamphier MS, Aizawa S, Mak TW, Taniguchi T (1994) Cellular commitment to oncogene-induced transformation or apoptosis is dependent on the transcription factor IRF-1. Cell 77:829–839
Tsujimura H, Tamura T, Kong HJ, Nishiyama A, Ishii KJ, Klinman DM, Ozato K (2004) Toll-like receptor 9 signaling activates NF-kappaB through IFN regulatory factor-8/IFN consensus sequence binding protein in dendritic cells. J Immunol 172:6820–6827
Tzeng YJ, Guhl E, Graessmann M, Graessmann A (1993) Breast cancer formation in transgenic animals induced by the whey acidic protein SV40 T antigen (WAP-SV-T) hybrid gene. Oncogene 8:1965–1971
Xia W, Lau YK, Hu MC, Li L, Johnston DA, Sheng S, El-Naggar A, Hung MC (2000) High tumoral maspin expression is associated with improved survival of patients with oral squamous cell carcinoma. Oncogene 19:2398–2403
Yin S, Li X, Meng Y, Finley RL Jr, Sakr W, Yang H, Reddy N, Sheng S (2005) Tumor suppressive maspin regulates cell response to oxidative stress by direct interaction with glutathione S-transferase. J Biol Chem 280(41):34985–34996
Yin S, Lockett J, Meng Y, Biliran H Jr, Blouse GE, Li X, Reddy N, Zhao Z, Lin X, Anagli J et al (2006) Maspin retards cell detachment via a novel interaction with the urokinase-type plasminogen activator/urokinase-type plasminogen activator receptor system. Cancer Res 66:4173–4181
Zhang M, Maass N, Magit D, Sager R (1997a) Transactivation through Ets and Ap1 transcription sites determines the expression of the tumor-suppressing gene maspin. Cell Growth Differ 8:179–186
Zhang M, Magit D, Botteri F, Shi Y, He K, Li M, Furth P, Sager R (1999) Maspin plays an important role in mammary gland development. Dev Biol 215:278–287

Zhang M, Magit D, Sager R (1997b) Expression of maspin in prostate cells is regulated by a positive ets element and a negative hormonal responsive element site recognized by androgen receptor. Proc Natl Acad Sci USA 94:5673–5678
Zhang M, Shi Y, Magit D, Furth PA, Sager R (2000a) Reduced mammary tumor progression in WAP-TAg/WAP-maspin bitransgenic mice. Oncogene 19:6053–6058
Zhang M, Volpert O, Shi YH, Bouck N (2000b) Maspin is an angiogenesis inhibitor. Nat Med 6:196–199
Zhang W, Shi HY, Zhang M (2005) Maspin overexpression modulates tumor cell apoptosis through the regulation of Bcl-2 family proteins. BMC Cancer 5:50
Zou Z, Anisowicz A, Hendrix MJ, Thor A, Neveu M, Sheng S, Rafidi K, Seftor E, Sager R (1994) Maspin, a serpin with tumor-suppressing activity in human mammary epithelial cells [see comments]. Science 263:526–529
Zou Z, Gao C, Nagaich AK, Connell T, Saito S, Moul JW, Seth P, Appella E, Srivastava S (2000) p53 regulates the expression of the tumor suppressor gene maspin. J Biol Chem 275:6051–6054

Chapter 18
Epigenetic Mouse Models

Cecilia Rosales and Manel Esteller

18.1 Epigenetics

Epigenetics is currently defined as "non-genetic inheritance" (Haig 2004) and means literally "beyond genetics." The word alludes to the study of the heritable changes in gene expression that cannot be explained by changes in DNA sequences (Bird 2002). Epigenetic modifications fall into two main categories: DNA methylation and histone modification (Bird 2002).

More than half of mammalian genes have "CpG islands" in promoter regions that can be differentially methylated. Epigenetic regulation is important not only for generating the diversity of tissues during development, but also for maintaining genomic stability and chromosomal integrity. It was initially useful as a defense mechanism against invasion by parasitic elements, permanently inactivating foreign genes in the host cell. This selective gene-silencing confers another important role on epigenetics in embryonic development and cell differentiation, and explains how all genes in an organism can be present in all cells although only a few are activated in particular cells and at particular times. Epigenetics is also responsible for the inactivation of the X chromosome in females and for gene imprinting. Dynamic changes in DNA methylation patterns and histone modification are known to occur during preimplantation and early postimplantation mouse development.

It is now widely accepted that cancer is, in part, an epigenetic disease, although epigenetic alterations are still viewed largely as a surrogate of genetic alterations. Studies of DNA methylation in tumor tissues have revealed at least as many epigenetic alterations as genetic alterations for any given gene (Feinberg et al. 2006).

C. Rosales • M. Esteller (✉)
Cancer Epigenetics Group, Spanish National Cancer Research Centre (CNIO), Madrid, Spain

Cancer Epigenetics and Biology Program (PEBC), Catalan Institute of Oncology (ICO), Barcelona, Catalonia, Spain
e-mail: mesteller@iconcologia.net

J.E. Green and T. Ried (eds.), *Genetically Engineered Mice for Cancer Research: Design, Analysis, Pathways, Validation and Pre-Clinical Testing*,
DOI 10.1007/978-0-387-69805-2_18, © Springer Science+Business Media, LLC 2012

18.1.1 DNA Methylation

The most widely studied epigenetic modification in humans to date has been the cytosine methylation of DNA. CpG dinucleotides are not uniformly distributed throughout vertebrate genomes but are concentrated in regions known as CpG islands, and are located mainly in the promoter region of many genes. In normal healthy cells, while repetitive genomic sequences are heavily methylated, most CpG islands remain unmethylated, thereby allowing genes to be expressed. However, in some cases, gene-promoter regions are methylated as part of normal development: DNA methylation is responsible for genomic imprinting, which means that one of the parental alleles of a gene is inactivated, resulting in monoallelic expression (Reik and Lewis 2005), X chromosome in females, and for tissue-specific gene expression (Baylin et al. 1998).

By contrast, in cancer cells the tumor genome becomes globally hypomethylated, due mainly to the generalized demethylation in the CpGs located in the body of the genes. Also, transcriptional silencing of tumor suppressor genes by CpG island promoter hypermethylation is the main process in the tumorigenic process, where it is a hallmark in cancer cells arising from tumor suppressor inactivation (Esteller 2007).

Methylated CpGs are usually recognized by a type of protein known as methyl binding domain (MBD), which is capable of recruiting histone modifier enzymes, such as histone deacetylases (HDACs) and histone methyltransferases (HMTs) (Sarraf and Stancheva 2004).

18.1.2 Histone Modifications

The nucleosome comprises a histone octamer, with two copies of each of the histones H2A, H2B, H3, and H4, wrapped by 147 bp of DNA, and a fifth histone, H1, which acts as a connector, stabilizing DNA–histone interactions (Ballestar and Esteller 2002).

The core histones that make up the nucleosome are subject to more than 100 post-translational modifications, including acetylation, methylation, phosphorylation, ubiquitination, and sumoylation (Bernstein et al. 2007). Posttranslational modifications of histone tails affect chromatin in two ways. In the first place, they alter the electrostatic charge of the histone, which leads to a change in the structure or the bond of the DNA. Secondly, these modifications can create binding sites for other proteins, recruiting specific complexes, both activators and repressors. Overall, histone acetylation and methylation are the best-known posttranslational modifications.

Histone acetyltransferases (HATs) are responsible for adding the acetyl group to the histone tail. This reaction can be reversed by the HDAC group of proteins, so the acetylation status of the histones is determined by the relative activities of the HATs and the HDACs. Histones can be methylated in lysine and arginine residues. Lysine can be mono-, di- or trimethylated, and arginine can be mono- or dimethylated. This modification is linked to transcriptional activation and repression. In general, heterochromatin is associated with methylation in lysines 9 and 27 of histone 3, and lysine 20 of histone 4; however, euchromatin is associated with methylation in lysines 4, 36, and 79 of histone 3 (Kouzarides 2007).

An appreciation of the varieties of these modifications gave rise to the hypothesis of the "histone code," which proposed that histone modifications constitute a pattern for interaction with specific factors (Ballestar and Esteller 2002). All these combinations allow the formation of different epigenetic states, such as gene activation rather than gene silencing, or, what amounts to the same, cellular proliferation in contrast to cellular differentiation.

This chapter considers the most important mouse models involving epigenetic genes, and discusses the main phenotypes and the application of these models in cancer and other diseases. It includes the genes controlling epigenetic modifications, the associated proteins and chromatin-remodeling complexes. Table 18.1 summarizes the disrupted genes and the corresponding phenotypes.

Table 18.1 Epigenetic mice models

Gene disrupted	Main phenotype
DNMT1	Embryonic lethal
DNMT2	No distinct phenotype
DNMT3a	Dead at 4 weeks of age
DNMT3b	Embryonic lethal
DNMT3L	Infertility
MBD1	No distinct phenotype
MBD2	No distinct phenotype
MBD3	Embryonic lethal
MBD4	Increase tumor predisposition
Mecp2	Rett syndrome phenotype
Suv39	Increase tumor predisposition
Ezh2	Embryonic lethal
Eed	Embryonic lethal
Suz12	Embryonic lethal
Riz1	Increase tumor predisposition
G9a	Embryonic lethal
GLP	Embryonic lethal
ESET	Embryonic lethal
Prmt1	Embryonic lethal
Prmt2	No distinct phenotype
Prmt3	No distinct phenotype
Prmt4	Perinatal lethal
MLL	Embryonic lethal
MLL2	Embryonic lethal
NSD1	Embryonic lethal
LSD1	Embryonic lethal
JMJD1A	Infertility
RBP2	No distinct phenotype
MSK1	No distinct phenotype

(continued)

Table 18.1 (continued)

Gene disrupted	Main phenotype
Gcn512	Embryonic lethal
PCAF	No distinct phenotype
P300	Embryonic lethal
CBP	Embryonic lethal
MORF	Perinatal lethal
SCR1	Hormone resistance
TIF2	Infertility
p/CIP	Female hypofertility
HDAC1	Embryonic lethal
HDAC2	Cardiac abnormalities
HDAC4	Perinatal lethal
HDAC5	Cardiac failure
HDAC7	Embryonic lethal
HDAC9	Cardiac failure
SIRT1	Perinatal lethal
SIRT2	No distinct phenotype
SIRT3	No distinct phenotype
Brm	Heavier mice
Brg1	Embryonic lethal
SNF5	Embryonic lethal
Srg3	Embryonic lethal

18.2 Mouse Models of Epigenetic Genes

18.2.1 DNA Methyltransferases

DNA methylation patterns are established and maintained by proteins known as DNA methyltransferases (DNMTs). To date, five DNMT genes have been described in mammals: DNMT1, DNMT2, DNMT3A, DNMT3B, and DNMT3L.

DNMT1 is the only one of these proteins that has a preference for hemimethylated DNA and is responsible for the maintenance of DNA methylation (Spada et al. 2007). It is strongly expressed in proliferating cells and is ubiquitous in all somatic tissues (Okano and Li 2002).

The first attempt to knock out DNMT1 in mice eliminated only part of exon 4, resulting in the weak expression of a truncated form of DNMT1. Under these conditions, the level of total DNA methylation of embryonal stem (ES) cells and homozygous embryos is reduced to about one-third of that in heterozygotes or wild types. Homozygous mutant embryos display severe stunting, developmental delay and death at mid-gestation (Li et al. 1992). These embryos show biallelic expression of several imprinted genes, and evidence of at least transient inactivation of all X chromosomes (Li et al. 1993; Beard et al. 1995). These findings strongly suggest that DNMT1 functions as a major maintenance methyltransferase *in vivo* and demonstrate that DNA methylation is essential for normal mammalian development.

The studies in mouse models are extremely interesting but puzzling: When the mouse deficient in DNA methylation owing to a defect in DNMT1 is crossed with the colon adenoma prone Min mouse (with a genetic defect in APC), the resulting mouse has fewer tumors (Laird et al. 1995); but another DNMT1 defective mouse may have an increased risk of lymphomas (Gaudet et al. 2003; Eden et al. 2003). This paradox is an important question that needs to be addressed in the near future.

The murine *DNMT3* family consists of two genes, DNMT3A and DNMT3B, which are strongly expressed in undifferentiated ES cells but downregulated after differentiation and expressed at low levels in adult somatic tissues (Okano et al. 1998).

To determine whether DNMT3A and DNMT3B also function in de novo methylation, Li's group developed a double-knockout (KO) mouse. Their double-homozygous embryos have a phenotype similar to that of the DNMT1 KO. The DNA methylation level of these mice is much lower than that of wild-type mice, but much higher than that of DNMT1 KO animals. This demonstrates that DNMT3A and DNMT3B are essential for de novo methylation in early embryogenesis, and that these two gene products are redundant with respect to this function (Okano et al. 1999).

DNMT3A KO mice develop to term and appear to be normal at birth, but most homozygotes are stunted and die at about 4 weeks of age (Okano et al. 1999).

In contrast, no *DNMT3B* KO mice have been recovered at birth. Instead, these animals suffer embryonic lethality between 9.5 and 16.5 days post-coitum (dpc), exhibiting multiple developmental defects, including growth impairment and rostral neural tube defects (Okano et al. 1999). The hypomethylation of centromeric minor satellite repeats that occurs in DNMT3B KO mice demonstrates that DNMT3B is required for methylation in these regions.

Approximately 40% of immunodeficiency–centromeric instability–facial anomalies (ICF) syndrome cases are due to mutations in the highly conserved catalytic domain of DNMT3B (Xu et al. 1999). ICF syndrome is an autosomal recessive condition in which affected individuals tend to succumb to recurrent infections and have unusual faces. Their chromosomes are characterized by decondensation and instability in the pericentromeric heterochromatin of chromosomes 1 and 16, and, of less importance, chromosome 9 (Tiepolo et al. 1979; Maraschio et al. 1988).

As the Dnmt3B-null allele results in embryonic lethality, Li's group have developed a mouse bearing a mutation in the catalytic domain that results in phenotypes reminiscent of ICF patients (Li et al. 1993). The gross anatomy of these animals is normal, although they have smaller bodies, shorter noses and wider nasal bridges, and they have a reduced absolute number of thymocytes (Ueda et al. 2006).

18.2.2 Methyl-Binding Domains

DNA methylation-mediated transcriptional silencing is achieved at least in part through an indirect mechanism in which a methyl-CpG binding protein specifically binds to methylated DNA to bring about transcriptional repression.

A family of four proteins, *MBD1–MBD4*, has been identified; three of these, MBD1–MBD3, are involved in transcriptional repression while MBD4 is a mismatch-repair protein (Hendrich et al. 2001).

MBD2 KO mice are viable, fertile and have a normal phenotype, implying that this gene is not required for murine development. These mice also have normal methylation levels. The only phenotypic peculiarity described in these animals is an abnormal maternal behavioral trait (Hendrich et al. 2001). Regarding to intestinal tumorigenesis, when these KO mice are crossed with Min mice, they show a direct relation between adenoma burden and dosage of the MBD2 allele, resulting in ten times fewer adenomas than in MBD2 wild-type controls (Sansom et al. 2003). This antitumoral activity of MBD2 was not seen when these mice were crossed with p53-null mice. In this case, loss of MBD2 has no effect neither in acceleration nor in reduction of lymphomagenesis (Sansom et al. 2005) Taking these findings together, the potential use of MBD2 in the clinic is very encouraging.

MBD3-deficient mice die at 8.5 dpc, showing signs of retardation and of being in the process of resorption, implying that the MBD3 gene is necessary for embryonic development (Hendrich et al. 2001).

Mammalian *MDB4* contains a methyl-CpG binding domain and can enzymatically remove thymine or uracil from a mismatched CpG site (Hendrich and Bird 1998; Hendrich et al. 1999). MBD4 is mutated in 26–43% of human colorectal tumors showing microsatellite instability (Riccio et al. 1999; Bader et al. 2000). Mice homozygous for this gene are viable and fertile, but show a deficit in DNA repair and increased incidence of tumor formation, suggesting that MBD4 plays a role in reducing inherited disease and cancer (Millar et al. 2002).

The *Mecp2* gene is mutated in most cases of human Rett syndrome (RTT), and several mouse models for this have been generated that exhibit phenotypic similarities to the disease (Jones et al. 1998; Nan et al. 1998). RTT is a severe mental retardation disorder mainly affecting females. It is chiefly characterized by developmental stagnation, ataxia, stereotyped hand-wringing motions and autism. Microcephaly and reduced neuron size have also been reported (Hagberg et al. 1983; Bauman et al. 1995). The first model described showed Mecp2 to be essential for embryonic development because these animals fail to gastrulate and consequently die between E8.5 and E12.5 (Tate et al. 1996). To overcome this problem of mortality, Chen et al. (2001) generated a conditional Mecp2 allele using the Cre–Lox recombination system and deleted most of the MBD domain. These animals appear healthy for the first weeks of life and are fertile, but develop abnormal behaviors at 5 weeks of age and die by the age of 10 weeks. Heterozygous females seem normal for the first 4 months, but subsequently begin to show symptoms, such as weight gain, reduced activity, and ataxic gait (Chen et al. 2001).

Even though Mecp2-null mice do not initially show this phenotype, they develop a stiff, uncoordinated gait and reduced spontaneous movement between 3 and 8 weeks of age. Subsequently, these animals develop hind-limb clasping and irregular breathing. The progression of symptoms leads to rapid weight loss and death at approximately 54 days. Heterozygous female mice may be the most appropriate model for human RTT.

Although Mecp2-heterozygous females initially show no symptoms and raise normal litters, they develop RTT-like symptoms at 4–12 months of age. Similar to what is seen in humans, the phenotype stabilizes, and the animals have an apparently normal life span. The mice often become obese, which is not a characteristic of the human disease (Guy et al. 2007). Unlike Mecp2-null mice, heterozygotes do not undergo rapid deterioration, raising the possibility that, like RTT, they are stable over the long term (Guy et al. 2001).

In addition to this, specific deletion of Mecp2 in the brain (by using a Nestin-cre; Mecp2 conditional allele) mimics the germ-line loss of Mecp2, indicating that Mecp2 is exclusively required for correct central nervous system function (Guy et al. 2001; Chen et al. 2001). Furthermore, deletion of Mecp2 in postmitotic neurons (using a calmodulin kinase promoter-driven Cre recombinase) also mirrors the RTT phenotype, although with later onset (Guy et al. 2001).

The brain-specific effects of Mecp2 deletion may be due to compensation for this lack by a different MBD. As MBD2-null mice are viable and fertile, a combination of Mecp2-null and MBD2-null mice has been developed. These double-mutant mice show the same onset of symptoms and mortality as single Mecp2-null mice, providing no evidence for a genetic interaction between Mecp2 and MBD2 (Guy et al. 2001).

Another mouse model expressing a truncated protein was generated by Shahbazian et al. (2002). In this case, they eliminated approximately the C-terminal third of the coding sequence and found that mutant male mice exhibit no apparent abnormalities until around 6 weeks of age. After this, they develop numerous neurological features reminiscent of RTT. Body tremors and motor abnormalities may first be noted at 2 months of age, and are progressive (Shahbazian et al. 2002). This model is the one that best resembles the RTT phenotype.

18.2.3 Histone Methylations

Histone methylation is another well-studied histone modification. It is catalyzed by a family of enzymes known as HMTs. The methyl group can be removed by the recently identified group of proteins called histone demethylases (HDMs). SET domain-containing proteins are potential HMTases, and the mouse genome encodes at least six H3K4 methyltransferases and at least seven H3K9 methyltransferases (Glaser et al. 2006).

Suv39 is an H3K9 methyltransferase associated predominantly with pericentric heterochromatin. Murine Suv39h HMTases are encoded by two loci, Suv39h1 and Suv39h2, which have overlapping expression profiles during embryogenesis, while in adult mice Suv39h2 is mainly expressed in testes (O'Carroll et al. 2000). Mice deficient in either Suv39h1 or Suv39h2 have normal viability and fertility and do not show any unusual phenotypic traits (Peters et al. 2001). Nevertheless, double-Suv39h-null mice have severely impaired viability and chromosomal instabilities that are associated with increased tumor risk, resulting from the greatly reduced level of H3K9 methylation. Furthermore, these animals are born at sub-Mendelian frequencies, and males and females are both growth-retarded and infertile.

Analysis of mouse embryogenesis indicates that development is normal until day 12.5, but in later stages Suv39h-null fetuses are smaller and show higher rates of resorption and prenatal lethality.

These mice have a predisposition to cancer, illustrated by their tendency to develop late-onset B cell lymphomas with a penetrance of nearly 33%. The tumor cells feature nonsegregated chromosomes that are linked via their acrocentric regions. These cancer cells resemble those in human slow-progressing non-Hodgkin lymphomas, raising the possibility that SUV39H1/2 is important in the development of this disease. On the basis of these results, it can be said that Suv39h-dependent H3K9 methylation is important for maintaining the structure at pericentric heterochromatin, which is required to protect genomic stability. If Suv39 is absent, the relaxed heterochromatin is prone to aberrant interactions and may induce chromosome missegregation (Peters et al. 2001). Transgenic mice that overexpress SUV39H1 HMTase have also been generated. These animals are growth-retarded, have a weak penetrance of skeletal transformations and are characterized by impaired erythroid differentiation (Czvitkovich et al. 2001).

G9a is the main mammalian H3K9 methyltransferase targeting euchromatic regions and is essential for murine embryogenesis. This enzyme dominantly regulates H3K9 mono- and dimethylation at euchromatic regions (Rice et al. 2003). G9a-deficient embryos display severe growth retardation and early lethality, and global H3K9 methylation is drastically decreased (Tachibana et al. 2002, 2005). Further studies of G9a have demonstrated an additional link between histone methylation and DNA methylation, showing that DNA methylation at the Prader–Willi imprinting center is lost in G9a KO ES cells (Xin et al. 2003).

GLP is the only related homolog of G9a, and is also important for H3K9 methylation of mouse euchromatin. The phenotypes of G9a- and GLP-null mice are identical in many respects, including embryonic lethality, drastic reduction of H3K9 mono- and dimethylation (but not of trimethylation), induction of Mage-a genes, and HP1 relocalization in ES cells. These findings indicate that G9a and GLP function cooperatively rather than redundantly to mediate methylation of euchromatin at H3K9 (Tachibana et al. 2005).

Polycomb group (PcG) proteins are conserved throughout evolution, forming multiprotein complexes that regulate transcription through the induction of chromatin changes (Jenuwein and Allis 2001). They are involved in heritable epigenetic regulation of gene expression during development (Orlando et al. 1998). Among these PcG genes are zeste 2 enhancer (EZH2) and its interacting partner embryonic ectoderm development (EED), which are essential for early development (Sewalt et al. 1998).

There are two mammalian homologs of zeste enhancer: EZH1, which is expressed predominantly in adult tissues, and EZH2, which is expressed primarily in early development (Laible et al. 1997). EZH2 and EED are part of a single multimeric complex that includes HDAC 1 and 2 (Tie et al. 2001; van der Vlag and Otte 1999).

EZH2 is a SET domain-containing protein associated with methylation of H3K27 (Strahl and Allis 2000). This gene is amplified in several neoplasias, leading to an increase in the rate of cell proliferation (Varambally et al. 2002). In mice, ezh2 is detected as a maternally inherited gene. The use of a conditional allele of ezh2 to deplete the oocyte of this maternal inheritance showed that it has a long-term effect, severely retarding growth of neonates despite rescue through embryonic transcription from the paternal allele (Erhardt et al. 2003). Ezh2-null mice suffer lethality at early embryonic stages (approximately 6.5 dpc) (O'Carroll et al. 2001).

The SET domain in *NSD1* possesses intrinsic HMT activity with specificity for Lys 36H3 and Lys 20H4. Members of the NSD family have all been implicated in human malignancy, which suggests that this subgroup of SET domain proteins has a key role in controlling cell growth and differentiation. NSD1 mutation has also been implicated in Sotos syndrome, a rare growth disorder also known as cerebral gigantism.

Mice heterozygous for the NSD1 mutation are viable and fertile. They have a normal growth rate, indicating that the Sotos phenotype in mice may be more subtle than in man. However, it is possible that the target genes in mouse differ from those in humans.

The NSD1 mutation is recessive embryonic lethal. NSD1-deficient mice are able to initiate mesoderm formation, but display a high incidence of apoptosis and fail to complete gastrulation, suggesting that it is a developmental regulatory gene essential for early postimplantation (Rayasam et al. 2003).

The mixed-lineage leukemia gene (MLL/HRX/ALL-1) codes for a large protein that shows homology with *Drosophila* trithorax. This gene is disrupted by chromosomal translocation in many human acute leukemias (ALL and AML) and the prognosis of these patients is poor (Yagi et al. 1998; Rubnitz et al. 1996). The translocation of MLL results in the loss of one functional copy and the generation of a chimeric fusion protein with potential dominant-negative or neomorphic activity. MLL is a positive regulator of Hox genes, which have been implicated in both axial skeleton patterning and hematopoietic development. It possesses a highly conserved SET domain by which it interacts with components of the SWI/SNF complex (Rozenblatt-Rosen et al. 1998; Kingston et al. 1996).

To assess its role in development, mutant mice were generated by disrupting exon 3B. MLL heterozygous mice have retarded growth and show hematopoietic abnormalities. These animals also display bidirectional homeotic transformations of the axial skeleton as well as sternal malformations (Hess et al. 1997).

MLL-null mice are embryonic lethal and show severe developmental abnormalities. The mesenchyme of arch tissue is hypocellular and shows evidence of apoptosis. Histological sections of the MLL-null embryos recovered up to E10.5 show abundant erythropoiesis, with pooling of erythroid cells in the coelomic cavity. The early lethality of these mice occurs before the time when the predominant site of hematopoiesis shifts from yolk sac to liver. This precludes assessment of the role of MLL in fetal liver or adult bone marrow hematopoiesis, and the role of MLL in lymphopoiesis (Yu et al. 1995, 1998; Hess et al. 1997).

Another group has generated mutant mice of MLL by replacing the region, including exons 12–14, because they consider the function of MLL to be dependent on the 3′ terminus of exon 14. Homozygous mutant mice die at E11.5–14.5, with edematous bodies and obvious bleeding under the skin. Histological examination reveals that there are fewer hematopoietic cells in the liver than in MLL Wt and heterozygous littermates although they are composed of erythroid, myeloid, monocytic, and megakaryocytic cells that differentiate normally (Yagi et al. 1998).

The difference in time of lethality between these two models could be because a truncated MLL protein is produced in the latter that retains some functions of MLL. However, it is clear that MLL is required for definitive hematopoiesis (Ernst et al. 2004a) and that it sustains expression of certain Hox genes (Yu et al. 1998; Ernst et al. 2004b).

Another model for this gene was generated by Ayton et al. (2001) in which truncation of exon 5 is lethal at the two-cell stage in homozygotes, and heterozygotes exhibit mild skeletal defects and some defects of neuroectodermal derivatives.

Mammals have a second trithorax homolog, *MLL2*, which methylates H3K4. Its loss provokes slow growth, increases apoptosis and retards development, leading to embryonic failure before E11.5. MLL-null embryos or ES cells show no decrease in global H3K4 dimethylation levels, indicating that MLL2 is not responsible for this modification in any major way (Glaser et al. 2006).

RIZ1 is an H3K9 methyltransferase frequently silenced in human cancer (Chadwick et al. 2000; Fang et al. 2000, 2001). Knockout mice that lack RIZ1 produce normal offspring with no abnormal gross phenotype. However, they are prone to developing B cell lymphomas and a broad spectrum of unusual tumors. These studies suggest that RIZ1 can be considered a tumor-susceptibility gene in mice (Steele-Perkins et al. 2001).

18.2.4 Histone Acetyltransferases

The acetylation of lysine residues located in the N-terminal tails of histone is one of the key features associated with active gene transcription. The addition of acetyl groups neutralizes the positive charge of lysine, which affects the interaction of the histone tails with DNA, RNA and protein. The acetyl group also provides a specific binding site for certain proteins via their bromodomain. Two classes of proteins determine the acetylation status of histones: HATs and HDACs (Gibbons 2005).

Mice heterozygous for *Gcn5l2* are obtained at expected rates, grow normally, reach sexual maturity at an age comparable to that of wild-type mice and show no abnormal phenotype. However, Gcn5l2-null mice are embryonic lethal (Yamauchi et al. 2000).

PCAF plays a role in transcriptional activation, cell-cycle arrest and cell differentiation in cultured cells by acetylating chromatin and transcription factors. Mice null for this gene are viable and fertile with no unusual phenotypic traits. The lack of obvious phenotypic differences in Pcaf KO mice suggests that Gcn5l2 functionally compensates for the missing Pcaf. In Pcaf-null mice, Gcn5l2 protein is drastically overproduced in lung and liver, whereas in wild-type mice Pcaf is dominantly expressed in these tissues (Xu et al. 2000; Yamauchi et al. 2000).

Mice heterozygous for Gcn 512 and null for Pcaf are morphologically normal, but mice null for both genes die at approximately 7.5 dpc, several days earlier than the Gcn512 single mutants. The combined loss of Pcaf and Gcn512 leads to more severe defects than those observed when Gcn512 alone is lost, which therefore means that the functions of Gcn512 and Pcaf must coincide before 6.5 dpc (Xu et al. 2000).

P300 and CBP make up the other family of HATs. Several lines of evidence indicate that deregulation of their activity can cause cellular transformation.

When a single p300 allele is inactivated, the resultant embryos suffer significantly reduced viability (up to 55% die in utero), although heterozygotes that do survive do not suffer from further p300 insufficiency after birth.

Animals lacking both p300 alleles die around mid-gestation despite the presence of normal quantities of highly homologous CBP. These embryos have pleiotropic defects in morphogenesis, cell differentiation, and proliferation.

All p300 mutants and some heterozygous embryos feature a severe open neural tube. A significant number of heterozygous embryos also have exencephaly, which is generally restricted to the cranial region (Yao et al. 1998).

CBP is mutated in Rubinstein–Taybi Syndrome (RTS), an autosomal dominant syndrome characterized by abnormal pattern formation, mental retardation, craniofacial malformations (including hypoplastic maxilla with narrow palate, downward-slanting palpebral fissures and large anterior fontanel), microcephaly, broad thumbs and toes. In addition, they also have a 350-fold greater risk of developing cancer, usually with a neural crest origin (Miller and Rubinstein 1995). Interestingly, affected individuals are heterozygous at the CBP locus, suggesting that the dose of the CBP gene is critically important in humans.

Three groups have reported CBP mutant mice: Tanaka generated a $CBP^{+/-}$ null mutant mouse in which amino acids 29–265 of CBP are replaced by a targeting vector using homologous recombination. However, those null mutants exhibit only growth retardation and skeletal abnormalities, such as a large anterior fontanel and distinct holes in the xiphoid process. Similar anomalies have also been reported for RTS patients. Furthermore, the frequency of these abnormalities was affected by the genetic background. However, some of the anomalies observed in RTS, such as cardiac anomalies, are not seen in CBP heterozygous mutant mice (Tanaka et al. 1997). Homozygous mutants die around E10.5–E12.5, as a result of massive hemorrhage caused by defective blood vessel formation in the central nervous system, and exhibit developmental retardation as well as delays in both primitive and definitive hematopoiesis. CBP-null embryos show defective neural tube closure and exencephaly, but do not display anomalous heart formation (Tanaka et al. 2000).

Yao et al. (1998) generated CBP-deficient mice by gene targeting. Although CBP-null mice in this study show defective neural tube closure similar to the p300-null mice, they feature no other deleterious characteristics. All KO embryos die by E10.5 (Yao et al. 1998). Oike et al. (1999a, b) generated *CBP*-mutant mice with truncated CBP protein, and found that heterozygotes show growth retardation with respect to weight and height, poor locomotor activity, long-term memory deficits, retarded osseous maturation, large anterior fontanel, hypoplastic maxilla with narrow palate, short nose and a broad nasal bridge, cardiac anomalies, skeletal abnormalities, and occasional seizures. However, they are fertile. The CBP mice generated in this study strongly mimic the clinical features of RTS (Oike et al. 1999a, b).

The homozygous CBP mutant is embryonic lethal with abnormalities of vascular network formation, especially abnormal development of endothelial cells which can cause the membranous-type ventricular septal defect. These animals express a truncated form of CBP protein, and die between E9.5 and E10.5 with defective neural tube closure. They appear pale and have a reduced number of erythroid cells and colony-forming cells in the yolk sac, suggesting defects in primitive hematopoiesis (Oike et al. 1999b). The differences between the two models can be explained by a haploinsufficiency mechanism in the mice of Tanaka et al., and by a dominant-negative mechanism in those of Oike et al.

No viable compound heterozygotes for p300 and CBP have been detected by the time of weaning. Indeed, like p300 and CBP single null embryos, the compound heterozygous embryos are also severely stunted compared with littermates and have open neural tube defects similar to those observed in p300 or CBP-null animals. Since both p300 and CBP-null embryos suffer early lethality, this in turn indicates that there is a significant p300 and cbp gene dosage requirement during embryogenesis (Yao et al. 1998).

18.2.5 Histone Deacetylases

In addition to aberrant promoter DNA hypermethylation, epigenetic gene silencing can also occur by aberrant targeting of HDACs to the gene promoter, which causes histone hypoacetylation. The typical gene inactivated by this process is the tumor suppressor p21. HDAC inhibitor compounds may achieve some of their antitumoral effects through reactivation of these tumor suppressor genes.

There are four families of HDACs: class I (HDAC1, 2, 3, and 8), class II (HDAC4, 5, 6, 7, 9, and 10), class III HDACs (Sirtuin1–7), and class IV (HDAC11). Those of classes I, II, and IV have a similar sequence and structure, but the Sirtuins have a different structural homology and use a different catalytic mechanism that is dependent on NAD^+ (Villar-Garea and Esteller 2004).

The class I HDACs are widely expressed and consist mainly of a catalytic domain. In contrast, the class II HDACs (HDAC4, 5, 7, and 9) display cell-type-restricted patterns of expression and contain an N-terminal extension that links them to specific transcription factors and confers responsiveness to a variety of signal transduction pathways, thereby connecting the genome with the extracellular environment (Verdin et al. 2003).

Targeted disruption of both *HDAC1* alleles results in embryonic lethality (presumably due to increased expression of cell-cycle inhibitors) before E10.5 due to severe proliferation defects and retardation in development.

HDAC1-null ES cells show reduced proliferation rates, significantly reduced overall deacetylase activity, hyperacetylation of a subset of histones H3 and H4 and concomitant changes in other histone modifications. Heterozygous animals are obtained at lower rates, but they are viable and fertile and appear to have a normal phenotype (Lagger et al. 2002).

HDAC2-deficient mice have been created from a gene-trap embryonic stem cell (ES) line. Loss of *HDAC2* results in partial perinatal lethality but surviving mice recover and appear overall normal after 2 months of age (Trivedi et al. 2007).

A study involving genetic inactivation of HDAC2 crossed with tumor prone APC-null mice revealed that the development of intestinal tumors due to mutation of APC is reduced in HDAC2-null mice, thereby implying that a single HDAC isoform acts during intestinal tumor development. The prominent phenotypic characteristic of these mice is their reduced body size. In any case, these animals have normal life spans, but females are subfertile for reasons that are currently unknown. The intestines of 1–4-month-old mice have reduced thickness and length (Zimmermann et al. 2007).

HDAC4 is expressed in prehypertrophic chondrocytes; disruption of the mouse HDAC4 gene by homologous recombination generates mice characterized by inappropriate chondrocyte hypertrophy leading to ectopic bone formation. Within days of their birth, homozygous mutants are readily identifiable by their "dome-shaped" heads and misshapen spines. Some mutants also display exencephaly, and none of them survives to weaning. Conversely, overexpression of HDAC4 in proliferating chondrocytes *in vivo* inhibits chondrocyte hypertrophy and differentiation (Vega et al. 2004).

Sirtuins comprise the class III family of HDACs. They are NAD^+-dependent and are involved in many cellular events, such as transcriptional silencing, chromatin remodeling, mitosis, and lifespan duration (Guarente 2000).

Mammalian cells contain seven sirtuins, SIRT1 to SIRT7. Despite being involved in a wide range of functions, their main role is to act as sensors of the redox state of the cell/organism. To date, only SIRT1, SIRT2, and SIRT3 have been identified as deacetylases while the rest are only mono-ADP ribosyltransferases (North et al. 2003; Haigis et al. 2006; Liszt et al. 2005). SIRT1-SIRT3 HDAC activity exhibits a preference for acetylated K16 of histone H4 and also for K9H3.

Mammalian sirtuin 1 (*SIRT1*) deacetylases not only histones (mainly K16H4 and K9H3) (Vaquero et al. 2004; Pruitt et al. 2006), but also p53 (Vaziri et al. 2001), p300 (Bouras et al. 2005) and many more genes. SIRT1 is expressed in most mammalian somatic and germ tissues (Afshar and Murnane 1999) and is upregulated in several mouse and human cancers (Chen et al. 2005; Yeung et al. 2004; Luo et al. 2001; Vaziri et al. 2001).

SIRT1 KO mice are smaller at birth and they subsequently develop more slowly than their littermates. Most of them die during the early postnatal period. The mutants also show defects in gametogenesis, sterility, chronic lung infection and atrophy of the pancreas. The latter two defects are probably responsible for the early postnatal lethality observed (McBurney et al. 2003).

SIRT3 is the only sirtuin clearly associated with lifespan increase; its levels are upregulated by caloric restriction and cold exposure in brown and white adipose tissue. SIRT3 KO mice show hyperacetylation of numerous mitochondrial proteins, suggesting that SIRT3 might be a global deacetylase in this organelle. However, the increase in mitochondrial lysine acetylation resulting from loss of SIRT3 does not produce any metabolic abnormalities, so these mice are indistinguishable from wild-type individuals, and are healthy and fertile (Lombard et al. 2007).

Cancer has been associated with global loss of monoacetylated K16H4 and trimethylated K20H4, which suggests that a sirtuin may be involved (Fraga et al. 2005). That SIRT1 and SIRT2 KO mice produce global H4K16 hyperacetylation and that SIRT1 is specifically upregulated in tumors suggest that this protein's action on chromatin could be involved in regulating cancer. By contrast, SIRT2 is downregulated in some cancers, indicating its role as a tumor suppressor (Vaquero et al. 2006).

18.2.6 SWI/SNF

Switch/sucrose nonfermentable (SWI/SNF) complexes use energy provided by ATP hydrolysis to break histone-DNA contacts and to slide nucleosomes to alternate positions on the same strand of DNA (Cote et al. 1994).

Mammalian SWI/SNF complexes consist of 9–12 subunits, with those from different tissues showing significant heterogeneity (Kim et al. 2001). These complexes have two catalytic subunits, encoded by brahma (Brm) (also known as SNF2 alpha or Smarca2) and brahma-related gene 1 (Brg1) (also known as SNF2 beta or Smarca4) that are present in two distinct MDa BAF complexes (Khavari et al. 1993).

Components of mammalian SWI/SNF complexes have been implicated in a variety of cellular processes, including gene activation and repression, development and differentiation, recombination and repair, and cell-cycle control (Guidi et al. 2001).

SWI/SNF complexes are evolutionarily conserved and involved in transcriptional activation of a considerable number of genes through chromatin remodeling (Klochendler-Yeivin et al. 2000).

Mouse *Brg1* locus was inactivated by homologous recombination in ES cells. In addition to being a recessive lethal, the Brg1 mutation confers exencephaly at embryonic day E16.5–18.5 in 15–30% of heterozygotes. Although the majority of Brg1 heterozygotes survive and are fertile, they are susceptible, with a latency of 14 months, to mammary tumors that arise because of haploinsufficiency. They exhibit genomic instability but not polyploidy, in contrast to SNF5 heterozygotes.

Brg1-null mice can only be recovered before implantation as blastocysts at E3.5 but not following implantation at E6.5 or later, indicating that Brg1 homozygotes die during the peri-implantation stage. Although Brg1-homozygous blastocysts appear to be morphologically normal, they fail to hatch from their zona pellucidae or survive *in vitro* (Bultman et al. 2000, 2008).

SNF5/INI1 is a component of both BRG1 and BRM ATP-dependent chromatin remodeling family SWI/SNF. Ini1 interacts with ALL-1, translocations of which are associated with several types of human acute leukemias. Three groups have reported mutant mice for this gene.

A loss-of-function mutation of the SNF5 gene has been generated in ES cells by replacing exons 1 and 2. Homozygous mice lacking SNF5 stop developing and die at the peri-implantation stage. 32% of heterozygous mice develop tumors by the age of 15 months. Tumors may be classified as undifferentiated sarcomas with variable rhabdoid features and are detected at different locations, most commonly at intracranial and paravertebral sites. In these tumors, the wild-type allele was lost, providing further evidence that SNF5 functions as a tumor suppressor gene in certain cell types (Klochendler-Yeivin et al. 2000).

SNF5 is widely expressed during embryogenesis with focal areas of high levels of expression in the mandibular portion of the first branchial arch and central nervous system. Homozygous KO mice generated by Roberts et al. (2000) suffered embryonic lethality by E7.5, before the onset of gastrulation and organogenesis, whereas an expected number of heterozygous mice was born and the offspring were normal in appearance, growth, and fertility. However at 5 weeks of age, heterozygotes develop tumors consistent with malignant rhabdoid tumor (MRT). The tumors behave aggressively, displaying rapid growth, epidermal ulceration, and hemorrhage. The majority of tumors arise in soft tissues derived from the first branchial arch (Roberts et al. 2000).

As mice deficient in either Brg1 or Snf5 die early in embryonic development, only haploinsufficient mice can be assessed for cancer occurrence. Using a reversibly inactivating conditional allele, loss of SNF5 results in a highly penetrant cancer predisposition. 100% of mice develop mature CD8 T cell lymphoma or rare rhabdoid tumors with a latency of 11 weeks (Roberts et al. 2002).

Guidi et al. (2001) developed SNF5-null embryos that die between 3.5 and 5.5 dpc, probably due to an inability of the blastocyst to hatch, implant in the uterus and continue development. Approximately 15% of SNF5-heterozygous mice have tumors, the majority of which are undifferentiated or poorly differentiated sarcomas. All of these tumors arise in the head and neck region, particularly in the soft tissue of the face (Guidi et al. 2001). Generation of mice with a lox conditional allele of SNF5 in the developing liver caused neonatal death due to severe hypoglycemia. Mutant animals fail to store glycogen and have impaired energetic metabolism. Close examination reveals that mutant mice are born normally, but die within 12 h of birth. The formation of hepatic epithelium is also affected. SNF5 is essential for the assembly of all types of epithelial cell–cell junctions. Nonetheless, neither growth retardation nor macroscopic liver defects occur at birth. The transcriptome analysis of SNF5 KO livers at E18.5 showed that genes involved in cell-cycle control are deregulated. The strongest affected gene was p21, but several genes involved in apoptosis were down-regulated in addition. Interestingly, most of these were proapoptotic genes, showing that SNF5 deletion leads to increased cell proliferation. This defect could be due to partial inactivation of the G1/S checkpoint (Gresh et al. 2005).

It is surprising that BRG1 and SNF5 are the only SWI/SNF-related subunits known to protect against cancer, considering that several other subunits interact with cancer-related proteins and are not expressed in tumor-derived cell lines. Some of these subunits have been mutated (Srg3, Baf180) or perturbed by RNAi (Baf60c) and found to confer embryonic lethality. Nevertheless, heterozygotes or partial KO individuals are not predisposed to cancer (Lickert et al. 2004; Wang et al. 2004).

18.3 Conclusion

The requirement for de novo and maintenance DNA methyltransferases, several HMTs, acetylases and some HDACs during mouse embryonic development indicates that epigenetic silencing of specific genes or of the genome as a whole is an essential process during mouse development. In recent years, we have seen the rapid growth in the number of diseases, including cancer, found to be caused by epigenetic alterations. Mutant mouse models, such as those summarized in this chapter, will be very useful for further elucidating the pathogenetic and molecular mechanisms underlying diseases with epigenetic contributions.

References

Afshar G, Murnane JP (1999) Characterization of a human gene with sequence homology to *Saccharomyces cerevisiae* SIR2. Gene 234(1):161–168

Ayton P, Sneddon SF, Palmer DB, Rosewell IR, Owen MJ, Young B, Presley R, Subramanian V (2001) Truncation of the Mll gene in exon 5 by gene targeting leads to early preimplantation lethality of homozygous embryos. Genesis 30(4):201–212

Bader S, Walker M, Harrison D (2000) Most microsatellite unstable sporadic colorectal carcinomas carry MBD4 mutations. Br J Cancer 83(12):1646–1649

Ballestar E, Esteller M (2002) The impact of chromatin in human cancer: linking DNA methylation to gene silencing. Carcinogenesis 23(7):1103–1109

Bauman ML, Kemper TL, Arin DM (1995) Pervasive neuroanatomic abnormalities of the brain in three cases of Rett's syndrome. Neurology 45(8):1581–1586

Baylin SB, Herman JG, Graff JR, Vertino PM, Issa JP (1998) Alterations in DNA methylation: a fundamental aspect of neoplasia. Adv Cancer Res 72:141–196

Beard C, Li E, Jaenisch R (1995) Loss of methylation activates Xist in somatic but not in embryonic cells. Genes Dev 9(19):2325–2334

Bernstein BE, Meissner A, Lander ES (2007) The mammalian epigenome. Cell 128(4):669–681

Bird A (2002) DNA methylation patterns and epigenetic memory. Genes Dev 16(1):6–21

Bouras T, Fu M, Sauve AA, Wang F, Quong AA, Perkins ND, Hay RT, Gu W, Pestell RG (2005) SIRT1 deacetylation and repression of p300 involves lysine residues 1020/1024 within the cell cycle regulatory domain 1. J Biol Chem 280(11):10264–10276

Brami-Cherrier K, Valjent E, Herve D, Darragh J, Corvol JC, Pages C, Arthur SJ, Girault JA, Caboche J (2005) Parsing molecular and behavioral effects of cocaine in mitogen- and stress-activated protein kinase-1-deficient mice. J Neurosci 25(49):11444–11454

Bultman S, Gebuhr T, Yee D, La Mantia C, Nicholson J, Gilliam A, Randazzo F, Metzger D, Chambon P, Crabtree G, Magnuson T (2000) A Brg1 null mutation in the mouse reveals functional differences among mammalian SWI/SNF complexes. Mol Cell 6(6):1287–1295

Bultman SJ, Herschkowitz JI, Godfrey V, Gebuhr TC, Yaniv M, Perou CM, Magnuson T (2008) Characterization of mammary tumors from Brg1 heterozygous mice. Oncogene 27(4): 460–468

Chadwick RB, Jiang GL, Bennington GA, Yuan B, Johnson CK, Stevens MW, Niemann TH, Peltomaki P, Huang S, de la Chapelle A (2000) Candidate tumor suppressor RIZ is frequently involved in colorectal carcinogenesis. Proc Natl Acad Sci USA 97(6):2662–2667

Chen RZ, Akbarian S, Tudor M, Jaenisch R (2001) Deficiency of methyl-CpG binding protein-2 in CNS neurons results in a Rett-like phenotype in mice. Nat Genet 27(3):327–331

Chen WY, Wang DH, Yen RC, Luo J, Gu W, Baylin SB (2005) Tumor suppressor HIC1 directly regulates SIRT1 to modulate p53-dependent DNA-damage responses. Cell 123(3):437–448

Cheng HL, Mostoslavsky R, Saito S, Manis JP, Gu Y, Patel P, Bronson R, Appella E, Alt FW, Chua KF (2003) Developmental defects and p53 hyperacetylation in Sir2 homolog (SIRT1)-deficient mice. Proc Natl Acad Sci USA 100(19):10794–10799

Cote J, Quinn J, Workman JL, Peterson CL (1994) Stimulation of GAL4 derivative binding to nucleosomal DNA by the yeast SWI/SNF complex. Science 265(5168):53–60

Czvitkovich S, Sauer S, Peters AH, Deiner E, Wolf A, Laible G, Opravil S, Beug H, Jenuwein T (2001) Over-expression of the SUV39H1 histone methyltransferase induces altered proliferation and differentiation in transgenic mice. Mech Dev 107(1–2):141–153

Eden A, Gaudet F, Waghmare A, Jaenisch R (2003) Chromosomal instability and tumors promoted by DNA hypomethylation. Science 300(5618):455

Erhardt S, Su IH, Schneider R, Barton S, Bannister AJ, Perez-Burgos L, Jenuwein T, Kouzarides T, Tarakhovsky A, Surani MA (2003) Consequences of the depletion of zygotic and embryonic enhancer of zeste 2 during preimplantation mouse development. Development 130(18):4235–4248

Ernst P, Fisher JK, Avery W, Wade S, Foy D, Korsmeyer SJ (2004a) Definitive hematopoiesis requires the mixed-lineage leukemia gene. Dev Cell 6(3):437–443

Ernst P, Mabon M, Davidson AJ, Zon LI, Korsmeyer SJ (2004b) An Mll-dependent Hox program drives hematopoietic progenitor expansion. Curr Biol 14(22):2063–2069

Esteller M (2007) Cancer epigenomics: DNA methylomes and histone-modification maps. Nat Rev Genet 8(4):286–298

Fang W, Piao Z, Simon D, Sheu JC, Huang S (2000) Mapping of a minimal deleted region in human hepatocellular carcinoma to 1p36.13-p36.23 and mutational analysis of the RIZ (PRDM2) gene localized to the region. Genes Chromosomes Cancer 28(3):269–275

Fang W, Piao Z, Buyse IM, Simon D, Sheu JC, Perucho M, Huang S (2001) Preferential loss of a polymorphic RIZ allele in human hepatocellular carcinoma. Br J Cancer 84(6):743–747

Feinberg AP, Ohlsson R, Henikoff S (2006) The epigenetic progenitor origin of human cancer. Nat Rev Genet 7(1):21–33

Fraga MF, Ballestar E, Villar-Garea A, Boix-Chornet M, Espada J, Schotta G, Bonaldi T, Haydon C, Ropero S, Petrie K, Iyer NG, Perez-Rosado A, Calvo E, Lopez JA, Cano A, Calasanz MJ, Colomer D, Piris MA, Ahn N, Imhof A, Caldas C, Jenuwein T, Esteller M (2005) Loss of acetylation at Lys16 and trimethylation at Lys20 of histone H4 is a common hallmark of human cancer. Nat Genet 37(4):391–400

Gaudet F, Hodgson JG, Eden A, Jackson-Grusby L, Dausman J, Gray JW, Leonhardt H, Jaenisch R (2003) Induction of tumors in mice by genomic hypomethylation. Science 300(5618): 489–492

Gibbons RJ (2005) Histone modifying and chromatin remodelling enzymes in cancer and dysplastic syndromes. Hum Mol Genet 14(Spec No 1):R85–R92

Glaser S, Schaft J, Lubitz S, Vintersten K, van der Hoeven F, Tufteland KR, Aasland R, Anastassiadis K, Ang SL, Stewart AF (2006) Multiple epigenetic maintenance factors implicated by the loss of Mll2 in mouse development. Development 133(8):1423–1432

Gresh L, Bourachot B, Reimann A, Guigas B, Fiette L, Garbay S, Muchardt C, Hue L, Pontoglio M, Yaniv M, Klochendler-Yeivin A (2005) The SWI/SNF chromatin-remodeling complex subunit SNF5 is essential for hepatocyte differentiation. EMBO J 24(18):3313–3324

Guarente L (2000) Sir2 links chromatin silencing, metabolism, and aging. Genes Dev 14(9):1021–1026

Guidi CJ, Sands AT, Zambrowicz BP, Turner TK, Demers DA, Webster W, Smith TW, Imbalzano AN, Jones SN (2001) Disruption of Ini1 leads to peri-implantation lethality and tumorigenesis in mice. Mol Cell Biol 21(10):3598–3603

Guy J, Hendrich B, Holmes M, Martin JE, Bird A (2001) A mouse Mecp2-null mutation causes neurological symptoms that mimic Rett syndrome. Nat Genet 27(3):322–326

Guy J, Gan J, Selfridge J, Cobb S, Bird A (2007) Reversal of neurological defects in a mouse model of Rett syndrome. Science 315(5815):1143–1147

Hagberg B, Aicardi J, Dias K, Ramos O (1983) A progressive syndrome of autism, dementia, ataxia, and loss of purposeful hand use in girls: Rett's syndrome: report of 35 cases. Ann Neurol 14(4):471–479

Haig D (2004) The (dual) origin of epigenetics. Cold Spring Harb Symp Quant Biol 69:67–70
Haigis MC, Mostoslavsky R, Haigis KM, Fahie K, Christodoulou DC, Murphy AJ, Valenzuela DM, Yancopoulos GD, Karow M, Blander G, Wolberger C, Prolla TA, Weindruch R, Alt FW, Guarente L (2006) SIRT4 inhibits glutamate dehydrogenase and opposes the effects of calorie restriction in pancreatic beta cells. Cell 126(5):941–954
Hendrich B, Bird A (1998) Identification and characterization of a family of mammalian methyl-CpG binding proteins. Mol Cell Biol 18(11):6538–6547
Hendrich B, Hardeland U, Ng HH, Jiricny J, Bird A (1999) The thymine glycosylase MBD4 can bind to the product of deamination at methylated CpG sites. Nature 401(6750):301–304
Hendrich B, Guy J, Ramsahoye B, Wilson VA, Bird A (2001) Closely related proteins MBD2 and MBD3 play distinctive but interacting roles in mouse development. Genes Dev 15(6):710–723
Hess JL, Yu BD, Li B, Hanson R, Korsmeyer SJ (1997) Defects in yolk sac hematopoiesis in Mll-null embryos. Blood 90(5):1799–1806
Jenuwein T, Allis CD (2001) Translating the histone code. Science 293(5532):1074–1080
Jones PL, Veenstra GJ, Wade PA, Vermaak D, Kass SU, Landsberger N, Strouboulis J, Wolffe AP (1998) Methylated DNA and MeCP2 recruit histone deacetylase to repress transcription. Nat Genet 19(2):187–191
Khavari PA, Peterson CL, Tamkun JW, Mendel DB, Crabtree GR (1993) BRG1 contains a conserved domain of the SWI2/SNF2 family necessary for normal mitotic growth and transcription. Nature 366(6451):170–174
Kim JK, Huh SO, Choi H, Lee KS, Shin D, Lee C, Nam JS, Kim H, Chung H, Lee HW, Park SD, Seong RH (2001) Srg3, a mouse homolog of yeast SWI3, is essential for early embryogenesis and involved in brain development. Mol Cell Biol 21(22):7787–7795
Kingston RE, Bunker CA, Imbalzano AN (1996) Repression and activation by multiprotein complexes that alter chromatin structure. Genes Dev 10(8):905–920
Klochendler-Yeivin A, Fiette L, Barra J, Muchardt C, Babinet C, Yaniv M (2000) The murine SNF5/INI1 chromatin remodeling factor is essential for embryonic development and tumor suppression. EMBO Rep 1(6):500–506
Kouzarides T (2007) Chromatin modifications and their function. Cell 128(4):693–705
Lagger G, O'Carroll D, Rembold M, Khier H, Tischler J, Weitzer G, Schuettengruber B, Hauser C, Brunmeir R, Jenuwein T, Seiser C (2002) Essential function of histone deacetylase 1 in proliferation control and CDK inhibitor repression. EMBO J 21(11):2672–2681
Laible G, Wolf A, Dorn R, Reuter G, Nislow C, Lebersorger A, Popkin D, Pillus L, Jenuwein T (1997) Mammalian homologues of the polycomb-group gene enhancer of zeste mediate gene silencing in *Drosophila* heterochromatin and at *S. cerevisiae* telomeres. EMBO J 16(11):3219–3232
Laird PW, Jackson-Grusby L, Fazeli A, Dickinson SL, Jung WE, Li E, Weinberg RA, Jaenisch R (1995) Suppression of intestinal neoplasia by DNA hypomethylation. Cell 81(2):197–205
Li E, Bestor TH, Jaenisch R (1992) Targeted mutation of the DNA methyltransferase gene results in embryonic lethality. Cell 69(6):915–926
Li E, Beard C, Forster AC, Bestor TH, Jaenisch R (1993) DNA methylation, genomic imprinting, and mammalian development. Cold Spring Harb Symp Quant Biol 58:297–305
Lickert H, Takeuchi JK, Von Both I, Walls JR, McAuliffe F, Adamson SL, Henkelman RM, Wrana JL, Rossant J, Bruneau BG (2004) Baf60c is essential for function of BAF chromatin remodelling complexes in heart development. Nature 432(7013):107–112
Liszt G, Ford E, Kurtev M, Guarente L (2005) Mouse Sir2 homolog SIRT6 is a nuclear ADP-ribosyltransferase. J Biol Chem 280(22):21313–21320
Lombard DB, Alt FW, Cheng HL, Bunkenborg J, Streeper RS, Mostoslavsky R, Kim J, Yancopoulos G, Valenzuela D, Murphy A, Yang Y, Chen Y, Hirschey MD, Bronson RT, Haigis M, Guarente LP, Farese RV Jr, Weissman S, Verdin E, Schwer B (2007) Mammalian Sir2 homolog SIRT3 regulates global mitochondrial lysine acetylation. Mol Cell Biol 27(24):8807–8814
Luo J, Nikolaev AY, Imai S, Chen D, Su F, Shiloh A, Guarente L, Gu W (2001) Negative control of p53 by Sir2alpha promotes cell survival under stress. Cell 107(2):137–148

Maraschio P, Zuffardi O, Dalla Fior T, Tiepolo L (1988) Immunodeficiency, centromeric heterochromatin instability of chromosomes 1, 9, and 16, and facial anomalies: the ICF syndrome. J Med Genet 25(3):173–180

McBurney MW, Yang X, Jardine K, Hixon M, Boekelheide K, Webb JR, Lansdorp PM, Lemieux M (2003) The mammalian SIR2alpha protein has a role in embryogenesis and gametogenesis. Mol Cell Biol 23(1):38–54

Millar CB, Guy J, Sansom OJ, Selfridge J, MacDougall E, Hendrich B, Keightley PD, Bishop SM, Clarke AR, Bird A (2002) Enhanced CpG mutability and tumorigenesis in MBD4-deficient mice. Science 297(5580):403–405

Miller RW, Rubinstein JH (1995) Tumors in Rubinstein-Taybi syndrome. Am J Med Genet 56(1):112–115

Nan X, Ng HH, Johnson CA, Laherty CD, Turner BM, Eisenman RN, Bird A (1998) Transcriptional repression by the methyl-CpG-binding protein MeCP2 involves a histone deacetylase complex. Nature 393(6683):386–389

North BJ, Marshall BL, Borra MT, Denu JM, Verdin E (2003) The human Sir2 ortholog, SIRT2, is an NAD^+-dependent tubulin deacetylase. Mol Cell 11(2):437–444

O'Carroll D, Scherthan H, Peters AH, Opravil S, Haynes AR, Laible G, Rea S, Schmid M, Lebersorger A, Jerratsch M, Sattler L, Mattei MG, Denny P, Brown SD, Schweizer D, Jenuwein T (2000) Isolation and characterization of Suv39h2, a second histone H3 methyltransferase gene that displays testis-specific expression. Mol Cell Biol 20(24):9423–9433

O'Carroll D, Erhardt S, Pagani M, Barton SC, Surani MA, Jenuwein T (2001) The polycomb-group gene Ezh2 is required for early mouse development. Mol Cell Biol 21(13):4330–4336

Oike Y, Hata A, Mamiya T, Kaname T, Noda Y, Suzuki M, Yasue H, Nabeshima T, Araki K, Yamamura K (1999a) Truncated CBP protein leads to classical Rubinstein-Taybi syndrome phenotypes in mice: implications for a dominant-negative mechanism. Hum Mol Genet 8(3):387–396

Oike Y, Takakura N, Hata A, Kaname T, Akizuki M, Yamaguchi Y, Yasue H, Araki K, Yamamura K, Suda T (1999b) Mice homozygous for a truncated form of CREB-binding protein exhibit defects in hematopoiesis and vasculo-angiogenesis. Blood 93(9):2771–2779

Okano M, Li E (2002) Genetic analyses of DNA methyltransferase genes in mouse model system. J Nutr 132(8 Suppl):2462S–2465S

Okano M, Xie S, Li E (1998) Dnmt2 is not required for de novo and maintenance methylation of viral DNA in embryonic stem cells. Nucleic Acids Res 26(11):2536–2540

Okano M, Bell DW, Haber DA, Li E (1999) DNA methyltransferases Dnmt3a and Dnmt3b are essential for de novo methylation and mammalian development. Cell 99(3):247–257

Orlando V, Jane EP, Chinwalla V, Harte PJ, Paro R (1998) Binding of trithorax and polycomb proteins to the bithorax complex: dynamic changes during early *Drosophila* embryogenesis. EMBO J 17(17):5141–5150

Peters AH, O'Carroll D, Scherthan H, Mechtler K, Sauer S, Schofer C, Weipoltshammer K, Pagani M, Lachner M, Kohlmaier A, Opravil S, Doyle M, Sibilia M, Jenuwein T (2001) Loss of the Suv39h histone methyltransferases impairs mammalian heterochromatin and genome stability. Cell 107(3):323–337

Pruitt K, Zinn RL, Ohm JE, McGarvey KM, Kang SH, Watkins DN, Herman JG, Baylin SB (2006) Inhibition of SIRT1 reactivates silenced cancer genes without loss of promoter DNA hypermethylation. PLoS Genet 2(3):e40

Rayasam GV, Wendling O, Angrand PO, Mark M, Niederreither K, Song L, Lerouge T, Hager GL, Chambon P, Losson R (2003) NSD1 is essential for early post-implantation development and has a catalytically active SET domain. EMBO J 22(12):3153–3163

Reik W, Lewis A (2005) Co-evolution of X-chromosome inactivation and imprinting in mammals. Nat Rev Genet 6(5):403–410

Riccio A, Aaltonen LA, Godwin AK, Loukola A, Percesepe A, Salovaara R, Masciullo V, Genuardi M, Paravatou-Petsotas M, Bassi DE, Ruggeri BA, Klein-Szanto AJ, Testa JR, Neri G, Bellacosa A (1999) The DNA repair gene MBD4 (MED1) is mutated in human carcinomas with microsatellite instability. Nat Genet 23(3):266–268

Rice JC, Briggs SD, Ueberheide B, Barber CM, Shabanowitz J, Hunt DF, Shinkai Y, Allis CD (2003) Histone methyltransferases direct different degrees of methylation to define distinct chromatin domains. Mol Cell 12(6):1591–1598
Roberts CW, Galusha SA, McMenamin ME, Fletcher CD, Orkin SH (2000) Haploinsufficiency of Snf5 (integrase interactor 1) predisposes to malignant rhabdoid tumors in mice. Proc Natl Acad Sci USA 97(25):13796–13800
Roberts CW, Leroux MM, Fleming MD, Orkin SH (2002) Highly penetrant, rapid tumorigenesis through conditional inversion of the tumor suppressor gene Snf5. Cancer Cell 2(5):415–425
Rozenblatt-Rosen O, Rozovskaia T, Burakov D, Sedkov Y, Tillib S, Blechman J, Nakamura T, Croce CM, Mazo A, Canaani E (1998) The C-terminal SET domains of ALL-1 and trithorax interact with the INI1 and SNR1 proteins, components of the SWI/SNF complex. Proc Natl Acad Sci USA 95(8):4152–4157
Rubnitz JE, Behm FG, Downing JR (1996) 11q23 rearrangements in acute leukemia. Leukemia 10(1):74–82
Sansom OJ, Berger J, Bishop SM, Hendrich B, Bird A, Clarke AR (2003) Deficiency of Mbd2 suppresses intestinal tumorigenesis. Nat Genet 34(2):145–147
Sansom OJ, Bishop SM, Bird A, Clarke AR (2005) MBD2 deficiency does not accelerate p53 mediated lymphomagenesis. Oncogene 24(14):2430–2432
Sarraf SA, Stancheva I (2004) Methyl-CpG binding protein MBD1 couples histone H3 methylation at lysine 9 by SETDB1 to DNA replication and chromatin assembly. Mol Cell 15(4):595–605
Sewalt RG, van der Vlag J, Gunster MJ, Hamer KM, den Blaauwen JL, Satijn DP, Hendrix T, van Driel R, Otte AP (1998) Characterization of interactions between the mammalian polycomb-group proteins Enx1/EZH2 and EED suggests the existence of different mammalian polycomb-group protein complexes. Mol Cell Biol 18(6):3586–3595
Shahbazian M, Young J, Yuva-Paylor L, Spencer C, Antalffy B, Noebels J, Armstrong D, Paylor R, Zoghbi H (2002) Mice with truncated MeCP2 recapitulate many Rett syndrome features and display hyperacetylation of histone H3. Neuron 35(2):243–254
Spada F, Haemmer A, Kuch D, Rothbauer U, Schermelleh L, Kremmer E, Carell T, Langst G, Leonhardt H (2007) DNMT1 but not its interaction with the replication machinery is required for maintenance of DNA methylation in human cells. J Cell Biol 176(5):565–571
Steele-Perkins G, Fang W, Yang XH, Van Gele M, Carling T, Gu J, Buyse IM, Fletcher JA, Liu J, Bronson R, Chadwick RB, de la Chapelle A, Zhang X, Speleman F, Huang S (2001) Tumor formation and inactivation of RIZ1, an Rb-binding member of a nuclear protein-methyltransferase superfamily. Genes Dev 15(17):2250–2262
Strahl BD, Allis CD (2000) The language of covalent histone modifications. Nature 403(6765):41–45
Tachibana M, Sugimoto K, Nozaki M, Ueda J, Ohta T, Ohki M, Fukuda M, Takeda N, Niida H, Kato H, Shinkai Y (2002) G9a histone methyltransferase plays a dominant role in euchromatic histone H3 lysine 9 methylation and is essential for early embryogenesis. Genes Dev 16(14):1779–1791
Tachibana M, Ueda J, Fukuda M, Takeda N, Ohta T, Iwanari H, Sakihama T, Kodama T, Hamakubo T, Shinkai Y (2005) Histone methyltransferases G9a and GLP form heteromeric complexes and are both crucial for methylation of euchromatin at H3-K9. Genes Dev 19(7):815–826
Tanaka Y, Naruse I, Maekawa T, Masuya H, Shiroishi T, Ishii S (1997) Abnormal skeletal patterning in embryos lacking a single Cbp allele: a partial similarity with Rubinstein–Taybi syndrome. Proc Natl Acad Sci USA 94(19):10215–10220
Tanaka Y, Naruse I, Hongo T, Xu M, Nakahata T, Maekawa T, Ishii S (2000) Extensive brain hemorrhage and embryonic lethality in a mouse null mutant of CREB-binding protein. Mech Dev 95(1–2):133–145
Tate P, Skarnes W, Bird A (1996) The methyl-CpG binding protein MeCP2 is essential for embryonic development in the mouse. Nat Genet 12(2):205–208

Tie F, Furuyama T, Prasad-Sinha J, Jane E, Harte PJ (2001) The *Drosophila* polycomb group proteins ESC and E(Z) are present in a complex containing the histone-binding protein p55 and the histone deacetylase RPD3. Development 128(2):275–286

Tiepolo L, Maraschio P, Gimelli G, Cuoco C, Gargani GF, Romano C (1979) Multibranched chromosomes 1, 9, and 16 in a patient with combined IgA and IgE deficiency. Hum Genet 51(2):127–137

Trivedi CM, Luo Y, Yin Z, Zhang M, Zhu W, Wang T, Floss T, Goettlicher M, Noppinger PR, Wurst W, Ferrari VA, Abrams CS, Gruber PJ, Epstein JA (2007) Hdac2 regulates the cardiac hypertrophic response by modulating Gsk3 beta activity. Nat Med 13(3):324–331

Ueda Y, Okano M, Williams C, Chen T, Georgopoulos K, Li E (2006) Roles for Dnmt3b in mammalian development: a mouse model for the ICF syndrome. Development 133(6):1183–1192

van der Vlag J, Otte AP (1999) Transcriptional repression mediated by the human polycomb-group protein EED involves histone deacetylation. Nat Genet 23(4):474–478

Vaquero A, Scher M, Lee D, Erdjument-Bromage H, Tempst P, Reinberg D (2004) Human SirT1 interacts with histone H1 and promotes formation of facultative heterochromatin. Mol Cell 16(1):93–105

Vaquero A, Scher MB, Lee DH, Sutton A, Cheng HL, Alt FW, Serrano L, Sternglanz R, Reinberg D (2006) SirT2 is a histone deacetylase with preference for histone H4 Lys 16 during mitosis. Genes Dev 20(10):1256–1261

Varambally S, Dhanasekaran SM, Zhou M, Barrette TR, Kumar-Sinha C, Sanda MG, Ghosh D, Pienta KJ, Sewalt RG, Otte AP, Rubin MA, Chinnaiyan AM (2002) The polycomb group protein EZH2 is involved in progression of prostate cancer. Nature 419(6907):624–629

Vaziri H, Dessain SK, Ng Eaton E, Imai SI, Frye RA, Pandita TK, Guarente L, Weinberg RA (2001) hSIR2(SIRT1) functions as an NAD-dependent p53 deacetylase. Cell 107(2):149–159

Vega RB, Matsuda K, Oh J, Barbosa AC, Yang X, Meadows E, McAnally J, Pomajzl C, Shelton JM, Richardson JA, Karsenty G, Olson EN (2004) Histone deacetylase 4 controls chondrocyte hypertrophy during skeletogenesis. Cell 119(4):555–566

Verdin E, Dequiedt F, Kasler HG (2003) Class II histone deacetylases: versatile regulators. Trends Genet 19(5):286–293

Villar-Garea A, Esteller M (2004) Histone deacetylase inhibitors: understanding a new wave of anticancer agents. Int J Cancer 112(2):171–178

Wang Z, Zhai W, Richardson JA, Olson EN, Meneses JJ, Firpo MT, Kang C, Skarnes WC, Tjian R (2004) Polybromo protein BAF180 functions in mammalian cardiac chamber maturation. Genes Dev 18(24):3106–3116

Xin Z, Tachibana M, Guggiari M, Heard E, Shinkai Y, Wagstaff J (2003) Role of histone methyltransferase G9a in CpG methylation of the Prader–Willi syndrome imprinting center. J Biol Chem 278(17):14996–15000

Xu GL, Bestor TH, Bourc'his D, Hsieh CL, Tommerup N, Bugge M, Hulten M, Qu X, Russo JJ, Viegas-Pequignot E (1999) Chromosome instability and immunodeficiency syndrome caused by mutations in a DNA methyltransferase gene. Nature 402(6758):187–191

Xu W, Edmondson DG, Evrard YA, Wakamiya M, Behringer RR, Roth SY (2000) Loss of Gcn5l2 leads to increased apoptosis and mesodermal defects during mouse development. Nat Genet 26(2):229–232

Yagi H, Deguchi K, Aono A, Tani Y, Kishimoto T, Komori T (1998) Growth disturbance in fetal liver hematopoiesis of Mll-mutant mice. Blood 92(1):108–117

Yamauchi T, Yamauchi J, Kuwata T, Tamura T, Yamashita T, Bae N, Westphal H, Ozato K, Nakatani Y (2000) Distinct but overlapping roles of histone acetylase PCAF and of the closely related PCAF-B/GCN5 in mouse embryogenesis. Proc Natl Acad Sci USA 97(21):11303–11306

Yao TP, Oh SP, Fuchs M, Zhou ND, Ch'ng LE, Newsome D, Bronson RT, Li E, Livingston DM, Eckner R (1998) Gene dosage-dependent embryonic development and proliferation defects in mice lacking the transcriptional integrator p300. Cell 93(3):361–372

Yeung F, Hoberg JE, Ramsey CS, Keller MD, Jones DR, Frye RA, Mayo MW (2004) Modulation of NF-kappaB-dependent transcription and cell survival by the SIRT1 deacetylase. EMBO J 23(12):2369–2380

Yu BD, Hess JL, Horning SE, Brown GA, Korsmeyer SJ (1995) Altered Hox expression and segmental identity in Mll-mutant mice. Nature 378(6556):505–508

Yu BD, Hanson RD, Hess JL, Horning SE, Korsmeyer SJ (1998) MLL, a mammalian trithorax-group gene, functions as a transcriptional maintenance factor in morphogenesis. Proc Natl Acad Sci USA 95(18):10632–10636

Zimmermann S, Kiefer F, Prudenziati M, Spiller C, Hansen J, Floss T, Wurst W, Minucci S, Gottlicher M (2007) Reduced body size and decreased intestinal tumor rates in HDAC2-mutant mice. Cancer Res 67(19):9047–9054

Chapter 19
Modeling Transforming Growth Factor-ß Signaling in Cancer

Veronica R. Placencio and Neil A. Bhowmick

19.1 Introduction

The components of the TGF-ß signal transduction pathway have multiple unique proteins as well as many that are common to other growth factor signaling mechanisms. The TGF-ß family of ligands includes those of the bone morphogenic protein, activin and inhibin. However, the TGF-ß isoforms 1, 2, and 3 are the principle interacting proteins of their cognate TGF-ß type receptors (Massague and Gomis 2006). The ligands can only signal once released from the latency associated peptide through proteolytic cleavage. The TGF-ß type I, II, and III receptors bind all three TGF-ß ligand isoforms (Massague and Gomis 2006). The TGF-ß type II receptor (TßRII) mediates the interaction with TGF-ß type I receptors for subsequent downstream intracellular signaling. Type III TGF-ß binding is especially critical for TGF-ß2 signaling. There are a number of TGF-ß type I receptors that are expressed in a tissue and cell type-dependent manner, that in some cases can mediate signals downstream of TGF-ß. Downstream of the receptor ligand complex, multiple parallel signaling pathways are activated in a context-dependent manner (Massague and Gomis 2006). The TGF-ß family of ligands uniquely uses the Smad family of proteins to regulate the cell cycle, development, differentiation, apoptosis, and cancer. The TGF-ß isoforms stimulate the transcription factors Smad2, Smad3, Smad4, and Smad7. Smad7 was originally identified as an inhibitory Smad protein that prevented the association of Smad2 and Smad3 to the TGF-ß type I receptor (Hayashi 1997). Smad7 has since been identified to have roles as a transcription fac-

V.R. Placencio
Departments of Cancer Biology, Urologic Surgery, Vanderbilt-Ingram Cancer Center, Vanderbilt University, Nashville, TN 37232-2765, USA

N.A. Bhowmick (✉)
Department of Medicine, Uro-Oncology Research Program, Cedars-Sinai Medical Center, 8750 Beverly Boulevard, Atrium 103, Los Angeles, CA 90048, USA
e-mail: bhowmickn@cshs.org

J.E. Green and T. Ried (eds.), *Genetically Engineered Mice for Cancer Research: Design, Analysis, Pathways, Validation and Pre-Clinical Testing*,
DOI 10.1007/978-0-387-69805-2_19, © Springer Science+Business Media, LLC 2012

tor cooperating in multiple signaling pathways and target proteosome degradation of signaling proteins (Zhang 2007). The other parallel downstream pathways of TGF-ß signaling include RhoA/ROCK, p38MAPK, JNK, and PI3kinase (Bierie and Moses 2006). As cells are responding to multiple external factors in the environment at all times, the interaction of TGF-ß pathways with that of integrins, hormones, growth factors, radiation, etc., is crucial and can only be evaluated thoroughly through mouse models.

In tissue samples from patients of different cancers, researchers have successfully identified increased or decreased amounts of TGF-ß signaling pathway components. Though these tissues only represent a snapshot in time, it demonstrates the relevance of TGF-ß signaling to cancer. It is up to future studies to incorporate events associated with aging [e.g., telomerase activity (Wong 2003)] in the context of TGF-ß signaling mutations. Mouse models enable a systems biology approach to study cancer for the progression of future therapeutics. As the sophistication of mouse models replicating the human condition increases, the understanding of TGF-ß signaling is realized. The mouse models described here are among those relevant to human cancers and future mouse model development.

19.1.1 Transgenic Mouse Models of TGF-β1, TGF-β2, and TGF-β3 Expression

TGF-ß signaling begins when a TGF-ß ligand binds the TGF-ß receptors forming the activated signaling complex. The ligands can result from many types of signaling, including autocrine, paracrine, or exocrine secretion from cells. We begin our overview with the mouse models expressing TGF-ß ligands. These were the first mouse models designed to study TGF-ß signaling.

The first ligand to be discovered, TGF-ß1, was disrupted in exon 6 using homologous recombination to study the loss of this ligand (Shull et al. 1992). At birth, the mice appeared normal. However, at approximately 20 days old, they appeared emaciated and had a multifocal mixed inflammatory response with tissue necrosis that led to organ failure and ultimately death. These observations alluded to the importance of TGF-ß1 in immune and inflammatory diseases. A similar independent study generated TGF-ß1 null mice, through homologous recombination to disrupt the open reading frame of exon 1 in TGF-ß1, was characterized with rapid emaciation by 2 weeks after birth and death by 3–4 weeks of age (Kulkarni et al. 1993). The mice had a massive inflammatory response with infiltration of immune cells, including lymphocytes and macrophages to many organs. Further understanding of the TGF-ß-mediated immune response led to crossing the TGF-ß1 null mice onto a background lacking $p21^{Cip1}$, which extended the lifespan of the mice (Kulkarni et al. 2002). $P21^{Cip1}$ plays a role in the survival of memory T cells, so the loss of these cells helps to overcome the severe immune response brought on from the loss of TGF-ß1. These studies suggested TGF-ß1 to have an important role in immune surveillance and development of autoimmunity. These studies also highlighted the cell-specific and downstream pathway-specific effects TGF-ß signaling disruption has on an organism.

The tissue-specific regulation of TGF-ß1 enabled better understanding of its role in cancer. The loss of TGF-ß1 was studied using a subcutaneous xenograft approach (Glick et al. 1994). Keratinocytes were initiated with the v-ras^{Ha} oncogene in the context of a targeted deletion of TGF-ß1. The v-ras^{Ha}/TGF-$ß1^{ko}$ keratinocytes were grafted under the skin of athymic mice with wild type or TGF-ß1 null dermal fibroblasts. The resulting grafts with the v-ras^{Ha} initiated TGF-ß1 null keratinocytes developed multifocal squamous cell carcinoma using either type of fibroblasts. Comparatively, control grafts with only initiated v-ras^{Ha} keratinocytes formed well-differentiated papillomas. These studies showed the importance of autocrine TGF-ß to suppress malignant progression.

Further evidence for the tumor suppressive role of TGF-ß was found in tissue-specific expression of a dominant-active mutant of TGF-ß1 (lacking interaction with the latency associated peptide) in the context of mammary tumor developing mouse models. The overexpression of TGF-α driven by the MMTV mammary promoter led to mammary ductal hyperplasia and development of mammary carcinoma. It is important to note that there are multiple MMTV promoter-driven mice that have varying specificity to the mammary gland, often the target tissue. There are multiple viable mouse models for breast cancer that include the MMTV-Neu, MMTV-TGF-α, and MMTV-polyoma virus middle T antigen (PyVmT) used as a context to understand signaling pathways of interest in general FVB strain mice. The overexpression of TGF-ß1 by the MMTV promoter alone had little effect on the tumorigenesis of the mammary gland. However, when crossed with MMTV-TGF-α mice, TGF-ß1 expression inhibited tumorigenesis of the epithelia (Pierce et al. 1995). In addition, treatment with 7,12-dimethylbenz [a] anthracene accelerated the onset of mammary carcinoma. MMTV-TGF-ß1 mice treated with 7,12-dimethylbenz[a]anthracene (DMBA) failed to develop mammary carcinoma. The treatment of organ cultures of wild-type lung, mammary gland, and prostate with TGF-ß1 have resulted in decreased branching morphogenesis. In each of these examples, TGF-ß1 signaling was tumor suppressive in the context of tumorigenic inducers.

There are contrasting mouse models suggesting the tumorigenic properties of TGF-ß1. A model was developed by expressing TGF-ß1 and the c-Neu (ErbB2) oncogene, both driven by the MMTV promoter. The resulting mouse had increased mammary tumor grade compared to the MMTV-c-Neu mice with local invasiveness and metastasis of the mammary adenocarcinoma (Muraoka et al. 2003). These studies demonstrate how TGF-ß can act to promote mammary carcinoma by increased activation of pathways, such as AKT and MAPK, through cooperative signaling with c-Neu. The cooperative signaling of TGF-ß with other growth factors is important as both tumor cells and tumor-associated stromal cells have elevated TGF-ß1 expression compared to their normal counterparts. Similarly, the mouse model with a knockout of both TGF-ß1 and $p21^{Cip1}$, had a longer lifespan than TGF-ß1 null mice as previously mentioned (Kulkarni et al. 2002), and developed spontaneous tumors in the gastrointestinal tract with no chronic underlying inflammatory process. The CDK inhibitor, $p21^{Cip1}$, often downregulated in cancer, is upregulated by

TGF-ß signaling. These dual transgenic mouse models suggest the role of autocrine TGF-ß1 signaling in cancer initiation and progression.

Much like the formerly discovered TGF-ß1 ligand, the loss of TGF-ß2 and TGF-ß3 results in many developmental defects within the mouse with the most severe observed in TGF-ß3 mice. Though these TGF-ß2 null mice are viable, they have cardiac, lung, craniofacial, limb, spinal column, eye, inner ear, and urogenital defects (Sanford et al. 1997). Notably, there did not seem to be phenotypic overlap compared to TGF-ß1 and TGF-ß3 null mice. TGF-ß3 null mice appear to have defects in epithelial–mesenchymal interactions (Kaartinen et al. 1995). These mice die within 20 h of birth and exhibit delayed pulmonary development and defective pathogenesis. The abnormal lung development resembled the human condition seen in premature infants at risk for developing bronchopulmonary dysplasia.

The individual ligand knockout mouse models demonstrated the importance of each in different developmental systems, which likely translates into effects within cancer models for the corresponding tissues. However, the deletion or overexpression of the individual ligands in a tissue-specific manner apparently affects a wide range of tissues, not only since they are secreted, but because TGF-ß impacts inflammatory cell recruitment and differentiation. Studies performed with these TGF-ß ligand models were important in making the foundation for subsequent mouse models that better delineated the cell-specific roles of TGF-ß signaling through altering TGF-ß receptor status.

19.1.2 Mouse Models Altering the TGF-β type I, II, and III Receptors

The receptors remain bound at the cell surface and make studying cell-specific TGF-ß signaling feasible. However, just as homozygous deletions of TGF-ß ligand gene expression were lethal, the knockout of the *Tgfbr1*, *Tgfbr2*, and *Tgfbr3* genes coding for the respective type I, II, and III TGF-ß receptors also resulted in lethal phenotypes. As a result, much of our understanding of TGF-ß receptor activity has largely come through the generation of transgenic dominant-negative receptors and conditional knockout mice. Since many cells express more than one isoform of the type I and III receptors, conditional knockouts of either receptor provide specific understanding of the receptor subtype disrupted. However, as there is one TßRII isoform recognizing all TGF-ß ligand isoforms that is necessary for all downstream activity, it has been the primary receptor gene targeted in understanding cell-specific TGF-ß regulation. We initially describe the few type I and type III receptor mouse models and delve into the specific influence TßRII mouse models have had on our understanding of TGF-ß signaling *in vivo.*

There is one reported conditional knockout of the TGF-ß type I receptor mouse model. These mice were generated by crossing a neurofilament-H driven Cre promoter with a *Tgfbr1* floxed background (Honjo et al. 2007). Spontaneous squamous cell carcinoma developed in the periorbital and perianal regions in approximately 35% of these mice. This study suggested that a loss of TßRI expression led to spontaneous tumor formation. Although the phenotype was observed with relatively low penetrance, the model has implications on the role of neuronal TGF-ß signaling on epithelial differentiation status. The location of the tumors around the eye and anus suggests a role for exogenous bacterial infiltration in the tumorigenic process. Neuronal signaling is well recognized to regulate inflammatory cell recruitment and immune surveillance. These studies would suggest that TGF-ß signaling supports immune-tolerance through neuronal cells.

The role of TGF-ß type III receptor in cancer initiation and progression is least understood. The *Tgfbr3* knockout mice with targeted recombination of exon 2 were embryonic lethal (Stenvers et al. 2003). The model by Stenvers et al. had myocardial thinning and a ventricular septal defect with lethality at embryonic day 13.5 thought to result from liver defects. A similar model of *Tgfbr3* deficient mice targeted deletion of exon 3 (Compton et al. 2007). These mice were also embryonic lethal, however, the cause of death was determined to result from reduced coronary vessels. Their reduced vessel size was unable to sustain the myocardium leading to death at embryonic day 14.5. Since uniquely TGF-ß2 requires TßRIII for downstream signaling and its knockout has not been shown to develop tumorigenic transformation, there was initially less enthusiasm to develop *Tgfbr3* conditional knockout mice for cancer studies. However, a basis for further study has emerged since TGF-ß2 ligand antisense therapeutics has shown efficacy in glioblastoma and gastric cancer (Maggard et al. 2001; Marzo et al. 1997; Nemunaitis et al. 2006). Models are currently being generated to determine the significance of TßRIII in breast cancer progression through conditional knockouts using the MMTV-Cre and MMTV-PyVmT models. But for now only effects on development have been determined using these mouse models. In the interim, heterozygous *Tgfbr3* knockout mice could perhaps be crossed with other mouse models known to develop cancer. Such mouse models would reveal haploinsufficiency of the gene in its role as a tumor suppressor or promoter. The results of studies on TßRIII would have implications on the role of TGF-ß2 expression in cancer and possibly shed light on the role of TßRIII on the other two TGF-ß ligand isoforms.

One of the earlier TßRII mouse model suggesting an effect on cancer progression was developed as a TGF-ß ligand antagonist uniquely overexpressing the extracellular domain of TßRII fused to the Fc domain of human IgG (SR2F) (Tang et al. 1998). The expression of SR2F in the mammary gland was driven by the MMTV promoter/enhancer in the context of the MMTV-Neu metastatic breast cancer model. The expression of SR2F antagonized breast cancer metastasis by presumably sequestering TGF-ß ligands (Yang et al. 2002). These metastasis regulatory roles of TGF-ß seem to be distinct from the MMTV-TGF-ß1/MMTV-TGF-α model discussed earlier that had reduced mammary tumor initiation with TGF-ß expression (Pierce et al. 1995). Further, the studies suggested a lack of deleterious systemic effects of long-term TGF-ß antagonist exposure and opened the doors for the testing of pharmaceutical antagonists of TGF-ß in patients. The use of neutralizing antibodies toward TGF-ß1 is shown to block TGF-ß signaling-based breast cancer mouse models and further inhibit metastatic progression (Muraoka-Cook et al. 2005). It is known that the overproduction of TGF-ß in some cancers can enhance blood vessel formation, increased protease activity, and altered immune responses. The strategies for blocking TGF-ß signaling include ligand antibodies, fusion proteins blocking the ligand from binding its receptor, and small-molecule inhibitors of the type I TGF-ß receptor. These TGF-ß antagonists show promise in mouse models of cancer, however, side effects in human trials need to be monitored carefully.

There are multiple mouse models developed altering the membrane-bound TßRII expression in mice to study cancer initiation and progression. A series of models were developed for cell-specific expression of dominant negative TßRII (DNIIR) constructs, lacking the cytoplasmic domain of the receptor. However, the DNIIR models may give different effects compared to transgenic knockout models of the *Tgfbr2* gene (Bhowmick et al. 2001b). *In vitro* studies suggest that lower expression level of DNIIR can inhibit the Smad signaling pathway, but not affect downstream signaling by the p38MAPK and PI3 kinase/AKT pathways. This is an important point to keep in mind when comparing effects between mouse models, as the different promoters and integration sites of the transgene can have varied expression of DNIIR.

Expressing DNIIR targeted to the mouse skin led to an increase in carcinoma incidence as well as susceptibility for carcinoma development (Amendt et al. 1998; Go et al. 1999). DNIIR expression was targeted to the basal cells and the follicular cells in the skin. Mouse skin was initiated with DMBA and promoted with 12-O-tetradecanolyphorbol-13-acetate (TPA). Mouse skin expressing DNIIR alone had little effect on skin development. Similar effects were seen in mouse models of DMBA-induced mammary and lung carcinomas expressing DNIIR (Bottinger et al. 1997). In these studies, MMTV-driven DNIIR expression (in multiple organs, including the mammary gland, salivary glands, lung, spleen, and testis) initiation with DMBA led to increased incidence of the tumors in the lung and breast. Again in these studies, little hyperplasia was reported in the time frame examined for either tissue in the absence of carcinogenic initiation. However, DNIIR expression driven by the mouse intestinal trefoil peptide (ITF)/TFF3 promoter induced spontaneous colitis (Hahm et al. 2001). Further progression of the ITF-DNIIR mice to colon cancer required intraperitoneal induction with azoxymethane led to colon carcinoma or H. pylori-induced stomach cancer (Hahm et al. 2002). These studies suggested that the loss of TGF-ß signaling in the epithelia of the skin, lung, breast, colon, and stomach supported tumorigenesis in the context of initiating carcinogens and significantly promoted tumor progression. Thus, the studies supported the tumor suppressor activities of TGF-ß.

Studies focussing on targeting the mammary and prostate glands for DNIIR expression provided *in vivo* evidence for the role of TGF-ß in tumor progression. An independent set of studies with transgenic MMTV-DNIIR mice exhibited the development of spontaneous ductal carcinoma *in situ* and focal incidence of adenocarcinoma following a long latency – nearly 2 years (Gorska et al. 2003). These MMTV-DNIIR mice developed precocious mammary gland differentiation during early stages of pregnancy and delayed mammary gland involution after lactation ceased. The subsequent cross of the MMTV-DNIIR mice with MMTV-TGF-α mice resulted in shorter tumor latency of less than 10 months and a suppression of tumor invasion compared to MMTV-TGF-α mice. Importantly, the MMTV-DNIIR and MMTV-DNIIR/MMTV-TGF-α mice had a loss of DNIIR transgene expression in areas of invasion by the carcinoma. An analogous prostate model of carcinogenesis was generated to study the effect of TGF-ß signaling on metastasis (Tu et al. 2003). Bigenic prostate epithelial-targeted mice expressing SV40 large T by the probasin promoter were crossed with mice expressing DNIIR by the metallothionein inducible promoter. The induction of the metallothionein sensitive promoter by zinc chloride (in the drinking water) was well suited to targeting the prostate since the divalent cation is enriched in the organ. The bigenic mice had more metastasis, to primarily the lung and liver, compared to mice expressing only SV40 large T antigen, however not statistically significant. The latency period for metastasis was significantly shorter in the DNIIR expressing mice. These studies implied the role of TGF-ß supporting DNA stability and inhibiting tumor initiation. The contribution of TGF-ß to tumor metastasis was less clear in these studies as primary tumor volume was also greater in the DNIIR expressing bigenic mice. Clearly, the genetic and tissue context of the expression of DNIIR is critical to the outcome of tumor progression. We are realizing metastasis is not only dictated by the potential for epithelial to mesenchymal transdifferentiation, but also mediated by collective invasion events (Gaggioli et al. 2007; Macpherson et al. 2007). While TGF-ß positively regulates mesenchymal transdifferentiation and motility (Bakin et al. 2000; Bhowmick et al. 2001a, b; Muraoka et al. 2002; Oft et al. 1998; Sahai et al. 2007), the role of TGF-ß in collective migration is not clear, and requires us to reevaluate the role of TGF-ß in metastasis of various cancers.

The overwhelming inflammation resulting from the knockout of the TGF-ß ligand isoforms and TßRII suggested an immune suppressor role for TGF-ß. So, the results of targeting of DNIIR to immune cells were anticipated to reveal the effect of TGF-ß on immune cells themselves. The CD4 promoter-mediated expression of DNIIR in helper T cells spontaneously developed immunity toward xenografted aggressive melanoma and thymoma cells that secrete TGF-ß (Gorelik and Flavell 2001). The tumors can apparently evade the immune system by expressing TGF-ß. These studies suggested the immune tolerance was mediated by TGF-ß responsiveness of T cells. We have more recently learned that the specific T cell population responsible for immune suppression are the T regulatory cells. The expression of FoxP3, the regulatory T cell determining gene, is principally upregulated by TGF-ß (Bettelli et al. 2007; Lehner 2008). In the context of an azoxymethane-induced colon tumorigenesis mouse model, another group targeted DNIIR expression in T cells using the CD2 promoter. In this study, colon tumorigenesis was accelerated with the expression of DNIIR. Apparently, T cells expressing DNIIR express IL-6 to activate the JAK-STAT pathway in tumor cells (Becker et al. 2004). The expression of a kinase-dead TßRII driven by the CD8 memory T cell promoter in mice, resulted in leukemia/lymphoma-like syndrome leading to death by 3–4 months of age (Lucas et al. 2004). This suggested that TGF-ß could inhibit the expansion of memory or antigen-experienced T cells. The disruption of TGF-ß signaling in the different populations of T cells resulted in clear differences in the final response to cancer. Thus, the targeting of TGF-ß antagonists to the appropriate T cell population can potentially inhibit the progression of specific cancers, but systemic inhibition of TGF-ß can promote the disease.

The conditional TßRII knockout mice were developed to ablate downstream signals resulting from TGF-ß in a cell-specific manner. A study looked at the effects of conditionally deleting TßRII expression in lymphocytes using interferon inducible Mx1-Cre (Leveen et al. 2002). They induced Cre expression with polyI:polyC led to a lethal inflammatory disease. The Mx1-Cre is targeted to lymphocytes and the liver. To show that the TßRII null lymphocytes from the bone marrow caused the lethal inflammation, and that the liver phenotype was not necessary for the observation, they performed bone marrow transplantations into wild-type mice. These studies show that bone marrow of TßRII null mice result in a similar inflammatory disorder compared to TGF-ß1 null mice.

A model of colon cancer was created by mating mice with an Apc mutation (activating ß-catenin, (Su et al. 1992)) to mice that were null for *Tgfbr2* in the intestinal epithelium (Munoz et al. 2006). The specific deletion for *Tgfbr2* was created by crossing Tgfbr2$^{floxE2/floxE2}$ mice with Villin-Cre mice. Then, these resulting mice were further crossed with Apc$^{1638N/wt}$ mice to generate the Apc$^{1638N/wt}$/Tgfbr2^{IEKO} mice that developed intestinal adenocarcinomas. Neither the Apc$^{1638N/wt}$ nor Tgfbr2^{IEKO} mice developed adenocarcinoma alone. These studies showed that the loss of TGF-ß signaling could promote malignancy and invasion in tumors initiated with Apc mutations, closely resembling human colon cancer. Another gastrointestinal model crossing Tgfbr2$^{floxE2/floxE2}$ mice with Fabpl$^{4xat-132}$-Cre mice conditionally knocked out *Tgfbr2* in the colon epithelium (Biswas et al. 2004). The treatment of Tgfbr2^{FabpKO} mice with azoxymethane resulted in accelerated induction of colon neoplasms. There was an increase in colon cancer incidence and pathological stage in Tgfbr2^{FabpKO} compared to Tgfbr2$^{floxE2/floxE2}$ azoxymethane-treated mice.

A mouse model of pancreatic ductal adenocarcinoma (PDAC) closely resembles human clinical and pathological conditions (Ijichi et al. 2006). A *Tgfbr2* knockout driven by a pancreas-specific promoter, pancreatic transcription factor-1a (Ptf1a) was crossed with Kras^{G12D} mice. Note that 100% of the bigenic mice resulted in well-differentiated PDAC within a much shorter time frame, having a median lifespan of less than 60 days, compared to control Kras^{G12D} mice that only progressed to intraepithelial neoplasia over a 1-year period. Also interesting was that the heterozygous *Tgfbr2* knockout-Kras^{G12D} mice also developed pancreatic adenocarcinoma over time and more frequent metastases compared to the homozygous Tgfbr2^{Ptf1aKO}-Kras^{G12D} mice. Mouse models, such as this one, allow the study of the human condition in a context that closely resembles the genetic mutations in actual disease making it desirable for future therapeutic studies.

An emerging pattern of epithelial knockout of *Tgfbr2* expression is the acceleration of cancer progression when coupled with an initiating oncogene or carcinogen. The conditional knockout of *Tgfbr2* in the mammary epithelium driven by MMTV-Cre (Tgfbr2^{MGKO}) in the context of MMTV-PyVmT mice was no different (Forrester et al. 2005). The Tgfbr2^{MGKO} mice resulted in lobular alveolar hyperplasia as the mammary gland developed. However, unlike the other models, the MMTV-PyVmT/Tgfbr^{MGKO} mice not only had shorter tumor latency, but there was also an increase in pulmonary metastases compared to MMTV-PyVmT mice. Unlike the previously discussed paper that showed reduced MMTV-PyVmT lung metastasis when treated with TGF-ß1 neutralizing antibody, the MMTV-PyVmT/Tgfbr^{MGKO} mice showed that TGF-ß acts as a tumor suppressor during tumor development as well as late stage tumorigenesis leading to metastasis. This demonstrated how the host response to a therapeutic agent may differ from that of the tumor response to the same therapeutic agent. We also need to keep in mind the heterogeneity of the host immune and stromal makeup since it is often the source of the uniqueness of therapy response in one individual, whether it is mouse or human. More specifically, these studies suggest that efficacy of systemic TGF-ß neutralization is not necessarily mediated through mammary epithelial cells which naturally emphasizes the importance of the tumor stroma.

A new direction in TGF-ß signaling was examined in a fibroblast conditional knockout mouse model of *Tgfbr2* (Bhowmick et al. 2004). Previous models had focused on the epithelial cells contributing to the initiation of cancer development; the importance of the stromal cells in transforming adjacent epithelia was not appreciated. A model was generated using floxed mice for *Tgfbr2* crossed to mice containing a fibroblast-specific protein-1 (S100A4) promoter-driven Cre (FSP-1-Cre). The resulting mice spontaneously developed squamous cell carcinoma and prostatic intraepithelial neoplasia (PIN) with 100% penetrance by 5–7 weeks of age. Due to the widespread expression of FSP-1, multiple mesenchymal cell types were affected and the mice died within 8 weeks of birth. Unfortunately, early lethality of the mice limited further determination if carcinoma would develop in other organs given enough time. Tissue rescue of the Tgfbr2^{fspKO} prostates as allografts in immunocompromised mice for 7 months resulted in the eventual development of prostate adenocarcinoma (Li et al. 2008). Together, the studies suggested that normal stromal cells play a critical role in preventing epithelial transformation with an emphasis on TGF-ß signaling as a master regulator of stromal–epithelial cross talk.

To further study cancer progression in the mammary and prostate glands of Tgfbr2^{fspKO}, we performed tissue recombination allografts in the renal capsule of immunocompromised mice to overcome the early lethality of the mice. The Tgfbr2^{fspKO} mammary glands had expansion of stromal fibroblasts and increased ductal epithelial cell turnover, but no evidence of epithelial hyperplasia (Cheng et al. 2005). However, the tissue recombination of Tgfbr2^{fspKO} stromal cells with PyVmT mammary carcinoma cells resulted in tumors that had increased tumor mass and invasiveness compared to PyVmT mammary carcinoma cells with Tgfbr2$^{floxE2/floxE2}$ mammary stromal cells. An analogous experiment using Tgfbr2^{fspKO} prostate stromal cells with SV40 large T-antigen transformed prostate epithelial organoids resulted in tissue recombinants that formed prostate adenocarcinoma. Of note, the SV40 large T-antigen expressing prostate epithelia of LADY mice primarily form PIN lesions (Kasper et al. 1998). Interestingly, the Tgfbr2^{fspKO}/T antigen prostates were refractory to androgen ablation (Placencio et al. 2008). These unexpected results led to the development of mouse models that enabled monitoring of potential paracrine cell signaling pathways.

The role of TGF-ß in androgen refractory prostate cancer initiation and progression was studied through a combination of mouse models. A conditional TGF-ß type II receptor knockout mouse model of the epithelia (Tgfbr2$^{NKX3.1KO}$) demonstrated that epithelial TGF-ß signaling was not necessary for proper androgen responsiveness after androgen ablation (Placencio et al. 2008). Castration of the previously generated conditional stromal fibroblast knockout in the prostate (Tgfbr2^{fspKO}) showed that stromal TGF-ß signaling was necessary for androgen responsiveness upon androgen ablation. To corroborate *in vitro* studies suggesting the role of canonical Wnt signaling activation by the knockout of TGF-ß signaling in prostate stromal cells, the Tgfbr2^{fspKO} mouse was crossed to a TOPGal (LEF/TCF4 transcriptional reporter) mouse to generate Tgfbr2^{fspKO}/TOPGal mice (Placencio et al. 2008). This allowed *in vivo* visualization of activated canonical Wnt signaling by way of ß-galactosidase detection in exclusively the prostatic epithelia of the proximal ducts at days 3–5 following castration. It was determined that TGF-ß signaling cooperates with Wnt signaling to allow prostate regression on the distal ducts of the prostate while maintaining prostate viability near the base of the prostate. The loss of stromal TGF-ß signaling resulted in constitutive Wnt signaling throughout the entire prostate epithelial compartment enabling prostate epithelial proliferation even in the absence of androgens. The use of multiple knockout mouse models combined with reporter mouse models revealed that prostate androgen sensitivity is dependent on stromal TGF-ß signaling. To determine if Wnt signaling was responsible for the observed adenocarcinoma phenotype observed in Tgfbr2^{fspKO}/T antigen tissue recombinant prostates, the host mice were treated with Wnt antagonists secreted frizzled-related protein-2. As hypothesized, inhibition of Wnt signaling restored the androgen responsiveness to allow the prostate to regress and also enabled a more differentiated prostate phenotype. Tissue recombination allografting closely resembles developmental processes and cancer development (Hayward 2002; Hayward et al. 1992, 1997). This approach allows further utilization of previously established mouse models that have early lethality. Since tissue recombination technology is also much faster and economical method of altering genes of interest than generating a transgenic mouse models, it can be used as a foundation for mouse model generation.

19.1.3 In Vivo Smad2, Smad3, and Smad4 Signaling

Studying the role of individual TGF-ß ligands and receptors lead to the study of downstream Smad proteins. The canonical TGF-ß signaling pathway involves Smad signaling directly downstream of the receptor complex activation. Conditional Smad2 mice were generated to overcome embryonic lethality resulting in complete knockout mice of the gene. The Smad2 conditional knockout mice were designed with a C-terminal truncation frameshift in exon 11 to delete the phosphorylation sites needed to function (Liu et al. 2004). The resulting homozygous Smad2^{3loxp} mice are viable, fertile, and healthy. Currently, these mice are being crossed with various Cre lines to identify tissue-specific roles of Smad2.

There were three Smad3 knockout mouse models developed, all of which were viable postnatally. The first model of Smad3 null mice targeted deletion of the gene by disruption of exon 2 (Zhu et al. 1998). These mice had high expression of Smad3 in the colon, and with the Smad3 null allele they develop colorectal adenocarcinoma between 4 and 6 months of age. They saw multiple lesions within the large intestine and found large lesions in the proximal region of the colon or the distal end of the bowel with smaller lesions dispersed in between the proximal and distal regions. Histologically tumors ranged from hyperplastic lesions and adenomatous polyps to cancerous lesions with progressive degrees of atypia and metastaser beyond the colon. Interestingly, this study compared the incidence of colon tumors between inbred 129/sv mice or hybrid mice with a mixed background of 129/sv-C57BL/6. It was demonstrated that the inbred 129/sv mice had a 100% penetrance, whereas hybrid mice had a penetrance rate of only 30%. The development of colon carcinoma in these mice was the first report of spontaneous cancer development resulting from an alteration of the TGF-ß pathway at the time (Zhu et al. 1998). As in other TGF-ß receptor models described, inflammation clearly had contributed to some of the neoplastic changes observed.

The next model used homologous recombination to delete exon 8 while introducing a stop codon after exon 7 of the Smad3 gene (Yang et al. 1999). These mice appeared similar to their control littermates at birth, but developed defects in their immune function and died at ages ranging from 1 to 8 months depending on the severity of their immune system defects. These defects include multifocal formation of pyogenic abscesses that are found in periorbital and periodontal areas or within the walls of the stomach and intestine. These mice also had constitutively activated T cells, enlarged lymph nodes, involuted thymus, and a small spleen. The intestinal inflammation that developed resulting from the autoimmunity was associated with colonic carcinoma in mice that lived past 6 months.

Another model of Smad3 null mice was generated by removing exon1 and part of the first intron to remove the ATG start site of transcription so the Smad3 gene would not be transcribed (Datto et al. 1999). Although these mice were viable, they are noticeably smaller and grew slower than their control littermates. One third of the affected mice had severe bends in their forepaw wrists, a hunchback appearance and rib cage malformations giving them an indentation at the base of their sternum. The phenotypes described in this Smad3 knockout model were similar to that described in the previous papers; however, there was no mention of neoplastic development or early death associated with the defects described. These differing results speak to both the importance of the location of the Smad3 deletion and the mouse strain differences among the three labs.

More recently, the exon 8 deleted Smad3 knockout model was closely studied in leukemogenesis (Wolfraim et al. 2004b). In this study, the mouse models had either one or two Smad3 alleles knocked out in combination with deletions of $p27^{Kip1}$. It was found that homozygous knockouts of Smad3 were unable to suppress T cell proliferation and mice missing one Smad3 allele were only able to suppress T cell proliferation half as much as mice with both functional Smad3 alleles. Further study showed that mice with a homozygous deletion of both Smad3 and $p27^{Kip1}$ had a high rate of embryonic death making study in these mice too difficult. However, mice with a heterozygous deletion of Smad3 and a homozygous deletion of $p27^{Kip1}$ were viable and 50% lived past 6 months of age. In these heterozygous mice, there was lymphocytic infiltration to many organs, lymphoid hyperplasia, nephropathy, and cardiomyopathy. Ten percent of these mice also developed T cell leukemia, which was characterized by lymphadenopathy, splenomegaly, thymic enlargement, and lymphoblasts in various organs (Wolfraim et al. 2004a). Thus, most prominently the loss of Smad3 and $p27^{Kip1}$ together promoting T cell leukemogenesis closely modeled pediatric T cell acute lymphoblastic leukemia (ALL).

The loss of Smad4 has long been identified in human colon cancer (Fink et al. 2001; Riggins et al. 1997). Accordingly, various mouse models have helped to give insight to the biology and relevance of Smad4 loss in colon cancer. Heterozygous mice for the Smad4 allele ($Smad4^{+/-}$) developed multiple polyps in their stomach and duodenum as they age (Xu et al. 2000; Takaku 1999). The polyps closely resemble those found in the human juvenile polyposis syndrome resulting from germ line mutation of Smad4. These features include abundant stroma and eosinophilic infiltrates. The loss of heterozygosity for Smad4 is thought to be an early step in the formation of polyps. Another mouse model introduced a Smad4 mutation into $Apc^{delta716}$ to generate a colon cancer model similar to human familial adenomatous polyposis (Oshima et al. 1997). This model generated malignant and invasive colon cancer within the mouse gastrointestinal tract.

A model studying pancreatic neoplasia was developed using the Smad4 null mouse model (Kojima et al. 2007). Modeling the human condition, a pancreatic promoter driven Cre (Pdx1-Cre) was used to activate $Kras^{G12D}$ and knockout Smad4 in this mouse model. The deletion of Smad4 caused a more rapid progression of pancreatic intraepithelial neoplasias (mPanIN), promoted intraductal papillary mucinous neoplasia and active fibrosis with high incidence, and led to occasional PDAC by 6 months. This shows that Smad4 helps to suppress pancreatic malignancy associated with activated $Kras^{G12D}$.

A similar study examining the progression of pancreatic cancer used two different mouse models (Bardeesy et al. 2006). The first model looked at the combination of a Smad4 knockout driven by Pdx1-Cre with activation of $Kras^{G12D}$. A second model recapitulated another common mutation with the Smad4 knockout, activation of $Kras^{G12D}$, and a heterozygous knockout of the INK4/ARF locus. They determined that the loss of Smad4 accelerated the progression in initiated neoplasms within the pancreas, derived from both the activation of $Kras^{G12D}$ or with the addition of the heterozygous knockout of the INK4/ARF locus.

Since both BMP and TGF-ß pathways converge on the common mediator Smad4, the observations with Smad4 knockout mice have added impact on more processes and not surprisingly result in embryonic lethality when knocked out. Conditional knockout of Smad4 in T cells lead to spontaneous epithelial cancer within the gastrointestinal tract, including colon, rectum, duodenum, stomach, and oral cavity (Kim et al. 2006). A conditional Smad4 knockout mouse was generated to study skin tumorigenesis (Qiao et al. 2006). MMTV-Cre mice, having expression in the skin, were crossed to Smad4 conditional mutant mice ($Smad4^{Co/Co}$). By 3–13 months, all mice developed spontaneous malignant skin tumors mostly classified as squamous cell carcinoma. These results supported the role of TGF-ß signaling in both the host and epithelia in suppressing tumor initiation.

19.1.4 Mouse Models Upregulating Smad7 Expression

Consistent with the known roles of Smad7, its overexpression in transgenic mice resulted in inhibition of TGF-ß and BMP signaling (He et al. 2002). However, in an inducible model for Smad7 expression in skin at levels comparable to those seen in pathologic conditions, Smad7 only partially reduced TGF-ß and BMP signaling activity. Interestingly, the relatively high Smad7 expression correlated with mice having decreased levels of ß-catenin (DasGupta and Fuchs 1999; DasGupta et al. 2002; He et al. 2002; Waikel et al. 2001). High Smad7 expression in the embryo inhibited hair follicle development and accelerated sebaceous gland development. Postnatal Smad7 expression blocked hair follicle entry into a new growth phase and resulted in hair follicle degeneration. The expected findings of tumorigenesis, like that just described for Smad4 conditional knockout mice were not observed. The rational for the finding was that ß-catenin protein levels were reduced in Smad7 overexpressing skin. The studies revealed that Smad7 complexes with both ß-catenin and the E3 ligase Smurf2. Smurf2 activity results in proteosome degradation of ß-catenin and loss of canonical Wnt signaling activity. Overexpression of Smurf2 in mice enhanced the phenotypes produced by Smad7 expression (He et al. 2002). Smurf1-deficient mice do not show any enhancement of TGF-ß signaling (Yamashita et al. 2005), probably because of the redundant functions of Smurf1 and 2 in TGF-β signaling. Normally, active TGF-ß receptor associates with the Smad7–Smurf2 complex to ubiquitinate and target proteasome and lysosomal degradation (Kavsak et al. 2000). However, the higher Smad7 threshold for ß-catenin suggests a potential therapeutic option for various cancers, including those of the prostate and colon that are reported to often be associated with Wnt/ß-catenin signaling upregulation.

As we understand TGF-ß signaling more, we uncover the amount of cross talk shared by selected downstream proteins that are activated as part of this signaling cascade. A targated approach to the individual smad proteins may be more informative in determing the effect of those proteins on the signaling pathway than knocking out a wide range of signals at the level of a TGF-β receptor. This will enable us to retain noncanonical signaling pathways that we do not quite understand yet, while picking apart the signaling resulting from the canonical signaling pathway proteins individually.

19.1.5 Perspectives

A recognizable gap in the field is the lack of tissue-specific stromal promoters for targeting genes of interest. Our understanding of the wild type, much less cancer stromal compartment is shallow. Many of the stromal promoters identified are common to many organs. The diversity of the immune cells in the not too distant past was only recognized as myeloid and lymphoid cells. Intense work has identified tissue-specific endothelial cell surface markers [Corti 2008]. Similarly, such markers will likely be identified for the otherwise fibroblastic stromal cells omnipresent in the majority of tissues. With the discovery of more specific promoters, Cre targeting can be enhanced to very specific cell types at different stages of development or cancer progression. For example, models knocking out *Tgfbr2* in the prostate stroma are significant; recently, a study of over 150 human prostate cancer patients revealed a similar loss in stromal TßRII expression in 70% of the tissues regardless of cancer grade (Li et al. 2008). However, the fact that the Tgfbr2^{fspKO} mice have early lethality detracts from the enthusiasm of using this as a cancer model. Inducible conditional knockouts of *Tgfbr2* targeted to the prostate stroma would possibly result in an otherwise healthy mouse with prostate epithelial malignancy. There is evidence to suggest that mutations in the stroma are not common, but epigenetic mechanism can easily account for differences in gene expression (Allinen et al. 2004). Genetic models in mice can try to mimic such epigenetic mutations. In the mean time, as we search for more specific promoters, mice that harbor inducible conditional Cre constructs can be invaluable tools.

There is still much to learn from the existing TGF-ß signaling mouse models. As new signaling reporter mouse lines become available, such as the TOPGal mice (DasGupta and Fuchs 1999), it will be important to cross these mice with existing mouse models altering TGF-ß signaling. Often, the effects of TGF-ß in an intended organ or cell type results in unexpected paracrine changes in the adjacent cells. An example of this was observed in the Tgfbr2^{fspKO} prostates, where the loss of TGF-ß signaling in the stroma resulted in epithelial canonical Wnt signaling (Li et al. 2008; Placencio et al. 2008). The importance of the signaling reporter mice are made even more important as we try to understand mechanisms of action of therapeutics. As we progress to mimic human cancers accurately in mouse models, we will most likely require more advanced techniques not only to target new and different signaling pathways (as has been the dogma in the field), but also to better understand how the impact a single alteration in one signaling pathway affects other signaling pathways. Although *in vitro* studies may be more efficient in addressing some of the questions of signaling cross talk, only the mouse models enable the understanding of the cell type and timing of the signaling during different stages of cancer progression. Based on what we have learned so far, TGF-ß plays a major role in the maintenance of tissue homeostasis and what still remains unclear is how the so called switch occurs that causes TGF-ß to also play a role in the promotion of cancer progression. The timing and mechanism of the "switch" can be revealed to us through careful observation of signaling reporter mouse lines in the context of cancer models (Bierie and Moses 2006). However, it is possible that there is no proverbial "switch" for the apparently dual opposing roles of TGF-ß in cancer progression. We recognize that there is naturally stromal heterogeneity as there is epithelial heterogeneity in a single tissue. In cancer, a specific epithelial population normally with a mutation enables it to compete out the other epithelial cell types. The term cancer stem cells has been used to describe a cell population that expands in cancer (Hendrix et al. 2007; Postovit et al. 2007). Analogously, certain reactive stromal populations may outcompete other stromal cells as the tumor and tumor stroma co-evolve to provide the complementary niche. Recent studies using zebrafish and chick embryonic models of human stem cells have shown the reversion of the metastatic phenotype of melanoma cells (Postovit et al. 2006; Topczewska et al. 2006). The studies revealed a convergence of embryonic and tumorigenic signaling pathways in both epithelial and stromal compartments (Hendrix et al. 2007). There are many mouse models interrogating TGF-ß signaling that were not discussed since their phenotypes were primarily impacting development. However, since the link between embryonic development and cancer are reemerging, the results of mouse models with developmental phenotypes may need to be revisited. Mouse models continue to provide us with a tool to study these mechanisms *in vivo*. However, to translate to the human condition, we have to be willing to examine all cell types in the tissue of interest as well as the role of immunity in solid tumors.

Acknowledgments We apologize to those whose work we could not cite. Our work is supported by grants from the NIH GM079879 (to VRP) and CA108646 (to NAB).

References

Allinen M, Beroukhim R, Cai L, Brennan C, Lahti-Domenici J, Huang H, Porter D, Hu M, Chin L, Richardson A et al (2004) Molecular characterization of the tumor microenvironment in breast cancer. Cancer Cell 6:17–32

Amendt C, Schirmacher P, Weber H, Blessing M (1998) Expression of a dominant negative type II TGF-beta receptor in mouse skin results in an increase in carcinoma incidence and an acceleration of carcinoma development. Oncogene 17:25–34

Bakin AV, Tomlinson AK, Bhowmick NA, Moses HL, Arteaga CL (2000) Phosphatidylinositol 3-kinase function is required for transforming growth factor beta-mediated epithelial to mesenchymal transition and cell migration. J Biol Chem 275:36803–36810

Bardeesy N, Cheng KH, Berger JH, Chu GC, Pahler J, Olson P, Hezel AF, Horner J, Lauwers GY, Hanahan D et al (2006) Smad4 is dispensable for normal pancreas development yet critical in progression and tumor biology of pancreas cancer. Genes Dev 20:3130–3146

Becker C, Fantini MC, Schramm C, Lehr HA, Wirtz S, Nikolaev A, Burg J, Strand S, Kiesslich R, Huber S et al (2004) TGF-beta suppresses tumor progression in colon cancer by inhibition of IL-6 trans-signaling. Immunity 21:491–501

Bettelli E, Oukka M, Kuchroo VK (2007) T(H)-17 cells in the circle of immunity and autoimmunity. Nat Immunol 8:345–350

Bhowmick NA, Ghiassi M, Bakin A, Aakre M, Lundquist CA, Engel ME, Arteaga CL, Moses HL (2001a) Transforming growth factor-beta1 mediates epithelial to mesenchymal transdifferentiation through a RhoA-dependent mechanism. Mol Biol Cell 12:27–36

Bhowmick NA, Zent R, Ghiassi M, McDonnell M, Moses HL (2001b) Integrin beta 1 signaling is necessary for transforming growth factor-beta activation of p38MAPK and epithelial plasticity. J Biol Chem 276:46707–46713

Bhowmick NA, Chytil A, Plieth D, Gorska AE, Dumont N, Shappell S, Washington MK, Neilson EG, Moses HL (2004) TGF-beta signaling in fibroblasts modulates the oncogenic potential of adjacent epithelia. Science 303:848–851

Bierie B, Moses HL (2006) TGF-beta and cancer. Cytokine Growth Factor Rev 17:29–40
Biswas S, Chytil A, Washington K, Romero-Gallo J, Gorska AE, Wirth PS, Gautam S, Moses HL, Grady WM (2004) Transforming growth factor beta receptor type II inactivation promotes the establishment and progression of colon cancer. Cancer Res 64:4687–4692
Bottinger EP, Jakubczak JL, Haines DC, Bagnall K, Wakefield LM (1997) Transgenic mice overexpressing a dominant-negative mutant type II transforming growth factor beta receptor show enhanced tumorigenesis in the mammary gland and lung in response to the carcinogen 7,12-dimethylbenz-[a]-anthracene. Cancer Res 57:5564–5570
Cheng N, Bhowmick NA, Chytil A, Gorksa AE, Brown KA, Muraoka R, Arteaga CL, Neilson EG, Hayward SW, Moses HL (2005) Loss of TGF-beta type II receptor in fibroblasts promotes mammary carcinoma growth and invasion through upregulation of TGF-alpha-, MSP- and HGF-mediated signaling networks. Oncogene 24:5053–5068
Compton LA, Potash DA, Brown CB, Barnett JV (2007) Coronary vessel development is dependent on the type III transforming growth factor beta receptor. Circ Res 101:784–791
Corti A, Curnis F, Arap W, and Pasqualini R. (2008). The neovasculature homing motif NGR: more than meets the eye. Blood 112:2628–2635
DasGupta R, Fuchs E (1999) Multiple roles for activated LEF/TCF transcription complexes during hair follicle development and differentiation. Development 126:4557–4568
DasGupta R, Rhee H, Fuchs E (2002) A developmental conundrum: a stabilized form of beta-catenin lacking the transcriptional activation domain triggers features of hair cell fate in epidermal cells and epidermal cell fate in hair follicle cells. J Cell Biol 158:331–344
Datto MB, Frederick JP, Pan L, Borton AJ, Zhuang Y, Wang XF (1999) Targeted disruption of Smad3 reveals an essential role in transforming growth factor beta-mediated signal transduction. Mol Cell Biol 19:2495–2504
Fink SP, Swinler SE, Lutterbaugh JD, Massague J, Thiagalingam S, Kinzler KW, Vogelstein B, Willson JK, Markowitz S (2001) Transforming growth factor-beta-induced growth inhibition in a Smad4 mutant colon adenoma cell line. Cancer Res 61:256–260
Forrester E, Chytil A, Bierie B, Aakre M, Gorska AE, Sharif-Afshar AR, Muller WJ, Moses HL (2005) Effect of conditional knockout of the type II TGF-beta receptor gene in mammary epithelia on mammary gland development and polyomavirus middle T antigen induced tumor formation and metastasis. Cancer Res 65:2296–2302
Gaggioli C, Hooper S, Hidalgo-Carcedo C, Grosse R, Marshall JF, Harrington K, Sahai E (2007) Fibroblast-led collective invasion of carcinoma cells with differing roles for RhoGTPases in leading and following cells. Nat Cell Biol 9:1392–1400
Glick AB, Lee MM, Darwiche N, Kulkarni AB, Karlsson S, Yuspa SH (1994) Targeted deletion of the TGF-beta 1 gene causes rapid progression to squamous cell carcinoma. Genes Dev 8:2429–2440
Go C, Li P, Wang XJ (1999) Blocking transforming growth factor beta signaling in transgenic epidermis accelerates chemical carcinogenesis: a mechanism associated with increased angiogenesis. Cancer Res 59:2861–2868
Gorelik L, Flavell RA (2001) Immune-mediated eradication of tumors through the blockade of transforming growth factor-beta signaling in T cells. Nat Med 7:1118–1122
Gorska AE, Jensen RA, Shyr Y, Aakre ME, Bhowmick NA, Moses HL (2003) Transgenic mice expressing a dominant-negative mutant type II transforming growth factor-beta receptor exhibit impaired mammary development and enhanced mammary tumor formation. Am J Pathol 163:1539–1549
Hahm KB, Im YH, Parks TW, Park SH, Markowitz S, Jung HY, Green J, Kim SJ (2001) Loss of transforming growth factor beta signalling in the intestine contributes to tissue injury in inflammatory bowel disease. Gut 49:190–198
Hahm KB, Lee KM, Kim YB, Hong WS, Lee WH, Han SU, Kim MW, Ahn BO, Oh TY, Lee MH et al (2002) Conditional loss of TGF-beta signalling leads to increased susceptibility to gastrointestinal carcinogenesis in mice. Aliment Pharmacol Ther 16(Suppl 2):115–127
Hayashi A, Kasahara T, Iwamoto K, Ishiwata M, Kametani M, Kakiuchi C, Furuichi T, and Kato T. (1997). The role of brain-derived neurotrophic factor (BDNF)-induced XBP1 splicing during brain development. J Biol Chem 282:34525–34534

Hayward SW (2002) Approaches to modeling stromal-epithelial interactions. J Urol 168:1165–1172
Hayward SW, Del Buono R, Deshpande N, Hall PA (1992) A functional model of adult human prostate epithelium. The role of androgens and stroma in architectural organisation and the maintenance of differentiated secretory function. J Cell Sci 102(Pt 2):361–372
Hayward SW, Rosen MA, Cunha GR (1997) Stromal-epithelial interactions in the normal and neoplastic prostate. Br J Urol 79(Suppl 2):18–26
He W, Li AG, Wang D, Han S, Zheng B, Goumans MJ, Ten Dijke P, Wang XJ (2002) Overexpression of Smad7 results in severe pathological alterations in multiple epithelial tissues. EMBO J 21:2580–2590
Hendrix MJ, Seftor EA, Seftor RE, Kasemeier-Kulesa J, Kulesa PM, Postovit LM (2007) Reprogramming metastatic tumour cells with embryonic microenvironments. Nature reviews 7:246–255
Honjo Y, Bian Y, Kawakami K, Molinolo A, Longenecker G, Boppana R, Larsson J, Karlsson S, Gutkind JS, Puri RK et al (2007) TGF-beta receptor I conditional knockout mice develop spontaneous squamous cell carcinoma. Cell Cycle 6:1360–1366
Ijichi H, Chytil A, Gorska AE, Aakre ME, Fujitani Y, Fujitani S, Wright CV, Moses HL (2006) Aggressive pancreatic ductal adenocarcinoma in mice caused by pancreas-specific blockade of transforming growth factor-beta signaling in cooperation with active Kras expression. Genes Dev 20:3147–3160
Kaartinen V, Voncken JW, Shuler C, Warburton D, Bu D, Heisterkamp N, Groffen J (1995) Abnormal lung development and cleft palate in mice lacking TGF-beta 3 indicates defects of epithelial-mesenchymal interaction. Nat Genet 11:415–421
Kasper S, Sheppard PC, Yan Y, Pettigrew N, Borowsky AD, Prins GS, Dodd JG, Duckworth ML, Matusik RJ (1998) Development, progression, and androgen-dependence of prostate tumors in probasin-large T antigen transgenic mice: a model for prostate cancer. Lab Invest 78:1–15
Kavsak P, Rasmussen RK, Causing CG, Bonni S, Zhu H, Thomsen GH, Wrana JL (2000) Smad7 binds to Smurf2 to form an E3 ubiquitin ligase that targets the TGF beta receptor for degradation. Mol Cell 6:1365–1375
Kim BG, Li C, Qiao W, Mamura M, Kasprzak B, Anver M, Wolfraim L, Hong S, Mushinski E, Potter M et al (2006) Smad4 signalling in T cells is required for suppression of gastrointestinal cancer. Nature 441:1015–1019
Kojima K, Vickers SM, Adsay NV, Jhala NC, Kim HG, Schoeb TR, Grizzle WE, Klug CA (2007) Inactivation of Smad4 accelerates Kras(G12D)-mediated pancreatic neoplasia. Cancer Res 67:8121–8130
Kulkarni AB, Huh CG, Becker D, Geiser A, Lyght M, Flanders KC, Roberts AB, Sporn MB, Ward JM, Karlsson S (1993) Transforming growth factor beta 1 null mutation in mice causes excessive inflammatory response and early death. Proc Natl Acad Sci USA 90:770–774
Kulkarni AB, Thyagarajan T, Letterio JJ (2002) Function of cytokines within the TGF-beta superfamily as determined from transgenic and gene knockout studies in mice. Curr Mol Med 2:303–327
Lehner T (2008) Special regulatory T cell review: the resurgence of the concept of contrasuppression in immunoregulation. Immunology 123:40–44
Leveen P, Larsson J, Ehinger M, Cilio CM, Sundler M, Sjostrand LJ, Holmdahl R, Karlsson S (2002) Induced disruption of the transforming growth factor beta type II receptor gene in mice causes a lethal inflammatory disorder that is transplantable. Blood 100:560–568
Li X, Placencio VR, Iturregui JM, Uwamariya C, Sharif-Afshar AR, Koyama T, Hayward SW, Bhowmick NA (2008) Prostate tumor progression is mediated by a paracrine TGF-ß/Wnt3a signaling axis. Oncogene 27:7118–7130
Liu Y, Festing MH, Hester M, Thompson JC, Weinstein M (2004) Generation of novel conditional and hypomorphic alleles of the Smad2 gene. Genesis 40:118–123
Lucas PJ, McNeil N, Hilgenfeld E, Choudhury B, Kim SJ, Eckhaus MA, Ried T, Gress RE (2004) Transforming growth factor-beta pathway serves as a primary tumor suppressor in CD8+ T cell tumorigenesis. Cancer Res 64:6524–6529
Macpherson IR, Hooper S, Serrels A, McGarry L, Ozanne BW, Harrington K, Frame MC, Sahai E, Brunton VG (2007) p120-catenin is required for the collective invasion of squamous cell carcinoma cells via a phosphorylation-independent mechanism. Oncogene 26:5214–5228

Maggard M, Meng L, Ke B, Allen R, Devgan L, Imagawa DK (2001) Antisense TGF-beta2 immunotherapy for hepatocellular carcinoma: treatment in a rat tumor model. Ann Surg Oncol 8:32–37
Marzo AL, Fitzpatrick DR, Robinson BW, Scott B (1997) Antisense oligonucleotides specific for transforming growth factor beta2 inhibit the growth of malignant mesothelioma both in vitro and in vivo. Cancer Res 57:3200–3207
Massague J, Gomis RR (2006) The logic of TGFbeta signaling. FEBS Lett 580:2811–2820
Munoz NM, Upton M, Rojas A, Washington MK, Lin L, Chytil A, Sozmen EG, Madison BB, Pozzi A, Moon RT et al (2006) Transforming growth factor beta receptor type II inactivation induces the malignant transformation of intestinal neoplasms initiated by Apc mutation. Cancer Res 66:9837–9844
Muraoka RS, Dumont N, Ritter CA, Dugger TC, Brantley DM, Chen J, Easterly E, Roebuck LR, Ryan S, Gotwals PJ et al (2002) Blockade of TGF-beta inhibits mammary tumor cell viability, migration, and metastases. J Clin Invest 109:1551–1559
Muraoka RS, Koh Y, Roebuck LR, Sanders ME, Brantley-Sieders D, Gorska AE, Moses HL, Arteaga CL (2003) Increased malignancy of Neu-induced mammary tumors overexpressing active transforming growth factor beta1. Mol Cell Biol 23:8691–8703
Muraoka-Cook RS, Dumont N, Arteaga CL (2005) Dual role of transforming growth factor beta in mammary tumorigenesis and metastatic progression. Clin Cancer Res 11:937s–943s
Nemunaitis J, Dillman RO, Schwarzenberger PO, Senzer N, Cunningham C, Cutler J, Tong A, Kumar P, Pappen B, Hamilton C et al (2006) Phase II study of belagenpumatucel-L, a transforming growth factor beta-2 antisense gene-modified allogeneic tumor cell vaccine in non-small-cell lung cancer. J Clin Oncol 24:4721–4730
Oft M, Heider KH, Beug H (1998) TGFbeta signaling is necessary for carcinoma cell invasiveness and metastasis. Curr Biol 8:1243–1252
Oshima H, Oshima M, Kobayashi M, Tsutsumi M, Taketo MM (1997) Morphological and molecular processes of polyp formation in Apc(delta716) knockout mice. Cancer Res 57:1644–1649
Pierce DF Jr, Gorska AE, Chytil A, Meise KS, Page DL, Coffey RJ Jr, Moses HL (1995) Mammary tumor suppression by transforming growth factor beta 1 transgene expression. Proc Natl Acad Sci USA 92:4254–4258
Placencio VR, Sharif-Afshar AR, Li X, Huang H, Uwamariya C, Neilson EG, Shen MM, Hayward SW, Matusik RJ, Bhowmick NA (2008) Stromal TGF-ß signaling mediates prostatic androgen response by paracrine Wnt activity. Cancer Res 68:4709–4718
Postovit LM, Seftor EA, Seftor RE, Hendrix MJ (2006) Influence of the microenvironment on melanoma cell fate determination and phenotype. Cancer Res 66:7833–7836
Postovit LM, Costa FF, Bischof JM, Seftor EA, Wen B, Seftor RE, Feinberg AP, Soares MB, Hendrix MJ (2007) The commonality of plasticity underlying multipotent tumor cells and embryonic stem cells. J Cell Biochem 101:908–917
Qiao W, Li AG, Owens P, Xu X, Wang XJ, Deng CX (2006) Hair follicle defects and squamous cell carcinoma formation in Smad4 conditional knockout mouse skin. Oncogene 25:207–217
Riggins GJ, Kinzler KW, Vogelstein B, Thiagalingam S (1997) Frequency of Smad gene mutations in human cancers. Cancer Res 57:2578–2580
Sahai E, Garcia-Medina R, Pouyssegur J, Vial E (2007) Smurf1 regulates tumor cell plasticity and motility through degradation of RhoA leading to localized inhibition of contractility. J Cell Biol 176:35–42
Sanford LP, Ormsby I, Gittenberger-de Groot AC, Sariola H, Friedman R, Boivin GP, Cardell EL, Doetschman T (1997) TGFbeta2 knockout mice have multiple developmental defects that are non-overlapping with other TGFbeta knockout phenotypes. Development 124:2659–2670
Shull MM, Ormsby I, Kier AB, Pawlowski S, Diebold RJ, Yin M, Allen R, Sidman C, Proetzel G, Calvin D et al (1992) Targeted disruption of the mouse transforming growth factor-beta 1 gene results in multifocal inflammatory disease. Nature 359:693–699
Stenvers KL, Tursky ML, Harder KW, Kountouri N, Amatayakul-Chantler S, Grail D, Small C, Weinberg RA, Sizeland AM, Zhu HJ (2003) Heart and liver defects and reduced transforming

growth factor beta2 sensitivity in transforming growth factor beta type III receptor-deficient embryos. Mol Cell Biol 23:4371–4385

Su LK, Kinzler KW, Vogelstein B, Preisinger AC, Moser AR, Luongo C, Gould KA, Dove WF (1992) Multiple intestinal neoplasia caused by a mutation in the murine homolog of the APC gene. Science 256:668–670

Takaku K, Miyoshi H, Matsunaga A, Oshima M, Sasaki N, and Taketo MM. (1999). Gastric and duodenal polyps in Smad4 (Dpc4) knockout mice. Cancer Res 59:6113–6117

Tang B, Bottinger EP, Jakowlew SB, Bagnall KM, Mariano J, Anver MR, Letterio JJ, Wakefield LM (1998) Transforming growth factor-beta1 is a new form of tumor suppressor with true haploid insufficiency. Nat Med 4:802–807

Topczewska JM, Postovit LM, Margaryan NV, Sam A, Hess AR, Wheaton WW, Nickoloff BJ, Topczewski J, Hendrix MJ (2006) Embryonic and tumorigenic pathways converge via Nodal signaling: role in melanoma aggressiveness. Nat Med 12:925–932

Tu WH, Thomas TZ, Masumori N, Bhowmick NA, Gorska AE, Shyr Y, Kasper S, Case T, Roberts RL, Shappell SB et al (2003) The loss of TGF-beta signaling promotes prostate cancer metastasis. Neoplasia 5:267–277

Waikel RL, Kawachi Y, Waikel PA, Wang XJ, Roop DR (2001) Deregulated expression of c-Myc depletes epidermal stem cells. Nat Genet 28:165–168

Wolfraim LA, Fernandez TM, Mamura M, Fuller WL, Kumar R, Cole DE, Byfield S, Felici A, Flanders KC, Walz TM et al (2004a) Loss of Smad3 in acute T-cell lymphoblastic leukemia. N Engl J Med 351:552–559

Wolfraim LA, Walz TM, James Z, Fernandez T, Letterio JJ (2004b) p21Cip1 and p27Kip1 act in synergy to alter the sensitivity of naive T cells to TGF-beta-mediated G1 arrest through modulation of IL-2 responsiveness. J Immunol 173:3093–3102

Wong KK, Maser RS, Bachoo RM, Menon J, Carrasco DR, Gu Y, Alt FW, DePinho RA (2003). Telomere dysfunction and Atm deficiency compromises organ homeostasis and accelerates ageing. Nature 421:643–648

Xu X, Brodie SG, Yang X, Im YH, Parks WT, Chen L, Zhou YX, Weinstein M, Kim SJ, Deng CX (2000) Haploid loss of the tumor suppressor Smad4/Dpc4 initiates gastric polyposis and cancer in mice. Oncogene 19:1868–1874

Yamashita M, Ying SX, Zhang GM, Li C, Cheng SY, Deng CX, Zhang YE (2005) Ubiquitin ligase Smurf1 controls osteoblast activity and bone homeostasis by targeting MEKK2 for degradation. Cell 121:101–113

Yang X, Letterio JJ, Lechleider RJ, Chen L, Hayman R, Gu H, Roberts AB, Deng C (1999) Targeted disruption of SMAD3 results in impaired mucosal immunity and diminished T cell responsiveness to TGF-beta. EMBO J 18:1280–1291

Yang YA, Dukhanina O, Tang B, Mamura M, Letterio JJ, MacGregor J, Patel SC, Khozin S, Liu ZY, Green J et al (2002) Lifetime exposure to a soluble TGF-beta antagonist protects mice against metastasis without adverse side effects. J Clin Invest 109:1607–1615

Zhang S, Fei T, Zhang L, Zhang R, Chen F, Ning Y, Han Y, Feng XH, Meng A, Chen YG (2007). Smad7 antagonizes transforming growth factor beta signaling in the nucleus by interfering with functional Smad-DNA complex formation. Mol Cell Biol 27:4488–4499

Zhu Y, Richardson JA, Parada LF, Graff JM (1998) Smad3 mutant mice develop metastatic colorectal cancer. Cell 94:703–714

Chapter 20
Modeling Stromal–Epithelial Interactions

Omar E. Franco, Douglas W. Strand, and Simon W. Hayward

20.1 Overview of Stromal–Epithelial Interaction

Reciprocal interactions between mesenchymal and epithelial cells are known to be essential during embryonic development. These interactions result in the coordinated development of organs in correct spatial orientation with other anatomical entities, and in the timely expression of functional activity consistent with the physiological demands of both intrauterine and postnatal life. Epithelial–mesenchymal interactions continue during adulthood, playing a homeostatic role in the maintenance of epithelial and stromal differentiation and growth-quiescence. One of the organs in which these phenomena have been well studied is the prostate. The prostate develops from the embryonic urogenital sinus under the influence of circulating androgens. Cunha and coworkers showed that during prostatic development, the urogenital mesenchyme (UGM) specifies urogenital epithelial (UGE) identity, induces epithelial bud formation, and promotes growth and differentiation of a secretory epithelium (Marker et al. 2003; Staack et al. 2003). The differentiated epithelium in turn induces the UGM not only to undergo smooth muscle differentiation, but also to govern the

O.E. Franco • D.W. Strand
Department of Urologic Surgery, Vanderbilt University Medical Center, Nashville, TN 37232-2765, USA

S.W. Hayward (✉)
Department of Urologic Surgery, Department of Cancer Biology, Vanderbilt-Ingram Cancer Center, Vanderbilt University Medical Center, Nashville, TN 37232-2765, USA

Department of Urologic Surgery, Vanderbilt University Medical Center, A-1302 Medical Center North, 1161 21st Avenue South, Nashville, TN 37232-2765, USA
e-mail: simon.hayward@vanderbilt.edu

J.E. Green and T. Ried (eds.), *Genetically Engineered Mice for Cancer Research: Design, Analysis, Pathways, Validation and Pre-Clinical Testing*,
DOI 10.1007/978-0-387-69805-2_20, © Springer Science+Business Media, LLC 2012

spatial patterning of the smooth muscle (Cunha et al. 1992a, b; Cunha et al. 1996; Hayward et al. 1996, 1998). Stromal changes are critical components of nonmalignant proliferative conditions, such as benign prostatic hyperplasia (BPH) and also play a role in the regulation of tumor progression (Chung 1995; Ronnov-Jessen et al. 1996; Hayward et al. 1997).

Disruption of the harmonic interplay between the stromal and epithelial tissue compartments during carcinogenesis led pathologist G. Barry Pierce to promulgate the concept that "Neoplasia is a caricature of differentiation" (Pierce et al. 1978). Stephen Paget in his 1889 paper "Distribution of secondary growths in cancer of the breast" introduced the concept of "*seed and soil*" in cancer metastasis, although his comments were preceded by those of Fuchs who observed that certain organs may be "more predisposed" because they could provide the proper environment (*soil*) for tumor cells (*seeds*) to grow (Fuchs 1882; Paget 1889). The practical consequences of this idea can also be observed during carcinogenesis and local tumor invasion. In a carcinoma, the *seeds* reside within the epithelium, whereas the composition of the *soil* (tumor stroma) is heterogeneous and complex. There are two major components of the stroma; the tumor extracellular matrix (ECM), which provides the connective-tissue framework of the tumor, and the cellular components, such as muscle, fat, fibroblasts, immune and inflammatory cells, nerves, and blood vessel cells. Because the constituents of the tumor stroma resemble that of the granulation tissue formed during wound healing, Hal Dvorak described a tumor as a "wound that never heals" (Dvorak 1986). "Activation" of the stromal fibroblasts and their conversion into myofibroblasts is accompanied by the secretion of growth factors and remodeling enzymes necessary for tissue repair (Tuxhorn et al. 2001). Stromal activation has been recognized as an early event followed by infiltration of blood vessels (angiogenesis) into the expanding tumor that favors tumor progression. The role of recruited inflammatory cells, such as macrophages or monocytes, was once thought to be protective but is now more controversial. In 1850, Rudolph Virchow described the positive effect of chronic inflammation in tumor-promotion. Examples of this association can be found in several malignancies, including stomach, colon, cervix, and prostate (Kornfeld et al. 1997; Castellsague et al. 2002; Nelson et al. 2002; Mantovani et al. 2008).

The process of carcinogenesis includes a series of changes in the interactions between the cells and tissues comprising the tumor (Fig. 20.1). In this respect, a tumor is a caricature of a developing organ with instructive and permissive signaling within and between the tissues, resulting in the formation of a complex, growing structure with metastatic capability. Studying these complex interactions requires appropriate models. Several groups have developed *in vitro* and *in vivo* models that allow the examination of putative regulators by manipulation of genes in different compartments. These are briefly discussed and contrasted to the higher level of complexity found in various types of models which utilize an *in vivo* environment.

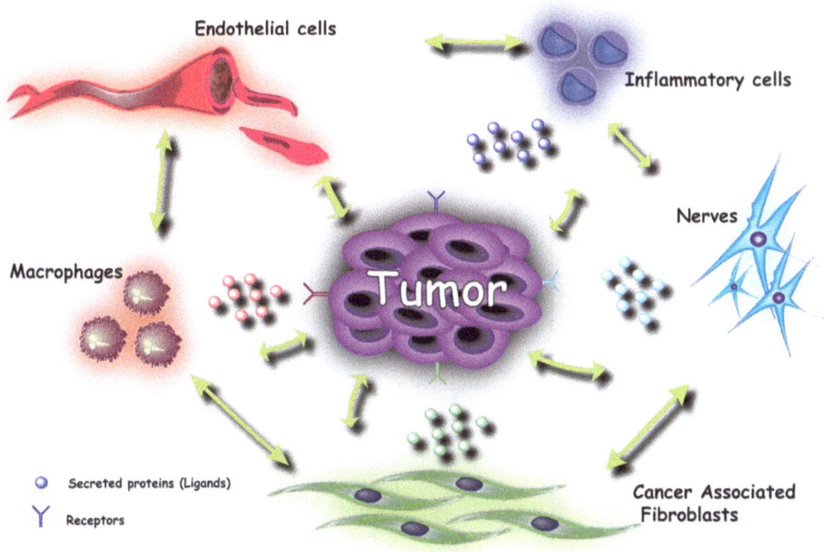

Fig. 20.1 Schematic representation of the crosstalk between cancer cells and host cells in the microenvironment of tumors. Studying the complex interactions in the vicinity of tumors between different cell types requires models that target different compartments. Any given cell in the microenvironment may have a specific role during cancer progression. For example, when fibroblasts (*green*) are "activated," they can secrete factors that attract inflammatory cells and promote cancer progression. It is important to understand that these are not simple two-way conversations; rather at any given time multiple cell populations are in communication and influencing each other

20.2 *In Vitro* Models of Stromal–Epithelial Interaction

Monolayer culture of cells. The availability of multiple cells lines, the simplicity of culturing cells and its reproducibility makes monolayer two-dimensional (2-D) culture a useful tool to quickly generate data on many cellular functions and provides explanations for some *in vivo* phenomena. Perhaps the utility of culture *in vitro* on plastic is that such cells can be transfected or transduced with putative stromal–epithelial paracrine regulators to directly assess molecular effects. Coculture of stromal and epithelial cells can be achieved using three different basic approaches (Fig. 20.2). The simplest method is the use of conditioned media in which secreted proteins from one cell support the growth of the same and other cell types. Although this one-way interaction can be informative of the type of response in the "target" cell (such as growth and differentiation), concomitant changes in the "feeder" cells can be overlooked. In the second approach, coculture using tissue culture inserts maintains a constant communication between stromal and epithelial cells supporting some aspects of cellular differentiation. The advantage of this method is that it can

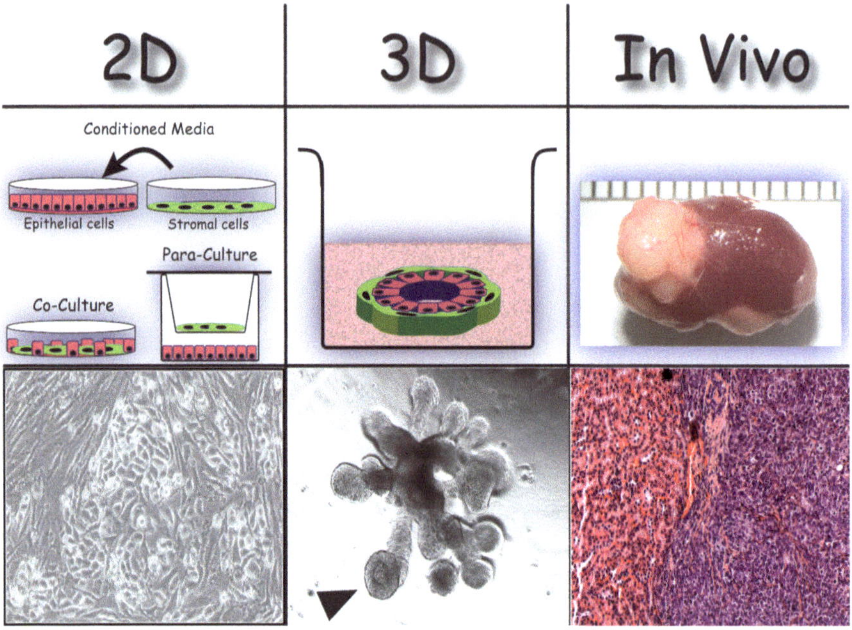

Fig. 20.2 Commonly used *in vitro* and *in vivo* techniques to study stromal–epithelial interactions. Cells cultured in two-dimensions (2-D) (*left panels*). Epithelial cells can be cultured in the presence of conditioned media (CM) from stromal cells, resulting in only a one-way communication. Alternatively, a "para-culture" system, using culture inserts allows two-way communication via soluble factors but no cell–cell contact. Co-culture of both cells in the same plate allows both two way communication and contact. BPH1 cells and normal human prostatic fibroblasts cultured in a Petri dish (*left bottom*). Embedding cells in a matrix (such as matrigel or collagen) recreates the three dimensional (3-D) organization which may mimic that present *in vivo* (shown stylistically in the upper middle panel). The lower central panel shows a phase contrast micrograph of human prostatic epithelial cells grown for 20 days in Matrigel. Note the coral-like branching and apparent lumen formation (*arrowhead*). *In vivo* techniques include subrenal capsule grafting of tissue recombinants. In the right hand panels, the gross and histologic appearance of recombinants composed of carcinoma associated fibroblasts + BPH-1 cells are shown. H&E staining shows invasive tumors (*right bottom*)

be manipulated to test the effects of putative modifiers of growth and differentiation between stromal and epithelial cells (Habib et al. 2000). The obvious disadvantage is the lack of direct contact between cells of different tissues. A third less widely used method consists of plating epithelial cells directly onto stromal cells. This approach has been further modified by the use of 3-D culture systems to coculture both stromal and epithelial cells within a collagen or Matrigel matrix.

Monolayer culture has many limitations. When cells are placed on plastic surface, they lose their normal cellular morphology and the expression of functional markers. Thus, cultured epithelial cells from different sources tend to look similar to each other, but different from the tissue of origin. Another major problem is the loss of polarity associated with the absence of the basal membrane component, which is critical for determining the cell orientation and morphology. By culturing in the presence of Matrigel, a natural substrate composed of reconstituted basal membrane from the Engelbreth-Holm-Swarm tumor, cell lines as well as primary cells from different organs can regain the formation of a differentiated phenotype (reviewed in Kleinman and Martin 2005).

Three-dimensional (3-D) culture systems. Researchers have become increasingly aware of the limitations of conventional 2-D tissue culture. This is mainly because *in vivo*, epithelial and stromal cells are colocalized in direct contact and are surrounded by ECM in an organized spatial configuration. In an attempt to mimic a more realistic micro- and local environment, a 3-D culture system offers a balanced alternative between a flask and an animal. This system allows manipulation in a more controlled manner of the cellular and noncellular components of the stromal–epithelial interaction. As 3-D cultures support cocultivation of multiple cell types, interactions between epithelial and stromal cells in normal and neoplastic development can be studied in real time. Perhaps the major advantage of 3-D cultures is the formation of glandular acinar structures with basal and luminal cell differentiation as seen during normal development or the distortion of cellular polarization and loss of normal cell architecture characteristic of tumor cells. Although 3-D cultures have been successful in cells from different organs, much of the information in regard to stromal–epithelia interaction has been obtained from studies in breast and prostate. Bissell's group has been able to recreate mammary gland acinus formation and production of milk proteins (Lee et al. 1984, 1985, 2007). They found that the nature and composition of the ECM was important in determining growth, differentiation, and survival of epithelial cells. The presence of laminin and collagen IV in the basement membrane (BM) could induce better expression of mammary-specific functions in epithelial cells compared to those containing collagen I or fibronectin (Roskelley et al. 1994). Laminin-rich BM favored polarization of epithelial cells allowing them to form structures resembling the normal mammary acinus that responds to hormones by secreting milk proteins into the lumen (Barcellos-Hoff et al. 1989). However, carcinoma-derived mammary epithelial cells were unable to polarize and instead form disorganized colonies that continue to grow. This assay has proven to be useful in distinguishing normal and cancer-derived breast epithelial cells (Petersen et al. 1992). More recently, experiments performed by Gudjonsen et al. have shown that the addition of myoepithelial cells derived from normal glands to the 3-D cultures could direct normal luminal polarity and recreate the epithelial bilayer that characterizes the breast acinus. In contrast, myoepithelial cells derived from 75% of breast cancer samples could not induce lumen formation due to the lack of laminin-1 production (Gudjonsson et al. 2002). Introduction of irradiated stromal cells demonstrated a proportional effect on tumor growth and invasiveness, characterized by changes in the ECM and activation of latent TGFß resulting in rapid remodeling of the microenvironment (Barcellos-Hoff and Ravani 2000).

Similar studies in the prostate have shown the importance of ECM components in acinar formation. The first description of prostate morphogenesis in 3-D culture came from the work done by Brinkmann. In his study, human prostatic epithelial cells grown on collagen formed branching structures that survived over 20 days in the presence of hepatocyte growth factor (HGF) (Brinkmann et al. 1995). Other paracrine factors can also control glandular morphogenesis. TGFβ inhibits acinar morphogenesis and increases the number of individual cells, whereas fibroblast growth factor 1 (FGF1) and HGF enhance the number of branch-like structures (Tyson et al. 2007). When prostate epithelial cells are grown on Matrigel, they can display an acinar phenotype. Also, the addition of human stromal-conditioned media results in glandular acinar structures with defined basal and luminal polarity (Webber et al. 1997). More recently, in order to maintain the 3-D organization present *in vivo*, Papini et al. developed a system in which small fragments (1-6 mm) of human neoplastic and nonneoplastic tissue were cultured in collagen. In this assay, proliferation of neoplastic cells and stromal cells was enhanced, and tissue architecture was preserved for about 3 weeks (Papini et al. 2004, 2007). Another attractive use of 3-D models is to study factors enhancing tumor progression and metastasis. In this regard, work by Wang using the rotary wall vessel, a NASA-designed 3-D cell culture model, explored the interaction between prostate cancer cells and bone. The rationale behind this technique is that the reduced gravity present in areas that are in contact with tumor cells like blood, lymphatic fluid and in bone marrow can be studied under microgravity-simulated conditions only present in outer space (Wang et al. 2005). Under these conditions, tumor cells proliferate faster.

Since 1988, 3-D cultures have also been used successfully with many undifferentiated embryonic stem (ES) cells (Bilozur and Hay 1988). It was found that ES cells could be cultured on Matrigel in the presence of mouse fibroblast conditioned media avoiding the classical use of a fibroblast feeder layer which raised concerns of contamination with endogenous mouse viruses (Xu et al. 2001). More recently, observations of the epigenetic effect of the human embryonic stem cell (hESC) microenvironment on tumor cells have shown that hESCs were able to moderate the behavior of aggressive melanoma cells resulting in a less invasive phenotype (Postovit et al. 2006).

Other traditional techniques to produce 3-D *in vitro* cancer models, such as spontaneous cell aggregation, liquid overlay, and spinner flasks are still widely used. For example, the human breast carcinoma cell MDA-MB-435 undergoes spontaneous homotypic cell aggregation, an important feature of metastatic cancer cells (Glinsky et al. 2000). In liquid overlay and spinner flask cultures, the inhibition of a meaningful contact with the culture vessel favors formation of spheroids, a useful model for studying solid tumors with an important component of desmoplasia (Sutherland et al. 1970; Kunz-Schughart et al. 2001). More sophisticated methods are now available. Microbarrier beads, that support the aggregation of cells which do not spontaneously aggregate or that are difficult to grow has provided many advantages to cell culture (Clark and Hirtenstein 1981; Bing et al. 1991). 3-D collagen scaffold-based cultures and more recently the use of biodegradable, pre-engineered scaffolds have been shown to closely resemble the natural ECM providing the physical/structural support and signaling to produce more organized and differentiated phenotypes (Jacquot et al. 1994; Tan et al. 2001). The aforementioned rotary cell culture system developed by NASA introduced a revolutionary concept in tissue engineering.

This method integrates cellular colocalization, stromal–epithelial–ECM interactions and very low shear forces which creates a more physiological environment in which more functional differentiated spheroids can be produced (Hammond and Hammond 2001). Several recent publications provide a more detailed explanation and in-depth exposition on the uses of 3-D models in tissue morphogenesis and cancer (Yamada and Cukierman 2007; Hebner et al. 2008).

Organ culture systems. Organ culture is a commonly used approach to study developmental biology making use of tissues harvested ex vivo (embryonic organs or intact tissue slices) and cultured *in vitro*. This "organ culture" technique has the advantage that the normal stromal/epithelial/ECM interactions are retained. It has been particularly effective for short-term cultures and for studies that examine the effects of steroid hormones and growth factors on development, or induction of angiogenesis and resistance to chemotherapeutic drugs in tumors (Hayward et al. 1992; Sugimura et al. 1996; Gahwiler et al. 1997; Lipschutz et al. 1997; Sakai et al. 2003). Disadvantages include the size of the specimen (<1 mm thick), which has to be thin enough to allow proper oxygenation and nutrition of the tissue interior, and another problem is the inability to separate endogenous from exogenous signaling.

3-D systems have an obvious advantage over routine 2-D cultures; however, they are not exact models of *in vivo* tissues. There are several major disadvantages of the 3-D cultures, such as variations in the ability to mimic *in vivo* tissue conditions, lack of proper vasculature and normal transport of small molecules, the absence of host immune response and other cell interactions. Such cultures at best represent static "snapshots" or short-term assessment of progression. As such, 3-D systems cannot completely replicate the complexities of the *in vivo* environment. Interpretation of results generated from such models must be considered in the light of these limitations; however, such results also provide a level of hypothesis testing that is likely to give rise to questions that require further validation *in vivo*.

20.3 *In Vivo* Models of Stromal–Epithelial Interaction

Modeling interactions between the stroma and the epithelium *in vivo* is challenging due to the lack of organ- and tissue-specific promoters from which to drive transgenes in the stroma. This is because, historically, much attention was focused on the epithelium leading to the development of tools specifically targeting this tissue. However, much less is known about the control of gene expression in the various cell types comprising the tumor microenvironment and the organ-specific targeting of stromally expressed genes. In this section, we describe historically how studying the tumor stroma evolved from first using environmentally induced cancer models, to the now widespread use of genetically engineered mice (GEM) with special emphasis in some of the genes that are exclusively expressed by the stroma. The usefulness of tissue rescue and recombination models is also addressed.

Each approach has advantages and disadvantages, and it is important to choose the system(s) that are most appropriate to address a particular question.

20.3.1 Carcinogen and Radiation-Induced Stromal Responses

A number of attempts have been made to address this issue using models in various organs. Bladder carcinogenesis is clearly a multistage process in which stromal–epithelial interactions can play a role. Early work by Hicks and coworkers established the idea that multiple doses of carcinogen were synergistic in producing tumors (Hicks and Wakefield 1972). They demonstrated that urothelium which had been "initiated" by exposure to a single (nontumorigenic) dose of a strong carcinogen could be promoted to form a tumor by chronic exposure to low levels of compounds which were not themselves tumorigenic (Hicks and Chowaniec 1977; Hicks 1980). Importantly, they also demonstrated that genetic insults specifically to the stromal cells of the bladder resulting from carcinogen treatment were able to elicit urothelial atypia and carcinogenesis (Hodges et al. 1977).

Sublethal DNA damage to fibroblasts through irradiation followed by the addition of nontransformed mammary epithelial cells into irradiated or nonirradiated cleared mammary fat pad in mice resulted in interesting observations. Only interactions with irradiated fibroblasts resulted in the formation of mammary carcinomas (Barcellos-Hoff and Ravani 2000). This study showed that stromal cells have an inductive role in the transformation of epithelial cells and, in this case, that TGFß was an important mediator in this phenomenon. More recently, similar experiments using irradiated fibroblasts recombined with pancreatic carcinoma cells resulted in an elevated incidence of more aggressive and invasive tumors compared to cells combined with nonirradiated fibroblasts (Ohuchida et al. 2004). In this instance, elevated expression of the HGF receptor, c-Met, and increased TGFß expressions were observed. These reports imply that alterations in stromal fibroblasts result in a paracrine transactivating mechanism in epithelial cells that enhances cancer progression.

20.3.2 Genetically Engineered Mouse Models

Introduction of the Cre-Lox recombination technique by Brian Sauer (patented by DuPont in 1980) has allowed researchers to genetically manipulate gene expression in mice (Sauer 2002). In this technique, expression of a Cre-recombinase causes genetic recombination of flox elements which can be engineered to either delete or activate a gene (Lakso et al. 1992; Frese and Tuveson 2007). This technology allows site-specific recombination of DNA, a valuable tool that was initially primarily used for *in vitro* DNA modification and is now widely used to create mammalian models of human diseases. Cre-Lox recombination in using tissue-specific promoters to drive expression of the Cre recombinase allows for spatially restricted expression to the compartment (tissue or organ) of interest. Promoters, such as collagen type 1 alpha 1 and 2, FSP-1/S100A, vimentin and alpha-smooth muscle actin, SM22/transgelin, and SMMHC have been used to target deletion or expression of osteoblasts, fibroblasts, myofibroblasts, and smooth muscle cells, major components of the stromal compartment (Kuhbandner et al. 2000; Miwa et al. 2000; Regan et al. 2000; Dacquin et al. 2002; Zhang et al. 2002; Bhowmick et al. 2004). A good example is FSP1-1, which is expressed in the development at embryonic day nine in precursor cells of the hematopoietic system, osteoblasts, and in the mesenchyme of nearly every tissue. The use of the FSP-1 promoter enables the study of the role of stromal cells in many tissues (Klingelhofer et al. 1997; Iwano et al. 2001, 2002). Notably, FSP-1 is only expressed in a fraction (30–50%) of the mature stromal cells. The disadvantage of stromal targeting using the FSP-1 promoter, as with other familiar stromal promoters, such as vimentin or SM22, is the lack of organ specificity. Alteration of genes that are essential in early development, as might be achieved using a combination of FSP1-Cre with a floxed gene can result in early mortality of the mouse model, often precluding analysis of the organ of interest. To circumvent this problem, new strategies using more restrictive expression of the transgene in fibroblasts have been developed. Zheng et al. generated a transgenic mouse in which the collagen type I alpha 2, a gene that is only expressed in fibroblasts that produce collagen type I, was linked to a Cre-ER fusion protein (Zheng et al. 2002). Treatment with tamoxifen causes nuclear translocation of the fusion protein resulting in Cre activity. This technique provided a tight regulation of the Cre activity allowing both temporal and tissue-specific control of gene regulation. Endogenous levels of estrogen present in pregnant female mice were not able to activate the transgene. The disadvantage of using tamoxifen during pregnancy is that it can be detrimental to the viability of the fetus at later stage and can lead to abortion. This unwanted effect could be avoided by the addition of an adjusted dose of tamoxifen at about E7.5, or concomitant administration of progesterone (Nakamura et al. 2006; Naiche and Papaioannou 2007). The tamoxifen-dependent construct is well suited for postnatal activation of Cre activity. Limitations of this system include the high-level expression of the ligand-binding domain of the estrogen receptor, which can be detrimental in some specific circumstances. For example, the receptor can bind the heat shock protein (Hsp)90 in the absence of ligand and also depletion of Hsp90 has been shown to have negative consequences (Kang et al. 1999). Perhaps the major limiting factor is that there are only a few Cre-ER models currently available for targeting the stromal compartment.

20.3.3 Targeting the Stroma

Transforming growth factor receptor Type II (TGFβRII). Several GEM models have been generated that develop prostatic intraepithelial neoplasia (PIN) without evidence of progression to invasive carcinoma during the course of the lifetime of a mouse.

Here, we describe a few that are relevant for studying the role of stromal-induced prostate carcinogenesis. In a report by Bhowmick and colleagues, generation of a conditional mesenchymal (stromal) transforming growth factor-beta receptor 2 (Tgfbr2) knockout mouse using FSP-1 Cre-mediated recombination resulted in prostatic intraepithelial neoplasia, a precursor of prostatic adenocarcinoma, and the appearance of invasive squamous cell carcinoma of the stomach by six weeks of age (Bhowmick et al. 2004). Reduction in the expression of p21 and p27 and an increase in the levels of c-Myc and the activating phosphorylation of the cognate HGF receptor, c-Met, were found to be associated with lesions in both sites. This study clearly demonstrates that loss of a critical signaling component in the stroma results in increased epithelial proliferation and in the promotion of invasive carcinomas in some tissues. Due to the early lethality of the Tgfbr2^{fspko} mouse model (about 6 weeks of age), studies have been focused on initiating factors involved in tumorigenesis and these mice have been less useful for studying subsequent tumor progression. However, the observations have been validated more recently in a study in which human mammary fibroblasts engineered to ectopically express HGF or TGFß1 alone or together induced human normal mammary epithelia to develop ductal carcinoma *in situ*, adenocarcinoma and poorly differentiated carcinoma when transplanted into cleared mammary fat pads of immuno-deficient mice (Kuperwasser et al. 2004).

Tumor protein p53. Tumor progression is a dynamic process in which each aberrant change has an impact on the biology of the tumor cell and its surroundings creating new selective pressures that affect the evolution of the cancer. Recently, several reports have identified mutations of tumor-suppressor genes, including p53, in the stromal compartment of advanced human carcinomas (Kurose et al. 2002; Matsumoto et al. 2003; Paterson et al. 2003; Fukino et al. 2004). Although oncogenic stress has previously been shown to induce p53 responses in epithelial cells, Hill and colleagues explored the nonautonomous loss of p53 in the stroma on tumor evolution (Hill et al. 2005). They showed in a mouse model of prostatic carcinoma that suppression of the Rb protein in the epithelia by expression of a fragment of the SV40 T antigen induces upregulation of p53 in stromal fibroblasts through a paracrine mechanism. This creates a selective pressure that results in subsequent loss of p53 in a subpopulation of stromal cells. Thus, an initial decrease in fibroblast proliferation is followed by a selective evolution of a highly proliferative p53 null subpopulation of stromal cells. These p53-deficient stromal cells nonautonomously increase the selective pressure against p53 in the epithelium. Similar to the prostate model used by Hill et al., complete loss of p53 in host fibroblasts accelerated proliferation of MCF7 human breast cancer cells when xenografted in SCID mice compared to wild-type control animals (Kiaris et al. 2005). Moreover, the tumor stroma in p53 heterozygous hosts showed p53 LOH, suggesting that the selection of p53-deficient fibroblasts is advantageous for tumor progression. It is not known whether this selection by epithelial cells occurs in all stromal cells or only in a specific lineage of cells bearing other mutations. Validation of these observations in other cancer models expands our understanding of the selective evolution of common stromal mutations relevant to human cancers.

Growth factor ligands and receptors. Multiple families of growth factors that have been implicated as paracrine mediators of stromal–epithelial interactions are altered in the tumor stroma. These include FGF, insulin growth factor (IGF), transforming growth factor alpha and beta (TGFα, TGFß) and HGF – all of which are predominantly stimulators of proliferation. Members of the FGF family have been identified as having essential roles in mediating mesenchymal–epithelial interactions in the prostate during development and carcinogenesis (Hayward et al. 2001; Thomson 2001). FGF ligands signal through four cognate high-affinity tyrosine kinase receptors, designated FGFR-1 to -4, leading to activation of multiple signaling transduction pathways, including the Erk, MAPK, and PI3 kinase pathways. Basal epithelial cells express FGFR-1 (IIIc isoform), FGFR-2 (IIIb isoform), and FGFR-3 (IIIb isoform) while FGFR-4 is expressed in luminal cells (Kwabi-Addo et al. 2004). The major source of FGFs in human prostate is the stroma. FGF2 (basic FGF), FGF7 (KGF), and FGF9 are found in biologically significant quantities in the prostatic stroma (Giri et al. 1999a, b). Although important in development, FGF10 is expressed at low levels in the adult, and FGF1, FGF5, and FGF8 have been detected at the mRNA level by RT-PCR (Ittman and Mansukhani 1997). Expression of other FGF family members was either undetected by traditional methods or has not been studied (Kwabi-Addo et al. 2004). Several studies have shown that multiple FGF ligands have elevated expression in prostate cancer and can act as a paracrine modulators of tumor progression. FGF1 is expressed in 80% of prostate cancers and can bind any type and isoform of the FGF receptors (Ornitz et al. 1996). Immunohistochemical analysis of FGF2 distribution has shown stromal staining in tissues from patients with localized prostate cancer, but tumor cell expression of FGF2 in more advanced patients. Studies with the TRAMP model of metastatic neuroendocrine prostate cancer revealed that when these animals were crossed with FGF2 knockout mice, there was an increase in the survival and a decrease in metastasis compared with mice bearing even one allele of the FGF2 gene (Polnaszek et al. 2003). Thus, it seems likely that FGF2 is acting as a paracrine factor expressed by stromal cells in early stages of malignant disease and then as the tumor advances switches to acting in an autocrine fashion favoring progression and metastasis of prostate cancer. Although FGF6, FGF7, and FGF8 expression has been demonstrated to be elevated in prostate cancer, their contribution and exact role remains unclear. A transgenic mouse model in which FGF8b was expressed in the prostatic epithelium resulted in the formation of prostatic intraepithelial neoplasia, suggesting that at least in this model overexpression of this ligand was insufficient to elicit full-blown carcinogenesis (Song et al. 2002). Memarzadeth et al. reported that overexpression of FGF10 in prostate stroma results in epithelial proliferation (Memarzadeh et al. 2007). Moreover, using tissue recombination techniques involving stromal cells overexpressing FGF10 recombined with epithelium expressing activated Akt (myristoylated Akt) resulted in cooperative effects and tumorigenesis. While interesting biologically, the relevance of these mouse studies is uncertain because FGF10 was found to be expressed at very low levels in human prostate cancer (Ropiquet et al. 2000; Abate-Shen and Shen 2007).

Tyrosine kinases have emerged as major potential therapeutic targets. Studies targeting the kinase domain of FGF receptors could be beneficial. Expression of FGF receptors and increased signaling are ubiquitous in prostate cancer. FGFR1 is upregulated in about 40% of poorly differentiated prostatic carcinomas. In a recent study, Acevedo et al. described the consequences that follow the activation of the FGFR1 receptor (Acevedo et al. 2007). They used a novel strategy for conditional and reversible activation of the FGFR1 cytoplasmic signaling by a chemical inducer of receptor heterodimerization. This genetically engineered mouse model was based on targeted, inducible FGFR1 (iFGFR1) activation, and designated juxtaposition of CID and kinase1 (JOCK1). FGFR1 expression and signaling was directed to epithelial cells under the control of the composite probasin promoter, ARR2PB (Zhang et al. 2000; Freeman et al. 2003). Biweekly treatment of the JOCK1 mice with the inducer for up to a year resulted in the occurrence of a wide spectrum of prostate malignancies from PIN to adenocarcinomas as well as metastases to lymph nodes and liver. Gene profiling revealed increased expression of members of the Wnt signaling pathway, including Fzd4, which is capable of inducing ß-catenin. Despite the long latency of tumorigenesis in this model, it is clear that paracrine actions of FGF signaling from stromal to epithelial tissues (which are mimicked here) may be critical for prostate tumorigenesis. More studies are needed to understand the roles of other members of the FGF family in this process.

Matrix metalloproteinases (MMPs). MMPs in cancer progression have been known to play a role in cancer progression for more than two decades (Liotta et al. 1980). Increased production of MMPs has been seen not only in tumor cells, but also in host-derived cells, such as fibroblasts, vascular endothelial cells, myofibroblasts and inflammatory cells, all residents of the tumor stroma. Degradation of the basement membrane, and thus a general structural loosening allowing invasion was thought to be the role played by these proteases. It is now clear that MMPs can not only cleave ECM proteins, but they can also proteolytically process growth factors, growth factor receptors, cytokines, chemokines, and precursor proteins to form biologically active fragments. Pathways that can be activated are FGFs, TGFß, IGF, VEGF, and integrins. (Agrez et al. 1994; Whitelock et al. 1996; Imai et al. 1997; Manes et al. 1997, 1999). The MMPs comprise a family of 23 enzymes; their expression in tumors varies among the types and subtypes of tumors. Studies in mouse models in which stromally derived MMPs are either upregulated or suppressed have shed some light on the role of these proteases during tumor progression. For example, in MMP-2 and MMP-9 knockout mice, a lower rate of tumor incidence and decreased growth of implanted neuroblastoma and ovarian cancer is seen compared with wild-type mice (Huang et al. 2002; Chantrain et al. 2004). The contribution of MMPs to tumor metastasis is also demonstrated by injection of melanoma cells into these mice with the resultant decreases in the number of metastatic nodules compared to the intact hosts (Itoh et al. 1999). The involvement of MMPs in the process of angiogenesis is well documented. For example, MMP-9 can cleave isoforms 121 and 165 of VEGF increasing its bioavailability in the vicinity of tumors. Tumors in mice deficient with MMP-2 and -9 exhibit decreased vascular density and size due to impaired endothelial cell invasion and inhibition of pericyte recruitment by endothelial cells (Huang et al. 2002; Chantrain et al. 2004). MMPs are also involved in vasculogenesis, which is important in the recruitment of endothelial precursor cells (EPC) to the tumor. EPC are mobilized from the bone marrow niche and directed to the tumor site by chemokines, such as stromal-derived factor-1 (SDF1/CXCL12) or the monocyte chemotactic protein-1 (MCP-1/CCL2) (Salcedo and Oppenheim 2003). MMP-9 expression is induced by SDF1 and causes the release of soluble sKitL, which enhances the recruitment of hematopoietic stem cells and EPC to the tumor (Lane et al. 2000; Heissig et al. 2002). Lack of MMP-9 by bone marrow-derived cells in knockout animals decreased colonization in xenotransplanted tumors (Jodele et al. 2005). Despite some pro-tumorigenic effects of MMPs, recent studies have also suggested protective and antitumorigenic effects. In particular, these arise from proteases expressed by infiltrating inflammatory cells, highlighting the importance of using immunocompetent models for studying tumor progression in the context of MMPs. Failed clinical trials using "promising" inhibitors of MMPs based on *in vitro* data emphasizes the need for a better understanding of the complex interactions of MMPs *in vivo* during cancer progression (Coussens et al. 2002) (see Chapter 21).

A d*isintegrin* a*nd* m*etalloproteinase protein (ADAM).* The family of ADAMs has key roles in cell–cell interactions because of their ability to cleave and release growth factors, cytokines, receptors, and other molecules from the membrane of the cell (Blobel 2005). Recent reports have linked some members of the ADAMs family to cell migration and to the control of various signals activated in several types of cancer. Of particular interest, ADAM12 has been reported to be useful for the identification of a subpopulation of α-smooth muscle actin-positive stromal cells adjacent to tumor cells. Evaluation of the expression of ADAM12 by *in situ* mRNA hybridization showed upregulation in stromal cells of three different mouse models of prostate (W^{10}), breast (MMTV-PyMT), and colon (Apc/Min/+) cancer (Arribas et al. 2006). Generation of W^{10} animals lacking ADAM12 (W10/ Adam12$^{-/-}$) resulted in smaller and better differentiated prostate cancer tumors compared to animals carrying one or both wild-type alleles. These results are in accordance with a recent study showing that "gain of function" by overexpression of ADAM12 in the PyMT mouse model of breast cancer accelerated tumor progression (Kveiborg et al. 2005). Further studies will be necessary to determine the contribution of the catalytic activity versus functions in cell–cell interaction during carcinogenesis.

Inflammatory cells and the immunity system. Other residents of the tumor microenvironment that cooperate with stromal-induced carcinogenesis are members of the immune system. The association between chronic inflammation and carcinogenesis is well established. Exemplary, studies have shown associations between gastric cancer and colonization of Helicobacter pylori, inflammatory bowel disease and colorectal cancer, cirrhosis and liver cancer, and familial pancreatitis and pancreatic cancer (Itzkowitz and Yio 2004; Houghton and Wang 2005; Jura et al. 2005; Seitz and Stickel 2006). The role of chronic inflammation in the development of gastric cancer has been elegantly shown using a mouse model of infection with *Helicobacter* spp (Eaton 1999). The study revealed that host T cell immunity was critical for the occurrence of gastric disease. In addition to the loss of parietal cells and chief cells in the glandular mucosa, chronic inflammation stimulated the recruitment of bone marrow-derived endothelial progenitor and myofibroblast cells similar to models of colitis and hepatitis (Forbes et al. 2004; Brittan et al. 2005). Another clear example of the critical contribution of inflammation when an oncogene is activated is illustrated in a model of pancreatic cancer. In this model, K-Ras oncogene expression was restricted to the acinar/centroacinar cells using a tet-off system allowing temporal control of expression. Induction of chronic pancreatitis following induction of K-Ras oncogene recapitulated all stages of human pancreatic ductal carcinoma (PDA) (Guerra et al. 2007).

Bone microenvironment. The roles of local cells at metastatic sites, such as bone, apparently play key roles in the biological regulation, establishment and growth of metastasis. Thus, it is not surprising that the cancer cells can display a preference to metastasize to a site that is permissive for the survival and growth of specific tumor cells. Current animal models of bone metastasis include spontaneous tumors (such as those that arise in cats and dogs), chemical induction, transgenic models, and xenografts of human tumors of cell lines into immunodeficient mice. Xenografted tissue or cells have been injected or implanted subcutaneously, intracardiac, orthotopic, or directly into the tibia or femur. Certain tumors have a predisposition to metastasize to bone. These include breast, prostate, and thyroid carcinoma, multiple myeloma, and renal cell carcinoma. Typically, bone metastases are osteolytic, meaning that, on balance, bone is resorbed (by osteoclasts) resulting in decreased mineral content. It is notable that the vast majority of metastases include both osteoblastic (new bone formation by osteoblasts) and osteolytic components. In the case of prostate cancer, the normal balance is reversed with some osteolytic but mostly osteoblastic lesions. Ultimately, both osteolytic and osteoblastic activity predispose to pathological fractures, pain, impaired mobility, spinal cord compression, and symptomatic hypercalcemia (Galasko 1986; Coleman 1997; Moul and Lipo 1999). In this regard, genetic manipulation of the metastatic microenvironment could be beneficial for modeling stromal–epithelial interactions at metastatic sites in animal models. Osteoblasts can be targeted using specific promoters, such as Col 1α1, FSP-1, or osteocalcin (Dacquin et al. 2002; Zhang et al. 2002; Bhowmick et al. 2004). Osteoclasts have been targeted with cathepsin K (CTSK) and tartrate-resistant acid phosphatase (TRAP) (Chiu et al. 2004). Promising results using the monoclonal antibody denosumab (AMG 162) have shown the role of the receptor activator of NFkB ligand (RANKL) in osteoclastogenesis and metastasis. These results suggest that RANKL could be a good candidate to study the intrinsic mechanism involved in osteoclast-tumor interaction. Current therapies for metastatic cancer are targeted at inhibiting osteolysis induced by tumor cells, but these have only palliative effects and do not eliminate the cancer. Thus, better understanding of the pathobiology in the bone stroma–tumor interaction is needed to continue developing improved therapies.

20.3.4 *Tissue Rescue*

Manipulation of critical genes that can impair viability, for example resulting in embryonic or neonatal lethality, of the transgenic or gene knockout mouse model is a major problem. In this case, it is often possible to "rescue" the organ of interest by dissection from the fetus or neonatal animal of the material or its embryonic precursor and subsequent growth under the renal capsule. The resultant organ can be used to examine the effects of the specific genetic change on organogenesis and carcinogenesis. Such tissues can also be used as a source of material with which to generate tissue recombinants for further study. There are many examples, but one that illustrates well this technique is the rescue of cloacal tissue from retinoblastoma (Rb) gene knockout fetuses. These animals die at about embryonic day 14 before the formation of the prostate. Rescued cloacal tissue before this crucial embryonic time period develops into Rb-deficient prostate, bladder, and other adjacent organs, depending on the efficiency of the dissection (Wang et al. 2000; Hayward et al. 2003). Similar tissue rescue experiments have also been performed to rescue Rb-deficient mammary gland (Robinson et al. 2001). These studies established that Rb is not required for prostate development but that Rb-deficient prostates were more sensitive to hormonal carcinogenesis creating a model that resembles prostate cancer (Wang et al. 2000; Hayward et al. 2003).

20.3.5 *Tissue Recombination*

Perhaps the most flexible *in vivo* model for studying stromal–epithelial, epithelial–stromal interactions is tissue recombination (TR) and grafting (Fig. 20.3). TR provides a good alternative to the labor- and time-intensive generation of transgenic mice for the *in vivo* study of gene function in a specific compartment. Tissue recombination is a widely used technique for examining different aspects of development. Briefly, tissue recombinants are made by mixing stromal or mesenchymal cells with epithelial cells within a matrix (most commonly type I collagen). These recombinants can then be grafted beneath the renal capsule of a young adult rodent host for a determined period of time (as long as a year). An overview of the renal capsule grafting procedure is available online in "Tools and Techniques in Mammary Gland Biology" at: (http://mammary.nih.gov/tools/mousework/cunha001/index.html). The renal capsule site is commonly chosen because the kidney has a rich vasculature that makes a suitable environment for the growth and survival of tissues.

TR has many advantages; in a large number of studies, the method has demonstrated faithful replication of key aspects of both development and carcinogenesis (Cunha et al. 1992a, b; Hayward 2002; Ishii et al. 2005). The technique is very plastic allowing for targeted modifications to all or some cells specifically within the epithelial and stromal tissues. Another advantage is that recombinants can be made using stromal and epithelial cells coming from the same (homotypic) or different

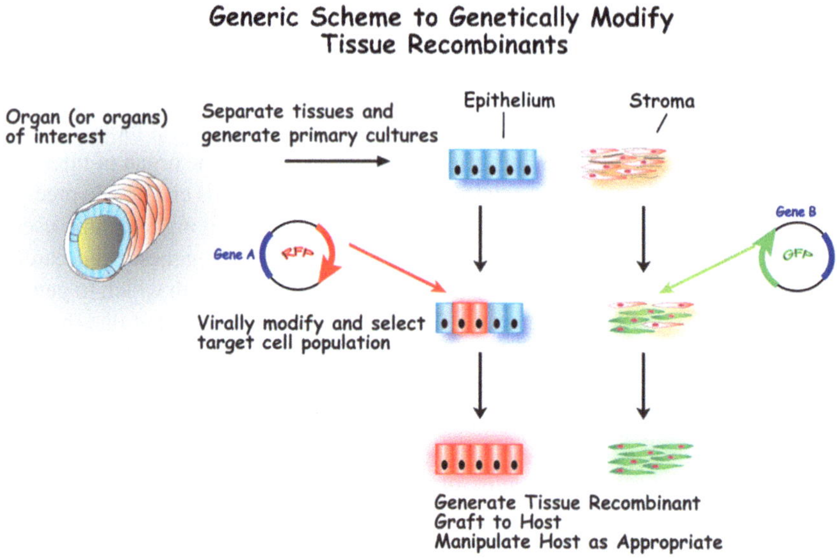

Fig. 20.3 Tissue recombination technique. Stylized illustration showing the generic technique to modify cells for recombination. Stromal and epithelial cells are separated from each other and isolated. The transgene (or shRNA) of interest is transduced into the cells that can be sorted using different color markers (or selected using biochemical selection as appropriate). Stromal and epithelial cells are then recombined and grafted under the renal capsule or in an orthotopic site of a mouse. This technique allows for recombination of cells from different tissues or species

(heterotypic) tissue as well as same (homospecific) or different (heterospecific) species. Disadvantages include the relative lack of metastasis when needed as a desired endpoint. Also recombinants are commonly grafted into immunocompromised mice (athymic or SCID) meaning that many effects of an intact immune system are lost. However, implantation into syngeneic hosts is also a possible alternative when appropriate cells are available.

TR can be used to explore questions relating to basic cellular lineage commitment. For example, recombinants composed of prostate mesenchymal and prostate epithelial cells give rise to the formation a prostatic glandular structure. Heterotypic recombinants composed of bladder epithelium (either embryonic or from an adult) and UGM also gives rise to prostatic tissue (Cunha et al. 1983; Neubauer et al. 1983). This suggests that the epithelium contains cells that are able to respond to the inductive mesenchyme by differentiating into different phenotypes. Whether this represents a true stem cell population or transdifferentiation needs to be defined.

Tissue recombinants have also been used to examine the tissue-specific roles of sex steroid hormones in the development of the urogenital system in males and females. Experiments using testicular feminized mice (Tfm) (these mice do not express a functional androgen receptor) demonstrated that androgens induce prostate gland morphogenesis through mesenchymal cells expressing androgen receptor, while expression of an epithelial androgen receptor was required for secretory function

(Donjacour and Cunha 1993). In contrast, combination of Tfm mesenchyme plus normal epithelia resulted in vaginal differentiation (Cunha and Chung 1981). This work demonstrated that control of epithelial proliferation and morphology is controlled via stromally located steroid receptors while control of differentiated function is a result of direct stimulation of epithelial receptors. An analogous system of control has since been demonstrated for estrogens and progesterone in the female genital tract (Buchanan et al. 1998a, b; Kurita et al. 2000, 2001). Such experiments would have been technically unfeasible without both TR as a method and the availability of transgenic mice as a tissue source, illustrating the combination of two approaches to solve an important question.

The increasing ratio of estrogen/androgen levels with age has been linked to the occurrence of prostate cancer. The use of testosterone and estradiol in various rat strains has been a classical method of inducing experimental prostatic tumors (Noble 1977, 1980; Wang and Wong 1998). It has been reported that administration of E2 in combination with testosterone to mice carrying recombinants of rat and mouse UGM and BPH1 cells stimulates malignant transformation and cancer progression (Wang et al. 2001b; Ricke et al. 2006). TR has allowed the dissection of target cell populations underlying the mechanism of carcinogenesis in such models (Ricke et al. 2007a, b).

Another benefit of TR is that *in vitro* genetic manipulation of the cells (stromal or epithelial) can be made before implantation. This approach was pioneered by Thompson to study malignant and benign disease progression in the prostate (Thompson et al. 1989). More recently, the use of lentiviral delivery systems allow either overexpression or downregulation of a gene of interest prior to grafting (Williams et al. 2005; Ao et al. 2006, 2007; He et al. 2007). With the addition of drug-inducible systems, genes can be controlled to study their contribution during different stages of cancer progression. Also, cells isolated from transgenic mice can be activated with adenovirus Cre (Ad-Cre) to knockout the expression of a floxed gene *in situ* prior to TR. An expanded use could also be the implantation of normal or tumor cells with genetic manipulation into transgenic mice to study the crosstalk between cells in the same or different compartments. A previously described role of the MMP-9 KO in metastasis in mice shows the potential use of this approach (Itoh et al. 1999; Huang et al. 2002).

Perhaps the best example of how the human tumor stroma can promote carcinogenesis of human initiated, but nontumorigenic epithelia, comes from tissue recombination experiments (Grossfeld et al. 1998; Olumi et al. 1999). In these experiments, fibroblasts from normal and cancer (carcinoma associated fibroblasts; CAF) regions of prostate cancer surgical specimens were isolated and recombined with the nontumorigenic SV40T-immortalized human prostatic epithelial cell line BPH-1 and grafted under the renal capsule of athymic mice (Hayward et al. 1995). Tissue recombinants composed of CAF formed large tumors composed of poorly differentiated carcinomas. This was in contrast to the absence of tumors when BPH-1 cells were grafted alone or in association with normal fibroblasts, or when CAFs were recombined with normal epithelium. Thus, fibroblasts in the vicinity of tumors can promote carcinogenesis in initiated but nontumorigenic epithelial cells.

In agreement with these observations, changes in gene expression in normal stromal cells have been shown to promote malignancy in neighboring unaltered epithelial cells (Nakamura et al. 1997; Barcellos-Hoff and Ravani 2000). We have recently demonstrated that the TGFβ pathway cooperates with the CXCL12/CXCR4 pathway in a paracrine manner to promote carcinogenesis (Ao et al. 2007).

The stroma can also play a role in the suppression of malignant transformation. There are a few reports suggesting that normal stromal cells have the capability to convert some malignant epithelia to morphologically benign lesions *in vitro* and *in vivo* (Cooper and Pinkus 1977; Hayashi and Cunha 1991).

In summary, the advantages of tissue recombination are: (1) cells can grow in a manageable *in vivo* environment; (2) Both, epithelial and stromal cells have normal interactions with each other (interactions that can be modified according to the experimental design); (3) flexibility of the system in which cells from a wide range of sources can be used; (4) cells isolated from transgenic mice can be used to study specific pathways; (5) genetically modified cells can be used.

20.3.6 Future Modeling Approaches

Transgenic mouse models represent a useful tool to evaluate the mechanistic fundamentals of malignancy. In these animals, tumors are derived from the manipulation of individual genes in specific (usually epithelial) tissues. Current models are limited in their ability to model stromal–epithelial interactions in tumor progression. In particular, the lack of organ-specific stromal promoters is a limiting factor. Moreover, the composition of both the inflammatory network and stromal compartments can be drastically different in parallel mouse and human organs. Due to our recent understanding that tumor progression is usually the result of an accumulation of genetic hits and functional changes in both epithelial and stromal tissues, the ability to manipulate multiple genes within the same animal remains a technological constraint. This limitation is currently overcome through the crossbreeding of various lineages; however, the generation and maintenance of multiple transgenic lineages is prohibitively time-consuming and costly for most complex experiments.

The discovery of genetic and epigenetic mutations in tumor stromal cells from clinical samples suggests that stroma co-evolves with cancer cells (Egeblad et al. 2005; Littlepage et al. 2005). However, the lack of organ-specific stromal promoters remains a technical problem for the use of transgenic animals in the study of stromal compartment contribution in tumor progression. Although transgenic mouse models have used promoter-regulated Cre-recombinase to target the stroma of multiple tissues, these commonly result in mosaic expression patterns (Bhowmick et al. 2004; Bardeesy et al. 2006). Accordingly, the components of the stromal compartment that are frequently heterogeneous in normal tissues and organs are underrepresented. By mutating the stroma in current autochthonous tumor mouse models, we can study stromal contribution during tumorigenesis and/or tumor progression.

Although the grafting of tissue recombinations of human cells into various mouse organs fails to incorporate a fully intact immune response, the ability to control gene expression in multiple tissues and incorporate multiple cell lineages makes tissue recombination an attractive alternative to generating and crossing of multiple transgenic mouse models. In addition, sophisticated and controlled scenarios using inducible genes allow for the mutation of different compartments at different time points in order to study the temporospatial contribution of each type of cell composing each tissue during cancer progression. This can be easily performed with cell lines but is much more challenging with transgenic mice.

It is well acknowledged that all models have their inherent advantages and limitations and should accordingly be utilized to address appropriate questions. The identification of genes and cells critical for tumor progression within the stroma, and the development of more sophisticated *in vivo* mouse models that mimic all aspects of human normal and tumor biology should prove invaluable for discovering new methodologies for the detection and treatment of cancer.

Acknowledgments The work of the authors is supported by funding through NIH grants U54 CA126505, U54 CA113007, R01 DK067049, by DOD-PCRP grant W81XWH-07-1-0479 and by the Frances Williams Preston Laboratories of the TJ Martell Foundation.

References

Abate-Shen C, Shen MM (2007) FGF signaling in prostate tumorigenesis – new insights into epithelial-stromal interactions. Cancer Cell 12:495–497

Acevedo VD, Gangula RD, Freeman KW, Li R, Zhang Y, Wang F, Ayala GE, Peterson LE, Ittmann M, Spencer DM (2007) Inducible FGFR-1 activation leads to irreversible prostate adenocarcinoma and an epithelial-to-mesenchymal transition. Cancer Cell 12:559–571

Agrez M, Chen A, Cone RI, Pytela R, Sheppard D (1994) The alpha v beta 6 integrin promotes proliferation of colon carcinoma cells through a unique region of the beta 6 cytoplasmic domain. J Cell Biol 127:547–556

Ao M, Franco OE, Park D, Raman D, Williams K, Hayward SW (2007) Cross-talk between paracrine-acting cytokine and chemokine pathways promotes malignancy in benign human prostatic epithelium. Cancer Res 67:4244–4253

Ao M, Williams K, Bhowmick NA, Hayward SW (2006) Transforming growth factor-beta promotes invasion in tumorigenic but not in nontumorigenic human prostatic epithelial cells. Cancer Res 66:8007–8016

Arribas J, Bech-Serra JJ, Santiago-Josefat B (2006) ADAMs, cell migration and cancer. Cancer Metastasis Rev 25:57–68

Barcellos-Hoff MH, Aggeler J, Ram TG, Bissell MJ (1989) Functional differentiation and alveolar morphogenesis of primary mammary cultures on reconstituted basement membrane. Development 105:223–235

Barcellos-Hoff MH, Ravani SA (2000) Irradiated mammary gland stroma promotes the expression of tumorigenic potential by unirradiated epithelial cells. Cancer Res 60:1254–1260

Bardeesy N, Cheng KH, Berger JH, Chu GC, Pahler J, Olson P, Hezel AF, Horner J, Lauwers GY, Hanahan D, DePinho RA (2006) Smad4 is dispensable for normal pancreas development yet critical in progression and tumor biology of pancreas cancer. Genes Dev 20:3130–3146

Bhowmick NA, Chytil A, Plieth D, Gorska AE, Dumont N, Shappell S, Washington MK, Neilson EG, Moses HL (2004) TGF-beta signaling in fibroblasts modulates the oncogenic potential of adjacent epithelia. Science 303:848–851

Bilozur ME, Hay ED (1988) Neural crest migration in 3D extracellular matrix utilizes laminin, fibronectin, or collagen. Dev Biol 125:19–33
Bing RJ, Binder T, Pataricza J, Kibira S, Narayan KS (1991) The use of microcarrier beads in the production of endothelium-derived relaxing factor by freshly harvested endothelial cells. Tissue Cell 23:151–159
Blobel CP (2005) ADAMs: key components in EGFR signalling and development. Nat Rev Mol Cell Biol 6:32–43
Brinkmann V, Foroutan H, Sachs M, Weidner KM, Birchmeier W (1995) Hepatocyte growth factor/scatter factor induces a variety of tissue-specific morphogenic programs in epithelial cells. J Cell Biol 131:1573–1586
Brittan M, Chance V, Elia G, Poulsom R, Alison MR, MacDonald TT, Wright NA (2005) A regenerative role for bone marrow following experimental colitis: contribution to neovasculogenesis and myofibroblasts. Gastroenterology 128:1984–1995
Buchanan DL, Kurita T, Taylor JA, Lubahn DL, Cunha GR, Cooke PS (1998a) Role of stromal and epithelial estrogen receptors in vaginal epithelial proliferation, stratification and cornification. Endocrinology 139:4345–4352
Buchanan DL, Setiawan T, Lubahn DL, Taylor JA, Kurita T, Cunha GR, Cooke PS (1998b) Tissue compartment-specific estrogen receptor participation in the mouse uterine epithelial secretory response. Endocrinology 140:484–491
Castellsague X, Bosch FX, Munoz N (2002) Environmental co-factors in HPV carcinogenesis. Virus Res 89:191–199
Chantrain CF, Shimada H, Jodele S, Groshen S, Ye W, Shalinsky DR, Werb Z, Coussens LM, DeClerck YA (2004) Stromal matrix metalloproteinase-9 regulates the vascular architecture in neuroblastoma by promoting pericyte recruitment. Cancer Res 64:1675–1686
Chiu WS, McManus JF, Notini AJ, Cassady AI, Zajac JD, Davey RA (2004) Transgenic mice that express Cre recombinase in osteoclasts. Genesis 39:178–185
Chung LW (1995) The role of stromal-epithelial interaction in normal and malignant growth. Cancer Surv 23:33–42
Clark JM, Hirtenstein MD (1981) Optimizing culture conditions for the production of animal cells in microcarrier culture. Ann N Y Acad Sci 369:33–46
Coleman RE (1997) Skeletal complications of malignancy. Cancer 80:1588–1594
Cooper M, Pinkus H (1977) Intrauterine transplantation of rat basal cell carcinoma as a model for reconversion of malignant to benign growth. Cancer Res 37:2544–2552
Coussens LM, Fingleton B, Matrisian LM (2002) Matrix metalloproteinase inhibitors and cancer: trials and tribulations. Science 295:2387–2392
Cunha GR, Alarid ET, Turner T, Donjacour AA, Boutin EL, Foster BA (1992a) Normal and abnormal development of the male urogenital tract: role of androgens, mesenchymal-epithelial interactions and growth factors. J Androl 13:465–475
Cunha GR, Battle E, Young P, Brody J, Donjacour A, Hayashi N, Kinbara H (1992b) Role of epithelial-mesenchymal interactions in the differentiation and spatial organization of visceral smooth muscle. Epithelial Cell Biol 1:76–83
Cunha GR, Chung LWK (1981) Stromal-epithelial interactions: I. Induction of prostatic phenotype in urothelium of testicular feminized (Tfm/y) mice. J. Steroid Biochem 14:1317–1321
Cunha GR, Fujii H, Neubauer BL, Shannon JM, Sawyer LM, Reese BA (1983) Epithelial-mesenchymal interactions in prostatic development. I. Morphological observations of prostatic induction by urogenital sinus mesenchyme in epithelium of the adult rodent urinary bladder. J Cell Biol 96:1662–1670
Cunha GR, Hayward SW, Dahiya R, Foster BA (1996) Smooth muscle-epithelial interactions in normal and neoplastic prostatic development. Acta Anat (Basel) 155:63–72
Dacquin R, Starbuck M, Schinke T, Karsenty G (2002) Mouse alpha1(I)-collagen promoter is the best known promoter to drive efficient Cre recombinase expression in osteoblast. Dev Dyn 224:245–251
Donjacour AA, Cunha GR (1993) Assessment of prostatic protein secretion in tissue recombinants made of urogenital sinus mesenchyme and urothelium from normal or androgen-insensitive mice. Endocrinology 131:2342–2350

Dvorak HF (1986) Tumors: wounds that do not heal. Similarities between tumor stroma generation and wound healing. N Engl J Med 315:1650–1659

Eaton KA (1999) Animal models of *Helicobacter gastritis*. Curr Top Microbiol Immunol 241:123–154

Egeblad M, Littlepage LE, Werb Z (2005) The fibroblastic coconspirator in cancer progression. Cold Spring Harb Symp Quant Biol 70:383–388

Forbes SJ, Russo FP, Rey V, Burra P, Rugge M, Wright NA, Alison MR (2004) A significant proportion of myofibroblasts are of bone marrow origin in human liver fibrosis. Gastroenterology 126:955–963

Freeman KW, Welm BE, Gangula RD, Rosen JM, Ittmann M, Greenberg NM, Spencer DM (2003) Inducible prostate intraepithelial neoplasia with reversible hyperplasia in conditional FGFR1-expressing mice. Cancer Res 63:8256–8263

Frese KK, Tuveson DA (2007) Maximizing mouse cancer models. Nat Rev Cancer 7:645–658

Fuchs E (1882) Das Sarkom des Uvealtractus. Graefe's Archiv für Ophthalmologie XII: 233

Fukino K, Shen L, Matsumoto S, Morrison CD, Mutter GL, Eng C (2004) Combined total genome loss of heterozygosity scan of breast cancer stroma and epithelium reveals multiplicity of stromal targets. Cancer Res 64:7231–7236

Gahwiler BH, Capogna M, Debanne D, McKinney RA, Thompson SM (1997) Organotypic slice cultures: a technique has come of age. Trends Neurosci 20:471–477

Galasko CS (1986) Skeletal metastases. Clin Orthop Relat Res:18–30

Giri D, Ropiquet F, Ittmann M (1999a) Alterations in expression of basic fibroblast growth factor (FGF) 2 and its receptor FGFR-1 in human prostate cancer. Clin Cancer Res 5:1063–1071

Giri D, Ropiquet F, Ittmann M (1999b) FGF9 is an autocrine and paracrine prostatic growth factor expressed by prostatic stromal cells. J Cell Physiol 180:53–60

Glinsky VV, Huflejt ME, Glinsky GV, Deutscher SL, Quinn TP (2000) Effects of Thomsen-Friedenreich antigen-specific peptide P-30 on beta-galactoside-mediated homotypic aggregation and adhesion to the endothelium of MDA-MB-435 human breast carcinoma cells. Cancer Res 60:2584–2588

Grossfeld GD, Hayward SW, Tlsty TD, Cunha GR (1998) The role of stroma in prostatic carcinogenesis. Endocr Relat Cancer 5:253–270

Gudjonsson T, Ronnov-Jessen L, Villadsen R, Rank F, Bissell MJ, Petersen OW (2002) Normal and tumor-derived myoepithelial cells differ in their ability to interact with luminal breast epithelial cells for polarity and basement membrane deposition. J Cell Sci 115:39–50

Guerra C, Schuhmacher AJ, Canamero M, Grippo PJ, Verdaguer L, Perez-Gallego L, Dubus P, Sandgren EP, Barbacid M (2007) Chronic pancreatitis is essential for induction of pancreatic ductal adenocarcinoma by K-Ras oncogenes in adult mice. Cancer Cell 11:291–302

Habib FK, Ross M, Bayne CW (2000) Development of a new in vitro model for the study of benign prostatic hyperplasia. Prostate Suppl 9:15–20

Hammond TG, Hammond JM (2001) Optimized suspension culture: the rotating-wall vessel. Am J Physiol Renal Physiol 281:F12–25

Hayashi N, Cunha GR (1991) Mesenchyme-induced changes in the neoplastic characteristics of the Dunning prostatic adenocarcinoma. Cancer Res 51:4924–4930

Hayward SW (2002) Approaches to modeling stromal-epithelial interactions. J Urol 168:1165–1172

Hayward SW, Cunha GR, Dahiya R (1996) Normal development and carcinogenesis of the prostate. A unifying hypothesis. Ann N Y Acad Sci 784:50–62

Hayward SW, Dahiya R, Cunha GR, Bartek J, Deshpande N, Narayan P (1995) Establishment and characterization of an immortalized but non-transformed human prostate epithelial cell line: BPH-1. In Vitro Cell Dev Biol Anim 31:14–24

Hayward SW, Del Buono R, Deshpande N, Hall PA (1992) A functional model of adult human prostate epithelium. The role of androgens and stroma in architectural organisation and the maintenance of differentiated secretory function. J Cell Sci 102(Pt 2):361–372

Hayward SW, Haughney PC, Rosen MA, Greulich KM, Weier HU, Dahiya R, Cunha GR (1998) Interactions between adult human prostatic epithelium and rat urogenital sinus mesenchyme in a tissue recombination model. Differentiation 63:131–140

Hayward SW, Rosen MA, Cunha GR (1997) Stromal-epithelial interactions in the normal and neoplastic prostate. Br J Urol 79(Suppl 2):18–26

Hayward SW, Wang Y, Cao M, Hom YK, Zhang B, Grossfeld GD, Sudilovsky D, Cunha GR (2001) Malignant transformation in a nontumorigenic human prostatic epithelial cell line. Cancer Res 61:8135–8142

Hayward SW, Wang Y, Day ML (2003) Rescue and isolation of Rb-deficient prostate epithelium by tissue recombination. Methods Mol Biol 218:17–33

He Y, Franco OE, Jiang M, Williams K, Love HD, Coleman IM, Nelson PS, Hayward SW (2007) Tissue-specific consequences of cyclin D1 overexpression in prostate cancer progression. Cancer Res 67:8188–8197

Hebner C, Weaver VM, Debnath J (2008) Modeling morphogenesis and oncogenesis in three-dimensional breast epithelial cultures. Annu Rev Pathol 3:313–339

Heissig B, Hattori K, Dias S, Friedrich M, Ferris B, Hackett NR, Crystal RG, Besmer P, Lyden D, Moore MA, Werb Z, Rafii S (2002) Recruitment of stem and progenitor cells from the bone marrow niche requires MMP-9 mediated release of kit-ligand. Cell 109:625–637

Hicks RM (1980) Initiation and promotion in the transitional epithelium of the rat bladder. Br J Cancer 41:504–505

Hicks RM, Chowaniec J (1977) The importance of synergy between weak carcinogens in the induction of bladder cancer in experimental animals and humans. Cancer Res 37:2943–2949

Hicks RM, Wakefield JSJ (1972) Rapid induction of bladder cancer in rats with N-methyl-N-nitrosourea. I. Histology. Chem Biol Interact 5:139–152

Hill R, Song Y, Cardiff RD, Van Dyke T (2005) Selective evolution of stromal mesenchyme with p53 loss in response to epithelial tumorigenesis. Cell 123:1001–1011

Hodges GM, Hicks RM, Spacey GD (1977) Epithelial-stromal interactions in normal and chemical carcinogen-treated adult bladder. Cancer Res 37:3720–3730

Houghton J, Wang TC (2005) *Helicobacter pylori* and gastric cancer: a new paradigm for inflammation-associated epithelial cancers. Gastroenterology 128:1567–1578

Huang S, Van Arsdall M, Tedjarati S, McCarty M, Wu W, Langley R, Fidler IJ (2002) Contributions of stromal metalloproteinase-9 to angiogenesis and growth of human ovarian carcinoma in mice. J Natl Cancer Inst 94:1134–1142

Imai K, Hiramatsu A, Fukushima D, Pierschbacher MD, Okada Y (1997) Degradation of decorin by matrix metalloproteinases: identification of the cleavage sites, kinetic analyses and transforming growth factor-beta1 release. Biochem J 322(Pt 3):809–814

Ishii K, Shappell SB, Matusik RJ, Hayward SW (2005) Use of tissue recombination to predict phenotypes of transgenic mouse models of prostate carcinoma. Lab Invest 85:1086–1103

Itoh T, Tanioka M, Matsuda H, Nishimoto H, Yoshioka T, Suzuki R, Uehira M (1999) Experimental metastasis is suppressed in MMP-9-deficient mice. Clin Exp Metastasis 17:177–181

Ittman M, Mansukhani A (1997) Expression of fibroblast growth factors (FGFs) and FGF receptors in human prostate. J Urol 157:351–356

Itzkowitz SH, Yio X (2004) Inflammation and cancer IV. Colorectal cancer in inflammatory bowel disease: the role of inflammation. Am J Physiol Gastrointest Liver Physiol 287:G7–17

Iwano M, Fischer A, Okada H, Plieth D, Xue C, Danoff TM, Neilson EG (2001) Conditional abatement of tissue fibrosis using nucleoside analogs to selectively corrupt DNA replication in transgenic fibroblasts. Mol Ther 3:149–159

Iwano M, Plieth D, Danoff TM, Xue C, Okada H, Neilson EG (2002) Evidence that fibroblasts derive from epithelium during tissue fibrosis. J Clin Invest 110:341–350

Jacquot J, Spilmont C, Burlet H, Fuchey C, Buisson AC, Tournier JM, Gaillard D, Puchelle E (1994) Glandular-like morphogenesis and secretory activity of human tracheal gland cells in a three-dimensional collagen gel matrix. J Cell Physiol 161:407–418

Jodele S, Chantrain CF, Blavier L, Lutzko C, Crooks GM, Shimada H, Coussens LM, Declerck YA (2005) The contribution of bone marrow-derived cells to the tumor vasculature in neuroblastoma is matrix metalloproteinase-9 dependent. Cancer Res 65:3200–3208

Jura N, Archer H, Bar-Sagi D (2005) Chronic pancreatitis, pancreatic adenocarcinoma and the black box in-between. Cell Res 15:72–77
Kang KI, Meng X, Devin-Leclerc J, Bouhouche I, Chadli A, Cadepond F, Baulieu EE, Catelli MG (1999) The molecular chaperone Hsp90 can negatively regulate the activity of a glucocorticosteroid-dependent promoter. Proc Natl Acad Sci USA 96:1439–1444
Kiaris H, Chatzistamou I, Trimis G, Frangou-Plemmenou M, Pafiti-Kondi A, Kalofoutis A (2005) Evidence for nonautonomous effect of p53 tumor suppressor in carcinogenesis. Cancer Res 65:1627–1630
Kleinman HK, Martin GR (2005) Matrigel: basement membrane matrix with biological activity. Semin Cancer Biol 15:378–386
Klingelhofer J, Ambartsumian NS, Lukanidin EM (1997) Expression of the metastasis-associated mts1 gene during mouse development. Dev Dyn 210:87–95
Kornfeld D, Ekbom A, Ihre T (1997) Is there an excess risk for colorectal cancer in patients with ulcerative colitis and concomitant primary sclerosing cholangitis? A population based study. Gut 41:522–525
Kuhbandner S, Brummer S, Metzger D, Chambon P, Hofmann F, Feil R (2000) Temporally controlled somatic mutagenesis in smooth muscle. Genesis 28:15–22
Kunz-Schughart LA, Heyder P, Schroeder J, Knuechel R (2001) A heterologous 3-D coculture model of breast tumor cells and fibroblasts to study tumor-associated fibroblast differentiation. Exp Cell Res 266:74–86
Kuperwasser C, Chavarria T, Wu M, Magrane G, Gray JW, Carey L, Richardson A, Weinberg RA (2004) Reconstruction of functionally normal and malignant human breast tissues in mice. Proc Natl Acad Sci USA 101:4966–4971
Kurita T, Lee KJ, Cooke PS, Lydon JP, Cunha GR (2000) Paracrine regulation of epithelial progesterone receptor and lactoferrin by progesterone in the mouse uterus. Biol Reprod 62:831–838
Kurita T, Wang YZ, Donjacour AA, Zhao C, Lydon JP, O'Malley BW, Isaacs JT, Dahiya R, Cunha GR (2001) Paracrine regulation of apoptosis by steroid hormones in the male and female reproductive system. Cell Death Differ 8:192–200
Kurose K, Gilley K, Matsumoto S, Watson PH, Zhou XP, Eng C (2002) Frequent somatic mutations in PTEN and TP53 are mutually exclusive in the stroma of breast carcinomas. Nat Genet 32:355–357
Kveiborg M, Frohlich C, Albrechtsen R, Tischler V, Dietrich N, Holck P, Kronqvist P, Rank F, Mercurio AM, Wewer UM (2005) A role for ADAM12 in breast tumor progression and stromal cell apoptosis. Cancer Res 65:4754–4761
Kwabi-Addo B, Ozen M, Ittmann M (2004) The role of fibroblast growth factors and their receptors in prostate cancer. Endocr Relat Cancer 11:709–724
Lakso M, Sauer B, Mosinger B Jr, Lee EJ, Manning RW, Yu SH, Mulder KL, Westphal H (1992) Targeted oncogene activation by site-specific recombination in transgenic mice. Proc Natl Acad Sci USA 89:6232–6236
Lane WJ, Dias S, Hattori K, Heissig B, Choy M, Rabbany SY, Wood J, Moore MA, Rafii S (2000) Stromal-derived factor 1-induced megakaryocyte migration and platelet production is dependent on matrix metalloproteinases. Blood 96:4152–4159
Lee EY, Lee WH, Kaetzel CS, Parry G, Bissell MJ (1985) Interaction of mouse mammary epithelial cells with collagen substrata: regulation of casein gene expression and secretion. Proc Natl Acad Sci USA 82:1419–1423
Lee EY, Parry G, Bissell MJ (1984) Modulation of secreted proteins of mouse mammary epithelial cells by the collagenous substrata. J Cell Biol 98:146–155
Lee GY, Kenny PA, Lee EH, Bissell MJ (2007) Three-dimensional culture models of normal and malignant breast epithelial cells. Nat Methods 4:359–365
Liotta LA, Tryggvason K, Garbisa S, Hart I, Foltz CM, Shafie S (1980) Metastatic potential correlates with enzymatic degradation of basement membrane collagen. Nature 284:67–68
Lipschutz JH, Foster BA, Cunha GR (1997) Differentiation of rat neonatal ventral prostates grown in a serum-free organ culture system. Prostate 32:35–42

Littlepage LE, Egeblad M, Werb Z (2005) Coevolution of cancer and stromal cellular responses. Cancer Cell 7:499–500

Manes S, Llorente M, Lacalle RA, Gomez-Mouton C, Kremer L, Mira E, Martinez AC (1999) The matrix metalloproteinase-9 regulates the insulin-like growth factor-triggered autocrine response in DU-145 carcinoma cells. J Biol Chem 274:6935–6945

Manes S, Mira E, Barbacid MM, Cipres A, Fernandez-Resa P, Buesa JM, Merida I, Aracil M, Marquez G, Martinez AC (1997) Identification of insulin-like growth factor-binding protein-1 as a potential physiological substrate for human stromelysin-3. J Biol Chem 272:25706–25712

Mantovani A, Romero P, Palucka AK, Marincola FM (2008) Tumour immunity: effector response to tumour and role of the microenvironment. Lancet 371:771–783

Marker PC, Donjacour AA, Dahiya R, Cunha GR (2003) Hormonal, cellular, and molecular control of prostatic development. Dev Biol 253:165–174

Matsumoto N, Yoshida T, Yamashita K, Numata Y, Okayasu I (2003) Possible alternative carcinogenesis pathway featuring microsatellite instability in colorectal cancer stroma. Br J Cancer 89:707–712

Memarzadeh S, Xin L, Mulholland DJ, Mansukhani A, Wu H, Teitell MA, Witte ON (2007) Enhanced paracrine FGF10 expression promotes formation of multifocal prostate adenocarcinoma and an increase in epithelial androgen receptor. Cancer Cell 12:572–585

Miwa T, Koyama T, Shirai M (2000) Muscle specific expression of Cre recombinase under two actin promoters in transgenic mice. Genesis 26:136–138

Moul JW, Lipo DR (1999) Prostate cancer in the late 1990s: hormone refractory disease options. Urol Nurs 19:125–131, quiz 132–123

Naiche LA, Papaioannou VE (2007) Tbx4 is not required for hindlimb identity or post-bud hindlimb outgrowth. Development 134:93–103

Nakamura E, Nguyen MT, Mackem S (2006) Kinetics of tamoxifen-regulated Cre activity in mice using a cartilage-specific CreER(T) to assay temporal activity windows along the proximodistal limb skeleton. Dev Dyn 235:2603–2612

Nakamura T, Matsumoto K, Kiritoshi A, Tano Y (1997) Induction of hepatocyte growth factor in fibroblasts by tumor-derived factors affects invasive growth of tumor cells: in vitro analysis of tumor-stromal interactions. Cancer Res 57:3305–3313

Nelson WG, DeWeese TL, DeMarzo AM (2002) The diet, prostate inflammation, and the development of prostate cancer. Cancer Metastasis Rev 21:3–16

Neubauer BL, Chung LWK, McCormick KA, Taguchi O, Thompson TC, Cunha GR (1983) Epithelial-mesenchymal interactions in prostatic development. II. Biochemical observations of prostatic induction by urogenital sinus mesenchyme in epithelium of the adult rodent urinary bladder. J Cell Biol 96:1671–1676

Noble RL (1977) Sex steroids as a cause of adenocarcinoma of the dorsal prostate in Nb rats, and their influence on the growth of transplants. Oncology 34:138–141

Noble RL (1980) Production of Nb rat carcinoma of the dorsal prostate and response of estrogen-dependent transplants to sex hormones and tamoxifen. Cancer Res 40:3547–3550

Ohuchida K, Mizumoto K, Murakami M, Qian LW, Sato N, Nagai E, Matsumoto K, Nakamura T, Tanaka M (2004) Radiation to stromal fibroblasts increases invasiveness of pancreatic cancer cells through tumor-stromal interactions. Cancer Res 64:3215–3222

Olumi AF, Grossfeld GD, Hayward SW, Carroll PR, Tlsty TD, Cunha GR (1999) Carcinoma-associated fibroblasts direct tumor progression of initiated human prostatic epithelium. Cancer Res 59:5002–5011

Ornitz DM, Xu J, Colvin JS, McEwen DG, MacArthur CA, Coulier F, Gao G, Goldfarb M (1996) Receptor specificity of the fibroblast growth factor family. J Biol Chem 271:15292–15297

Paget S (1889) The distribution of secondary growths in cancer of the breast. Lancet 1:571–573

Papini S, Rosellini A, Campani D, DeMatteis A, Selli C, Revoltella RP (2004) Selective growth of epithelial basal cells from human prostate in a three-dimensional organ culture. Prostate 59:383–392

Papini S, Rosellini A, De Matteis A, Campani D, Selli C, Caporali A, Bettuzzi S, Revoltella RP (2007) Establishment of an organotypic in vitro culture system and its relevance to the

characterization of human prostate epithelial cancer cells and their stromal interactions. Pathol Res Pract 203:209–216
Paterson RF, Ulbright TM, MacLennan GT, Zhang S, Pan CX, Sweeney CJ, Moore CR, Foster RS, Koch MO, Eble JN, Cheng L (2003) Molecular genetic alterations in the laser-capture-microdissected stroma adjacent to bladder carcinoma. Cancer 98:1830–1836
Petersen OW, Ronnov-Jessen L, Howlett AR, Bissell MJ (1992) Interaction with basement membrane serves to rapidly distinguish growth and differentiation pattern of normal and malignant human breast epithelial cells. Proc Natl Acad Sci USA 89:9064–9068
Pierce G, Shikes R, Fink L (1978) Cancer: a problem of developmental biology. Q Rev Biol 54:183–184
Polnaszek N, Kwabi-Addo B, Peterson LE, Ozen M, Greenberg NM, Ortega S, Basilico C, Ittmann M (2003) Fibroblast growth factor 2 promotes tumor progression in an autochthonous mouse model of prostate cancer. Cancer Res 63:5754–5760
Postovit LM, Seftor EA, Seftor RE, Hendrix MJ (2006) A three-dimensional model to study the epigenetic effects induced by the microenvironment of human embryonic stem cells. Stem Cells 24:501–505
Regan CP, Manabe I, Owens GK (2000) Development of a smooth muscle-targeted cre recombinase mouse reveals novel insights regarding smooth muscle myosin heavy chain promoter regulation. Circ Res 87:363–369
Ricke WA, Ishii K, Ricke EA, Simko J, Wang Y, Hayward SW, Cunha GR (2006) Steroid hormones stimulate human prostate cancer progression and metastasis. Int J Cancer 118:2123–2131
Ricke WA, McPherson SJ, Bianco JJ, Cunha GR, Wang Y, Risbridger GP (2007a) Prostatic hormonal carcinogenesis is mediated by in situ estrogen production and estrogen receptor alpha signaling. FASEB J 22(5):1512–20
Ricke WA, Wang Y, Cunha GR (2007b) Steroid hormones and carcinogenesis of the prostate: the role of estrogens. Differentiation 75:871–882
Robinson GW, Wagner KU, Hennighausen L (2001) Functional mammary gland development and oncogene-induced tumor formation are not affected by the absence of the retinoblastoma gene. Oncogene 20:7115–7119
Ronnov-Jessen L, Petersen OW, Bissell MJ (1996) Cellular changes involved in conversion of normal to malignant breast: importance of the stromal reaction. Physiol Rev 76:69–125
Ropiquet F, Giri D, Kwabi-Addo B, Schmidt K, Ittmann M (2000) FGF-10 is expressed at low levels in the human prostate. Prostate 44:334–338
Roskelley CD, Desprez PY, Bissell MJ (1994) Extracellular matrix-dependent tissue-specific gene expression in mammary epithelial cells requires both physical and biochemical signal transduction. Proc Natl Acad Sci USA 91:12378–12382
Sakai T, Larsen M, Yamada KM (2003) Fibronectin requirement in branching morphogenesis. Nature 423:876–881
Salcedo R, Oppenheim JJ (2003) Role of chemokines in angiogenesis: CXCL12/SDF-1 and CXCR4 interaction, a key regulator of endothelial cell responses. Microcirculation 10:359–370
Sauer B (2002) Cre/lox: one more step in the taming of the genome. Endocrine 19:221–228
Seitz HK, Stickel F (2006) Risk factors and mechanisms of hepatocarcinogenesis with special emphasis on alcohol and oxidative stress. Biol Chem 387:349–360
Song Z, Wu X, Powell WC, Cardiff RD, Cohen MB, Tin RT, Matusik RJ, Miller GJ, Roy-Burman P (2002) Fibroblast growth factor 8 isoform B overexpression in prostate epithelium: a new mouse model for prostatic intraepithelial neoplasia. Cancer Res 62:5096–5105
Staack A, Donjacour AA, Brody J, Cunha GR, Carroll P (2003) Mouse urogenital development: a practical approach. Differentiation 71:402–413
Sugimura Y, Foster BA, Hom YK, Lipschutz JH, Rubin JS, Finch PW, Aaronson SA, Hayashi N, Kawamura J, Cunha GR (1996) Keratinocyte growth factor (KGF) can replace testosterone in the ductal branching morphogenesis of the rat ventral prostate. Int J Dev Biol 40:941–951
Sutherland RM, Inch WR, McCredie JA, Kruuv J (1970) A multi-component radiation survival curve using an in vitro tumour model. Int J Radiat Biol Relat Stud Phys Chem Med 18:491–495

Tan W, Krishnaraj R, Desai TA (2001) Evaluation of nanostructured composite collagen – chitosan matrices for tissue engineering. Tissue Eng 7:203–210
Thompson TC, Southgate J, Kitchener G, Land H (1989) Multistage carcinogenesis induced by ras and myc oncogenes in a reconstituted organ. Cell 56:917–930
Thomson AA (2001) Role of androgens and fibroblast growth factors in prostatic development. Reproduction 121:187–195
Tuxhorn JA, Ayala GE, Rowley DR (2001) Reactive stroma in prostate cancer progression. J Urol 166:2472–2483
Tyson DR, Inokuchi J, Tsunoda T, Lau A, Ornstein DK (2007) Culture requirements of prostatic epithelial cell lines for acinar morphogenesis and lumen formation in vitro: role of extracellular calcium. Prostate 67:1601–1613
Wang R, Xu J, Juliette L, Castilleja A, Love J, Sung SY, Zhau HE, Goodwin TJ, Chung LW (2005) Three-dimensional co-culture models to study prostate cancer growth, progression, and metastasis to bone. Semin Cancer Biol 15:353–364
Wang Y, Hayward SW, Donjacour AA, Young P, Jacks T, Sage J, Dahiya R, Cardiff RD, Day ML, Cunha GR (2000) Sex hormone-induced carcinogenesis in Rb-deficient prostate tissue. Cancer research 60:6008–6017
Wang YZ, Wong YC (1998) Sex hormone-induced prostatic carcinogenesis in the noble rat: the role of insulin-like growth factor-I (IGF-I) and vascular endothelial growth factor (VEGF) in the development of prostate cancer. Prostate 35:165–177
Webber MM, Bello D, Kleinman HK, Hoffman MP (1997) Acinar differentiation by non-malignant immortalized human prostatic epithelial cells and its loss by malignant cells. Carcinogenesis 18:1225–1231
Whitelock JM, Murdoch AD, Iozzo RV, Underwood PA (1996) The degradation of human endothelial cell-derived perlecan and release of bound basic fibroblast growth factor by stromelysin, collagenase, plasmin, and heparanases. J Biol Chem 271:10079–10086
Williams K, Fernandez S, Stien X, Ishii K, Love HD, Lau YF, Roberts RL, Hayward SW (2005) Unopposed c-MYC expression in benign prostatic epithelium causes a cancer phenotype. Prostate 63:369–384
Xu C, Inokuma MS, Denham J, Golds K, Kundu P, Gold JD, Carpenter MK (2001) Feeder-free growth of undifferentiated human embryonic stem cells. Nat Biotechnol 19:971–974
Yamada KM, Cukierman E (2007) Modeling tissue morphogenesis and cancer in 3D. Cell 130:601–610
Zhang J, Thomas TZ, Kasper S, Matusik RJ (2000) A small composite probasin promoter confers high levels of prostate-specific gene expression through regulation by androgens and glucocorticoids in vitro and in vivo. Endocrinology 141:4698–4710
Zhang M, Xuan S, Bouxsein ML, von Stechow D, Akeno N, Faugere MC, Malluche H, Zhao G, Rosen CJ, Efstratiadis A, Clemens TL (2002) Osteoblast-specific knockout of the insulin-like growth factor (IGF) receptor gene reveals an essential role of IGF signaling in bone matrix mineralization. J Biol Chem 277:44005–44012
Zheng B, Zhang Z, Black CM, de Crombrugghe B, Denton CP (2002) Ligand-dependent genetic recombination in fibroblasts: a potentially powerful technique for investigating gene function in fibrosis. Am J Pathol 160:1609–1617

Chapter 21
Utilizing Mouse Models of Human Cancer for Assessing Immune Modulation of Cancer Development

Karin E. de Visser and Lisa M. Coussens

21.1 Introduction

For the past 40 years, cancer research has predominantly focussed on delineating significant genetic alterations underlying enhanced cell survival and hyperproliferation of neoplastic tumor cells. Insights gained from these elegant studies have aided our understanding of important cell cycle and signal transduction pathways in both neoplastic and tumor-associated stromal cells that regulate human cancer development (Hanahan and Weinberg 2000; Nelson and Bissell 2006). Clinical translation of these insights has led to the development of novel anticancer-targeted therapeutics, some of which extend patient survival (Sawyers 2004). However, what has also emerged is the realization that some tumors develop resistance to targeted therapeutics similarly to what has been previously observed for cytotoxic drugs largely due to acquisition of additional genetic and/or epigenetic alterations that limit therapeutic efficacy (Klein et al. 2005), and in some cases lead to emergence of more aggressive disease (De Giorgi et al. 2007; Druker 2004; Jain 2008; Persano et al. 2007). Realization of this therapeutic dilemma implies that whether targeting neoplastic or tumor-associated stromal cells with single-agent therapies, therapeutic efficacy may be enhanced by employing combinatorial strategies, where multiple critical programs

K.E. de Visser
Department of Molecular Biology, The Netherlands Cancer Institute, Plesmanlaan 121, 1066 CX Amsterdam, The Netherlands

L.M. Coussens (✉)
Department of Pathology, University of California, San Francisco, 513 Parnassus Ave. HSW-450C, San Francisco, CA 94143, USA

Helen Diller Family Comprehensive Cancer Center, University of California, San Francisco, 513 Parnassus Ave. HSW-450C, San Francisco, CA 94143-0502, USA
e-mail: Lisa.Coussens@ucsf.edu

J.E. Green and T. Ried (eds.), *Genetically Engineered Mice for Cancer Research: Design, Analysis, Pathways, Validation and Pre-Clinical Testing*,
DOI 10.1007/978-0-387-69805-2_21,

essential for tumor survival are crippled (Albini and Sporn 2007; Joyce 2005). That said, there is a growing awareness that tissues containing either *initiated* preneoplastic or premalignant cells require co-option of multiple stromal-derived pro-tumor programs in order to favor the development of an invasive *tumor* – these include programs regulating angiogenic, lymphangiogenic, tissue remodeling, and immune function (Balkwill et al. 2005; Cha and DuBois 2007; de Visser et al. 2006; Jain 2005, 2008; Karin 2006; Nelson and Bissell 2006).

One important host-derived program that regulates pro-tumor programming of tissues is the immune system (Balkwill et al. 2005; de Visser et al. 2006). Accumulating evidence indicates that cells of the immune system play critical roles as regulators of malignancy due to their differential abilities to regulate either or both pro- and/or antitumor immunity (de Visser et al. 2006; Dunn et al. 2004). Examination of genetically modified or chemically induced mouse models (Tuveson and Jacks 2002; Van Dyke and Jacks 2002; Yuspa 2000) susceptible to de novo tumor development, where selective components of the immune system have been deleted or modified, have provided important clues regarding the functional significance of specific immune response programs as regulators of pro- versus antitumor immunity. Herein, we provide an overview of several mouse cancer models, where immune response has been manipulated and thus revealed significant control points that regulate tumor progression.

21.2 Activation of Immune Response

The immune system can be divided into two subsets, adaptive and innate. In order to provide optimal protection against invading pathogens while simultaneously maintaining tolerance toward self-antigens, both subsets of the immune system are intimately associated (Luster 2002; Murphy et al. 2000). The adaptive immune system is composed of $CD4^+$ (helper), $CD8^+$ (cytotoxic T lymphocytes, CTLs) and $FoxP3^+$ regulatory T lymphocytes, B lymphocytes, and antibodies (humoral immunity). Cells of the adaptive immune system are characterized by antigen-specificity and memory formation. T and B lymphocytes express unique, highly diverse, somatically generated antigen-specific receptors, T cell receptors (TCRs) or B cell receptors (BCRs), respectively (Goldrath and Bevan 1999; Tonegawa 1983). Tremendously diverse B and T lymphocyte repertoires are generated that ensure recognition of an almost unlimited array of pathogenic antigens (Goldrath and Bevan 1999; Tonegawa 1983). Fully activated T lymphocytes protect against invaders by cytotoxic killing of cells expressing the antigen of specificity ($CD8^+$ T cells), by producing cytokine production, and by enlisting the help of B cells ($CD4^+$ T cells). Regulatory T (T_{reg}) cells suppress effector functions of CTLs and play important roles in preventing autoimmune disease and exacerbated immunity against infections (Colombo and Piconese 2007; Zou 2006). B lymphocytes exert their main effector function by secreting antibodies with the same antigen specificity as their BCR (Janeway et al. 2001; McHeyzer-Williams 2003).

The innate arm of the immune system is composed of monocytes, macrophages, granulocytes (neutrophils, basophils, and eosinophils), dendritic cells (DCs), mast cells, natural killer (NK) cells and soluble complement components. In contrast to adaptive immune cells, innate leukocytes are relatively nonspecific and not intrinsically affected by prior contact with infectious agents. Innate cells express germ-line encoded pattern-recognition Toll-like receptors (TLRs) with which they recognize conserved molecular patterns found on microorganisms, but not in self-tissue (Akira and Takeda 2004; Janeway and Medzhitov 2002). Innate immune cells can also be activated via crosslinking of Fc receptors by antibodies and immune complexes or via crosslinking of complement receptors by components of the activated complement cascade.

Acute inflammatory responses to invading pathogens or tissue injury represent a multistep process initiated by activation of resident innate immune cells that release preformed and newly synthesized pro-inflammatory mediators, followed by recruitment and activation of other immune cells from the periphery, leading to subsequent elimination of damaged cells and invading organisms (Janeway and Medzhitov 2002; Walport 2001a, b; Zlotnik and Yoshie 2000). Acute activation of the innate immune system not only forms the first line of immune defense against invading pathogens, but is also necessary for efficient activation of the highly specific adaptive immune system (Belardelli and Ferrantini 2002; Carroll 2004; Janeway and Medzhitov 2002). The kinetics of primary adaptive immune responses are slower than innate responses, largely due to the need for clonal expansion of antigen-specific lymphocytes that is required for generation of sufficient antigen-specific T and/or B lymphocytes (McHeyzer-Williams 2003; Sprent and Surh 2002); however, upon activation, a subset of lymphocytes differentiate into long-lived memory cells, thus forming heightened states of immune reactivity used for future contact with the same antigen (Sprent and Surh 2002). Thus, the interplay between innate and adaptive immunity in acute states results in the removal of invading pathogens and damaged cells, resolution of tissue damage and reestablishment of tissue integrity and homeostasis. However, as is now clear from multiple experimental and clinical studies, alteration of immune response in chronically damaged tissues can underlie disease pathogenesis, including cancer (de Visser et al. 2006).

21.2.1 Antitumor Immunity

For decades, it was believed that the immune system predominantly exerted an antitumor protective role against solid tumor development (Dunn et al. 2002; Ostrand-Rosenberg 2008; Qin and Blankenstein 2004; Swann and Smyth 2007). Indeed, numerous clinical observations have revealed that under certain circumstances some adaptive lymphocytes limit tumor development (Ostrand-Rosenberg 2008; Swann and Smyth 2007). For example, patients with suppressed adaptive immunity, e.g., HIV+ AIDS patients or recipients of organ transplantation, typically exhibit increased relative risk (RR) for viral-associated (Boshoff and Weiss 2002) and carcinogen-induced malignancies (de Visser et al. 2006). In support of this, the presence of infiltrating T lymphocytes in human colon adenocarcinomas was recently revealed to correlate with better clinical prognosis (Galon et al. 2006). However, under other circumstances, the presence of adaptive immune cells is not sufficient to protect tissues from incipient neoplasia. Although most sporadic cancers typically contain tumor-specific antigens that induce spontaneous antitumor T cell responses, antigenic tumors frequently progress and limit life-span. Similarly, activation of antitumor T cell responses by vaccination strategies also rarely culminates in tumor eradication (Choudhury et al. 2006; Tabi and Man 2006). Moreover, whereas patients with suppressed adaptive immunity exhibit increased RR of viral-associated malignancies, incidence of nonviral epithelial malignancies such as breast and prostate cancer is not increased, and in some instances is decreased with RR < 1.0 (Buell et al. 2005; de Visser et al. 2006; Grulich et al. 2007; Peto 2001).

Why do tumors evade adaptive immune antitumor responses? One plausible explanation is that neoplastic microenvironments favor polarized chronic pro-tumorigenic immunity as opposed to those representing acute antitumor immune states (Balkwill et al. 2005; Johansson et al. 2008; Qin and Blankenstein 2004; Zou 2005). Clinical data indicate that "immune status" in healthy individuals as compared to cancer patients is distinct, where in the later populations T cells are functionally impaired (Finke et al. 1999). In addition, accumulation of chronically activated inflammatory monocytes and T_{reg} cells can indirectly contribute to cancer development via suppression of antitumor adaptive responses, allowing tumor escape from immunosurveillance through direct cell–cell contact with neoplastic cells, by the production of immunosuppressive mediators, as well as by the production of pro-angiogenic and pro-tissue remodeling enzymes that favor tumor development (Almand et al. 2001; Kusmartsev and Gabrilovich 2006; Serafini et al. 2004; van Kempen et al. 2006). Other underlying mechanisms for the lack of tumor protection by adaptive lymphocytes have been described, including but not limited to the failure of T cell homing to tumors (Buckanovich et al. 2008; Garbe et al. 2006), induction of T cell tolerance (Willimsky and Blankenstein 2005), the presence of local immunosuppressive networks (Zou 2005), and collaboration of adaptive immune cells with pro-tumor inflammatory responses (de Visser et al. 2005). Thus, while adaptive leukocytes might protect tissues against some types of malignancy, such as viral-associated cancers, they do not sufficiently protect tissues from all cancer-inducing assaults, but instead, in some neoplastic microenvironments, the balance between innate and adaptive immunity often favors pro-tumor programming of tissues (Johansson et al. 2008).

21.2.2 Inflammation and Cancer

Over the last decade, a paradigm shift has occurred regarding our awareness of the significant interplay between fibroblasts, immune cells and vascular cells with neoplastic cells capable of developing into tumors (Bhowmick and Moses 2005).

Solid tumors are frequently characterized by the presence of chronically activated innate immune cells whose presence often correlates with unfavorable clinical prognosis (Imada et al. 2000; Leek et al. 1996). Although correlative, these observations led to the hypothesis that innate immune cells might exacerbate disease progression. A convincing link between innate cells and malignancy emerged from epidemiological studies revealing that tissues containing infiltrates of chronically activated myeloid subsets (inflammation) exhibit increased RR for cancer development (Balkwill et al. 2005). The feasibility of therapeutically targeting chronic inflammation to prevent or "treat" cancer has been demonstrated by long-term usage of nonsteroidal anti-inflammatory drugs, such as aspirin or selective COX-2 inhibitors, where significant reduction of cancer recurrence has been revealed (Arber et al. 2006; Bertagnolli et al. 2006; Dannenberg and Subbaramaiah 2003; Gupta and Dubois 2001).

Innate immune cells directly and indirectly potentiate cancer risk through the diversity of bioactive mediators they deliver to neoplastic tissues. Leukocytes are loaded with chemokines, cytokines, cytotoxic mediators, including reactive oxygen species, serine-, cysteine-, and metallo-proteases, membrane-perforating agents, and soluble mediators of cell killing, such as tumor necrosis factor-alpha (TNF-α), interleukins (IL), and interferons (de Visser et al. 2006). Individually, these molecules are known mediators of acute inflammation and trigger innate cell recruitment and/or activation, tissue remodeling and angiogenesis, and together, create a microenvironment favoring cell proliferation, genomic instability, and tissue remodeling, all critical programs necessary for solid tumor development. Thus, whereas the historical viewpoint was that host immunity exerted a protective role with regards to cancer development (Qin and Blankenstein 2004), it is now clear that subsets of chronically activated innate cells instead promote growth and/or facilitate survival of neoplastic cells.

21.2.3 Mouse Models to Study Significance of Immune Cells as Regulators of Solid Tumor Development

Xenograft, transgenic, and chemically induced mouse models of human cancer have aided our understanding of the multitude of roles played by various immune cells and their soluble mediators during cancer development (Tables 21.1 and 21.2). Of these and by far the most simplistic, tumor transplantation studies using xenograft approaches have historically provided comparative analyses, where tumor growth is followed after inoculation of a bolus of malignant tumor cells in immune-deficient versus immune-proficient animals. While much was learned from these early approaches, it is clear that xenograft models neither resemble nor mimic the complexities of de novo cancer development. Bolus injection of tumor cells is followed by massive tumor cell death prior to engagement of angiogenic programs and establishment of a favorable microenvironment; thus, early release of tumor antigens would be expected to trigger acute adaptive immune responses that are not

Table 21.1 Mouse genotypes harboring homozygous null deletions in genes encoding regulators of adaptive immunity

Immune deficiency	Genotype	Phenotype	References
Lymphocytes	RAG-1$^{-/-}$	Deficient in T, B, and NK T cells	Mombaerts et al. (1992)
	RAG-2$^{-/-}$	Deficient in T, B, and NK T cells	Shinkai et al. (1992)
	CD4$^{-/-}$	Deficient in CD4^{+} T cells	Rahemtulla et al. (1991)
	CD8$^{-/-}$	Deficient in CD8^{+} T cells	Fung-Leung et al. (1991)
	TCR$\beta^{-/-}$	Deficient in $\alpha\beta$ T cells	Mombaerts et al. (1993)
	TCR$\delta^{-/-}$	Deficient in $\gamma\delta$ T cells	Itohara et al. (1993), Mombaerts et al. (1993)
	JH	Deficient in B lymphocytes	Chen et al. (1993)
	*mu*MT	Deficient in B lymphocytes	Kitamura et al. (1991)
	Jα281$^{-/-}$	Deficient in NK T cells	Cui et al. (1997)
	Athymic nude mice	T cell deficiency (incomplete)	Ikehara et al. (1984)
	SCID	T and B cell deficiency (incomplete)	Bosma et al. (1983)
	Beige	NK cell deficiency	Beck and Henney (1981)
Effector molecules	Perforin$^{-/-}$	Deficient in perforin	Kagi et al. (1994)
	Fasl/lpr	Point mutation Fas	Takahashi et al. (1994)
	Fasl/gld	Point mutation FasL	Takahashi et al. (1994)
	Gzm A$^{-/-}$	Granzyme A deficient mice	Ebnet et al. (1995)
	Gzm B$^{-/-}$	Granzyme B deficient mice	Heusel et al. (1994)
Antigen presentation	MHC II$^{-/-}$	Deficient in MHC class II expression	Kontgen et al. (1993)
	Beta-2 micro-globulin$^{-/-}$	Deficient in MHC class I expression	Koller et al. (1990)
	TAP-1$^{-/-}$	Deficient in peptide transport and MHC-I restricted antigen presentation	Van Kaer et al. (1992)
Co-stimulation	CD28$^{-/-}$	Deficient in CD28	Shahinian et al. (1993)
	CD80 (B7-1) -/-	Deficient in B7.1	Freeman et al. (1993)
	CD86 (B7-2) -/-	Deficient in B7.2	Borriello et al. (1997)
	CTLA4$^{-/-}$	Deficient in CTLA4	Chambers et al. (1997)
	4-1BB$^{-/-}$	Deficient in 4-1BB (CD137)	Kwon et al. (2002)

Table 21.2 Mouse genotypes harboring genetic modifications in genes encoding regulators of innate immunity

Immune deficiency	Genotype	Phenotype	References
Leukocytes	KIT^{W}/KIT^{Wv}	Deficient in mast cells	Kitamura et al. (1978)
	$KIT^{W\text{-}sh}/KIT^{W\text{-}sh}$	Deficient in mast cells	Duttlinger et al. (1993), Grimbaldeston et al. (2005)
	$CSF1^{op}/CSF1^{op}$	Deficient in colony-stimulating factor 1, resulting in many defects, including macrophage deficiency	Marks and Lane (1976)
	CD11b-DTR	Transgenic expression of the human diphtheria toxin receptor (DTR) in $CD11b^{+}$ cells, allowing conditional ablation of macrophages by treatment with diphtheria toxin (DT)	Duffield et al. (2005)
	CD11c-DTR	Transgenic expression of the human DTR in $CD11c^{+}$ cells, allowing conditional ablation of $CD11c^{+}$ DCs by treatment with DT	Jung et al. (2002)
Effector molecules and receptors	$FcR\gamma^{-/-}$	Deficient in the common gamma subunit of the immunoglobulin Fc receptor	Takai et al. (1994)
	$Fc\gamma RIII^{-/-}$	Deficient in the alpha chain of Fc gamma receptor III	Hazenbos et al. (1996)
	C1q, C2, C3, C4, and C5-deficient mice	Deficient in complement proteins	Wessels et al. (1995), Botto et al. (1998), Taylor et al. (1998)
	Cr1, Cr2, and C5aR-deficient mice	Deficient in complement protein receptors	Hopken et al. (1996), Molina et al.(1996)
	$COX1^{-/-}$ ($Ptgs1^{-/-}$)	Deficient in cyclooxygenase-1	Langenbach et al. (1995)
	$COX2^{-/-}$ ($Ptgs2^{-/-}$)	Deficient in cyclooxygenase-2	Dinchuk et al. (1995), Morham et al. (1995)
Toll-like receptors	$MyD88^{-/-}$	Deficient in the TLR intracellular adaptor molecule myeloid differentiation factor 88 (MyD88), affecting TLR, IL-1R, and IL-18R signaling	Adachi et al. (1998)
	TLR1, TLR2, TLR3, TLR4, TLR5, TLR6 TLR7, and TLR9 deficiency	Deficient for members of the TLRfamily	Alexopoulou et al. (2001), Feuillet et al. (2006), Hemmi et al. (2000, 2002), Hoshino et al. (1999), Takeuchi et al. (1999, 2001, 2002)

characteristically observed during the progressive events accompanying de novo tumor development (Yu et al. 2006). With the awareness that adaptive and innate leukocyte bioactivity are intimately associated, it seems unreasonable to examine immune modulation of cancer development in such an artificial setting.

Two-stage models of chemically induced cancer development provide a good alternative to xenograft models since tissues evolve through multiple stages, including initiation, promotion, and malignant conversion stages, that mimic many aspects of de novo cancer formation (Hennings et al. 1993; Yuspa 2000). Two-stage models of skin, colon, and liver carcinogenesis have been broadly exploited to reveal functional involvement of immunity as a regulator of multistage carcinogenesis (Greten et al. 2004; Moore et al. 1999).

A more elegant and clinically relevant approach is to elucidate importance of immune mediators utilizing genetically engineered de novo mouse cancer models harboring transgenic expression of oncogenes (or tumor suppressor genes) under tissue and/or cell type-specific control (Frese and Tuveson 2007; Hanahan et al. 2007; Jonkers and Derksen 2007; Kim et al. 2005; Tuveson and Jacks 2002; Van Dyke and Jacks 2002). Intercrossings between these onco-mice and immune-modified mice have led to surprising and paradigm shifting insights regarding the significant role played by the immune system during cancer development, with the caveat that entire tissues typically express high levels of dominant oncogenes that result in rapid and multifocal tumor formation. Conditional mouse models, where tumors arise stochastically in otherwise normal tissue recapitulate human cancer and metastasis formation more rigorously.

In the paragraphs below, we have focussed on recent studies that have provided compelling insights and shaped our fundamental understanding of how cancer development is regulated by either pro- or antitumor immunity. Given the overwhelming number and diversity of such studies, it is an impossible task to provide an exhaustive list of all relevant studies; thus, we have focussed on those studies utilizing spontaneous tumor models. For additional information and more detail on this topic, we refer the reader to published studies summarized in the accompanying tables and to excellent reviews (Albini and Sporn 2007; Balkwill et al. 2005; Croci et al. 2007; de Visser et al. 2006; Dunn et al. 2004; Swann and Smyth 2007).

21.2.4 Cancer Development in Mice Deficient in Components of Adaptive Immunity

As shown in Table 21.1, many mouse strains have been developed that alter regulation and/or the presence of adaptive immune mediators, e.g., mice lacking selective adaptive immune cells, adaptive effector molecules, antigen presentation, or co-stimulatory molecules. The functional significance of spontaneous adaptive immune responses during de novo cancer development has intrigued many tumor biologists and immunologists for several decades, and is still not fully understood and a controversial topic. Whereas some studies have provided convincing data supporting the concept that the immune system exerts a protective role against certain tumor types, other compelling studies do not support this so-called immunosurveillance concept (de Visser et al. 2006; Qin and Blankenstein 2004). In Table 21.3, studies are summarized in which the effect of deficiency on components of the adaptive immune system was assessed during de novo tumor formation. As these studies indicate, the role of adaptive leukocytes is paradoxical; tumor progression is enhanced in some settings, whereas in others, suppression by adaptive immunity is achieved, thus supporting the notion that malignant outcome is etiology-, context-, and organ-dependent.

Initial mechanistic studies that supported the concept of antitumor immunity provided by adaptive lymphocytes were performed with the 3-methylcholanthrene (MCA)-induced carcinogenesis model (Dunn et al. 2006). It was revealed that RAG-2-deficient mice lacking all mature T and B lymphocytes, exhibit a heightened susceptibility to chemically induced tumorigenesis (Shankaran et al. 2001). Mice with other adaptive immune-deficiencies, e.g., perforin-deficient mice, TCRαβ-deficient mice, STAT1-deficient mice, IFNγR-deficient mice, and IL12p40-deficient mice were also reported to have increased susceptibility to MCA-induced tumor formation (Table 21.3). Based on these studies, it was concluded that adaptive immunity exerted the so-called immunoediting effect, e.g., spontaneous adaptive immune responses elicited by developing tumors recognize and eliminate sarcoma cells; however, after an equilibrium phase, immunosurveillance results in inadvertent selection of tumor escape variants that ultimately develop into full-blown sarcomas with reduced immunogenicity (Dunn et al. 2002).

More recent examination of tumor-prone mice not initiated by exposure to MCA, however, indicates that carcinomas may be less prone to the process of cancer immunoediting (Table 21.3) (Casanovas et al. 2005; de Visser et al. 2005; Garbe et al. 2006; Willimsky and Blankenstein 2005). For example, the classical two-stage skin carcinogenesis model using DMBA followed by TPA appears not to be inhibited by immunosurveillance mechanisms, since tumor incidence is unaltered in perforin-deficient animals (van den Broek et al. 1996), and is instead even reduced in TNFα$^{-/-}$ (Moore et al. 1999), and in αβ T cell-deficient mice (Girardi et al. 2001) (Table 21.3). Blankenstein and colleagues utilized a mouse cancer model based on rare spontaneous activation of a dormant oncogene and found that tumors grew progressively, despite the presence of spontaneous humoral and cellular immune responses with specificity for the tumor (Willimsky and Blankenstein 2005). These sporadic tumors did not lose their intrinsic immunogenicity, as they were rejected after transplantation in immune-competent mice, and thus were not affected by adaptive immunity (Willimsky and Blankenstein 2005). Likewise, genetic elimination of the entire T and B cell compartment in a transgenic mouse model of neuroendocrine pancreatic islet cell carcinomas, e.g., RIP1-Tag2 mice, failed to alter tumor development (Casanovas et al. 2005), indicating that malignant cells are not eliminated in these tissues by adaptive lymphocytes. In contrast, we previously identified a novel role for the adaptive immune system during de novo carcinoma formation, where the absence of T and B lymphocytes in a transgenic mouse model of squamous carcinogenesis, e.g., K14-HPV16 mice, protected against spontaneous tumor

Table 21.3 Immune-deficient mouse models and de novo tumor formation

Tumor model	Immune deficiency	Malignant outcome	References
Adaptive immunity and cancer			
MCA-induced carcinogenesis	RAG-$2^{-/-}$	↑	Shankaran et al. (2001)
	TCR$\beta^{-/-}$	↑	Girardi et al. (2001)
	Perforin$^{-/-}$	↑	van den Broek et al. (1996)
	J$\alpha 281^{-/-}$	↑	Smyth et al. (2000, 2001)
DMBA/TPA-induced skin carcinogenesis	Perforin$^{-/-}$	nc	van den Broek et al. (1996)
	TCR$\beta^{-/-}$	↓	Girardi et al. (2001)
	TCR$\delta^{-/-}$	↑	Girardi et al. (2001, 2003)
DMBA/phorbol 12-myristate 13-acetate-induced skin carcinogenesis	CD$8^{-/-}$	↓	Roberts et al. (2007)
RIP-TAG (pancreatic islet carcinogenesis)	RAG-$1^{-/-}$	nc	Casanovas et al. (2005)
K14-HPV16 (squamous cell carcinoma of skin and cervix)	RAG-$1^{-/-}$, CD$4^{-/-}$/CD$8^{-/-}$	↓	de Visser et al. (2005)
	CD$4^{-/-}$	↓	Daniel et al. (2003)
	CD$8^{-/-}$	nc	Daniel et al. (2003)
BALB/c Her2/neu (mammary carcinomas)	Perforin$^{-/-}$	↑	Street et al. (2007)
Innate immunity and cancer			
K14-HPV16 (squamous cell carcinoma of skin)	KIT^{W}/KIT^{Wv}	↓	Coussens et al. (1999)
	C$3^{-/-}$	nc	de Visser et al. (2004)
p*Ins-myc*ER^{TAM} ;RIP7-bcl-x$_L$ (pancreatic β-cell tumors)	KIT^{W}/KIT^{Wv}	↓	Soucek et al. (2007)
APC(Min/+) (intestinal tumors)	KIT$^{W\text{-}sh}$/KIT$^{W\text{-}sh}$	↑	Sinnamon et al. (2008)
1,2-dimethylhydrazine-induced intestinal tumors	KIT^{W}/KIT^{Wv}	↓	Wedemeyer and Galli (2005)
MMTV–PyMT (mammary tumors)	CSF1^{op}/CSF1^{op}	↓	Lin et al. (2001)
MCA-induced carcinogenesis	MyD$88^{-/-}$	↓	Swann et al. (2008)
DMBA/TPA-induced carcinogenesis	MyD$88^{-/-}$	↓	Swann et al. (2008)
Mdr$2^{-/-}$ (hepatocellular carcinoma)	Hepatocyte-specific IκB-super-repressor	↓	Pikarsky et al. (2004)
Azoxymethane/dextran sulfate-induced colon cancer	Enterocyte-specific deletion of IKKβ	↓	Greten et al. (2004)
	Myeloid-specific deletion of IKKβ	↓	Greten et al. (2004)
TRAMP (prostate cancer)	Inactive IKKα	↓	Luo et al. (2007)
Cytokines, chemokines, immune receptors, and other (soluble) immune mediators of cancer development			
DMBA/TPA-induced carcinogenesis	TNF$\alpha^{-/-}$	↓	Moore et al. (1999)
	TNF-R$1^{-/-}$ and TNF-R$2^{-/-}$	↓	Arnott et al. (2004)
	IL-23p$19^{-/-}$	↓	Langowski et al. (2006)
	IL-12p$35^{-/-}$	↑	Langowski et al. (2006)

(continued)

Table 21.3 (continued)

Tumor model	Immune deficiency	Malignant outcome	References
MCA-induced carcinogenesis	IFNγR$^{-/-}$	↑	Kaplan et al. (1998), Shankaran et al. (2001)
	STAT1$^{-/-}$	↑	Kaplan et al. (1998), Shankaran et al. (2001)
	IL-12p40$^{-/-}$	↑	Smyth et al. (2000)
Infectious agents-induced carcinogenesis	GM-CSF$^{-/-}$/ IFNγ$^{-/-}$	↑	Enzler et al. (2003)
APC$^{\Delta716}$ (intestinal tumors)	COX-2$^{-/-}$	↓	Oshima et al. (1996)
APC (Min/+) (intestinal tumors)	COX-1$^{-/-}$ or COX-2$^{-/-}$	↓	Chulada et al. (2000)
MMTV–NDL (mammary tumors)	COX-2$^{-/-}$	↓	Howe et al. (2005)
DMBA/TPA-induced carcinogenesis	COX-1$^{-/-}$ or COX-2$^{-/-}$	↓	Tiano et al. (2002)
DMBA-induced carcinogenesis	TLR4$^{-/-}$	↑	Yusuf et al. (2008)
Azoxymethane and dextran sodium sulfate-induced colon carcinogenesis	TLR4$^{-/-}$	↓	Fukata et al. (2007)

nc, no change in progression to malignancy; ↑, enhanced progression to malignancy; ↓, decreased progression to malignancy

formation and arrested neoplastic progression at an early benign hyperplastic stage (de Visser et al. 2005). Adoptive transfer of educated B lymphocytes or serum from congenic HPV16 mice rescued the defect and reinstated pro-tumor programs, including productive infiltration of neoplastic tissue by inflammatory monocytes and mast cells, and the development of angiogenic vasculature, both necessary components of precursor lesions that enable malignant conversion, thus supporting a role for the humoral component of adaptive immunity as a critical determinant of malignant progression (de Visser et al. 2005). In addition to the concept of immunoediting in which the adaptive immune system "sculpts" developing tumors, these studies indicate existence of alternative programs, where a spontaneously developing tumor sculpts, avoids, or even harnesses adaptive immune responsiveness to its own advantage. The degree to which these programs are tissue-, organ-, or oncogene-specific remains to be evaluated.

21.2.5 Mice Deficient in Components of Innate Immunity

As shown in Table 21.2 a multitude of genetically engineered strains of mice have been developed to examine innate immune functions during cancer development and include mice lacking selective innate immune cells, innate effector molecules, signal transduction components, or surface receptors. Excellent models for mast cell deficiency exist; however, models completely deficient for macrophages, granulocytes, and eosinophils have not been developed. The majority of tumor studies in which innate immune cell populations or responding pathways have been genetically eliminated have revealed that chronic engagement of innate immunity promotes tumorigenesis (Table 21.3). Using K14-HPV16 mice (Arbeit et al. 1994; Coussens et al. 1996), we reported that transgenic oncogene expression alone is not sufficient for complete cancer development; instead, additional signals provided by mast cells and inflammatory monocytes are required for full neoplastic progression (Coussens et al. 1999; de Visser et al. 2005). Likewise, mast cells are essential for tumor development in experimental mouse models of de novo pancreatic islet carcinogenesis driven by the *myc* proto-oncogene (Soucek et al. 2007), in two-stage carcinogenesis of skin (Wedemeyer and Galli 2005) and for polyp development in colon (Gounaris et al. 2007). Lin et al. reported that genetic elimination of colony stimulating factor-1 in the MMTV–PyMT mouse model of mammary carcinogenesis inhibited macrophage recruitment into mammary tissue that resulted in delayed late-stage adenocarcinoma development and attenuated pulmonary metastasis formation (2001). Other studies have similarly revealed a causal link between chronic presence of activated myeloid cells and cancer progression (Bergers et al. 2000; Coussens et al. 1999, 2000; De Palma et al. 2005; Fischer et al. 2007; Giraudo et al. 2004; Greten et al. 2004; Guerra et al. 2007; Nozawa et al. 2006; Pikarsky et al. 2004).

The availability of a growing number of de novo carcinogenesis and immune-modified mouse models has allowed investigators to take steps toward understanding the complex mechanisms underlying the relationship between chronic inflammation and cancer. Since chronic inflammation is a complex and dynamic process involving multiple cell types and soluble mediators, it is not surprising that diverse mechanisms have been identified, whereby inflammation promotes malignancy (Table 21.3) (Balkwill et al. 2005; Coussens et al. 2000; Greten et al. 2004; Lin et al. 2001; Pikarsky et al. 2004; Soucek et al. 2007; Wyckoff et al. 2004). Detailed characterization of tumor formation in the presence and absence of specific components of the innate immune system revealed that chronic inflammation contributes to tumor formation through induction of DNA damage or paracrine stimulation of proliferation of neoplastic cells, and indirectly through activation of angiogenesis, tissue remodeling, and suppression of antitumor adaptive immune responses (de Visser et al. 2006). Mechanistic studies have already dissected some of the underlying pathways through which the immune system exerts these tumor-modulating effects. For instance, chronic inflammation is frequently associated with activation of the pro-inflammatory NF-κB signaling pathway (Karin 2006). Experimental studies have revealed that activation of NF-κB signaling pathways in both neoplastic and immune cells is critical for the development of inflammation-associated cancers (Table 21.3) (Greten et al. 2004; Luo et al. 2007; Pikarsky et al. 2004). Inflammation also results in upregulation of cyclooxygenase type 2 (COX-2), a key enzyme in the synthesis of prostaglandins from arachidonic acid (Turini and DuBois 2002) that plays an important role in neoplasia (Table 21.3). COX-2-deficient mice exhibit reduced susceptibility to colon and mammary carcinomas, and chemically induced skin cancer (Table 21.3). Likewise, immune cell-derived growth factors and cytokines can modulate proliferation, migration, and survival of initiated epithelial cells. Mice deficient for TNFα or TNFα receptors display decreased susceptibility to DMBA/TPA-induced skin carcinogenesis, whereas IFNγ unresponsiveness renders mice more susceptible to MCA-induced cancer formation (Table 21.3).

These and many other reports have revealed that some immune cell subsets, through continuous production of a rich array of growth factors, pro-angiogenic mediators, and proteolytic enzymes, are significant modulators of neoplastic progression and malignant outcome. A more thorough understanding into the temporal and spatial involvement and underlying mechanisms of individual immune cell populations and inflammatory mediators during the development of distinct types of sporadic cancers will help determine where tractable targets for anticancer therapy lie.

21.2.6 Approaches for Manipulating Immunity During Cancer Development

In some circumstances, it is advantageous to use nongenetic approaches to manipulate immune function in cancer-prone animals. Intercrossing of genetically complex mouse models or backcrossing onto congenic strains with immune-deficient/modified mice is time-consuming and can be cost-prohibitive. Moreover, some mechanistic evaluations require temporal manipulation of immunity, e.g., during early neoplastic development versus later when mice are tumor bearing. Such investigations cannot be achieved by studying cancer development directly in conventional immune-deficient animals since these mice lack immune components during development and throughout their entire adult life.

Alternative approaches to circumvent these barriers can be utilized by generating bone marrow chimeric mice or mice receiving congenic subpopulations of leukocytes following adoptive transfer. Chimeric mice are straightforward to generate, where donor bone marrow is harvested and transplanted into lethally irradiated congenic recipient tumor-prone mice. In these studies, it is critical to control for irradiation-induced changes in tumor progression by including control bone marrow chimeras transplanted with wild-type bone marrow. Adoptive transfer of isolated subpopulations of leukocytes from wild-type or immune-modified mice can be given intraperitoneally to congenic recipients at any stage of neoplastic development. Temporal manipulation of immune cell populations can also be achieved using depleting or neutralizing antibodies, by pharmacological methods, or by using mouse models that allow conditional depletion of immune cell subsets. For example, macrophages can be temporarily depleted by injecting liposomal clodronate (Van Rooijen and Sanders 1994) and mast cells can be stabilized by the treatment with chromolyn (Soucek et al. 2007). Conditional ablation of particular immune cell populations can also be achieved in mice by transgenic expression of a high affinity diphtheria toxin receptor (DTR) under the control of an immune cell-specific promoter (Bennett and Clausen 2007), where administration of diphtheria toxin to mice results in ablation of selective immune cells expressing the DTR.

The combined use of immune-deficient mouse models and adoptive transfer techniques can be useful in further dissecting underlying mechanisms. For example, to elucidate the immune cell type responsible for reduced squamous carcinoma development in K14-HPV16/RAG-$1^{-/-}$ mice, we performed adoptive transfer studies in which B lymphocytes or serum isolated from control K14-HPV16 mice were transferred into K14-HPV16/RAG-$1^{-/-}$ cohorts. These studies revealed that educated B cells or serum from donor K14-HPV16 mice restored hallmarks of premalignant progression, indicating that B lymphocytes and humoral immunity were necessary for tumor progression (de Visser et al. 2005).

21.3 Conclusions

The availability and diversity of mouse models of human cancer that develop organ-specific neoplasms has increased significantly in the past 10 years. These, coupled with mouse strains where components of the immune system have been manipulated or deleted, have provided novel insights into how immune cells and immune function significantly contributes to cancer development. Importantly, these studies have revealed immune cell populations that when appropriately activated are capable of exerting either pro- or antitumor immune function. The hope for the future is that these insights will aid in our therapeutic armament with which to develop anti-cancer drugs whose efficacy will either diminish cancer development or stabilize premalignant or dormant disease thus that life-span and the quality of life for patients with cancer can be significantly increased.

Acknowledgments The authors thank all those that have contributed to work discussed here but are not referenced due to space constraints. The authors acknowledge support by grants from the National Institutes of Health (NIH)/NCI (R01CA130980, R01CA132566, R01CA140943, P50CA58207), and the Department of Defense (W81XWH-06-1-0416, PR080717 to L.M.C., and grants from the Dutch Cancer Society (2006-3715 and 2011-5004), the Netherlands Organization for Scientific Research (VIDI 917.96.307), and the Association for International Cancer Research (AICR 11-0677) to K.E.d.V.

References

Adachi O, Kawai T, Takeda K, Matsumoto M, Tsutsui H, Sakagami M, Nakanishi K, Akira S (1998) Targeted disruption of the MyD88 gene results in loss of IL-1- and IL-18-mediated function. Immunity 9:143–150

Akira S, Takeda K (2004) Toll-like receptor signalling. Nat Rev Immunol 4:499–511

Albini A, Sporn MB (2007) The tumour microenvironment as a target for chemoprevention. Nat Rev Cancer 7:139–147

Alexopoulou L, Holt AC, Medzhitov R, Flavell RA (2001) Recognition of double-stranded RNA and activation of NF-kappaB by Toll-like receptor 3. Nature 413:732–738

Almand B, Clark JI, Nikitina E, van Beynen J, English NR, Knight SC, Carbone DP, Gabrilovich DI (2001) Increased production of immature myeloid cells in cancer patients: a mechanism of immunosuppression in cancer. J Immunol 166:678–689

Arbeit JM, Munger K, Howley PM, Hanahan D (1994) Progressive squamous epithelial neoplasia in K14-human papillomavirus type 16 transgenic mice. J Virol 68:4358–4368
Arber N, Eagle CJ, Spicak J, Racz I, Dite P, Hajer J, Zavoral M, Lechuga MJ, Gerletti P, Tang J et al (2006) Celecoxib for the prevention of colorectal adenomatous polyps. N Engl J Med 355:885–895
Arnott CH, Scott KA, Moore RJ, Robinson SC, Thompson RG, Balkwill FR (2004) Expression of both TNF-alpha receptor subtypes is essential for optimal skin tumour development. Oncogene 23:1902–1910
Balkwill F, Charles KA, Mantovani A (2005) Smoldering and polarized inflammation in the initiation and promotion of malignant disease. Cancer Cell 7:211–217
Beck BN, Henney CS (1981) An analysis of the natural killer cell defect in beige mice. Cell Immunol 61:343–352
Belardelli F, Ferrantini M (2002) Cytokines as a link between innate and adaptive antitumor immunity. Trends Immunol 23:201–208
Bennett CL, Clausen BE (2007) DC ablation in mice: promises, pitfalls, and challenges. Trends Immunol 28:525–531
Bergers G, Brekken R, McMahon G, Vu TH, Itoh T, Tamaki K, Tanzawa K, Thorpe P, Itohara S, Werb Z et al (2000) Matrix metalloproteinase-9 triggers the angiogenic switch during carcinogenesis. Nat Cell Biol 2:737–744
Bertagnolli MM, Eagle CJ, Zauber AG, Redston M, Solomon SD, Kim K, Tang J, Rosenstein RB, Wittes J, Corle D et al (2006) Celecoxib for the prevention of sporadic colorectal adenomas. N Engl J Med 355:873–884
Bhowmick NA, Moses HL (2005) Tumor-stroma interactions. Curr Opin Genet Dev 15:97–101
Borriello F, Sethna MP, Boyd SD, Schweitzer AN, Tivol EA, Jacoby D, Strom TB, Simpson EM, Freeman GJ, Sharpe AH (1997) B7-1 and B7-2 have overlapping, critical roles in immunoglobulin class switching and germinal center formation. Immunity 6:303–313
Boshoff C, Weiss R (2002) AIDS-related malignancies. Nat Rev Cancer 2:373–382
Bosma GC, Custer RP, Bosma MJ (1983) A severe combined immunodeficiency mutation in the mouse. Nature 301:527–530
Botto M, Dell'Agnola C, Bygrave AE, Thompson EM, Cook HT, Petry F, Loos M, Pandolfi PP, Walport MJ (1998) Homozygous C1q deficiency causes glomerulonephritis associated with multiple apoptotic bodies. Nat Genet 19:56–59
Buckanovich RJ, Facciabene A, Kim S, Benencia F, Sasaroli D, Balint K, Katsaros D, O'Brien-Jenkins A, Gimotty PA, Coukos G (2008) Endothelin B receptor mediates the endothelial barrier to T cell homing to tumors and disables immune therapy. Nat Med 14:28–36
Buell JF, Gross TG, Woodle ES (2005) Malignancy after transplantation. Transplantation 80:S254–S264
Carroll MC (2004) The complement system in regulation of adaptive immunity. Nat Immunol 5:981–986
Casanovas O, Hicklin DJ, Bergers G, Hanahan D (2005) Drug resistance by evasion of antiangiogenic targeting of VEGF signaling in late-stage pancreatic islet tumors. Cancer Cell 8:299–309
Cha YI, DuBois RN (2007) NSAIDs and cancer prevention: targets downstream of COX-2. Annu Rev Med 58:239–252
Chambers CA, Cado D, Truong T, Allison JP (1997) Thymocyte development is normal in CTLA-4-deficient mice. Proc Natl Acad Sci USA 94:9296–9301
Chen J, Trounstine M, Alt FW, Young F, Kurahara C, Loring JF, Huszar D (1993) Immunoglobulin gene rearrangement in B cell deficient mice generated by targeted deletion of the JH locus. Int Immunol 5:647–656
Choudhury A, Mosolits S, Kokhaei P, Hansson L, Palma M, Mellstedt H (2006) Clinical results of vaccine therapy for cancer: learning from history for improving the future. Adv Cancer Res 95:147–202
Chulada PC, Thompson MB, Mahler JF, Doyle CM, Gaul BW, Lee C, Tiano HF, Morham SG, Smithies O, Langenbach R (2000) Genetic disruption of Ptgs-1, as well as Ptgs-2, reduces intestinal tumorigenesis in Min mice. Cancer Res 60:4705–4708
Colombo MP, Piconese S (2007) Regulatory T-cell inhibition versus depletion: the right choice in cancer immunotherapy. Nat Rev Cancer 7:880–887

Coussens LM, Hanahan D, Arbeit JM (1996) Genetic predisposition and parameters of malignant progression in K14- HPV16 transgenic mice. Am J Pathol 149:1899–1917

Coussens LM, Raymond WW, Bergers G, Laig-Webster M, Behrendtsen O, Werb Z, Caughey GH, Hanahan D (1999) Inflammatory mast cells up-regulate angiogenesis during squamous epithelial carcinogenesis. Genes Dev 13:1382–1397

Coussens LM, Tinkle CL, Hanahan D, Werb Z (2000) MMP-9 supplied by bone marrow-derived cells contributes to skin carcinogenesis. Cell 103:481–490

Croci DO, Zacarias Fluck MF, Rico MJ, Matar P, Rabinovich GA, Scharovsky OG (2007) Dynamic cross-talk between tumor and immune cells in orchestrating the immunosuppressive network at the tumor microenvironment. Cancer Immunol Immunother 56:1687–1700

Cui J, Shin T, Kawano T, Sato H, Kondo E, Toura I, Kaneko Y, Koseki H, Kanno M, Taniguchi M (1997) Requirement for Valpha14 NKT cells in IL-12-mediated rejection of tumors. Science 278:1623–1626

Daniel D, Meyer-Morse N, Bergsland EK, Dehne K, Coussens LM, Hanahan D (2003) Immune enhancement of skin carcinogenesis by CD4+ T cells. J Exp Med 197:1017–1028

Dannenberg A, Subbaramaiah K (2003) Targeting cyclooxygenase-2 in human neoplasia: rationale and promise. Cancer Cell 4:431–436

De Giorgi U, Pupi A, Turrisi G, Montenora I, Morini S, Fayyaz M, De Simone M, Fiorentini G (2007) Critical update and emerging trends in imatinib treatment for gastrointestinal stromal tumor. Rev Recent Clin Trials 2:43–48

De Palma M, Venneri MA, Galli R, Sergi Sergi L, Politi LS, Sampaolesi M, Naldini L (2005) Tie2 identifies a hematopoietic lineage of proangiogenic monocytes required for tumor vessel formation and a mesenchymal population of pericyte progenitors. Cancer Cell 8:211–226

de Visser KE, Korets LV, Coussens LM (2004) Early neoplastic progression is complement independent. Neoplasia 6:768–776

de Visser KE, Korets LV, Coussens LM (2005) De novo carcinogenesis promoted by chronic inflammation is B lymphocyte dependent. Cancer Cell 7:411–423

de Visser KE, Eichten A, Coussens LM (2006) Paradoxical roles of the immune system during cancer development. Nat Rev Cancer 6:24–37

Dinchuk JE, Car BD, Focht RJ, Johnston JJ, Jaffee BD, Covington MB, Contel NR, Eng VM, Collins RJ, Czerniak PM et al (1995) Renal abnormalities and an altered inflammatory response in mice lacking cyclooxygenase II. Nature 378:406–409

Druker BJ (2004) Imatinib as a paradigm of targeted therapies. Adv Cancer Res 91:1–30

Duffield JS, Forbes SJ, Constandinou CM, Clay S, Partolina M, Vuthoori S, Wu S, Lang R, Iredale JP (2005) Selective depletion of macrophages reveals distinct, opposing roles during liver injury and repair. J Clin Invest 115:56–65

Dunn GP, Bruce AT, Ikeda H, Old LJ, Schreiber RD (2002) Cancer immunoediting: from immunosurveillance to tumor escape. Nat Immunol 3:991–998

Dunn GP, Old LJ, Schreiber RD (2004) The immunobiology of cancer immunosurveillance and immunoediting. Immunity 21:137–148

Dunn GP, Koebel CM, Schreiber RD (2006) Interferons, immunity and cancer immunoediting. Nat Rev Immunol 6:836–848

Duttlinger R, Manova K, Chu TY, Gyssler C, Zelenetz AD, Bachvarova RF, Besmer P (1993) W-sash affects positive and negative elements controlling c-kit expression: ectopic c-kit expression at sites of kit-ligand expression affects melanogenesis. Development 118:705–717

Ebnet K, Hausmann M, Lehmann-Grube F, Mullbacher A, Kopf M, Lamers M, Simon MM (1995) Granzyme A-deficient mice retain potent cell-mediated cytotoxicity. EMBO J 14:4230–4239

Enzler T, Gillessen S, Manis JP, Ferguson D, Fleming J, Alt FW, Mihm M, Dranoff G (2003) Deficiencies of GM-CSF and interferon gamma link inflammation and cancer. J Exp Med 197:1213–1219

Feuillet V, Medjane S, Mondor I, Demaria O, Pagni PP, Galan JE, Flavell RA, Alexopoulou L (2006) Involvement of Toll-like receptor 5 in the recognition of flagellated bacteria. Proc Natl Acad Sci USA 103:12487–12492

Finke J, Ferrone S, Frey A, Mufson A, Ochoa A (1999) Where have all the T cells gone? Mechanisms of immune evasion by tumors. Immunol Today 20:158–160

Fischer C, Jonckx B, Mazzone M, Zacchigna S, Loges S, Pattarini L, Chorianopoulos E, Liesenborghs L, Koch M, De Mol M et al (2007) Anti-PlGF inhibits growth of VEGF(R)-inhibitor-resistant tumors without affecting healthy vessels. Cell 131:463–475

Freeman GJ, Borriello F, Hodes RJ, Reiser H, Hathcock KS, Laszlo G, McKnight AJ, Kim J, Du L, Lombard DB et al (1993) Uncovering of functional alternative CTLA-4 counter-receptor in B7-deficient mice. Science 262:907–909

Frese KK, Tuveson DA (2007) Maximizing mouse cancer models. Nat Rev Cancer 7:645–658

Fukata M, Chen A, Vamadevan AS, Cohen J, Breglio K, Krishnareddy S, Hsu D, Xu R, Harpaz N, Dannenberg AJ et al (2007) Toll-like receptor-4 promotes the development of colitis-associated colorectal tumors. Gastroenterology 133:1869–1881

Fung-Leung WP, Schilham MW, Rahemtulla A, Kundig TM, Vollenweider M, Potter J, van Ewijk W, Mak TW (1991) CD8 is needed for development of cytotoxic T cells but not helper T cells. Cell 65:443–449

Galon J, Costes A, Sanchez-Cabo F, Kirilovsky A, Mlecnik B, Lagorce-Pages C, Tosolini M, Camus M, Berger A, Wind P et al (2006) Type, density, and location of immune cells within human colorectal tumors predict clinical outcome. Science 313:1960–1964

Garbe AI, Vermeer B, Gamrekelashvili J, von Wasielewski R, Greten FR, Westendorf AM, Buer J, Schmid RM, Manns MP, Korangy F et al (2006) Genetically induced pancreatic adenocarcinoma is highly immunogenic and causes spontaneous tumor-specific immune responses. Cancer Res 66:508–516

Girardi M, Oppenheim DE, Steele CR, Lewis JM, Glusac E, Filler R, Hobby P, Sutton B, Tigelaar RE, Hayday AC (2001) Regulation of cutaneous malignancy by gammadelta T cells. Science 294:605–609

Girardi M, Glusac E, Filler RB, Roberts SJ, Propperova I, Lewis J, Tigelaar RE, Hayday AC (2003) The distinct contributions of murine T cell receptor (TCR)gammadelta+ and TCRalphabeta+ T cells to different stages of chemically induced skin cancer. J Exp Med 198:747–755

Giraudo E, Inoue M, Hanahan D (2004) An amino-bisphosphonate targets MMP-9-expressing macrophages and angiogenesis to impair cervical carcinogenesis. J Clin Invest 114:623–633

Goldrath AW, Bevan MJ (1999) Selecting and maintaining a diverse T-cell repertoire. Nature 402:255–262

Gounaris E, Erdman SE, Restaino C, Gurish MF, Friend DS, Gounari F, Lee DM, Zhang G, Glickman JN, Shin K et al (2007) Mast cells are an essential hematopoietic component for polyp development. Proc Natl Acad Sci USA 104:19977–19982

Greten FR, Eckmann L, Greten TF, Park JM, Li ZW, Egan LJ, Kagnoff MF, Karin M (2004) IKKbeta links inflammation and tumorigenesis in a mouse model of colitis-associated cancer. Cell 118:285–296

Grimbaldeston MA, Chen CC, Piliponsky AM, Tsai M, Tam SY, Galli SJ (2005) Mast cell-deficient W-sash c-kit mutant Kit W-sh/W-sh mice as a model for investigating mast cell biology in vivo. Am J Pathol 167:835–848

Grulich AE, van Leeuwen MT, Falster MO, Vajdic CM (2007) Incidence of cancers in people with HIV/AIDS compared with immunosuppressed transplant recipients: a meta-analysis. Lancet 370:59–67

Guerra C, Schuhmacher AJ, Canamero M, Grippo PJ, Verdaguer L, Perez-Gallego L, Dubus P, Sandgren EP, Barbacid M (2007) Chronic pancreatitis is essential for induction of pancreatic ductal adenocarcinoma by K-Ras oncogenes in adult mice. Cancer Cell 11:291–302

Gupta RA, Dubois RN (2001) Colorectal cancer prevention and treatment by inhibition of cyclooxygenase-2. Nat Rev Cancer 1:11–21

Hanahan D, Weinberg RA (2000) The hallmarks of cancer. Cell 100:57–70

Hanahan D, Wagner EF, Palmiter RD (2007) The origins of oncomice: a history of the first transgenic mice genetically engineered to develop cancer. Genes Dev 21:2258–2270

Hazenbos WL, Gessner JE, Hofhuis FM, Kuipers H, Meyer D, Heijnen IA, Schmidt RE, Sandor M, Capel PJ, Daeron M et al (1996) Impaired IgG-dependent anaphylaxis and Arthus reaction in Fc gamma RIII (CD16) deficient mice. Immunity 5:181–188
Hemmi H, Takeuchi O, Kawai T, Kaisho T, Sato S, Sanjo H, Matsumoto M, Hoshino K, Wagner H, Takeda K et al (2000) A Toll-like receptor recognizes bacterial DNA. Nature 408:740–745
Hemmi H, Kaisho T, Takeuchi O, Sato S, Sanjo H, Hoshino K, Horiuchi T, Tomizawa H, Takeda K, Akira S (2002) Small anti-viral compounds activate immune cells via the TLR7 MyD88-dependent signaling pathway. Nat Immunol 3:196–200
Hennings H, Glick AB, Greenhalgh DA, Morgan DL, Strickland JE, Tennenbaum T, Yuspa SH (1993) Critical aspects of initiation, promotion, and progression in multistage epidermal carcinogenesis. Proc Soc Exp Biol Med 202:1–8
Heusel JW, Wesselschmidt RL, Shresta S, Russell JH, Ley TJ (1994) Cytotoxic lymphocytes require granzyme B for the rapid induction of DNA fragmentation and apoptosis in allogeneic target cells. Cell 76:977–987
Hopken UE, Lu B, Gerard NP, Gerard C (1996) The C5a chemoattractant receptor mediates mucosal defence to infection. Nature 383:86–89
Hoshino K, Takeuchi O, Kawai T, Sanjo H, Ogawa T, Takeda Y, Takeda K, Akira S (1999) Cutting edge: Toll-like receptor 4 (TLR4)-deficient mice are hyporesponsive to lipopolysaccharide: evidence for TLR4 as the Lps gene product. J Immunol 162:3749–3752
Howe LR, Chang SH, Tolle KC, Dillon R, Young LJ, Cardiff RD, Newman RA, Yang P, Thaler HT, Muller WJ et al (2005) HER2/neu-induced mammary tumorigenesis and angiogenesis are reduced in cyclooxygenase-2 knockout mice. Cancer Res 65:10113–10119
Ikehara S, Pahwa RN, Fernandes G, Hansen CT, Good RA (1984) Functional T cells in athymic nude mice. Proc Natl Acad Sci USA 81:886–888
Imada A, Shijubo N, Kojima H, Abe S (2000) Mast cells correlate with angiogenesis and poor outcome in stage I lung adenocarcinoma. Eur Respir J 15:1087–1093
Itohara S, Mombaerts P, Lafaille J, Iacomini J, Nelson A, Clarke AR, Hooper ML, Farr A, Tonegawa S (1993) T cell receptor delta gene mutant mice: independent generation of alpha beta T cells and programmed rearrangements of gamma delta TCR genes. Cell 72:337–348
Jain RK (2005) Normalization of tumor vasculature: an emerging concept in antiangiogenic therapy. Science 307:58–62
Jain RK (2008) Lessons from multidisciplinary translational trials on anti-angiogenic therapy of cancer. Nat Rev Cancer 8:309–316
Janeway CA Jr, Medzhitov R (2002) Innate immune recognition. Annu Rev Immunol 20:197–216
Janeway CA, Travers P, Walport M, Shlomchik M (2001) Immunobiology, 5th edn. Garland Publishing, New York
Johansson M, Denardo DG, Coussens LM (2008) Polarized immune responses differentially regulate cancer development. Immunol Rev 222:145–154
Jonkers J, Derksen PW (2007) Modeling metastatic breast cancer in mice. J Mammary Gland Biol Neoplasia 12:191–203
Joyce JA (2005) Therapeutic targeting of the tumor microenvironment. Cancer Cell 7:513–520
Jung S, Unutmaz D, Wong P, Sano G, De los Santos K, Sparwasser T, Wu S, Vuthoori S, Ko K, Zavala F et al (2002) In vivo depletion of CD11c(+) dendritic cells abrogates priming of CD8(+) T cells by exogenous cell-associated antigens. Immunity 17:211–220
Kagi D, Ledermann B, Burki K, Seiler P, Odermatt B, Olsen KJ, Podack ER, Zinkernagel RM, Hengartner H (1994) Cytotoxicity mediated by T cells and natural killer cells is greatly impaired in perforin-deficient mice. Nature 369:31–37
Kaplan DH, Shankaran V, Dighe AS, Stockert E, Aguet M, Old LJ, Schreiber RD (1998) Demonstration of an interferon gamma-dependent tumor surveillance system in immunocompetent mice. Proc Natl Acad Sci USA 95:7556–7561
Karin M (2006) NF-kappaB and cancer: mechanisms and targets. Mol Carcinog 45:355–361
Kim CF, Jackson EL, Kirsch DG, Grimm J, Shaw AT, Lane K, Kissil J, Olive KP, Sweet-Cordero A, Weissleder R et al (2005) Mouse models of human non-small-cell lung cancer: raising the bar. Cold Spring Harb Symp Quant Biol 70:241–250

Kitamura Y, Go S, Hatanaka K (1978) Decrease of mast cells in W/Wv mice and their increase by bone marrow transplantation. Blood 52:447–452

Kitamura D, Roes J, Kuhn R, Rajewsky K (1991) A B cell-deficient mouse by targeted disruption of the membrane exon of the immunoglobulin mu chain gene. Nature 350:423–426

Klein S, McCormick F, Levitzki A (2005) Killing time for cancer cells. Nat Rev Cancer 5:573–580

Koller BH, Marrack P, Kappler JW, Smithies O (1990) Normal development of mice deficient in beta 2M, MHC class I proteins, and CD8+ T cells. Science 248:1227–1230

Kontgen F, Suss G, Stewart C, Steinmetz M, Bluethmann H (1993) Targeted disruption of the MHC class II Aa gene in C57BL/6 mice. Int Immunol 5:957–964

Kusmartsev S, Gabrilovich DI (2006) Effect of tumor-derived cytokines and growth factors on differentiation and immune suppressive features of myeloid cells in cancer. Cancer Metastasis Rev 25:323–331

Kwon BS, Hurtado JC, Lee ZH, Kwack KB, Seo SK, Choi BK, Koller BH, Wolisi G, Broxmeyer HE, Vinay DS (2002) Immune responses in 4-1BB (CD137)-deficient mice. J Immunol 168:5483–5490

Langenbach R, Morham SG, Tiano HF, Loftin CD, Ghanayem BI, Chulada PC, Mahler JF, Lee CA, Goulding EH, Kluckman KD et al (1995) Prostaglandin synthase 1 gene disruption in mice reduces arachidonic acid-induced inflammation and indomethacin-induced gastric ulceration. Cell 83:483–492

Langowski JL, Zhang X, Wu L, Mattson JD, Chen T, Smith K, Basham B, McClanahan T, Kastelein RA, Oft M (2006) IL-23 promotes tumour incidence and growth. Nature 442:461–465

Leek RD, Lewis CE, Whitehouse R, Greenall M, Clarke J, Harris AL (1996) Association of macrophage infiltration with angiogenesis and prognosis in invasive breast carcinoma. Cancer Res 56:4625–4629

Lin EY, Nguyen AV, Russell RG, Pollard JW (2001) Colony-stimulating factor 1 promotes progression of mammary tumors to malignancy. J Exp Med 193:727–740

Luo JL, Tan W, Ricono JM, Korchynskyi O, Zhang M, Gonias SL, Cheresh DA, Karin M (2007) Nuclear cytokine-activated IKKalpha controls prostate cancer metastasis by repressing Maspin. Nature 446:690–694

Luster AD (2002) The role of chemokines in linking innate and adaptive immunity. Curr Opin Immunol 14:129–135

Marks SC Jr, Lane PW (1976) Osteopetrosis, a new recessive skeletal mutation on chromosome 12 of the mouse. J Hered 67:11–18

McHeyzer-Williams MG (2003) B cells as effectors. Curr Opin Immunol 15:354–361

Molina H, Holers VM, Li B, Fung Y, Mariathasan S, Goellner J, Strauss-Schoenberger J, Karr RW, Chaplin DD (1996) Markedly impaired humoral immune response in mice deficient in complement receptors 1 and 2. Proc Natl Acad Sci USA 93:3357–3361

Mombaerts P, Iacomini J, Johnson RS, Herrup K, Tonegawa S, Papaioannou VE (1992) RAG-1-deficient mice have no mature B and T lymphocytes. Cell 68:869–877

Mombaerts P, Arnoldi J, Russ F, Tonegawa S, Kaufmann SH (1993) Different roles of alpha beta and gamma delta T cells in immunity against an intracellular bacterial pathogen. Nature 365:53–56

Moore RJ, Owens DM, Stamp G, Arnott C, Burke F, East N, Holdsworth H, Turner L, Rollins B, Pasparakis M et al (1999) Mice deficient in tumor necrosis factor-alpha are resistant to skin carcinogenesis. Nat Med 5:828–831

Morham SG, Langenbach R, Loftin CD, Tiano HF, Vouloumanos N, Jennette JC, Mahler JF, Kluckman KD, Ledford A, Lee CA et al (1995) Prostaglandin synthase 2 gene disruption causes severe renal pathology in the mouse. Cell 83:473–482

Murphy JE, Robert C, Kupper TS (2000) Interleukin-1 and cutaneous inflammation: a crucial link between innate and acquired immunity. J Invest Dermatol 114:602–608

Nelson CM, Bissell MJ (2006) Of extracellular matrix, scaffolds, and signaling: tissue architecture regulates development, homeostasis, and cancer. Annu Rev Cell Dev Biol 22:287–309

Nozawa H, Chiu C, Hanahan D (2006) Infiltrating neutrophils mediate the initial angiogenic switch in a mouse model of multistage carcinogenesis. Proc Natl Acad Sci USA 103:12493–12498

Oshima M, Dinchuk JE, Kargman SL, Oshima H, Hancock B, Kwong E, Trzaskos JM, Evans JF, Taketo MM (1996) Suppression of intestinal polyposis in Apc delta716 knockout mice by inhibition of cyclooxygenase 2 (COX-2). Cell 87:803–809

Ostrand-Rosenberg S (2008) Immune surveillance: a balance between protumor and antitumor immunity. Curr Opin Genet Dev 18:11–18

Persano L, Crescenzi M, Indraccolo S (2007) Anti-angiogenic gene therapy of cancer: current status and future prospects. Mol Aspects Med 28:87–114

Peto J (2001) Cancer epidemiology in the last century and the next decade. Nature 411:390–395

Pikarsky E, Porat RM, Stein I, Abramovitch R, Amit S, Kasem S, Gutkovich-Pyest E, Urieli-Shoval S, Galun E, Ben-Neriah Y (2004) NF-kappaB functions as a tumour promoter in inflammation-associated cancer. Nature 431:461–466

Qin Z, Blankenstein T (2004) A cancer immunosurveillance controversy. Nat Immunol 5:3–4

Rahemtulla A, Fung-Leung WP, Schilham MW, Kundig TM, Sambhara SR, Narendran A, Arabian A, Wakeham A, Paige CJ, Zinkernagel RM et al (1991) Normal development and function of CD8+ cells but markedly decreased helper cell activity in mice lacking CD4. Nature 353:180–184

Roberts SJ, Ng BY, Filler RB, Lewis J, Glusac EJ, Hayday AC, Tigelaar RE, Girardi M (2007) Characterizing tumor-promoting T cells in chemically induced cutaneous carcinogenesis. Proc Natl Acad Sci USA 104:6770–6775

Sawyers C (2004) Targeted cancer therapy. Nature 432:294–297

Serafini P, De Santo C, Marigo I, Cingarlini S, Dolcetti L, Gallina G, Zanovello P, Bronte V (2004) Derangement of immune responses by myeloid suppressor cells. Cancer Immunol Immunother 53:64–72

Shahinian A, Pfeffer K, Lee KP, Kundig TM, Kishihara K, Wakeham A, Kawai K, Ohashi PS, Thompson CB, Mak TW (1993) Differential T cell costimulatory requirements in CD28-deficient mice. Science 261:609–612

Shankaran V, Ikeda H, Bruce AT, White JM, Swanson PE, Old LJ, Schreiber RD (2001) IFNgamma and lymphocytes prevent primary tumour development and shape tumour immunogenicity. Nature 410:1107–1111

Shinkai Y, Rathbun G, Lam KP, Oltz EM, Stewart V, Mendelsohn M, Charron J, Datta M, Young F, Stall AM et al (1992) RAG-2-deficient mice lack mature lymphocytes owing to inability to initiate V(D)J rearrangement. Cell 68:855–867

Sinnamon MJ, Carter KJ, Sims LP, Lafleur B, Fingleton B, Matrisian LM (2008) A protective role for mast cells in intestinal tumorigenesis. Carcinogenesis 29(4):880–886

Smyth MJ, Thia KY, Street SE, Cretney E, Trapani JA, Taniguchi M, Kawano T, Pelikan SB, Crowe NY, Godfrey DI (2000) Differential tumor surveillance by natural killer (NK) and NKT cells. J Exp Med 191:661–668

Smyth MJ, Crowe NY, Godfrey DI (2001) NK cells and NKT cells collaborate in host protection from methylcholanthrene-induced fibrosarcoma. Int Immunol 13:459–463

Soucek L, Lawlor ER, Soto D, Shchors K, Swigart LB, Evan GI (2007) Mast cells are required for angiogenesis and macroscopic expansion of Myc-induced pancreatic islet tumors. Nat Med 13:1211–1218

Sprent J, Surh CD (2002) T cell memory. Annu Rev Immunol 20:551–579

Street SE, Zerafa N, Iezzi M, Westwood JA, Stagg J, Musiani P, Smyth MJ (2007) Host perforin reduces tumor number but does not increase survival in oncogene-driven mammary adenocarcinoma. Cancer Res 67:5454–5460

Swann JB, Smyth MJ (2007) Immune surveillance of tumors. J Clin Invest 117:1137–1146

Swann JB, Vesely MD, Silva A, Sharkey J, Akira S, Schreiber RD, Smyth MJ (2008) Demonstration of inflammation-induced cancer and cancer immunoediting during primary tumorigenesis. Proc Natl Acad Sci USA 105:652–656

Tabi Z, Man S (2006) Challenges for cancer vaccine development. Adv Drug Deliv Rev 58:902–915

Takahashi T, Tanaka M, Brannan CI, Jenkins NA, Copeland NG, Suda T, Nagata S (1994) Generalized lymphoproliferative disease in mice, caused by a point mutation in the Fas ligand. Cell 76:969–976

Takai T, Li M, Sylvestre D, Clynes R, Ravetch JV (1994) FcR gamma chain deletion results in pleiotrophic effector cell defects. Cell 76:519–529

Takeuchi O, Hoshino K, Kawai T, Sanjo H, Takada H, Ogawa T, Takeda K, Akira S (1999) Differential roles of TLR2 and TLR4 in recognition of gram-negative and gram-positive bacterial cell wall components. Immunity 11:443–451
Takeuchi O, Kawai T, Muhlradt PF, Morr M, Radolf JD, Zychlinsky A, Takeda K, Akira S (2001) Discrimination of bacterial lipoproteins by Toll-like receptor 6. Int Immunol 13:933–940
Takeuchi O, Sato S, Horiuchi T, Hoshino K, Takeda K, Dong Z, Modlin RL, Akira S (2002) Cutting edge: role of Toll-like receptor 1 in mediating immune response to microbial lipoproteins. J Immunol 169:10–14
Taylor PR, Nash JT, Theodoridis E, Bygrave AE, Walport MJ, Botto M (1998) A targeted disruption of the murine complement factor B gene resulting in loss of expression of three genes in close proximity, factor B, C2, and D17H6S45. J Biol Chem 273:1699–1704
Tiano HF, Loftin CD, Akunda J, Lee CA, Spalding J, Sessoms A, Dunson DB, Rogan EG, Morham SG, Smart RC et al (2002) Deficiency of either cyclooxygenase (COX)-1 or COX-2 alters epidermal differentiation and reduces mouse skin tumorigenesis. Cancer Res 62:3395–3401
Tonegawa S (1983) Somatic generation of antibody diversity. Nature 302:575–581
Turini ME, DuBois RN (2002) Cyclooxygenase-2: a therapeutic target. Annu Rev Med 53:35–37
Tuveson DA, Jacks T (2002) Technologically advanced cancer modeling in mice. Curr Opin Genet Dev 12:105–110
van den Broek ME, Kagi D, Ossendorp F, Toes R, Vamvakas S, Lutz WK, Melief CJ, Zinkernagel RM, Hengartner H (1996) Decreased tumor surveillance in perforin-deficient mice. J Exp Med 184:1781–1790
Van Dyke T, Jacks T (2002) Cancer modeling in the modern era: progress and challenges. Cell 108:135–144
Van Kaer L, Ashton-Rickardt PG, Ploegh HL, Tonegawa S (1992) TAP1 mutant mice are deficient in antigen presentation, surface class I molecules, and CD4-8+ T cells. Cell 71:1205–1214
van Kempen LC, de Visser KE, Coussens LM (2006) Inflammation, proteases and cancer. Eur J Cancer 42:728–734
Van Rooijen N, Sanders A (1994) Liposome mediated depletion of macrophages: mechanism of action, preparation of liposomes and applications. J Immunol Methods 174:83–93
Walport MJ (2001a) Complement. First of two parts. N Engl J Med 344:1058–1066
Walport MJ (2001b) Complement. Second of two parts. N Engl J Med 344:1140–1144
Wedemeyer J, Galli SJ (2005) Decreased susceptibility of mast cell-deficient Kit(W)/Kit(W-v) mice to the development of 1, 2-dimethylhydrazine-induced intestinal tumors. Lab Invest 85:388–396
Wessels MR, Butko P, Ma M, Warren HB, Lage AL, Carroll MC (1995) Studies of group B streptococcal infection in mice deficient in complement component C3 or C4 demonstrate an essential role for complement in both innate and acquired immunity. Proc Natl Acad Sci USA 92:11490–11494
Willimsky G, Blankenstein T (2005) Sporadic immunogenic tumours avoid destruction by inducing T-cell tolerance. Nature 437:141–146
Wyckoff J, Wang W, Lin EY, Wang Y, Pixley F, Stanley ER, Graf T, Pollard JW, Segall J, Condeelis J (2004) A paracrine loop between tumor cells and macrophages is required for tumor cell migration in mammary tumors. Cancer Res 64:7022–7029
Yu P, Rowley DA, Fu YX, Schreiber H (2006) The role of stroma in immune recognition and destruction of well-established solid tumors. Curr Opin Immunol 18:226–231
Yuspa SH (2000) Overview of carcinogenesis: past, present and future. Carcinogenesis 21:341–344
Yusuf N, Nasti TH, Long JA, Naseemuddin M, Lucas AP, Xu H, Elmets CA (2008) Protective role of Toll-like receptor 4 during the initiation stage of cutaneous chemical carcinogenesis. Cancer Res 68:615–622
Zlotnik A, Yoshie O (2000) Chemokines: a new classification system and their role in immunity. Immunity 12:121–127
Zou W (2005) Immunosuppressive networks in the tumour environment and their therapeutic relevance. Nat Rev Cancer 5:263–274
Zou W (2006) Regulatory T cells, tumour immunity and immunotherapy. Nat Rev Immunol 6:295–307

Chapter 22
Transplanted Tumor Models for Preclinical Drug Testing and the Potential Benefit of Genetically Engineered Mouse Models

Melinda Hollingshead, Michelle Ahalt, and Sergio Alcoser

22.1 Historical Perspectives on Transplanted Rodent Models in Preclinical Testing

Rodents were actively bred in research laboratories in the nineteenth century for various studies, many of which were designed to understand diseases and genetics. These efforts, advanced by the activities of "fancy" mouse breeding clubs in late nineteenth century England (Fancy Mouse Breeders Club 2008), led to the production of a variety of inbred mouse strains (Beck et al. 2000). The first fully inbred strain (DBA), developed in 1909, was followed over the next few years by several others (C57BL, C3H, CBA, and BALB/c) (Staats 1966). This resulted in the identification of strains with higher than average incidences of neoplasia thereby providing the research community with early tools to study neoplasia (Slye 1915; McCampbell 1909). An obvious offshoot of these studies was the pursuit of methods and means to treat and/or prevent cancer using spontaneous mouse tumors (Murphy et al. 1923; Strong 1927; Hammett 1936). These inbred strains and their tumors, originating both as spontaneous and as experimentally induced tumors (chemical or radiation) (Nettleship 1943; Carruthers and Suntzeff 1944; Greene and Murphy 1945), provided a means to reproducibly, serially passage tumors in mice (Ludford 1933; Bittner 1934; Barrett 1940; Cloudman 1946).

In 1955, the National Cancer Institute (NCI) initiated a program that became the leading public screening effort for anticancer drug development. This screen consisted of a large-scale *in vivo* testing scheme using many of the transplanted rodent tumor models developed in the first half of the twentieth century. The goal was to identify

M. Hollingshead (✉) • M. Ahalt • S. Alcoser
Developmental Therapeutics Program, NCI-Frederick, 1050 Boyles St, Building 1052/239, Frederick, MD 21702, USA
e-mail: hollingm@mail.nih.gov

J.E. Green and T. Ried (eds.), *Genetically Engineered Mice for Cancer Research: Design, Analysis, Pathways, Validation and Pre-Clinical Testing*,
DOI 10.1007/978-0-387-69805-2_22,

and develop antitumor compounds based on the hypothesis that the transplanted tumor models correlated with clinical outcomes (Talmadge et al. 2007). Initially, the screen consisted of three mouse tumors (Sarcoma 180, Carcinoma 755, and Leukemia 1210), which were selected as sensitive, predictive models based on retrospective analysis of the Gellhorn-Hirshberg report (Schepartz 1977). Subsequent data analyses induced several changes to the screening panel composition between 1955 and 1985. These panels included the well-known B16 melanoma, Lewis lung, and Walker 256 tumors as well as others. Ultimately, the murine P388 leukemia, demonstrated to be sensitive to most classes of clinically effective drugs (Venditti 1981), served as a prescreen and compounds active against P388 were referred for additional evaluations in other transplanted tumor models (Schepartz 1977; Frei 1982). These tumors were maintained by serial mouse-to-mouse transplantation either as intraperitoneal ascites tumors (e.g., P388, L1210, Sarcoma 180) or as subcutaneous tumors (e.g., Lewis lung, Carcinoma 755). For drug studies, the tumor-bearing mice were randomized into groups and treated with the test compounds, primarily by intraperitoneal injection, on one of the several treatment schedules (Goldin et al. 1977). The progressive testing plan was designed to provide an increasingly more rigorous challenge at each subsequent step by using more resistant tumors (Goldin et al. 1977). At the height of this drug testing modality, over 15,000 compounds were screened annually in multiple dedicated laboratories (Driscoll 1984; Frei 1982). This effort was discontinued in 1985 due to concerns that rodent tumors were selecting agents with activity in human leukemias, but with little activity in solid tumors. In addition, there was a growing belief that compounds could be identified by *in vitro* screening techniques prior to conducting mouse studies (Schuh 2004; Brown 1997; Suggitt and Bibby 2005; Schein and Scheffler 2006).

While these traditional models have been described as suboptimal and nonpredictive (Corbett et al. 1987), they are the means by which many of the early anticancer agents were identified and advanced to human clinical trials. That is, many of the currently used clinical agents were initially identified as active using transplantable rodent tumors (Johnson et al. 2001; Driscoll 1984; Staquet et al. 1983). In fact, eight (cisplatin, cyclophosphamide, cytarabine, doxorubicin, methotrexate, melphalan, mitomycin, and vincristine) of the 16 commercially available drugs discovered between 1955 and 1984 were identified through the NCI rodent tumor model screen (Driscoll 1984).

22.2 Development of the Human Tumor Xenograft

While spontaneous and induced rodent tumors were the primary tumor models in the first half of the twentieth century, the identification of the immunocompromised athymic nude mouse in 1966 induced a shift in research models (Flanagan 1966). In the homozygous state, these mice lack a functional thymus, which results in an absence of T cell function while maintaining other specific and nonspecific immune functions (Rygaard 1969; Gershwin 1977). Shortly after the athymic mouse was described, it was reported that skin and tumors from human and veterinary species could be successfully grafted onto these mice (Manning et al. 1973; Rygaard and Povlsen 1969). Additional studies demonstrated that human tumor xenografts not only grew progressively and were transplantable mouse-to-mouse, but they also showed that the tumors appeared stable during serial passage (Povlsen et al. 1975; Lozzio et al. 1976; Shimosato et al. 1976). While xenotransplantation onto athymic nude mice was possible, difficulties in maintaining immunocompromised mouse colonies due to adventitious mouse viruses (e.g., mouse hepatitis virus, Sendai virus) had to be resolved for the technology to be broadly utilized (Osieka et al. 1977). Alleviating these husbandry issues (Eaton et al. 1975; Guflino et al. 1976) allowed more laboratories to use human tumor xenografts. This revolutionized the drug screening process by providing what was presumed to be a more clinically relevant model. As mouse husbandry issues diminished, additional laboratories worked with immunocompromised mice and literally hundreds of transplantable human tumor models were developed. As part of the shift to what was hypothesized to be a more relevant tumor model, the NCI drug screening effort adopted three human tumor xenografts (breast MX-1, lung LX-1, and colon CX-1 xenografts) as part of its testing paradigm in 1975 (Driscoll 1984). However, in 1985 the NCI mouse tumor screening program was abolished and resources were redirected to establish an *in vitro* drug selection program. This program established the NCI 60 human tumor cell line screen (Brown 1997; Monks et al. 1991). As a component of this effort, NCI developed human tumor xenografts for each of the 60 cell lines (Plowman et al. 1997; Stinson et al. 1992). By the early 1990s, NCI had established a panel of over 40 tumor xenograft models for drug screening (Plowman et al. 1997). This panel continues to serve the NCI drug development effort.

Indeed, while the athymic nude mouse provided a major step forward for human tumor xenotransplantation, it did not universally support all human tumor xenografts. In 1983, the severe combined immunodeficient (SCID) mouse was identified and found to be more immunosuppressed than the athymic mouse (Bosma et al. 1983). SCID mice do not have functional T or B cells due to a loss of V(D)J recombinase activity, but they have an otherwise intact immune system (Weiss and Shannon 2004). This mouse offers an alternative for grafting human tumors that grow poorly in athymic mice. In fact, the SCID mouse proved particularly useful for generating human leukemia and lymphoma xenografts (Schmidtwolf et al. 1991; Liu et al. 1996). Other immunocompromised mice that are used in xenograft models include the NIH III (nu-xid-bg) (Kamel-Reid and Dick 1988), the SCID.bg (Croy and Chapeau 1990), and the NOD.SCID (Prochazka et al. 1992) strains. The nude (nu) autosomal recessive mutation produces a severe depletion of T-lymphocytes due to thymic dysgenesis. The xid mutation results in defects in B cell function, while the beige (bg) mutation results in impaired macrophage chemotaxis and motility and a natural killer (NK) cell deficiency. Finally, the nonobese diabetic (NOD) mutation produces a defect in NK cell activity. The NOD.SCID mouse model has been especially useful for the cancer stem cell research community as the preferred host for the isolation and characterization of tumor stem cells (Singh et al. 2004).

In addition to spontaneous mutations, mice can also be selectively depleted of immune functions through chemical (e.g., cyclosporine, cyclophosphamide), radiation, and antibody treatments (e.g., anti-asialo GM1 antibody to decrease NK activity) or by genetic engineering (e.g., NOD/LtSz-Rag1null mice) (Shultz et al. 2000). These chemically and genetically immunocompromised mice also serve as transplanted tumor hosts.

Thousands of publications describe the use of subcutaneous human tumor xenografts to assess the potential efficacy of new anticancer compounds. In fact, a literature search (conducted in August 2008) seeking papers that report xenograft data for the 175 approved antineoplastic agents yielded 2,783 references (Biological Testing Branch 2008). The general protocol for these studies is to implant tumors into a large cohort of mice, randomize mice into experimental groups, and monitor subsequent tumor growth. While there are variations on the theme (e.g., treatment start time, specific tumor implant site), the basic methodology involves collecting serial tumor measurements for each experimental animal and comparing the tumor status of the treated animals to the control animals. The most common means to monitor tumor growth is to measure tumor dimensions (length and width) using calipers (Plowman et al. 1997). These values are then used to calculate the tumor weight or volume with one of several formulas. Commonly used formulas are $[(\text{length} \times \text{width}^2)/2]$ and $[(\pi/6) \times d]$, where d is the mean tumor diameter (Plowman et al. 1997; Houghton et al. 2007). Endpoints, calculated from the tumor weights, provide a basis for assessing the effect of treatment on tumor growth. There is no single set of endpoints applied by all research groups, but common ones include: percent test/control tumor weights (%T/C) calculated at each tumor measurement time point, tumor growth delay, net log cell kill, median days to a specified tumor size, tumor regressions, tumor free animals, and statistical comparisons between the treated and control tumor weights (Plowman et al. 1997; Alley et al. 2004; Houghton et al. 2007). It is important to remember that these are not all clinically relevant endpoints, as human responses are defined in several ways that cannot be directly compared to preclinical testing outcomes (e.g., time to disease progression, objective response rates, surrogate markers, and quality of life parameters) (Flaherty and O'Dwyer 2004).

The use of transplanted models in efficacy testing has expanded to include the evaluation of the impact of treatment on cellular targets. For these types of studies, tumors or tumor biopsies are collected and analyzed for target expression following administration of one or more doses of the experimental agent. This allows the mechanism of action defined in laboratory studies to be confirmed in the complex tumor microenvironment. Furthermore, this approach allows assessment of the capacity for a molecularly targeted anticancer agent to reach and modulate the target(s) without having to demonstrate tumor growth inhibitory activity (Kinders et al. 2007). For example, Dudkin et al. (2001) demonstrated the capacity of the rapamycin analog CCI-779 to modulate its target, mTOR, by analyzing resected tumor tissue. In this case, they correlated the molecular effects of CCI-779 with the growth inhibitory activity in sensitive and insensitive xenograft models. More recently, this type of approach has been approved by the Food and

Drug Administration for use in human clinical trials, provided the test agent meets specific requirements. As of 2008, preclinical and clinical trials have not provided the critical mass necessary to evaluate the predictive value of preclinical models for this approach. However, it is worth noting that the Phase 0 trial of the poly-ADP ribose polymerase (PARP) inhibitor, ABT-888, demonstrated excellent correlation between target modulation in the preclinical models and that detected in the patients (Kinders et al. 2007).

The advantages of the subcutaneous xenograft model for drug evaluations include (1) progressive tumor growth that can be repeatedly assessed over the course of the experiment allowing for impact on tumor growth relative to the time of drug administration to be correlated; (2) xenografts provide a vascularized model to evaluate toxicity, antitumor activity, and compound bioavailability (Burchill 2006); (3) time required to conduct most xenograft studies is measured in weeks allowing for experimental time frames to be minimized and space for rodent housing does not become limiting; (4) large databases exist for comparing results obtained with experimental agents to those seen with clinically approved agents or to agents with a known mechanism of action; (5) mastery of the subcutaneous xenograft techniques is within the skills of most researchers (Bibby 2004); and (6) early comparisons of subcutaneous xenograft results with clinical outcomes suggest a positive correlation between activity in these models and activity in the clinic (Winograd et al. 1987; Giovanella et al. 1983; Schold and Bigner 1983; Osieka 1984).

While advantages exist with subcutaneous xenograft models, they also have shortcomings. An obvious concern is the absence of a normal immune system since drug evaluations occur in the absence of assistance from immune surveillance. A second well-described deficiency of the subcutaneous xenograft is its failure to reproduce the microenvironment of the primary growth site of common human cancers (Bibby 2004; Talmadge et al. 2007). For this reason, orthotopic (the grafting of cells/tissues into their tissue of origin) sites of implantation for use in preclinical studies were developed in the late 1980s (Naito et al. 1987) to more closely model human neoplasms (Talmadge et al. 2007; Bibby 2004). Certain prostate tumor cell lines (LnCap) actually require orthotopic implantation for tumor development since they are unable to reproducibly grow in the subcutaneous compartment (Stephenson et al. 1992). In addition, the sensitivity of xenograft tumors to a compound may be affected by the tumor's location (Killion et al. 1998). Some examples include the response of KM12 human colon xenograft tumors to doxorubicin, the response of pancreatic tumor xenografts to 17-DMAG, as well as the response of small cell lung tumor xenografts to cisplatin and mitomycin C (Kung 2007; Fidler et al. 1994; Borgel et al. 2003). In each of these cases, orthotopically implanted tumors are more sensitive than subcutaneous tumors.

Another concern with subcutaneous xenograft models is their failure to metastasize from the primary implant site to expected distant sites (e.g., lung, brain, bone marrow) as occurs in human patients (Bibby 2004). Models that accurately recapitulate the metastatic process are extremely beneficial for understanding the biological process as well as providing a means for assessing the antimetastatic potential of new compounds (Céspedes et al. 2006). Orthotopically implanted tumors are reported to provide better metastatic models than subcutaneously implanted tumors (Talmadge et al. 2007; Bibby 2004); however, many of the orthotopically transplanted tumors do not produce easily visualized metastatic lesions. The availability of highly sensitive imaging systems (e.g., bioluminescence, MRI, PET) provides an opportunity to identify microscopic disease in transplanted tumor models (see Chapter 11). This should improve detection since micrometastases from xenografted tumors can be more easily visualized with these technologies than with classical histological evaluations.

There is literature suggesting that because orthotopically implanted tumors are better able to recapitulate human tumor behavior (Talmadge et al. 2007; Bibby 2004), they should be routinely used for drug evaluations. However, it is important to note that (1) orthotopic models require technically skilled operators as the tumors generally require surgical implantation; (2) the endpoints are more difficult to monitor since the tumor is not apparent and thus must be visualized with imaging technologies (e.g., MRI, PET, ultrasound); (3) there is inadequate data correlating the outcome of drug testing in orthotopic models with human clinical trials; and (4) the technical challenges of these models often mean less mice are used per experimental group which can adversely affect the statistical power of the study (Teicher 2006).

Let us now consider the predictive value of the preclinical transplanted tumor models for human clinical antitumor activity. As early as 1981, data comparing the outcome of preclinical models to clinical responses suggested that compounds active in several preclinical models were more likely to be active in man than those active in only one or two models (Goldin et al. 1981). A comprehensive independent statistical review of the data generated by the NCI was reported by Staquet et al. (1983). While their goal was to develop a recommendation as to the best preclinical screening strategy, a second outcome of their studies was the demonstrable correlation between clinical activity and the number of preclinical models in which the agent was active. For example, 12 of 15 compounds that were active in four or more preclinical models were active in at least one human clinical trial. However, only 21 of 45 compounds found active in less than four preclinical models were active clinically (Staquet et al. 1983). More recently, Johnson et al. (2001) reported a similar conclusion based upon clinical results obtained with 39 compounds for which preclinical drug testing data was available. Voskoglou-Nomikos et al. (2003) reported the results of a retrospective study of NCI-generated data and their conclusion was that human tumor xenografts are superior to murine allografts for predicting clinical activity. While one can argue the individual value of a particular xenograft study, the data analyzed by several independent groups over the course of many years suggests that the greater the percentage of preclinical models that are positive, the greater the chance of clinical activity. This is logical if we consider that each xenograft or allograft, even though it is conducted with hundreds of mice, actually represents a single tumor. So, the presence of activity in four of eight xenograft models is no different than a clinical study in which four of eight patients respond to the therapeutic agent. Therefore, if clinical activity is defined by a response in at least 20% of patients, then the preclinical correlate would require evaluation of at least ten tumor models with activity seen in at least two.

The data suggests that there is a positive relationship between the preclinical and clinical outcomes. However, this requires diligence in preclinical modeling. Oftentimes, inadequate study designs and the subsequent overinterpretation of the resulting data increase the ambiguity surrounding the use of mouse models because of false positive results. Prior to initiating and interpreting any preclinical animal experiment, the following should be incorporated into the study design (1) review of tumor and compound sensitivity data; (2) select the proper model expressing the purported drug target; (3) define the control animals; (4) select appropriate animal group sizes to properly power the experiment; (5) define an effective drug vehicle; (6) develop and implement a randomization protocol; (7) design dosing schedules that are justified based upon the mechanism proposed for the therapeutic being assessed; and (8) define endpoints before initiating the experiment (Hollingshead 2008). In conjunction with these considerations, it is imperative that the resulting data cannot be overinterpreted as this leads to the referral of compounds for clinical trial that have little possibility for achieving success.

22.3 Transgenic Mouse Tumor Models in Preclinical Drug Testing

While multiple publications demonstrate that xenograft tumors used in preclinical drug development have predictive value, there are many who conclude that these models fail to accurately predict clinical activity. The growing sophistication of transgenic mouse technologies has led proponents to hypothesize that these mice will be more predictive of clinical anticancer activity than traditional xeno- and allogeneic transplant models (Weiss and Shannon 2004). Many of the tumors occurring in genetically engineered mice (GEM) recapitulate the characteristics observed in human tumors, including progression from benign hyperplastic lesions into aggressive tumors which mimic human histopathology. Currently, hundreds of GEM models predisposed to tumor development exist (Mouse Models of Human Cancer Consortium 2008). GEM models offer the advantages of known genetic alterations in an immunocompetent host that can be targeted by drug therapies directed at that target in a host that has genetically matched normal and tumor tissues so differential toxicity can be accurately assessed. On the downside, many of the GEM models do not develop metastatic tumors, yet metastatic disease is the leading cause of human cancer deaths. Secondly, some models require crossing of multiple GEM backgrounds to induce the desired tumor occurrence. This can be costly due to the numbers of animals required to generate adequate experimental numbers and the resulting space constraints. Furthermore, many of the germ line models have variable penetrance and lengthy tumor latencies which make testing drugs complicated since study animals with varying tumor burdens are difficult to randomize or standardize. To minimize the impact of these issues on the early preclinical drug screening application of GEM models, Varticovski et al. (2007) applied tumor transplant methods to tumors harvested from GEM models. This allows traditional drug testing paradigms to be used while taking advantage of the unique features of the GEM models, including growth in immunocompetent mice and a known genetic perturbation causing the neoplasia. In addition, this offers the opportunity to evaluate new therapeutics in transplanted GEM tumors as a screening tool and then to apply the positive compounds to the original GEM mouse, where issues of drug scheduling, drug resistance, and multidrug cycling can be explored.

22.4 General Difficulties with Preclinical Mouse Models in Drug Testing Programs

An ideal preclinical mouse model would have a single primary tumor arising in the adult mouse that recapitulates the histology and gene expression profile of a specific human cancer subtype. This tumor could be followed through the stages of tumor development and metastasis, and it would respond to therapeutics in a manner that mimics the human clinical experience. However, there are limitations in extrapolating data to humans due to interspecies differences. For example, cancer develops over years and decades in humans, but only in weeks to months in mice. Furthermore, drugs developed against cancers in mice do so in a controlled environment, and they may not behave the same in the human physiology.

It can take in excess of 10 years and more than 500 million dollars in resources for a compound to make it from concept to Phase III clinical trials (Kelland 2004; DiMasi et al. 2003). These costs coupled with the failure in human clinical trials of many promising anticancer compounds (Kerbel 2003; Sharpless and DePinho 2006), insures that the search for innovative means to identify effective therapeutic compounds remains a priority. While mouse models have been and continue to be extensively used in drug screening efforts, general issues associated with their use remain. For example, how do we extrapolate mouse data to the human patient, what preclinical experimental design best mimics that which is possible in humans, and how should the preclinical data be interpreted to avoid false negatives while reducing false positives? Whether it is establishing a maximum tolerated dose (for which the mouse MTD often exceeds the human MTD), selecting an appropriate vehicle, or developing a therapeutic dosing schedule, there are distinct disparities between the mouse and the human that limit our ability to clearly predict the efficacy and therapeutic index of a compound in humans (Sausville and Hollingshead 2004; Kerbel 2003).

Ensuring that data is reproducible, statistically significant, and the resulting conclusions justified will alleviate many of the issues resulting from negative results in human clinical trials. The important contributions of mouse models to preclinical testing programs cannot be overstated. Nevertheless, improved strategies and models are needed to enhance identification and to predict the efficacy and potential toxicity of anticancer compounds. While it is currently unclear what role each will play as the technologies evolve, it is promising that advancing technologies will continue to serve as a catalyst to integrate more predictive, clinically relevant, and targeted technologies into preclinical screening programs to provide the millions of people diagnosed with cancer with safe, effective therapeutic treatments.

References

Alley MC, Hollingshead MG et al (2004) Human tumor xenograft models in NCI drug development. In: Teicher BA, Andrews PA (eds) Anticancer drug development guide: preclinical screening, clinical trials, and approval, 2nd edn. Humana, Totowa, NJ

American Cancer Society (2008) Cancer facts and figures. http://www.cancer.org. Accessed 25 Aug 2008

Barrett MK (1940) The influence of genetic constitution upon the induction of resistance to transplantable mouse tumors. J Natl Cancer Inst 1:387–393

Beck JA, Lloyd S et al (2000) Genealogies of mouse inbred strains. Nat Genet 24:23–25

Bibby MC (2004) Orthotopic models of cancer for preclinical drug evaluation: advantages and disadvantages. Eur J Cancer 40:852–857

Biological Testing Branch (2008) http://dtp.nci.nih.gov/branches/btb/btb_index.html. Accessed 25 Aug 2008

Bittner JJ (1934) Linkage in transplantable tumours. J Genet 29(1):17–27

Borgel SD, Carter JP et al (2003) The impact of tumor location on the activity of 17-DMAG (NSC 707545), a water soluble geldanamycin analog. Clin Cancer Res 9(16):6215S

Bosma GC, Custer RP, Bosma MJ (1983) A severe combined immunodeficiency mutation in the mouse. Nature 301(5900):527–530

Brown JM (1997) NCI's anticancer drug screening program may not be selecting for clinically active compounds. Oncol Res 9:213–215

Burchill SA (2006) What do, can and should we learn from models to evaluate potential anticancer agents? Future Med 2(2):201–211

Carruthers C, Suntzeff V (1944) Chemical studies on the transformation of mouse epidermis by methylcholanthrene to squamous cell carcinoma. J Biol Chem 155(2):459–464

Céspedes MV, Casanova I et al (2006) Mouse models in oncogenesis and cancer therapy. Clin Transl Oncol 8(5):318–329

Cloudman AM (1946) A study of the organophilic tendencies of 2 transplantable mouse tumors. Cancer Res 6(9):503

Corbett TH, Valeriote FA, Baker LH (1987) Is the P388 murine tumor model no longer adequate as a drug discovery model? Invest New Drugs 5:3–20

Croy BA, Chapeau C (1990) Evaluation of the pregnancy immunotrophism hypothesis by assessment of the reproductive performance of young adult mice of genotype scid/scid.bg/bg. J Reprod Fertil 88(1):231–239

DiMasi JA, Hansen RW, Grabowski HG (2003) The price of innovation: New estimates of drug development costs. J Health Econ 22(20):151–185

Driscoll JS (1984) The preclinical new drug research program of the National Cancer Institute. Cancer Treat Rep 68(1)

Dudkin L, Dilling MB et al (2001) Biochemical correlates of mTOR inhibition by the rapamycin ester CCI-779 and tumor growth inhibition. Clin Cancer Res 7(6):1758–1764

Eaton GJ, Outzen HC et al (1975) Husbandry of the "Nude" mouse in conventional and germ-free environments. Lab Anim Sci 5:309–314

Fancy Mouse Breeders Club (2008) http://www.nationalmouseclub.co.uk/history.html. Accessed 25 Aug 2008

Fidler IJ, Wilmanns C et al (1994) Modulation of tumor cell response to chemotherapy by the organ environment. Cancer Metastasis Rev 13(2):209–222

Flaherty KT, O'Dwyer PJ (2004) Conventional design and novel strategies in the era of targeted therapies. In: Teicher BA, Andrews PA (eds) Anticancer drug development guide: preclinical screening, clinical trials, and approval, 2nd edn. Humana, Totowa, NJ

Flanagan SP (1966) 'Nude', a new hairless gene with pleiotropic effects in the mouse. Genet Res 8:295–309

Frei E III (1982) The national cancer chemotherapy program. Science 217:600–606

Gershwin ME (1977) DiGeorge syndrome: congenital thymic hypoplasia. Animal model: congenitally athymic (nude) mouse. Am J Pathol 89(3):809–812

Giovanella BC, Stehlin JS Jr et al (1983) Correlation between response to chemotherapy of human tumors in patients and in nude mice. Cancer 52(7):1146–1152
Goldin A, Venditti JM, Carter SK (1977) Screening at the National Cancer Institute. Natl Cancer Inst Monogr 45:37–48
Goldin A, Venditti JM et al (1981) Current results of the screening program at the Division of Cancer Treatment, National Cancer Institute. Eur J Cancer 17:129–142
Greene HSN, Murphy ED (1945) The heterologous transplantation of mouse and rat tumors. Cancer Res 5(5):269–282
Guflino PM, Ediger RD et al (1976) Guide for the care and use of the nude (thymus deficient) mouse in biomedical research. ILAR News 19:M3–M20
Hammett FS (1936) Effect of cystine disulfoxide on spontaneous tumors of the mouse. Science 83:108–109
Hollingshead MG (2008) Antitumor efficacy testing in rodents. J Natl Cancer Inst 100(21):1500–1510
Houghton PJ, Morton CL et al (2007) The pediatric preclinical testing program: description of models and early testing results. Pediatr Blood Cancer 49:928–940
Johnson JI, Decker S et al (2001) Relationships between drug activity in NCI preclinical in vitro and in vivo models and early clinical trials. Br J Cancer 84(10):1424–1431
Kamel-Reid BA, Dick JE (1988) Engraftment of immune-deficient mice with human hematopoietic stem cells. Science 242(4886):1706–1709
Kelland LR (2004) "Of mice and men:" values and liabilities of the athymic nude mouse model in anticancer drug development. Eur J Cancer 40:827–836
Kerbel RS (2003) Human tumor xenografts as predictive preclinical models for anticancer drug activity in humans: getter than commonly perceived – but they can be improved. Cancer Biol Ther 2(4):S134–S139
Killion JJ, Radinsky R, Fidler IJ (1998) Orthotopic models are necessary to predict therapy of transplantable tumors in mice. Cancer Metastasis Rev 17(3):279–284
Kinders R, Parchment RE et al (2007) Phase 0 clinical trials in cancer drug development: from FDA guidance to clinical practice. Mol Interv 7(6):325–334
Kung AL (2007) Practices and pitfalls of mouse cancer models in drug discovery. Adv Cancer Res 96:191–212
Liu CN, Lambert JM et al (1996) Cure of multidrug-resistant human B-cell lymphoma xenografts by combinations of anti-B4-blocked ricin and chemotherapeutic drugs. Blood 87(9):3892–3898
Lozzio BB, Lozzi CB, Machado E (1976) Human myelogenous (Ph+) leukemia cell line: transplantation into athymic mice. J Natl Cancer Inst 56(3):627–629
Ludford RJ (1933) Differences in the growth of transplantable tumours in plasma and serum culture media. Proc R Soc Lond B 112(776):250–263
Manning DD, Reed ND, Shaffer CF (1973) Maintenance of skin xenografts of widely divergent phylogenetic origin on congenitally athymic (nude) mice. J Exp Med 138(2):488–494
McCampbell EF (1909) Malignant tumors in mice with a report of a spontaneous adeno-carcinoma in a house mouse (mus musculus). J Med Res 20(113):261–273
Monks A, Scudiero D et al (1991) Feasibility of a high-flux anticancer drug screen using a diverse panel of cultured human tumor cell lines. J Natl Cancer Inst 83(11):757–766
Mouse Models of Human Cancer Consortium (2008) http://emice.nci.nih.gov. Accessed 25 Aug 2008
Murphy JB, Maisin J, Sturm E (1923) Local resistance to spontaneous mouse cancer induced by X-rays. J Exp Med 38:45–653
Naito S, Giavazzi R et al (1987) Growth and Metastatic behavior of human tumor cells implanted into nude and beige nude mice. Clin Exp Metastasis 5(2):135–146
Nettleship A (1943) Study of a spontaneous mouse rhabdomyosarcoma. J Natl Cancer Inst 3(5):563–568
Osieka R (1984) Primary and acquired resistance to antineoplastic chemotherapy. A preclinical and clinical study. Cancer 54(6):1168–1174
Osieka R, Houchens DP et al (1977) Chemotherapy of human colon cancer xenografts in athymic nude mice. Cancer 40(5):2640–2650

Plowman J, Dykes DJ et al (1997) Human tumor xenograft models in NCI drug development. In: Teicher BA (ed) Anticancer drug development guide: preclinical screening, clinical trials, and approval, 1st edn. Humana, Totowa, NJ
Povlsen CO, Visfeldt J et al (1975) Growth patterns and chromosome constitutions of human malignant tumours after long-term serial transplantation in nude mice. Acta Pathol Microbiol Scand 83(6):709–716
Prochazka M, Gaskins HR et al (1992) The nonobese diabetic scid mouse: model for spontaneous thymomagenesis associated with immunodeficiency. Proc Natl Acad Sci USA 89(8):3290–3294
Rygaard J (1969) Immunobiology of the mouse mutant "Nude". Preliminary investigations. Acta Pathol Microbiol Scand 77(4):761–762
Rygaard J, Povlsen CO (1969) Heterotransplantation of a human malignant tumour to "Nude" mice. Acta Pathol Microbiol Scand 77(4):758–760
Sausville EA, Hollingshead MG (2004) Mouse models in cancer drug discovery and development. In: Figg WD, McLeod HL (eds) Handbook of anticancer pharmacokinetics and pharmacodynamics. Humana, Totowa, NJ
Schein PS, Scheffler B (2006) Barriers to efficient development of cancer therapeutics. Clin Cancer Res 12:3243–3248
Schepartz SA (1977) Antitumor screening procedures of the National Cancer Institute. Jpn J Antibiot 30:35–40
Schmidtwolf IGH, Negrin RS et al (1991) Use of a SCID mouse human lymphoma model to evaluate cytokine-induced killer-cells with potent antitumor cell-activity. J Exp Med 174(1):139–149
Schold SC Jr, Bigner DD (1983) Treatment of five subcutaneous human glioma tumor lines in athymic mice with carmustine, procarbazine, and mithramycin. Cancer Treat Rep 67(9):811–819
Schuh JC (2004) Trials, tribulations, and trends in tumor modeling in mice. Toxicol Pathol 32:53–66
Sharpless NE, Depinho RA (2006) The mighty mouse: genetically engineered mouse models in cancer drug development. Nat Rev Drug Discov 5(9):741–754
Shimosato Y, Kameya T et al (1976) Transplantation of human tumors in nude mice. J Natl Cancer Inst 56(6):1251–1260
Shultz LD, Lang PA et al (2000) NOD/LtSz-Rag1null mice: an immunodeficient and radioresistant model for engraftment of human hematolymphoid cells, HIV infection, and adoptive transfer of NOD mouse diabetogenic T cells. J Immunol 164(5):2496–2507
Singh SK, Hawkins C et al (2004) Identification of human brain tumour initiating cells. Nature 432(7015):396–401
Slye M (1915) The incidence and inheritability of spontaneous cancer in mice (third report). J Med Res 32(1):159–200
Staats J (1966) The laboratory mouse. In: Green EL (ed) Biology of the laboratory mouse. McGraw-Hill, New York
Staquet MJ, Byar DP et al (1983) Clinical predictivity of transplantable tumor system in the selection of new drugs for solid tumors: rationale for a three-stage strategy. Cancer Treat Rep 67(9):753–765
Stephenson RA, Dinney CP et al (1992) Metastatic model for human prostate cancer using orthotopic implantation in nude mice. J Natl Cancer Inst 84(12):951–957
Stinson SF, Alley MC et al (1992) Morphological and immunocytochemical characteristics of human tumor cell lines for use in a disease-oriented anticancer drug screen. Anticancer Res 12(4):1035–1053
Strong LC (1927) Studies on the effect of potassium alum-hydrochloric acid solutions on the growth and fate of neoplastic tissue. 1. Effect on a slow growing adeno carcinoma of the mouse. Proc Natl Acad Sci USA 13:141–145
Suggitt M, Bibby MC (2005) 50 years of preclinical anticancer drug screening: empirical to target-driven approaches. Clin Cancer Res 11:971–981
Talmadge JE, Singh RK et al (2007) Murine models to evaluate novel and conventional therapeutic strategies for cancer. Am J Pathol 170(3):793–804
Teicher BA (2006) Tumor models for efficacy determination. Mol Cancer Ther 5(10):2435–2443

Varticovski L, Hollingshead MG et al (2007) Accelerated preclinical testing using transplanted tumors from genetically engineered mouse breast cancer models. Clin Cancer Res 13(7):2168–2177

Venditti JM (1981) Preclinical drug development: rationale and methods. Semin Oncol 8:349–361

Voskoglou-Nomikos T, Pater JL, Seymour L (2003) Clinical predictive value of the in vitro cell line, human xenograft and mouse allograft preclinical cancer models. Clin Cancer Res 9:4227–4239

Weiss B, Shannon K (2004) Preclinical trials in mouse cancer models. In: Holland EC (ed) Mouse models of human cancer. Wiley, Hoboken, NJ

Winograd B, Boven E et al (1987) Human tumor xenografts in the nude mouse and their value as test models in anticancer drug development (review). In Vivo 1(1):1–13

Chapter 23
The Development and Use of Genetically Tractable Preclinical Mouse Models

Michael T. Hemann

23.1 Background

Until recently, "preclinical mouse models" invariably referred to mouse xenograft studies (Sharpless and Depinho 2006). Even today, the majority of mouse-based cancer therapy studies performed in either academia or industry use xenograft approaches to assess therapeutic efficacy. In these studies, large numbers of cultured human tumor cells or primary human tumor material are injected into immunocompromised recipient mice. Tumors grown at nonnative sites, most frequently subcutaneously in the rear flank, are then monitored for growth kinetics following drug administration. The disseminated use of these models can be roughly distilled to two major advantages. First, human tumor material, as opposed to mouse tumors, can be treated. Second, these tumors can be rapidly generated and readily monitored over the course of therapy. These models have had significant value in cancer drug development, most appreciably in the examination of drug pharmacokinetics and pharmacodynamics (Johnson et al. 2001a). They have also been effectively used in lead compound optimization, identifying drug variants among a set with the highest efficacy and lowest toxicity (Besterman 1996; Traxler et al. 2001). However, in identifying effective anticancer therapies that translate to clinical success, xenograft models have been an unmitigated failure. Notable examples of the dubious history of these models include studies on farnesyltransferase inhibitors (Gibbs and Oliff 1997; Macdonald et al. 2005), peroxisome proliferator-activated receptor-γ agonists (Sarraf et al. 1998; Kulke et al. 2002) and angiogenesis inhibitors (Boehm et al. 1997; Twombly 2002), all of which showed potent anticancer activity in xenograft models but failed to produce significant clinical efficacy as single agents.

M.T. Hemann (✉)
The Koch Institute for Integrative Cancer Research at MIT,
700 Main Street, Cambridge, MA 02139, USA
e-mail: hemann@mit.edu

J.E. Green and T. Ried (eds.), *Genetically Engineered Mice for Cancer Research: Design, Analysis, Pathways, Validation and Pre-Clinical Testing*,
DOI 10.1007/978-0-387-69805-2_23,

The failure of xenograft models can be attributed to a variety of factors. First, tumor xenografts fail to recapitulate the normal tumor microenvironment. Primary tumors exist as complex mixtures of tumor cells, neighboring normal tissue, stromal cells, supporting vasculature and infiltrating immune cells, all of which may provide crucial pro-survival signals that counteract the effect of cancer therapies. Second, xenografts are primarily derived from cultured tumor cell lines. Such cell lines are inefficiently generated from primary tumors and successful growth in culture likely requires the acquisition of additional mutations. For example, the incidence of p53 mutation in primary lymphomas ranges from 15 to 20% (Klumb et al. 2003), but 60–80% of Burkitt's cell lines express mutant p53 (Cherney et al. 1997). Additionally, tumor growth following subcutaneous injection may require additional adaptation. Thus, the tumor that arises in a xenograft model may be profoundly different than its antecedent human tumor. Finally, xenografts are derived from clonal outgrowths of human tumors. Consequently, resulting xenografts consist of homogenous mixtures of cells that may lack the developmental precursors or cancer stem cells that have been postulated to underlie long-term drug resistance.

The development of genetically engineered mouse models (GEMMs) over the last two decades has yielded preclinical systems that lack the myriad deficiencies of xenograft models. Notably, tumors can be treated in their native microenvironment and developmental context. However, the use of such models has been hampered by some of the same limitations as human clinical trials, including long tumor latency and high tumor variability. This chapter describes recent adaptations in the development and use of GEMMs that have fundamentally advanced our ability to model human chemotherapeutic response in mouse preclinical systems. A particular focus is placed on the adaptation of tractable genetic systems to mouse models, systems that, for the first time, allow forward loss-of-function genetic experiments to be performed in mammalian systems.

23.2 What Makes a Good Therapy Model?

If xenograft models represent "poor" preclinical systems, what represents a "good" model for evaluating therapeutic response? There may not be a single answer to this question, as researchers may have disparate objectives for preclinical systems. For example, while most researchers use preclinical models to evaluate drug efficacy, these systems can also be engineered to evaluate mechanisms of drug resistance and general determinants of therapeutic response. Additionally, tractable genetic models can be used to identify strategies to sensitize tumors to existing therapies.

Nevertheless, there are certain common requirements for effective preclinical systems. First, tumor generation needs to be rapid. Tumors that arise over the course of years preclude the examination of complex tumor genotypes or the use of multiple treatment regimens, and, consequently, they encounter problems associated with conventional clinical problems. Second, tumorigenesis should occur in a physiologically relevant microenvironment. After all, the essential purpose of a preclinical model is to provide a therapeutic context that better approximates the human clinical setting. Lastly, therapeutic response should be readily observable. This final issue is far from trivial, as different tumor types require distinct imaging modalities. While certain tumor types can be monitored by visual inspection, palpation, or simple fluorescence or bioluminescence imaging, other tumors require imaging equipment currently in human clinical use. These include ultrasound and microCT systems, as well as functional imaging equipment like PET or MRI systems (reviewed in Beckmann et al. 2007) (see Chapter 11). Even when available, these systems may not represent experimentally feasible approaches to monitoring therapeutic response, given the extensive time required to monitor individual animals. Thus, a "good" preclinical model is one in which tumors form rapidly in the appropriate microenvironment in a manner that is readily observable.

Other standards for "good" preclinical models are subject to debate and their importance can vary with the specific objectives of the experimental system. One common objective for a mouse cancer model is that the mouse tumor closely recapitulates the pathology of its human counterpart. This includes not only the end stage of the primary tumor, but also various stages of neoplastic progression and metastasis – allowing one to investigate the effect of cancer stage on therapeutic response. Another emerging standard for "good" preclinical models is that tumor-initiating lesions are induced in adult, as opposed to developing, tissues. Additionally, tumors should arise from genetic lesions present in a subset of cells, as opposed to an entire organ system or the entire mouse. In other words, cancer-promoting mutations should appear as they do during human tumor progression. Finally, given the enormous complexity of cancer genotypes, there is increasing desire for preclinical models that can rapidly accommodate the introduction of diverse genetic lesions. The mouse models described below all fit some, but not all, of these criteria for "good" preclinical systems (Fig. 23.1). Importantly, effective preclinical systems will likely be as varied as the cancers they attempt to model, and it may require the combined use of several distinct systems to effectively predict human therapeutic response.

23.3 Directed Oncogene Expression and Therapy

While gene knockout studies have taught us a great deal about tumor progression, the models that have best defined the potential of GEMMs as preclinical systems are those in which activated oncogenes are expressed under the control of tissue-specific promoters. Several of these models, including the *Eμ-myc* mouse (Adams et al. 1985) (which will be discussed later in this chapter), the Rip-Tag mouse (Hanahan 1985), and mice expressing MMTV-directed oncogenes (Bouchard et al. 1989), were generated over three decades ago. Only in the last decade have these models been used as the basis for preclinical studies.

The great value of these cancer models as preclinical systems arises from several key characteristics. First, tumors appear with complete penetrance. Second,

Genetically Tractable GEMMs

GEMM	Directed Oncogene	Chimeric	Inducible	Transplant	Screening
Example	Rip-Tag	Latent *K-ras*^{G12D}	Reversible RNAi in Liver	*iBMK*	Transplant /RNAi Screening
Adv.	Accurately model disease progression	Models somatic induction of mutations	Can evaluate putative drug targets	Complex genetic alterations can be examined	Forward genetic screens can be performed
Disadv.	Lesions present in all cells in a given organ	Limited to small numbers of alterations	Limited to small numbers of alterations	Tumors are not in their native developmental context	Transplants are not autochthonous tumors

Fig. 23.1 A diagram showing the spectrum of tractable mouse models in current preclinical use. No single model is appropriate for all applications, so the ideal preclinical program would incorporate several of these models

tumor development occurs via a defined set of premalignant and early neoplastic stages. Consequently, tumor development can be temporally linked to defined premalignant stages, and therapy can be applied at various stages during malignant growth – a process that can be extrapolated to early intervention in human cancers. Finally, these tumors arise in autochthonous locations and can, therefore, faithfully recapitulate critical components of tumor-microenvironment and tumor–drug interactions.

A classic "directed oncogene" mouse model is the transgenic mouse model for prostate and mammary cancer or TRAMP mouse (Greenberg et al. 1995). This mouse expresses the simian virus 40 early region transforming sequence (SV40 T antigen), functionally inactivating the tumor suppressors p53 and Rb, under the regulatory control of the rat prostatic steroid binding protein promotor sequence. Male transgenic mice develop prostatic intraepithelial neoplasia (PIN) early in life that progresses to adenocarcinoma at 6–7 months, with occasional accompanying metastasis. These mice have been the subject of innumerable preclinical experiments, ranging from the use of genotoxic chemotherapeutics to angiogenesis inhibitors to hormone-based therapies and/or physical castration (Kasper and Smith 2004; Klein 2005). Additionally, the TRAMP model has been a mouse model of choice for cancer immunotherapy experiments, due to the continued interest in using prostate-specific antigens to direct a cancer-specific T cell repertoire (Foster et al. 1997). More recently, additional prostate models have been developed and used as preclinical systems. These include tissue-specific Pten and p53-deficient mice (Chen et al. 2005), which faithfully recapitulate all of the human premaligant stages of prostate cancer, as well as tissue-specific Pten and Nkx3.11-deficient mice (Abate-Shen et al. 2003; Kinkade et al. 2008). Recent studies in the latter mice have identified possible synergy between inhibition of the mTOR and map kinase pathways in treating prostate cancer.

Another well-established "directed oncogene" model driven by the expression of SV40 T antigen is the Rip-Tag model for pancreatic neuroendocrine ('Islet cell') tumorigenesis (Hanahan 1985). In this model, SV40 Large T antigen is expressed under the control of the rat insulin promotor and is, consequently, expressed exclusively in the beta cells of the endocrine pancreas in transgenic mice. In transgenic mice, approximately 70% of pancreatic islet cells become hyperplastic and 4–10% of these cells undergo an "angiogenic switch." Because tumor neovascularization occurs in all transgenic mice at a defined age, the Rip-Tag mouse has been the mouse model of choice to examine the efficacy angiogenesis inhibitors. Not surprisingly, these drugs, which include AGM-1470, angiostatin, BB-94, and endostatin, show best efficacy when administered concurrently with the angiogenic switch (Bergers et al. 1999; Bergers and Hanahan 2002). Additionally, because many of these tumors adapt to the presence of angiogenesis inhibitors by becoming more invasive, this model has also been employed to examine inhibitors of tumor microenvironment remodeling proteins, like matrix metalloproteinases (Bergers et al. 2000). Thus, this work highlights the ability of GEMMs to serve as powerful preclinical systems for examining drug efficacy during specific stages of tumor development. Additionally, therapy experiments in this model have begun to identify mechanisms by which tumors evade the action of anti-angiogenic drugs (Bergers and Hanahan 2008). Of note, neither the TRAMP nor RIP-TAG malignancies are "curable" following treatment with targeted are conventional chemotherapeutics. This will represent a common theme among nearly all of the autochthonous tumor models discussed in this chapter – a property that decisively distinguishes these systems from xenograft models.

One drawback of the TRAMP and RIP-TAG models as preclinical systems is that they express an artificial oncogene. SV40 Large T antigen is obviously not expressed in human cancers and the relevance of this lesion to tumor development driven by actual mouse or human oncogenes remains unclear. Additionally, while the loss of *Rb* and *p53* are both critical alterations in human cancer, the inactivation of both of these tumor suppressor genes (with the possible exception of HPV-mediated inactivation of Rb and p53) is neither an initiating nor a universal lesion in human cancer. Thus, therapy experiments in these models invariably occur on a genetic background, that of *p53*-deficiency, known to impact therapeutic response.

In contrast to the TRAMP and RIP-TAG models, a mouse model that faithfully recapitulates the lesions seen in its human counterpart is the *K-ras;p53*$^{-/-}$ model of pancreatic ductal adenocarcinoma (Hingorani et al. 2003). In this mouse, activated mutant *K-ras(G12D)* expression is repressed in most tissues due to the presence of a "floxed" stop cassette in the *K-ras* transgene. Pancreas-specific expression results from the expression of Cre under the control of the tissue-specific *pdx-1* promotor. Directed *K-ras* expression in a *p53*-deficient background results in ductal lesions similar to those seen in pancreatic intraepithelial neoplasias – precursor lesions to pancreatic cancer. Some of these lesions spontaneously progress to invasive and metastatic adenocarcinomas. Importantly, this model has been used extensively as a preclinical system, ranging from the development of ultrasound imaging of developing tumors to the analysis of tumor biomarkers before and subsequent to therapy (Olive and Tuveson 2006). Additionally, this model has been used to study the action and efficacy of gemcitabine, a frontline therapy for human pancreatic adenocarcinoma (Cook et al. 2008).

"Directed oncogene" models in the hematopoietic system also generally express initiating lesions that precisely recapitulate established alterations in human leukemias. A classic example of the use of these models as preclinical systems are mice engineered to express fusion proteins characteristic of acute promyelocytic leukemia (APL) (He et al. 2001). APL fusion proteins almost universally involve the retinoic acid receptor α (*RARα*) gene, and the vast majority of APLs fuse the *RARα* gene with *PML*, although a small subset of these leukemias involve other fusion partners. Mice expressing *PML-RARα* under the control of a *Cathepsin G* promotor develop APL at a 10–30% penetrance, and like their human counterpart, these leukemias are highly sensitive to retinoic acid. Conversely, mice expressing another APL fusion protein, PLZF-RARα, develop leukemias at complete penetrance that are resistant to retinoic acid. Interestingly, these tumors can be sensitized to the effects of retinoic acid through the use histone deacetylase inhibitors. Thus, by accurately modeling the precise lesions present in a give malignancy, these mice can serve as preclinical models to predict drug efficacy as well as systems to explore the potential of combination therapeutic regimens to overcome intrinsic drug resistance.

23.4 Chimeric Systems

A potential limitation in extrapolating results from preclinical experiments in GEMMs to human cancer is the extent to which the stochastic nature of human tumor development is faithfully recapitulated in the mouse. Specifically, aside from inherited cancer predisposition syndromes, tumor development in humans generally initiates from mutations in a small set of cells on a genetically unperturbed background. Conversely, cancer-prone mice expressing deregulated oncogenes or lacking tumor suppressor genes generally carry these alterations either in all cells of the animal or every cell of a given type. This distinction raises several potential concerns: First, mouse tumors are frequently polyclonal, arising independently from multiple progenitor cells. Thus, effective cancer therapies need to be able to target a "mixed" tumor that comprises several cell populations at various stages of tumor development. Second, if all tissue stem cells carry the cancer-predisposing alteration, de novo tumor development can presumably occur at any point – even subsequent to cancer treatment. Thus, it would be difficult to differentiate between tumor relapse and new tumor development. Finally, tumor therapy in GEMMs frequently occurs on a background in which the predisposing lesion is present throughout a given organ system. Thus, drug–microenvironment interactions may impact the overall response of a given tumor or mouse to a drug.

Numerous approaches have been taken to produce "chimeric" preclinical mouse models, in which tumor development occurs on an essentially normal background. One notable example is the use of retroviral infection in combination with adoptive transfer in the murine hematopoietic system (Pear et al. 1998; Schmitt et al. 2000). Using this approach, hematopoietic stem cells (HSCs) are infected with a cancer-promoting oncogene and used to reconstitute the hematopoietic system of a lethally irradiated recipient mouse. If the infection efficiency is kept low, then the majority of the transferred HSCs will remain wild type. Thus, tumor development and treatment in the recipient mouse occurs in the presence of an essentially normal hematopoietic system. Importantly, this approach can also be used to rapidly interrogate the impact of a series of oncogene variants on therapeutic outcome. However, tumor development may not be clonal in this context. For example, potent oncogenes like *Bcr/Abl*, which promote leukemia development at high efficiency and short latency, produce polyclonal malignancies in recipient mice even when transduced at low levels in HSC cultures (Pear et al. 1998).

The production of clonal, chimeric malignancies has proven more feasible in several well-established models of epithelial cancer. For example, two mouse models of Ras-induced lung adenocarcinoma generated in the Jacks lab develop clonal malignancies on an otherwise normal background (Jackson et al. 2001; Johnson et al. 2001b). In one model, a "latent" *K-ras* allele undergoes spontaneous recombination to activate mutant *ras* and promote tumor development. In another model, adenovirally encoded Cre recombinase, delivered through the nasal cavity into the lungs, promotes the formation of an active *K-ras* mutant allele that drives tumorigenesis. While multiple tumors are formed in the lungs of each of these mice following K-*ras* activation, individual tumor clones can still be distinguished. This is particularly important in the context of therapy, in that it allows one to track the response of single tumors following therapy – a process that facilitates comparison with comparable human data and highlights clonal variability in drug response between tumors. Interestingly, recent therapy experiments have shown that the tumor response to conventional chemotherapeutics in both of these models parallels the dismal response seen in human lung adenocarcinoma (T. Oliver and T. Jacks, personal communication). Specifically, tumor regression is seen following cisplatin treatment, but treated tumors rapidly relapse and grow more aggressively than control tumors – resulting in no change in overall survival. Here, again, an autochthonous tumor model recapitulates the profound therapeutic resistance of its human counterpart.

Chimeric models are essential for the analysis of tumor development and therapeutic response in certain systems. For example, disseminated expression of a deregulated oncogene in the brain would be embryonic lethal. One strategy used to produce chimeric lesions in the brain has been to express and avian viral receptor (TVA) under the control of a variety of tissue-specific promoters, including a brain-specific promotor (Holland et al. 1998; Holland and Varmus 1998). Delivery of the cancer-promoting oncogene is then mediated by local delivery of an avian retrovirus (RCAS) expressing the gene of interest. Here, only a small subset of proliferating cells is infected, resulting in tumor development amid an essentially normal surrounding brain. Tumors produced by RCAS-mediated delivery of activated *Akt* and *K-ras* have been used to test the efficacy of an mTOR inhibitor, CCI-799, as a glioblastoma chemotherapy (Hu et al. 2005).

The use of chimeric models as preclinical systems is particularly important in modeling the therapeutic response of sporadic malignancies. However, in modeling the treatment of familial cancers, the predisposing cancer alteration may constitute a critical component of the overall therapeutic response – even when that alteration is present in nontumor tissue. A particularly striking example of the relevance of mutations in support tissue to therapeutic response comes from a model of neurofibromatosis type 1 (Yang et al. 2008). In this model, the development of plexiform neurofibromas from Schwann cells is dependent upon the complete absence of NF1 in Schwann cells and *NF1* heterozygosity in surrounding mast cells. Interestingly, the use of Imatinib, a c-kit inhibitor, shows significant efficacy in these tumors, and this efficacy is due to the action of Imatinib on surrounding *NF1* heterozygous mast cells, rather than the neurofibroma itself. Thus, effective preclinical models need to incorporate the appropriate stromal as well as tumor genotype.

23.5 Inducible Systems and Cancer Therapy

The preclinical models described thus far in this chapter examine the use of conventional or targeted therapeutics on genetically engineered tumors. However, one of the great recent advances in molecular biology is the ability to use genetic tools to recapitulate the putative effects of drug target inhibition in diverse cellular contexts. This approach allows one to establish the efficacy of inhibiting a specific protein or pathway prior to any lengthy drug discovery effort or in the absence of any pharmacokinetic or pharmacodynamic barriers. Thus, the identification of cancer drug targets is an emerging and critically important role for preclinical mouse models of cancer.

This type of approach has generally focused on the requirement for the persistent expression of a "driving" oncogene following tumor initiation. The first such experiment to be performed examined the requirement for ongoing $H\text{-}Ras^{V12}$ expression in established tumor (Chin et al. 1999). Using a doxycycline-regulated system, $H\text{-}Ras^{V12}$ expression was silenced in primary melanomas. In all cases, tumors regressed and were barely detectable at 2 weeks following doxycycline withdrawal. Thus, ongoing $H\text{-}Ras^{V12}$ expression is required for tumor maintenance. Interestingly, all tumors relapsed following $H\text{-}Ras^{V12}$ reexpression, indicating the Ras-independent survival of minimal residual disease following tumor regression.

Similar experiments were also performed to examine the requirement for ongoing *c-myc* expression in tumor maintenance. For example, ubiquitous expression of *c-myc* in mice with a doxycycline-regulated system resulted in the development of either T cell hyperplasia or lymphoma, and both of these conditions were efficiently reversed for as long as 30 weeks following the termination of *myc* expression (Jain et al. 2002). Interestingly, a small percentage of these mice developed osteogenic sarcomas. In these mice, *myc* inactivation resulted in the differentiation of tumor cells into mature osteocytes, and subsequent reactivation of *myc* in these differentiated cells induced apoptosis. Thus, the transient inactivation of *myc* can lead to sustained tumor regression, and even tumor elimination upon *myc* reactivation. The impact of *myc* silencing can, however, vary significantly depending upon tumor type. For example, inactivation of *myc* in pancreatic beta cell tumors or breast adenocarcinoma results in tumor regression and the reappearance of normal organ histology (Pelengaris et al. 2002; Boxer et al. 2004). However, reactivation of *myc* in the breast results in the rapid restoration of neoplastic phenotypes, and breast adenocarcinomas can recur even in the absence of restored *myc* expression (Boxer et al. 2004).

Both *ras* and *myc* have been referred to as "undruggable" targets given the lack of effective inhibitors despite massive drug discovery efforts. Thus, it has been impossible to compare phenotypes associated with genetic inactivation of these mutant oncogenes with targeted drug-based inhibition of these proteins. However, this is not the case for all preclinical mouse models expressing inducible oncogenes. For example, mice and humans overexpressing an activated form of the epidermal growth factor receptor (EGFR) develop lung adenocarcinoma (Pao et al. 2004; Politi et al. 2006), and activated EGFR can be targeted with drugs that bind its ATP binding pocket, including erlotinib and gefitinib. Experiments using erlotinib and an inducible allele of mutant EGFR show that both chemical inhibition and silencing of EGFR yield a similar phenotype, including progressive depletion of lung adenocarcinoma cells and a return of normal lung histology (Politi et al. 2006). Thus, genetic ablation of an oncogene can predict the consequence of its pharmacological inhibition.

Recent experiments have extended the use of inducible alleles to the study of tumor suppressor genes and tumor maintenance. Specifically, three different approaches in three different tumor models have been used to test whether the absence of *p53* is required for the maintenance of *p53*-deficient tumors. First, an estrogen receptor (ER)-fused p53 protein was rendered transcriptionally active in p53-deficient T cell lymphomas via addition of tamoxifen (Martins et al. 2006). Second, Cre-induced recombination was used to activate a *p53* allele carrying a "floxed" stop cassette in *p53*-deficient sarcomas (Ventura et al. 2007). Third, an inducible RNA interference (RNAi) system was used to silence, and then subsequently activates, p53 in p53-deficent hepatocellular carcinoma (Xue et al. 2007). In all cases, p53 reintroduction led to sustained tumor regression, suggesting that p53 loss is required continually during tumor growth. Interestingly, the precise mechanism of tumor regression varied in each case, ranging from apoptotic cell death to senescence and clearance by tumor-associated immune cells. Thus, the "therapeutic" reintroduction of p53 is effective in multiple cancers and demonstrates the potential reversibility of tumor suppressive function tumors.

In all cases discussed thus far, elimination of an initiating oncogenic or tumor suppressive lesion resulted in tumor regression, suggesting that this may be a universal feature of cancer-promoting alterations. However, this is not the case. For example, analysis of murine malignancies bearing an inactive *Brca2* tumor suppressor gene has shown that these tumors can reactivate Brca2, through deletion of a domain containing the inactivating mutation, to acquire resistance to poly ADP ribose polymerase (PARP) inhibitors (Edwards et al. 2008). Thus, not only is the absence of Brca2 dispensable for tumor maintenance, but its reactivation may be necessary to circumvent the action of certain targeted therapeutics.

Perhaps the most critical role of inducible mouse models as preclinical systems is the ability of these models to identify critical nodes of action downstream of highly pleiotropic oncogenes. In addition to being "non-targetable," oncogenes like *ras* and *myc* activate countless effector pathways, any of which could be essential for their action in a given tumor. Thus, mechanisms to identify critical pathways downstream of these broadly acting factors are essential for the identification of better drug targets. A definitive example of the use of preclinical mouse models to address this problem is the generation of a conditional *p110β* knockout mouse (Jia et al. 2008). Inactivation of the Pten tumor suppressor activates numerous PI3 kinases, including p110β , leading to the subsequent induction of numerous downstream signaling pathways – any combination of which could be critical for Pten's tumor suppressor ability. Interestingly, conditional ablation of *p110β* in adult mice completely inhibited the development of prostate cancer in Pten-deficient, tumor-prone mice. Thus, p110β inhibition represents a primary tumor suppressive function of Pten in this context. As such, this work clearly establishes p110β as a meaningful drug target. Additionally, while *p110β* deficiency is embryonic lethal, this study suggests that *p110β* loss is tolerated in adult animals. Thus, this kind of preclinical model can simultaneously define novel drug targets and provide initial data concerning drug toxicity.

A similar set of experiments, while not performed in an inducible model, involved the investigation of the cyclin-dependent kinases Cdk4 and Cdk6 (Malumbres et al. 2004). These proteins have long been thought to be essential regulators of cell cycle progression. Consequently, they have been the target of numerous drug discovery efforts. Surprisingly, embryos lacking *cdk4/cdk6* are viable, and cells derived from these embryos can proliferate indefinitely in culture. Additionally, these cells immortalize like wild-type cells. Here, again, mouse models can be used to predict the efficacy of pharmacological inhibition of specific cancer drug targets.

23.6 Transplant Systems

Data from recent tumor genome sequencing efforts suggest that individual cancers contain at least six causative genetic lesions and that the specific identity of these alterations varies significantly between tumors (Ding et al. 2008; Jones et al. 2008). Thus, any attempt to model the diversity of alterations present in human cancers in the context of preclinical studies requires genetically tractable systems. One strategy for rapidly constructing tumors with complex genotypes is to manipulate tumor cells ex vivo and then transplant these tumors into recipient mice. This approach differs from tumor xenografts in several important ways. First, tumor transplants are performed in syngeneic, immunocompetent recipient mice, so tumor therapy can be performed in the presence of a functional immune system. Second, tumors are initially generated from defined genetic lesions, so they lack the inherent variability – prior to genetic manipulation – of human tumors. Finally, in many cases the tumor transplant can be reintroduced into its native microenvironment.

One approach to ex vivo cell manipulation has been to isolate embryonic epithelial cells and immortalize them through standard genetic manipulation, including the introduction of dominant negative alleles of *p53* and viral oncogenes, like adenoviral *E1A* (Degenhardt and White 2006). The resulting cell lines are not transformed, but are highly predisposed, such that the addition of any number of additional alterations can promote tumor development. Importantly, a wide variety of cell lines, including breast, kidney, ovarian, and prostate, have been developed using such an approach. A great strength of this system is the ability to correlate the presence of particular transforming alterations with ultimate therapeutic response. For example, it has been shown that suppression of the BH3-only protein Bim provides resistance to paclitaxel in tumors produced from transformed embryonic kidney cells (Tan et al. 2005). Additionally, Bim induction can sensitize *ras*-driven tumors to the proteasome inhibitor bortezomib.

The mouse hematopoietic system has proven a particularly tractable context for studying chemotherapeutic response. For example, putative mediators of drug response or resistance can be retrovirally introduced into established lymphomas, which can then be transplanted into syngeneic recipient mice. This approach can be combined with the use of lymphomas deficient for a particular tumor suppressor in a way that lymphomas with multiple genotypes are rapidly produced (Schmitt et al. 2000). Moreover, the rapid ex vivo modification of preestablished lymphomas by retroviral insertion allow the generation of "matched pairs" of tumors differing in a single defined lesion, i.e., aliquots of the same primary malignancy with and without a gene of interest. These matched pairs can be systemically introduced into recipient mice for treatment studies. Importantly, unlike the use of transformed embryonic epithelial cells or tumor xenografts, hematopoietic tumors home to the appropriate anatomical location following tail vein injection, directly recapitulating the tumor burden of the donor mouse. Thus, tumors can be treated in their native microenvironment. Of note, the hematopoietic system is not the only organ system that can be reconstituted with genetically modified cells. For example, cultured embryonic hepatocytes can be used to reconstitute a liver that has been chemically or surgically ablated (Zender et al. 2006). Similarly, cultured murine mammary epithelial cells can give rise to normal ductal structures following injection into a mammary fat pad (Vargo-Gogola and Rosen 2007). While preliminary, these data suggest that numerous anatomical systems may be similarly amenable to these tractable genetic approaches.

Retroviral transduction systems, such as the one described above have generally been limited to gene overexpression studies and, until recently, the examination of loss-of-function phenotypes has been limited to using HSCs derived from available knockout mice. Recent work has shown that RNAi can stably suppress gene expression in stem cells and mature lymphomas (Hemann et al. 2003, 2004). Thus, RNAi can be used as a tool to study gene function *in vivo*. This work has led to the adaptation of loss-of-function genetic approaches to tractable chemotherapy models in the hematopoietic system.

This combination of retrovirus-mediated oncogene activation and RNAi-mediated gene suppression has been used most extensively to examine the genetics of therapeutic response in the *Eμ-myc* model. This mouse carries a transgene that places *myc* under the control of an immunoglobulin enhancer, recapitulating the characteristic lesion in human Burkitt's lymphoma (Adams et al. 1985). Transgenic mice develop B cell lymphomas at 4–6 months, and these tumors can be cultured, infected, and transplanted into recipient mice for treatment studies. Early studies in this mouse showed that overexpression of *bcl2*, suppression of *p53,* and deletion of the *Ink4a/Arf* locus all promote resistance to genotoxic chemotherapies (Schmitt and Lowe 2001; Schmitt et al. 2002). Furthermore, these studies established tumor senescence as a *bona fide* response to DNA damaging chemotherapy in tumors unable to undergo apoptosis.

The *Eμ-myc* has also been used as a preclinical system to examine the efficacy of combination chemotherapies. For example, tumors generated following retroviral introduction of activated Akt were treated with the front-line chemotherapeutic doxorubicin and/or the mTOR inhibitor rapamycin (Wendel et al. 2004). While tumors treated with either drug alone showed limited tumor regression, tumors treated with both drugs showed a sustained remission. Interestingly, in control tumors, the combination of both drugs performed worse than doxorubicin alone, highlighting the importance of tumor genotype in therapeutic outcome.

More recently, the *Eμ-myc* model has been used to rapidly examine putative modulators of therapeutic outcome. For example, cell culture studies involving topoisomerase II suppression suggested that suppression of this protein may promote resistance to doxorubicin and sensitivity to the topoisomerase I poison camptothecin (Burgess et al. 2008). This result was confirmed in *Eμ-myc* lymphomas, highlighting the power of this system to rapidly examine the effect of diverse genetic alterations on chemotherapeutic response in a relevant *in vivo* setting. Additional studies examining checkpoint kinases in this model showed that suppression of either Chk2 or ATM kinase promotes resistance to genotoxic drugs. However, inactivation of either of these proteins on a p53-deficient background results in drug sensitivity. This result has subsequently been confirmed in human epithelial cancers, further establishing the value of this system as a preclinical model.

Other models of hematopoietic malignancy have similarly used ex vivo approaches to examine the determinants of therapeutic response. For example, experiments in which a *Bcr-Abl* transgene was introduced into *p53*-deficient HSCs showed that chronic myelogenous leukemias that are deficient for p53 develop resistance to the targeted therapy imatinib (Wendel et al. 2006). Similarly, *Bcr-Abl*-induced acute lymphocytic leukemias that are deficient for the Arf tumor suppressor are resistant to imatinib (Williams et al. 2006). Thus, combinatorial genetics can rapidly evaluate putative mechanisms of drug resistance.

23.7 RNA Interference (RNAi) Genetic Screens

Forward genetic screens in the mouse hematopoietic system have proven to be highly effective strategies for identifying novel and cooperating cancer genes. For example, Moloney-based retroviral insertional mutagenesis screens first characterized the potent oncogenes *bmi-1* (van Lohuizen et al. 1991) and *pim1-3* (van Lohuizen et al. 1989). Subsequently, these screens have been performed on numerous sensitized backgrounds, revealing significant insight into the relationship between specific oncogenic and tumor suppressor alterations. The more recent development of mice expressing active transposons has similarly facilitated the identification of novel oncogenes (Collier et al. 2005; Dupuy et al. 2005). Unfortunately, these screens have not yet been used to explore the genetics of therapeutic response and drug resistance. One major reason for this is that insertional mutagenesis screens are generally limited to the identification of oncogenes, and tumor suppressors are thought to play a very significant role in modulating the response to cancer therapies. Another reason is that these screens are only suitable as positive selection screens and cannot be used to identify alterations that sensitize tumors to a given chemotherapy.

Given the advent of RNAi technology, genetic screens can be performed to examine loss-of-function phenotypes – including changes in therapeutic response (Fig. 23.2). Moreover, our group has recently adapted this technology to carry out therapy-based genetic screens in hematopoietic malignancies *in vivo*. In these experiments, we used ex vivo infection of lymphoma cells with large libraries of short hairpin RNA (shRNAs), followed by tail vein injection, to produce tumors expressing a diverse set of genetic lesions. This process was contingent upon overcoming several major technical issues, as will its adaptation to other tumor models. First, shRNAs must function effectively at single copy following retroviral infection (Dickins et al. 2005). Second, a large number of cells must contribute to a given tumor following injection. In the case of the *Eμ-myc* model, nearly 1,000 lymphoma cells help give rise to a resulting lymphoma following transplantation – placing the "screenable" number of shRNAs at approximately 1,000 per mouse (C. Meacham and M. Hemann, unpublished data). Finally, systems need to be available to track changes in the representation of shRNAs following administration of a given therapy. In our case, we used high-throughput sequencing to track shRNA representation. Briefly, common primers can be used to amplify shRNAs or associated "bar codes." The PCR products can be sequenced using emerging chip-based sequencing technology. This approach can provide quantitative identification of multiple millions of PCR products per read, allowing one to examine the representation of whole-genome shRNA libraries. Importantly, this approach can be used to track positive selection (if a given shRNA promotes drug resistance and is enriched following treatment) or negative

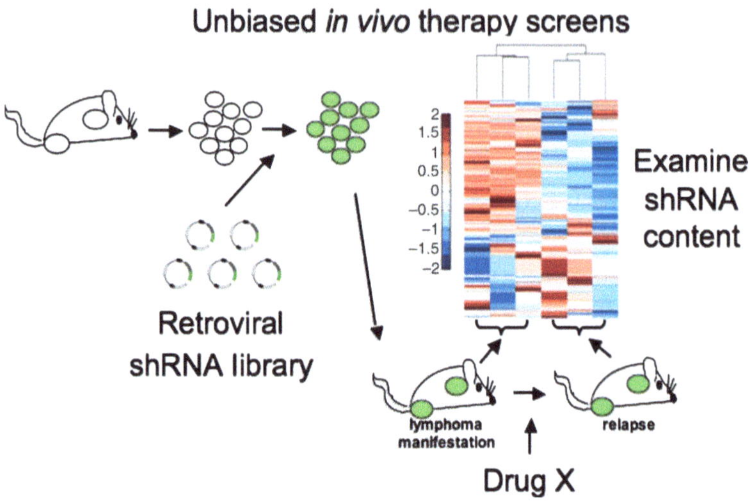

Fig. 23.2 A diagram showing the strategy for using *in vivo* RNAi to perform screens for genes involved in chemotherapeutic response in physiologically relevant contexts. This representational approach can identify both positive and negative regulators of therapeutic outcome

selection (if a given shRNA promotes drug sensitivity and is depleted from the library following treatment).

Using this system, we have been able to screen for modulators of therapeutic response in a physiologically relevant context. Importantly, we also now have the potential to screen for *in vivo* phenotypes that correlate with drug resistance, including tumor metastasis or dissemination into the CNS. Whether other tumor models will be similarly amenable to this kind of screening approach remains unclear. However, preliminary studies in at least one model of hepatocellular carcinoma suggest that screening approaches can be used for mouse epithelial cancers. In these experiments, up to 50 shRNAs could be screened for contributions to tumor development in the liver of an individual mouse.

Importantly, these tumor transplantation/screening approaches have inherent limitations. First, while transplanted tumors may home to appropriate sites, they are not autochthonous tumors. Thus, the response of a tumor in its appropriate developmental context to a given anticancer therapy may vary from that seen in tumor transplants. Second, if tumors relapse due to the differentiation and outgrowth of "tumor stem cells," then tumor transplants will not accurately recapitulate the response of native tumors if they fail to include similar stem cell populations. One solution to this problem would be to directly introduce retroviruses expressing a diverse mixture of shRNAs into autochthonus tumors during development. Given the number of tumor models that show a precise temporal progression of premalignant stages, it should be possible to time the introduction of a pool of retroviruses with the peak of tumor cell proliferation. This approach could combine the strengths of highly accurate cancer models with genetically tractable systems for modifying tumor genotype.

23.8 Conclusions

There is little doubt that preliminary studies using GEMMs as preclinical systems have yielded profoundly different results relative to xenograft-based systems. Put simply, cancers produced from GEMMs better recapitulate the therapeutic challenges associated with treating human cancers and, consequently, their disseminated use should not be saddled with the burden of overcoming years of painful or uninformative experiences with xenograft systems. That is not to argue that these models should not be critically evaluated or that the future use of these models is not without considerable challenges – both at the level of the individual academic researcher and at the level of preclinical efforts at pharmaceutical companies.

In the case of academic research laboratories, there has been a strange hesitance to examine newly developed cancer models in therapeutic contexts. In fact, there are still vastly more laboratories involved in generating new mouse models of cancer than there are evaluating the preclinical value of such systems. This clearly needs to change. If we are to argue that these systems have value as cancer models, then we need to test this hypothesis in the most stringent manner possible. This is particularly true as the prevailing mission of cancer GEMMs makes the inevitable transition from cancer gene discovery efforts to preclinical studies.

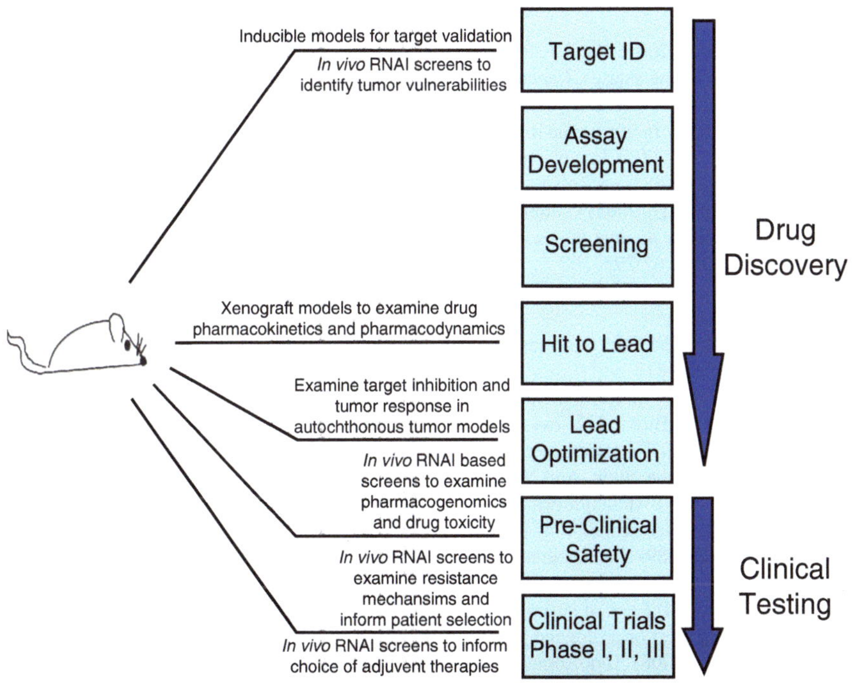

Fig. 23.3 GEMMs can provide valuable information at multiple stages of drug development. This includes the ability to inform clinical trials and late-stage drug development

There are perhaps even greater challenges facing the use of GEMMs as cancer preclinical systems in the pharmaceutical industry. While the "Oncomouse" patent has been a substantial barrier for the disseminated use of GEMMs in industry (Hanahan et al. 2007) (see Chapter 28), it is apparent that additional barriers exist. The most significant of these is the common industry view of where mouse models can contribute during the drug design process. Specifically, mouse models are used, when they are used, in a narrow window between cell-based efficacy studies and human clinical trials. The work described in this chapter suggests that mouse models might be used throughout the drug development process (Fig. 23.3). This not only applies to stages of initial target identification and toxicity studies, but also includes end-stage processes. Mouse preclinical models have the potential to identify common mechanisms of drug resistance, allowing for the informed identification of patient cohorts most likely to benefit from the drug. Additionally, drug combinations can be examined in preclinical systems, enabling physicians to make better choices of adjuvant therapies during clinical trials. In fact, it is the use of mouse models in these late-stage processes of drug development that holds the most potential for substantially reducing drug costs and fundamentally improving the process of pharmaceutical development.

References

Abate-Shen C et al (2003) Nkx3.1; Pten mutant mice develop invasive prostate adenocarcinoma and lymph node metastases. Cancer Res 63:3886–3890

Adams JM et al (1985) The c-myc oncogene driven by immunoglobulin enhancers induces lymphoid malignancy in transgenic mice. Nature 318:533–538

Beckmann N et al (2007) In vivo mouse imaging and spectroscopy in drug discovery. NMR Biomed 20:154–185

Bergers G, Hanahan D (2002) Combining antiangiogenic agents with metronomic chemotherapy enhances efficacy against late-stage pancreatic islet carcinomas in mice. Cold Spring Harb Symp Quant Biol 67:293–300

Bergers G, Hanahan D (2008) Modes of resistance to anti-angiogenic therapy. Nat Rev Cancer 8:592–603

Bergers G et al (1999) Effects of angiogenesis inhibitors on multistage carcinogenesis in mice. Science 284:808–812

Bergers G et al (2000) Matrix metalloproteinase-9 triggers the angiogenic switch during carcinogenesis. Nat Cell Biol 2:737–744

Besterman JM (1996) Topoisomerase I inhibition by the camptothecin analog Gl147211C. From the laboratory to the clinic. Ann N Y Acad Sci 803:202–209

Boehm T et al (1997) Antiangiogenic therapy of experimental cancer does not induce acquired drug resistance. Nature 390:404–407

Bouchard L et al (1989) Stochastic appearance of mammary tumors in transgenic mice carrying the MMTV/c-neu oncogene. Cell 57:931–936

Boxer RB et al (2004) Lack of sustained regression of c-MYC-induced mammary adenocarcinomas following brief or prolonged MYC inactivation. Cancer Cell 6:577–586

Burgess DJ et al (2008) Topoisomerase levels determine chemotherapy response in vitro and in vivo. Proc Natl Acad Sci USA 105:9053–9058

Chen Z et al (2005) Crucial role of p53-dependent cellular senescence in suppression of Pten-deficient tumorigenesis. Nature 436:725–730

Cherney BW et al (1997) Role of the p53 tumor suppressor gene in the tumorigenicity of Burkitt's lymphoma cells. Cancer Res 57:2508–2515
Chin L et al (1999) Essential role for oncogenic Ras in tumour maintenance. Nature 400:468–472
Collier LS et al (2005) Cancer gene discovery in solid tumours using transposon-based somatic mutagenesis in the mouse. Nature 436:272–276
Cook N et al (2008) K-Ras-driven pancreatic cancer mouse model for anticancer inhibitor analyses. Methods Enzymol 439:73–85
Degenhardt K, White E (2006) A mouse model system to genetically dissect the molecular mechanisms regulating tumorigenesis. Clin Cancer Res 12:5298–5304
Dickins RA et al (2005) Probing tumor phenotypes using stable and regulated synthetic microRNA precursors. Nat Genet 37:1289–1295
Ding L et al (2008) Somatic mutations affect key pathways in lung adenocarcinoma. Nature 455:1069–1075
Dupuy AJ et al (2005) Mammalian mutagenesis using a highly mobile somatic Sleeping Beauty transposon system. Nature 436:221–226
Edwards SL et al (2008) Resistance to therapy caused by intragenic deletion in BRCA2. Nature 451:1111–1115
Foster BA et al (1997) Characterization of prostatic epithelial cell lines derived from transgenic adenocarcinoma of the mouse prostate (TRAMP) model. Cancer Res 57:3325–3330
Gibbs JB, Oliff A (1997) The potential of farnesyltransferase inhibitors as cancer chemotherapeutics. Annu Rev Pharmacol Toxicol 37:143–166
Greenberg NM, DeMayo F, Finegold MJ, Medina D, Tilley WD, Aspinall JO, Cunha GR, Donjacour AA, Matusik RJ, Rosen JM (1995) Prostate cancer in a transgenic mouse. Proc Natl Acad Sci USA 92(8):3439–3443
Hanahan D (1985) Heritable formation of pancreatic beta-cell tumours in transgenic mice expressing recombinant insulin/simian virus 40 oncogenes. Nature 315:115–122
Hanahan D, Wagner EF, Palmiter RD (2007) The origins of oncomice: a history of the first transgenic mice genetically engineered to develop cancer. Genes Dev 21:2258–2270
He LZ et al (2001) Histone deacetylase inhibitors induce remission in transgenic models of therapy-resistant acute promyelocytic leukemia. J Clin Invest 108:1321–1330
Hemann MT et al (2003) An epi-allelic series of p53 hypomorphs created by stable RNAi produces distinct tumor phenotypes in vivo. Nat Genet 33:396–400
Hemann MT et al (2004) Suppression of tumorigenesis by the p53 target PUMA. Proc Natl Acad Sci USA 101:9333–9338
Hingorani SR et al (2003) Preinvasive and invasive ductal pancreatic cancer and its early detection in the mouse. Cancer Cell 4:437–450
Holland EC, Varmus HE (1998) Basic fibroblast growth factor induces cell migration and proliferation after glia-specific gene transfer in mice. Proc Natl Acad Sci USA 95:1218–1223
Holland EC et al (1998) A constitutively active epidermal growth factor receptor cooperates with disruption of G1 cell-cycle arrest pathways to induce glioma-like lesions in mice. Genes Dev 12:3675–3685
Hu X et al (2005) mTOR promotes survival and astrocytic characteristics induced by Pten/AKT signaling in glioblastoma. Neoplasia 7:356–368
Jackson EL et al (2001) Analysis of lung tumor initiation and progression using conditional expression of oncogenic K-ras. Genes Dev 15:3243–3248
Jain M et al (2002) Sustained loss of a neoplastic phenotype by brief inactivation of MYC. Science 297:102–104
Jia S et al (2008) Essential roles of PI(3)K-p110beta in cell growth, metabolism and tumorigenesis. Nature 454:776–779
Johnson JI et al (2001a) Relationships between drug activity in NCI preclinical in vitro and in vivo models and early clinical trials. Br J Cancer 84:1424–1431
Johnson L et al (2001b) Somatic activation of the K-ras oncogene causes early onset lung cancer in mice. Nature 410:1111–1116

Jones S et al (2008) Core signaling pathways in human pancreatic cancers revealed by global genomic analyses. Science 321:1801–1806
Kasper S, Smith JA Jr (2004) Genetically modified mice and their use in developing therapeutic strategies for prostate cancer. J Urol 172:12–19
Kinkade CW et al (2008) Targeting AKT/mTOR and ERK MAPK signaling inhibits hormone-refractory prostate cancer in a preclinical mouse model. J Clin Invest 118:3051–3064
Klein RD (2005) The use of genetically engineered mouse models of prostate cancer for nutrition and cancer chemoprevention research. Mutat Res 576:111–119
Klumb CE et al (2003) DNA sequence profile of TP53 gene mutations in childhood B-cell non-Hodgkin's lymphomas: prognostic implications. Eur J Haematol 71:81–90
Kulke MH et al (2002) A phase II study of troglitazone, an activator of the PPARgamma receptor, in patients with chemotherapy-resistant metastatic colorectal cancer. Cancer J 8:395–399
Macdonald JS et al (2005) A phase II study of farnesyltransferase inhibitor R115777 in pancreatic cancer: a Southwest oncology group (SWOG 9924) study. Invest New Drugs 23:485–487
Malumbres M et al (2004) Mammalian cells cycle without the D-type cyclin-dependent kinases Cdk4 and Cdk6. Cell 118:493–504
Martins CP, Brown-Swigart L, Evan GI (2006) Modeling the therapeutic efficacy of p53 restoration in tumors. Cell 127:1323–1334
Olive KP, Tuveson DA (2006) The use of targeted mouse models for preclinical testing of novel cancer therapeutics. Clin Cancer Res 12:5277–5287
Pao W et al (2004) EGF receptor gene mutations are common in lung cancers from "never smokers" and are associated with sensitivity of tumors to gefitinib and erlotinib. Proc Natl Acad Sci USA 101:13306–13311
Pear WS et al (1998) Efficient and rapid induction of a chronic myelogenous leukemia-like myeloproliferative disease in mice receiving P210 bcr/abl-transduced bone marrow. Blood 92:3780–3792
Pelengaris S, Khan M, Evan GI (2002) Suppression of Myc-induced apoptosis in beta cells exposes multiple oncogenic properties of Myc and triggers carcinogenic progression. Cell 109:321–334
Politi K et al (2006) Lung adenocarcinomas induced in mice by mutant EGF receptors found in human lung cancers respond to a tyrosine kinase inhibitor or to down-regulation of the receptors. Genes Dev 20:1496–1510
Sarraf P et al (1998) Differentiation and reversal of malignant changes in colon cancer through PPARgamma. Nat Med 4:1046–1052
Schmitt CA, Lowe SW (2001) Bcl-2 mediates chemoresistance in matched pairs of primary E(mu)-myc lymphomas in vivo. Blood Cells Mol Dis 27:206–216
Schmitt CA, Rosenthal CT, Lowe SW (2000) Genetic analysis of chemoresistance in primary murine lymphomas. Nat Med 6:1029–1035
Schmitt CA et al (2002) A senescence program controlled by p53 and p16INK4a contributes to the outcome of cancer therapy. Cell 109:335–346
Sharpless NE, Depinho RA (2006) The mighty mouse: genetically engineered mouse models in cancer drug development. Nat Rev Drug Discov 5:741–754
Tan TT et al (2005) Key roles of BIM-driven apoptosis in epithelial tumors and rational chemotherapy. Cancer Cell 7:227–238
Traxler P et al (2001) Tyrosine kinase inhibitors: from rational design to clinical trials. Med Res Rev 21:499–512
Twombly R (2002) First clinical trials of endostatin yield lukewarm results. J Natl Cancer Inst 94:1520–1521
van Lohuizen M, Breuer M, Berns A (1989) N-myc is frequently activated by proviral insertion in MuLV-induced T cell lymphomas. EMBO J 8:133–136
van Lohuizen M et al (1991) Identification of cooperating oncogenes in E mu-myc transgenic mice by provirus tagging. Cell 65:737–752
Vargo-Gogola T, Rosen JM (2007) Modelling breast cancer: one size does not fit all. Nat Rev Cancer 7:659–672

Ventura A et al (2007) Restoration of p53 function leads to tumour regression in vivo. Nature 445:661–665

Wendel HG et al (2004) Survival signalling by Akt and eIF4E in oncogenesis and cancer therapy. Nature 428:332–337

Wendel HG et al (2006) Loss of p53 impedes the antileukemic response to BCR-ABL inhibition. Proc Natl Acad Sci USA 103:7444–7449

Williams RT, Roussel MF, Sherr CJ (2006) Arf gene loss enhances oncogenicity and limits imatinib response in mouse models of Bcr-Abl-induced acute lymphoblastic leukemia. Proc Natl Acad Sci USA 103:6688–6693

Xue W et al (2007) Senescence and tumour clearance is triggered by p53 restoration in murine liver carcinomas. Nature 445:656–660

Yang FC et al (2008) Nf1-dependent tumors require a microenvironment containing $Nf1^{+/-}$ and c-kit-dependent bone marrow. Cell 135:437–448

Zender L et al (2006) Identification and validation of oncogenes in liver cancer using an integrative oncogenomic approach. Cell 125:1253–1267

Chapter 24
Animal Models for Breast Cancer Prevention Research

Chunyu Wang and Powel H. Brown

24.1 Introduction

Breast cancer is the most commonly diagnosed cancer and the second leading cause for cancer-related death for women in the USA, with over 182,460 projected new cases and 40,480 deaths in 2008 (Jemal et al. 2008). Breast cancer is a multistep disease involving multiple genetic and environmental factors. Animal models provide an invaluable tool in understanding the complexity of breast cancer. Understanding the genetic pathways is a prerequisite for the development of pharmacological treatment and chemoprevention of breast cancer. Over the past decades, the studies on animal models of breast cancer have provided in-depth insight into molecular pathways of normal mammary gland development and breast tumorigenesis.

There has recently been significant progress in the chemoprevention of breast cancer. Selected estrogen receptor modulators (SERMs) and aromatase inhibitors were successfully used to prevent the development of breast cancer in animal models and in clinical trials. However, these drugs do not prevent all breast cancers. Tamoxifen and raloxifene prevent the development of estrogen receptor (ER)-positive breast cancers but have no effect on the development of ER-negative breast cancer. In addition, the use of these drugs has been limited due to their toxicity. Thus, there is an urgent need of novel strategies to develop tolerable anti-estrogens and to prevent ER-negative breast cancers (Li and Brown 2007, 2009). We and others investigated the cancer preventive ability of several classes of drugs for the prevention of ER-negative breast cancer in preclinical models and clinical trials. Results from these studies demonstrated that many drugs, such as rexinoids and tyrosine

---

C. Wang • P.H. Brown (✉)
Department of Clinical Cancer Prevention, The University of Texas MD Anderson Cancer Center, Houston, TX 77030, USA
e-mail: phbrown@mdanderson.org

J.E. Green and T. Ried (eds.), *Genetically Engineered Mice for Cancer Research: Design, Analysis, Pathways, Validation and Pre-Clinical Testing*,
DOI 10.1007/978-0-387-69805-2_24,

kinase inhibitors, can prevent ER-negative breast cancer in transgenic mouse models. Other promising agents are also being tested in early phase cancer prevention trials.

In this article, we review the animal models of breast cancer and the current status of breast cancer prevention studies. We anticipate that the generation of animal models for breast cancer research will provide new tools for understanding the molecular mechanisms of mammary tumorigenesis. Using these models, more effective chemopreventive drugs and intervention strategies will be developed for the prevention of human breast cancer in the future.

24.2 Animal Models of Breast Cancer

24.2.1 Traditional Models: Spontaneous or Carcinogenesis Rat Models

Among all the experimental animal models for mammary carcinogenesis, the induction of breast tumors in rats by chemical carcinogens is one of the earliest models. Many strains of rodents develop spontaneous tumors, and respond to a variety of chemical carcinogens and radiation with the development of either hormone-dependent or hormone-independent mammary tumors. These models have many features that make them attractive, e.g., easy induction and reliability, organ site specificity, histopathological characteristics, and hormone responsiveness.

The two most widely used carcinogen-induced systems for the study of mammary tumorigenesis are the tumors induced by 12-dimethylbenz(a)anthracene (DMBA) (Huggins et al. 1961; Rogers and Lee 1986) in the Sprague-Dawley (SD) rats and by N-methylnitrosourea (NMU) (Huggins et al. 1981; Gusterson and Williams 1981) in the SD or Fischer 344 rats. DMBA when given by gavage in a single dose of 2.5–20 mg, or NMU when given by intravenous or subcutaneous injection in a single dose of 25–50 mg/kg body weight, induces tumors with latencies that range between 8 and 21 weeks with final tumor incidences close to 100% (Rogers and Lee 1986). The tumor latency in chemically induced tumors is inversely correlated with dosage of the drug. Tumor histology is also influenced by dosage of the drug. In SD rats given a single NMU dose of 10 mg/kg body weight, 42% of tumors were malignant, whereas 86–94% were malignant at doses from 35 to 50 mg/kg (McCormick et al. 1981).

These carcinogen-induced rat mammary tumors are generally hormone-dependent for both induction and growth. The influence of the endocrine system on chemically-induced carcinogenesis is reviewed by Welsch (1985). Ovariectomy causes tumor regression in about 80% of DMBA-treated SD rats (McCormick et al. 1982). The DMBA and NMU-induced mammary tumor models are useful for the assessment of antiestrogen prevention of breast cancer. In both models, tamoxifen and other antiestrogens given after carcinogen exposure but before tumor formation can delay tumor appearance (Russo and Russo 1996).

Continuous treatment with E2 delivered through release from subcutaneous pellet implants can induce the development of mammary cancers in female August-Copenhagen-Irish (ACI) rats. The ACI strain differs from all other inbred rat strains in that those rats develop mammary tumors at high incidence when treated with the synthetic estrogen diethylstilbestrol (DES) (Dunning et al. 1948; Shull et al. 1997; Shellabarger et al. 1980; Rothschild et al. 1987). Although ACI rats develop mammary tumors when chronically treated with estrogen, spontaneous mammary tumors are rare. ACI rats develop relatively few mammary tumors when treated with chemical carcinogens. The ACI rat strain is an animal model for evaluating the role of estrogens in the etiology of human breast cancer. The ACI rat is a unique model of human breast cancer in that mammary cancers are induced by estrogen without carcinogens, irradiation, xenografts, or transgenic manipulations. Naturally occurring estrogen 17β-estradiol (E2) is able to induce the development of mammary tumors in ovary-intact and ovariectomized ACI rats. E2 rapidly induces the development of mammary carcinomas that often display invasive characteristics (Shull et al. 1997). The tumors express ERs and progesterone receptor (PR). Concomitant treatment with the anti-estrogen tamoxifen completely prevents E2 induction of mammary tumors in female ACI rats (Li et al. 2002). In a congenic variant of the ACI rat (ACI.COPEpt2), all rats with estradiol implants developed mammary cancers in about 6 months. Tamoxifen treatment suppresses the growth of these tumors. Thus, this model is highly relevant to hormone-responsive human breast cancers. Based on biochemical and morphological characteristics, this model is most representative of hormone-responsive human breast cancers (Ruhlen et al. 2009).

24.2.2 Transgenic Mouse Models

Mammary tumorigenesis is thought to involve multiple genetic events. The majority of genetic changes can be divided into two categories, gain-of-function of proto-oncogenes, which are involved in cell growth, proliferation and survival, and loss-of-function of tumor suppressor genes, which are involved in preventing cell growth or promoting apoptosis. Transgenic mouse models are especially useful in deciphering how specific genetic changes affect breast cancer tumorigenesis since these transgenic mouse models can genetically recapitulate important molecular features of particular subtypes of human breast cancer (reviewed in Shen and Brown 2005). Table 24.1 presents a list of commonly used mouse models.

24.2.2.1 First Generation: Oncogene Transgenic Mouse Models (Hanahan et al. 2007)

The first transgenic mouse with breast cancer was reported in 1984 by Stewart et al. (1984). This landmark paper established an entirely new research area to identify genetic pathways that control breast cancer. Transgenic mice were generated by pronuclear injection of the c-myc oncogene expression cassettes into fertilized

Table 24.1 Summary of animal models of mammary tumorigenesis

Model	Promoter	Transgene	Strain and species	Time to tumor development	Percent animal with tumor	ER status	Tumor metastasis target	References
DMBA	–	–	SD rat	8–21 weeks	100	+	Lung	Huggins et al. (1961), Rogers and Lee (1986)
NMU	–	–	SD/Fischer 344 rat	8–21 weeks	100	+	Lung	Huggins et al. (1981), Gusterson and Williams (1981)
c-myc	MMTV/WAP	c-myc	C57BL/6J mouse	4–10 months	80–100	–	Lung	Stewart et al. (1984), Pelengaris et al. (2002), Grushko et al. (2004), Schoenenberger et al. (1988), Sandgren et al. (1995), Rose-Hellekant and Sandgren (2000)
Ras	MMTV/WAP	Ha-ras	C57BL/6J mouse	5 weeks to 6 months	50	–	Liver and lung	Malaney and Daly (2001), Huang et al. (1981), Andres et al. 1987, Nielsen et al. (1991, 1992, 1995), Sinn et al. (1987)
SV40	C3(1)	SV40 large tag	FVB mouse	3–6 months	100	±	Lung, liver, adrenal and heart	Maroulakou et al. (1994), Shibata et al. (1996), Green et al. (2000)
ErbB2	MMTV/WAP	neu/ErbB2	FVB mouse	5–10 months	80–100	–	Lung	Muller et al. (1988), Bouchard et al. v1989, Guy et al. (1992)
Wnt1	MMTV	Wnt1	FVB mouse	3–6 months	100	±	Lymph node and lung	Nusse and Varmus (1982), Nusse et al. (1984), Tsukamoto et al. (1988)
IRS-1/2	MMTV	IRS-1/2	FVB/N mouse	12 months	100	–	Lung	Dearth et al. (2006)
AIB1	MMTV	AIB1	FVB/N mouse	16 months	50	±	Lung, bone and kidney	Torres-Arzayus et al. (2004)

Brca1 KO	–	Mutant Brca1	C57BL/6 mouse	10–13 months	<20	±	–	Ludwig et al. (1997, 2001), Hakem et al. (1996), Shen et al. (1998), Gowen et al. (1996), Liu et al. (1996), Xu et al. (1999)
Brca2 KO	–	Mutant Brca2	C57BL/6 mouse	>1 year	<20	±	–	Connor et al. (1997), Bennett et al. (2000)
p53 KO	–	Mutant p53	C57BL/6 mouse	1–2 years	90–100	±	Lung and liver	Donehower et al. (1992), Li et al. (1998), Lin et al. (2004)
RCAS	MMTV	TVA	FVB/N mouse	5–25 weeks	100	+	Lung	Du et al. (2006), Du and Li (2007)

mouse eggs. MYC encodes a multifunctional, nuclear phosphoprotein. MYC is frequently amplified and overexpressed in breast cancer (Pelengaris et al. 2002; Grushko et al. 2004). They produced transgenic mice that carry a normal mouse c-myc gene in which the c-myc promoter region has been replaced by a hormonally inducible mouse mammary tumor virus (MMTV) promoter. The female transgenic mice spontaneously developed mammary adenocarcinomas during early pregnancy. Both the tumors and the breast tissue of these animals expressed the transgene. The constitutionally deregulated c-myc gene can act as a heritable predisposing factor accelerating the development of breast adenocarcinoma. Schoenenberger et al. generated WAP-c-myc transgenic mice (1988). WAP-c-myc tumors were poorly differentiated, solid carcinomas with a minority of adenocarcinomas (Sandgren et al. 1995). About 20% of the tumor-bearing WAP-c-myc transgenic mice displayed grossly visible lung metastases (Rose-Hellekant and Sandgren 2000).

Rat sarcoma (RAS) proteins are cytoplasmic guanine-5'-triphosphate (GTP)-binding proteins that function as critical signal mediators to couple upstream receptor tyrosine kinases (RTKs) to downstream serine/threonine kinases, such as mitogen-activated protein kinases (MAPKs). The Harvey (Ha)-ras protein is expressed and activated in mammary tumors (reviewed in Malaney and Daly 2001). Huang et al. (1981) showed that when the MMTV long terminal repeat (LTR) was fused to the Ras oncogene, its expression was regulated by steroid hormones in cultured mammalian cells. Ras driven by WAP or MMTV promoters is sufficient to induce tumorigenesis and metastatic mammary tumors in transgenic mouse models (Andres et al. 1987; Nielsen et al. 1991, 1992, 1995). Sinn et al. (1987) mated separate strains of transgenic mice that carry either the v-Ha-ras or the c-myc gene driven by the MMTV promoter. The co-expression of both oncogenes in the hybrid mice results in a dramatic and synergistic increase in tumor formation.

Simian virus 40 (SV40) large T- and small t-antigen (Ag) are the products of the early region of SV40 which are involved in cellular transformation (Moens et al. 1997). Large T antigen (Tag) has been well characterized and used as a transforming agent in many cell types. SV40 large Tag can induce transformation and tumorigenesis by binding to and inactivating the tumor suppressor genes p53 (Mietz et al. 1992) and retinoblastoma (Rb) (Dyson et al. 1989). Initially, the SV40 large Tag transgene was driven by the rat prostatic steroid-binding protein C3(1) promoter in order to direct expression of large Tag to the prostate epithelium of transgenic mice (Maroulakou et al. 1994). Most male transgenic mice develop prostatic hyperplasia in early life that progresses to adenoma or adenocarcinoma while female transgenic mice develop mammary hyperplasia by 3 months of age with subsequent development of mammary adenocarcinoma by 6 months of age in 100% of the animals (Shibata et al. 1996). The mammary carcinomas are poorly differentiated high-grade tumors that are histologically similar to ER-negative, metaplastic human breast cancer (reviewed in Green et al. 2000). Estrogen is not able to activate C3(1) transcription. Although mammary tumors that develop in the C3(1)-SV40 Tag mice appear to be estrogen responsive at early stages, the invasive carcinomas are estrogen independent and do not express the estrogen receptor (Green et al. 2000).

The EGF ligand/receptor family has been implicated in breast cancer genetics (reviewed in Hynes and Stern 1994; Olayioye et al. 2000; Yarden 2001).

This family is composed of four RTKs (EGFR/HER/erbB1, erbB2/HER2/neu, erbB3, and erbB4) (Menard et al. 2004) which can bind to the ligands (ligands of the EGF family), including EGF, amphiregulin (AREG), heregulin (NRG1/HRG), and transforming growth factor α (TGFα). Many of these ligands/receptors are implicated in breast cancers, particularly erbB2/HER2/neu (Eccles 2001). ErbB2 is amplified and/or overexpressed in up to 30% of breast cancers (Slamon et al. 1987). HER2/neu amplification/overexpression can induce breast tumors and can produce aggressive characteristics on these breast cancers (Lohrisch and Piccart 2001). Patients whose tumors overexpress HER2 have a worse prognosis than those with tumors having normal HER2/neu expression (Ravdin and Chamness 1995; Andrulis et al. 1998; Piccart et al. 2001). Direct evidence implicating ErbB2 as an oncogene is derived from studies of transgenic mice expressing an activated Neu (the rat homologue of ErbB2). MMTV-driven overexpression of the neu/ErbB2 oncogene resulted in the formation of mammary tumors which are similar to well-differentiated breast adenocarcinoma in human (Muller et al. 1988; Bouchard et al. 1989). These mice develop mammary tumors both in female and male transgene carriers. The tumors can also metastasize to the lung (Guy et al. 1992).

The Wingless-type (Wnt) family is one of the most important families of intercellular signaling factors in animals (reviewed in Howe and Brown 2004; Turashvili et al. 2006). There are 19 Wnt genes in mammals. All encode cysteine-rich secreted glycoproteins. Wnt signals through intracellular protein β-catenin. Elevated β-catenin levels, a hallmark of active Wnt signaling, can be seen in a majority of breast tumors (Ryo et al. 2001). There are also several studies describing overexpression of Wnts in breast cancer relative to normal tissue (Ugolini et al. 2001; Huguet et al. 1994; Bui et al. 1997; Dale et al. 1996). The Wnt/β-catenin pathways have been shown to regulate cell fate in development and also affect proliferation, apoptosis, and differentiation. Wnt1, originally named int1, was the first identified protooncogene activated by viral insertion in naturally occurring mouse mammary tumors (Nusse and Varmus 1982; Nusse et al. 1984). Tsukamoto et al. (1988) generated MMTV-Wnt1 transgenic mice, which express Wnt1 at high levels in mammary and salivary glands of male and female mice and in male reproductive organs. The mammary glands are grossly hyperplastic compared with those of nontransgenic littermates. Mammary and salivary adenocarcinomas also occur in these animals. Transgenic expression of Wnt1 causes extensive ductal hyperplasia early in life and mammary adenocarcinomas in approximately 50% of the female transgenic mice by 6 months of age. Metastasis to the lung and proximal lymph nodes is rare at the time tumors are detected but frequent after the removal of the primary neoplasm. The potent mitogenic effect mediated by Wnt-1 expression does not require estrogen stimulation (Li et al. 2000). The tumors that develop in MMTV-Wnt1 mice are heterogeneous and have ER-positive and/or ER-negative cells.

Insulin receptor substrate (IRS)-1 and IRS-2 are adaptor proteins in the insulin-like growth factor I (IGF-I)/IGF-I receptor (IGF-IR) pathway (Dearth et al. 2007). Both IRS-1 and IRS-2 are oncogenes and can induce transformation and metastasis *in vitro* and *in vivo*. In breast cancers, IRSs have unique functions, with IRS-1 being mainly involved in cell proliferation and survival, whereas IRS-2 has a clear role in regulating cell migration and metastasis (Chan and Lee 2008; Gibson et al. 2007). Overexpression of the functional IRS protein was sufficient to induce a mitogenic response to insulin (D'Ambrosio et al. 1995). Both MMTV-HA-IRS-1 and MMTV-HA-IRS-2 transgenic mice developed progressive mammary gland hyperplasia, tumors, and metastasis. In these mice, preneoplastic ductal hyperplasia appeared as early as 24 weeks of age, and multiple mammary gland lesions by 1 year of age (Dearth et al. 2006).

Amplified in breast cancer 1 (AIB1)/steroid receptor coactivator-3 (SRC-3) is a member of the p160 family of nuclear receptor co-regulators (Xu and Li 2003). The AIB1/SRC-3 gene is amplified in up to 10% of human breast cancers and is overexpressed in 30% of breast cancer cell lines (List et al. 2001; Gnanapragasam et al. 2001). In breast cancer, high levels of AIB1/SRC-3 and the growth factor receptor HER2/neu predict resistance to endocrine therapy and poor outcome. Preclinical breast cancer models have been generated by overexpressing AIB1/SRC-3 in breast epithelium (Avivar et al. 2006; Torres-Arzayus et al. 2004). Transgenic female mice display mammary hyperplasia at the onset of puberty, with enhanced proliferation of primary mammary epithelial cultures and augmented levels of cyclin D1 and E-cadherin (Avivar et al. 2006). AIB1/SRC-3 overexpression leads to increased levels of (IGF)-I, suggesting that its mechanism of action involves the establishment of an autocrine IGF-I loop (Torres-Arzayus et al. 2004). AIB1/SRC-3 is required for HER2/neu oncogenic activity and for the phosphorylation and activation of the HER2/neu receptor (Fereshteh et al. 2008).

24.2.2.2 Second Generation: Tumor Suppressor Genes Knockout Mouse Models (Abate-Shen et al. 2008)

Breast cancer can arise from loss-of-function mutations or epigenetic silencing of tumor suppressor genes (reviewed in Widschwendter and Jones 2002). Studies of human familial breast cancer showed that many breast cancers can be predisposed by germline mutation of several tumor suppressor genes, such as BRCA1, BRCA2, and p53.

The tumor suppressor gene TP53 is mutated in approximately 50% of primary human breast cancers (Allred et al. 1994; Elledge and Allred 1994). Donehower et al. introduced a null mutation into the p53 gene by homologous recombination in murine embryonic stem cells (1992). p53 knockout mice are prone to spontaneous development of a variety of neoplasms. However, although mice homozygous for null p53 do develop a diverse array of tumors, mammary tumors are only rarely observed. This may be because the absence of p53 alone is not sufficient to transform mammary cells, or that early death of p53 null mice from thymic lymphomas may obscure other tumor phenotypes that would develop later. In order to overcome these limitations, several approaches have been used. Transgenic mice have been generated that carry a mutant p53 (172R-H) under the control of the whey acidic protein (WAP) promoter (Li et al. 1998; Medina et al. 2002). These mice display lower tumor incidence, but exhibit increased mammary tumor incidence in response to the carcinogen DMBA (Medina et al. 2002). Another way to assess the role of p53 loss on mammary tumorigenesis has been to use mammary transplant techniques. In the study by Jerry et al., p53-null mammary epithelium was transplanted into cleared mammary fat pads of wild-type p53 Bagg albino (BALB)/c hosts to allow long-term analysis of mammary tumor phenotypes (Jerry et al. 2000). This study showed that ER-positive and ER-negative mammary tumors develop in the p53-null transplanted mammary epithelial within 1 year. Similar results were obtained using the Cre/lox system to generate mammary specific knockout of p53. Lin et al. reported that somatic mutations of p53 in mouse mammary epithelial cells using the Cre/loxP system leads to ER-positive and -negative tumors, and that these tumors have a high rate of metastasis (2004). This tumor system reproduces many important features of high-grade ER-positive and ER-negative human breast cancers and provides a tool for studying the origins of ER-positive and -negative breast tumors in mice (Lin et al. 2004).

BRCA1 and BRCA2 are well-known breast cancer susceptibility genes (reviewed in Fackenthal and Olopade 2007; Lee and Boyer 2001; Narod and Foulkes 2004), whose inactivation is the most common cause of familial breast cancer. BRCA1/2 are large nuclear phosphoproteins with critical functions in DNA repair. Germline mutations in BRCA1 have been detected in up to 50% of familial breast cancers (Paterson 1998). BRCA1 tumors show a uniform tumor type of high-grade invasive ductal carcinomas and are usually ER and HER2/neu negative (Chappuis et al. 2000; Phillips 2000). Several groups have generated Brca1 (Ludwig et al. 1997; Hakem et al. 1996; Shen et al. 1998; Gowen et al. 1996; Liu et al. 1996; Ludwig et al. 2001; Xu et al. 1999) and Brca2 (Connor et al. 1997; Bennett et al. 2000) knockout mice with mutations in different portions of the genes to recapitulate the effects of these genetic lesions in mouse models. None of these heterozygous mutants showed a strong tumor predisposing phenotype; and most homozygous mice displayed embryonic lethality (Hohenstein et al. 2001; Suzuki et al. 1997; Sharan et al. 1997). Therefore, mammary tissue-specific knockout models were made using the Cre/lox system. Cre-mediated mutation of the Brca1 gene in mouse mammary epithelial cells caused increased apoptosis and abnormal ductal development, with mammary tumor formation after long latency (Xu et al. 1999; Deng 2002; Weaver et al. 2002).

24.2.2.3 Modern Multigene Transgenic Models

As discussed above, transgenic and knockout mice are valuable tools for breast cancer research. However, these mouse models have drawbacks, such as embryonic lethality and lack of tissue specificity. To circumvent the drawbacks, conditional knockout methods have been developed. Tissue-specific knockouts have been generated using the Cre/lox recombination system (Sauer 1998; Kuhn et al. 1995). Currently, several tissue-specific promoters have been used for generation of Cre-recombinase mice. Studies using tissue-specific Cre expression have shown that transgenic mice with WAP-Cre but not MMTV-Cre can be used as a powerful tool to study gene function in development and tumorigenesis in the mammary gland (Wagner et al. 1997).

Another limitation of the several tissue-specific promoter-driven Cre-recombinase transgenes is that expression is influenced by hormones. The WAP-Cre and MMTV-Cre must be subject to multiple cycles of pregnancy/lactation to induce expression. To circumvent these limitations, transgenic mice were created expressing Cre under the control of the cytokeratin-14 (K14) promoter (Jonkers et al. 2001). Cre-recombinase activity was found in skin, salivary glands and in luminal epithelial cells and myoepithelial cells of virgin mammary glands.

24.2.3 Other Transgene Delivery Models

Traditionally, mouse models of breast cancer have been generated by introducing genetic alterations in the entire mammary epithelium using transgenic or knockout approaches.

However, recently other systems for delivering oncogenes have been developed. Du et al. (2006) adapted the avian leukosis virus RCAS (replication-competent avian sarcoma-leukosis virus LTR splice acceptor)-mediated somatic gene transfer technique to introduce oncogenes into mammary cells. The transgene is avian subgroup A receptor gene (TVA, a member of the low density lipoprotein family and a cell surface receptor for subgroup A avian leukosis virus), which is under the control of the MMTV promoter. In contrast to other methods, the RCAS-TVA model allows the introduction of genes into a small subset of somatic mammary cells in developmentally normal mammary glands. This new method allows the testing of the carcinogenic potential of many candidate oncogenes *in vivo* without the need to create individual transgenic lines. Moreover, the progression of mammary tumor in this RCAS-TVA model mimics the naturally occurring human breast cancer in that a small number of ductal epithelial cells express oncogenic protein. Thus, this mouse model closely recapitulates evolution of human breast cancer, which may help to understand human breast cancer initiation and progression (Du and Li 2007). This model has also been successfully used to study cancers other than breast cancer (Mayr et al. 2008).

24.3 Use of Animal Models for Breast Cancer Prevention

Modern breast cancer animal models mimic the situation of the human disease more accurately than early models. These models have been used for basic, translational, and preclinical studies. In addition to helping to understand breast carcinogenesis, the modern models have been used to test novel chemopreventive agents and to define the mechanisms of these chemopreventive drugs (Shen and Brown 2003; Uray and Brown 2006; Brown and Lippman 2000). Targeting various gene regulation and signal transduction pathways becomes a promising strategy for breast cancer prevention (William et al. 2009). These targets are illustrated in Fig. 24.1.

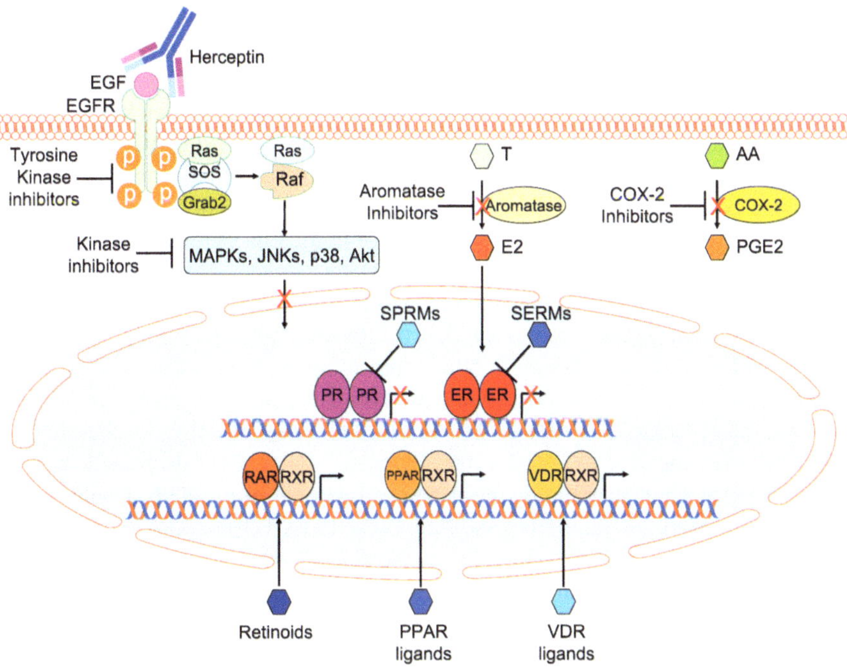

Fig. 24.1 Targeting various gene regulation and signal transduction pathways for the prevention of breast cancer. *EGF* epidermal growth factor, *EGFR* epidermal growth factor receptor, *MAPK* mitogen-activated protein kinase, *JNK* Jun N-terminal kinase, *T* testosterone, *E2* estradiol, *AA* arachidonic acid, *COX-2* cyclooxygenase-2, *PGE2* prostaglandin E2, *PR* progesterone receptor, *ER* estrogen receptor, *SERM* selective estrogen receptor modulator, *SPRM* selective progesterone receptor modulator, *RAR* retinoic acid receptor, *RXR* retinoic X receptor, *PPAR* peroxisome proliferator-activated receptor, *VDR* vitamin D receptor

24.3.1 Antiestrogens or Selective Estrogen Receptor Modulators (SERMs)

Blocking estrogen signaling using SERMs has been shown to prevent mammary cancer in animals and humans (Fabian and Kimler 2005; Ariazi et al. 2006). The best studied chemopreventive agents are tamoxifen and raloxifene (reviewed in Jordan and Morrow 1999).

Tamoxifen can prevent rat mammary carcinogenesis induced by DMBA (Jordan and Allen 1980), NMU (Gottardis and Jordan 1987), and ionizing radiation (Welsch et al. 1981). In addition, long-term treatment with tamoxifen prevents spontaneous carcinogenesis in mice infected with MMTV (Jordan et al. 1991). Tamoxifen has also proven to be effective for the chemoprevention of breast cancer in women at increased risk (reviewed in Jordan 2007; Chlebowski et al. 1999). Results from the National Surgical Adjuvant Breast and Bowel Project (NSABP) prevention trial showed that tamoxifen suppresses the development of ER-positive breast cancers by about 50% (Fisher et al. 1998), but has no effect on ER-negative breast cancer (Wickerham et al. 2009). Results from the International Breast Cancer Intervention Study (IBIS)-I trial found that tamoxifen reduced the risk of invasive ER-positive tumors by 31% in women at increased risk for breast cancer (Cuzick et al. 2007; Cuzick et al. 2002). However, tamoxifen is associated with side effects (Rohatgi et al. 2002). The agonist effects of tamoxifen on the uterus and liver, which result in an increased incidence of uterine cancer and thromboembolic phenomena, keep tamoxifen from being an ideal SERM and have limited its use in high-risk women (Bergman et al. 2000).

Raloxifene belongs to the benzothiophene class of SERMs. Raloxifene has similar effects as tamoxifen in mouse models (O'Regan et al. 2002). Raloxifene has a greater estrogen agonist activity in bone (preventing bone loss) and reduced estrogen agonist activity in the uterus (without causing uterine hypertrophy) in ovariectomized rats (Black et al. 1994). In the Multiple Outcomes of Raloxifene Evaluation (MORE) trial, the risk of invasive breast cancer significantly decreased in postmenopausal women (Cummings et al. 1999; Lippman et al. 2006; Cauley et al. 2001). In the Study of Tamoxifen and Raloxifene (STAR) trial, raloxifene is shown to be as effective as tamoxifen in reducing the risk of invasive breast cancer and has a lower risk of adverse events, such as uterine cancer, thromboemboli and cataract (Vogel et al. 2006; Vogel 2009).

Arzoxifene (LY353381) is a novel third-generation SERM that is a potent estrogen antagonist in mammary and uterine tissue while acting as an estrogen agonist to maintain bone density and lower serum cholesterol (Sporn 2004). Arzoxifene is a highly effective agent for the prevention of mammary cancer induced in the rat by the carcinogen NMU and is significantly more potent than raloxifene (Suh et al. 2001). Arzoxifene has been tested in early phase cancer prevention trials (Fabian et al. 2004). Acolbifene (EM-652, or its active metabolite, EM-800) is a fourth-generation SERM of the benzopyrans class. Acolbifene not only prevented estrogen-induced tumor growth, but also reduced tumor size in the tamoxifen-resistant ZR75-1 xenograft mouse model (Roy et al. 2003). Lasofoxifene is another fourth-generation SERM (Gennari 2005). The Postmenopausal Evaluation and Risk-reduction with Lasofoxifene (PEARL) trial showed that lasofoxifene reduces the incidence of ER-positive breast cancer and has a favorable benefit-risk profile for the prevention of clinical fractures in postmenopausal women with osteoporosis (LaCroix et al. 2009).

24.3.2 Progesterone Antagonists

Result from hormone replacement therapy (HRT) demonstrated that combined use of estrogen plus progestin has been associated with an elevation in breast cancer risk in postmenopausal women (Rossouw et al. 2002). Progesterone regulates target gene expression via two progesterone receptors: PR-A and PR-B (Graham et al. 1996). PRs play an important role in mammary tumorigenesis (Lange 2008; Liang et al. 2007). Various studies were conducted to assess progesterone receptor antagonists and selective progesterone receptor modulators (SPRMs) with respect to PR agonistic and antagonistic activities in breast cancer (Klijn et al. 2000). There are many SPRMs developed, such as mifepristone (RU 486), asoprisnil (J 867), onapristone (ZK 98299), ulipristal (CDB 2914), Proellex (CDB 4124), ORG 33628, and ORG 31710 (Benagiano et al. 2008a, b, c; Spitz 2006).

Mifepristone (RU 486) (DiPierri 1994) is the most widely used progesterone receptor antagonist. Mifepristone treatment increased the tumor latency period in DMBA-induced rat mammary tumor model (Bakker et al. 1987). Combined treatment with mifepristone and tamoxifen caused additive tumor growth inhibitory effects (Bakker et al. 1989, 1990). In the nude mice transplantation model, the addition of mifepristone was also effective in delaying or preventing tumor relapse from the antiestrogenic effect of tamoxifen (El Etreby and Liang 1998). Poole et al. showed that progesterone antagonist mifepristone prevented mammary tumorigenesis using the Brca1/p53-deficient mouse model (2006). Onapristone has also been shown to have potent anti-breast tumor activity in DMBA-and NMU-induced rat mammary tumors (Michna et al. 1989a, b; Schneider et al. 1990).

24.3.3 Aromatase Inhibitors

Aromatase, a member of the cytochrome P450 superfamily, is the enzyme responsible for a key step in the biosynthesis of estrogens. Overexpression of aromatase in mammary glands of transgenic mice results in hyperplasia and morphological abnormalities (Tekmal et al. 1996), and changes the expression of genes involved in apoptosis, cell cycle, and growth (Kirma et al. 2001). Aromatase overexpression also increases the expression of both estrogen and progesterone receptors. Aromatase inhibitors can prevent aromatase-mediated conversion of androstenedione and testosterone to estrone and estradiol, thus lowering estradiol concentrations in tissue and depriving breast tumor cells of the growth promoting effects of estrogens. Many drugs are currently available which can nearly completely inhibit aromatase activity (Brueggemeier et al. 2005; Campos 2004; Santen et al. 1999; Cocconi 1994; Masamura et al. 1995).

In mouse models, 4-hydroxy-f-androstene-3,17-dione (4-OH-A) can inhibit the conversion of 4-androstene-3,17-dione to estrogens. 4-OH-A treatment of rats with estrogen-dependent breast tumors induced by DMBA caused 80% of the tumors to regress significantly after 4 weeks, 42% of these tumors regressed completely (Brodie et al. 1977). In a transgenic mouse model, overexpression of aromatase results in increased tissue estrogenic activity and induction of hyperplastic and dysplastic lesions in mammary glands (Kirma et al. 2001). In prevention studies, these aromatase-induced changes in the mammary glands can be abrogated with very low concentrations of letrozole, without affecting normal physiology (Luthra et al. 2003). Another aromatase inhibitor, vorozole (R-83842), significantly decreases NMU-induced mammary tumor incidence while simultaneously decreasing tumor multiplicity (Lubet et al. 1994, 1998).

Clinical studies have demonstrated the effectiveness of aromatase inhibitors for the treatment of postmenopausal women with early and advanced breast cancer (Cuzick 2005, 2008a; Dunn and Ryan 2009). Anastrozole and letrozole were more effective than tamoxifen in postmenopausal women with advanced breast cancer. Anastrozole also has a more favorable side-effect profile compared to tamoxifen with fewer gynecological problems and vascular events (Bonneterre et al. 2000; Nabholtz et al. 2000). Data from the Arimidex, Tamoxifen, Alone or in Combination (ATAC) adjuvant trial indicate that the aromatase inhibitor anastrozole is more effective than tamoxifen in reducing recurrence and preventing new mammary tumors (Baum et al. 2002, 2003; Howell et al. 2005). Recently, a phase III breast cancer prevention trial, IBIS-II, has been launched aiming to recruit 6,000 high-risk postmenopausal women comparing anastrozole against placebo (Cuzick 2003, 2008b). Another phase III breast cancer prevention trial is currently underway with aromatase inhibitor exemestane. In addition, a National Cancer Institute of Canada Clinical Trials Group Mammary Prevention 3 (MAP.3) trial has been launched which compares exemestane to placebo for breast cancer risk reduction in 4,560 postmenopausal women (Goss et al. 2007; Richardson et al. 2007). These three important trials should determine whether these aromatase inhibitors are able to prevent the development of breast cancer in high-risk women.

24.3.4 Retinoids/Rexinoids

SERMs and aromatase inhibitors prevent only ER-positive breast cancers. Therefore, there is an urgent need for preventive options for ER-negative breast cancers. Thus, other agents are being studied. Retinoids, which are vitamin A analogues, have been shown to prevent head and neck tumors and have been tested in animals and humans for breast cancer prevention (Yang et al. 1999; Lippman and Lotan 2000). Retinoids bind to their receptors, i.e., retinoic acid receptors (RARs) and retinoid X receptors (RXRs) (Lala et al. 1996). In contrast to RAR, RXR proteins can form heterodimers with different partners, including thyroid hormone receptors (TR), vitamin D receptor (VDR), peroxisome proliferator-activated receptors (PPARs), and a number of orphan receptors (Chambon 1996). Binding of retinoids/rexinoids to these receptors leads to regulation of several cellular processes, including growth, differentiation, and apoptosis.

The naturally occurring retinoid 9-cis-retinoic acid (9cRA) prevents the development of NMU-induced mammary tumors in rats (Anzano et al. 1994), and suppresses ER-negative mammary tumor development in the C3(1)-SV40 Tag transgenic mouse model (Wu et al. 2000). However, in human clinical trials, 9cRA has been found to have significant toxicity, including skin changes, liver toxicity, cracking of the lips, and headaches (Miller et al. 1996). Increased understanding of retinoid receptor biology has facilitated the design of synthetic ligands with increased selectivity and hence decreased toxicity (reviewed in Howe 2007). Ligands selective for the RXR subclass have been developed, termed rexinoids, which have been found to be equally or more effective at preventing mammary tumors with substantially diminished toxicity. These RXR-selective retinoids are now being tested for their ability to prevent breast cancer in humans. Compounds of particular interest include bexarotene and LG100268.

Bexarotene (LGD1069) suppresses ER-negative mammary tumorigenesis in C3(1)-SV40 Tag transgenic mice (Wu et al. 2002a) and MMTV-erbB2 mice (Wu et al. 2002b), with minimal toxicity compared with RAR-selective retinoids. Studies using the p53-null mammary epithelium transplantation model also showed that bexarotene decreased the incidence of mammary tumor development in these mice (Medina et al. 2009). Other studies in the MMTV-erbB2 mice model and in the p53-null mice model have shown that the second-generation rexinoid LG100268 is even more effective at preventing ER-negative mammary tumors (Liby et al. 2008; Li et al. 2007, 2008). A novel rexinoid NRX194204 not only delayed the development of ER-negative mammary tumors using the MMTV-erbB2 mouse model in prevention studies, but also caused marked tumor regression or growth arrest in animals with established tumors (Liby et al. 2007). This class of preventive drugs has very promising potential in preventing ER-positive and ER-negative breast cancers. Bexarotene is now being tested in human clinical trials for the prevention of breast cancer in high-risk women. Recently, the results of a phase II prevention trial using bexarotene were presented (Brown et al. 2007). This study showed a reduction in cyclin D1 expression in women taking bexarotene.

24.3.5 Peroxisome Proliferator-Activated Receptor (PPAR) γ Ligands

PPAR γ is a PPAR subfamily member with three different isoforms (Issemann and Green 1990). The natural ligands of PPARs include long-chain polyunsaturated fatty acids, eicosanoid derivatives, and oxidized lipids (Issemann and Green 1990). PPARγ ligands have anticancer activity against a wide variety of cancers *in vitro* and *in vivo* (Koeffler 2003), including breast cancers (Baranova 2008). PPARs can form heterodimers with RXRs (Issemann et al. 1993; Serghides and Kain 2005; Palmer et al. 1994; Gearing et al. 1993). Rexinoids can potentiate the antiproliferative and apoptotic responses of breast cancer cell lines to PPAR ligands (Crowe and Chandraratna 2004).

Troglitazone (TGZ) is a synthetic antidiabetic drug, a high affinity-specific ligand of PPARγ (Lehmann et al. 1995). TGZ causes inhibition of proliferation in cultured breast cancer cells and inhibition of MCF7 xenograft tumor growth in immuno-deficient mice. Combined administration of TGZ and all-trans retinoic acid (ATRA) causes prominent apoptosis of the tumors without toxic effects (Elstner et al. 1998). TGZ in combination with rexinoids also prevented DMBA-induced preneoplastic mammary lesions in the BALB/c mouse model (Mehta et al. 2000a). A new ligand for PPARγ, GW7845, significantly reduced tumor incidence, tumor number, and tumor weight in NMU- or DMBA-induced mammary tumorigenesis rat models (Suh et al. 1999; Yin et al. 2005). These agents are promising cancer preventive drugs, particularly in combination with retinoids or rexinoids.

24.3.6 VDR Ligands

Vitamin D is a small lipophilic molecule which easily penetrates the cell membrane and binds to VDR. VDR is a nuclear hormone transcription factor that dimerizes with RXR proteins. VDR–RXR heterodimers bind to target DNA sequences to regulate target gene expression (McCullough et al. 2009). An increasing body of research supports the hypothesis that vitamin D has protective effects against the development of cancer, including breast cancer (Ingraham et al. 2008; Bertone-Johnson 2009; Colston 2008; Welsh 2007). Epidemiological studies indicated that high vitamin D intake and plasma concentration is inversely associated with breast cancer risk (Shin et al. 2002; Abbas et al. 2009; Bertone-Johnson et al. 2005; Freedman et al. 2007; Lowe et al. 2005). The active metabolite of vitamin D, 1α, 25-dihydroxyvitamin D3 (1,25 (OH)2 D3) has been shown to be able to reduce the growth and induce differentiation of human breast adenocarcinoma cells regardless of their sex steroid dependency (Chouvet et al. 1986; Bortman et al. 2002). However, recent clinical trials have not shown a cancer preventive effect of vitamin D supplementation (Chlebowski et al. 2008; Rohan et al. 2009).

Naturally occurring vitamin D can induce hypercalcemia *in vivo*, which has hindered its clinical use. Recently, the development of noncalcemic vitamin D analogues has renewed interest in targeting VDR for cancer prevention (Mehta et al. 2000b; Verlinden et al. 2000). Vitamin D analogues EB1089 and CB1093 inhibited the growth of breast cancer cells and prevented the anti-apoptotic effects of IGF-I (Xie et al. 1999). 22-oxacalcitriol (OCT) significantly suppressed the growth of tumors in the athymic mice MDA-MB-231 xenograft model (Matsumoto et al. 1999). 1α-hydroxy-24-ethylcholecalciferol [1α(OH)D(5)] can prevent preneoplastic mammary lesion development in the DMBA-induced mouse model (Mehta et al. 1997), and 1α(OH) D(5)treatment showed significant reduction of tumor incidence and multiplicity in the NMU carcinogenesis rat model (Mehta et al. 2003). Preclinical studies have shown that the vitamin D analogue EB 1089 has significantly less calcemic activity and is well-tolerated in animals (Gulliford et al. 1998). In an ER-negative xenograft model, Gemini vitamin D compounds significantly suppressed mammary tumor growth without hypercalcemia toxicity (Lee et al. 2008). These less toxic vitamin D analogues may provide promising agents to prevent ER-negative breast cancers.

24.3.7 Cyclooxygenase-2 Inhibitors

Cyclooxygenase-2 (COX-2) is an enzyme involved in prostaglandin synthesis from arachidonic acid (AA) that plays a vital role in inflammation. In breast cancer patients, elevated expression of COX-2 protein is associated with unfavorable distant disease-free survival, large tumor size, high histologic grade, and negative hormone receptor status (Ristimaki et al. 2002). Transgenic mice with overexpression of the human COX-2 gene in the mammary glands expressed reduced levels of the proapoptotic proteins Bcl-2-associated X protein (Bax) and B-cell lymphoma-extra large (Bcl-XL) and an increase in the anti-apoptotic protein Bcl-2. Enhanced COX-2 expression is sufficient to induce mammary gland tumorigenesis due to decreased apoptosis of mammary epithelial cells (Liu et al. 2001). Conversely, knocking out COX-2 significantly decreases HER2/neu-induced mammary tumor formation in mice (Howe et al. 2005).

Epidemiological studies provided clinical evidence of a chemopreventive effect of aspirin (Bosetti et al. 2006, 2009). Previous studies showed a chemopreventive effect for aspirin and other nonsteroidal anti-inflammatory drugs (NSAIDs) on colorectal cancer and perhaps other cancer types (Cuzick et al. 2009). The COX-2 inhibitor celecoxib is a nonsteroidal anti-inflammatory drug and is a potential candidate for the prevention of hormonally nonresponsive breast cancer (Chow et al. 2005; Arun and Goss 2004). Administration of celecoxib to rats fed a high fat diet rich in n-6 polyunsaturated fatty acids suppresses the promotion of mammary tumorigenesis induced by NMU (Lu et al. 2002). Celecoxib (1,500 ppm) also produced striking reductions in the incidence, multiplicity, and volume of breast tumors in DBMA-induced rats (Harris et al. 2000; Jang et al. 2002). Celecoxib (500 ppm) significantly reduced the incidence of mammary tumors in MMTV/neu mice (Howe et al. 2002; Lanza-Jacoby et al. 2003) and in a human xenograft model of breast cancer (Barnes et al. 2007). Celecoxib treatment also caused significant reduction in mammary tumor burden associated with increased tumor cell apoptosis, decreased proliferation, and suppressed angiogenesis in the MMTV-polyoma middle T mouse mammary tumor model (Basu et al. 2004). These results led to the development of several phase II clinical breast cancer prevention trials testing the effect of celecoxib in women at risk of breast cancer. However, several of these trials were stopped early because of concerns about the toxicity of COX-2 inhibitors. Thus, while the COX-2 pathway is a promising pathway for breast cancer prevention, less toxic drugs are needed before this pathway can be effectively targeted for breast cancer prevention.

24.3.8 Vaccination: Anti-erbB2 Antibody

Vaccination against selected oncoantigens is a promising option for breast cancer prevention (Lollini et al. 2006). Esserman et al. immunized neu-transgenic mice with a vaccine consisting of the extracellular domain of the HER2/neu oncogene product, p185 (1999). Immunized mice developed neu-specific humoral immune responses. The subsequent development of mammary tumors was significantly lower and vaccine treatment was associated with a significant increase in median survival (Esserman et al. 1999). Trastuzumab (Herceptin) is an anti-erbB2 antibody that has shown an impressive ability to shrink metastatic erbB2-positive breast cancers, and reduce the recurrence of early erbB2-positive breast cancer by 50%. Trastuzumab is a mouse–human chimeric monoclonal antibody that acts by blocking the HER2/neu receptor and abolishing its function, and can also interact with human immune cells to affect antibody dependent cell-mediated cytotoxicity (reviewed in Pegram et al. 2000; Dillman 1999). The success of trastuzumab in treating breast cancer has raised the possibility that an anti-erbB2 vaccine could be developed to prevent HER2-positive breast cancer (Hortobagyi 2001). In a phase I/II study, trastuzumab and an HER2/neu vaccine proved to be associated with minimal toxicity and results in prolonged antigen-specific immune responses (Salazar and Disis 2005).

24.3.9 Small-Molecule Epidermal Growth Factor Receptor Tyrosine Kinase Inhibitors

Uncontrolled epidermal growth factor receptor (EGFR) activity has been implicated in many aspects of breast tumor growth, including proliferative, angiogenic, invasive, and metastatic aspects (Woodburn 1999). Mutations in the ATP-binding site of the intracellular domain of the EGFR disable ligand-induced responses (Chen et al. 1987). Many synthetic compounds of different chemical classes which may inhibit EGFR kinase activity have been identified (Janmaat and Giaccone 2003). These molecules generally differ in their abilities to bind the EGFR ATP-binding pocket, either reversibly or irreversibly (Noonberg and Benz 2000). Some of these agents have been investigated in preclinical and clinical studies. Gefitinib (ZD1839, Iressa) is an orally active, selective, and reversible EGFR tyrosine kinase inhibitor (Wakeling et al. 2002). Gefitinib exhibited antitumor activity in the HER-2/neu overexpression ER-negative xenograft (Moulder et al. 2001) and in the MMTV-erbB2 transgenic mouse model (Lu et al. 2003). Lapatinib (GW572016) is an orally active small molecule that functions as a dual kinase inhibitor and reversibly inhibits ErbB1/ErbB2 tyrosine kinases (Nelson and Dolder 2006; Moy and Goss 2006), leading to inhibition of MAPK and PI3K signaling in ErbB1/ErbB2-overexpressing tumor cells and xenografts (Xia et al. 2002; Rusnak et al. 2001). Lapatinib suppressed the development of ER-negative and ErbB2-positive invasive mammary tumors in MMTV-erbB2 mice and reduced the numbers of premalignant lesions in the mammary glands (Strecker et al. 2009). These results led to the development of a clinical trial testing lapatinib's ability to reduce the growth of noninvasive breast cancer. This trial called the Ductal Carcinoma *In Situ* (DCIS) Lapatinib trial (LAPIS) is ongoing (http://clinicaltrials.gov/ct2/show/NCT00570453). The results of this clinical trial will provide critical information to develop tyrosine kinase inhibitors as agents for the prevention of breast cancer.

24.3.10 Combination Prevention

Carcinogenesis is considered to be a multistep process that involves the activation of multiple complex signal transduction pathways. Thus, it may be necessary to combine cancer preventive drugs that have different targets to effectively prevent breast cancer. The SERM arzoxifene and the rexinoid LG100268 have been shown to totally prevent NMU-induced carcinogenesis in the rat model (Suh et al. 2002; Rendi et al. 2004). The rexinoid LG100268 and the SERMs arzoxifene and acolbifene were also extremely effective in preventing ER-negative breast cancer in the MMTV-neu mice (Liby et al. 2006). Combination of submaximal doses of the selective COX-2 inhibitor celecoxib and the RXR-selective retinoid LGD1069 was substantially more effective than either drug administered individually in the MMTV-neu transgenic mouse model (Brown et al. 2008). Growth of the tumors in an intratumoral aromatase model was inhibited by both anastrozole and fulvestrant compared with the control tumors (Macedo et al. 2008).

24.4 Conclusions and Future Directions

The generation of animal models for breast cancer research has provided many new tools for understanding the molecular and cellular mechanisms of mammary tumorigenesis. These models are invaluable tools for breast cancer prevention research. Since chemoprevention was first introduced by Sporn et al. (Sporn 1976; Sporn et al. 1976), there has been significant progress in using vitamins or drugs to prevent cancer. Using these animal models for breast cancer, many important chemopreventive drugs and intervention strategies have been developed. Several drugs tested in these preclinical models are now being studied in human clinical trials. It is expected that the available carcinogen-induced and genetically engineered animal models will be used even more frequently in the future to test and develop cancer preventive drugs. Such models are invaluable to evaluate the cancer preventive activity of many drugs singly or in combination, and in evaluating the toxicities of these agents and combinations of agents.

References

Abate-Shen C et al (2008) The untapped potential of genetically engineered mouse models in chemoprevention research: opportunities and challenges. Cancer Prev Res (Phila) 1(3): 161–166

Abbas S, Chang-Claude J, Linseisen J (2009) Plasma 25-hydroxyvitamin D and premenopausal breast cancer risk in a German case–control study. Int J Cancer 124(1):250–255

Allred DC et al (1994) The p53 tumor-suppressor gene in human breast cancer. Cancer Treat Res 71:63–77

Andres AC et al (1987) Ha-ras oncogene expression directed by a milk protein gene promoter: tissue specificity, hormonal regulation, and tumor induction in transgenic mice. Proc Natl Acad Sci USA 84(5):1299–1303

Andrulis IL et al (1998) neu/erbB-2 amplification identifies a poor-prognosis group of women with node-negative breast cancer. Toronto Breast Cancer Study Group. J Clin Oncol 16(4):1340–1349

Anzano MA et al (1994) Prevention of breast cancer in the rat with 9-cis-retinoic acid as a single agent and in combination with tamoxifen. Cancer Res 54(17):4614–4617

Ariazi EA et al (2006) Estrogen receptors as therapeutic targets in breast cancer. Curr Top Med Chem 6(3):181–202
Arun B, Goss P (2004) The role of COX-2 inhibition in breast cancer treatment and prevention. Semin Oncol 31(2 Suppl 7):22–29
Avivar A et al (2006) Moderate overexpression of AIB1 triggers pre-neoplastic changes in mammary epithelium. FEBS Lett 580(22):5222–5226
Bakker GH et al (1987) Comparison of the actions of the antiprogestin mifepristone (RU486), the progestin megestrol acetate, the LHRH analog buserelin, and ovariectomy in treatment of rat mammary tumors. Cancer Treat Rep 71(11):1021–1027
Bakker GH et al (1989) Endocrine and antitumor effects of combined treatment with an antiprogestin and antiestrogen or luteinizing hormone-releasing hormone agonist in female rats bearing mammary tumors. Endocrinology 125(3):1593–1598
Bakker GH et al (1990) Treatment of breast cancer with different antiprogestins: preclinical and clinical studies. J Steroid Biochem Mol Biol 37(6):789–794
Baranova A (2008) PPAR ligands as potential modifiers of breast carcinoma outcomes. PPAR Res 2008:230893
Barnes NL et al (2007) Cyclooxygenase-2 inhibition: effects on tumour growth, cell cycling and lymphangiogenesis in a xenograft model of breast cancer. Br J Cancer 96(4):575–582
Basu GD et al (2004) Cyclooxygenase-2 inhibitor induces apoptosis in breast cancer cells in an in vivo model of spontaneous metastatic breast cancer. Mol Cancer Res 2(11):632–642
Baum M et al (2002) Anastrozole alone or in combination with tamoxifen versus tamoxifen alone for adjuvant treatment of postmenopausal women with early breast cancer: first results of the ATAC randomised trial. Lancet 359(9324):2131–2139
Baum M et al (2003) Anastrozole alone or in combination with tamoxifen versus tamoxifen alone for adjuvant treatment of postmenopausal women with early-stage breast cancer: results of the ATAC (Arimidex, Tamoxifen Alone or in Combination) trial efficacy and safety update analyses. Cancer 98(9):1802–1810
Benagiano G, Bastianelli C, Farris M (2008a) Selective progesterone receptor modulators 3: use in oncology, endocrinology and psychiatry. Expert Opin Pharmacother 9(14):2487–2496
Benagiano G, Bastianelli C, Farris M (2008b) Selective progesterone receptor modulators 2: use in reproductive medicine. Expert Opin Pharmacother 9(14):2473–2485
Benagiano G, Bastianelli C, Farris M (2008c) Selective progesterone receptor modulators 1: use during pregnancy. Expert Opin Pharmacother 9(14):2459–2472
Bennett LM et al (2000) BRCA2-null embryonic survival is prolonged on the BALB/c genetic background. Mol Carcinog 28(3):174–183
Bergman L et al (2000) Risk and prognosis of endometrial cancer after tamoxifen for breast cancer. Comprehensive Cancer Centres' ALERT Group. Assessment of Liver and Endometrial cancer Risk following Tamoxifen. Lancet 356(9233):881–887
Bertone-Johnson ER (2009) Vitamin D and breast cancer. Ann Epidemiol 19(7):462–467
Bertone-Johnson ER et al (2005) Plasma 25-hydroxyvitamin D and 1,25-dihydroxyvitamin D and risk of breast cancer. Cancer Epidemiol Biomarkers Prev 14(8):1991–1997
Black LJ et al (1994) Raloxifene (LY139481 HCI) prevents bone loss and reduces serum cholesterol without causing uterine hypertrophy in ovariectomized rats. J Clin Invest 93(1):63–69
Bonneterre J et al (2000) Anastrozole versus tamoxifen as first-line therapy for advanced breast cancer in 668 postmenopausal women: results of the Tamoxifen or Arimidex Randomized Group Efficacy and Tolerability study. J Clin Oncol 18(22):3748–3757
Bortman P et al (2002) Antiproliferative effects of 1,25-dihydroxyvitamin D3 on breast cells: a mini review. Braz J Med Biol Res 35(1):1–9
Bosetti C, Gallus S, La Vecchia C (2006) Aspirin and cancer risk: an updated quantitative review to 2005. Cancer Causes Control 17(7):871–888
Bosetti C, Gallus S, La Vecchia C (2009) Aspirin and cancer risk: a summary review to 2007. Recent Results Cancer Res 181:231–251
Bouchard L et al (1989) Stochastic appearance of mammary tumors in transgenic mice carrying the MMTV/c-neu oncogene. Cell 57(6):931–936

Brodie AM et al (1977) The effect of an aromatase inhibitor, 4-hydroxy-4-androstene-3,17-dione, on estrogen-dependent processes in reproduction and breast cancer. Endocrinology 100(6):1684–1695

Brown PH, Lippman SM (2000) Chemoprevention of breast cancer. Breast Cancer Res Treat 62(1):1–17

Brown P et al (2007) Prevention of breast cancer using rexinoids: results of a phase II biomarker modulation trial using bexarotene in women at high risk of breast cancer. Breast Cancer Res Treat 106(Suppl):181, #4046

Brown PH et al (2008) Combination chemoprevention of HER2/neu-induced breast cancer using a cyclooxygenase-2 inhibitor and a retinoid X receptor-selective retinoid. Cancer Prev Res (Phila) 1(3):208–214

Brueggemeier RW, Hackett JC, Diaz-Cruz ES (2005) Aromatase inhibitors in the treatment of breast cancer. Endocr Rev 26(3):331–345

Bui TD et al (1997) A novel human Wnt gene, WNT10B, maps to 12q13 and is expressed in human breast carcinomas. Oncogene 14(10):1249–1253

Campos SM (2004) Aromatase inhibitors for breast cancer in postmenopausal women. Oncologist 9(2):126–136

Cauley JA et al (2001) Continued breast cancer risk reduction in postmenopausal women treated with raloxifene: 4-year results from the MORE trial. Multiple outcomes of raloxifene evaluation. Breast Cancer Res Treat 65(2):125–134

Chambon P (1996) A decade of molecular biology of retinoic acid receptors. FASEB J 10(9):940–954

Chan BT, Lee AV (2008) Insulin receptor substrates (IRSs) and breast tumorigenesis. J Mammary Gland Biol Neoplasia 13(4):415–422

Chappuis PO, Nethercot V, Foulkes WD (2000) Clinico-pathological characteristics of BRCA1- and BRCA2-related breast cancer. Semin Surg Oncol 18(4):287–295

Chen WS et al (1987) Requirement for intrinsic protein tyrosine kinase in the immediate and late actions of the EGF receptor. Nature 328(6133):820–823

Chlebowski RT et al (1999) American Society of Clinical Oncology technology assessment on breast cancer risk reduction strategies: tamoxifen and raloxifene. J Clin Oncol 17(6):1939–1955

Chlebowski RT et al (2008) Calcium plus vitamin D supplementation and the risk of breast cancer. J Natl Cancer Inst 100(22):1581–1591

Chouvet C et al (1986) 1,25-Dihydroxyvitamin D3 inhibitory effect on the growth of two human breast cancer cell lines (MCF-7, BT-20). J Steroid Biochem 24(1):373–376

Chow LW, Loo WT, Toi M (2005) Current directions for COX-2 inhibition in breast cancer. Biomed Pharmacother 59(Suppl 2):S281–S284

Cocconi G (1994) First generation aromatase inhibitors–aminoglutethimide and testololactone. Breast Cancer Res Treat 30(1):57–80

Colston KW (2008) Vitamin D and breast cancer risk. Best Pract Res Clin Endocrinol Metab 22(4):587–599

Connor F et al (1997) Tumorigenesis and a DNA repair defect in mice with a truncating Brca2 mutation. Nat Genet 17(4):423–430

Crowe DL, Chandraratna RA (2004) A retinoid X receptor (RXR)-selective retinoid reveals that RXR-alpha is potentially a therapeutic target in breast cancer cell lines, and that it potentiates antiproliferative and apoptotic responses to peroxisome proliferator-activated receptor ligands. Breast Cancer Res 6(5):R546–R555

Cummings SR et al (1999) The effect of raloxifene on risk of breast cancer in postmenopausal women: results from the MORE randomized trial. Multiple outcomes of raloxifene evaluation. JAMA 281(23):2189–2197

Cuzick J (2003) Aromatase inhibitors in prevention – data from the ATAC (arimidex, tamoxifen alone or in combination) trial and the design of IBIS-II (the second International Breast Cancer Intervention Study). Recent Results Cancer Res 163:96–103;discussion 264–266

Cuzick J (2005) Aromatase inhibitors for breast cancer prevention. J Clin Oncol 23(8): 1636–1643

Cuzick J (2008a) Chemoprevention of breast cancer. Breast Cancer 15(1):10–16

Cuzick J (2008b) IBIS II: a breast cancer prevention trial in postmenopausal women using the aromatase inhibitor anastrozole. Expert Rev Anticancer Ther 8(9):1377–1385

Cuzick J et al (2002) First results from the International Breast Cancer Intervention Study (IBIS-I): a randomised prevention trial. Lancet 360(9336):817–824

Cuzick J et al (2007) Long-term results of tamoxifen prophylaxis for breast cancer–96-month follow-up of the randomized IBIS-I trial. J Natl Cancer Inst 99(4):272–282

Cuzick J et al (2009) Aspirin and non-steroidal anti-inflammatory drugs for cancer prevention: an international consensus statement. Lancet Oncol 10(5):501–507

Dale TC et al (1996) Compartment switching of WNT-2 expression in human breast tumors. Cancer Res 56(19):4320–4323

D'Ambrosio C et al (1995) Transforming potential of the insulin receptor substrate 1. Cell Growth Differ 6(5):557–562

Dearth RK et al (2006) Mammary tumorigenesis and metastasis caused by overexpression of insulin receptor substrate 1 (IRS-1) or IRS-2. Mol Cell Biol 26(24):9302–9314

Dearth RK et al (2007) Oncogenic transformation by the signaling adaptor proteins insulin receptor substrate (IRS)-1 and IRS-2. Cell Cycle 6(6):705–713

Deng CX (2002) Tumor formation in Brca1 conditional mutant mice. Environ Mol Mutagen 39(2–3):171–177

Dillman RO (1999) Perceptions of herceptin: a monoclonal antibody for the treatment of breast cancer. Cancer Biother Radiopharm 14(1):5–10

DiPierri D (1994) RU 486, mifepristone: a review of a controversial drug. Nurse Pract 19(6):59–61

Donehower LA et al (1992) Mice deficient for p53 are developmentally normal but susceptible to spontaneous tumours. Nature 356(6366):215–221

Du Z, Li Y (2007) RCAS-TVA in the mammary gland: an in vivo oncogene screen and a high fidelity model for breast transformation? Cell Cycle 6(7):823–826

Du Z et al (2006) Introduction of oncogenes into mammary glands in vivo with an avian retroviral vector initiates and promotes carcinogenesis in mouse models. Proc Natl Acad Sci USA 103(46):17396–17401

Dunn BK, Ryan A (2009) Phase 3 trials of aromatase inhibitors for breast cancer prevention: following in the path of the selective estrogen receptor modulators. Ann N Y Acad Sci 1155:141–161

Dunning WF, Curtis MR, Segaloff A (1948) Strain differences in response to diethylstilbestrol and the induction of mammary gland, adrenal and bladder cancer in the rat. J Mich State Med Soc 47(3):305

Dyson N et al (1989) The cellular 107K protein that binds to adenovirus E1A also associates with the large T antigens of SV40 and JC virus. Cell 58(2):249–255

Eccles SA (2001) The role of c-erbB-2/HER2/neu in breast cancer progression and metastasis. J Mammary Gland Biol Neoplasia 6(4):393–406

El Etreby MF, Liang Y (1998) Effect of antiprogestins and tamoxifen on growth inhibition of MCF-7 human breast cancer cells in nude mice. Breast Cancer Res Treat 49(2):109–117

Elledge RM, Allred DC (1994) The p53 tumor suppressor gene in breast cancer. Breast Cancer Res Treat 32(1):39–47

Elstner E et al (1998) Ligands for peroxisome proliferator-activated receptorgamma and retinoic acid receptor inhibit growth and induce apoptosis of human breast cancer cells in vitro and in BNX mice. Proc Natl Acad Sci USA 95(15):8806–8811

Esserman LJ et al (1999) Vaccination with the extracellular domain of p185neu prevents mammary tumor development in neu transgenic mice. Cancer Immunol Immunother 47(6):337–342

Fabian CJ, Kimler BF (2005) Selective estrogen-receptor modulators for primary prevention of breast cancer. J Clin Oncol 23(8):1644–1655

Fabian CJ et al (2004) Breast cancer chemoprevention phase I evaluation of biomarker modulation by arzoxifene, a third generation selective estrogen receptor modulator. Clin Cancer Res 10(16):5403–5417

Fackenthal JD, Olopade OI (2007) Breast cancer risk associated with BRCA1 and BRCA2 in diverse populations. Nat Rev Cancer 7(12):937–948
Fereshteh MP et al (2008) The nuclear receptor coactivator amplified in breast cancer-1 is required for Neu (ErbB2/HER2) activation, signaling, and mammary tumorigenesis in mice. Cancer Res 68(10):3697–3706
Fisher B et al (1998) Tamoxifen for prevention of breast cancer: report of the National Surgical Adjuvant Breast and Bowel Project P-1 Study. J Natl Cancer Inst 90(18):1371–1388
Freedman DM et al (2007) Prospective study of serum vitamin D and cancer mortality in the United States. J Natl Cancer Inst 99(21):1594–1602
Gearing KL et al (1993) Interaction of the peroxisome-proliferator-activated receptor and retinoid X receptor. Proc Natl Acad Sci USA 90(4):1440–1444
Gennari L (2005) Lasofoxifene (Pfizer). Curr Opin Investig Drugs 6(10):1067–1078
Gibson SL, Ma Z, Shaw LM (2007) Divergent roles for IRS-1 and IRS-2 in breast cancer metastasis. Cell Cycle 6(6):631–637
Gnanapragasam VJ et al (2001) Expression of RAC 3, a steroid hormone receptor co-activator in prostate cancer. Br J Cancer 85(12):1928–1936
Goss PE et al (2007) National Cancer Institute of Canada Clinical Trials Group MAP.3 Trial: evaluation of exemestane to prevent breast cancer in postmenopausal women. Clin Breast Cancer 7(11):895–900
Gottardis MM, Jordan VC (1987) Antitumor actions of keoxifene and tamoxifen in the N-nitrosomethylurea-induced rat mammary carcinoma model. Cancer Res 47(15):4020–4024
Gowen LC et al (1996) Brca1 deficiency results in early embryonic lethality characterized by neuroepithelial abnormalities. Nat Genet 12(2):191–194
Graham JD et al (1996) Progesterone receptor A and B protein expression in human breast cancer. J Steroid Biochem Mol Biol 56(1–6 Spec No):93–98
Green JE et al (2000) The C3(1)/SV40 T-antigen transgenic mouse model of mammary cancer: ductal epithelial cell targeting with multistage progression to carcinoma. Oncogene 19(8):1020–1027
Grushko TA et al (2004) MYC is amplified in BRCA1-associated breast cancers. Clin Cancer Res 10(2):499–507
Gulliford T et al (1998) A phase I study of the vitamin D analogue EB 1089 in patients with advanced breast and colorectal cancer. Br J Cancer 78(1):6–13
Gusterson BA, Williams JC (1981) N-nitrosomethylurea-induced rat mammary tumours as models of human breast cancer. J R Soc Med 74(1):56–59
Guy CT et al (1992) Expression of the neu protooncogene in the mammary epithelium of transgenic mice induces metastatic disease. Proc Natl Acad Sci USA 89(22):10578–10582
Hakem R et al (1996) The tumor suppressor gene Brca1 is required for embryonic cellular proliferation in the mouse. Cell 85(7):1009–1023
Hanahan D, Wagner EF, Palmiter RD (2007) The origins of oncomice: a history of the first transgenic mice genetically engineered to develop cancer. Genes Dev 21(18):2258–2270
Harris RE et al (2000) Chemoprevention of breast cancer in rats by celecoxib, a cyclooxygenase 2 inhibitor. Cancer Res 60(8):2101–2103
Hohenstein P et al (2001) A targeted mouse Brca1 mutation removing the last BRCT repeat results in apoptosis and embryonic lethality at the headfold stage. Oncogene 20(20):2544–2550
Hortobagyi GN (2001) Overview of treatment results with trastuzumab (Herceptin) in metastatic breast cancer. Semin Oncol 28(6 Suppl 18):43–47
Howe LR (2007) Rexinoids and breast cancer prevention. Clin Cancer Res 13(20):5983–5987
Howe LR, Brown AM (2004) Wnt signaling and breast cancer. Cancer Biol Ther 3(1):36–41
Howe LR et al (2002) Celecoxib, a selective cyclooxygenase 2 inhibitor, protects against human epidermal growth factor receptor 2 (HER-2)/neu-induced breast cancer. Cancer Res 62(19):5405–5407
Howe LR et al (2005) HER2/neu-induced mammary tumorigenesis and angiogenesis are reduced in cyclooxygenase-2 knockout mice. Cancer Res 65(21):10113–10119
Howell A et al (2005) Results of the ATAC (arimidex, tamoxifen, alone or in combination) trial after completion of 5 years' adjuvant treatment for breast cancer. Lancet 365(9453):60–62

Huang AL et al (1981) Glucocorticoid regulation of the Ha-MuSV p21 gene conferred by sequences from mouse mammary tumor virus. Cell 27(2 Pt 1):245–255
Huggins C, Grand LC, Brillantes FP (1961) Mammary cancer induced by a single feeding of polymucular hydrocarbons, and its suppression. Nature 189:204–207
Huggins CB, Ueda N, Wiessler M (1981) N-Nitroso-N-methylurea elicits mammary cancer in resistant and sensitive rat strains. Proc Natl Acad Sci USA 78(2):1185–1188
Huguet EL et al (1994) Differential expression of human Wnt genes 2, 3, 4, and 7B in human breast cell lines and normal and disease states of human breast tissue. Cancer Res 54(10):2615–2621
Hynes NE, Stern DF (1994) The biology of erbB-2/neu/HER-2 and its role in cancer. Biochim Biophys Acta 1198(2–3):165–184
Ingraham BA, Bragdon B, Nohe A (2008) Molecular basis of the potential of vitamin D to prevent cancer. Curr Med Res Opin 24(1):139–149
Issemann I, Green S (1990) Activation of a member of the steroid hormone receptor superfamily by peroxisome proliferators. Nature 347(6294):645–650
Issemann I et al (1993) The retinoid X receptor enhances the function of the peroxisome proliferator activated receptor. Biochimie 75(3–4):251–256
Jang TJ et al (2002) Chemopreventive effect of celecoxib and expression of cyclooxygenase-1 and cyclooxygenase-2 on chemically-induced rat mammary tumours. Int J Exp Pathol 83(4):173–182
Janmaat ML, Giaccone G (2003) Small-molecule epidermal growth factor receptor tyrosine kinase inhibitors. Oncologist 8(6):576–586
Jemal A et al (2008) Cancer statistics, 2008. CA Cancer J Clin 58(2):71–96
Jerry DJ et al (2000) A mammary-specific model demonstrates the role of the p53 tumor suppressor gene in tumor development. Oncogene 19(8):1052–1058
Jonkers J et al (2001) Synergistic tumor suppressor activity of BRCA2 and p53 in a conditional mouse model for breast cancer. Nat Genet 29(4):418–425
Jordan VC (2007) Chemoprevention of breast cancer with selective oestrogen-receptor modulators. Nat Rev Cancer 7(1):46–53
Jordan VC, Allen KE (1980) Evaluation of the antitumour activity of the non-steroidal antioestrogen monohydroxytamoxifen in the DMBA-induced rat mammary carcinoma model. Eur J Cancer 16(2):239–251
Jordan VC, Morrow M (1999) Tamoxifen, raloxifene, and the prevention of breast cancer. Endocr Rev 20(3):253–278
Jordan VC, Lababidi MK, Langan-Fahey S (1991) Suppression of mouse mammary tumorigenesis by long-term tamoxifen therapy. J Natl Cancer Inst 83(7):492–496
Kirma N et al (2001) Overexpression of aromatase leads to hyperplasia and changes in the expression of genes involved in apoptosis, cell cycle, growth, and tumor suppressor functions in the mammary glands of transgenic mice. Cancer Res 61(5):1910–1918
Klijn JG, Setyono-Han B, Foekens JA (2000) Progesterone antagonists and progesterone receptor modulators in the treatment of breast cancer. Steroids 65(10–11):825–830
Koeffler HP (2003) Peroxisome proliferator-activated receptor gamma and cancers. Clin Cancer Res 9(1):1–9
Kuhn R et al (1995) Inducible gene targeting in mice. Science 269(5229):1427–1429
LaCroix AZ et al (2009) Effects of 5 years of treatment with lasofoxifene on incidence of breast cancer in older women. Cancer Res 69(Suppl 2):11
Lala DS et al (1996) Activation of specific RXR heterodimers by an antagonist of RXR homodimers. Nature 383(6599):450–453
Lange CA (2008) Challenges to defining a role for progesterone in breast cancer. Steroids 73(9–10):914–921
Lanza-Jacoby S et al (2003) The cyclooxygenase-2 inhibitor, celecoxib, prevents the development of mammary tumors in Her-2/neu mice. Cancer Epidemiol Biomarkers Prev 12(12):1486–1491
Lee WH, Boyer TG (2001) BRCA1 and BRCA2 in breast cancer. Lancet 358(Suppl):S5
Lee HJ et al (2008) Gemini vitamin D analogues inhibit estrogen receptor-positive and estrogen receptor-negative mammary tumorigenesis without hypercalcemic toxicity. Cancer Prev Res (Phila) 1(6):476–484

Lehmann JM et al (1995) An antidiabetic thiazolidinedione is a high affinity ligand for peroxisome proliferator-activated receptor gamma (PPAR gamma). J Biol Chem 270(22):12953–12956
Li Y, Brown PH (2007) Translational approaches for the prevention of estrogen receptor-negative breast cancer. Eur J Cancer Prev 16(3):203–215
Li Y, Brown PH (2009) Prevention of ER-negative breast cancer. Recent Results Cancer Res 181:121–134
Li B et al (1998) A transgenic mouse model for mammary carcinogenesis. Oncogene 16(8):997–1007
Li Y, Hively WP, Varmus HE (2000) Use of MMTV-Wnt-1 transgenic mice for studying the genetic basis of breast cancer. Oncogene 19(8):1002–1009
Li SA et al (2002) Prevention of solely estrogen-induced mammary tumors in female aci rats by tamoxifen: evidence for estrogen receptor mediation. J Endocrinol 175(2):297–305
Li Y et al (2007) The Rexinoid LG100268 prevents the development of preinvasive and invasive estrogen receptor negative tumors in MMTV-erbB2 mice. Clin Cancer Res 13(20):6224–6231
Li Y et al (2008) The rexinoid, bexarotene, prevents the development of premalignant lesions in MMTV-erbB2 mice. Br J Cancer 98(8):1380–1388
Liang Y et al (2007) Progestin-dependent progression of human breast tumor xenografts: a novel model for evaluating antitumor therapeutics. Cancer Res 67(20):9929–9936
Liby K et al (2006) The combination of the rexinoid, LG100268, and a selective estrogen receptor modulator, either arzoxifene or acolbifene, synergizes in the prevention and treatment of mammary tumors in an estrogen receptor-negative model of breast cancer. Clin Cancer Res 12(19): 5902–5909
Liby K et al (2007) A new rexinoid, NRX194204, prevents carcinogenesis in both the lung and mammary gland. Clin Cancer Res 13(20):6237–6243
Liby K et al (2008) Prevention and treatment of experimental estrogen receptor-negative mammary carcinogenesis by the synthetic triterpenoid CDDO-methyl Ester and the rexinoid LG100268. Clin Cancer Res 14(14):4556–4563
Lin SC et al (2004) Somatic mutation of p53 leads to estrogen receptor alpha-positive and -negative mouse mammary tumors with high frequency of metastasis. Cancer Res 64(10): 3525–3532
Lippman SM, Lotan R (2000) Advances in the development of retinoids as chemopreventive agents. J Nutr 130(2S Suppl):479S–482S
Lippman ME et al (2006) Effect of raloxifene on the incidence of invasive breast cancer in postmenopausal women with osteoporosis categorized by breast cancer risk. Clin Cancer Res 12(17):5242–5247
List HJ et al (2001) Expression of the nuclear coactivator AIB1 in normal and malignant breast tissue. Breast Cancer Res Treat 68(1):21–28
Liu CY et al (1996) Inactivation of the mouse Brca1 gene leads to failure in the morphogenesis of the egg cylinder in early postimplantation development. Genes Dev 10(14):1835–1843
Liu CH et al (2001) Overexpression of cyclooxygenase-2 is sufficient to induce tumorigenesis in transgenic mice. J Biol Chem 276(21):18563–18569
Lohrisch C, Piccart M (2001) An overview of HER2. Semin Oncol 28(6 Suppl 18):3–11
Lollini PL et al (2006) Vaccines for tumour prevention. Nat Rev Cancer 6(3):204–216
Lowe LC et al (2005) Plasma 25-hydroxy vitamin D concentrations, vitamin D receptor genotype and breast cancer risk in a UK Caucasian population. Eur J Cancer 41(8):1164–1169
Lu S et al (2002) Cyclooxygenase-2 inhibitor celecoxib inhibits promotion of mammary tumorigenesis in rats fed a high fat diet rich in n-6 polyunsaturated fatty acids. Cancer Lett 184(1):7–12
Lu C et al (2003) Effect of epidermal growth factor receptor inhibitor on development of estrogen receptor-negative mammary tumors. J Natl Cancer Inst 95(24):1825–1833
Lubet RA et al (1994) Chemopreventive effects of the aromatase inhibitors vorozole (R-83842) and 4-hydroxyandrostenedione in the methylnitrosourea (MNU)-induced mammary tumor model in Sprague-Dawley rats. Carcinogenesis 15(12):2775–2780
Lubet RA et al (1998) Chemopreventive effects of the aromatase inhibitor vorozole (R 83842) in the methylnitrosourea-induced mammary cancer model. Carcinogenesis 19(8):1345–1351

Ludwig T et al (1997) Targeted mutations of breast cancer susceptibility gene homologs in mice: lethal phenotypes of Brca1, Brca2, Brca1/Brca2, Brca1/p53, and Brca2/p53 nullizygous embryos. Genes Dev 11(10):1226–1241
Ludwig T et al (2001) Tumorigenesis in mice carrying a truncating Brca1 mutation. Genes Dev 15(10):1188–1193
Luthra R et al (2003) Use of letrozole as a chemopreventive agent in aromatase overexpressing transgenic mice. J Steroid Biochem Mol Biol 86(3–5):461–467
Macedo LF et al (2008) Combination of anastrozole with fulvestrant in the intratumoral aromatase xenograft model. Cancer Res 68(9):3516–3522
Malaney S, Daly RJ (2001) The ras signaling pathway in mammary tumorigenesis and metastasis. J Mammary Gland Biol Neoplasia 6(1):101–113
Maroulakou IG et al (1994) Prostate and mammary adenocarcinoma in transgenic mice carrying a rat C3(1) simian virus 40 large tumor antigen fusion gene. Proc Natl Acad Sci USA 91(23): 11236–11240
Masamura S et al (1995) Aromatase inhibitor development for treatment of breast cancer. Breast Cancer Res Treat 33(1):19–26
Matsumoto H et al (1999) Antitumor effect of 22-oxacalcitriol on estrogen receptor-negative MDA-MB-231 tumors in athymic mice. Oncol Rep 6(2):349–352
Mayr U et al (2008) RCAS-mediated retroviral gene delivery: a versatile tool for the study of gene function in a mouse model of pancreatic cancer. Hum Gene Ther 19(9):896–906
McCormick DL et al (1981) Lifetime dose-response relationships for mammary tumor induction by a single administration of N-methyl-N-nitrosourea. Cancer Res 41(5):1690–1694
McCormick DL et al (1982) Enhanced inhibition of mammary carcinogenesis by combined treatment with N-(4-hydroxyphenyl)retinamide and ovariectomy. Cancer Res 42(2):508–512
McCullough ML, Bostick RM, Mayo TL (2009) Vitamin d gene pathway polymorphisms and risk of colorectal, breast, and prostate cancer. Annu Rev Nutr 29:111–132
Medina D et al (2002) Environmental carcinogens and p53 tumor-suppressor gene interactions in a transgenic mouse model for mammary carcinogenesis. Environ Mol Mutagen 39(2–3):178–183
Medina D et al (2009) Prevention of tumorigenesis in p53-null mammary epithelium by rexinoid bexarotene, tyrosine kinase inhibitor gefitinib, and celecoxib. Cancer Prev Res (Phila) 2(2):168–174
Mehta RG et al (1997) Prevention of preneoplastic mammary lesion development by a novel vitamin D analogue, 1alpha-hydroxyvitamin D5. J Natl Cancer Inst 89(3):212–218
Mehta RG et al (2000a) A ligand of peroxisome proliferator-activated receptor gamma, retinoids, and prevention of preneoplastic mammary lesions. J Natl Cancer Inst 92(5):418–423
Mehta RR et al (2000b) Differentiation of human breast carcinoma cells by a novel vitamin D analog: 1alpha-hydroxyvitamin D5. Int J Oncol 16(1):65–73
Mehta RG et al (2003) Chemoprevention of mammary carcinogenesis by 1alpha-hydroxyvitamin D5, a synthetic analog of Vitamin D. Mutat Res 523–524:253–264
Menard S et al (2004) Role of HER2/neu in tumor progression and therapy. Cell Mol Life Sci 61(23):2965–2978
Michna H et al (1989a) Antitumor activity of the antiprogestins ZK 98.299 and RU 38.486 in hormone dependent rat and mouse mammary tumors: mechanistic studies. Breast Cancer Res Treat 14(3):275–288
Michna H et al (1989b) The antitumor mechanism of progesterone antagonists is a receptor mediated antiproliferative effect by induction of terminal cell death. J Steroid Biochem 34(1–6):447–453
Mietz JA et al (1992) The transcriptional transactivation function of wild-type p53 is inhibited by SV40 large T-antigen and by HPV-16 E6 oncoprotein. EMBO J 11(13):5013–5020
Miller VA et al (1996) Initial clinical trial of the retinoid receptor pan agonist 9-cis retinoic acid. Clin Cancer Res 2(3):471–475
Moens U et al (1997) Mechanisms of transcriptional regulation of cellular genes by SV40 large T- and small T-antigens. Virus Genes 15(2):135–154

Moulder SL et al (2001) Epidermal growth factor receptor (HER1) tyrosine kinase inhibitor ZD1839 (Iressa) inhibits HER2/neu (erbB2)-overexpressing breast cancer cells in vitro and in vivo. Cancer Res 61(24):8887–8895
Moy B, Goss PE (2006) Lapatinib: current status and future directions in breast cancer. Oncologist 11(10):1047–1057
Muller WJ et al (1988) Single-step induction of mammary adenocarcinoma in transgenic mice bearing the activated c-neu oncogene. Cell 54(1):105–115
Nabholtz JM et al (2000) Anastrozole is superior to tamoxifen as first-line therapy for advanced breast cancer in postmenopausal women: results of a North American multicenter randomized trial. Arimidex Study Group. J Clin Oncol 18(22):3758–3767
Narod SA, Foulkes WD (2004) BRCA1 and BRCA2: 1994 and beyond. Nat Rev Cancer 4(9):665–676
Nelson MH, Dolder CR (2006) Lapatinib: a novel dual tyrosine kinase inhibitor with activity in solid tumors. Ann Pharmacother 40(2):261–269
Nielsen LL et al (1991) Histopathology of salivary and mammary gland tumors in transgenic mice expressing a human Ha-ras oncogene. Cancer Res 51(14):3762–3767
Nielsen LL, Gurnani M, Tyler RD (1992) Evaluation of the wap-ras transgenic mouse as a model system for testing anticancer drugs. Cancer Res 52(13):3733–3738
Nielsen LL et al (1995) In wap-ras transgenic mice, tumor phenotype but not cyclophosphamide-sensitivity is affected by genetic background. Anticancer Res 15(2):385–392
Noonberg SB, Benz CC (2000) Tyrosine kinase inhibitors targeted to the epidermal growth factor receptor subfamily: role as anticancer agents. Drugs 59(4):753–767
Nusse R, Varmus HE (1982) Many tumors induced by the mouse mammary tumor virus contain a provirus integrated in the same region of the host genome. Cell 31(1):99–109
Nusse R et al (1984) Mode of proviral activation of a putative mammary oncogene (int-1) on mouse chromosome 15. Nature 307(5947):131–136
Olayioye MA et al (2000) The ErbB signaling network: receptor heterodimerization in development and cancer. EMBO J 19(13):3159–3167
O'Regan RM et al (2002) Effects of raloxifene after tamoxifen on breast and endometrial tumor growth in athymic mice. J Natl Cancer Inst 94(4):274–283
Palmer CN et al (1994) Interaction of the peroxisome proliferator-activated receptor alpha with the retinoid X receptor alpha unmasks a cryptic peroxisome proliferator response element that overlaps an ARP-1-binding site in the CYP4A6 promoter. J Biol Chem 269(27):18083–18089
Paterson JW (1998) BRCA1: a review of structure and putative functions. Dis Markers 13(4): 261–274
Pegram MD, Konecny G, Slamon DJ (2000) The molecular and cellular biology of HER2/neu gene amplification/overexpression and the clinical development of herceptin (trastuzumab) therapy for breast cancer. Cancer Treat Res 103:57–75
Pelengaris S, Khan M, Evan G (2002) c-MYC: more than just a matter of life and death. Nat Rev Cancer 2(10):764–776
Phillips KA (2000) Immunophenotypic and pathologic differences between BRCA1 and BRCA2 hereditary breast cancers. J Clin Oncol 18(21 Suppl):107S–112S
Piccart M et al (2001) The predictive value of HER2 in breast cancer. Oncology 61(Suppl 2): 73–82
Poole AJ et al (2006) Prevention of Brca1-mediated mammary tumorigenesis in mice by a progesterone antagonist. Science 314(5804):1467–1470
Ravdin PM, Chamness GC (1995) The c-erbB-2 proto-oncogene as a prognostic and predictive marker in breast cancer: a paradigm for the development of other macromolecular markers–a review. Gene 159(1):19–27
Rendi MH et al (2004) The selective estrogen receptor modulator arzoxifene and the rexinoid LG100268 cooperate to promote transforming growth factor beta-dependent apoptosis in breast cancer. Cancer Res 64(10):3566–3571
Richardson H et al (2007) The National Cancer Institute of Canada Clinical Trials Group MAP.3 trial: an international breast cancer prevention trial. Curr Oncol 14(3):89–96

Ristimaki A et al (2002) Prognostic significance of elevated cyclooxygenase-2 expression in breast cancer. Cancer Res 62(3):632–635
Rogers AE, Lee SY (1986) Chemically-induced mammary gland tumors in rats: modulation by dietary fat. Prog Clin Biol Res 222:255–282
Rohan TE et al (2009) A randomized controlled trial of calcium plus vitamin D supplementation and risk of benign proliferative breast disease. Breast Cancer Res Treat 116(2):339–350
Rohatgi N, Blau R, Lower EE (2002) Raloxifene is associated with less side effects than tamoxifen in women with early breast cancer: a questionnaire study from one physician's practice. J Womens Health Gend Based Med 11(3):291–301
Rose-Hellekant TA, Sandgren EP (2000) Transforming growth factor alpha- and c-myc-induced mammary carcinogenesis in transgenic mice. Oncogene 19(8):1092–1096
Rossouw JE et al (2002) Risks and benefits of estrogen plus progestin in healthy postmenopausal women: principal results From the Women's Health Initiative randomized controlled trial. JAMA 288(3):321–333
Rothschild TC et al (1987) Transplacental effects of diethylstilbestrol on mammary development and tumorigenesis in female ACI rats. Cancer Res 47(16):4508–4516
Roy J et al (2003) A novel pure SERM achieves complete regression of the majority of human breast cancer tumors in nude mice. Breast Cancer Res Treat 81(3):223–229
Ruhlen RL et al (2009) Tamoxifen induces regression of estradiol-induced mammary cancer in the ACI.COP-Ept2 rat model. Breast Cancer Res Treat 117(3):517–524
Rusnak DW et al (2001) The effects of the novel, reversible epidermal growth factor receptor/ ErbB-2 tyrosine kinase inhibitor, GW2016, on the growth of human normal and tumor-derived cell lines in vitro and in vivo. Mol Cancer Ther 1(2):85–94
Russo J, Russo IH (1996) Experimentally induced mammary tumors in rats. Breast Cancer Res Treat 39(1):7–20
Ryo A et al (2001) Pin1 regulates turnover and subcellular localization of beta-catenin by inhibiting its interaction with APC. Nat Cell Biol 3(9):793–801
Salazar LG, Disis ML (2005) Cancer vaccines: the role of tumor burden in tipping the scale toward vaccine efficacy. J Clin Oncol 23(30):7397–7398
Sandgren EP et al (1995) Inhibition of mammary gland involution is associated with transforming growth factor alpha but not c-myc-induced tumorigenesis in transgenic mice. Cancer Res 55(17):3915–3927
Santen RJ et al (1999) The potential of aromatase inhibitors in breast cancer prevention. Endocr Relat Cancer 6(2):235–243
Sauer B (1998) Inducible gene targeting in mice using the Cre/lox system. Methods 14(4):381–392
Schneider MR et al (1990) Antitumor activity and mechanism of action of different antiprogestins in experimental breast cancer models. J Steroid Biochem Mol Biol 37(6):783–787
Schoenenberger CA et al (1988) Targeted c-myc gene expression in mammary glands of transgenic mice induces mammary tumours with constitutive milk protein gene transcription. EMBO J 7(1):169–175
Serghides L, Kain KC (2005) Peroxisome proliferator-activated receptor gamma and retinoid X receptor agonists have minimal effects on the interaction of endothelial cells with Plasmodium falciparum-infected erythrocytes. Infect Immun 73(2):1209–1213
Sharan SK et al (1997) Embryonic lethality and radiation hypersensitivity mediated by Rad51 in mice lacking Brca2. Nature 386(6627):804–810
Shellabarger CJ et al (1980) Interaction of dimethylbenzanthracene and diethylstilbestrol on mammary adenocarcinoma formation in female ACI rats. Cancer Res 40(6):1808–1811
Shen Q, Brown PH (2003) Novel agents for the prevention of breast cancer: targeting transcription factors and signal transduction pathways. J Mammary Gland Biol Neoplasia 8(1):45–73
Shen Q, Brown PH (2005) Transgenic mouse models for the prevention of breast cancer. Mutat Res 576(1–2):93–110
Shen SX et al (1998) A targeted disruption of the murine Brca1 gene causes gamma-irradiation hypersensitivity and genetic instability. Oncogene 17(24):3115–3124

Shibata MA et al (1996) Progression of prostatic intraepithelial neoplasia to invasive carcinoma in C3(1)/SV40 large T antigen transgenic mice: histopathological and molecular biological alterations. Cancer Res 56(21):4894–4903

Shin MH et al (2002) Intake of dairy products, calcium, and vitamin d and risk of breast cancer. J Natl Cancer Inst 94(17):1301–1311

Shull JD et al (1997) Ovary-intact, but not ovariectomized female ACI rats treated with 17beta-estradiol rapidly develop mammary carcinoma. Carcinogenesis 18(8):1595–1601

Sinn E et al (1987) Coexpression of MMTV/v-Ha-ras and MMTV/c-myc genes in transgenic mice: synergistic action of oncogenes in vivo. Cell 49(4):465–475

Slamon DJ et al (1987) Human breast cancer: correlation of relapse and survival with amplification of the HER-2/neu oncogene. Science 235(4785):177–182

Spitz IM (2006) Progesterone receptor antagonists. Curr Opin Investig Drugs 7(10):882–890

Sporn MB (1976) Approaches to prevention of epithelial cancer during the preneoplastic period. Cancer Res 36(7 Pt 2):2699–2702

Sporn MB (2004) Arzoxifene: a promising new selective estrogen receptor modulator for clinical chemoprevention of breast cancer. Clin Cancer Res 10(16):5313–5315

Sporn MB et al (1976) Prevention of chemical carcinogenesis by vitamin A and its synthetic analogs (retinoids). Fed Proc 35(6):1332–1338

Stewart TA, Pattengale PK, Leder P (1984) Spontaneous mammary adenocarcinomas in transgenic mice that carry and express MTV/myc fusion genes. Cell 38(3):627–637

Strecker TE et al (2009) Effect of lapatinib on the development of estrogen receptor-negative mammary tumors in mice. J Natl Cancer Inst 101(2):107–113

Suh N et al (1999) A new ligand for the peroxisome proliferator-activated receptor-gamma (PPAR-gamma), GW7845, inhibits rat mammary carcinogenesis. Cancer Res 59(22):5671–5673

Suh N et al (2001) Arzoxifene, a new selective estrogen receptor modulator for chemoprevention of experimental breast cancer. Cancer Res 61(23):8412–8415

Suh N et al (2002) Prevention and treatment of experimental breast cancer with the combination of a new selective estrogen receptor modulator, arzoxifene, and a new rexinoid, LG 100268. Clin Cancer Res 8(10):3270–3275

Suzuki A et al (1997) Brca2 is required for embryonic cellular proliferation in the mouse. Genes Dev 11(10):1242–1252

Tekmal RR et al (1996) Overexpression of int-5/aromatase in mammary glands of transgenic mice results in the induction of hyperplasia and nuclear abnormalities. Cancer Res 56(14):3180–3185

Torres-Arzayus MI et al (2004) High tumor incidence and activation of the PI3K/AKT pathway in transgenic mice define AIB1 as an oncogene. Cancer Cell 6(3):263–274

Tsukamoto AS et al (1988) Expression of the int-1 gene in transgenic mice is associated with mammary gland hyperplasia and adenocarcinomas in male and female mice. Cell 55(4):619–625

Turashvili G et al (2006) Wnt signaling pathway in mammary gland development and carcinogenesis. Pathobiology 73(5):213–223

Ugolini F et al (2001) WNT pathway and mammary carcinogenesis: loss of expression of candidate tumor suppressor gene SFRP1 in most invasive carcinomas except of the medullary type. Oncogene 20(41):5810–5817

Uray IP, Brown PH (2006) Prevention of breast cancer: current state of the science and future opportunities. Expert Opin Investig Drugs 15(12):1583–1600

Verlinden L et al (2000) Two novel 14-Epi-analogues of 1,25-dihydroxyvitamin D3 inhibit the growth of human breast cancer cells in vitro and in vivo. Cancer Res 60(10):2673–2679

Vogel VG (2009) The NSABP Study of Tamoxifen and Raloxifene (STAR) trial. Expert Rev Anticancer Ther 9(1):51–60

Vogel VG et al (2006) Effects of tamoxifen vs raloxifene on the risk of developing invasive breast cancer and other disease outcomes: the NSABP Study of Tamoxifen and Raloxifene (STAR) P-2 trial. JAMA 295(23):2727–2741

Wagner KU et al (1997) Cre-mediated gene deletion in the mammary gland. Nucleic Acids Res 25(21):4323–4330

Wakeling AE et al (2002) ZD1839 (Iressa): an orally active inhibitor of epidermal growth factor signaling with potential for cancer therapy. Cancer Res 62(20):5749–5754
Weaver Z et al (2002) Mammary tumors in mice conditionally mutant for Brca1 exhibit gross genomic instability and centrosome amplification yet display a recurring distribution of genomic imbalances that is similar to human breast cancer. Oncogene 21(33):5097–5107
Welsch CW (1985) Host factors affecting the growth of carcinogen-induced rat mammary carcinomas: a review and tribute to Charles Brenton Huggins. Cancer Res 45(8):3415–3443
Welsch CW et al (1981) Effect of an estrogen antagonist (tamoxifen) on the initiation and progression of gamma-irradiation-induced mammary tumors in female Sprague-Dawley rats. Eur J Cancer Clin Oncol 17(12):1255–1258
Welsh J (2007) Vitamin D and prevention of breast cancer. Acta Pharmacol Sin 28(9):1373–1382
Wickerham DL et al (2009) The use of tamoxifen and raloxifene for the prevention of breast cancer. Recent Results Cancer Res 181:113–119
Widschwendter M, Jones PA (2002) DNA methylation and breast carcinogenesis. Oncogene 21(35):5462–5482
William WN Jr et al (2009) Molecular targets for cancer chemoprevention. Nat Rev Drug Discov 8(3):213–225
Woodburn JR (1999) The epidermal growth factor receptor and its inhibition in cancer therapy. Pharmacol Ther 82(2–3):241–250
Wu K et al (2000) 9-cis-Retinoic acid suppresses mammary tumorigenesis in C3(1)-simian virus 40T antigen-transgenic mice. Clin Cancer Res 6(9):3696–3704
Wu K et al (2002a) Suppression of mammary tumorigenesis in transgenic mice by the RXR-selective retinoid, LGD1069. Cancer Epidemiol Biomarkers Prev 11(5):467–474
Wu K et al (2002b) The retinoid X receptor-selective retinoid, LGD1069, prevents the development of estrogen receptor-negative mammary tumors in transgenic mice. Cancer Res 62(22):6376–6380
Xia W et al (2002) Anti-tumor activity of GW572016: a dual tyrosine kinase inhibitor blocks EGF activation of EGFR/erbB2 and downstream Erk1/2 and AKT pathways. Oncogene 21(41): 6255–6263
Xie SP, Pirianov G, Colston KW (1999) Vitamin D analogues suppress IGF-I signalling and promote apoptosis in breast cancer cells. Eur J Cancer 35(12):1717–1723
Xu J, Li Q (2003) Review of the in vivo functions of the p160 steroid receptor coactivator family. Mol Endocrinol 17(9):1681–1692
Xu X et al (1999) Conditional mutation of Brca1 in mammary epithelial cells results in blunted ductal morphogenesis and tumour formation. Nat Genet 22(1):37–43
Yang LM et al (1999) Role of retinoid receptors in the prevention and treatment of breast cancer. J Mammary Gland Biol Neoplasia 4(4):377–388
Yarden Y (2001) Biology of HER2 and its importance in breast cancer. Oncology 61(Suppl 2): 1–13
Yin Y et al (2005) Peroxisome proliferator-activated receptor delta and gamma agonists differentially alter tumor differentiation and progression during mammary carcinogenesis. Cancer Res 65(9):3950–3957

Chapter 25
Oncogene Addiction: Mouse Models and Clinical Relevance for Molecularly Targeted Therapies

James V. Alvarez, Elizabeth S. Yeh, Yi Feng, and Lewis A. Chodosh

Cancer results from the dysregulation of pathways controlling the growth, proliferation, differentiation, and survival of tumor cells, as well as fundamental alterations in the manner in which cells interact with their microenvironment (Hanahan and Weinberg 2000). Several lines of evidence suggest that these alterations are due to the accumulation of multiple mutations in oncogenes and tumor suppressor genes that disrupt their normal function or regulation. These mutations provide a selective advantage to the cells in which they occur, leading to their expansion and clinical manifestation as a tumor.

The frequency of cancer increases exponentially with age, consistent with our understanding that the development of this disease requires multiple independent mutations (Renan 1993). In an analogous manner, analyses of several tumor types

J.V. Alvarez • E.S. Yeh • Y. Feng
Department of Cancer Biology, Raymond and Ruth Perelman School of Medicine at the University of Pennsylvania, Philadelphia, PA 19104, USA

Abramson Family Cancer Research Institute, Raymond and Ruth Perelman School of Medicine at the University of Pennsylvania, Philadelphia, PA 19104-6160, USA

L.A. Chodosh (✉)
Department of Cancer Biology, Raymond and Ruth Perelman School of Medicine at the University of Pennsylvania, Philadelphia, PA 19104, USA

Department of Cell and Developmental Biology, Raymond and Ruth Perelman School of Medicine at the University of Pennsylvania, Philadelphia, PA 19104, USA

Department of Medicine, Raymond and Ruth Perelman School of Medicine at the University of Pennsylvania, Philadelphia, PA 19104, USA

Abramson Family Cancer Research Institute, Raymond and Ruth Perelman School of Medicine at the University of Pennsylvania, Philadelphia, PA 19104-6160, USA
e-mail: chodosh@mail.med.upenn.edu

J.E. Green and T. Ried (eds.), *Genetically Engineered Mice for Cancer Research: Design, Analysis, Pathways, Validation and Pre-Clinical Testing*,
DOI 10.1007/978-0-387-69805-2_25, © Springer Science+Business Media, LLC 2012

suggest that cancers arise through one or more premalignant intermediates. In some cases, some of the molecular aberrations that accompany the transition from normal tissue to premalignancy, and from premalignancy to malignancy, have been identified (Burstein et al. 2004; Vogelstein and Kinzler 1993). In most cases, however, the specific characteristics of premalignant intermediates, the particular mutations involved in their generation and progression, and the order in which they occur remain speculative. Experimental models of transformation also support the multistep nature of tumorigenesis. Mouse models of cancer characteristically exhibit tumor latencies and clonality consistent with the requirement of several genetic events for tumor formation. Thus, a preponderance of evidence argues that multiple genetic events are required for tumorigenesis and, as a corollary, that clinically evident tumors contain multiple dysregulated pathways that contribute to their growth.

In light of the multiple mutations present within human cancer cells, considerable skepticism existed regarding the proposition that targeted therapies designed to inhibit a single oncogene might achieve a therapeutic effect. Rather, it seemed likely that just as activation of multiple oncogenic pathways is required to form a tumor, so might inhibition of several of these proteins be required to reverse the tumorigenic phenotype. Furthermore, the genetic instability that typically accompanies tumor progression, as well as the considerable genetic heterogeneity of cells within a given tumor, were considered likely to render cells resistant to inhibition of a single oncogene that may have contributed to tumor development. More troublesome still, loss of tumor suppressor genes would not be predicted to be amenable to this targeted inhibitory approach.

Despite these theoretical considerations, accumulating evidence suggests that many cancers do in fact remain dependent upon one or more dominant oncogenic pathways for their growth and maintenance. For example, pharmacologic studies using targeted therapies in human leukemias and some solid cancers have shown a striking dependence for tumor growth on a single activated oncogene. Paralleling clinical observations, a number of mouse models have provided compelling genetic evidence that tumors arising in a variety of organs and driven by a variety of oncogenes frequently remain dependent upon those oncogenes for their maintenance and growth. This phenomenon, known as oncogene dependence or oncogene addiction (Weinstein 2002), has significant implications for the development and use of targeted therapies to treat human cancers.

In this chapter, we summarize evidence suggesting that oncogene addiction is a clinically important phenomenon. We then highlight the genetically engineered mouse models that have been used to study oncogene addiction, and the insights these have provided into the molecular and cellular mechanisms underlying this process. Importantly, in patients as well as mouse models, tumors almost invariably escape their dependence upon a single pathway and recur, often following an indeterminate period of dormancy. We describe additional mouse models that recapitulate clinically relevant aspects of tumor dormancy and relapse that are essential to our understanding of the mechanisms that underlie the ability of human cancer cells to escape treatment. Finally, we speculate on the possibility that a unifying explanation for observations related to oncogene addiction in mice and humans may be provided by the concept of cancer stem cells. These cells – for cancers in which they exist – may be resistant to oncogene inhibition and capable of surviving in a dormant state before acquiring the ability to resume growth in a manner independent of the oncogene that initiated primary tumor growth. The implications of the cancer stem cell hypothesis for oncogene addiction and tumor recurrence are discussed.

25.1 Clinical Evidence for Oncogene Addiction

The observation that targeting a single growth-promoting pathway can be an effective strategy for treating human cancer was recognized prior to the pharmacological targeting of oncogenic proteins. For example, the use of tamoxifen to treat ER+ breast cancer was widespread long before the introduction of small-molecule inhibitors to block oncogenic pathways (Jensen and Jordan 2003). However, clinical experience using targeted therapies that inhibit oncogenic kinases has provided the most compelling evidence for oncogene addiction in human cancers.

25.1.1 BCR-ABL in Chronic Myelogenous Leukemia

Perhaps the most striking evidence that oncogene addiction occurs in human cancers has come from studies of the oncogene BCR-ABL and the disease it causes, chronic myelogenous leukemia (CML). Originally identified as the gene product of the chromosomal translocation producing the Philadelphia chromosome, BCR-ABL was the first oncogene to be effectively inhibited by a targeted molecular therapy (Druker 2002). Imatinib mesylate (Gleevec) is a potent inhibitor of BCR-ABL tyrosine kinase activity, and its initial testing in BCR-ABL-positive CML patients represented a landmark test of the principle of targeted therapies. The results from these trials, initially conducted in the late 1990s, demonstrated that imatinib was remarkably effective in treating CML with >95% of patients exhibiting a complete hematologic response within 12 months (Sherbenou and Druker 2007). Overall 5-year survival for patients receiving imatinib as initial therapy was 89%, compared to 70% for patients receiving the prior standard of care, interferon-α plus cytarabine. Thus, imatinib constituted the first instance in which a therapy targeting a single activated oncogene induced tumor regression and prolonged patient survival.

Beyond its surprising initial success in treating CML, some of the most valuable insights garnered from clinical experience with imatinib have related to the mechanisms of resistance to targeted therapies. Early on it was noted that imatinib has more limited efficacy against the more advanced stage of CML referred to as blast crisis, wherein hematologic responses were observed in only 52% of patients (Druker et al. 2001; Sawyers et al. 2002). Tumor cells in many of these patients exhibited secondary chromosomal abnormalities consistent with the outgrowth of resistant clones. Furthermore, a fraction of patients treated with imatinib in chronic phase ultimately relapse and become resistant to imatinib. A critical insight into this process came from the finding that resistant clones commonly harbored either BCR-ABL amplification or mutations in the BCR-ABL protein that conferred resistance to inhibition by imatinib (Gorre et al. 2001). Thus, in this context most instances of relapse occur through a BCR-ABL-dependent pathway. Beyond identifying mechanisms by which tumors can become resistant to targeted therapy, these findings further emphasized the importance of oncogene addiction as, even in relapsed tumors, BCR-ABL signaling plays a critical role in driving malignancy.

25.1.2 EGFR in Non-small-Cell Lung Cancers

A second instance in which targeted therapies against a single oncogene have proven effective in treating cancer is inhibition of the epidermal growth factor receptor (EGFR) in non-small-cell lung cancers (NSCLC). NSCLC is the leading cause of death from cancer worldwide as tobacco use is widespread and cytotoxic chemotherapy induces responses in only ~25% of patients, and even these tend to be transient (Sequist and Lynch 2008). Given that EGFR is frequently overexpressed in NSCLC (Rusch et al. 1993), EGFR inhibitors were tested as therapeutic agents in this disease. Two small-molecule inhibitors of EGFR, gefitinib (Iressa) and erlotinib (Tarceva), have been developed and tested in clinical trials. Initial trials with each drug were disappointing, showing no improvements in response rates, 1-year overall survival, or median survival compared to standard therapy alone (Sequist and Lynch 2008). However, it soon became apparent that a small subset of NSCLC patients exhibited robust responses to EGFR inhibitors, and the molecular determinant of this response was the presence of activating mutations in EGFR itself (Lynch et al. 2004; Paez et al. 2004; Pao et al. 2004). Both retrospective analyses and prospective trials have demonstrated response rates of ~78% in patients whose tumors harbor mutant alleles of EGFR (Sequist et al. 2007). Thus, EGFR inhibition in NSCLC represents another proof-of-principle that inhibition of a single oncogene can elicit a dramatic clinical response, further supporting the existence of oncogene addiction in human cancers.

Analogous to imatinib treatment of CML, NSCLC that initially respond to gefitinib or erlotinib treatment can relapse through the development of resistance to EGFR inhibition. A substantial fraction of resistance can be attributed to mutations in EGFR that render it insensitive to inhibition by gefitinib or erlotinib (Kobayashi et al. 2005; Pao et al. 2005). Interestingly, the mechanism of resistance of mutant proteins appears to be similar between BCR-ABL and EGFR, with mutation of analogous residues in the kinase domains of the two proteins. An alternative route of resistance appears to occur through amplification of the receptor tyrosine kinase, c-Met, which has been identified in ~20% of resistant tumors (Engelman et al. 2007). Thus, both EGFR-dependent and EGFR-independent pathways for resistance to small-molecule inhibitors of EGFR are observed in NSCLC.

25.1.3 HER2/Neu in Breast Cancer

The EGFR family member HER2 (neu, ErbB2) is amplified and overexpressed in approximately one-third of human breast cancers and its amplification identifies a particularly aggressive form of the disease that is associated with a poor prognosis. Trastuzumab is a humanized monoclonal antibody that targets HER2, and has significant antitumor activity against HER2-positive breast cancers. Initially used for metastatic disease, trastuzumab is now also used as first-line therapy, and it provides survival benefits for patients with both early stage and metastatic disease.

Initial response rates for HER2-positive patients with metastatic breast cancer treated with trastuzumab are between 30 and 70%, depending on the context (Slamon et al. 2001; Vogel et al. 2002). However, as with the other targeted therapies, initial response to trastuzumab in the setting of advanced disease is invariably followed by disease progression, usually within 1 year, indicating that tumors have acquired resistance to trastuzumab. One mechanism by which HER2-positive tumors become resistant to trastuzumab is through upregulation of the PI3K-AKT pathway, either by deletion of PTEN or mutational activation of PI3K (Berns et al. 2007; Nagata et al. 2004). Thus, as with EGFR inhibitors, HER2-positive tumors can escape their dependence on this oncogenic pathway by HER2-independent pathways.

25.1.4 Other Oncogene-Dependent Human Cancers

Several other tumor types exist for which clinical evidence suggests dependence upon a single oncogenic pathway. Gastrointestinal stromal tumors (GIST) frequently harbor activating mutations in either KIT or PDGFRA; each of these kinases is inhibited by imatinib and the treatment of GIST tumors with imatinib leads to clinical responses in upward of three-quarters of patients (Demetri et al. 2002; Fletcher and Rubin 2007; Verweij et al. 2003). Similarly, some patients with chronic eosinophilic leukemia (CEL) harbor constitutively activated PDGFRα resulting from a fusion with the protein FIP1L. These tumors also show a marked response to imatinib, with nearly all patients whose tumors express the fusion protein exhibiting a complete hematological and molecular response (Pardanani and Tefferi 2004).

25.1.5 Clinical Synthesis

A wealth of clinical experience, derived primarily from the use of small-molecule inhibitors and antibodies directed against protein kinases, has provided important insights into the consequences of targeted inhibition of oncogenic pathways in human cancers. This experience suggests that a variety of human cancers – including tumors of hematologic, epithelial, and mesenchymal cell origin – can exhibit an exquisite dependence upon the continued activity of a single oncogene for their maintenance and growth.

While the specifics vary among tumor types, a general picture emerges of the role that a dominantly active oncogene may play in the development and maintenance of tumors. First, activation of an oncogene through mutation, amplification, or overexpression is a necessary step in the development of many tumors. However, as discussed above, this lone event is not sufficient for tumorigenesis, but instead requires the collaboration of additional genetic alterations.

Once a tumor has formed, inhibiting the function of a dominant oncogenic pathway to which a tumor is addicted may halt tumor growth or induce tumor regression (Fig. 25.1). A subset of tumors are initially resistant to the effects of oncogene inhibition, a phenomenon termed primary resistance. In cases in which tumors do regress, a period in which tumor growth – or the tumor itself – is not clinically evident is almost invariably followed by tumor escape and recurrence. Recurrent tumors can acquire resistance through both oncogene-dependent and oncogene-independent mechanisms. Recurrent tumors that have developed resistance through mutations in the targeted oncogene that render it insensitive to inhibition are likely still oncogene-addicted, and secondary inhibitors that overcome this resistance may be therapeutically useful. Alternately, recurrent tumors that have acquired resistance through secondary pathways may become addicted to a different oncogenic pathway, and inhibition of these secondary pathways may have therapeutic utility.

The simplicity of the above model belies the many questions concerning oncogene addiction that remain unanswered. What accounts for the substantial differences observed in the clinical responses of different tumors to oncogene inhibition? To what extent do cell type-specific and oncogene-specific differences play an important role? What are the cellular mechanisms responsible for tumor regression? Are there genetic determinants and modifiers of tumor regression, and can these be identified prospectively? Given that tumor relapse is almost universally observed, residual tumor cells must persist following tumor regression – possibly in a dormant state. What is the biologic state of these cells and how do they acquire the ability to reinitiate malignant growth? What are the molecular pathways that underlie tumor recurrence? What dictates whether tumors will recur in a manner dependent or independent of the oncogenic pathway targeted? These and other critical questions about the mechanisms underlying oncogene addiction and escape have yet to be addressed.

Unfortunately, these mechanistic questions are extremely difficult to answer using human patient material. As such, the development of mouse models that recapitulate key features of oncogene addiction as it is observed in the clinic has been invaluable. Below we highlight examples of these models with special emphasis on the insights that they have provided into the cellular mechanisms of oncogene addiction, tumor regression, tumor dormancy, and oncogene-dependent and independent recurrence. Finally, we consider whether the cancer stem-cell hypothesis might provide a unifying explanation for clinical and experimental observations made in this area of targeted therapies.

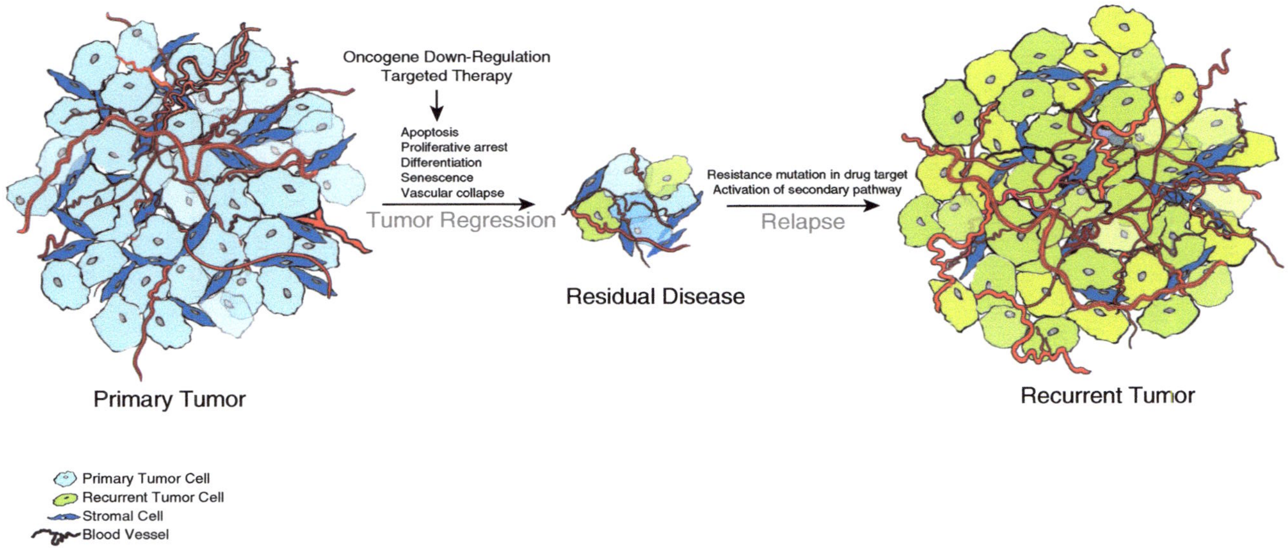

Fig. 25.1 Oncogene addiction and escape. A significant fraction of tumors are dependent upon the continued activity of an initiating oncogene for their maintenance and growth. Upon loss of oncogenic activity – either through experimental down-regulation of the oncogene or through its inhibition with a drug – tumors regress, often to a clinically undetectable state. This regression occurs through some combination of proliferative arrest, apoptosis, differentiation, senescence, and vascular collapse. However, in both mice and humans, this regression is short-lived and is followed by tumor relapse. Recurrent tumors can arise through mutation of the drug target or through activation of secondary oncogenic pathways

25.2 Genetically Engineered Mouse Models of Oncogene Addiction

25.2.1 Conditional Mouse Models to Study Oncogene Addiction

Transgenic and knockout mice have provided important insights into the role of oncogenes and tumor suppressors in the initiation and progression of cancer. However, determining whether established tumors remain dependent upon the continued activity of an initiating oncogene requires the ability to experimentally downregulate the oncogene following tumor formation. The generation of mouse models with inducible oncogene expression has provided the technology essential for addressing this issue.

Several systems have been developed to permit the conditional expression of oncogenes in mice. The two most commonly used are the tetracycline-regulated system and the tamoxifen-regulated system. In the tetracycline system, levels of the transgene can be reversibly regulated by the tetracycline analog doxycycline; in the tamoxifen system, the stability or activity of the protein product of the transgene can be reversibly regulated by 4-OH tamoxifen. In both cases, the use of tissue-specific promoters allows for transgene expression in selected cell types. Thus, these systems allow for inducible, tissue-specific, and reversible expression of transgenes in mice.

The ability to express an oncogene to induce tumor formation, and then downregulate that oncogene once a tumor has formed, has allowed for a direct test of the concept of oncogene addiction *in vivo*. Indeed, a number of studies have demonstrated that oncogene addiction is a common feature of oncogene-induced tumors, irrespective of the initiating oncogene or the tissue type in which the tumor formed (Table 25.1). For instance, tumors driven by Myc, neu, Ras, Wnt1, BCR-ABL, EGFR, and PIK3CA each exhibit dependence upon the initiating oncogene, as do tumors arising in the lung, mammary gland, melanocytes, hematopoietic system, skin, pancreatic β-cells, liver, and bone cells (Boxer et al. 2004; Chin et al. 1999; D'Cruz et al. 2001; Engelman et al. 2008; Felsher and Bishop 1999; Flores et al. 2004; Gunther et al. 2003; Huettner et al. 2000; Jain et al. 2002; Moody et al. 2002; Pelengaris et al. 2002, 1999; Politi et al. 2006; Sarkisian et al. 2007; Shachaf et al. 2004; Tran et al. 2008). Thus, oncogene addiction is a general phenomenon of tumors and is not specific for a particular oncogene or tissue type. These studies have provided important insights into the molecular and cellular mechanisms of oncogene addiction.

25.2.2 Addiction to Proliferative Signals

Tumors typically exhibit inappropriately elevated rates of cellular proliferation compared to the normal tissues from which they arise. These proliferative signals originate from activated oncogenic signaling pathways; logically, inhibiting these

Table 25.1 Evidence for oncogene addiction derived from inducible mouse models and the use of targeted therapies in humans

Oncogene	Tissue	Dependence in mice	Mechanism of dependence	Recurrence in mice (mechanism)	Targeted therapies	Dependence in humans	Resistance in humans (mechanism)	References
HER2/neu	Mammary gland	Yes	Apoptosis, proliferative arrest	Yes (Snail up-regulation/ EMT)	Trastuzumab	Yes	Yes (PTEN mutation)	Moody et al. (2002)
MYC	Mammary gland	Partial (~50% of tumors)	Apoptosis, proliferative arrest	Yes (K-Ras mutation)	None	N/A	N/A	Boxer et al. (2004), D'Cruz et al. (2001)
	Liver	Yes	Apoptosis, proliferative arrest, differentia-tion, senescence	nd				Shachaf et al. (2004), Wu et al. (2007)
	Bone	Yes	Proliferative arrest, differentiation, senescence	nd				Jain et al. (2002), Wu et al. (2007)
	Skin	Yes	Vascular collapse	nd				Flores et al. (2004), Pelengaris et al. (1999)
	Hematopoietic system	Yes	Apoptosis, proliferative arrest, senescence	Yes (nd)				Felsher and Bishop (1999), Wu et al. (2007)
	Pancreatic b cell	Yes	Vascular collapse, differentiation	nd				Pelengaris et al. (2002)
	Lung	Partial (incomplete regression)	nd	nd				Tran et al. (2008)
BCR-ABL	Hematopoietic system	Yes	Apoptosis	nd	Imatinib	Yes	Yes (BCR-ABL mutation, amplification)	Huettner et al. (2000)
EGFR	Lung	Yes	Apoptosis, proliferative arrest	nd	Gefitinib, erlotinib	Yes	Yes (EGFR mutation, Met amplification)	Ji et al. (2006), Politi et al. (2006)

(continued)

Table 25.1 (continued)

Oncogene	Tissue	Dependence in mice	Mechanism of dependence	Recurrence in mice (mechanism)	Targeted therapies	Dependence in humans	Resistance in humans (mechanism)	References
Ras	Melanoma	Yes	Apoptosis, vascular collapse	nd	None			Chin et al. (1999)
	Lung	Yes	Apoptosis					Fisher et al. (2001), Tran et al. (2008)
	Mammary gland	Yes	Apoptosis, proliferative arrest					Podsypanina et al. (2008)
	Hematopoietic system	Yes	nd					Tran et al. (2008)
Wnt	Mammary gland	Yes	Apoptosis, proliferative arrest	Yes (p53 loss, p16/p19 loss)	None	N/A	N/A	Gunther et al. (2003) #2057, Debies et al. (2008) #5321
PIK3CA	Lung	Yes	nd		None	ns	ns	Engelman et al. (2008)

signaling pathways in established tumors would be predicted to result in reduced cellular proliferation. Nearly every study that has examined tumor cell proliferation following oncogene down-regulation has confirmed this prediction.

For example, mice in which MYC is conditionally expressed in hematopoietic cells (Eμ-tTA;tetO-MYC) develop T-cell lymphoma and, less frequently, acute myeloid leukemia (AML) (Felsher and Bishop 1999). Upon deinduction of MYC by administration of doxycycline, 90% of these cancers regress with 80% of those mice reported to maintain sustained regression. Cancer cells appeared to be cleared following MYC deinduction as monitored by gross morphological observation and Gallium imaging. Clearance of cancer cells is accompanied by a decrease in BrdU-positive cells and an accumulation of cells in the G1 phase of the cell cycle (Felsher and Bishop 1999). In another study, activation of c-Myc in the skin by treating mice bearing a Myc-ER fusion protein with 4-OH tamoxifen induced papillomatosis. These lesions contained high levels of proliferating keratinocytes, and down-regulating Myc activity led to a profound loss of proliferating cells in the suprabasal layer (Pelengaris et al. 1999). A similar decrease in proliferation was observed following Myc deinduction in Myc-driven liver and bone tumors (Jain et al. 2002; Shachaf et al. 2004). These studies indicate that in diverse cell types sustained Myc activity is required for the maintenance of proliferative signals, even in established tumors.

The loss of proliferation following oncogene inactivation is not specific to Myc. Expression of HER2/neu in the mammary gland leads to the formation of invasive carcinomas, and down-regulation of neu induces tumor regression in >90% of mice. Our laboratory has determined that as early as 2 days following neu deinduction, mammary tumors exhibit a near complete loss of BrdU incorporation indicative of proliferative arrest (Moody et al. 2002). This suggests that neu signaling is required for the maintenance of pro-proliferative signals. Similarly, our laboratory has found that oncogene down-regulation in mammary tumors driven by Myc, Wnt1, and H-Ras also leads to rapid and dramatic decreases in tumor cell proliferation ((D'Cruz et al. 2001; Gunther et al. 2003; Sarkisian et al. 2007) and unpublished data).

Cellular senescence is a specific form of growth arrest characterized by stereotypic morphological changes, marker expression, and signaling pathway activation. As opposed to quiescence, or arrest in the G0/G1 phase of the cell cycle, senescence is generally thought to be irreversible. In addition, oncogene-induced senescence has been shown to be a tumor suppressive mechanism (Braig et al. 2005; Collado et al. 2005; Michaloglou et al. 2005; Sarkisian et al. 2007). Interestingly, down-regulation of Myc in Myc-driven lymphomas, osteosarcomas, and liver tumors has been shown to induce cellular senescence, and this senescence response appears to be required for these tumor types to exhibit sustained tumor regression (Wu et al. 2007).

The above studies indicate that sustained oncogenic signaling is almost invariably required for the continued proliferation of tumor cells, and suggest that the addiction of tumor cells to dominant oncogenic pathways is at least partly explained through their addiction to pro-proliferative signals.

25.2.3 Addiction to Oncogenic Survival Signals

The loss of tumor cell proliferation alone is unlikely to account for the rapid tumor regression commonly observed following oncogene down-regulation. A second cellular consequence almost universally observed following oncogene down-regulation is the induction of apoptosis. One of the first demonstrations that disruption of oncogenic signaling leads to apoptosis was a study of BCR-ABL-induced leukemia (Huettner et al. 2000). Mice with conditional expression of the p210 isoform of *BCR-ABL1* develop acute B-cell leukemia. Administration of doxycycline to animals to down-regulate p210 resulted in a marked decrease in circulating white blood cells within 48 h and complete regression within 5 days. Additionally, no blasts remained in the peripheral blood. This dramatic reversal of phenotype was accompanied by an increase in apoptosis in leukemic animals within 20 h of doxycycline treatment. This finding is consistent with previous observations that *BCR-ABL1* expressing cells are protected from apoptosis (Bedi et al. 1994; McGahon et al. 1994).

Loss of Myc expression in Myc-dependent tumors has also been shown to induce apoptosis. For instance, down-regulation of Myc in Myc-driven T-cell lymphomas, liver tumors, and β-cell tumors elicits tumor cell apoptosis (Felsher and Bishop 1999; Pelengaris et al. 2002; Shachaf et al. 2004). In contrast, down-regulation of Myc in established papillomas and osteogenic sarcomas induces regression in the absence of apoptosis, suggesting that the consequence of loss of Myc signaling on cell survival is cell type-specific (Flores et al. 2004; Jain et al. 2002). Finally, down-regulation of oncogenic Ras in melanomas (Chin et al. 1999) as well as activated neu in mammary adenocarcinomas (Moody et al. 2002) also leads to a dramatic induction of apoptosis.

Together these observations indicate that the continued activity of oncogenes is required to maintain tumor cell survival. This suggests that during the course of transformation tumor cells sustain irreversible alterations in the homeostatic control of survival. Thus, rather than reverting to a nontransformed but viable state upon cessation of oncogenic signaling, tumor cells undergo apoptosis. The molecular mechanisms underlying this altered homeostasis are not well understood, but an intriguing proposed mechanism is discussed below.

25.2.4 Addiction to Oncogenic Dedifferentiation Signals

In addition to proliferation and apoptosis, cellular differentiation is also critical for achieving normal tissue homeostasis. Consistent with this, loss of differentiation is a cardinal feature of many tumors. If oncogenic signals are required to maintain tumor cells in an undifferentiated state, then it is possible that oncogene down-regulation could lead to tumor cell differentiation. Indeed, tumor cell differentiation has been observed in various contexts following oncogene deinduction.

As described above, loss of Myc expression in established osteogenic sarcomas, hepatocellular cancers, or β-cell tumors induces tumor regression (Jain et al. 2002; Pelengaris et al. 2002; Shachaf et al. 2004). Histologic examination of these tumor cells following Myc down-regulation reveals that these cells exhibit features suggestive of terminal differentiation. For instance, osteogenic sarcomas in which Myc has been down-regulated *in vivo* form calcified masses reminiscent of bone, and osteogenic sarcoma cells in which Myc is down-regulated *in vitro* appear to differentiate into mature osteocytes (Jain et al. 2002). Similarly, after Myc is turned off in liver tumors, cells adopt normal liver cell morphology and markers indicative of differentiated hepatocytes and biliary cells (Shachaf et al. 2004). Finally, pancreatic β-cell tumors driven by Myc undergo redifferentiation following Myc down-regulation, as evidenced by the reacquisition of insulin expression (Pelengaris et al. 2002). These data suggest that the intrinsic differentiation program of cells, which is disrupted by oncogenic signaling during the course of transformation, can be restored when that oncogenic signaling is down-regulated.

25.2.5 Addiction to Angiogenesis

In addition to the tumor cell-autonomous mechanisms described above, evidence also exists that oncogene down-regulation can induce tumor regression through effects on the tumor stroma. Specifically, a number of studies have found that the tumor vasculature, establishment of which is a necessary step in the development of most solid tumors, requires continued oncogenic signaling for its maintenance (Chin et al. 1999; Giuriato et al. 2006; Pelengaris et al. 2002, 1999). Loss of oncogenic signaling in these models leads to vascular collapse, which in turn promotes tumor regression due to limited nutrient and oxygen delivery. For instance, in Ras-driven melanomas, endothelial cells lining blood vessels undergo apoptosis as early as 24 h following Ras down-regulation, and this is coincident with a loss in cells staining positive for endothelial markers (Chin et al. 1999). Notably, vascular collapse was not prevented by constitutive expression in tumor cells of the pro-angiogenic factor VEGF, suggesting that Ras-mediated angiogenesis is mediated by additional proteins.

A number of studies have indicated that Myc expression is also required for the maintenance of tumor vasculature. Myc-driven lymphomas, skin, and β-cell tumors each possess a robust tumor vasculature that rapidly collapses following Myc down-regulation (Giuriato et al. 2006; Pelengaris et al. 2002, 1999). Again, this collapse is associated with apoptosis of endothelial cells lining blood vessels.

While it is unlikely that the loss of tumor-associated vasculature alone accounts for tumor regression following the down-regulation of oncogenic signaling, the studies described above indicate that vascular collapse may be a common consequence of the targeted inhibition of dominant oncogenic pathways. As such, they suggest the intriguing hypothesis that oncogene addiction may partly be due to the active maintenance of a permissive stroma.

25.2.6 Addiction to Cooperating Oncogenes

As described above, tumorigenesis required multiple independent mutations in both oncogenes and tumor suppressors. Consistent with this, it is well documented that oncogenes often act cooperatively to induce tumorigenesis, such that certain combinations of oncogenes promote transformation more potently and more rapidly than either oncogene alone. This raises the intriguing question of whether tumors arising from the concerted action of multiple oncogenes remain dependent upon the continued activity of one or both oncogenes that induced transformation.

One of the first pairs of oncogenes shown to cooperate, Myc and Ras, synergistically induce transformation of cells *in vitro* and *in vivo* (Land et al. 1983; Sinn et al. 1987). Moreover, mouse mammary tumors induced by Myc frequently harbor spontaneous activating point mutations in Kras2, suggesting that Ras pathway activation represents a preferred secondary pathway for Myc-induced tumorigenesis (D'Cruz et al. 2001). Notably, while down-regulation of Myc in established mammary tumors typically results in dramatic tumor regression, down-regulation of Myc in tumors bearing activated alleles of *Kras2* does not. This suggests that tumors induced by one oncogenic pathway may escape their dependence on the initiating oncogene by becoming addicted to a second oncogenic pathway, and represents an example of Myc-independent tumor resistance.

A recent study tested the consequences of down-regulating Myc alone, Ras alone, or both Myc and Ras in mammary tumors induced by the concerted action of these two oncogenes (Podsypanina et al. 2008). Consistent with previous results, co-expression of Myc and Ras caused tumors to form more quickly than expression of either oncogene alone. Deinduction of both oncogenes led to tumor regression, as expected. Interestingly, deinduction of either oncogene alone led to substantial tumor regression, suggesting that both cooperating oncogenes are required to maintain tumorigenesis and that tumors can regress despite the continued activation of a major oncogenic pathway. However, whereas Ras down-regulation resulted in prolonged tumor regression, Myc down-regulation elicited only short-lived tumor regression and, consistent with prior findings in mice with Myc tumors bearing spontaneous *Kras2* mutations, only a subset of tumors exhibited any regression (Boxer et al. 2004; D'Cruz et al. 2001; Jang et al. 2006; Podsypanina et al. 2008). Thus, mammary tumors arising from the concerted action of Myc and Ras may exhibit a greater dependence on Ras signaling than Myc signaling, and loss of both oncogenes has the greatest effect on tumor regression.

Another study examined the dependence of lymphomas and lung tumors on continued signaling from Myc and Ras (Tran et al. 2008). As expected, lymphomas driven by Myc or Ras alone exhibited a requirement for the continued expression of the inducing oncogene for tumor growth. Furthermore, Myc and Ras cooperated in the formation of lymphomas, and down-regulation of both genes led to complete tumor regression, indicating that these tumors are dependent upon both oncogenes. Lung tumors, in contrast, behaved quite differently. First, Myc-induced lung tumors did not remain dependent upon continued Myc signaling, as they failed to fully regress following Myc down-regulation. Second, Myc and Ras did not cooperate to induce lung tumors; the latency of tumor formation in mice expressing Myc and Ras was similar to that in mice expressing Ras alone. Nonetheless, combined inactivation of Myc and Ras in Myc/Ras-induced lung tumors led to more pronounced tumor regression than inactivation of Myc alone in Myc-driven tumors. This suggests that the presence of activated Ras signaling influences the dependence of lung tumors on sustained Myc activity, as observed in mouse mammary tumors (D'Cruz et al. 2001). These results emphasize that oncogene cooperation and addiction are tissue and context-specific and highlight the importance of conditional mouse models for preclinical validation of anti-oncogenic therapies.

25.2.7 Molecular Mechanisms of Oncogene Addiction

The cellular consequences of oncogene down-regulation described above – growth arrest, senescence, apoptosis, differentiation, and vascular collapse – provide a logical explanation for how tumors regress. However, the molecular mechanisms that account for oncogene addiction are less clear. In particular, the reasons why a tumor cell may exhibit exquisite dependence upon a single oncogenic signaling pathway, whereas the normal cell from which it derived does not, remain a mystery. Stated differently, it is plausible that inhibiting a dominant oncogenic pathway within an established tumor might have cytostatic effects such that the tumor would stop growing but would remain clinically evident. However, what is instead observed following oncogene down-regulation is tumor regression. This suggests that the molecular circuitry of the cell becomes altered during neoplastic transformation in a manner that renders transformed cells dependent upon the transforming oncogene.

One mechanism that has been proposed to underlie this alteration is differential signal attenuation (Sharma et al. 2006a, b). This model posits that oncogenic signaling elicits both pro-survival and pro-apoptotic cascades, but that these cascades exhibit differential decay following oncogene down-regulation. If the pro-survival pathway decays more rapidly than the pro-apoptotic pathway, this would result in a temporal window during which pro-apoptotic pathways would be dominant over pro-survival pathways, leading to cell death. This mechanism has been identified in several systems, including BCR-ABL-, Src-, and EGFR-transformed cells (Sharma et al. 2006b). In each of these cell types, loss of activation of the ERK, Akt, and STAT3 and STAT5 pro-survival pathways occurs rapidly, and the pro-apoptotic p38 pathway becomes activated after a delay. While this model provides an explanation for the cell death that occurs following oncogene down-regulation, it does not address differentiation, senescence, growth arrest, or vascular collapse, each of which has been shown to be an important mediator of tumor regression. Nonetheless, this study provides important clues to one potential molecular mechanism underlying the cell death commonly observed following oncogene inhibition.

25.3 Mouse Models of Tumor Recurrence

The mouse models described above have provided compelling evidence that, consistent with clinical experience, tumors can be dependent upon a single oncogene and down-regulation of that oncogene can induce tumor regression. However, clinical experience is also clear that in patients this type of tumor regression is transient; prolonged remission is the exception, whereas relapse of resistant tumors is the rule. Understanding the pathways by which tumors recur is therefore of paramount importance to improving the success with which human cancers are treated, particularly with targeted therapies.

Conditional mouse models of tumorigenesis have begun to address this issue. As in the clinic, full tumor regression observed following deinduction of several oncogenes in the mouse is followed, at variable intervals, by tumor regrowth. This has provided insight into molecular pathways that contribute to tumor recurrence. One such pathway has been identified through work in our laboratory on neu-induced tumorigenesis (Moody et al. 2005). Mammary tumors induced by activated neu fully regress following neu down-regulation, but ultimately recur with >90% penetrance; recurrent tumors are independent of the neu transgene that initiated primary tumor growth. Molecular and histological analysis has revealed that recurrent tumors undergo an epithelial-to-mesenchymal transition (EMT) and that the transcriptional repressor Snail – which is known to induce EMT – is spontaneously up-regulated in recurrences. Consistent with a functional role in this process, Snail is sufficient to promote tumor recurrence in mouse models systems and high Snail expression predicts an increased risk of recurrence in women with breast cancer. Thus, Snail resides in a pathway by which neu-driven tumors – and likely other breast cancer subtypes as well – escape oncogene addiction and relapse (Moody et al. 2005).

Studies of Wnt-driven tumors have provided evidence that the ARF-p53 tumor suppressor pathway also regulates recurrence. Expression of Wnt1 in the mammary gland results in the formation of adenocarcinomas that are dependent upon continued Wnt signaling, as down-regulation of the transgene leads to tumor regression (Gunther et al. 2003). As with neu-induced mammary tumors, however, regression is followed by Wnt1-independent tumor regrowth. The latency and frequency with which tumors recur is also markedly influenced by the p53 status of the primary tumors: Wnt tumors arising in a *p53* heterozygous background recur more readily and have commonly undergone *p53* loss of heterozygosity. The same acceleration in Wnt tumor recurrence is observed in mice lacking the INK4a/ARF locus (Debies et al. 2008). Careful dissection of the two gene products encoded by this locus has revealed that ARF, but not INK4a, is responsible for inhibition of tumor recurrence. Thus, the ARF-p53 pathway is critical for suppressing the tendency of Wnt-induced tumors to recur.

One possible explanation for this observation is that oncogene down-regulation unmasks a p53-dependent apoptotic program. In the absence of p53, cells may be able to survive oncogene down-regulation and eventually acquire the secondary mutations necessary to reinitiate growth. Alternatively, p53 loss may be required to tolerate the molecular events that drive tumor regrowth, such as activation of a second oncogenic pathway. Distinguishing between these possibilities will require temporal dissection of the requirement for p53 in the process of tumor regression and recurrence.

The ability of Myc-induced tumors to undergo Myc-independent recurrence is dependent upon the tissue type in which it is expressed. As described above, Myc down-regulation in Myc-driven lymphomas leads to tumor regression, which is followed in ~10% of cases by Myc-independent recurrences (Felsher and Bishop 1999; Karlsson et al. 2003). Recurrent tumors acquire additional genetic alterations, as evidenced by the detection of novel chromosomal abnormalities. While the specific genetic events that drive Myc-independent recurrences have yet to be identified, these observed chromosomal abnormalities offer a starting point for dissecting these alternative pathways. By comparison, nearly 50% of Myc-driven mammary adenocarcinomas recur and, while some recurrences are Myc-dependent, most are Myc-independent (Boxer et al. 2004).

While studies on tumor recurrence using mouse models are still in their infancy, several general themes have emerged. First, tumor recurrence is common following oncogene down-regulation, a finding consistent with clinical experience with cancer patients. There also seems to be variation in the ability of tumors driven by different oncogenes to recur: neu-induced tumors in the mammary gland invariably recur; Wnt-induced tumors in the mammary gland recur at low levels unless there is concurrent loss of one p53 allele; and Myc-induced tumors recur at a rate that depends on the tissue type in which it is expressed. This final point – the strong tissue type-dependence of a tumor's response to oncogene down-regulation – may provide insights into why some tumors recur and others do not. The cellular program uncovered by oncogene down-regulation, whether apoptosis, differentiation, or quiescence, is undoubtedly governed by the cell type in which it occurs. A better understanding of these programs, and of the genes that control them, may allow us to better predict the consequences of targeted therapies in humans.

25.4 Oncogene Dependence and Cancer Stem Cells: A Unifying Hypothesis?

The initial response of tumors to therapeutic inhibition or experimental down-regulation of an oncogenic pathway, the persistence of residual tumor cells in a presumed dormant state, and the eventual reemergence of a recurrent tumor following a period of latency, raise a number of interesting questions about the biology of cells that mediate recurrence. These cells must be able to survive oncogene inhibition, persist in a dormant state for an extended period of time and, finally, undergo additional changes to reinitiate malignant growth. Given that cancer stem cells have been proposed to possess a number of these properties, it is possible that such cells may contribute to tumor recurrence. Nevertheless, the extent to which cancer stem cells exist in different solid tumor type is unknown.

The existence of cancer stem cells has been demonstrated conclusively for some malignancies, including acute myelogenous leukemia, but has been only postulated for others (Park et al. 2009; Visvader and Lindeman 2008; Wang and Dick 2005). Notwithstanding specific evidence for or against the presence of stem cells in different tumor types, cancer stem cells have been postulated to possess certain common properties. These include the ability to sustain or reinitiate tumor growth; the ability to generate heterogeneity within a tumor; and the ability to self-renew. In addition, several studies have demonstrated that cancer stem cells may be relatively resistant to cytotoxic therapy and, by analogy with normal tissue stem cells, may be quiescent. It is tempting to speculate that these properties may account for the behavior of tumors in response to inhibition of a dominant oncogenic pathway.

The following model suggests itself: first, cancer stem cells may be relatively resistant to oncogene down-regulation owing to the presence of strong intrinsic survival pathways. Thus, while the bulk of the tumor regresses, cancer stem cells may be able to survive and persist as residual neoplastic cells. Next, these cells may be able to survive in a dormant state for extended periods of time, in which in the absence of oncogenic signaling they are unable to sustain tumor growth. Finally, these cells may accumulate additional genetic or epigenetic alterations that allow for the reinitiation of growth. This model, while speculative, has several important clinical implications. For example, it suggests that targeting quiescent cancer stem cells that are undetectable following targeted therapy may prevent tumor relapse, and ultimately lead to more effective treatments (Jones et al. 2004). The mouse models described above hold promise for testing this model and examining the role of cancer stem cells in mediating tumor recurrence.

25.5 Summary

The use of molecularly targeted therapies in patients has demonstrated conclusively that a wide variety of cancers may become addicted to a single dominant oncogenic pathway. Mouse models that permit reversible oncogene expression in specific tissues have confirmed this finding, and have provided important insights into the mechanisms of oncogene addiction. Going forward, these models have the potential to explore and guide the use of molecularly targeted therapies in patients. Specifically, studies of tumor recurrence in conditional mouse models may help identify secondary oncogenic pathways that mediate resistance to specific targeted therapies in patients. In this manner, mouse models for oncogene addiction may directly impact the treatment of human cancers by identifying novel secondary oncogenic pathways as candidates for therapeutic inhibition.

References

Bedi A, Zehnbauer BA, Barber JP, Sharkis SJ, Jones RJ (1994) Inhibition of apoptosis by BCR-ABL in chronic myeloid leukemia. Blood 83:2038–2044

Berns K, Horlings HM, Hennessy BT, Madiredjo M, Hijmans EM, Beelen K, Linn SC, Gonzalez-Angulo AM, Stemke-Hale K, Hauptmann M et al (2007) A functional genetic approach identifies the PI3K pathway as a major determinant of trastuzumab resistance in breast cancer. Cancer Cell 12:395–402

Boxer RB, Jang JW, Sintasath L, Chodosh LA (2004) Lack of sustained regression of c-MYC-induced mammary adenocarcinomas following brief or prolonged MYC inactivation. Cancer Cell 6:577–586
Braig M, Lee S, Loddenkemper C, Rudolph C, Peters AH, Schlegelberger B, Stein H, Dorken B, Jenuwein T, Schmitt CA (2005) Oncogene-induced senescence as an initial barrier in lymphoma development. Nature 436:660–665
Burstein HJ, Polyak K, Wong JS, Lester SC, Kaelin CM (2004) Ductal carcinoma in situ of the breast. N Engl J Med 350:1430–1441
Chin L, Tam A, Pomerantz J, Wong M, Holash J, Bardeesy N, Shen Q, O'Hagan R, Pantginis J, Zhou H et al (1999) Essential role for oncogenic Ras in tumour maintenance. Nature 400:468–472
Collado M, Gil J, Efeyan A, Guerra C, Schuhmacher AJ, Barradas M, Benguria A, Zaballos A, Flores JM, Barbacid M et al (2005) Tumour biology: senescence in premalignant tumours. Nature 436:642
D'Cruz CM, Gunther EJ, Boxer RB, Hartman JL, Sintasath L, Moody SE, Cox JD, Ha SI, Belka GK, Golant A et al (2001) c-MYC induces mammary tumorigenesis by means of a preferred pathway involving spontaneous Kras2 mutations. Nat Med 7:235–239
Debies MT, Gestl SA, Mathers JL, Mikse OR, Leonard TL, Moody SE, Chodosh LA, Cardiff RD, Gunther EJ (2008) Tumor escape in a Wnt1-dependent mouse breast cancer model is enabled by p19Arf/p53 pathway lesions but not p16 Ink4a loss. J Clin Invest 118:51–63
Demetri GD, von Mehren M, Blanke CD, Van den Abbeele AD, Eisenberg B, Roberts PJ, Heinrich MC, Tuveson DA, Singer S, Janicek M et al (2002) Efficacy and safety of imatinib mesylate in advanced gastrointestinal stromal tumors. N Engl J Med 347:472–480
Druker BJ (2002) Perspectives on the development of a molecularly targeted agent. Cancer Cell 1:31–36
Druker BJ, Sawyers CL, Kantarjian H, Resta DJ, Reese SF, Ford JM, Capdeville R, Talpaz M (2001) Activity of a specific inhibitor of the BCR-ABL tyrosine kinase in the blast crisis of chronic myeloid leukemia and acute lymphoblastic leukemia with the Philadelphia chromosome. N Engl J Med 344:1038–1042
Engelman JA, Chen L, Tan X, Crosby K, Guimaraes AR, Upadhyay R, Maira M, McNamara K, Perera SA, Song Y et al (2008) Effective use of PI3K and MEK inhibitors to treat mutant Kras G12D and PIK3CA H1047R murine lung cancers. Nat Med 14:1351–1356
Engelman JA, Zejnullahu K, Mitsudomi T, Song Y, Hyland C, Park JO, Lindeman N, Gale CM, Zhao X, Christensen J et al (2007) MET amplification leads to gefitinib resistance in lung cancer by activating ERBB3 signaling. Science 316:1039–1043
Felsher DW, Bishop JM (1999) Reversible tumorigenesis by MYC in hematopoietic lineages. Mol Cell 4:199–207
Fletcher JA, Rubin BP (2007) KIT mutations in GIST. Curr Opin Genet Dev 17:3–7
Flores I, Murphy DJ, Swigart LB, Knies U, Evan GI (2004) Defining the temporal requirements for Myc in the progression and maintenance of skin neoplasia. Oncogene 23:5923–5930
Giuriato S, Ryeom S, Fan AC, Bachireddy P, Lynch RC, Rioth MJ, van Riggelen J, Kopelman AM, Passegue E, Tang F et al (2006) Sustained regression of tumors upon MYC inactivation requires p53 or thrombospondin-1 to reverse the angiogenic switch. Proc Natl Acad Sci USA 103:16266–16271
Gorre ME, Mohammed M, Ellwood K, Hsu N, Paquette R, Rao PN, Sawyers CL (2001) Clinical resistance to STI-571 cancer therapy caused by BCR-ABL gene mutation or amplification. Science 293:876–880
Gunther EJ, Moody SE, Belka GK, Hahn KT, Innocent N, Dugan KD, Cardiff RD, Chodosh LA (2003) Impact of p53 loss on reversal and recurrence of conditional Wnt-induced tumorigenesis. Genes Dev 17:488–501
Hanahan D, Weinberg RA (2000) The hallmarks of cancer. Cell 100:57–70
Huettner CS, Zhang P, Van Etten RA, Tenen DG (2000) Reversibility of acute B-cell leukaemia induced by BCR-ABL1. Nat Genet 24:57–60
Jain M, Arvanitis C, Chu K, Dewey W, Leonhardt E, Trinh M, Sundberg CD, Bishop JM, Felsher DW (2002) Sustained loss of a neoplastic phenotype by brief inactivation of MYC. Science 297:102–104
Jang JW, Boxer RB, Chodosh LA (2006) Isoform-specific ras activation and oncogene dependence during MYC- and Wnt-induced mammary tumorigenesis. Mol Cell Biol 26:8109–8121

Jensen EV, Jordan VC (2003) The estrogen receptor: a model for molecular medicine. Clin Cancer Res 9:1980–1989

Jones RJ, Matsui WH, Smith BD (2004) Cancer stem cells: are we missing the target? J Natl Cancer Inst 96:583–585

Karlsson A, Giuriato S, Tang F, Fung-Weier J, Levan G, Felsher DW (2003) Genomically complex lymphomas undergo sustained tumor regression upon MYC inactivation unless they acquire novel chromosomal translocations. Blood 101:2797–2803

Kobayashi S, Boggon TJ, Dayaram T, Janne PA, Kocher O, Meyerson M, Johnson BE, Eck MJ, Tenen DG, Halmos B (2005) EGFR mutation and resistance of non-small-cell lung cancer to gefitinib. N Engl J Med 352:786–792

Land H, Parada LF, Weinberg RA (1983) Tumorigenic conversion of primary embryo fibroblasts requires at least two cooperating oncogenes. Nature 304:596–602

Lynch TJ, Bell DW, Sordella R, Gurubhagavatula S, Okimoto RA, Brannigan BW, Harris PL, Haserlat SM, Supko JG, Haluska FG et al (2004) Activating mutations in the epidermal growth factor receptor underlying responsiveness of non-small-cell lung cancer to gefitinib. N Engl J Med 350:2129–2139

McGahon A, Bissonnette R, Schmitt M, Cotter KM, Green DR, Cotter TG (1994) BCR-ABL maintains resistance of chronic myelogenous leukemia cells to apoptotic cell death. Blood 83:1179–1187

Michaloglou C, Vredeveld LC, Soengas MS, Denoyelle C, Kuilman T, van der Horst CM, Majoor DM, Shay JW, Mooi WJ, Peeper DS (2005) BRAFE600-associated senescence-like cell cycle arrest of human naevi. Nature 436:720–724

Moody SE, Perez D, Pan TC, Sarkisian CJ, Portocarrero CP, Sterner CJ, Notorfrancesco KL, Cardiff RD, Chodosh LA (2005) The transcriptional repressor Snail promotes mammary tumor recurrence. Cancer Cell 8:197–209

Moody SE, Sarkisian CJ, Hahn KT, Gunther EJ, Pickup S, Dugan KD, Innocent N, Cardiff RD, Schnall MD, Chodosh LA (2002) Conditional activation of Neu in the mammary epithelium of transgenic mice results in reversible pulmonary metastasis. Cancer Cell 2:451–461

Nagata Y, Lan KH, Zhou X, Tan M, Esteva FJ, Sahin AA, Klos KS, Li P, Monia BP, Nguyen NT et al (2004) PTEN activation contributes to tumor inhibition by trastuzumab, and loss of PTEN predicts trastuzumab resistance in patients. Cancer Cell 6:117–127

Paez JG, Janne PA, Lee JC, Tracy S, Greulich H, Gabriel S, Herman P, Kaye FJ, Lindeman N, Boggon TJ et al (2004) EGFR mutations in lung cancer: correlation with clinical response to gefitinib therapy. Science 304:1497–1500

Pao W, Miller V, Zakowski M, Doherty J, Politi K, Sarkaria I, Singh B, Heelan R, Rusch V, Fulton L et al (2004) EGF receptor gene mutations are common in lung cancers from "never smokers" and are associated with sensitivity of tumors to gefitinib and erlotinib. Proc Natl Acad Sci USA 101:13306–13311

Pao W, Miller VA, Politi KA, Riely GJ, Somwar R, Zakowski MF, Kris MG, Varmus H (2005) Acquired resistance of lung adenocarcinomas to gefitinib or erlotinib is associated with a second mutation in the EGFR kinase domain. PLoS Med 2:e73

Pardanani A, Tefferi A (2004) Imatinib therapy for hypereosinophilic syndrome and eosinophilia-associated myeloproliferative disorders. Leuk Res 28(Suppl 1):S47–52

Park CY, Tseng D, Weissman IL (2009) Cancer stem cell-directed therapies: recent data from the laboratory and clinic. Mol Ther 17:219–230

Pelengaris S, Khan M, Evan GI (2002) Suppression of Myc-induced apoptosis in beta cells exposes multiple oncogenic properties of Myc and triggers carcinogenic progression. Cell 109:321–334

Pelengaris S, Littlewood T, Khan M, Elia G, Evan G (1999) Reversible activation of c-Myc in skin: induction of a complex neoplastic phenotype by a single oncogenic lesion. Mol Cell 3:565–577

Podsypanina K, Politi K, Beverly LJ, Varmus HE (2008) Oncogene cooperation in tumor maintenance and tumor recurrence in mouse mammary tumors induced by Myc and mutant Kras. Proc Natl Acad Sci USA 105:5242–5247

Politi K, Zakowski MF, Fan PD, Schonfeld EA, Pao W, Varmus HE (2006) Lung adenocarcinomas induced in mice by mutant EGF receptors found in human lung cancers respond to a tyrosine kinase inhibitor or to down-regulation of the receptors. Genes Dev 20:1496–1510

Renan MJ (1993) How many mutations are required for tumorigenesis? Implications from human cancer data. Mol Carcinog 7:139–146

Rusch V, Baselga J, Cordon-Cardo C, Orazem J, Zaman M, Hoda S, McIntosh J, Kurie J, Dmitrovsky E (1993) Differential expression of the epidermal growth factor receptor and its ligands in primary non-small cell lung cancers and adjacent benign lung. Cancer Res 53:2379–2385

Sarkisian CJ, Keister BA, Stairs DB, Boxer RB, Moody SE, Chodosh LA (2007) Dose-dependent oncogene-induced senescence in vivo and its evasion during mammary tumorigenesis. Nat Cell Biol 9:493–505

Sawyers CL, Hochhaus A, Feldman E, Goldman JM, Miller CB, Ottmann OG, Schiffer CA, Talpaz M, Guilhot F, Deininger MW et al (2002) Imatinib induces hematologic and cytogenetic responses in patients with chronic myelogenous leukemia in myeloid blast crisis: results of a phase II study. Blood 99:3530–3539

Sequist LV, Bell DW, Lynch TJ, Haber DA (2007) Molecular predictors of response to epidermal growth factor receptor antagonists in non-small-cell lung cancer. J Clin Oncol 25:587–595

Sequist LV, Lynch TJ (2008) EGFR tyrosine kinase inhibitors in lung cancer: an evolving story. Annu Rev Med 59:429–442

Shachaf CM, Kopelman AM, Arvanitis C, Karlsson A, Beer S, Mandl S, Bachmann MH, Borowsky AD, Ruebner B, Cardiff RD et al (2004) MYC inactivation uncovers pluripotent differentiation and tumour dormancy in hepatocellular cancer. Nature 431:1112–1117

Sharma SV, Fischbach MA, Haber DA, Settleman J (2006a) "Oncogenic shock": explaining oncogene addiction through differential signal attenuation. Clin Cancer Res 12:4392s–4395s

Sharma SV, Gajowniczek P, Way IP, Lee DY, Jiang J, Yuza Y, Classon M, Haber DA, Settleman J (2006b) A common signaling cascade may underlie "addiction" to the Src, BCR-ABL, and EGF receptor oncogenes. Cancer Cell 10:425–435

Sherbenou DW, Druker BJ (2007) Applying the discovery of the Philadelphia chromosome. J Clin Invest 117:2067–2074

Sinn E, Muller W, Pattengale P, Tepler I, Wallace R, Leder P (1987) Coexpression of MMTV/v-Ha-ras and MMTV/c-myc genes in transgenic mice: synergistic action of oncogenes in vivo. Cell 49:465–475

Slamon DJ, Leyland-Jones B, Shak S, Fuchs H, Paton V, Bajamonde A, Fleming T, Eiermann W, Wolter J, Pegram M, Baselga J, Norton L (2001) Use of chemotherapy plus a monoclonal antibody against HER2 for metastatic breast cancer that overexpresses HER2. N Engl J Med 344:783–792

Tran TP, Fan AC, Bendapudi PK, Koh S, Komatsubara K, Chen J, Horng G, Bellovin DI, Giuriato S, Wang CS et al (2008) Combined Inactivation of MYC and K-Ras oncogenes reverses tumorigenesis in lung adenocarcinomas and lymphomas. PLoS One 3:e2125

Verweij J, van Oosterom A, Blay JY, Judson I, Rodenhuis S, van der Graaf W, Radford J, Le Cesne A, Hogendoorn PC, di Paola ED et al (2003) Imatinib mesylate (STI-571 Glivec, Gleevec) is an active agent for gastrointestinal stromal tumours, but does not yield responses in other soft-tissue sarcomas that are unselected for a molecular target. Results from an EORTC Soft Tissue and Bone Sarcoma Group phase II study. Eur J Cancer 39:2006–2011

Visvader JE, Lindeman GJ (2008) Cancer stem cells in solid tumours: accumulating evidence and unresolved questions. Nat Rev Cancer 8:755–768

Vogel CL, Cobleigh MA, Tripathy D, Gutheil JC, Harris LN, Fehrenbacher L, Slamon DJ, Murphy M, Novotny WF, Burchmore M et al (2002) Efficacy and safety of trastuzumab as a single agent in first-line treatment of HER2-overexpressing metastatic breast cancer. J Clin Oncol 20:719–726

Vogelstein B, Kinzler KW (1993) The multistep nature of cancer. Trends Genet 9:138–141

Wang JC, Dick JE (2005) Cancer stem cells: lessons from leukemia. Trends Cell Biol 15:494–501

Weinstein IB (2002) Cancer. Addiction to oncogenes – the Achilles heal of cancer. Science 297:63–64

Wu CH, van Riggelen J, Yetil A, Fan AC, Bachireddy P, Felsher DW (2007) Cellular senescence is an important mechanism of tumor regression upon c-Myc inactivation. Proc Natl Acad Sci USA 104:13028–13033

Chapter 26
Mouse Models in Preclinical Drug Development: Applications to CNS Models

Eletha Carbajal and Eric C. Holland

26.1 Introduction

Primary tumors of the CNS can be divided into several groups; the most common in adults being the gliomas and in children the medulloblastomas (Kleihues et al. 2002). Gliomas are thought to arise from cells at some point of differentiation along glial lineage leading to astrocytes and oligodendrocytes. In the normal brain, the primary function of astrocytes is to maintain homeostasis of neuronal extracellular milieu and protect the neurons by establishing the blood–brain barrier during development. Oligodendrocytes provide the CNS axons with myelin sheaths that insulate the neurons and thereby accelerating the action potential transduction. By contrast, medulloblastomas appear to arise from cells of the external granule layer in the developing cerebellum that give rise to the internal granule layer neurons. These tumors are the most common tumors found in children. Gliomas are relatively radiation-resistant and the clinical outcome for patients with the most aggressive and common of the gliomas, Glioblastoma Multiforme or GBM, has been poor and essentially unchanged in the last 50 years with the median survival averaging 12–14 months (CBTRUS. Statistical report: primary brain tumors in the USA). Medulloblastomas on the other hand are much more responsive than GBMs to radiation and chemotherapy with a 70% cure rate. An important step toward improving existing treatments and discovering new ones comes from creating genetically and histologically accurate mouse models that can be used as a representative system of human tumors in order to study the biological and mechanistic causes of brain cancer and the way that these tumor interact with their microenvironment. Furthermore, accurate mouse models of these tumors are helping us to identify novel targets and therapies for clinical testing (Hambardzumyan et al. 2007b).

E. Carbajal • E.C. Holland (✉)
Program in Cancer Biology and Genetics, Memorial Sloan Kettering Cancer Center, New York, NY 10021, USA
e-mail: hollande@mskcc.org

J.E. Green and T. Ried (eds.), *Genetically Engineered Mice for Cancer Research: Design, Analysis, Pathways, Validation and Pre-Clinical Testing*,
DOI 10.1007/978-0-387-69805-2_26,

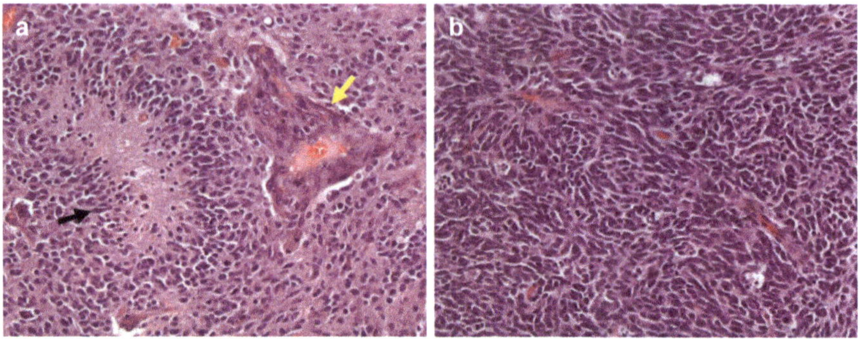

Fig. 26.1 Histology of murine PDGF-induced GBM and SHH induced Medulloblastoma. (**a**) H&E staining of GBM, *black arrow* represents pseudopalisading necrosis and *yellow arrow* represents microvascular proliferation. (**b**) H&E staining for SHH driven-medulloblastoma

26.2 Histology of Gliomas and Medulloblastomas

Gliomas are classified as astrocytic, oligodendrocytic, or mixed based on their histology, including expression of certain markers and morphologic similarities to normal cells of the respective type (Kleihues et al. 2002). As seen in Fig. 26.1, gliomas are histologically characterized by diffuse invasion into surrounding brain structures, microvascular proliferation, pseudopalisading necrosis, and expression of glial markers. These tumors generally progress over time, increasing in grade as the tumor cell population acquires additional mutations. Upon recurrence, tumors commonly increase in grade and may not show histologic similarity to the original resected tumor. Medulloblastomas are small blue cell tumors with expression of neuronal markers that usually form in the cerebellum. In addition, they are often characterized by groups of tumor cells arranged in a circle around a fibrillary center (rosette formation). Medulloblastomas are classified as desmoplastic when they contain large-cytoplasm cell islands in the field of more typical medulloblastoma cells, or large cell when they contain large pleiomorphic cells.

26.3 Signaling Characteristics of the Two Tumor Types

GBMs are the most malignant astrocytomas and frequently contain several genetic alterations that result in the disregulation of specific signaling pathways. The pathways activated by these mutations contain components that are frequently drug-able and therefore these pathways are reasonable therapeutic targets. The most frequent genetic abnormalities typically result in the stimulation of RAS and AKT pathways (Holland et al. 2000). Gliomas often produce growth factors such as platelet-derived growth factor (PDGF) A and B (Hermanson et al. 1992), epidermal growth factor (EGF) (Ekstrand et al. 1994), transforming growth factor-alpha (TGF-a)

(Nister et al. 1988), and insulin-like growth factor I (IGF-I) (Sara et al. 1986; Trojan et al. 1992), making autocrine or paracrine stimulation of the corresponding growth factor receptors likely (Guha et al. 1995; Hermanson et al. 1992; Nister et al. 1988). The corresponding receptors, PDGFRa, -b, EGFR, and IGFR are also often overexpressed, sometimes due to amplification (Ekstrand et al. 1994). Downstream of these receptors, RAS activity is elevated in all human GBMs and AKT is elevated in approximately 75% of GBMs. Although AKT activity is elevated in the majority of human GBMs (Holland et al. 2000), no gain-of-function mutations of AKT have been found thus far. However, there are at least two additional routes to achieving AKT activation other than activation of the tyrosine kinase receptors upstream. Loss of PTEN is frequent (Li et al. 1997), and recent modeling studies have shown that loss of PTEN is similar to activated Akt in gliomagenesis (Hu et al. 2005). Further, activating mutations of PI3 kinase have been shown in these tumors as well (Gallia et al. 2006).

The different histologic types of medulloblastomas are likely to be driven by different oncogenic pathways. In the desmoplastic subset of these tumors, the sonic hedgehog (SHH) pathway is hyperactive either by loss of PTC function (Johnson et al. 1996) or activating mutations in Smo or suppressor of fused (Taylor et al. 2002). This signaling characteristic accounts for approximately 30% of these tumors. The pathways that lead to the formation of the other medulloblastoma subtypes are less well understood although loss of *p53* and amplification of *Myc* is common in the anaplastic variants of the tumor (Zindy et al. 2007).

Many pharmaceutical companies are currently developing compounds that target the Ras pathway at either Raf or Erk; the PI3 kinase pathway by targeting PI3K, Akt, TOR kinase, or mTOR; and the SHH pathway by targeting SMO. Few if any of these compounds are being developed specifically for the brain and frequently the compound either does not get into the brain or its brain penetrance is unknown. Nonetheless, this collection of compounds may comprise a few drugs that will ultimately be useful in brain tumors in people. Accurate mouse models of these tumors will hopefully provide tools for sorting through the available drugs for efficacy and penetration of the blood–brain barrier.

26.4 Models of Brain Tumors

While researchers use several different animal species to model cancer biology, mice are the most common cancer-modeling organisms because of their high degree of genetic homology, relatively low maintenance costs, and the availability of sophisticated genetic and testing tools (Fomchenko and Holland 2006). Mouse models provide the necessary tumor–host interaction that is difficult or impossible to model *in vitro*. Furthermore, mouse models can be used to address issues in drug development such as the determination of toxicity and proof of *in vivo* antitumor effectiveness. Mouse models not only allow us to identify mutations or pathways involved in tumor initiation, therefore, furthering our understanding of the microenvironment and etiology of brain cancer, but they also help identify whether specific oncogenic signaling pathways are necessary or sufficient for tumor induction or maintenance of the tumor phenotype (Hu and Holland 2005). Lastly, mouse models faithfully recapitulating human disease provide an opportunity for testing novel targets for potential therapies (Hambardzumyan et al. 2007b).

26.5 Xenograft Models

The simplest and most commonly used xenografts are subcutaneous injections of a tumor cell line creating a tumor in the flank. The primary advantages of using subcutaneous xenografts include the induction of a large number of synchronized, predictable, and rapidly forming tumors that are easily observed and can be treated once tumors reach an optimal size (Becher and Holland 2006; Sausville and Burger 2006). However, the standard cell lines used for these studies have been passaged for many generations in cell culture, they commonly acquire additional mutations during the culture times and often fail to induce formation of the histologically appropriate vasularization and rarely recapitulate tumor-of-origin phenotype (Finkelstein et al. 1994). Also, immunocompromised mice do not show a typical immune response toward tumor cells and can produce false positives during drug trials. Orthotopic transplantation into the brain gives a more realistic environment for the tumors but the standard cell lines used typically do not give the diffuse invasion, pseudopalisading necrosis, and microvascular proliferation that are the histologic definition of these tumors in patients. Some of these shortcomings have been overcome by using tumor lines that have never been grown on ex vivo plastic dishes but rather passaged directly into the mouse and then serially from mouse to mouse. These xenograft lines replicate the diffuse invasion of the human tumors but generally lack the pseudopalisading necrosis or vascular proliferation (Giannini et al. 2005). These characteristics are seen in some of the genetically engineered mouse models (GEMs) (Becher and Holland 2006).

26.6 Genetically Engineered Models

Recently, GEMs of cancer have made significant contributions to our understanding of the molecular pathways causally linked to tumor initiation, progression, and metastasis (Hambardzumyan et al. 2007b). GEMs have allowed us to gain a clearer understanding of the role of the tumor microenvironment while at the same time expanding our knowledge of the mechanistic roles oncogenes and tumor suppressor genes play in this process. Tumors induced genetically in mouse models not only can recapitulate the histology and genetics of human tumors, but also demonstrate the causal role of specific genetic abnormalities in tumor formation (Uhrbom et al. 2005). The two general categories of GEM brain tumor models are those produced by germline modification strategies and those induced by somatic cell-gene transfer.

26.7 GEM Produced by Germline Modification Strategies for Brain Tumors

With the introduction of gene-targeting technology in mouse embryonic stem cells, germline modification strategies created lines of mice with oncogene-bearing transgenic gain-of-function mice, and tumor suppressor knockout or loss-of-function mice. In transgenic mice, every cell in the body has the genetic construct and will express the transgene if the promoter is active in it (Grisendi and Pandolfi 2004). For glioma and medulloblastoma models, these promoters are Glial Fibrillary Acidic Protein (GFAP) promoter and Nestin promoter; both are normally active in glial cells and are also stem/progenitor cell markers (Fomchenko and Holland 2006). Good examples of this strategy for gliomas are the expression of vErbB from the S100 promoter, giving rise to oligodendrogliomas (Ohgaki et al. 2006) and expression of mutant active HRas from the GFAP promoter giving rise to astrocytic tumors (Shannon et al. 2005).

In contrast to the gain of function transgenic lines, knockout mice show a loss of gene function in all cells resulting in functional consequences in cell types that would normally express that gene, or in cell types derived from precursors that normally express the gene during development. This is seen in the knockout of PTC that elevates SHH signaling which is lethal as a homozygote, but as a heterozygote gives rise to medulloblastomas among other tumor types (Goodrich et al. 1997).

More sophisticated strategies using combined conditional alleles allow limiting the deletion of tumor suppressors in specific cell types such as the combined loss of NF1 (resulting in elevation of Ras activity) and p53 driven by cre expression from the Krox20 promoter that is active in Schwann cell progenitors (Zhu et al. 2002). Therefore, combined deletion of NF1 and p53 results in the activation of Ras and loss of p53 specifically in cell destined to form Schwann cells, the glial cell equivalent of the peripheral nervous system. Later, it was shown that loss of p53 and activation to the Ras pathway via NF1 inactivation in CNS cells is sufficient to cause malignant astrocytoma formation with 100% penetrance (Zhu et al. 2005).

Although using GEM strains with germline modifications allows us to better understand the cooperative effects of different genetic alterations, these methods create genetic alterations in large numbers of cells which are not sufficient to transform a given cell. If the genetic alterations in the cells were sufficient to convert a normal cell to a cancerous one, the entire organ (in this case the brain) would be tumor and there would not be a mouse to study. Since these mice exist and tumors only arise in a certain percentage of them, it is likely that tumor initiation requires secondary mutations. As a result, germline models are useful in identifying mutations that contribute to tumorigenesis but are insufficient for tumor induction.

26.8 Somatic Cell Gene Transfer

Somatic cell gene transfer using retroviral vectors is a complementary modeling system to germline modification. The design differs from germline modification strategies in that targeted gene deletion or overexpression cannot be inherited and is presented in a targeted cell population rather than in the whole mouse. At the same time, somatic cells that were genetically modified and transformed by viral vectors can pass on acquired mutation to daughter cells. Murine retroviral vectors can be used with any mouse line but have no cell type specificity. By contrast, the RCAS/tv-a system allows cell-type specificity but requires specific transgenic recipient mouse lines to achieve gene transfer. RCAS (Replication Competent ALV-Splice Acceptor) vectors are based on the subgroup A avian leucosis virus (Petropoulos and Hughes 1991). The subgroup A receptor (Tv-a) is not normally expressed in mammalian cells, therefore the infection with RCAS is very low. However, when mammalian cells are genetically modified to express the tv-a under the control of a cell or tissue-specific promoter, the cells become infectable by these vectors and gene transfer is limited to those cells that express the tv-a transgene (Federspiel et al. 1994). As an example, the Gtv-a mouse line used in the RCAS/tv-a system expresses tv-a under the control of the GFAP promoter, and the Ntv-a line expresses tv-a under the control of the nestin promoter (Holland et al. 1998). These can be infected with RCAS vectors to transfer genes into astrocytes and stem cells in the brain leading to tumors in areas such as the cortex, thalamus, brain stem, and cerebellum.

Furthermore, the RCAS/tv-a system allows for lineage tracing with an ability to genetically tag the originally infected cells to identify the progeny of the cell-of-origin, as well as study the use of oncogene combinations.

26.9 Somatic Cell Gene Transfer Models of Brain Tumors

PDGF was first shown to be gliomagenic using MMLV-based vector transfer of the PDGF-B gene. Infection of MMLV-producing cells into neonatal mouse brain, induced malignant gliomas characterized by PDGFB/PDGFRα co-expression (Uhrbom et al. 1998). Additional loss of the tumor suppressor gene P53 in the same mouse model has resulted in increased tumor frequency and increased latency (Hesselager et al. 2003).

The RCAS/tv-a system has also been used to enforce expression of PDGF-B by retroviral infection of tv-a expressing neural progenitor cells and primarily results in oligodendrogliomas. Retroviral PDGFB transfer to nestin-expressing progenitor cells of *ink4a/arf* null mice in the same system results in higher grade gliomas with more astrocytic character, which confirms the ability of PDGFB to cooperate with tumor suppressor loss during gliomagenesis. The tumor cell structures that define low-grade gliomas are recreated by this modeling system (Dai et al. 2001). The tumor cells migrate along white matter tracks, surround neurons and blood vessels and accumulate at the edge of the brain in the subpial space. The structures that define gliomas are the way that these tumor cells interact with their microenvironment and “stroma” of the brain. This model is therefore an excellent experimental system to understand the interactions between the tumor cell and its microenvironment. Both high and low grade tumors occur in this model, and continued PDGF signaling is required for tumor maintenance. By elevating levels of PDGF-B expression, it is possible to generate glial tumors of higher grade and increased vascularization, correlated with significant recruitment of vascular smooth muscle progenitor cells (Shih et al. 2004; Shih and Holland 2006b).

GBM modeling using the RCAS/tv-a system and constitutive expression of K-Ras and Akt in nestin-expressing progenitor cells implicate a causal relationship between these pathways and GBM formation resulting in 26% tumor incidence at 12 weeks. However, single or combined expression of activated K-Ras or Akt in differentiated astrocytes appeared insufficient to induce tumors. This information suggests that both cell differentiation status and activation of signaling pathways are critical for GBM formation.

Modeling medulloblastoma with RCAS/tv-a can be achieved by transferring SHH to nestin–expressing cells in the developing cerebellum resulting in 15% incidence of medulloblastomas within 3 months (Goodrich et al. 1997). Co-infection of these nestin progenitor cells with RCAS vectors expressing Akt, N-myc, IGF2, or BCL-2 increases the incidence of medulloblastomas (Hallahan et al. 2004; Rao et al. 2003). In addition, the RCAS/tv-a system was used to generate a minor subtypes of medulloblastomas. The medulloblastoma subtype with extensive nodularity can be recreated by overexpressing SHH with loss of PTEN in nestin expressing progenitor cells from the external granular layer (EGL) (Hambardzumyan et al. 2007a). Additionally, anaplastic medulloblastomas can be achieved by infecting $p53^{-/-}$ GFAP-expressing cells with RCAS-cMyc and RCAS-beta catenin (Momota et al. 2008).

26.10 Stem Cells in These Tumor Types

In recent years, evidence has been mounting in support of a role for stem-like cells in cancer formation. A subset of cells in gliomas and medulloblastomas have a significantly higher ability to initiate tumors when transplanted into recipient animals than the bulk of the tumor cells (Singh et al. 2004). These cells express stem cell markers and have the capacity to self-renew and are relatively resistant to therapy (Bao et al. 2006). These cells occupy specific regions of the tumors referred to as stem cell niches which in the case of medulloblastomas are perivascular (Calabrese et al. 2007) and in the case of gliomas are both perivascular and in the region of the pseudopalisading necroses. These stem cell niches contain multiple cell types that help promote the stem cell characteristics of these cells by several signaling pathways including Notch and SHH that are potentially therapeutic targets.

26.11 Notch

Notch1–4 in mammals is a family of transmembrane proteins that regulate cell-to-cell signaling which is important in cell communication. Notch ligands, which are also transmembrane proteins (Delta-like 1, Delta-like 3, Delta-like 4, and Jagged 1-2 in mammals) are expressed on adjacent cells. Notch signaling appears to affect both tumorigenesis and stem cell development. Notch signaling has many functions including inhibition of neurogenesis, increasing neural stem cell proliferation, and regulating cell fate decisions throughout glial and neuronal development (Morrison et al. 2000; Nye et al. 1994). In the oligodendrocyte lineage, Notch activation has been shown to suppress terminal oligodendrocyte differentiation and also to support the specification of oligodendrocyte precursors from the initial stem cell pool (Genoud et al. 2002; Wang et al. 1998).

Notch signaling has been implicated in various steps throughout tumorigenesis. The Ras pathway collaborates with Notch signaling to maintain and establish a neoplastic phenotype (Fitzgerald et al. 2000; Weijzen et al. 2002). In CNS tumors, Notch signaling components were found to be deregulated in meningiomas and medulloblastomas (Cuevas et al. 2005). Notch receptors and ligands are expressed in human GBMs and medulloblastomas (Fan et al. 2006). In Kras-induced gliomas in mice, Notch appeared responsible for the transcription of specific gene targets, particularly that of nestin, a neural progenitor marker. The activation of the nestin promoter by Notch was also seen in culture where Notch directly acts to generate lesions along the subventricular zone (SVZ). These lesions proliferate and express nestin, suggesting that they may be early precursor lesions to tumorigenesis (Shih and Holland 2006a). The central role for Notch signaling in medulloblastomas has been suggested by cultured and xenograft models where notch is cleaved and notch targets are expressed and gamma secretase inhibitors have shown to block these effects and cell growth (De Strooper et al. 1999).

26.12 Sonic Hedgehog (SHH)

As noted above, about a third of human medulloblastomas have mutations that activate the SHH signaling pathway and the causal relationship between SHH and medulloblastoma formation has been demonstrated in mice by deletion of the SHH receptor Patched or by enforced SHH activation (Marino 2005). The connection between SHH signaling and the brain is well established. PTCH is expressed in the developing cerebellum by neuronal precursors in the EGL while SHH is produced by Purkinje neurons that lie beneath the EGL (Dahmane and Ruiz i Altaba 1999; Wechsler-Reya and Scott 1999). *In vitro*, SHH has been shown to be a potent mitogen for EGL precursors. Blocking of SHH signaling *in vivo* leads to hypoplastic cerebellae in which granule neurons are greatly reduced or absent. These factors that lead to differentiation and migration of the postmitotic neurons are not clear at this time. The glioma-associated oncogene homologue Gli was first identified as a gene amplified in gliomas (Kinzler et al. 1987). The product of this gene was later found to be a component of the SHH signaling pathway. SHH signaling regulates multiple aspects of CNS development, controlling both cell proliferation and cell differentiation. SHH is required for the differentiation of floor plate cells and ventral neurons in the early neural tube. SHH also regulates proliferation of granule cell precursors in the developing cerebellum and contributes to oligodendrocyte specification in the mammalian forebrain (Nery et al. 2001; Tekki-Kessaris et al. 2001). Furthermore, SHH is necessary and sufficient for oligodendrocyte precursor production in cortical neuroepithelial cultures, and drives proliferation in CNS precursor cells (Rowitch et al. 1999). Several small molecule inhibitors of SMO are being developed for cancer therapeutics as a means of blocking SHH signaling to gli (Romer and Curran 2005).

26.13 Blood–Brain Barrier

One aspect of treating tumors in the brain that is a unique challenge is the blood-brain barrier that prevents many drugs of achieving adequate levels within this organ site. The normal blood-brain barrier is composed of endothelial cells, astrocytes, and pericytes. It functions to prevent certain large molecules from penetrating the brain parenchyma and is the leading reason for low efficacy of drug delivery to tumors (Szakacs et al. 2006). A major component of the blood-brain barrier (BBB) is the activity of various efflux transporters (ABCB1, ABCC1, ABCC2, and ABCC4) that pump compounds from the brain back into the bloodstream. In theory, inhibiting these transporters could bring down drug resistance and allow efficient killing of tumor cells. However, most of the developed ABC inhibitors have so far failed to significantly improve the results of chemotherapy in patients.

A most intriguing property of ABC transporters is their high level of expression in stem cells, and the loss of their expression in mature cells. So far, three ABC genes have been identified in tumor stem cells: ABCB1, (MDRI), ABCCI ABCC1 (MRP1), and ABCG2 ABCG2 (BCRP1) (Dean et al. 2005). The drug-resistance property of stem cells, as conferred by these ABC transporters, is extremely useful in the isolation and analysis of stem cells. While most cells accumulate the fluorescent dyes Hoeschst 33342, stem cells do not since the dye is effluxed by ABC pumps. The latter property allows the sorting of stem cells as a population referred to as the side population (SP) (Challen and Little 2006).

26.14 Radiation Therapy

Radiation therapy is a critical component of the standard of care for patients with CNS tumors, partly because the BBB is not an issue with radiation. Medulloblastomas respond much better to radiation than do gliomas (Mayer et al. 1976). Even with the relatively poor response in gliomas, radiation is the standard of care due to the lack of adequate alternatives. Any new treatments will need to be given with radiation there up front or at relapse. The biology of radiation response is not fully known at this time, especially the differences in response between the tumor bulk and the stem cells within the tumor.

The term stem cell was coined in the context of clonogenic cells surviving radiation that were able to repopulate the spleen (Fisher and Puck 1956). In the gut, cells with a relatively low baseline proliferation rate were found to be relatively resistant to radiation and respond to it with cell division giving rise to cells that repopulate the crypt (Hornsey 1973; Marshman et al. 2002). Radiation therapy works by creating double strand breaks in the DNA and it is the standard therapy for both gliomas and medulloblastomas following surgical resection. In both humans and mouse models, medulloblastomas are far more sensitive to radiation therapy than are gliomas. Stem cells existing within p53 wild-type medulloblastomas in mice are resistant to radiation therapy, initially responding to radiation treatment by cell cycle arrest rather than cell death. These arrested cells reenter the cell cycle and repopulate the tumor. By contrast, many cells of gliomas are resistant to radiation and behave similarly to the stem cells of medulloblastomas (Hambardzumyan et al. 2008). It is likely that both p53 and the PI3 kinase pathways are critical for the response of cells in these tumors to radiation, the PI3 kinase pathway is a potential therapeutic target in this regard.

26.15 Preclinical Trials in GEM Models of Brain Tumors

With the exception of GBM, most primary CNS tumors are relatively rare, hampering informative trials in patients. Several advantages exist in using murine models in trials such as the ability to overcome the limited cohort sizes and to histologically analyze the effect of a given therapy at various points during treatment. Generally speaking, the larger the effect seen relative to the background noise of the population the fewer mice are needed to achieve significance. Several techniques allow for a reduction in mouse numbers by following individual mice overtime.

26.16 Imaging

The use of imaging technology is crucial for identifying mice with tumors so that they can be enrolled in preclinical trials (Ross et al. 2004). Recent advances in preclinical imaging technologies also make it possible to establish end-points for tumor-bearing mice in preclinical studies. Proliferation status can be monitored by imaging the mice at specified intervals of time throughout the study. Preclinical imaging allows the mouse to be used as its own control. Not only do these imaging techniques allow the mouse to be its own control, but they also allow us to produce statistically significant results with fewer mice than standard comparisons between populations of mice.

26.17 Magnetic Resonance Imaging

Anatomic imaging techniques are used in preclinical trials as they parallel the standard image analysis in humans. Magnetic resonance imaging (MRI) reveals similarities between the characteristics of GEM tumors and those seen in humans such as contrast enhancement on T1 images in regions of high-grade histology, and T2 signal in regions of edema and tumor cell invasion (Koutcher et al. 2002). The major drawbacks of MRI scans as routine metrics for preclinical trials are the costs and availability of the scanning equipment. MRI can be used for detecting tumors in mice based upon the appearance of imaging characteristics and their pattern of enhancement that can be useful for detecting mice that are suitable candidates for preclinical trials. MRI imaging also allows monitoring of tumor size for stratification, measurements of growth rates, and treatment response (McConville et al. 2007). By contrast, CT scans used in humans with intracranial pathology do not show sufficient contrast with intracranial tumors to make this modality useful in the murine setting.

26.18 Positron Emission Tomography

GEMs with intracranial tumors can also be detected and monitored noninvasively with the use of positron emission tomography (PET) imaging. The most common PET ligand used clinically is 18F-fluorodeoxyglucose (FDG). This ligand is taken up by cells with high glucose metabolism and trapped in the cell by the enzymatic action of hexokinase. Other PET ligands exist such as the nucleic acid analog 18F-deoxyfluorothymidine (FLT) (Chen et al. 2005) that is an indirect measure of cell proliferation, and ACBC that is an amino acid analog, which is a readout of amino acid uptake and anabolism (Bradbury et al. 2008). The resolution of PET is relatively low but newer instruments are being developed that allow the fusion of PET images with MRI or CT providing the anatomical landmarks. Not only can the PET image report on the presence of a tumor and potentially provide a readout for therapeutic cell killing, but the glucose and amino acid uptake of a tumor may correlate with the activity of critical pathways such as the PI3 kinase pathway that is not only critical for the biology of the tumor but also is a therapeutic target as well. It remains to be seen if PET imaging will provide a quantitative molecular readout of therapeutic blockade of this pathway.

26.19 Bioluminescence Imaging

The easiest imaging technology for preclinical trial studies is bioluminescence imaging (BLI). BLI technology visualizes the conversion of chemical energy within luciferin substrates into visible light by luciferase enzymes. Although the wavelength

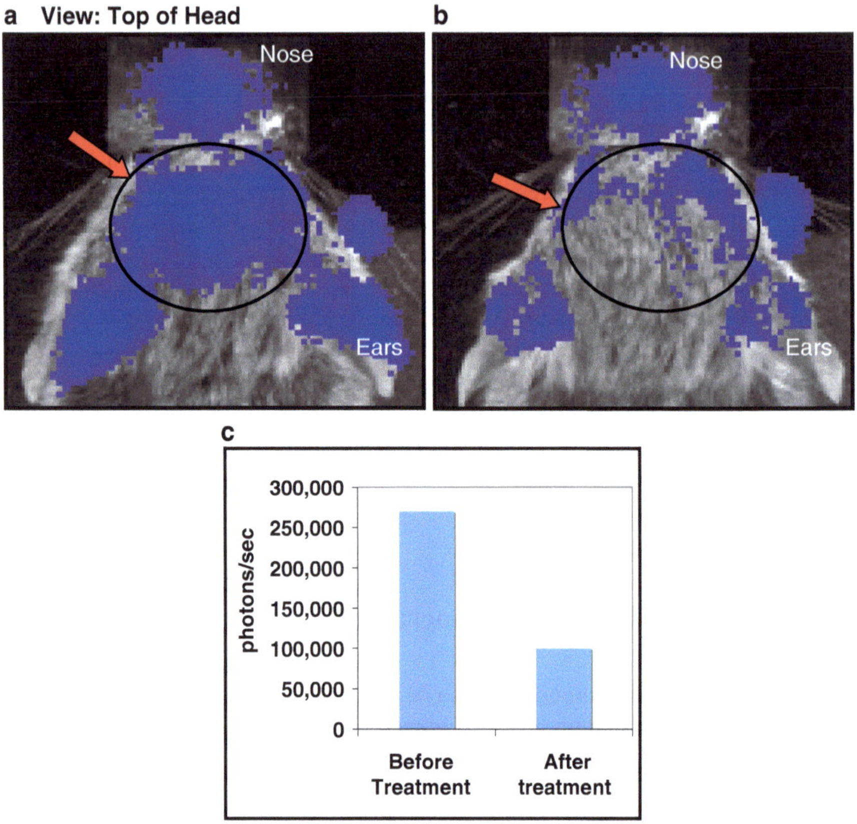

Fig. 26.2 Bioluminescence imaging of Gli-luc mouse with Glioma before and after treatment. (**a**) Image illustrates Gli-luc mouse with Glioma before treatment with drug. Light coming from nose and ears represents background light emitted from skin. *Red arrows* indicate light emitted from tumor before and after treatment. (**b**) Gli-luc mouse with Glioma after treatment with drug. (**c**) Bar graph demonstrating bioluminescence signal from glioma in panel (**a**) and (**b**) before and after treatment

of light produced by luciferin is attenuated to some extent by passing through tissue, tumors in the brain are easily visualized noninvasively *in vivo* with this strategy and is an established method this to measure cell numbers and proliferation in grafted tumors *in vivo* (Momota and Holland 2005). Recently, the use of *in vivo* BLI has shown to be more sensitive than any other *in vivo* imaging methods presently available with low background and deep tissue penetration (Contag et al. 1998; Edinger et al. 1999) (see Chapter 11). There are many examples of BLI as a method for following implanted intracranial xenografts or metastatic cells to the brain (Dinca et al. 2007). In these cases, the host animal does not express luciferase and the cell lines express luciferase from a constitutive promoter resulting in a correlation between light output and tumor cell number. In addition, transgenic mice have been developed in which the light output correlates with a specific biologic process or pathway activity.

One example is a reporter line that expresses the luciferase gene under the control of an E2F1 promoter. When luciferase mice are crossed with genetically modified mouse models of human cancer, the proliferation of the tumor cells can be monitored with proportional light production with regards to both tumor mass and tumor volume (Uhrbom et al. 2004). As seen in Fig. 26.2, a second example is one where the SHH signaling pathway (specifically Gli activity as seen by a synthetic promoter with 8 Gli binding sites) is seen to be activated in many tumor types including medulloblastomas and gliomas (Becher et al. 2008). Although there are several useful advantages to BLI and other optical imaging, strategies do not translate to use in humans.

26.20 Biomarkers

There are several biomarkers in addition to imaging that can indicate the presence of an intracranial tumor or its response to treatment. For example, extreme weight loss or lack of grooming can indicate the presence of a CNS tumor. Oftentimes, mice with advanced tumors become noticeably erratic or sluggish and sedentary, or spin in their cages. Serum markers such as YKL-40 are currently being developed for gliomas in humans (Hormigo et al. 2006). However, the use of serum markers in mice is not currently available.

26.21 Histology

All of the above noninvasive measures listed above are only surrogates for the real effect that the therapy is having on the tumor. Further, because of cellular heterogeneity of the tumors the detailed effect that a given therapy has on specific cell types can only be seen by histologic analysis. Furthermore, histology is needed to interpret all imaging studies. Immunohistochemical staining for proliferation markers such as PCNA (Fig. 26.3) or Ki67 are quite effective and report the number of cells in cycle at the moment of sacrifice. BrDU labeling prior to sacrifice is a direct measure of which cells were in cycle prior to therapy, or during therapy just prior to sacrifice. Cell death is identified by morphology of cells stained with H&E and by immunohistochemical staining for cleaved caspase 3 or TUNEL. Immunoflorescence staining for the above markers allow double or triple staining with other antibodies to determine the identity of the cells in question.

26.22 Published Trials in GEM Models

Several trials of small molecule inhibitors have been published using GEM models of CNS tumors. The first was that of a trial of the SMO inhibitor HhAntag treating Medulloblastomas from $PTC^{+/-}$ mice in a $p53^{-/-}$ background (Romer et al. 2004).

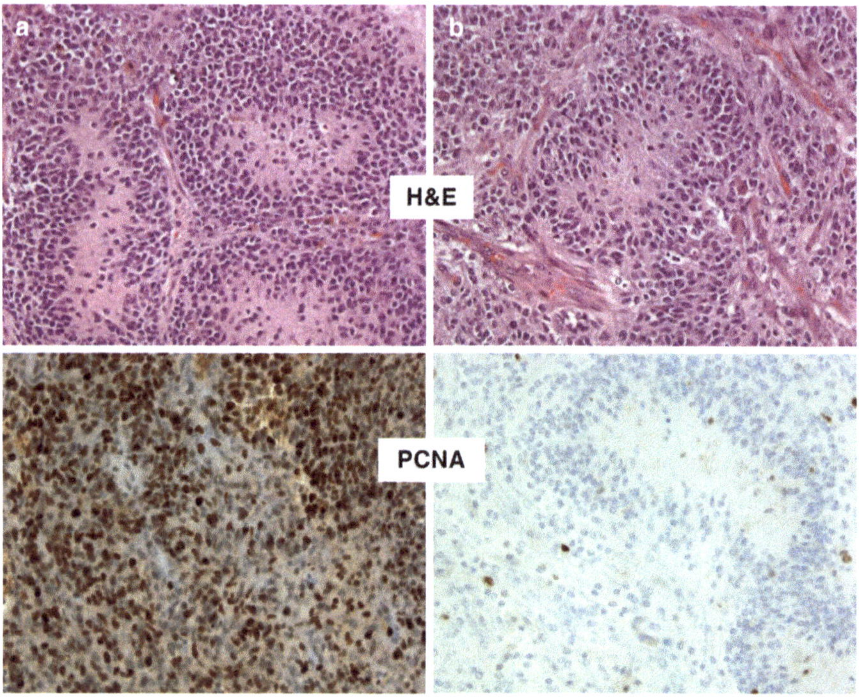

Fig. 26.3 H&E and PCNA for PDGF-induced GBM before and after treatment. (**a**) H&E and PCNA staining of untreated PDGF-induced GBM, showing high proliferation (*brown nuclei are positive*). (**b**) H&E and PCNA staining of PDGF-induced GBM treated with CCI779 for 8 days, showing inhibition of proliferation corresponding to low PCNA

These mice develop tumors by 12 weeks of age and in this study the mice were enrolled on study without imaging. The authors found no medulloblastomas in the treated cohort and interpreted the results that this drug irradicated these tumors within a treatment window of 2 weeks. A second study using several mouse models of medulloblastomas showed that p53 status mattered greatly to the response of both the tumor bulk and stem cell compartment in response to irradiation. In this study, wild-type p53 tumors responded within 6 h to 2 Gy radiation with the tumor bulk undergoing apoptosis while the cells in the perivascular niche undergo cell cycle arrest and eventually re-enter the cell cycle and repopulate the tumor. The cells of the perivascular niche elevate Akt signaling in the process of surviving the treatment and blockade of this pathway with perifosine prior to irradiation lead to reduced survival of cells in that niche (Hambardzumyan et al. 2008). Several trials have been published on somatic cell gene transfer models of gliomas using the PDGF-induced oligodendroglioma model. First, blockade of the PDGF receptor with PTK787 lead to arrest of cycling cells in the tumor and loss of high-grade structures and contrast enhancement on MRI scans in these tumors with resultant lower grade tumors (Shih et al. 2004). A similar effect was seen with blockade of mTOR using the rapamycin analog CCI779 (Uhrbom et al. 2004). Finally, blockade of Akt signaling with perifosine leads to a reduction in proliferation (Momota et al. 2005) of low-grade but not high-grade gliomas. Combining perifosine with CCI779 lead to a much more substantial and uniform effect on high-grade gliomas.

26.23 Examples of Drugs from Mice to Man

Based on the results of the mouse experiments, several trials in patients with gliomas have been initiated. In one case, perifosine as a single agent was tested in humans with anaplastic gliomas and GBM as a phase I/II study (Andrew Lassman, personal communication). In another case, rapamycin was tested in glioblastoma patients (Cloughesy et al. 2008). Finally, the combination of perifosine and CCI779 is currently in trials for malignant gliomas based on the mouse modeling data. Hopefully, the results in medulloblastomas will result in trials in children with this disease either of inhibitors of the SHH signaling pathway in the appropriate tumor subtype or blockade of Akt signaling in association with radiation treatment.

26.24 Conclusion

The current WHO classification lists 124 different distinct types of CNS tumors. The majority of these tumors are very rare, so rare that meaningful trials will never be possible. If models of many of these tumor types were created and validated to the degree that they have been for gliomas and medulloblastomas, these models could be used as surrogates in trials for the patients with these tumors. However, it remains to be seen whether these improved modeling systems will be more predictive than previous systems for testing drugs prior to testing them in patients.

References

Bao S, Wu Q, McLendon RE, Hao Y, Shi Q, Hjelmeland AB, Dewhirst MW, Bigner DD, Rich JN (2006) Glioma stem cells promote radioresistance by preferential activation of the DNA damage response. Nature 444(7120):756–760

Becher OJ, Holland EC (2006) Genetically engineered models have advantages over xenografts for preclinical studies. Cancer Res 66(7):3355–3358, discussion 3358–3359

Becher OJ, Hambardzumyan D, Fomchenko EI, Momota H, Mainwaring L, Bleau AM, Katz AM, Edgar MA, Kenney AM, Cordon-Cardo C, Blasberg RG, Holland EC (2008) Gli activity correlates with tumor grade in platelet-derived growth factor-induced gliomas. Cancer Res 68(7):2241–2249

Bradbury MS, Hambardzumyan D, Zanzonico PB, Schwartz J, Cai S, Burnazi EM, Longo V, Larson SM, Holland EC (2008) Dynamic small-animal pet imaging of tumor proliferation with 3′-deoxy-3′-18F-fluorothymidine in a genetically engineered mouse model of high-grade gliomas. J Nucl Med 49(3):422–429

Calabrese C, Poppleton H, Kocak M, Hogg TL, Fuller C, Hamner B, Oh EY, Gaber MW, Finklestein D, Allen M, Frank A, Bayazitov IT, Zakharenko SS, Gajjar A, Davidoff A, Gilbertson RJ (2007) A perivascular niche for brain tumor stem cells. Cancer Cell 11(1):69–82
CBTRUS. Statistical report: primary brain tumors in the United States, 1997–2001 years data collected. Chicago: Central Brain Tumor Registry of the United States; 2004–2005. http://www.cbtrus.org
Challen GA, Little MH (2006) A side order of stem cells: the SP phenotype. Stem Cells 24(1):3–12
Chen W, Cloughesy T, Kamdar N, Satyamurthy N, Bergsneider M, Liau L, Mischel P, Czernin J, Phelps ME, Silverman DH (2005) Imaging proliferation in brain tumors with 18F-FLT PET: comparison with 18F-FDG. J Nucl Med 46(6):945–952
Cloughesy TF, Yoshimoto K, Nghiemphu P, Brown K, Dang J, Zhu S, Hsueh T, Chen Y, Wang W, Youngkin D, Liau L, Martin N, Becker D, Bergsneider M, Lai A, Green R, Oglesby T, Koleto M, Trent J, Horvath S, Mischel PS, Mellinghoff IK, Sawyers CL (2008) Antitumor activity of rapamycin in a phase I trial for patients with recurrent PTEN-deficient glioblastoma. PLoS Med 5(1):e8
Contag PR, Olomu IN, Stevenson DK, Contag CH (1998) Bioluminescent indicators in living mammals. Nat Med 4(2):245–247
Cuevas IC, Slocum AL, Jun P, Costello JF, Bollen AW, Riggins GJ, McDermott MW, Lal A (2005) Meningioma transcript profiles reveal deregulated Notch signaling pathway. Cancer Res 65(12):5070–5075
Dahmane N, Ruiz i Altaba A (1999) Sonic hedgehog regulates the growth and patterning of the cerebellum. Development 126(14):3089–3100
Dai C, Celestino JC, Okada Y, Louis DN, Fuller GN, Holland EC (2001) PDGF autocrine stimulation dedifferentiates cultured astrocytes and induces oligodendrogliomas and oligoastrocytomas from neural progenitors and astrocytes in vivo. Genes Dev 15(15):1913–1925
De Strooper B, Annaert W, Cupers P, Saftig P, Craessaerts K, Mumm JS, Schroeter EH, Schrijvers V, Wolfe MS, Ray WJ, Goate A, Kopan R (1999) A presenilin-1-dependent gamma-secretase-like protease mediates release of Notch intracellular domain. Nature 398(6727):518–522
Dean M, Fojo T, Bates S (2005) Tumour stem cells and drug resistance. Nat Rev Cancer 5(4):275–284
Dinca EB, Sarkaria JN, Schroeder MA, Carlson BL, Voicu R, Gupta N, Berger MS, James CD (2007) Bioluminescence monitoring of intracranial glioblastoma xenograft: response to primary and salvage temozolomide therapy. J Neurosurg 107(3):610–616
Edinger M, Sweeney TJ, Tucker AA, Olomu AB, Negrin RS, Contag CH (1999) Noninvasive assessment of tumor cell proliferation in animal models. Neoplasia 1(4):303–310
Ekstrand AJ, Longo N, Hamid ML, Olson JJ, Liu L, Collins VP, James CD (1994) Functional characterization of an EGF receptor with a truncated extracellular domain expressed in glioblastomas with EGFR gene amplification. Oncogene 9(8):2313–2320
Fan X, Matsui W, Khaki L, Stearns D, Chun J, Li YM, Eberhart CG (2006) Notch pathway inhibition depletes stem-like cells and blocks engraftment in embryonal brain tumors. Cancer Res 66(15):7445–7452
Federspiel MJ, Bates P, Young JA, Varmus HE, Hughes SH (1994) A system for tissue-specific gene targeting: transgenic mice susceptible to subgroup A avian leukosis virus-based retroviral vectors. Proc Natl Acad Sci USA 91(23):11241–11245
Finkelstein SD, Black P, Nowak TP, Hand CM, Christensen S, Finch PW (1994) Histological characteristics and expression of acidic and basic fibroblast growth factor genes in intracerebral xenogeneic transplants of human glioma cells. Neurosurgery 34(1):136–143
Fisher HW, Puck TT (1956) On the functions of X-irradiated "Feeder" cells in supporting growth of single mammalian cells. Proc Natl Acad Sci USA 42(12):900–906
Fitzgerald K, Harrington A, Leder P (2000) Ras pathway signals are required for notch-mediated oncogenesis. Oncogene 19(37):4191–4198
Fomchenko EI, Holland EC (2006) Mouse models of brain tumors and their applications in preclinical trials. Clin Cancer Res 12(18):5288–5297
Gallia GL, Rand V, Siu IM, Eberhart CG, James CD, Marie SK, Oba-Shinjo SM, Carlotti CG, Caballero OL, Simpson AJ, Brock MV, Massion PP, Carson BS Sr, Riggins GJ (2006) PIK3CA gene mutations in pediatric and adult glioblastoma multiforme. Mol Cancer Res 4(10):709–714

Genoud S, Lappe-Siefke C, Goebbels S, Radtke F, Aguet M, Scherer SS, Suter U, Nave KA, Mantei N (2002) Notch1 control of oligodendrocyte differentiation in the spinal cord. J Cell Biol 158(4):709–718

Giannini C, Sarkaria JN, Saito A, Uhm JH, Galanis E, Carlson BL, Schroeder MA, James CD (2005) Patient tumor EGFR and PDGFRA gene amplifications retained in an invasive intracranial xenograft model of glioblastoma multiforme. Neuro Oncol 7(2):164–176

Goodrich LV, Milenkovic L, Higgins KM, Scott MP (1997) Altered neural cell fates and medulloblastoma in mouse patched mutants. Science 277(5329):1109–1113

Grisendi S, Pandolfi PP (2004) Germline modifications strategies. In: Holland EC (ed) Mouse models of human cancer. Wiley, p 43–65

Guha A, Dashner K, Black PM, Wagner JA, Stiles CD (1995) Expression of PDGF and PDGF receptors in human astrocytoma operation specimens supports the existence of an autocrine loop. Int J Cancer 60(2):168–173

Hallahan AR, Pritchard JI, Hansen S, Benson M, Stoeck J, Hatton BA, Russell TL, Ellenbogen RG, Bernstein ID, Beachy PA, Olson JM (2004) The SmoA1 mouse model reveals that notch signaling is critical for the growth and survival of sonic hedgehog-induced medulloblastomas. Cancer Res 64(21):7794–7800

Hambardzumyan D, Becher OJ, Rosenblum M, Manova-Todorova K, Holland EC (2007a) P53 and PTEN dependent radiation response in medulloblastoma cell types in vivo. Genes Dev 22(4):436–448

Hambardzumyan D, Lyustikman Y, Holland EC (2007b) An update on mouse brain tumor models in cancer drug advice. Expert Opin 2(11):1435–1451

Hambardzumyan D, Becher OJ, Rosenblum MK, Pandolfi PP, Manova-Todorova K, Holland EC (2008) PI3K pathway regulates survival of cancer stem cells residing in the perivascular niche following radiation in medulloblastoma in vivo. Genes Dev 22(4):436–448

Hermanson M, Funa K, Hartman M, Claesson-Welsh L, Heldin CH, Westermark B, Nister M (1992) Platelet-derived growth factor and its receptors in human glioma tissue: expression of messenger RNA and protein suggests the presence of autocrine and paracrine loops. Cancer Res 52(11):3213–3219

Hesselager G, Uhrbom L, Westermark B, Nister M (2003) Complementary effects of platelet-derived growth factor autocrine stimulation and p53 or Ink4a-Arf deletion in a mouse glioma model. Cancer Res 63(15):4305–4309

Holland EC, Hively WP, DePinho RA, Varmus HE (1998) A constitutively active epidermal growth factor receptor cooperates with disruption of G1 cell-cycle arrest pathways to induce glioma-like lesions in mice. Genes Dev 12(23):3675–3685

Holland EC, Celestino J, Dai C, Schaefer L, Sawaya RE, Fuller GN (2000) Combined activation of Ras and Akt in neural progenitors induces glioblastoma formation in mice. Nat Genet 25(1):55–57

Hormigo A, Gu B, Karimi S, Riedel E, Panageas KS, Edgar MA, Tanwar MK, Rao JS, Fleisher M, DeAngelis LM, Holland EC (2006) YKL-40 and matrix metalloproteinase-9 as potential serum biomarkers for patients with high-grade gliomas. Clin Cancer Res 12(19):5698–5704

Hornsey S (1973) The radiosensitivity of the intestine. Strahlenschutz Forsch Prax 13:78–88

Hu X, Holland EC (2005) Applications of mouse glioma models in preclinical trials. Mutat Res 576(1–2):54–65

Hu X, Pandolfi PP, Li Y, Koutcher JA, Rosenblum M, Holland EC (2005) mTOR promotes survival and astrocytic characteristics induced by Pten/AKT signaling in glioblastoma. Neoplasia 7(4):356–368

Johnson RL, Rothman AL, Xie J, Goodrich LV, Bare JW, Bonifas JM, Quinn AG, Myers RM, Cox DR, Epstein EH Jr, Scott MP (1996) Human homolog of patched, a candidate gene for the basal cell nevus syndrome. Science 272(5268):1668–1671

Kinzler KW, Bigner SH, Bigner DD, Trent JM, Law ML, O'Brien SJ, Wong AJ, Vogelstein B (1987) Identification of an amplified, highly expressed gene in a human glioma. Science 236(4797):70–73

Kleihues P, Louis DN, Scheithauer BW, Rorke LB, Reifenberger G, Burger PC, Cavenee WK (2002) The WHO classification of tumors of the nervous system. J Neuropathol Exp Neurol 61(3):215–225, discussion 226–219

Koutcher JA, Hu X, Xu S, Gade TP, Leeds N, Zhou XJ, Zagzag D, Holland EC (2002) MRI of mouse models for gliomas shows similarities to humans and can be used to identify mice for preclinical trials. Neoplasia 4(6):480–485

Li J, Yen C, Liaw D, Podsypanina K, Bose S, Wang SI, Puc J, Miliaresis C, Rodgers L, McCombie R, Bigner SH, Giovanella BC, Ittmann M, Tycko B, Hibshoosh H, Wigler MH, Parsons R (1997) PTEN, a putative protein tyrosine phosphatase gene mutated in human brain, breast, and prostate cancer. Science 275(5308):1943–1947

Marino S (2005) Medulloblastoma: developmental mechanisms out of control. Trends Mol Med 11(1):17–22

Marshman E, Booth C, Potten CS (2002) The intestinal epithelial stem cell. Bioessays 24(1):91–98

Mayer EG, Boone ML, Aristizabal SA (1976) Role of radiation therapy in the management of neoplasms of the central nervous system. Adv Neurol 15:201–220

McConville P, Hambardzumyan D, Moody JB, Leopold WR, Kreger AR, Woolliscroft MJ, Rehemtulla A, Ross BD, Holland EC (2007) Magnetic resonance imaging determination of tumor grade and early response to temozolomide in a genetically engineered mouse model of glioma. Clin Cancer Res 13(10):2897–2904

Momota H, Holland EC (2005) Bioluminescence technology for imaging cell proliferation. Curr Opin Biotechnol 16(6):681–686

Momota H, Nerio E, Holland EC (2005) Perifosine inhibits multiple signaling pathways in glial progenitors and cooperates with temozolomide to arrest cell proliferation in gliomas in vivo. Cancer Res 65(16):7429–7435

Momota H, Shih AH, Edgar MA, Holland EC (2008) c-Myc and β-catenin cooperate with loss of p53 to generate multiple members of the primitive neuroectodermal tumor (PTEN) family in mice. Oncogene 27(32):4392–4401

Morrison SJ, Perez SE, Qiao Z, Verdi JM, Hicks C, Weinmaster G, Anderson DJ (2000) Transient Notch activation initiates an irreversible switch from neurogenesis to gliogenesis by neural crest stem cells. Cell 101(5):499–510

Nery S, Wichterle H, Fishell G (2001) Sonic hedgehog contributes to oligodendrocyte specification in the mammalian forebrain. Development 128(4):527–540

Nister M, Libermann TA, Betsholtz C, Pettersson M, Claesson-Welsh L, Heldin CH, Schlessinger J, Westermark B (1988) Expression of messenger RNAs for platelet-derived growth factor and transforming growth factor-alpha and their receptors in human malignant glioma cell lines. Cancer Res 48(14):3910–3918

Nye JS, Kopan R, Axel R (1994) An activated Notch suppresses neurogenesis and myogenesis but not gliogenesis in mammalian cells. Development 120(9):2421–2430

Ohgaki H, Kita D, Favereaux A, Huang H, Homma T, Dessen P, Weiss WA, Kleihues P, Heppner FL (2006) Brain tumors in S100beta-v-erbB transgenic rats. J Neuropathol Exp Neurol 65(12):1111–1117

Petropoulos CJ, Hughes SH (1991) Replication-competent retrovirus vectors for the transfer and expression of gene cassettes in avian cells. J Virol 65(7):3728–3737

Rao G, Pedone CA, Coffin CM, Holland EC, Fults DW (2003) c-Myc enhances sonic hedgehog-induced medulloblastoma formation from nestinexpressing neural progenitors in mice. Neoplasia 5(3):198–204

Romer J, Curran T (2005) Targeting medulloblastoma: small-molecule inhibitors of the Sonic Hedgehog pathway as potential cancer therapeutics. Cancer Res 65(12):4975–4978

Romer JT, Kimura H, Magdaleno S, Sasai K, Fuller C, Baines H, Connelly M, Stewart CF, Gould S, Rubin LL, Curran T (2004) Suppression of the Shh pathway using a small molecule inhibitor eliminates medulloblastoma in Ptc1(+/–)p53(–/–) mice. Cancer Cell 6(3):229–240

Ross BD, Chenevert TL, Moffat BA, Rehemtulla A, Hall DE (2004) Use of magnetic resonance imaging for evaluation of treatment response. In: Holland EC (ed) Mouse models of human cancer. Wiley

Rowitch DH, St. Jacques B, Lee SM, Flax JD, Snyder EY, McMahon AP (1999) Sonic hedgehog regulates proliferation and inhibits differentiation of CNS precursor cells. J Neurosci 19(20):8954–8965

Sara VR, Prisell P, Sjogren B, Persson L, Boethius J, Enberg G (1986) Enhancement of insulin-like growth factor 2 receptors in glioblastoma. Cancer Lett 32(3):229–234

Sausville EA, Burger AM (2006) Contributions of human tumor xenografts to anticancer drug development. Cancer Res 66(7):3351–3354, discussion 3354

Shannon P, Sabha N, Lau N, Kamnasaran D, Gutmann DH, Guha A (2005) Pathological and molecular progression of astrocytomas in a GFAP:12 VHa-Ras mouse astrocytoma model. Am J Pathol 167(3):859–867

Shih AH, Holland EC (2006a) Notch signaling enhances nestin expression in gliomas. Neoplasia 8(12):1072–1082

Shih AH, Holland EC (2006b) Platelet-derived growth factor (PDGF) and glial tumorigenesis. Cancer Lett 232(2):139–147

Shih AH, Dai C, Hu X, Rosenblum MK, Koutcher JA, Holland EC (2004) Dose-dependent effects of platelet-derived growth factor-B on glial tumorigenesis. Cancer Res 64(14):4783–4789

Singh SK, Hawkins C, Clarke ID, Squire JA, Bayani J, Hide T, Henkelman RM, Cusimano MD, Dirks PB (2004) Identification of human brain tumour initiating cells. Nature 432(7015):396–401

Szakacs G, Paterson JK, Ludwig JA, Booth-Genthe C, Gottesman MM (2006) Targeting multidrug resistance in cancer. Nat Rev Drug Discov 5(3):219–234

Taylor MD, Liu L, Raffel C, Hui CC, Mainprize TG, Zhang X, Agatep R, Chiappa S, Gao L, Lowrance A, Hao A, Goldstein AM, Stavrou T, Scherer SW, Dura WT, Wainwright B, Squire JA, Rutka JT, Hogg D (2002) Mutations in SUFU predispose to medulloblastoma. Nat Genet 31(3):306–310

Tekki-Kessaris N, Woodruff R, Hall AC, Gaffield W, Kimura S, Stiles CD, Rowitch DH, Richardson WD (2001) Hedgehog-dependent oligodendrocyte lineage specification in the telencephalon. Development 128(13):2545–2554

Trojan J, Blossey BK, Johnson TR, Rudin SD, Tykocinski M, Ilan J, Ilan J (1992) Loss of tumorigenicity of rat glioblastoma directed by episome-based antisense cDNA transcription of insulin-like growth factor I. Proc Natl Acad Sci USA 89(11):4874–4878

Uhrbom L, Hesselager G, Nister M, Westermark B (1998) Induction of brain tumors in mice using a recombinant platelet-derived growth factor B-chain retrovirus. Cancer Res 58(23):5275–5279

Uhrbom L, Nerio E, Holland EC (2004) Dissecting tumor maintenance requirements using bioluminescence imaging of cell proliferation in a mouse glioma model. Nat Med 10(11):1257–1260

Uhrbom L, Kastemar M, Johansson FK, Westermark B, Holland EC (2005) Cell type-specific tumor suppression by Ink4a and Arf in Kras-induced mouse gliomagenesis. Cancer Res 65(6):2065–2069

Wang S, Sdrulla AD, diSibio G, Bush G, Nofziger D, Hicks C, Weinmaster G, Barres BA (1998) Notch receptor activation inhibits oligodendrocyte differentiation. Neuron 21(1):63–75

Wechsler-Reya RJ, Scott MP (1999) Control of neuronal precursor proliferation in the cerebellum by Sonic Hedgehog. Neuron 22(1):103–114

Weijzen S, Rizzo P, Braid M, Vaishnav R, Jonkheer SM, Zlobin A, Osborne BA, Gottipati S, Aster JC, Hahn WC, Rudolf M, Siziopikou K, Kast WM, Miele L (2002) Activation of Notch-1 signaling maintains the neoplastic phenotype in human Ras-transformed cells. Nat Med 8(9):979–986

Wetmore C, Eberhart DE, Curran T (2001) Loss of p53 but not ARF accelerates medulloblastoma in mice heterozygous for patched. Cancer Res 61(2):513–516

Zhu Y, Ghosh P, Charnay P, Burns DK, Parada LF (2002) Neurofibromas in NF1: Schwann cell origin and role of tumor environment. Science 296(5569):920–922

Zhu Y, Guignard F, Zhao D, Liu L, Burns DK, Mason RP, Messing A, Parada LF (2005) Early inactivation of p53 tumor suppressor gene cooperating with NF1 loss induces malignant astrocytoma. Cancer Cell 8(2):119–130

Zindy F, Uziel T, Ayrault O, Calabrese C, Valentine M, Rehg JE, Gilbertson RJ, Sherr CJ, Roussel MF (2007) Genetic alterations in mouse medulloblastomas and generation of tumors de novo from primary cerebellar granule neuron precursors. Cancer Res 67(6):2676–2684

Chapter 27
Mouse Models of Human Cancer: Role in Preclinical Testing and Personalized Medicine

Alice Hawley Berger and Pier Paolo Pandolfi

27.1 Introduction

In the last few decades, information about the human genome and the genetic landscape of human cancer has led to therapeutic strategies guided by specific genetic knowledge and executed with molecularly targeted drugs. The era of personalized medicine for cancer treatment is coming of age. Genetic knowledge leads to identification of potential drug targets that in turn is used to develop molecularly targeted therapies for cancer (see Chapter 28). Central to the pursuit of this approach are faithful mouse models of human cancer. As we discuss in this chapter, these models not only serve to define the critical and causal molecular events required for cancer, but they also provide powerful systems for preclinical testing of cancer drugs.

Such a personalized medicine strategy has already been used to great success for cancer therapy. In each success story, mouse models have proven to be informative and predictive models for targeted therapy in humans. For example, acute promyelocytic leukemia (APL) is caused by defined chromosomal translocations between the *RARα* gene and variable genes, most commonly *PML*. The identification of these translocations and the recognition of the role of *RARα* in the pathogenesis of APL has allowed for molecular diagnosis of the disease and targeted therapies towards disease eradication. Studies in mouse models led to the adoption of combined retinoic acid (RA) and arsenic trioxide (Ar_2O_3) as the standard of care for APL, as we discuss in detail below.

A.H. Berger • P.P. Pandolfi (✉)
Cancer Genetics Program, Beth Israel Deaconess Cancer Center, Departments of Medicine and Pathology, Beth Israel Deaconess Medical Center, Harvard Medical School, Boston, MA, USA

Weill Graduate School of Medical Sciences, Cornell University, New York, NY, USA
e-mail: ppandolf@bidmc.harvard.edu

J.E. Green and T. Ried (eds.), *Genetically Engineered Mice for Cancer Research: Design, Analysis, Pathways, Validation and Pre-Clinical Testing*,
DOI 10.1007/978-0-387-69805-2_27, © Springer Science+Business Media, LLC 2012

Another classic targeted therapy success story is the discovery of the BCR-ABL fusion oncogene in chronic myelogenous leukemia (CML). The recognition of BCR-ABL as the causal molecular event in CML led to the development of small-molecule ABL inhibitors, such as imatinib, dasatinib, and nilotinib, that are extremely effective in the treatment of CML (Druker 2008). Once again, mouse models have been invaluable to optimize therapeutic strategies for CML eradication (Ito et al. 2008; Peng and Li 2010).

In cases of drugs developed to target a cancer without initial consideration of the genetic makeup of the tumors, subsequent elucidation of the molecular targets and interacting signaling pathways has provided an understanding of why some patients respond to that particular drug and other patients do not. In the case of estrogen inhibitors like tamoxifen for the treatment of breast cancer, it is now known that such a treatment is only effective in estrogen receptor- (ER-) positive tumors and not ER-negative tumors (Jordan 2008). Similarly, the EGFR inhibitors erlotinib and gefitinib were used to treat lung adenocarcinoma with limited success before it was recognized that drug-responsive tumors harbored activating *EGFR* mutations, whereas tumors with *de novo* resistance did not have *EGFR* mutations and instead harbored other mutations, such as *KRAS* (Lynch et al. 2004; Paez et al. 2004; Pao et al. 2004, 2005b). Mouse models of oncogenic *KRAS*- and *EGFR*-driven lung tumorigenesis faithfully recapitulate both the phenotype and the therapeutic response of human tumors with *KRAS* or *EGFR* mutations. These models are now being used to develop novel treatment modalities for the management of human lung cancer, as we describe in this chapter.

These examples demonstrate the power of personalized medicine and reinforce the importance of moving forward with similar approaches to treat other cancers and diseases. Critical to this pursuit will be the integrated use of sophisticated *in vitro* and *in vivo* experimental approaches to identify new drug targets, analyze molecular pathways, and interrogate biological effects of drugs on cells with different defined genetic alterations.

Mouse models have been central to the advent of personalized medicine in the last decade and they continue to serve crucial roles in preclinical testing of cancer therapies and advancement of genetic and biological knowledge. Mouse models facilitate the development of molecularly targeted therapies in many ways (Fig. 27.1). First, these models allow the study of critical signaling pathways that are involved in tumorigenesis. Crosses of different genetically modified mouse strains can demonstrate cooperativity or antagonism of various pathways and identify key signaling molecules that are necessary for tumor formation and therefore represent potential targets for drug development. Second, mouse models allow the study of the progression of preneoplastic and neoplastic lesions. In humans, it is difficult to determine if benign lesions represent a precursor to a malignant lesion, or if the two lesions are two distinct entities with differing pathological histories. Using mouse models, investigators can study cancer from initiation in a defined manner. Such studies have shown that many cancers progress in a step-wise fashion from preneoplastic hyperplastic tissue to dysplastic tissue to invasive, cancerous tissue and finally to metastatic disease (Vogelstein and Kinzler 1993; Wu and Pandolfi 2001). Third, mouse models allow preclinical testing of drugs for *in vivo*

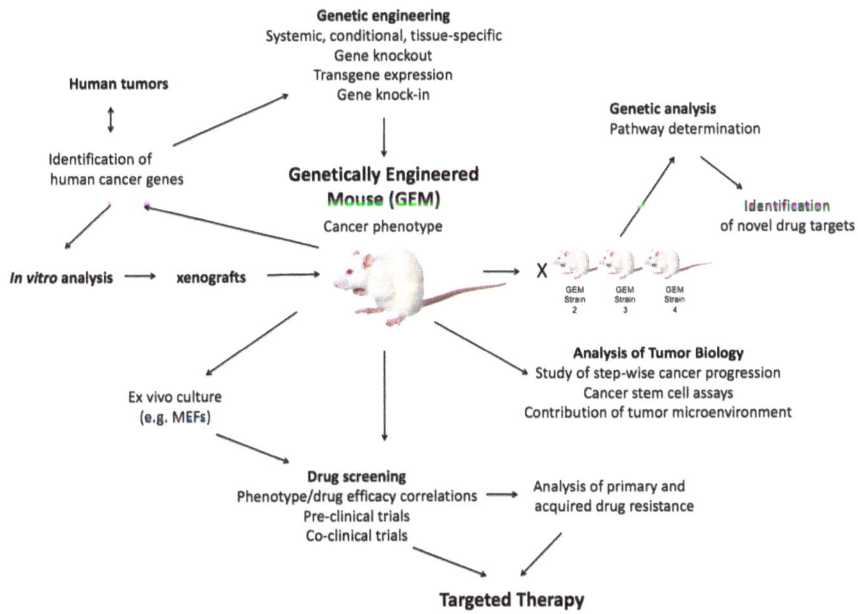

Fig. 27.1 Mouse modeling of human cancer for personalized medicine. Genetically engineered mouse (GEM) strains can faithfully model human cancers, such as APL, prostate cancer, and lung cancer, as discussed in the text. These GEM lines can be bred to other GEM lines as shown for genetic analysis and determination of molecular signaling pathways. This knowledge can in turn be used to identify novel drug targets. At the same time, these models can be used to predict human cancer genes and study the cell biological processes underlying tumor development. Importantly, these models act as subjects for preclinical or "co-clinical" testing of drug candidates, allowing the analysis of genotype/phenotype/drug response relationships. Complex drug response scenarios, such as acquired resistance, are also faithfully modeled in the mouse, allowing prediction of mechanisms of acquired resistance in human cancer

efficacy in tumors with a defined and controlled genetic make-up. Drugs can not only be tested on established tumors, but also tested prior to cancer initiation for the development of chemoprevention strategies. Cells can be harvested from tumor-bearing mice for ex vivo screening of drug efficacy. Moreover, acquired resistance to molecular therapy can be modeled and analyzed *in vivo*, allowing prediction of human mechanisms of resistance and development of alternative strategies to combat resistance. Finally, mouse models allow elucidation of the complex cell biological processes that lead to cancer development and progression, including the contribution of the tumor microenvironment (Joyce 2005) and the role of cancer stem cells in tumorigenesis and disease relapse (Al-Hajj and Clarke 2004; Jordan et al. 2006; Reya et al. 2001). Although these topics are outside the scope of this chapter, the utility of mouse modeling for cancer stem cell research has been reviewed (Cheng et al. 2010; Dick 2009; Perez-Caro and Sanchez-Garcia 2006) including a chapter in this book (see Chapter 12).

Here, we discuss murine models of leukemia, prostate cancer, and lung cancer and their use in preclinical testing. We illustrate through these specific examples the power of mouse modeling in the dissection of the genetic requirements of particular cancers and the preclinical testing of molecularly targeted therapies.

27.2 Modeling APL and Preclinical Testing of Novel Therapeutic Modalities

APL, the M3 subtype of the French-American-British (FAB) classification of acute myelogenous leukemia (AML), accounts for 10–15% of all AML cases (Warrell 1996). APL is characterized by the presence of specific chromosomal translocations and a unique and exquisite sensitivity of APL blasts to all trans retinoic acid (RA). During the pathogenesis of APL, an accumulation in the bone marrow of promyelocytic-like blast cells leads to intravascular coagulation and a hemorrhagic tendency resulting in significant mortality if left untreated (Warrell et al. 1993).

APL is invariably associated with reciprocal chromosomal translocations involving the *RARα* gene on chromosome 17 and one of the five variable "X" genes (Piazza et al. 2001; Redner 2002). RARα is a transcription factor belonging to the retinoic acid receptor (RAR) superfamily of hormone receptor transcription factors. RAR family members dimerize with retinoid X-receptor (RXR) family nuclear receptors to together form RA-inducible transcriptional activators. Most commonly, patients exhibit a chromosomal translocation, t(15;17), that leads to fusion of the *RARα* gene on chromosome 17 to the *PML* gene on chromosome 15 (Alcalay et al. 1991; Borrow et al. 1990; de The et al. 1990, 1991; Goddard et al. 1991; Kakizuka et al. 1991; Longo et al. 1990; Pandolfi et al. 1991). In other cases, the *RARα* gene is involved in a reciprocal translocation with one of the four other genes, PLZF, NPM, NUMA, or STAT5b (Arnould et al. 1999; Chen et al. 1993; Hummel et al. 1999; Redner et al. 1996; Wells et al. 1997). Each of these *RARα* translocations results in the formation of reciprocal fusion genes *X-RARα* and *RARα-X*, both of which are typically expressed in APL blasts (Alcalay et al. 1992).

The PML-RARα mutant protein retains the ability to bind to RA and DNA, and to heterodimerize with RXRs. It is therefore thought to act as a dominant negative *RARα* mutant that interferes with the normal function of RXRs and other nuclear receptors (Kastner et al. 1992; Perez et al. 1993). X-RARα proteins can also heterodimerize with the wild-type X protein because all of the X proteins contain oligomerization domains that are retained in the X-RARα fusions (Piazza et al. 2001). The additional translocation product, RARα-X, contains the transactivation domain of RARα coupled to the C-terminus of the X protein. Because *RARα* was consistently mutated in all APL cases, whereas the X genes were variable, disruption of *RARα* function initially was assumed to be the critical event for APL pathogenesis, and the relevance of the translocation partner X gene to disease development was unclear. Mouse models were pivotal in determining if one or both of the *X-RARα* and *RARα-X* fusion genes were necessary and/or sufficient for APL development

(Carver and Pandolfi 2006; Piazza et al. 2001). In fact, such mouse models may be the only bona fide assay for assessing the causality of genetic mutations and translocations in cancer initiation and progression. These models were also critical for determining the best treatment strategies for management or eradication of APL.

Tissue-specific transgenic mouse models have been constructed expressing *PML-RARα, PLZF-RARα, NPM-RARα,* or *NUMA-RARα* in the hematopoietic compartment (Brown et al. 1997; Cheng et al. 1999; Grisolano et al. 1997; He et al. 1997, 1998; Sukhai et al. 2004). These *X-RARα* transgenic models develop abnormal hematopoietic phenotypes with similarities and differences observed in the phenotypes induced by the different fusion genes. *PML-RARα* transgenic mice develop leukemia phenotypically similar to human APL after a long latency and with a penetrance of 10–30% (Grisolano et al. 1997; He et al. 1997). A long latency and incomplete penetrance suggests that *PML-RARα* is necessary but not sufficient for APL development and other genetic events are required to initiate the full disease phenotype (Piazza et al. 2001). Such additional events could include the expression of the reciprocal *RARα-X* fusion gene. Indeed, simultaneous expression of *RARα-PML* and *PML-RARα* increased the penetrance of the APL phenotype, indicating that the *RARα-PML* fusion acts as a tumor modifier. However, expression of *RARα-PML* alone is not sufficient to induce leukemia (Piazza et al. 2001; Pollock et al. 1999).

In contrast to *PML-RARα* mice, *PLZF-RARα* mice develop a myeloproliferative disorder resembling CML (Cheng et al. 1999; He et al. 1998). Strikingly, this phenotype is converted to APL when *PLZF-RARα* is co-expressed with the reciprocal translocation product, *RARα-PLZF*, thus proving that the *RARα-X* fusion is not simply an innocent bystander but instead directly contributes to the APL phenotype (He et al. 2000).

These faithful mouse models not only recapitulate the disease phenotype with respect to morphological features, but also with respect to drug sensitivity (Table 27.1). The varying sensitivity of leukemias in these different murine APL models closely mimics what is seen in the human scenario (Table 27.1). Treatment of human APL with RA induces remission in approximately 85% of APL patients, including virtually all patients with a confirmed molecular diagnosis of a *PML-RARα* translocation (Warrell 1996). Similarly, RA treatment of APL in *PML-RARα* transgenic mice can induce complete disease remission (Grisolano et al. 1997; He et al. 1997; Rego et al. 2000). In contrast, human APL with the atypical t(11;17) translocation resulting in the *PLZF/RARα* fusions exhibits resistance to RA (Licht et al. 1995), and RA is ineffective in treating leukemia found in transgenic *PLZF-RARα* mice (He et al. 1998; Rego et al. 2000) or *PLZF-RARα/RARα-PLZF* mice (He et al. 2000). The close conservation between the human disease and these murine models in their differential response to RA indicates that these models reproduce critical features of the human disease and therefore these models can reliably improve our understanding of human APL.

Preclinical testing of various combination therapies in APL murine models also demonstrated that combination RA and Ar_2O_3 treatment could be beneficial for human APL patients harboring the *PML-RARα* fusion. In *PML-RARα* transgenic

Table 27.1 Sensitivity of APL subtypes to therapeutic intervention in human patients and mouse models

	Response to RA		Response to RA+Ar_2O_3		Response to RA+HDACi		
Translocation	Human	Mouse	Human	Mouse	Human	Mouse	References
PML-RARα	+	+	+	+	+	+	Brown et al. (1997), Grisolano et al. (1997), He et al. (1997), (1998), (2001), Hu et al. (2009), Warrell et al. (1998)
PLZF-RARα	–	–	–	–	+	+	Cheng et al. (1999), He et al. (1998), (2001), Licht et al. (1995), Petti et al. (2002)
NPM-RARα	+	+	*N.D.*	*N.D.*	*N.D.*	*N.D.*	Cheng et al. (1999), Hummel et al. (1999)
STAT5B-RARα	–	*N.D.*	*N.D.*	*N.D.*	*N.D.*	*N.D.*	Arnould et al. (1999)
NUMA-RARα	+	+	*N.D.*	*N.D.*	*N.D.*	*N.D.*	Sukhai et al. (2004), Wells et al. (1997)

Summary of sensitivity of APL subtypes harboring the indicated translocations to therapeutic strategies, including retinoic acid (RA), arsenic trioxide (Ar_2O_3) and histone deacetylase inhibitors (HDACi). "+" complete remission achieved, "–" partial response or no response, "*N.D.*" no data

mice, combination RA/Ar_2O_3 showed additional benefit compared to RA alone (Rego et al. 2000). These data proved decisive and instructive for human clinical care. Previously, misleading *in vitro* findings suggested that RA and Ar_2O_3 would antagonize the positive effects of each respective drug. Instead, the murine data demonstrated a potential benefit of combination therapy that has now become the standard of care and can induce complete durable remission in the vast majority of patients with the *PML-RARα* translocation (Hu et al. 2009). Unfortunately, even combination RA/Ar_2O_3 treatment could not induce complete remission *PLZF-RARα* mice, suggesting that human APL patients with this subtype of APL will not benefit from this therapeutic strategy (Rego et al. 2000).

Further studies in *PML-RARα* and *PLZF-RARα* transgenic mice demonstrated that the APL in both mouse models responds to combination therapy of RA plus histone deacetylase (HDAC) inhibitors (He et al. 2001). This is the first treatment in mice to induce complete remission of APL initiated by *PLZF-RARα*. Moreover, these findings led to clinical trials that demonstrated the combination of HDAC inhibitors with RA is effective to treat not only t(15;17) APL (Warrell et al. 1998), but also for the treatment of t(11;17) APL harboring *PLZF-RARα* (Petti et al. 2002). Thus, the knowledge gained from murine preclinical studies can influence the development of clinical trials and allow clinicians to tailor a patient's treatment based on defined molecular criteria. Importantly, clinical trials alone would have been insufficient to unravel the complexities of APL harboring *PLZF-RARα* because this translocation is so rare that only a very small number of cases have ever been reported. This is a problem that will be encountered more and more in the future as we accept that cancers of similar histology are not one disease but many at the genetic level.

27.3 Prostate Cancer Murine Models and Implications for Therapy

Phosphatase and tensin homolog (*PTEN*) is a tumor suppressor gene mutated in many human cancers, including prostate cancer, where approximately 70% of human primary prostate cancers show loss or mutation of at least one allele of *PTEN* (Gray et al. 1998). The importance of *PTEN* loss in the initiation and progression of prostate cancer has been confirmed by the study of a variety of *in vivo* mouse models of *PTEN* loss. Like human prostate cancer, prostate cancer progression in the mouse follows a sequential progression from epithelial hyperplasia, to prostatic intraepithelial neoplasia (PIN) to high-grade invasive PIN and then to invasive adenocarcinoma. Complete knockout of *PTEN* in the mouse results in embryonic lethality as a result of severe cephalic and caudal patterning defects (Di Cristofano et al. 1998; Podsypanina et al. 1999; Stambolic et al. 1998; Suzuki et al. 1998). Due to the embryonic lethality of these mice, investigators have turned to the analysis of *PTEN* heterozygous (+/−) mice, tissue-specific deletion of *PTEN*, or analysis of $PTEN^{-/-}$ murine embryonic fibroblasts (MEFs) to study the function of *PTEN* as a tumor suppressor.

$PTEN^{+/-}$ mice are susceptible to cancers in multiple tissues, including the prostate, hematopoietic system, thyroid, mammary gland, and endometrium (Di Cristofano et al. 1998; Podsypanina et al. 1999; Suzuki et al. 1998). Tissue-specific conditional *PTEN* loss has been studied using Cre-loxP systems in dozens of tissues, including the prostate (Backman et al. 2004; Ma et al. 2005b; Trotman et al. 2003; Wang et al. 2003), brain (Backman et al. 2001; Fraser et al. 2004; Lesche et al. 2002; Marino et al. 2002; Xiao et al. 2005) and blood (Suzuki et al. 2001; Yilmaz et al. 2006; Zhang et al. 2006). Studies of *PTEN* loss in the prostate have not only confirmed the causal role of *PTEN* loss in prostate tumorigenesis, but also have demonstrated the critical importance of *PTEN* dose to normal cellular function. In the prostate, loss of *PTEN* leads to tumorigenesis with the penetrance and latency of the phenotype varying with the level of *PTEN* gene dosage. The effect of *PTEN* dose has been demonstrated using an allelic series of *PTEN* mutants (Trotman et al. 2003). Heterozygous loss of *PTEN* results in high-grade PIN with incomplete penetrance (Di Cristofano et al. 1998; Trotman et al. 2003). A further reduction of *PTEN* dose (below heterozygous level) further enhances tumorigenesis, with mice heterozygous for a *PTEN* hypomorphic allele (corresponding to a final gene dosage of approximately 30%) developing invasive prostate cancer, in contrast to PIN, after long latency (Trotman et al. 2003).

Interestingly, complete loss of *PTEN* is not always more tumorigenic than partial loss of *PTEN* because complete loss of *PTEN* initiates a *PTEN* loss-induced cellular senescence (PICS) through a p53-dependent fail-safe mechanism (Chen et al. 2005). Cellular senescence, an irreversible cell cycle arrest, is also induced by hyperactive oncogenic signaling in a process known as oncogene-induced senescence (OIS). OIS can serve to restrict the growth of a tumor, and this process is involved in limiting the growth of benign dysplastic nevi (skin moles) into cancerous tumors (Michaloglou et al. 2005). Since complete, but not partial, *PTEN* loss induces cellular senescence, partial loss of *PTEN* can be more oncogenic than complete loss by inducing proliferation without triggering PICS. However, if the tumor suppressor p53 is inactivated in the tumors, the cellular senescence is overcome and in that molecular context complete *PTEN* loss is more oncogenic than partial *PTEN* loss. In line with this notion, mice with compound tissue-specific loss of *PTEN* and p53 in the mouse prostate rapidly develop a lethal, locally aggressive prostate cancer with a shorter latency than that found in mice lacking only *PTEN* (Chen et al. 2005). These observations in the mouse may explain why heterozygote *PTEN* mutation/loss typically occurs in early stage tumors, whereas later stage tumors exhibit complete *PTEN* loss (Di Cristofano and Pandolfi 2000). Complete *PTEN* loss would provide an additional selective advantage to the tumor only in the context of p53 mutation, and mutation of p53 generally, but not always, occurs as a late event in prostate cancer.

The generation of mouse models of prostate cancer has facilitated study of the *in vivo* signaling networks involved in cellular senescence and tumorigenesis, and this knowledge can in turn be used to select protein targets for therapeutic intervention. For example, genetic analysis using *PTEN* conditional knockout mice has shown that Akt1 and PDK1 are required for tumor formation following *PTEN* loss (Bayascas et al. 2005; Chen et al. 2006). Such experiments have confirmed the epistatic relationships in the PTEN-PDK1-AKT pathway (Fig. 27.2). Interestingly, conditional deletion of the PI3K catalytic subunit p110β in conditional *PTEN* knockout mice was sufficient to block *PTEN* loss-induced prostate tumorigenesis (Jia et al. 2008). In contrast, deletion of p110α had no effect on tumor progression in the same model. This finding demonstrates that tumorigenesis initiated by *PTEN* loss specifically requires p110β and therefore identifies p110β as a relevant drug target for the treatment of prostate cancer with *PTEN* loss or mutation. Selective p110β inhibitors (Knight et al. 2006) have been developed and could be tested in these models to confirm the efficacy of such a strategy for prostate cancer therapy.

Mouse modeling has also indicated the importance of the mTOR pathway in tumorigenesis initiated by loss of *PTEN* (Fig. 27.2). Mammalian target of rapamycin (mTOR) is a protein kinase that acts as a master sensor and integrator of energy and nutrient signaling (Wullschleger et al. 2006). mTOR exists in two distinct complexes, mTORC1 and mTORC2. Of these two complexes, only mTORC1 is inhibited by the allosteric inhibitor rapamycin. mTOR is intimately involved in a signaling loop with the AKT pathway. AKT signaling activates mTOR in the mTORC1 complex while mTORC2-complexed mTOR phosphorylates and activates AKT on S473. AKT-mediated mTORC1 activation involves phosphorylation and inhibition of the TSC1/TSC2 complex. This in turn releases a TSC-dependent inhibition of the small GTPase Rheb, which then can activate mTORC1 (Fig. 27.2).

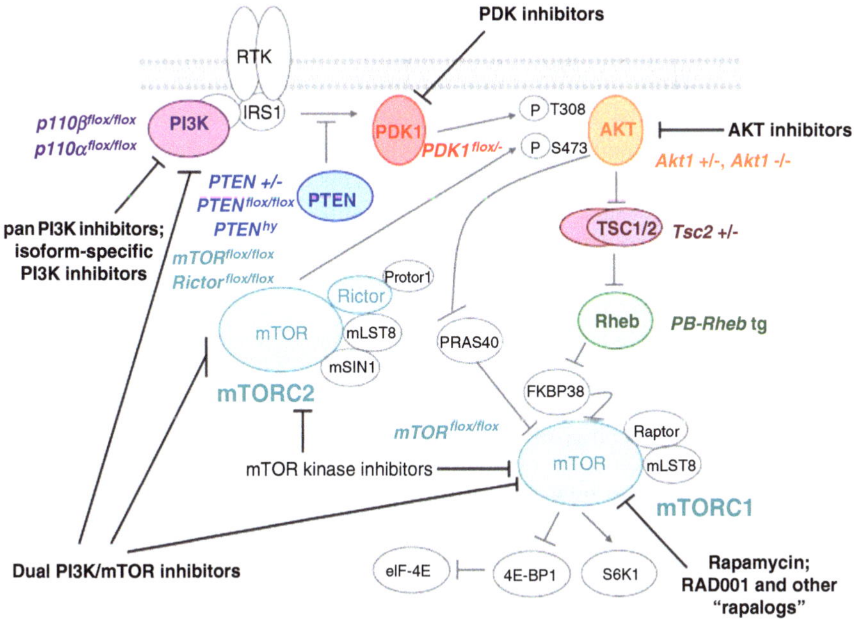

Fig. 27.2 Genetics and therapeutic modulation of the PTEN/AKT/mTOR pathway in prostate cancer. Proteins modulated in mouse models discussed in the text are shown in *color*. Mouse models are shown in *bold italic* in the same shade of color as the corresponding protein. Opportunities for therapeutic modulation are shown in *bold black text*.

The involvement of mTOR signaling in *PTEN*-loss-induced tumorigenesis is clearly evidenced by murine genetic studies, with implications for the treatment of prostate cancers with *PTEN* loss. Heterozygosity for *TSC2*, a negative regulator of the mTOR pathway, cooperates with *PTEN* haploinsufficiency in prostate cancer (Ma et al. 2005a; Manning et al. 2005). Consistently, overexpression of Rheb, a negative regulator of the TSC1/2 complex, can also cooperate with *PTEN* loss-induced cancer (Nardella et al. 2008). On the other hand, deletion or knockout of key positive mTOR signaling components, such as mTOR (Nardella et al. 2009) or the mTORC2-component Rictor (Guertin et al. 2009), can inhibit prostate tumorigenesis initiated by *PTEN* loss, thus confirming the critical role of the mTOR pathway downstream of *PTEN*. Importantly, these results demonstrate that kinase inhibitors of mTOR that completely abrogate its activity in both the mTORC1 and mTORC2 complexes could be of significant therapeutic value in prostate tumors harboring deletion or mutation of *PTEN* (Fig. 27.2).

Genetically engineered mouse models (GEMs) can not only be used for genetic analysis, but also for derivation of cell lines for *in vitro* biochemical and cell biology characterization of the cellular response to therapy. Cultivation of primary murine cells harvested from GEMs provides an ideal ex vivo system to study the consequences of the genetic manipulation on response to therapeutic agents. For example, in the case of the APL models, bone marrow cells could be harvested and studied *in vitro*. In the case of knockout or transgenic mice, MEFs are a useful system for target identification, drug screening, and analysis of the mechanism of drug action. Another useful feature of MEFs is that they can often be harvested from the embryos of genetically modified mouse strains harboring mutations that cause embryonic lethality. Unless the embryonic lethality occurs at a very early stage, embryos can still be harvested and used to derive cell lines. This feature allows investigators to use MEFs as a tool to study the genetic modification despite the limitation that the mice cannot be generated. Moreover, these MEFs or other cells harvested from GEMs can be transplanted into nude mice or syngeneic mice to allow further *in vivo* testing of therapeutic strategies.

PTEN$^{-/-}$ MEFs have been studied extensively and have aided in the determination of PTEN function. Recently, studies in *PTEN*$^{-/-}$ MEFs, in tandem with the analysis of human tumors, led to the identification of a novel feedback pathway that is activated upon inhibition of the mTORC1 complex with the drug rapamycin (Carracedo et al. 2008). Such inhibition was already known to elicit activation of AKT via upregulation of RTKs, such as PDGFR (Zhang et al. 2003, 2007) or adaptors such as IRS1 (Harrington et al. 2004; Shah et al. 2004). This finding could partially explain why mTORC1 inhibitors, such as rapamycin, have performed poorer than expected in clinical trials. Now, studies in *PTEN*$^{-/-}$ MEFs showed that mTOR inhibition also leads to activation of Erk via an additional feedback mechanism involving S6K, PI3K, and Ras (Carracedo et al. 2008). This phenomenon was then investigated *in vivo* in GEMs and human samples. Mice with prostate cancer caused by conditional inactivation of *PTEN* were treated with RAD001, an mTORC1 inhibitor derived from rapamycin (Carracedo et al. 2008). Prostate tissue from animals treated with rapamycin showed a marked increase in phospho-Erk staining compared to tissue from vehicle-treated control animals (Carracedo et al. 2008). In human breast cancer samples, RAD001 treatment was associated with activation of ERK in a dosage schedule-dependent manner, suggestive of a direct involvement of RAD001 in activation of Erk signaling (Carracedo et al. 2008). Because Erk activation is generally oncogenic, this Erk-related feedback mechanism further explains why rapamycin and rapamycin analogs have typically performed more poorly as anticancer agents than originally expected. Furthermore, this study shows the power of combinatorial analysis of human samples, mouse models, and ex vivo cell culture.

Importantly, these studies suggest that combination therapy with rapamycin analogs and inhibitors of the PI3K/AKT and MAPK/ERK pathways could be more effective than single agent treatment with rapamycin alone (Carracedo et al. 2008; Guertin and Sabatini 2007). Indeed, combination therapy using rapamycin and PD0325901, a MAPK inhibitor, could inhibit prostate tumor growth in a mouse model of prostate cancer caused by combined *PTEN* and *Nkx3.1* heterozygosity (Kinkade et al. 2008). To date, many dual PI3K/mTOR inhibitors have been developed, such as PI-103 (Raynaud et al. 2007) and NVP-BEZ235 (Maira et al. 2008), the latter of which is under clinical development at Novartis. Studies using such inhibitors have demonstrated that dual PI3K/mTOR inhibition may be effective for glioma treatment (Fan et al. 2006) and lung cancer treatment (Engelman et al. 2008), as we discuss in detail below.

27.4 Mouse Modeling of Lung Cancer and Targeted Therapy

Lung cancer mouse models provide another example of the power of mouse modeling for lung cancer genetic analysis and preclinical testing. Genetic analysis of human lung adenocarcinoma specimens has identified regions of recurrent amplification, loss, and mutation (Chitale et al. 2009; Ding et al. 2008; Weir et al. 2007). Many of the putative oncogenes and tumor suppressors in human lung cancer have been validated through the use of murine models, including oncogenes such as *EGFR* and *KRAS*, and tumor suppressors such as p53, LKB1, and recently, *DOK* family genes (Berger et al. 2010; Fisher et al. 2001; Jackson et al. 2001; Ji et al. 2006, 2007; Johnson et al. 2001; Politi et al. 2006).

KRAS and *EGFR* are two of the most frequently mutated oncogenes in human lung adenocarcinoma, and each can be found mutated in approximately 15–30% of lung adenocarcinoma cases (Ding et al. 2008; Shigematsu and Gazdar 2006). Mutations in *KRAS* and *EGFR* are mutually exclusive and do not occur in the same tumors. *EGFR* mutations were first identified in 2004, when it was recognized that patients exhibiting sensitivity to the *EGFR* inhibitors erlotinib and gefitinib harbored activating mutations in *EGFR* (Lynch et al. 2004; Paez et al. 2004; Pao et al. 2004). Later, it was discovered that an additional secondary mutation of EGFR (T790M) was associated with acquired resistance to *EGFR* inhibitors (Pao et al. 2005a). These findings both implicated *EGFR* as a human lung cancer oncogene and confirmed the functional relevance of *EGFR* to the cancer maintenance. Nevertheless, it was unknown if *EGFR* mutation was sufficient to initiate tumorigenesis or if it was a late event that the tumors eventually grew to require. Furthermore, human studies are not amenable to study the process of tumor formation and the relevant pathways involved. Thus, mouse models have been useful to address these issues as well as to identify novel drug targets for therapy of gefitinib- and erlotinib-resistant tumors, to analyze the mechanism of drug resistance, and to test novel molecular therapies.

Mouse models have been constructed that overexpress lung cancer oncogenes, such as KRAS, EGFR, and PI3K subunits in the murine lung (Engelman et al. 2008; Fisher et al. 2001; Jackson et al. 2001; Ji et al. 2006; Johnson et al. 2001; Politi et al. 2006). Generally, overexpression of such genes in the mouse embryo leads to embryonic lethality or neonatal lethality just after birth, when the lung function becomes critical. For this reason, investigators typically use a method of temporal control of gene expression. Such methods commonly include a tamoxifen-inducible ER-regulated transgene, a doxycycline-inducible tetracycline-response-element-regulated transgene (Fig. 27.3), or "lox-stop-lox" systems where expression of Cre induces excision of a stop codon in front of the oncogene, leading to oncogene expression. These studies have confirmed the causal nature of these mutations to cancer initiation and progression and the response to different therapies, as we describe below for the case of *EGFR*.

Models of EGFR-induced lung tumorigenesis utilize a bitransgenic system (Fig. 27.3) (Ji et al. 2006; Politi et al. 2006). The first transgene drives expression of the reverse tetracycline transactivator (rtTA) under control of the lung-specific

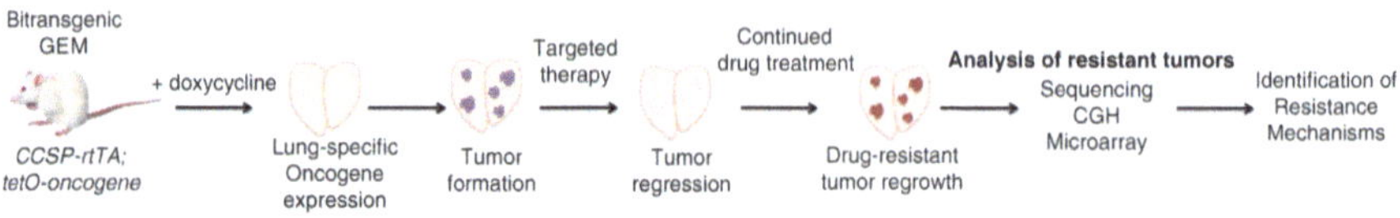

Fig. 27.3 Modeling lung cancer and acquired resistance to lung cancer therapy. Schematic of lung cancer modeling strategy currently employed for modeling of oncogenic EGFR (Ji et al. 2006; Politi et al. 2006) and analysis of acquired resistance to EGFR inhibitors (Politi et al. 2010). The bitransgenic GEM contains one transgene expressing the reverse tetracycline transactivator (rtTA) under the control of the lung-specific Clara cell secretory protein (CCSP) promoter. The second transgene expresses the oncogene under the control of tet operator regulatory sequences. Oncogene expression is induced by doxycycline treatment, ultimately leading to tumor formation. Treatment with a targeted molecular therapy (e.g., small-molecule EGFR inhibitors) will cause tumor regression. However, under prolonged treatment resistant tumors can emerge. Genomic analysis of these tumors can then be used to identify resistance mechanisms as in Politi et al. (2010)

CCSP promoter (Fig. 27.3). A second transgene carries tet-responsive elements (TRE) that drive expression of mutated forms of human *EGFR* found in human cancer, such as the L858R point mutant and a small in-frame deletion mutant (DEL). Without any treatment, rtTA suppresses expression of the oncogene, and animals remain phenotypically normal. Upon doxycycline administration, the rtTA is inhibited and the oncogene is expressed (Fig. 27.3), leading eventually to the development of either diffuse bronchioalveolar carcinoma (BAC) or invasive adenocarcinomas (Ji et al. 2006; Politi et al. 2006). Similar to human tumors with EGFR mutation, lung tumors found in EGFR bitransgenic mice were sensitive to treatment with the EGFR-targeting drugs erlotinib, HKI-272, and cetuximab. These studies demonstrate that expression of mutant, oncogenic EGFR is sufficient to initiate and maintain lung tumorigenesis and tumors are dependent on continued EGFR signaling for tumor maintenance.

The ability to temporally and spatially control oncogene expression in carefully controlled mouse models is a powerful tool for understanding cancer progression. Not only can one control when to turn on the gene, but the investigator can also shut off transgene expression in the case of tamoxifen and doxycycline-inducible models. Using this approach, it has been shown that *KRAS* and *EGFR* lung-specific transgenic mice require continuous expression of the transgene for tumor maintenance (Fisher et al. 2001; Ji et al. 2006; Politi et al. 2006). This corroborates the finding in humans that EGFR inhibitors induce regression of the tumors and confirms that the effectiveness of EGFR inhibitors in human tumors reflects a direct on-target effect of the drug. In the case of KRAS, a protein that is largely considered "undruggable," these data provide a proof-of-principle that inhibition of KRAS function could indeed be a successful therapeutic strategy and therefore further efforts should be devoted to finding such a drug.

Another benefit of mouse models is the enhanced ability for pathway analysis and genetic analysis of response to treatment modalities, as we have already illustrated in the case of APL and prostate cancer. Such studies enable identification of new biomarkers or novel protein targets for therapy. Although some may argue that the mouse is too different from humans to be clinically predictive, in fact studies of *KRAS* lung-specific transgenic mice have shown that the gene expression profile of *KRAS* mutant murine lung tumors correlates well with the gene expression profile of *KRAS* mutant human tumors (Sweet-Cordero et al. 2005). Furthermore, studies of drug resistance in mouse models recapitulate the mechanisms of resistance found in human tumors (see below).

Similar to the genetic analysis described above to dissect the PTEN/AKT/MTOR pathway in prostate cancer, genetic studies utilizing mouse models of lung cancer can demonstrate which downstream pathway effectors are required for tumorigenesis. Such proteins are potential targets for therapeutic intervention in the analogous human cancer. For example, one model of *KRAS*-induced lung tumorigenesis is the lox-stop-lox KRAS G12D model. These mice express the oncogenic form of *KRAS* after Cre-mediated excision of the stop codon in front of the oncogene sequence and rapidly develop adenomas and adenocarcinomas in the lung (Tuveson et al. 2004). Simultaneous Cre-mediated inactivation of the *Rac1* gene, a small GTPase that is activated by Ras, inhibits tumor formation and any tumors that did form had failed to inactivate *Rac1* (Kissil et al. 2007). Therefore, *Rac1* is required for *KRAS*-induced lung tumorigenesis and *Rac1* or other *Rac1*-effectors are potential targets for therapy of human lung tumors harboring *KRAS* mutations. Dozens of other interactors of the RAS pathway have also been elucidated or validated in this manner.

Like the studies with $PTEN^{-/-}$ MEFs discussed above, MEFs from mouse models of lung cancer have proven useful for analysis of the signaling pathways activated by the expressed oncogene. For example, studies with lox-stop-lox *KRAS* G12D MEFs showed that MAPK and PI3K inhibition with the inhibitors U0216 and LY294002 could inhibit the phenotypic changes accompanied by *KRAS* expression in MEFs (Tuveson et al. 2004). These preliminary studies have in turn led to *in vivo* analysis of the efficacy of combined PI3K and MAPK inhibition that are now forming the basis for clinical trials in human patients (see below).

Mouse models serve as preclinical subjects and can be used to test novel EGFR inhibitors for target validation and efficacy (Arteaga 2006). At the time of the discovery of *EGFR* mutations, it was recognized that EGFR inhibitors, such as erlotinib and gefitinib, were effective in the tumors that harbored *EGFR* mutations and not in others, such as those harboring *KRAS* mutations (Lynch et al. 2004; Paez et al. 2004; Pao et al. 2004, 2005b). Since mouse models of EGFR-driven tumorigenesis have been created and faithfully mimic the human disease, these mouse models have become critical players in the preclinical testing of novel-targeted therapies. EGFR-transgenic models have already confirmed the *in vivo* efficacy of the EGFR-targeted drugs, such as erlotinib (a reversible small-molecule inhibitor), cetuximab (a monoclonal antibody), and HKI-272 (an irreversible small-molecule inhibitor) (Ji et al. 2006; Li et al. 2007; Politi et al. 2006).

Unfortunately, a significant problem in the current management of patients with *EGFR* mutation is the issue of acquired resistance (Fig. 27.3). Although the majority of patients harboring *EGFR* mutations respond to EGFR inhibitors initially

(Lynch et al. 2004; Paez et al. 2004; Pao et al. 2004), acquired resistance can occur through mechanisms such as MET amplification (Engelman et al. 2007b) or a secondary mutation, T790M (Kobayashi et al. 2005; Pao et al. 2005a), that increases the affinity of ATP binding to EGFR (Yun et al. 2008). Studies in mice confirm that lung tumors expressing EGFR L858R in combination with T790M are not sensitive to the EGFR inhibitor erlotinib (Li et al. 2007; Regales et al. 2007). However, these mice have been used to test other therapeutic strategies, and have shown that tumors with dual L858R-T790M mutation respond to combination treatment of HKI-272, an irreversible EGFR inhibitor, and rapamycin (Li et al. 2007). HKI-272 is now in Phase 2 clinical trials (Wong 2007), although the results of the murine study suggest that combination therapy may be more effective than single agent treatment for the management of *EGFR* mutant tumors.

Additional irreversible inhibitors of EGFR, such as PF00299804 and BIBW2922, are also in development and have shown promise in preclinical testing in EGFR L858R-T790M transgenic mice or xenograft studies (Engelman et al. 2007a; Regales et al. 2009). Finally, a new class of irreversible inhibitors has been developed that selectively targets T790M mutant EGFR and shows efficacy for the treatment of lung tumors found in transgenic mice harboring L858R-T790M or DEL-T790M compound EGFR mutations (Zhou et al. 2009). These studies are just a few of the numerous other examples, whereby mouse modeling has informed the development and interpretation of clinical trials.

Importantly, mouse models not only recapitulate the response to therapy of genetically similar human tumors, but these models also demonstrate the same mechanisms of drug resistance found in human cancer (Fig. 27.3). Studies of acquired resistance in murine models of EGFR-induced lung tumorigenesis have shown that these murine tumors, like human cancer, can develop resistance via mutation to T790M or amplification of MET (Politi et al. 2010). Therefore, these models may be useful in identifying additional, currently unknown, mechanisms of acquired resistance to EGFR inhibitors (Fig. 27.3).

Although *EGFR* mutant tumors respond well to EGFR inhibitors, *KRAS* mutant lung tumors have proven more problematic for therapeutic intervention and display primary resistance to EGFR inhibitors (Pao et al. 2005b). Recently, however, Engelman et al. demonstrated that combined treatment of murine *KRAS*-mutant lung tumors with the dual PI3K/mTOR inhibitor NVP-BEZ235 and the MEK inhibitor ARRY-142886 was effective to shrink *RAS*-drive lung tumors *in vivo* (Engelman et al. 2008) These results demonstrate that combined PI3K/mTOR/MEK inhibition could be a viable strategy for the treatment of human lung tumors harboring *KRAS* mutations. Once again, knowledge gained from preclinical testing in mouse models of cancer is informing and shaping drug development and clinical trials of human-targeted therapies.

The success of EGFR inhibitors for targeted therapy of lung cancer and the knowledge we have gained from mouse models of this disease will likely be extended now for the treatment of ALK-positive human lung cancers. Anaplastic lymphoma kinase (ALK) is a tyrosine kinase frequently mutated or translocated in hematopoietic malignancies and found translocated with EML4 in lung cancer

(Soda et al. 2007). Small-molecule ALK inhibitors are already in clinical development (Solomon et al. 2009) and have shown efficacy in a mouse model of EML4-ALK-driven lung cancer (Soda et al. 2008). Thus, such faithful mouse models could predict the genetic basis underlying acquired resistance to ALK inhibitors, and on the basis of this genetic information inform how to best combine ALK inhibitors with other anticancer drugs toward disease eradication.

27.5 Conclusion and Future Directions

The examples above represent only a tiny fraction of the preclinical studies carried out in mouse models of each disease. For each scenario we described, there are literally dozens of other studies performed with different drugs or mouse models with different genetic alterations. Unfortunately, due to space limitations it is impossible to catalog the vast number of preclinical trials that are influencing the development of personalized medicine for cancer therapy. Instead, we simply highlighted a few examples for differing tumors types that became paradigmatic or are coming to fruition right now for human cancer therapy on the basis of preclinical efforts.

The realization of genetic-based and personalized medicine as an attainable and important part of cancer therapy is changing the way preclinical trials and mouse models interface with translational medicine to bring about safe and effective treatments. "Co-clinical trials" that are performed in mice in parallel with human clinical trials may represent the future of mouse modeling of targeted therapy (Caterina et al. 2011). These "co-clinical" trials would be performed in mouse models with genetically defined alterations similar or identical to those found in the human cancers treated in the clinical trials. Biological, pharmacological, and clinical information, such as tumor-specific somatic variations upon treatment, drug response on various regimens, and gene expression or proteomic profiling could be performed in the mouse samples in tandem with the human samples. The controlled environment and genetic background of the mouse models, as well as the large number of subjects available, would allow rapid identification of predictive prognostic factors for therapeutic response that could then be analyzed in the human patient. Such studies accelerate rapid advances in targeted therapy development, patient stratification based on genetic criteria, and drug approval, all of which are critical to bring effective personalized medicine into the oncology clinic as rapidly as possible.

References

Al-Hajj M, Clarke MF (2004) Self-renewal and solid tumor stem cells. Oncogene 23:7274–7282

Alcalay M, Zangrilli D, Fagioli M, Pandolfi PP, Mencarelli A, Lo Coco F, Biondi A, Grignani F, Pelicci PG (1992) Expression pattern of the RAR alpha-PML fusion gene in acute promyelocytic leukemia. Proc Natl Acad Sci USA 89:4840–4844

Alcalay M, Zangrilli D, Pandolfi PP, Longo L, Mencarelli A, Giacomucci A, Rocchi M, Biondi A, Rambaldi A, Lo Coco F et al (1991) Translocation breakpoint of acute promyelocytic leukemia lies within the retinoic acid receptor alpha locus. Proc Natl Acad Sci USA 88:1977–1981

Arnould C, Philippe C, Bourdon V, Gr goire MJ, Berger R, Jonveaux P (1999) The signal transducer and activator of transcription STAT5b gene is a new partner of retinoic acid receptor alpha in acute promyelocytic-like leukaemia. Hum Mol Genet 8:1741–1749

Arteaga CL (2006) EGF receptor mutations in lung cancer: from humans to mice and maybe back to humans. Cancer Cell 9:421–423

Backman SA, Ghazarian D, So K, Sanchez O, Wagner KU, Hennighausen L, Suzuki A, Tsao MS, Chapman WB, Stambolic V et al (2004) Early onset of neoplasia in the prostate and skin of mice with tissue-specific deletion of Pten. Proc Natl Acad Sci USA 101:1725 1730

Backman SA, Stambolic V, Suzuki A, Haight J, Elia A, Pretorius J, Tsao MS, Shannon P, Bolon B, Ivy GO et al (2001) Deletion of Pten in mouse brain causes seizures, ataxia and defects in soma size resembling Lhermitte-Duclos disease. Nat Genet 29:396–403

Bayascas JR, Leslie NR, Parsons R, Fleming S, Alessi DR (2005) Hypomorphic mutation of PDK1 suppresses tumorigenesis in PTEN(+/–) mice. Curr Biol 15:1839–1846

Berger AH, Niki M, Morotti A, Taylor BS, Socci ND, Viale A, Brennan C, Szoke J, Motoi N, Rothman PB et al (2010) Identification of DOK genes as lung tumor suppressors. Nat Genet 42(3):216–223

Borrow J, Goddard AD, Sheer D, Solomon E (1990) Molecular analysis of acute promyelocytic leukemia breakpoint cluster region on chromosome 17. Science 249:1577–1580

Brown D, Kogan S, Lagasse E, Weissman I, Alcalay M, Pelicci PG, Atwater S, Bishop JM (1997) A PMLRARalpha transgene initiates murine acute promyelocytic leukemia. Proc Natl Acad Sci USA 94:2551–2556

Carracedo A, Ma L, Teruya-Feldstein J, Rojo F, Salmena L, Alimonti A, Egia A, Sasaki AT, Thomas G, Kozma SC et al (2008) Inhibition of mTORC1 leads to MAPK pathway activation through a PI3K-dependent feedback loop in human cancer. J Clin Invest 118:3065–3074

Carver BS, Pandolfi PP (2006) Mouse modeling in oncologic preclinical and translational research. Clin Cancer Res 12:5305–5311

Caterina N, Andrea L, Akash P, Lewis CC, Pier PP Cancer Discovery July 2011 1:108–116; doi:10.1158/2159-8290.CD-11-0061

Chen ML, Xu PZ, Peng XD, Chen WS, Guzman G, Yang X, Di Cristofano A, Pandolfi PP, Hay N (2006) The deficiency of Akt1 is sufficient to suppress tumor development in Pten+/– mice. Genes Dev 20:1569–1574

Chen Z, Brand NJ, Chen A, Chen SJ, Tong JH, Wang ZY, Waxman S, Zelent A (1993) Fusion between a novel Kruppel-like zinc finger gene and the retinoic acid receptor-alpha locus due to a variant t(11;17) translocation associated with acute promyelocytic leukaemia. EMBO J 12:1161–1167

Chen Z, Trotman LC, Shaffer D, Lin HK, Dotan ZA, Niki M, Koutcher JA, Scher HI, Ludwig T, Gerald W et al (2005) Crucial role of p53-dependent cellular senescence in suppression of Pten-deficient tumorigenesis. Nature 436:725–730

Cheng GX, Zhu XH, Men XQ, Wang L, Huang QH, Jin XL, Xiong SM, Zhu J, Guo WM, Chen JQ et al (1999) Distinct leukemia phenotypes in transgenic mice and different corepressor interactions generated by promyelocytic leukemia variant fusion genes PLZF-RARalpha and NPM-RARalpha. Proc Natl Acad Sci USA 96:6318–6323

Cheng L, Ramesh AV, Flesken-Nikitin A, Choi J, Nikitin AY (2010) Mouse models for cancer stem cell research. Toxicol Pathol 38:62–71

Chitale D, Gong Y, Taylor BS, Broderick S, Brennan C, Somwar R, Golas B, Wang L, Motoi N, Szoke J et al (2009) An integrated genomic analysis of lung cancer reveals loss of DUSP4 in EGFR-mutant tumors. Oncogene 28:2773–2783

de The H, Chomienne C, Lanotte M, Degos L, Dejean A (1990) The t(15;17) translocation of acute promyelocytic leukaemia fuses the retinoic acid receptor alpha gene to a novel transcribed locus. Nature 347:558–561

de The H, Lavau C, Marchio A, Chomienne C, Degos L, Dejean A (1991) The PML-RAR alpha fusion mRNA generated by the t(15;17) translocation in acute promyelocytic leukemia encodes a functionally altered RAR. Cell 66:675–684

Di Cristofano A, Pandolfi PP (2000) The multiple roles of PTEN in tumor suppression. Cell 100:387–390
Di Cristofano A, Pesce B, Cordon-Cardo C, Pandolfi PP (1998) Pten is essential for embryonic development and tumour suppression. Nat Genet 19:348–355
Dick JE (2009) Looking ahead in cancer stem cell research. Nat Biotechnol 27:44–46
Ding L, Getz G, Wheeler DA, Mardis ER, McLellan MD, Cibulskis K, Sougnez C, Greulich H, Muzny DM, Morgan MB et al (2008) Somatic mutations affect key pathways in lung adenocarcinoma. Nature 455:1069–1075
Druker BJ (2008) Translation of the Philadelphia chromosome into therapy for CML. Blood 112:4808–4817
Engelman JA, Chen L, Tan X, Crosby K, Guimaraes AR, Upadhyay R, Maira M, McNamara K, Perera SA, Song Y et al (2008) Effective use of PI3K and MEK inhibitors to treat mutant Kras G12D and PIK3CA H1047R murine lung cancers. Nat Med 14:1351–1356
Engelman JA, Zejnullahu K, Gale CM, Lifshits E, Gonzales AJ, Shimamura T, Zhao F, Vincent PW, Naumov GN, Bradner JE et al (2007a) PF00299804, an irreversible pan-ERBB inhibitor, is effective in lung cancer models with EGFR and ERBB2 mutations that are resistant to gefitinib. Cancer Res 67:11924–11932
Engelman JA, Zejnullahu K, Mitsudomi T, Song Y, Hyland C, Park JO, Lindeman N, Gale CM, Zhao X, Christensen J et al (2007b) MET amplification leads to gefitinib resistance in lung cancer by activating ERBB3 signaling. Science 316:1039–1043
Fan QW, Knight ZA, Goldenberg DD, Yu W, Mostov KE, Stokoe D, Shokat KM, Weiss WA (2006) A dual PI3 kinase/mTOR inhibitor reveals emergent efficacy in glioma. Cancer Cell 9:341–349
Fisher GH, Wellen SL, Klimstra D, Lenczowski JM, Tichelaar JW, Lizak MJ, Whitsett JA, Koretsky A, Varmus HE (2001) Induction and apoptotic regression of lung adenocarcinomas by regulation of a K-Ras transgene in the presence and absence of tumor suppressor genes. Genes Dev 15:3249–3262
Fraser MM, Zhu X, Kwon CH, Uhlmann EJ, Gutmann DH, Baker SJ (2004) Pten loss causes hypertrophy and increased proliferation of astrocytes in vivo. Cancer Res 64:7773–7779
Goddard AD, Borrow J, Freemont PS, Solomon E (1991) Characterization of a zinc finger gene disrupted by the t(15;17) in acute promyelocytic leukemia. Science 254:1371–1374
Gray IC, Stewart LM, Phillips SM, Hamilton JA, Gray NE, Watson GJ, Spurr NK, Snary D (1998) Mutation and expression analysis of the putative prostate tumour-suppressor gene PTEN. Br J Cancer 78:1296–1300
Grisolano JL, Wesselschmidt RL, Pelicci PG, Ley TJ (1997) Altered myeloid development and acute leukemia in transgenic mice expressing PML-RAR alpha under control of cathepsin G regulatory sequences. Blood 89:376–387
Guertin DA, Sabatini DM (2007) Defining the role of mTOR in cancer. Cancer Cell 12:9–22
Guertin DA, Stevens DM, Saitoh M, Kinkel S, Crosby K, Sheen JH, Mullholland DJ, Magnuson MA, Wu H, Sabatini DM (2009) mTOR complex 2 is required for the development of prostate cancer induced by Pten loss in mice. Cancer Cell 15:148–159
Harrington LS, Findlay GM, Gray A, Tolkacheva T, Wigfield S, Rebholz H, Barnett J, Leslie NR, Cheng S, Shepherd PR et al (2004) The TSC1-2 tumor suppressor controls insulin-PI3K signaling via regulation of IRS proteins. J Cell Biol 166:213–223
He LZ, Bhaumik M, Tribioli C, Rego EM, Ivins S, Zelent A, Pandolfi PP (2000) Two critical hits for promyelocytic leukemia. Mol Cell 6:1131–1141
He LZ, Guidez F, Tribioli C, Peruzzi D, Ruthardt M, Zelent A, Pandolfi PP (1998) Distinct interactions of PML-RARalpha and PLZF-RARalpha with co-repressors determine differential responses to RA in APL. Nat Genet 18:126–135
He LZ, Tolentino T, Grayson P, Zhong S, Warrell RP Jr, Rifkind RA, Marks PA, Richon VM, Pandolfi PP (2001) Histone deacetylase inhibitors induce remission in transgenic models of therapy-resistant acute promyelocytic leukemia. J Clin Invest 108:1321–1330
He LZ, Tribioli C, Rivi R, Peruzzi D, Pelicci PG, Soares V, Cattoretti G, Pandolfi PP (1997) Acute leukemia with promyelocytic features in PML/RARalpha transgenic mice. Proc Natl Acad Sci USA 94:5302–5307

Hu J, Liu YF, Wu CF, Xu F, Shen ZX, Zhu YM, Li JM, Tang W, Zhao WL, Wu W et al (2009) Long-term efficacy and safety of all-trans retinoic acid/arsenic trioxide-based therapy in newly diagnosed acute promyelocytic leukemia. Proc Natl Acad Sci USA 106:3342–3347
Hummel JL, Wells RA, Dube ID, Licht JD, Kamel-Reid S (1999) Deregulation of NPM and PLZF in a variant t(5;17) case of acute promyelocytic leukemia. Oncogene 18:633–641
Ito K, Bernardi R, Morotti A, Matsuoka S, Saglio G, Ikeda Y, Rosenblatt J, Avigan DE, Teruya-Feldstein J, Pandolfi PP (2008) PML targeting eradicates quiescent leukaemia-initiating cells. Nature 453:1072–1078
Jackson EL, Willis N, Mercer K, Bronson RT, Crowley D, Montoya R, Jacks T, Tuveson DA (2001) Analysis of lung tumor initiation and progression using conditional expression of oncogenic K-ras. Genes Dev 15:3243–3248
Ji H, Li D, Chen L, Shimamura T, Kobayashi S, McNamara K, Mahmood U, Mitchell A, Sun Y, Al-Hashem R et al (2006) The impact of human EGFR kinase domain mutations on lung tumorigenesis and in vivo sensitivity to EGFR-targeted therapies. Cancer Cell 9:485–495
Ji H, Ramsey MR, Hayes DN, Fan C, McNamara K, Kozlowski P, Torrice C, Wu MC, Shimamura T, Perera SA et al (2007) LKB1 modulates lung cancer differentiation and metastasis. Nature 448:807–810
Jia S, Liu Z, Zhang S, Liu P, Zhang L, Lee SH, Zhang J, Signoretti S, Loda M, Roberts TM et al (2008) Essential roles of PI(3)K-p110beta in cell growth, metabolism and tumorigenesis. Nature 454:776–779
Johnson L, Mercer K, Greenbaum D, Bronson RT, Crowley D, Tuveson DA, Jacks T (2001) Somatic activation of the K-ras oncogene causes early onset lung cancer in mice. Nature 410:1111–1116
Jordan CT, Guzman ML, Noble M (2006) Cancer stem cells. N Engl J Med 355:1253–1261
Jordan VC (2008) Tamoxifen: catalyst for the change to targeted therapy. Eur J Cancer 44:30–38
Joyce JA (2005) Therapeutic targeting of the tumor microenvironment. Cancer Cell 7:513–520
Kakizuka A, Miller WH Jr, Umesono K, Warrell RP Jr, Frankel SR, Murty VV, Dmitrovsky E, Evans RM (1991) Chromosomal translocation t(15;17) in human acute promyelocytic leukemia fuses RAR alpha with a novel putative transcription factor, PML. Cell 66:663–674
Kastner P, Perez A, Lutz Y, Rochette-Egly C, Gaub MP, Durand B, Lanotte M, Berger R, Chambon P (1992) Structure, localization and transcriptional properties of two classes of retinoic acid receptor alpha fusion proteins in acute promyelocytic leukemia (APL): structural similarities with a new family of oncoproteins. EMBO J 11:629–642
Kinkade CW, Castillo-Martin M, Puzio-Kuter A, Yan J, Foster TH, Gao H, Sun Y, Ouyang X, Gerald WL, Cordon-Cardo C et al (2008) Targeting AKT/mTOR and ERK MAPK signaling inhibits hormone-refractory prostate cancer in a preclinical mouse model. J Clin Invest 118:3051–3064
Kissil JL, Walmsley MJ, Hanlon L, Haigis KM, Bender Kim CF, Sweet-Cordero A, Eckman MS, Tuveson DA, Capobianco AJ, Tybulewicz VL et al (2007) Requirement for Rac1 in a K-ras induced lung cancer in the mouse. Cancer Res 67:8089–8094
Knight ZA, Gonzalez B, Feldman ME, Zunder ER, Goldenberg DD, Williams O, Loewith R, Stokoe D, Balla A, Toth B et al (2006) A pharmacological map of the PI3-K family defines a role for p110alpha in insulin signaling. Cell 125:733–747
Kobayashi S, Boggon TJ, Dayaram T, Janne PA, Kocher O, Meyerson M, Johnson BE, Eck MJ, Tenen DG, Halmos B (2005) EGFR mutation and resistance of non-small-cell lung cancer to gefitinib. N Engl J Med 352:786–792
Lesche R, Groszer M, Gao J, Wang Y, Messing A, Sun H, Liu X, Wu H (2002) Cre/loxP-mediated inactivation of the murine Pten tumor suppressor gene. Genesis 32:148–149
Li D, Shimamura T, Ji H, Chen L, Haringsma HJ, McNamara K, Liang MC, Perera SA, Zaghlul S, Borgman CL et al (2007) Bronchial and peripheral murine lung carcinomas induced by T790M-L858R mutant EGFR respond to HKI-272 and rapamycin combination therapy. Cancer Cell 12:81–93
Licht JD, Chomienne C, Goy A, Chen A, Scott AA, Head DR, Michaux JL, Wu Y, DeBlasio A, Miller WH Jr et al (1995) Clinical and molecular characterization of a rare syndrome of acute promyelocytic leukemia associated with translocation (11;17). Blood 85:1083–1094

Longo L, Pandolfi PP, Biondi A, Rambaldi A, Mencarelli A, Lo Coco F, Diverio D, Pegoraro L, Avanzi G, Tabilio A et al (1990) Rearrangements and aberrant expression of the retinoic acid receptor alpha gene in acute promyelocytic leukemias. J Exp Med 172:1571–1575

Lynch TJ, Bell DW, Sordella R, Gurubhagavatula S, Okimoto RA, Brannigan BW, Harris PL, Haserlat SM, Supko JG, Haluska FG et al (2004) Activating mutations in the epidermal growth factor receptor underlying responsiveness of non-small-cell lung cancer to gefitinib. N Engl J Med 350:2129–2139

Ma L, Teruya-Feldstein J, Behrendt N, Chen Z, Noda T, Hino O, Cordon-Cardo C, Pandolfi PP (2005a) Genetic analysis of Pten and Tsc2 functional interactions in the mouse reveals asymmetrical haploinsufficiency in tumor suppression. Genes Dev 19:1779–1786

Ma X, Ziel-van der Made AC, Autar B, van der Korput HA, Vermeij M, van Duijn P, Cleutjens KB, de Krijger R, Krimpenfort P, Berns A et al (2005b) Targeted biallelic inactivation of Pten in the mouse prostate leads to prostate cancer accompanied by increased epithelial cell proliferation but not by reduced apoptosis. Cancer Res 65:5730–5739

Maira SM, Stauffer F, Brueggen J, Furet P, Schnell C, Fritsch C, Brachmann S, Chene P, De Pover A, Schoemaker K et al (2008) Identification and characterization of NVP-BEZ235, a new orally available dual phosphatidylinositol 3-kinase/mammalian target of rapamycin inhibitor with potent in vivo antitumor activity. Mol Cancer Ther 7:1851–1863

Manning BD, Logsdon MN, Lipovsky AI, Abbott D, Kwiatkowski DJ, Cantley LC (2005) Feedback inhibition of Akt signaling limits the growth of tumors lacking Tsc2. Genes Dev 19:1773–1778

Marino S, Krimpenfort P, Leung C, van der Korput HA, Trapman J, Camenisch I, Berns A, Brandner S (2002) PTEN is essential for cell migration but not for fate determination and tumourigenesis in the cerebellum. Development 129:3513–3522

Michaloglou C, Vredeveld LC, Soengas MS, Denoyelle C, Kuilman T, van der Horst CM, Majoor DM, Shay JW, Mooi WJ, Peeper DS (2005) BRAFE600-associated senescence-like cell cycle arrest of human naevi. Nature 436:720–724

Nardella C, Carracedo A, Alimonti A, Hobbs RM, Clohessy JG, Chen Z, Egia A, Fornari A, Fiorentino M, Loda M et al (2009) Differential requirement of mTOR in postmitotic tissues and tumorigenesis. Sci Signal 2:ra2

Nardella C, Chen Z, Salmena L, Carracedo A, Alimonti A, Egia A, Carver B, Gerald W, Cordon-Cardo C, Pandolfi PP (2008) Aberrant Rheb-mediated mTORC1 activation and Pten haploinsufficiency are cooperative oncogenic events. Genes Dev 22:2172–2177

Paez JG, Janne PA, Lee JC, Tracy S, Greulich H, Gabriel S, Herman P, Kaye FJ, Lindeman N, Boggon TJ et al (2004) EGFR mutations in lung cancer: correlation with clinical response to gefitinib therapy. Science 304:1497–1500

Pandolfi PP, Grignani F, Alcalay M, Mencarelli A, Biondi A, LoCoco F, Pelicci PG (1991) Structure and origin of the acute promyelocytic leukemia myl/RAR alpha cDNA and characterization of its retinoid-binding and transactivation properties. Oncogene 6:1285–1292

Pao W, Miller V, Zakowski M, Doherty J, Politi K, Sarkaria I, Singh B, Heelan R, Rusch V, Fulton L et al (2004) EGF receptor gene mutations are common in lung cancers from "never smokers" and are associated with sensitivity of tumors to gefitinib and erlotinib. Proc Natl Acad Sci USA 101:13306–13311

Pao W, Miller VA, Politi KA, Riely GJ, Somwar R, Zakowski MF, Kris MG, Varmus H (2005a) Acquired resistance of lung adenocarcinomas to gefitinib or erlotinib is associated with a second mutation in the EGFR kinase domain. PLoS Med 2:e73

Pao W, Wang TY, Riely GJ, Miller VA, Pan Q, Ladanyi M, Zakowski MF, Heelan RT, Kris MG, Varmus HE (2005b) KRAS mutations and primary resistance of lung adenocarcinomas to gefitinib or erlotinib. PLoS Med 2:e17

Peng C, Li S (2010) CML mouse model in translational research. Methods Mol Biol 602:253–266

Perez A, Kastner P, Sethi S, Lutz Y, Reibel C, Chambon P (1993) PMLRAR homodimers: distinct DNA binding properties and heteromeric interactions with RXR. EMBO J 12:3171–3182

Perez-Caro M, Sanchez-Garcia I (2006) Killing time for cancer stem cells (CSC): discovery and development of selective CSC inhibitors. Curr Med Chem 13:1719–1725

Petti MC, Fazi F, Gentile M, Diverio D, De Fabritiis P, De Propris MS, Fiorini R, Spiriti MA, Padula F, Pelicci PG et al (2002) Complete remission through blast cell differentiation in PLZF/RARalpha-positive acute promyelocytic leukemia: in vitro and in vivo studies. Blood 100:1065–1067

Piazza F, Gurrieri C, Pandolfi PP (2001) The theory of APL. Oncogene 20:7216–7222

Podsypanina K, Ellenson LH, Nemes A, Gu J, Tamura M, Yamada KM, Cordon-Cardo C, Catoretti G, Fisher PE, Parsons R (1999) Mutation of Pten/Mmac1 in mice causes neoplasia in multiple organ systems. Proc Natl Acad Sci USA 96:1563–1568

Politi K, Fan PD, Shen R, Zakowski M, Varmus H (2010) Erlotinib resistance in mouse models of epidermal growth factor receptor-induced lung adenocarcinoma. Dis Model Mech 3:111–119

Politi K, Zakowski MF, Fan PD, Schonfeld EA, Pao W, Varmus HE (2006) Lung adenocarcinomas induced in mice by mutant EGF receptors found in human lung cancers respond to a tyrosine kinase inhibitor or to down-regulation of the receptors. Genes Dev 20:1496–1510

Pollock JL, Westervelt P, Kurichety AK, Pelicci PG, Grisolano JL, Ley TJ (1999) A bcr-3 isoform of RARalpha-PML potentiates the development of PML-RARalpha-driven acute promyelocytic leukemia. Proc Natl Acad Sci USA 96:15103–15108

Raynaud FI, Eccles S, Clarke PA, Hayes A, Nutley B, Alix S, Henley A, Di-Stefano F, Ahmad Z, Guillard S et al (2007) Pharmacologic characterization of a potent inhibitor of class I phosphatidylinositide 3-kinases. Cancer Res 67:5840–5850

Redner RL (2002) Variations on a theme: the alternate translocations in APL. Leukemia 16:1927–1932

Redner RL, Rush EA, Faas S, Rudert WA, Corey SJ (1996) The t(5;17) variant of acute promyelocytic leukemia expresses a nucleophosmin-retinoic acid receptor fusion. Blood 87:882–886

Regales L, Balak MN, Gong Y, Politi K, Sawai A, Le C, Koutcher JA, Solit DB, Rosen N, Zakowski MF et al (2007) Development of new mouse lung tumor models expressing EGFR T790M mutants associated with clinical resistance to kinase inhibitors. PLoS One 2:e810

Regales L, Gong Y, Shen R, de Stanchina E, Vivanco I, Goel A, Koutcher JA, Spassova M, Ouerfelli O, Mellinghoff IK et al (2009) Dual targeting of EGFR can overcome a major drug resistance mutation in mouse models of EGFR mutant lung cancer. J Clin Invest 119:3000–3010

Rego EM, He LZ, Warrell RP Jr, Wang ZG, Pandolfi PP (2000) Retinoic acid (RA) and As2O3 treatment in transgenic models of acute promyelocytic leukemia (APL) unravel the distinct nature of the leukemogenic process induced by the PML-RARalpha and PLZF-RARalpha oncoproteins. Proc Natl Acad Sci USA 97:10173–10178

Reya T, Morrison SJ, Clarke MF, Weissman IL (2001) Stem cells, cancer, and cancer stem cells. Nature 414:105–111

Shah OJ, Wang Z, Hunter T (2004) Inappropriate activation of the TSC/Rheb/mTOR/S6K cassette induces IRS1/2 depletion, insulin resistance, and cell survival deficiencies. Curr Biol 14:1650–1656

Shigematsu H, Gazdar AF (2006) Somatic mutations of epidermal growth factor receptor signaling pathway in lung cancers. Int J Cancer 118:257–262

Soda M, Choi YL, Enomoto M, Takada S, Yamashita Y, Ishikawa S, Fujiwara S, Watanabe H, Kurashina K, Hatanaka H et al (2007) Identification of the transforming EML4-ALK fusion gene in non-small-cell lung cancer. Nature 448:561–566

Soda M, Takada S, Takeuchi K, Choi YL, Enomoto M, Ueno T, Haruta H, Hamada T, Yamashita Y, Ishikawa Y et al (2008) A mouse model for EML4-ALK-positive lung cancer. Proc Natl Acad Sci USA 105:19893–19897

Solomon B, Varella-Garcia M, Camidge DR (2009) ALK gene rearrangements: a new therapeutic target in a molecularly defined subset of non-small cell lung cancer. J Thorac Oncol 4:1450–1454

Stambolic V, Suzuki A, de la Pompa JL, Brothers GM, Mirtsos C, Sasaki T, Ruland J, Penninger JM, Siderovski DP, Mak TW (1998) Negative regulation of PKB/Akt-dependent cell survival by the tumor suppressor PTEN. Cell 95:29–39

Sukhai MA, Wu X, Xuan Y, Zhang T, Reis PP, Dube K, Rego EM, Bhaumik M, Bailey DJ, Wells RA et al (2004) Myeloid leukemia with promyelocytic features in transgenic mice expressing hCG-NuMA-RARalpha. Oncogene 23:665–678

Suzuki A, de la Pompa JL, Stambolic V, Elia AJ, Sasaki T, del Barco Barrantes I, Ho A, Wakeham A, Itie A, Khoo W et al (1998) High cancer susceptibility and embryonic lethality associated with mutation of the PTEN tumor suppressor gene in mice. Curr Biol 8:1169–1178

Suzuki A, Yamaguchi MT, Ohteki T, Sasaki T, Kaisho T, Kimura Y, Yoshida R, Wakeham A, Higuchi T, Fukumoto M et al (2001) T cell-specific loss of Pten leads to defects in central and peripheral tolerance. Immunity 14:523–534

Sweet-Cordero A, Mukherjee S, Subramanian A, You H, Roix JJ, Ladd-Acosta C, Mesirov J, Golub TR, Jacks T (2005) An oncogenic KRAS2 expression signature identified by cross-species gene-expression analysis. Nat Genet 37:48–55

Trotman LC, Niki M, Dotan ZA, Koutcher JA, Di Cristofano A, Xiao A, Khoo AS, Roy-Burman P, Greenberg NM, Van Dyke T et al (2003) Pten dose dictates cancer progression in the prostate. PLoS Biol 1:E59

Tuveson DA, Shaw AT, Willis NA, Silver DP, Jackson EL, Chang S, Mercer KL, Grochow R, Hock H, Crowley D et al (2004) Endogenous oncogenic K-ras(G12D) stimulates proliferation and widespread neoplastic and developmental defects. Cancer Cell 5:375–387

Vogelstein B, Kinzler KW (1993) The multistep nature of cancer. Trends Genet 9:138–141

Wang S, Gao J, Lei Q, Rozengurt N, Pritchard C, Jiao J, Thomas GV, Li G, Roy-Burman P, Nelson PS et al (2003) Prostate-specific deletion of the murine Pten tumor suppressor gene leads to metastatic prostate cancer. Cancer Cell 4:209–221

Warrell RP Jr (1996) Pathogenesis and management of acute promyelocytic leukemia. Annu Rev Med 47:555–565

Warrell RP Jr, de The H, Wang ZY, Degos L (1993) Acute promyelocytic leukemia. N Engl J Med 329:177–189

Warrell RP Jr, He LZ, Richon V, Calleja E, Pandolfi PP (1998) Therapeutic targeting of transcription in acute promyelocytic leukemia by use of an inhibitor of histone deacetylase. J Natl Cancer Inst 90:1621–1625

Weir BA, Woo MS, Getz G, Perner S, Ding L, Beroukhim R, Lin WM, Province MA, Kraja A, Johnson LA et al (2007) Characterizing the cancer genome in lung adenocarcinoma. Nature 450:893–898

Wells RA, Catzavelos C, Kamel-Reid S (1997) Fusion of retinoic acid receptor alpha to NuMA, the nuclear mitotic apparatus protein, by a variant translocation in acute promyelocytic leukaemia. Nat Genet 17:109–113

Wong KK (2007) HKI-272 in non small cell lung cancer. Clin Cancer Res 13:s4593–s4596

Wu X, Pandolfi PP (2001) Mouse models for multistep tumorigenesis. Trends Cell Biol 11:S2–S9

Wullschleger S, Loewith R, Hall MN (2006) TOR signaling in growth and metabolism. Cell 124:471–484

Xiao A, Yin C, Yang C, Di Cristofano A, Pandolfi PP, Van Dyke T (2005) Somatic induction of Pten loss in a preclinical astrocytoma model reveals major roles in disease progression and avenues for target discovery and validation. Cancer Res 65:5172–5180

Yilmaz OH, Valdez R, Theisen BK, Guo W, Ferguson DO, Wu H, Morrison SJ (2006) Pten dependence distinguishes haematopoietic stem cells from leukaemia-initiating cells. Nature 441:475–482

Yun CH, Mengwasser KE, Toms AV, Woo MS, Greulich H, Wong KK, Meyerson M, Eck MJ (2008) The T790M mutation in EGFR kinase causes drug resistance by increasing the affinity for ATP. Proc Natl Acad Sci USA 105:2070–2075

Zhang H, Bajraszewski N, Wu E, Wang H, Moseman AP, Dabora SL, Griffin JD, Kwiatkowski DJ (2007) PDGFRs are critical for PI3K/Akt activation and negatively regulated by mTOR. J Clin Invest 117:730–738

Zhang H, Cicchetti G, Onda H, Koon HB, Asrican K, Bajraszewski N, Vazquez F, Carpenter CL, Kwiatkowski DJ (2003) Loss of Tsc1/Tsc2 activates mTOR and disrupts PI3K-Akt signaling through downregulation of PDGFR. J Clin Invest 112:1223–1233

Zhang J, Grindley JC, Yin T, Jayasinghe S, He XC, Ross JT, Haug JS, Rupp D, Porter-Westpfahl KS, Wiedemann LM et al (2006) PTEN maintains haematopoietic stem cells and acts in lineage choice and leukaemia prevention. Nature 441:518–522

Zhou W, Ercan D, Chen L, Yun CH, Li D, Capelletti M, Cortot AB, Chirieac L, Iacob RE, Padera R et al (2009) Novel mutant-selective EGFR kinase inhibitors against EGFR T790M. Nature 462:1070–1074

Chapter 28
Mighty, But How Useful? The Emerging Role of Genetically Engineered Mice in Cancer Drug Discovery and Development

Reinhard Ebner, Jeffrey W. Strovel, Stephen K. Horrigan, and Kenneth C. Carter

28.1 Introduction

The focus of this chapter is on the use of genetically engineered mouse models (GEMMs) during cancer drug discovery and development: past, present, and future. From the beginning of history, humans have been seeking antidotes to disease through the ingestion of therapeutic substances. In recent centuries, scientists and governmental agencies have increasingly sought to investigate, define, and regulate therapeutic substances based on empirical testing in both humans and animals. A central challenge to this approach is balancing evidence that suggests that a candidate drug or other medical treatment is effective and provides clinical benefit versus data that suggests that the same treatment causes adverse side effects (Drews 2000; Bolon and Galbreath 2002; Bolon 2004; Roberts et al. 2004; DeVita and Chu 2008). During the past 50 years, the use of animal models for human disease has become central to the investigation of potential therapeutics and is now a definitive early-stage hurdle to the scientific and regulatory acceptance of a new drug candidate. GEMMs are a relatively new addition to the ever-increasing panoply of animal models that are used to search for new, more effective treatments for cancer. Because of the perceived potential of these models to enhance drug development and provide critical clues for superior treatment of disease, GEMMs have garnered an exceptional amount of attention and biomedical research funding (Van Dyke and Jacks 2002; Weiss and Shannon 2003; O'Hagan et al. 2005; Olive and Tuveson 2006; Sharpless and Depinho 2006; Singh and Johnson 2006; Frese and Tuveson 2007;

R. Ebner (✉)
National Cancer Institute, Section of Cancer Genomics, National Institutes of Health, Bethesda, MD 20892, USA
e-mail: reinhard.ebner@yahoo.com

J.W. Strovel • S.K. Horrigan • K.C. Carter
Noble Life Sciences, Inc., 22 Firstfield Road, Gaithersburg, MD 20878, USA

J.E. Green and T. Ried (eds.), *Genetically Engineered Mice for Cancer Research: Design, Analysis, Pathways, Validation and Pre-Clinical Testing*,
DOI 10.1007/978-0-387-69805-2_28,

Talmadge et al. 2007; Walrath et al. 2010). Here, we examine a range of current and potential uses for GEMM in cancer drug discovery and examine to what degree these models have, thus far, contributed to develop better medicines and what they may contribute in the future.

The proper exploration of this topic requires a contextual review of the historical dependence on animal models during drug development, the use of which now serves as a routine requirement for all investigational new drug (IND) data packages presented for regulatory approval to the FDA and other regulatory agencies. Following the discovery of a compound or biomolecule that is believed to have potential as a therapeutic agent, the typical next steps almost always include testing for efficacy and safety of the molecule in various *in vitro* and *in vivo* models. For small molecular weight organic compounds (small molecules) of the type that makes up the vast majority of currently marketed drugs, and in some cases for biomolecules, the drug development process typically involves repeated cycles of testing both *in vitro* and *in vivo* during which multiple related analogues of a molecule are tested. For cancer drug candidates, potential efficacy is often tested *in vitro* by examining the effect of the drug on a range of appropriate human cell lines and *in vivo* by thorough animal model testing. There are several different cancer animal models used for drug efficacy testing, the most common being mouse/human xenografts in which human cancer cells of an appropriate tissue and/or genetic subtype are injected into a standard immune compromised mouse and allowed to form a tumor. Candidate drugs are typically tested for the ability to delay or reduce tumor burden in these models. Xenograft models have been a near-requirement for cancer drug testing for many years, but, as reviewed elsewhere, they are far from an ideal model for many reasons, including their notorious lack of predictive value for human response, particularly for targeted drugs that work by mechanisms other than broad-acting cytotoxicity (Rosenberg and Bortner 1999; Kerbel 2003; Richmond and Su 2008; Gopinathan and Tuveson 2008). Increasingly, scientists in academic, federal, and academic laboratories have turned to a variety of alternatives to xenograft models, including GEMMS, in the hope that alternative models will provide greater insights into the mechanisms of tumor biology and the potential of candidate human therapeutics. The ability to manipulate the murine genome with ever increasing sophistication has generated great expectations for their success in this regard. Indeed, GEMMs have shown remarkable power to faithfully recapitulate some key aspects of human tumorigenesis and therapy response, but their use in drug discovery and development has remained limited. There are several reasons for this, but one important factor is that specific GEMMs remain both expensive and cumbersome to produce and maintain, and have limited applicability outside of a very narrow disease niche. These economic and practical considerations, combined with legal issues of complex licensing requirements, often preclude their widespread use. In addition, the historical inertia and preference for established traditional models by granting and regulatory agencies, such as FDA and NIH, may play a role in the limited use of GEMMs in drug discovery and development. That said, GEMMs have begun to have an impact on the biomedical industry in the areas of target and pathway validation, disease history elucidation, and the discovery and refinement of pharmacodynamic and toxicity biomarkers as reviewed below.

28.2 Disease Models in Drug Development: Historical Perspective

The use of animals in basic and applied life sciences research has a tradition dating back at least two millennia (Fox and Tucker 1984), and the evaluation of the effects of new therapies in animals prior to human testing is legally required in most jurisdictions (Schein and Scheffler 2006). However, these requirements were put in place predominantly for the prevention of detriment to human subjects, i.e., for the early detection of unwanted effects of novel agents, not for the discovery and proof of desired effects; however, animal testing for the latter has become routine in drug development. As depicted in Fig. 28.1, cancer drug discovery follows a somewhat routine process in which traditional mouse models play a central role; also depicted are a few of the roles that GEMMs are beginning to play in this process. Live mammals, especially rodents, will continue to be needed for toxicity screening of new agents for the foreseeable future, irrespective of which model systems or data were used to justify any therapeutic development campaign (Greaves et al. 2004). In other words, the need to avoid adverse reactions in the clinic is not in itself an argument for efficacy screening *in vivo*, although examples exist where unexpected systemic as well as organ-specific adverse activities have been first observed in live rodents, including GEMMs (Bolon 2004; Roberts et al. 2004). Pointing out the difference between therapeutic action discovery and toxicity testing – and the necessity of the latter irrespective of the former – is important in any discussion of *in vivo* model systems.

The choice of species used for adverse effect prevention has been influenced by many historical and empirical factors, including increasingly specific regulatory guidelines that have been put in place over several decades. For example, chronic toxicity testing in the USA today must be carried out in two mammalian species, one rodent – most frequently rats – and one nonrodent – most often dogs

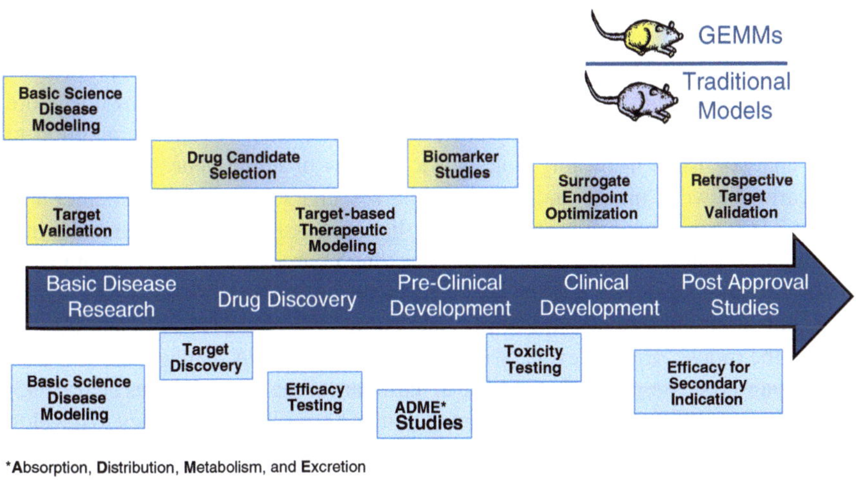

Fig. 28.1 Selected uses for traditional and GEMMs in cancer drug discovery and development

(http://www.regulations.gov/search/Regs/contentStreamer?objectId...pdf). Other species are sometimes used in cases where there is reason to believe that they more closely resemble certain physiological characteristics in specific organs, such as pigs, guinea pigs, hamsters, rabbits, sheep, and some nonhuman primates (Greaves et al. 2004). For cancer drugs, the selection of animal species for toxicity testing was guided in large part by systematic studies carried out in the mid-1960s demonstrating significant correlation between toxic effects observed in the mouse, rat, hamster, dog, and monkey, and that seen in humans (Freireich et al. 1966).

In contrast, a long list of organisms has been used as models of mechanism or efficacy for the discovery of potential anticancer agents, including the primary model organisms of genetics: *Escherichia coli*, Drosophila, mouse, zebrafish, *Xenopus*, and *Caenorhabditis elegans*. Going further back in history, models have included a large number of other organisms, including a variety of more exotic species such as the armadillo and the housefly (Storrs 1971; Mitlin and Baroody 1958). Over the past one and a half centuries, the laboratory mouse has grown into the undisputed primary *in vivo* model for anticancer testing. Aside from tradition and historical coincidence, this is due to a number of advantageous traits, including their small size, short gestation period, and modest housing requirements, which – combined with the availability of well-characterized inbred strains – makes mice an attractive species to breed and maintain. More importantly, though, more advanced genetic manipulation techniques are available for the mouse than any other mammalian species, and only a year after the first publication of the human genome sequence (Venter et al. 2001; Lander et al. 2001) the draft sequence of the mouse genome became available (Waterston et al. 2002). From this, we now know that murine chromosomes harbor syntenic regions for approximately 40% of the human genome, and that there is significant sequence homology between 80% or more of their gene products (de Jong and Maina 2010; Dennis 2006). Despite the many apparent similarities in organ systems and physiology between the two species, there are also many differences, both obvious and hidden, that should be kept in mind. These include body size, longevity, reproduction, olfaction, behavior, metabolism, and immunity – not to mention lifestyle, nutrition, exposure, and cognitive as well as immune memory. Many of these differences affect therapeutic agent behavior and response; for example, mice have an almost ten times faster heart beat and have been shown to more rapidly clear certain drugs (Jorge-Nebert et al. 2007) than humans. In addition to these characteristics in overall biology, the process of tumorigenesis in particular also varies between mouse and man in some key characteristics. The smaller size (about 3,000 times on average) and shorter life span (ca. 30–50 times) results in a significantly smaller number of mitoses occurring in an individual – 10^{11} compared to 10^{16} by some estimates. Yet, the lifetime cancer risk for both organisms is roughly the same, about 30% at the end of life, indicating greater cancer susceptibility per time in the mouse. Higher levels of oxidative and other DNA damage and less efficient repair processes are believed to be at least part of the reason (Ames et al. 1993; Adelman et al. 1988). It is also important that mice have a higher tendency to develop lymphomas or sarcomas, humans are more prone to carcinomas. The typical chromosomal aberrations, including aneuploidy and nonreciprocal translocations typical of most human solid tumors, are rare in mice. Interestingly, it has long been known that murine cells are easier to immortalize than human cell types, which more readily display replicative crises and senescence.

Although many oncogenes and tumor suppressor genes function in quite similar ways in the two species, the disease specifics are often different; for example, murine knockouts of the Rb gene product show pituitary as opposed to retinal tumors, and those of p53 more often develop soft tissue sarcomas rather than breast epithelial tumors. Much of this is believed to be a result of strikingly different telomerase systems. Murine telomeres are on average five times longer, and telomerase is expressed in most adult mouse tissues; in fact, telomerase-deficient mice show a more human-like malignancy spectrum. However, simply humanizing oncogenic components does not always lead to the appearance of human pathology in the mouse, and when transplanted into the mouse, human tumor tissue often undergoes changes in physiology, tissue architecture, organ specificity, latency, and malignancy.

28.2.1 Tumor Models for Compound Screening

The disease relevance of model systems that are employed in the preclinical testing of novel therapeutics has been the subject of much debate. This is true in all disease areas but has been particularly pronounced in the oncology arena (Hanauske 2000; Kamb 2005; Burchill 2006; Garber 2006; Dankort et al. 2009), where there are a number of critical issues that remain hot topics of debate regarding model choices for drug screening. These include:

1. *Phenotypic- versus target-driven screens.* As knowledge of tumor biology has grown over the past century, the field of anticancer agent testing has followed a discontinuous path of trends, favoring either phenotypic or mechanistic approaches (DeVita and Chu 2008). The earliest anticancer treatments were developed in a highly empirical fashion, guided by little mechanistic knowledge of the pathogenic progress. In pace with gradual improvements in the understanding of the molecular aspects of tumor formation, this was followed by efforts to identify molecularly targeted agents. Target-directed approaches have dominated novel anticancer agent research for the past two decades. Only recently, recognizing the daunting complexities of the processes shaping a cell's response behavior, hypothesis-free phenotypic drug screens have again found a place in the drug discovery arena (Root et al. 2003; Stockwell 2004; Bol and Ebner 2006). However, more often than not, most of these programs become target-directed later on as well, since even in the case of compounds initially derived from contemporary chemical genomics and systems biology initiatives, at least some knowledge on molecular target and mechanism of action is gained by the time they reach the later stages of preclinical testing. The influence of all of these factors is reflected closely in the history of the major drug testing programs of the last several decades, as described in more detail below.

Table 28.1 Preclinical drug testing platforms

Cell-free	Cell-based	3D cell colony assays	Live animal
• Enzymatic • Competitive protein binding • Kinetic	• Tumor cell lines • Primary tumor material • Lines with inducible vectors or transgenes • Gene expression • Protein expression • Cell death • High content assays (protein translocation)	• Soft agar • 3D culture • Tumor spheroids (homotypic and heterotypic) • Invasion	• Diffusion chamber • Hollow fibers • Murine tumors (spontaneous and induced) • Transplantable murine tumors (syngeneic and allogeneic) • Xenografts (heterotopic and orthotopic) • Primary tissue xenografts • GEMM

2. *Unique toxicity threshold for cancer drugs.* Levels of both systemic as well as organ-specific toxicities that would be deemed intolerable in most other disease areas are often considered acceptable for a cancer drug, which in some cases has positioned whole animal testing later in the development process (Olive and Tuveson 2006).
3. *Disappointing rate of improvement for cancer treatments.* As measured by its high ranking among the leading causes of death in the general population and massive societal efforts to find cures, advances in cancer therapy have been disappointing when compared to progress in other diseases. Epidemiological records have been taken in the USA for the past eight decades, during which overall mortality by the other major causes of death – infectious diseases in the earlier parts of the twentieth century, metabolic and cardiovascular diseases since – has decreased significantly (Kola and Landis 2004; Jemal et al. 2006). But 5-year survival rates for the main cancers have improved only marginally over the same time period, although the number has decreased modestly starting in 2003, for the first time since 1930. Cancer surpassed cardiovascular disease and stroke as the primary cause of death in people under 85 years of age in the early 2000s (Jemal et al. 2008). While this is in part attributable to an aging population and improved diagnostic methods, the lack of disease models with sufficient power to predict clinical behavior of candidate therapeutic agents is frequently cited as one important reason for our slow progress in the "war on cancer" (Leaf 2004).
4. *Individual cell-based nature of cancer etiology.* Finally, since cancer is generally recognized as a genetic disease originating from a single cell, or from clones of a few initiating cells, cell-based testing has always played a special role, and debates as to the relative impact of *in vitro* versus *in vivo* platforms continue to this day (Sharma et al. 2010).

Conceptually, preclinical testing platforms for new cancer drugs can most easily be categorized by their complexity and divided into four main types – (a) cell-free biochemical assays, (b) cell-based screens, (c) Clonogenic and organotypic (3D) assays, and (d) testing involving live animals. These are reviewed briefly below and outlined in Table 28.1.

1. *Cell-free biochemical assays.* Cell-free assays, in which collections of compounds are tested for their effect on an isolated target molecule, typically a pure protein in a test tube, have played an increasingly important role in primary screening for anticancer compounds over the past 20–30 years. This has been driven by: (a) the discovery of molecular targets, cellular structures, and pathways involved in the neoplastic disease process starting in the second half of the twentieth century, and (b) dramatic advancements in combinatorial chemistry, liquid handling and robotic technologies, and in the capture and handling of large datasets. The majority of targeted synthetic small-molecule compounds that have entered the clinic in recent decades have initially been identified in cell-free assays.
2. *Cell-based screens.* Assays based on the behavior of cells grown in culture have formed a mainstay of standard cancer drug screening for a much longer time period than cell-free biochemical assays. The culturing of animal cells became a routine procedure in the middle of the twentieth century, and the first human tumor cell line derived from a patient was grown in culture in 1951. Since then, thousands of immortalized cancer cell lines have been isolated and described, many of which have been maintained since. An important element of standardization was introduced to both the academic and private drug screening landscapes with the establishment of the NCI60 panel platform developed at the National Cancer Institute in the years 1986–1990; until today, cell lines of this panel are used more frequently than any others (Shoemaker 2006), although there has been much debate as to the physiological relatedness of these long-cultured lines to their tumors of origin. While there is little doubt that fast growing subclones have been selected from these lines, many of them appear to have reached a remarkably stable metabolic state and cytogenetic profiles (Macville et al. 1999; Roschke et al. 2002; Chen et al. 2006; Haddad and Yee 2008).

   Largely for practical and economic reasons, primary cells taken directly from individual patients – that have not been subjected to such selection in culture – have been used less frequently for drug screening, although some such efforts have continued over many years and have shown promise. In fact, several impressive examples exist where serial rounds of adaptive screening, with cells from patients that had acquired resistance to prior therapies, have successfully guided therapeutic drug selection (Fichtner et al. 2008; Rubio-Viqueira et al. 2006). Conceptually, such adaptive screening is comparable to the standard monitoring for adaptive resistance, and subsequent periodic changing of therapeutic agent, in many patients in the clinic today. But it provides the additional advantage of proactive preselection of treatment agents that are empirically deemed promising. This clinical course of action is of course only useful if screens can be completed by the time resistance is observed and new agents are needed. Moreover, whether individual patient-tailored screening initiatives can be economically viable outside the research hospital setting has yet to be seen.
3. *Three dimensional cell colony assays.* A large number of 3D culture systems have been developed in recent years due to an increasing appreciation for physiological consequences of tumor microenvironment and three dimensional tumor architecture, but only a few 3D culture systems have ever been used in drug testing.

The two most widely known methods are: soft agar colonies and multicellular spheroids.

Soft agar colonies based on a cellular ability to form anchorage-independent colonies have long been used as one of the basic tests of tumorigenic cell potential and are still frequently employed. Several older as well as some recent studies found such assays to predict clinical effectiveness of chemotherapeutics with impressive accuracy (Fiebig et al. 1987, 1990, 2004), but drug response is sometimes limited to only a subset of tumors, even from the same patient. Absent a clear demonstration of patient benefit, combined with the practical hurdles of poor tissue availability and low plating efficiency of many solid tumors, these assays were never widely accepted. However, with the tremendous advances in ultra-high throughput screening, small sample handling and sensitive detection methods that have been made recently, it may well be that this so-called clonogenic human tumor stem cell assay (HTSC) deserves renewed attention.

The most well-characterized and developed 3D culture technique is the multicellular tumor spheroid (MCTS) system, in which cultured tumor cells are allowed to form unattached spheroids. Tumor spheroids have been shown to simulate many pathophysiological and histological aspects of the *in vivo* growth of human tumors, and of their response to various therapies, and have been used in therapeutic programs modeling metabolic and chemical gradients, drug tissue penetration, tumor hypoxia, cell-to-cell and cell–matrix interactions and therapy resistance (Kunz-Schughart et al. 2004). While no large numbers of agents have yet been tested in any of these systems, industry-applicable large scale and high-throughput screening protocols have been developed (Friedrich et al. 2009). On the whole, 3D culture techniques for oncology drug screening are still in their infancy and have not nearly reached a level of importance or attention that some organ or tissue culture systems have reached in other pharmaceutical disciplines, such as the cardiovascular, musculoskeletal, nervous system, and metabolic disease areas or in wound healing and regenerative medicine.

In short, because of their particular advantages, platforms using primary cells or cell lines – in 2D or 3D – will likely be an important cancer drug screening approach for many years to come (Sharma et al. 2010; Balis 2002).

4. *Testing involving live animals*. The overwhelming majority of *in vivo* anticancer testing systems use the mouse. These mice either harbor endogenously grown tumors or xenografted tumors transplanted from an external source. Two of the simpler models using live mice are the diffusion chamber and the hollow fiber assay (HFA); although the term "animal culture" may be more appropriate a description for these systems than mouse models. Except in some basic research programs, diffusion chambers are rarely used in cancer screens today. The HFA has been developed and used routinely at the NCI, but less so elsewhere (Sausville and Feigal 1999; Suggitt and Bibby 2005).

   Endogenous mouse tumors, formed spontaneously or induced by radiation or carcinogenic agents, were used frequently in the first half of the last century, typically using murine leukemia. It is not surprising that these efforts produced a number of approved antileukemic compounds, several of which are still in clinical use. However, in more recent decades, endogenous mouse models have been used less frequently, particularly after the development of mouse/human xenograft models which allow for the direct examination of the behavior of tumors formed by human cancer cells in an animal model.

The most prominent *in vivo* model by far is the human tumor xenograft system in which human tumor cells, most often a culture from one of the established cancer cell lines, is allowed to grow in an immunocompromised but otherwise normal murine host (Morton and Houghton 2007). Injection of the human tumor cells typically occurs at an easily accessible site, most often subcutaneously, or, less frequently, intravenously or intraperitoneally. Orthotopic xenografts, transplanted into the organ environment from which the tumor was derived, have shown some specific advantages, although they are technically more challenging; to our knowledge, they have not been employed in any of the earlier stages of a larger therapeutic screening campaign.

The most technically straightforward and thus most popular xenograft methods use established tumor lines, e.g., members of the NCI60 panel. The bulk, if not all, cancer chemotherapeutic agents today have been tested in at least one xenograft system. As in the case of cell-based models, xenografts of primary tumor tissue from cancer patients have also been used in various therapeutic agent studies, with a number of promising results (Jin et al. 2010; Dong et al. 2010). However, since plating efficiency of primary tumor cells is sometimes low, and considering the regulatory and clinical hurdles of biopsy material procurement, the use of patient cell xenografts is not widespread.

28.2.2 Genetically Engineered Mouse Models

Starting with the generation of transgenic mice harboring external oncogenes, and knockout mice with targeted deletion of genomic loci, in the early and late 1980s, respectively, genetically engineered mice have been added as the newest addition to the *in vivo* arsenal of cancer testing models. The excitement generated by these technological milestones, combined with the wealth of new knowledge about tumorigenesis triggered several intense waves of further investigation and refinement of GEMMs. A large number of disparate GEMMs have been generated within the past two decades that can be categorized into a few basic types that here we call constitutive, conditional, chimeric, and complex. These categories are defined below and some common examples in each category are outlined in Table 28.2.

Constitutive. GEMMs carry heritable manipulations in all cells of the organism and in all stages of development. Using methods largely developed during the early years of GEMM development, knockout or transgenic clones for most, if not all, murine genes have been created, although it will be many years until they are exhaustively characterized.

Table 28.2 Categories of GEMMs

Category	Types of GEMM	Example of use in drug development
Constitutive	Germ line (e.g., APC^{min}) (Su et al. 1992)	Chemoprevention of aspirin, curcumin, and celecoxib in colon cancer (Corpet and Pierre 2005)
Conditional	Promoter driven (Cre-lox; Flp-frt) (Indra et al. 1999; Schmidt-Supprian et al. 2007)	HRAS and MYC oncogene dependence in melanoma (Chin et al. 1999), lymphoma (Felsher and Bishop 1999), breast (D'Cruz et al. 2001), and liver cancer (Jain et al. 2002)
Chimeric	Embryonic Stem Cell (EGFR, $BRAF^{V600E}$) (Zhou et al. 2010b)	Prediction of clinical activity for EGFR inhibitor (AV-412) in EGFR-L858R and KRAS-G12V models (Zhou et al. 2010b)
Complex	Human in mouse transplantation	Herceptin efficacy (Wu et al. 2009); MLL cell characterization (Barabé et al. 2007)

Conditional. Conditional models carry genomic changes that can be turned on or off externally, allowing temporal (inducible) and/or anatomic (tissue-specific) control of the altered gene.

Chimeric. Chimeric models are GEMMs with tissue of mixed species or mixed strain origin, including various types of mosaic, nongermline or transplanted systems. Most typically, they are generated by implantation of engineered embryonic stem cells into preimplantation embryos.

Complex. GEMMs are those complexes involving multiple heterogeneous components, including mouse-in-mouse or human-in-mouse systems, sometimes following genetic manipulation ex vivo, including advanced models of increasingly sophisticated levels of engineering, often comprising parts from various genetic or species backgrounds, such as immunocompromised and further engineered mice used as hosts for transplantation of tissue from distinctly engineered animals (mouse in mouse), or, of human tissue or cell lines following ex vivo manipulations (human in mouse). Into this category also fall model animals with significant levels of genomic humanization, containing for example not only human drug target genes, but also genes for entire components of the human adaptive or humoral immune system (Traggiai et al. 2004; Shultz et al. 2007) or for enzymes of the human detoxification apparatus (Powley et al. 2009).

Another way sometimes used to categorize genetically engineered animals is by contrasting transgenic models expressing genes in nonphysiological situations (e.g., from a foreign promoter) with endogenous systems, whose genome has been manipulated, but without the use of foreign strain or species elements (e.g., knockout or mutants expressed from their native promoter) (Frese and Tuveson 2007). Since the models called complex here virtually always contain material derived from different individuals, strains or species, they are generally "transgenic" by that definition.

28.2.3 Comparative Predictiveness

The clinical predictive value of preclinical model systems has been the subject of frequent and passionate debate. However, there have been few systematic studies that would provide definitive answers – i.e., comparing model readouts with known clinical activities – or even systematic scientific literature reviews that directly address this question in the form of a quasi meta-analysis. Furthermore, most such studies are dated and largely limited to the armamentarium of classical generally cytotoxic agents. Nevertheless, a few observations are worth noting. For one, while it is often stated that *in vivo* platforms overall are somewhat more predictive than *in vitro* assays, this is usually not based on evidence but assumption. In fact, studies demonstrating greater predictive clinical value of tumor cell line panels compared to xenografts have been reported (Voskoglou-Nomikos et al. 2003). A frequently cited analysis carried out at NCI compared cell line panels, the hollow fiber assay, and xenografts with known phase II or later clinical data (Johnson et al. 2001; Newell 2001). The HFA was found to be a good predictor of activity in xenografts but not of clinical activity. One of the primary positive conclusions was that the higher the number of xenografts for which a drug or drug candidate displays activity, the greater the chance of observing clinical activity. However, the same has long known to be true for small versus large panels of cell lines as well. This study is also notable in that is has been quoted repeatedly as a reference both in support of or against the predictiveness of mainstream cancer therapeutic screening processes (Kamb 2005; Sager and Lengauer 2003).

Much effort has also been devoted to define the relative strengths and weaknesses of GEMMs and xenografts (Rosenberg and Bortner 1999; Kerbel 2003; Richmond and Su 2008; Gopinathan and Tuveson 2008), even to the point of one journal publishing an eclipsing series of back-to-back opinion articles (Becher and Holland 2006; Sausville and Burger 2006). Even though much of this controversy was published in or around 2006, most of the main arguments remain relevant. For example, in virtually all of the cases where a GEMM provided a correct or explanatory observation that was missed in xenografts, the observations were made in retrospective studies, often long after a drug had been on the market and with much of its biology known. On the other hand, the argument – often used by xenograft defenders – that most drugs in clinical use today have gone through at least one xenograft – results from the fact that such testing was usually dictated by regulatory, not scientific considerations. Whether valuable therapeutic agents have failed to reach the clinic because they lacked activity in mice, either GEMMs or xenografts (but would have worked in humans), is impossible to know.

In summary, the comparative predictiveness of different models remains a subject with no straightforward synoptic answer. Almost all of the above models have contributed positively to the prioritization of compounds that were later approved, but some models, including GEMMs have, to date, garnered few well-documented predictive examples of use.

With definitive answers lacking, debates about the disease relevance of drug screening models are unlikely to subside anytime soon. However, there are common themes that can be pointed out: one of these is the notion that in cases where a novel therapeutic agent targets cells that are genetically identical or very similar to cancer patient cells, predictions of clinical efficacy are often very good (Kamb 2005; Kamb et al. 2007) – sometimes largely irrespective of model platform. A recent example of this is the remarkable response rate in a phase I trial of patients with oncogenic BRAF mutations to the B-RAF kinase inhibitor PLKX4032, which preclinically had been shown to selectively block the RAF/MEK/ERK pathway in BRAF mutant cells and to cause regression of BRAF mutant xenografts (Dankort et al. 2007). The durability of response has yet to be determined.

28.2.4 Summary of Historical Perspective

As illustrated above, the disease relevance of drug screening models is more hotly debated in oncology than in most other therapeutic areas and, historically, the anti-cancer agent discovery enterprise has not taken a straightforward course from using simple to increasingly complex models. Rather, choice of testing systems is, to this day, guided largely by available technology and growing knowledge about the disease. During the late nineteenth and early twentieth century, gradual advances in cancer treatment were achieved predominantly through surgery and radiation. However, progress from both neared a plateau in the second half of the last century, and most progress since has come from chemotherapy (DeVita and Chu 2008). The main historical phases of the latter can be divided into three main, distinct but overlapping periods, which may be called compound-centered, disease-oriented, and target and mechanism focused. These phases are reflected in the primary screening paradigms of the National Cancer Institute of the USA that have periodically changed over time; the most recent preclinical testing guidelines are reflected in a 2006 fact sheet (NCI 2006). No updates have been posted since. There may be a number of reasons for this, both scientific and political; but two contributing factors are undoubtedly the increased activity level on molecularly targeted therapies, for which more specialized test models had already been developed, and the increased availability and use of GEMMs specifically constructed for particular therapeutics under development.

For the foreseeable future, debate will continue concerning the optimal preclinical model for oncology. A major challenge in this regard is the alignment of measureable endpoints in the model versus the clinical situation (Table 28.3). Faithful representation of as many physiological, histological, genetic, and pharmacological disease parameters as possible is generally considered desirable, with qualities not related to the disease kept to a minimum. In addition, high penetrance (translating into sample homogeneity) and short latency (translating into testing speed) of *in vivo* cancer models are not only valuable but necessary because of the significant financial costs associated with program delays. By definition, the more faithfully a model is

Table 28.3 Endpoints: model versus clinical

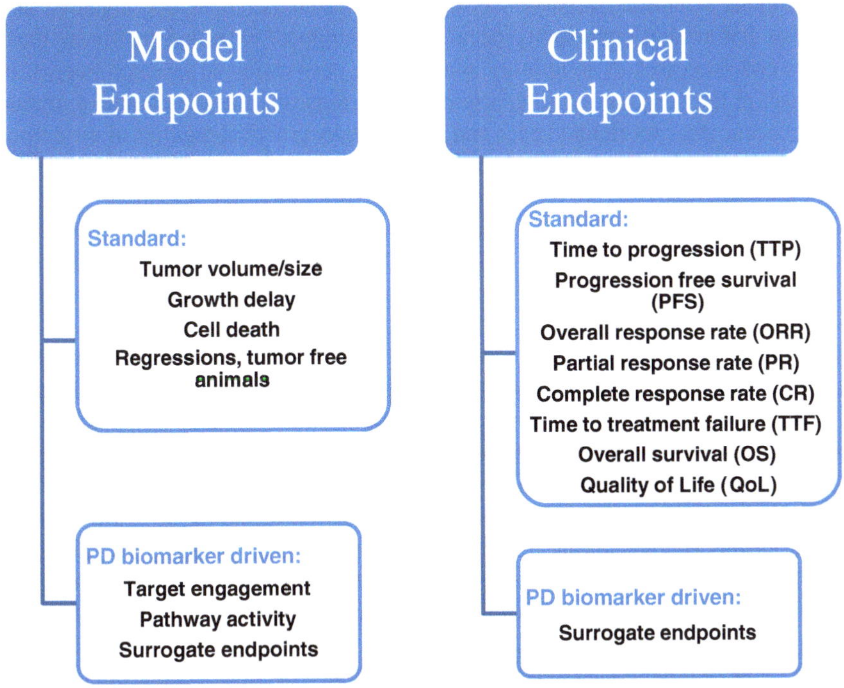

able to recapitulate all phases and aspects of a human disease, the more valuable it is as a platform for drug discovery and development. However, it is impossible to replicate all aspects in a single model, and it is not clear that this is needed in the testing of experimental therapeutics at different stages. There is a broad range of human disease characteristics that can be modeled in GEMMs, but in light of the practical realities of industrial drug development it is clear that currently the cost, legal hurdles, and other factors remain as serious roadblocks to their routine implementation in corporate drug development initiatives.

28.3 Realities of Industrial Drug Discovery

When deciding on disease models in the industrial environment, a number of economic, practical, temporal, legal, and regulatory considerations come into play that usually have little or no part in the decision making in a purely academic setting. In addition, scientific decision making at drug discovery companies is often influenced by perceived needs of the organization and trends in the business community and with potential pharmaceutical partners.

Scientific tradition and practices in academia reinforce the wisdom and autonomous scientific decision making of individual principal investigators, and success, failure and opportunity occur on the tempo of extended multiyear grant cycles and processes of submitting and revising peer reviewed research papers. In contrast, in the corporate setting, scientific decisions, such as the choice of animal model occur at an accelerated pace and are driven by practical realities of the corporate environment, such as quarterly budget decisions and by more subtle demands of a sociological matrix in the business world in which individual scientists, no matter how brilliant, accomplished, or qualified, rarely have the power to independently drive scientific strategy. Instead, a surprisingly broad set of people influence scientific decisions, including scientific or medical advisors, investors, members of the board of directors, multidepartment review committees, and corporate executives; it is also often the case that these individuals do not have appropriate credentials to understand the scientific decisions they are influencing and are often more comfortable with traditional methods regardless of the scientific logic of a new approach, such as a GEMM. Finally, during the past several years, the institutional instability of pharmaceutical and biotechnology companies, driven by the economic downturn, has added an additional, less than ideal influence on scientific decision making. In the ideal corporate setting, the choice of which animal model to use would be driven by a logical decision-making process that results in a choice that is scientifically relevant and provides meaningful results rapidly and within budget. In addition to human and sociological imperfections that stand in the way of this, there are also a range of other concrete practical hurdles.

Issues of legal freedom to operate, licensing and patent protection are particularly convoluted concerning the use of GEMMs. However, there is a growing infrastructure providing legal and practical assistance in sorting through these issues that is more readily available than just a few years ago (Deftos and Harvard 2001; Marshall 2002; Blaug et al. 2004). Starting in 2000, agreements between patent holders and federal agencies have greatly simplified the situation for academia. While significant statutory and regulatory hurdles remain for industry, technology adoption will become more widespread as a result of "safe harbor" provisions (Sharpless and Depinho 2006; Raubicheck et al. 2003) for research tools and the pending expiration of some key patents. While the first so-called Oncomouse patent, covering the derivation of cell cultures from transgenic animals (US 5,087,571) has recently expired, protection on using transgenic mice for testing remains in place until 2016 or later (US 5,925,803). Nevertheless, it is without question that more biotechnology firms would already have used GEMM technology if they could afford the licensing fees.

Another major practical consideration for drug discovery is that the models for therapeutic screening should ideally be simple and easy to use, give unambiguous results, and must provide actionable information, i.e., data to support the often-called go/no go-decision-making process. But when it comes to infrastructure and resource constraints, some common challenges are less a function of public versus private settings, but one of institutional sizes, funding and stage of development.

However, for small companies, these issues can be a matter of institutional life and death. Only large pharmaceutical corporations, universities, or national laboratories and centers can afford the luxury of maintaining extensive panels of *in vivo* test platforms. This has been true for some time, even prior to the advent of GEMMs. For example, while small biotech firms engaged in the anticancer field typically employ a few xenograft types, lead compounds in large companies are often tested through panels of dozens of different models. This is a remarkable fact for several reasons, one being that an unfortunate consequence of this has been that because it is often possible to find at least one xenograft model in which a compound is active (Francia and Kerbel 2010), cases where therapeutic campaigns have been pushed forward on the basis of observations made in one out of many models are not unusual. In this context, the incorporation of expensive and sophisticated GEMMs may only increase such disparity between large and small institutions.

This also leads to the intriguing observation that there has not been a clear positive correlation between number and sophistication of models used and the success of a therapeutic development program, and sometimes the correlation is negative. Yet in recent years, it has been the smaller innovative companies from whence a significant percentage of novel therapeutic entities have originated (Kola and Landis 2004; Hughes 2009, 2010). This is particularly apparent in oncology compared to other disease areas particularly with recently approved anticancer biologicals.

A final factor that influences decisions about using GEMMs are the ethical considerations associated with the use of whole live mammalian species (de Dios 2002; Workman et al. 2010). These can sometimes pose greater complications for studies carried out in for-profit settings compared to academic or government labs. This is the case even though the number of experimental and control animals used in most therapeutic testing programs is typically smaller than those required for answering basic biology questions. Ultimately, as a result of public perception and policy debate, the number of animals in medical, consumer product, and public health testing had been measurably reduced over the course of the last two decades. However, this trend has stalled in recent years and is likely to be reversed, in large part owing to the growing number of use areas for mouse models due to the increasing power of GEM techniques. Moreover, in light of the various national and international genome-scale mouse phenotyping initiatives now underway the number of animals required may in fact rise drastically (Beckers et al. 2009).

Even with all the complications noted above, one scenario that would be likely to provide a boost to routine adoption of GEMMs in drug discovery campaigns would be a case in which a promising therapeutic activity is discovered in such a model that would have been missed in all other model platforms. However, to our knowledge, such examples are still lacking. An important reason for this, in addition to the hurdles reviewed above, is the fact that the genetics of mice and their correlation to predicting the behavior of therapeutics in humans remains a complicated issue that has only become more convoluted by an explosion of recent studies and data concerning mouse genomic expression and genetic knockouts. In 2003, a systematic study of the phenotypes of mice with knockouts of the targets (where known) for the then 100 best-selling drugs found that they correlated well with drug efficacy (Zambrowicz and Sands 2003), and a subsequent analysis of the major pipeline drugs of the ten largest pharmaceutical companies made similar observations (Zambrowicz 2003). These retrospective analyses may, however, have carried some bias in that the generation of knockout animals could have been more likely in cases, where a therapeutic rationale already existed from a substantial body of physiological knowledge. In addition, only four of the drugs were anticancer agents. Historically, about half of the constitutive germ line knockouts that by now have been generated for almost all murine genes have shown unexpected, no, or even paradoxical phenotypes. This is especially relevant for oncology since genetic lesions are usually found only in the cancer cells not in all cells, and many key cancer genes encode oncofetal signaling molecules – receptors, and growth and transcription factors with crucial roles during embryonic and early development that often differ from their adult function (Ebner and Ried 2007). In contrast to widespread views from a decade ago, we now know that drastic, early lethal phenotypes of target mutants do not discount therapeutic potential of a target. The early embryonic lethality of knockouts for the targets of recent blockbuster anticancer monoclonal antibodies (Miettinen et al. 1995; Fong and Kakar 2009; Shalaby et al. 1995) illustrates this well. The title of one of the above-mentioned studies ended with the question "will they model the next 100?", expressing a then much-anticipated hope. However, this prediction has clearly not come true in the intervening years. Even though correlations between *in vivo* characteristics of target mutations and drug action were eventually found for many novel drugs and drug candidates, the majority of GEMM studies on the target of newly approved agents since were retrospective or entirely unrelated to the therapeutic development project. It is also worth noting that, because they are labor-intensive and time-consuming, some large corporations have excluded oncology assays from their otherwise large-scale and genome-wide mouse phenotyping programs (Sacca et al. 2010).

To be clear, it is indisputable that many and sometimes long sought new answers in basic cancer biology were obtained from GEMM studies, along with a myriad path-breaking new questions. Yet on the whole, the promise of an industry-reshaping and productivity-boosting impact of GEMMs on anticancer therapeutic development, frequently announced during the last two decades, has yet to become a reality, at least in terms of efficacy prediction for new molecular entities. On the other hand, these models have already made a mark in some of the later stages of the drug discovery process, for example in target validation, resistance modeling and explanation of unanticipated toxicities, oncogene addiction and tumor maintenance modeling, patient stratification, and simulating combination therapies. The area where GEMMs have had the most visible influence on both biotechnology and pharmaceutical oncology has been in the discovery, preclinical and co-clinical validation, and refinement of molecular biomarkers – of disease status, compound action and pharmacokinetics, therapy response and prognosis.

28.4 Biomarkers

The use of GEMMs in the discovery, characterization, and validation of biomarkers holds great promise for enabling the more effective use of therapeutics in the clinical setting. This is particularly true for the early stages of drug development, where animal models are a significant and necessary source of information about both biological and pharmacologic and toxicologic activity of therapeutic molecules. Biomarkers for these uses can be broadly defined as markers that measure biological processes and are used to monitor disease or therapy, and include molecular markers measuring, for example, glucose, DNA, RNA, or protein or phenotypic tests, measuring more generalized effects using techniques, such as traditional stains for tracking tissue pathology or MRI or other imaging technologies.

While animal models have been extensively used in the preclinical testing of therapeutics for both disease activity and for ADME and toxicology studies, the use in developing and validating markers of drug activity and sensitivity for making drug development decisions or for clinical trial design has only recently become commonplace. There are several potential roles for GEMMs for biomarker use in drug discovery, including correlating target inhibition with mechanistic results (how much/long do I have to inhibit my target to have the desired effect?), pharmacodynamic (how much/how long do I have to give drug to inhibit target?), and resistance (what determines drug sensitivity?). These are properties that cannot be easily evaluated in early clinical trials because of the lack of both number and homogeneity of patients. Animal models, by contrast, make it possible to quickly and easily manipulate conditions of each assay, and to readily monitor the biomarkers in the context of multiple therapeutic interventions and conditions. This has the potential to facilitate selection of those biomarkers that have the greatest promise for robust performance, and importantly for providing a model system for characterizing and validating the biomarker before entry into testing within clinical trials.

In practice, GEMMs have both potential advantages and disadvantages for use in biomarker studies (Van Dyke and Jacks 2002; Frese and Tuveson 2007; Jonkers and Berns 2002). The most significant disadvantage is that deleterious mutations in mice often result in a tissue distribution significantly different than is found in human tumors; they are therefore unlikely to be of great use when modeling the human condition and have the potential to modulate biomarkers that are not representative of the human disease. The advent of newer technologies for generating conditional genetic disruptions may help alleviate some of these concerns (Sharpless and Depinho 2006; Talmadge et al. 2007; Walrath et al. 2010). The potential difference in reagents for detecting the biomarkers in mouse models and human cells can also cause difficulties when attempting to translate biomarkers to the clinical setting (Mandrekar and Sargent 2009; Simon 2010; Tan et al. 2009; Alymani et al. 2010). This is particularly true when antibody-based reagents are used as they have a significant potential to have species-specific performance differences.

In some circumstances, GEMMs do provide advantages in biomarker studies as has been demonstrated for different aspects of drug development including: preclinical selection and qualification of candidate therapeutics, identification and validation of clinical biomarkers, identification and validation of companion biomarkers, and identification of surrogate endpoints for drug optimization.

1. *Preclinical selection and qualification of candidate therapeutics.* Incorporation of biomarkers in preclinical selection and qualification of candidate therapeutics is now frequently seen. The ability to monitor downstream pathway activity and to determine the extent of drug exposure necessary for target modulation and phenotypic outcome is an important part of candidate selection. An example of a GEMM employed in this regard is the use of HER2/INK4a-driven breast tumors to evaluate gamma secretase inhibitors and identify molecular and imaging biomarkers of compound activity (Watters et al. 2009; Efferson et al. 2010), or the evaluation of inhibitors of the Ras/Raf signaling pathway and downstream biomarkers of pathway inhibition (Karreth and Tuveson 2009).
2. *Identification and validation of clinical biomarkers.* For the validation of clinical biomarkers, particular predictive biomarkers have been identified from a series of cancer types by monitoring changes in either gene expression or the serum proteome. Transcriptional changes from a large number of different GEMMs have been studied, including p53 and Kras, that induce changes similar to human tumors, and can be used to define biomarkers for specific cancer pathways (García-Escudero et al. 2010). Several studies have been done comparing the serum proteome of mice carrying representative defects for the particular cancer type and confirming these observations in human clinical samples. For example, the K-rasLSLG12D/+ model of pancreatic and lung cancer and pancreatic insulinomas in the RIP1-Tag2 model provided platforms for the identification of serum biomarkers correlated with human disease (Pitteri et al. 2009; Faca et al. 2008). Studies of this type are necessary because the proteomic profiling of human serum for biomarker discovery is challenging due to genetic and environmental heterogeneity among patients and their tumors. GEMMs, which have homogeneous background and controlled environmental conditions, allow the more robust and relevant markers to be selected and manipulated in an animal model system to confirm the robustness of the identified markers (Singh and Johnson 2006; Weiss 2003; Firestone 2010).
3. *Identification of surrogate endpoints for drug optimization.* GEMMs have also been found useful for tagging of cells with bioluminescent molecules to enable the monitoring of specific cell types or cellular states (Singh et al. 2010). Imaging systems then allow the response of these cells to be monitored in response to therapeutic intervention or environmental changes (Maggi and Ciana 2005). This has been used recently to identify markers of tumor stem cells and monitor their fate during development or treatment (Haegebarth and Clevers 2009).

The availability of biomarker data from preclinical models serves as an avenue for validation and corroboration of those identified in early clinical trials (Walrath et al. 2010). This allows candidate clinical biomarkers to be tested in relevant animal models not only for performance characteristics, but also for the response to intervention by candidate therapeutics and generation of statistically relevant data in clinical trials. GEMMs that develop relevant disease models are ideal for the selection and validation of a "fit-for-purpose" companion biomarker strategy allowing the robust qualification of the markers before clinical trials begin. For example, modeling the resistance of EGFR inhibitors to mutations in the Ras signaling pathway has helped guide clinical care to the testing and selection of patients that are Ras wild type for treatment with therapies targeting EGFR (Singh et al. 2010).

Mechanistic studies that rely on the use of tissue or tumors with known genetic mutations are well suited to be carried out in GEMMs. For example, antibodies directed against HER2 were shown to be effective in tumors overexpressing HER2, leading to a clinical strategy which enriched patients based on a molecular biomarker (Singh and Johnson 2006). BRCA1 and BRCA2 defective cells were shown to be particularly sensitive to preclinical models designed for PARP inhibitors, leading to successful clinical trials in patient populations enriched for these molecular markers (Rottenberg et al. 2010). The Min mouse, which recapitulates the APC mutation found in familial adenomatous coli and the majority of sporadic colon cancers, has been used frequently for the identification of biomarkers and testing of candidate therapeutics (McCarthy 2009; Uronis and Threadgill 2009) (see Chapter 15).

The use of GEMMs, in particular conditional and chimeric models, holds potential for the identification and characterization of biomarkers for diagnostic assays for cancer detection or relapse, and importantly for pharmacodynamic monitoring during the development and use of novel-targeted therapeutics that cannot be easily accomplished in the clinical setting. The advantages of using GEMM for biomarker discovery and validation over less expensive more simplistic xenograft models must be evaluated on a case-by-case basis, with more extensive use in biomarker characterization awaiting the further evolution of GEMMs that more accurately recapitulate human tumors, and are more functionally accessible.

28.5 Success Stories

A large number of clinical outcomes have been modeled closely in GEM systems. Although the majority of these studies were retrospective and did not directly impact preclinical decision points or clinical trial strategy, there are now a number of cases of prospective information from preclinical and co-clinical studies that were valuable for dosing, timing, patient selection, and combination therapy. For example, using a transgenic mosaic model, the treatment of mice harboring alterations common in human acute myelocytic leukemia was used to accurately predict chemotherapy response in patients (McCarthy 2009; Zuber et al. 2009). In another example, the treatment of chimeric mouse tumor models transgenic for activating EGFR or KRAS mutations – as a model of Kras-driven lung adenocarcinoma – with a receptor tyrosine kinase inhibitor showed mutation-specific differences in sensitivity that correlated closely with clinical observations, revealing patient subtype-specific pathway signaling status and offering guidance for future treatment options (Zhou et al. 2010a; Cheng et al. 2010).

Work with transgenic mouse models of acute promyelocytic leukemia expressing defined retinoic acid receptor mutants or fusion proteins has helped explain the mechanism and the differential sensitivity of patient subsets to differentiation therapy with retinoic acid (Rego et al. 2000; Curing 2009; Nasr et al. 2008) and has prompted the use of combined arsenic and RA treatment for APL patients (Lallemand-Breitenbach et al. 1999).The necessity and value of GEMMs of human hematopoietic malignancies is sometimes put into question, since with most human leukemia a single, relatively homogeneous cell population is ultimately targeted in treatment. However, a substantial number of such "co-clinical" trials, including several on solid tumors, are now underway (Zielinska et al. 2011).

A highly informative comparative study published earlier this year sought to determine whether the outcome of phase 3 clinical trials could be improved with the use of properly designed GEMMs (Francia and Kerbel 2010; Singh et al. 2010). Here, two conditional mouse models designed to simulate Kras-driven nonsmall-cell lung carcinoma and pancreas adenocarcinoma, respectively, were treated with standard chemotherapeutic agents alone or in combination with inhibitors of EGFR and VEGFR. Comparing their results with data from several recent phase 3 trials, the authors found close correlation between trial results and results from many, though not all, of their models. In parallel, response to treatment with the same agents or combinations was also tested in xenografts and was largely consistent. While a clear advantage of the more expensive GEMMs could not yet be demonstrated, only a limited number of xenograft types and treatment modalities were used.

This last study is remarkable also in another important aspect: the authors monitored survival and progression-free survival, two very common outcome criteria determined in clinical trials, in their mouse models. The measurements taken in clinical studies, for example tumor shrinkage or growth delay or surrogate molecular markers, have more often than not been very different from those that are taken during clinical trials (Table 3), and it is very likely that some of the predictive weaknesses of the standard preclinical models could be ameliorated by including multiple testing parameters, in addition to ensuring that clinically relevant parameters like drug dose, delivery, and pharmacokinetics are kept as close as possible between model and patient.

28.6 Summary and Outlook

As noted in many places in this chapter and book, GEMMs hold great promise to further our understanding of human tumors and for converting that understanding into better medicines. However, for the reasons reviewed above, it seems unlikely that GEMM will transform drug discovery and development as quickly as some optimistic reports have predicted because of the complicated nature of the disease and the procedural and regulatory requirements of the drug discovery process. However, this point of view must be tempered with the fact that corporations often do not publish preclinical experimental data on emerging drug candidate and when they do so it is often years after the fact. It is likely that there are important GEMM experiments that are or have been done in a corporate setting that are not reviewed here simply because they are not published. Also, there are already some public examples in which the use of GEMMs have influenced acceptance or understanding of an already established drug (Heyer et al. 2010), giving hope that these models will become more routinely useful in earlier stages of drug development in the future.

As with many other emerging technologies, the adoption and use of GEMMs is influenced considerably by a range of factors. A nontrivial challenge to the widespread effective use of GEMMs is the fact that intellectual property ownership and licensing issues can stand in the way of the adoption of a particular model. Also, the NIH and FDA will continue to wield considerable influence over the type of models that are developed and how they are used for basic early stage analysis of drugs and their mechanisms of action and as part of the approval process. In recent years, beginning with the Human Genome Project in the 1990s, the NIH has embraced various "big science" projects that entail centrally directed efforts to solve particular scientific questions that are perceived to be answerable only through a major centrally coordinated federally funded effort. It is conceivable that GEMMs could be the subject of, or part of, a massive federal initiative to develop better tools for cancer drug discovery and development. In fact, there have been some efforts in this direction (Marks 2009); to date, there has been no effective program to make models readily available outside of federally funded laboratories. It would likely be a great boon to the improvement of cancer medicines if such models and service based on those models were to be provided to industry through a true public–private partnership. Relevant to this question, we note that the vast majority of new medicines for all indications, including cancer, for the past 40 years have been discovered and developed in the corporate setting (DeVita and Chu 2008), a trend that we believe is likely to continue. In this context, we believe the recent trend at NIH to fund drug discovery and development work (Minig et al. 2010), would be better directed toward creation of better tools for testing therapeutics, including GEMMs, data, and basic scientific discoveries that would be made readily available to all researchers in academic, federal, and industrial settings.

Whether the setting is academia, federal laboratory, or industry and regardless of whence the tools are developed and supplied, the central challenge of developing better medicines for cancer remains a profound lack of screening and preclinical modeling tools that provide accurate reliable predictions about the behavior of drugs in the clinic. In the current era of molecularly targeted anticancer drug development, it is astounding that still only one of every ten drugs that is assessed in Phase I clinical trials becomes approved for use by the FDA (Kamb et al. 2007; Caskey 2007). This high rate of attrition is due mainly to lack of efficacy, not to intolerable toxicity as was prevalent in the era of cytotoxic drug development less than 20 years ago. Though we have seen much progress in the field with regards to the development of less toxic, targeted agents, this change in the drug development paradigm has led to failures of drugs later in the drug development cycle. This has led to an ever escalating cost to reach an NDA (Kola 2008). Central to this problem is the dearth of *in vivo* models that accurately reflect clinical cancer types and subtypes and that accurately predict the activity of therapeutics in clinical trials.

Ideally, preclinical tumor models, including GEMMs, would exactly mimic the properties that contribute to the initiation, progression, and drug response in human cancer. Most importantly, the ideal mouse model would provide clear and easily interpretable data describing pharmacokinetic and pharmacodynamic properties, which are paramount in defining drug dosing, scheduling, and toxicity in preclinical studies. Current models only weakly predict these properties, and significantly better models are unlikely to be generated in the very near future. For this reason, it is vital that researchers understand the limitations and true utility of their models for drug development. We argue above that mouse/human xenograft models are of limited value for the prediction of clinical efficacy (at least as typically used in most settings), but these models can be quite useful for tumor PK and PD studies; in this regard, GEMM models may be too costly and have other limitations for routine use for preclinical drug efficacy testing, but their development will likely offer excellent tools in other ways. For example, greater use of GEMMs to generate models of novel targets, clinically relevant genetic lesions, and for chemoprevention modeling and in mimicking human pharmacogenetic variation should be encouraged. The modeling of combination therapies is another promising but as yet underused application. In addition, the ex vivo screening of GEMM-derived material containing humanized tissue or organs will become more frequent.

Genetically engineered species other than the mouse will also find growing application in therapeutic development; this trend has already generated transgenic pigs and rats for testing (Zan et al. 2003) but will likely extend to other species that have long been considered better models of organ-specific responses – dogs for modeling cardiotoxicity or nephrotoxicity, for example. Finally, standardization of systems for the acquisition, use and interpretation of GEMMs and unified guidelines for preclinical applications would almost certainly enhance power and acceptance of models.

A particularly promising future use of GEMMs is in the development of new tools for the genetic characterization of cancer coupled with functional characterization of drugs and drug targets. This use has already yielded much new insight into the pathogenesis of cancer. For example, recent studies using humanized mice have lead to a better understanding of how immunotherapies and humanized antibodies function in the treatment of cancer. Though improvements, such as the inclusion humanized MHC genes and cytokines, are needed to fully realize the potential of the humanized mouse model, it is likely to become important for preclinical development of biotherapeutics in the very near future. Currently, we are also witnessing a shift in clinical trial design from a pathology-based process to biomarker trials using biomarkers such as mutant Kras and BRCA1 in order to identify patient populations more likely to respond to targeted therapeutics. These breakthroughs, in large part, are due to the concomitant and rapid advances in our knowledge of cancer biology and continued development of new mouse models, including GEMMs, that allow the mechanism of drugs to be elucidated at the molecular level and targeted to individual patients. We have entered the personalized medicine era and only our continued progress in understanding cancer and anticancer therapeutics has the potential to allow the mouse model to become a bridge spanning the gap between targeted drug discovery and development in the laboratory and targeted patient populations in the clinic. In this era, a refocusing of the enormous modeling power of GEMMs – from answering basic questions in tumor biology to therapeutic questions – may open up exciting therapeutic development opportunities.

References

Adelman R, Saul RL, Ames BN (1988) Oxidative damage to DNA: relation to species metabolic rate and life span. Proc Natl Acad Sci USA 85(8):2706–2708

Alymani NA, Smith MD, Williams DJ, Petty RD (2010) Predictive biomarkers for personalised anti-cancer drug use: discovery to clinical implementation. Eur J Cancer 46(5):869–879

Ames BN, Shigenaga MK, Gold LS (1993) DNA lesions, inducible DNA repair, and cell division: three key factors in mutagenesis and carcinogenesis. Environ Health Perspect 101(Suppl 5): 35–44

Balis FM (2002) Evolution of anticancer drug discovery and the role of cell-based screening. J Natl Cancer Inst 94(2):78–79

Barabé F, Kennedy JA, Hope KJ, Dick JE (2007) Modeling the initiation and progression of human acute leukemia in mice. Science 316(5824):600–604

Becher OJ, Holland EC (2006) Genetically engineered models have advantages over xenografts for preclinical studies. Cancer Res 66(7):3355–3358, discussion 3358–9

Beckers J, Wurst W, de Angelis MH (2009) Towards better mouse models: enhanced genotypes, systemic phenotyping and envirotype modelling. Nat Rev Genet 10(6):371–380

Blaug S, Chien C, Shuster MJ (2004) Managing innovation: university-industry partnerships and the licensing of the Harvard mouse. Nat Biotechnol 22(6):761–764

Bol D, Ebner R (2006) Gene expression profiling in the discovery, optimization and development of novel drugs: one universal screening platform. Pharmacogenomics 7(2):227–235

Bolon B (2004) Genetically engineered animals in drug discovery and development: a maturing resource for toxicologic research. Basic Clin Pharmacol Toxicol 95(4):154–161

Bolon B, Galbreath E (2002) Use of genetically engineered mice in drug discovery and development: wielding Occam's razor to prune the product portfolio. Int J Toxicol 21(1):55–64

Burchill SA (2006) What do, can and should we learn from models to evaluate potential anticancer agents? Future Oncol 2(2):201–211

Caskey CT (2007) The drug development crisis: efficiency and safety. Annu Rev Med 58:1–16

Chen Q, Watson JT, Marengo SR et al (2006) Gene expression in the LNCaP human prostate cancer progression model: progression associated expression in vitro corresponds to expression changes associated with prostate cancer progression in vivo. Cancer Lett 244(2):274–288

Cheng L, Ramesh AV, Flesken-Nikitin A, Choi J, Nikitin AY (2010) Mouse models for cancer stem cell research. Toxicol Pathol 38(1):62–71

Chin L, Tam A, Pomerantz J et al (1999) Essential role for oncogenic Ras in tumour maintenance. Nature 400(6743):468–472

Corpet DE, Pierre F (2005) How good are rodent models of carcinogenesis in predicting efficacy in humans? A systematic review and meta-analysis of colon chemoprevention in rats, mice and men. Eur J Cancer 41(13):1911–1922

Curing KSC (2009) Curing APL: differentiation or destruction? Cancer Cell 15(1):7–8

D'Cruz CM, Gunther EJ, Boxer RB et al (2001) c-MYC induces mammary tumorigenesis by means of a preferred pathway involving spontaneous Kras2 mutations. Nat Med 7(2): 235–239

Dankort D, Filenova E, Collado M, Serrano M, Jones K, McMahon M (2007) A new mouse model to explore the initiation, progression, and therapy of BRAFV600E-induced lung tumors. Genes Dev 21(4):379–384

Dankort D, Curley DP, Cartlidge RA et al (2009) Braf(V600E) cooperates with Pten loss to induce metastatic melanoma. Nat Genet 41(5):544–552
de Dios ES (2002) The use of animal models in cancer research. Clin Transl Oncol 4(2):55–58
de Jong M, Maina T (2010) Of mice and humans: are they the same? – implications in cancer translational research. J Nucl Med 51(4):501–504
Deftos LJ, Harvard V (2001) Canada: the myc mouse that still squeaks in the maze of biopatent law. Acad Med 76(7):684–692
Dennis C (2006) Cancer: off by a whisker. Nature 442(7104):739–741
DeVita VT, Chu E (2008) A history of cancer chemotherapy. Cancer Res 68(21):8643–8653
Dong X, Guan J, English JC et al (2010) Patient-derived first generation xenografts of non-small cell lung cancers: promising tools for predicting drug responses for personalized chemotherapy. Clin Cancer Res 16(5):1442–1451
Drews J (2000) Quo vadis, biotech? (Part 1). Drug Discov Today 5(12):547–553
Ebner R, Ried T (2007) Editorial. Drug Discov Today Dis Mech 4(4):259–260
Efferson CL, Winkelmann CT, Ware C et al (2010) Downregulation of Notch pathway by a gamma-secretase inhibitor attenuates AKT/mammalian target of rapamycin signaling and glucose uptake in an ERBB2 transgenic breast cancer model. Cancer Res 70(6):2476–2484
Faca VM, Song KS, Wang H et al (2008) A mouse to human search for plasma proteome changes associated with pancreatic tumor development. PLoS Med 5(6):e123
NCI Fact Sheet[Internet] (2006) http://www.cancer.gov/cancertopics/factsheet/nci/drugdiscovery
Felsher DW, Bishop JM (1999) Reversible tumorigenesis by MYC in hematopoietic lineages. Mol Cell 4(2):199–207
Fichtner I, Rolff J, Soong R et al (2008) Establishment of patient-derived non-small cell lung cancer xenografts as models for the identification of predictive biomarkers. Clin Cancer Res 14(20):6456–6468
Fiebig HH, Schmid JR, Bieser W, Henss H, Lohr GW (1987) Colony assay with human tumor xenografts, murine tumors and human bone marrow. Potential for anticancer drug development. Eur J Cancer Clin Oncol 23(7):937–948
Fiebig HH, Berger DP, Winterhalter BR, Plowman J (1990) In vitro and in vivo evaluation of US-NCI compounds in human tumor xenografts. Cancer Treat Rev 17(2–3):109–117
Fiebig HH, Maier A, Burger AM (2004) Clonogenic assay with established human tumour xenografts: correlation of in vitro to in vivo activity as a basis for anticancer drug discovery. Eur J Cancer 40(6):802–820
Firestone B (2010) The challenge of selecting the "right" in vivo oncology pharmacology model. Curr Opin Pharmacol 10(4):391–396
Fong MY, Kakar SS (2009) Ovarian cancer mouse models: a summary of current models and their limitations. J Ovarian Res 2(1):12
Fox RR, Tucker FS (1984) Atropine esterase status of laboratory mice. Lab Anim Sci 34(4): 381–382
Francia G, Kerbel RS (2010) Raising the bar for cancer therapy models. Nat Biotechnol 28(6): 561–562
Freireich EJ, Gehan EA, Rall DP, Schmidt LH, Skipper HE (1966) Quantitative comparison of toxicity of anticancer agents in mouse, rat, hamster, dog, monkey, and man. Cancer Chemother Rep 50(4):219–244
Frese KK, Tuveson DA (2007) Maximizing mouse cancer models. Nat Rev Cancer 7(9):645–658
Friedrich J, Seidel C, Ebner R, Kunz-Schughart LA (2009) Spheroid-based drug screen: considerations and practical approach. Nat Protoc 4(3):309–324
Garber K (2006) Realistic rodents? Debate grows over new mouse models of cancer. J Natl Cancer Inst 98(17):1176–1178
García-Escudero R, Martínez-Cruz AB, Santos M et al (2010) Gene expression profiling of mouse p53-deficient epidermal carcinoma defines molecular determinants of human cancer malignancy. Mol Cancer 9:193
Gopinathan A, Tuveson DA (2008) The use of GEM models for experimental cancer therapeutics. Dis Model Mech 1(2–3):83–86

Greaves P, Williams A, Eve M (2004) First dose of potential new medicines to humans: how animals help. Nat Rev Drug Discov 3(3):226–236
Haddad TC, Yee D (2008) Of mice and (wo)men: is this any way to test a new drug? J Clin Oncol 26(6):830–832
Haegebarth A, Clevers H (2009) Wnt signaling, lgr5, and stem cells in the intestine and skin. Am J Pathol 174(3):715–721
Hanauske A (2000) In vitro and in vivo predictive tests. In: Bast RC, Kufe DW, Pollock RE (eds) Cancer medicine, 5th edn. BC Decker, Hamilton, ON
Heyer J, Kwong LN, Lowe SW, Chin L (2010) Non-germline genetically engineered mouse models for translational cancer research. Nat Rev Cancer 10(7):470–480
Hughes B (2009) 2008 FDA drug approvals. Nat Rev Drug Discov 8(2):93–96
Hughes B (2010) 2009 FDA drug approvals. Nat Rev Drug Discov 9(2):89–92
Indra AK, Warot X, Brocard J et al (1999) Temporally-controlled site-specific mutagenesis in the basal layer of the epidermis: comparison of the recombinase activity of the tamoxifen-inducible Cre-ER(T) and Cre-ER(T2) recombinases. Nucleic Acids Res 27(22):4324–4327
Jain M, Arvanitis C, Chu K et al (2002) Sustained loss of a neoplastic phenotype by brief inactivation of MYC. Science 297(5578):102–104
Jemal A, Siegel R, Ward E et al (2006) Cancer statistics, 2006. CA Cancer J Clin 56(2):106–130
Jemal A, Siegel R, Ward E et al (2008) Cancer statistics, 2008. CA Cancer J Clin 58(2):71–96
Jin K, Teng L, Shen Y, He K, Xu Z, Li G (2010) Patient-derived human tumour tissue xenografts in immunodeficient mice: a systematic review. Clin Transl Oncol 12(7):473–480
Johnson JI, Decker S, Zaharevitz D et al (2001) Relationships between drug activity in NCI preclinical in vitro and in vivo models and early clinical trials. Br J Cancer 84(10):1424–1431
Jonkers J, Berns A (2002) Conditional mouse models of sporadic cancer. Nat Rev Cancer 2(4): 251–265
Jorge-Nebert LF, Derkenne S, Nebert DW (2007) Chapter 16. Drugs and the Mouse: Pharmacology, Pharmacogenetics, and Pharmacogenomics. In: Fox JG, Muriel T, Davisson MT, Fred W, Quimby FW, Stephen W, Barthold SW, Christian E, Newcomer CE and Smith AL. The Mouse in Biomedical Research (Second Edition) History, Wild Mice, and Genetics, Elsevier Inc
Kamb A (2005) What's wrong with our cancer models? Nat Rev Drug Discov 4(2):161–165
Kamb A, Wee S, Lengauer C (2007) Why is cancer drug discovery so difficult? Nat Rev Drug Discov 6(2):115–120
Karreth FA, Tuveson DA (2009) Modelling oncogenic Ras/Raf signalling in the mouse. Curr Opin Genet Dev 19(1):4–11
Kerbel RS (2003) Human tumor xenografts as predictive preclinical models for anticancer drug activity in humans: better than commonly perceived-but they can be improved. Cancer Biol Ther 2(4 Suppl 1):S134–S139
Kola I (2008) The state of innovation in drug development. Clin Pharmacol Ther 83(2):227–230
Kola I, Landis J (2004) Can the pharmaceutical industry reduce attrition rates? Nat Rev Drug Discov 3(8):711–715
Kunz-Schughart LA, Freyer JP, Hofstaedter F, Ebner R (2004) The use of 3-D cultures for high-throughput screening: the multicellular spheroid model. J Biomol Screen 9(4):273–285
Lallemand-Breitenbach V, Guillemin MC, Janin A et al (1999) Retinoic acid and arsenic synergize to eradicate leukemic cells in a mouse model of acute promyelocytic leukemia. J Exp Med 189(7):1043–1052
Lander ES, Linton LM, Birren B et al (2001) Initial sequencing and analysis of the human genome. Nature 409(6822):860–921
Leaf C (2004) Why we're losing the war on cancer (and how to win it). Fortune 149(6):76–82, 84–6, 88 passim
Macville M, Schröck E, Padilla-Nash H et al (1999) Comprehensive and definitive molecular cytogenetic characterization of HeLa cells by spectral karyotyping. Cancer Res 59(1):141–150
Maggi A, Ciana P (2005) Reporter mice and drug discovery and development. Nat Rev Drug Discov 4(3):249–255

Mandrekar SJ, Sargent DJ (2009) Clinical trial designs for predictive biomarker validation: one size does not fit all. J Biopharm Stat 19(3):530–542
Marks C (2009) Mouse Models of Human Cancers Consortium (MMHCC) from the NCI. Dis Model Mech 2(3–4):111
Marshall E (2002) Dupont ups ante on use of Harvard's oncomouse. Science 296:1212–1213
McCarthy N (2009) Mouse models: closer than you think. Nat Rev Cancer 9(6):382–383
Miettinen PJ, Berger JE, Meneses J et al (1995) Epithelial immaturity and multiorgan failure in mice lacking epidermal growth factor receptor. Nature 376(6538):337–341
Minig L, Trimble EL, Birrer MJ, Kim KY, Takebe N, Abrams JS (2010) NIH and NCI support for development of novel therapeutics in gynecologic cancer: a user's guide. Gynecol Oncol 116(2):177–180
Mitlin N, Baroody AM (1958) Use of the housefly as screening agent for tumorinhibiting agents. Cancer Res 18(6):708–710
Morton CL, Houghton PJ (2007) Establishment of human tumor xenografts in immunodeficient mice. Nat Protoc 2(2):247–250
Nasr R, Guillemin M-C, Ferhi O et al (2008) Eradication of acute promyelocytic leukemia-initiating cells through PML-RARA degradation. Nat Med 14(12):1333–1342
Newell DR (2001) Flasks, fibres and flanks – pre-clinical tumour models for predicting clinical antitumour activity. Br J Cancer 84(10):1289–1290
O'Hagan RC, Wu M, Rideout WM, Zhou Y, Heyer J (2005) Genetically engineered mouse models of human cancer for drug discovery and development. The oncogenomics handbook. Humana, Totowa, NJ, pp 247–261
Olive KP, Tuveson DA (2006) The use of targeted mouse models for preclinical testing of novel cancer therapeutics. Clin Cancer Res 12(18):5277–5287
Pitteri SJ, JeBailey L, Faça VM et al (2009) Integrated proteomic analysis of human cancer cells and plasma from tumor bearing mice for ovarian cancer biomarker discovery. PLoS One 4(11):7916
Powley MW, Frederick CB, Sistare FD, DeGeorge JJ (2009) Safety assessment of drug metabolites: implications of regulatory guidance and potential application of genetically engineered mouse models that express human P450s. Chem Res Toxicol 22(2):257–262
Raubicheck C, White BS, Kowalski TJ, Brown DG, Leahy A, Fekete P (2003) Integra v. Merck: a mixed bag for research tool patents. Nat Biotechnol 21(9):1099–1101
Rego EM, He LZ, Warrell RP, Wang ZG, Pandolfi PP (2000) Retinoic acid (RA) and As_2O_3 treatment in transgenic models of acute promyelocytic leukemia (APL) unravel the distinct nature of the leukemogenic process induced by the PML-RARalpha and PLZF-RARalpha oncoproteins. Proc Natl Acad Sci USA 97(18):10173–10178
Richmond A, Su Y (2008) Mouse xenograft models vs. GEM models for human cancer therapeutics. Dis Model Mech 1(2–3):78–82
Roberts TG, Goulart BH, Squitieri L et al (2004) Trends in the risks and benefits to patients with cancer participating in phase 1 clinical trials. JAMA 292(17):2130–2140
Root DE, Flaherty SP, Kelley BP, Stockwell BR (2003) Biological mechanism profiling using an annotated compound library. Chem Biol 10(9):881–892
Roschke AV, Stover K, Tonon G, Schäffer AA, Kirsch IR (2002) Stable karyotypes in epithelial cancer cell lines despite high rates of ongoing structural and numerical chromosomal instability. Neoplasia 4(1):19–31
Rosenberg MP, Bortner D (1999) Why transgenic and knockout animal models should be used (for drug efficacy studies in cancer). Cancer Metastasis Rev 17(3):295–299
Rottenberg S, Pajic M, Jonkers J (2010) Studying drug resistance using genetically engineered mouse models for breast cancer. Methods Mol Biol 59:633–645
Rubio-Viqueira B, Jimeno A, Cusatis G et al (2006) An in vivo platform for translational drug development in pancreatic cancer. Clin Cancer Res 12(15):4652–4661
Sacca R, Engle SJ, Qin W, Stock JL, McNeish JD (2010) Genetically engineered mouse models in drug discovery research. Methods Mol Biol 60:237–254

Sager J, Lengauer C (2003) New paradigms for cancer drug discovery? Cancer Biol Ther 2(4): 452–455
Sausville EA, Burger AM (2006) Contributions of human tumor xenografts to anticancer drug development. Cancer Res 66(7):3351–3354, discussion 3354
Sausville EA, Feigal E (1999) Evolving approaches to cancer drug discovery and development at the National Cancer Institute, USA. Ann Oncol 10(11):1287–1291
Schein PS, Scheffler B (2006) Barriers to efficient development of cancer therapeutics. Clin Cancer Res 12(11 Pt 1):3243–3248
Schmidt-Supprian M, Wunderlich FT, Rajewsky K (2007) Excision of the Frt-flanked neo (R) cassette from the CD19cre knock-in transgene reduces Cre-mediated recombination. Transgenic Res 16(5):657–660
Shalaby F, Rossant J, Yamaguchi TP et al (1995) Failure of blood-island formation and vasculogenesis in Flk-1-deficient mice. Nature 376(6535):62–66
Sharma SV, Haber DA, Settleman J (2010) Cell line-based platforms to evaluate the therapeutic efficacy of candidate anticancer agents. Nat Rev Cancer 10(4):241–253
Sharpless NE, Depinho RA (2006) The mighty mouse: genetically engineered mouse models in cancer drug development. Nat Rev Drug Discov 5(9):741–754
Shoemaker RH (2006) The NCI60 human tumour cell line anticancer drug screen. Nat Rev Cancer 6(10):813–823
Shultz LD, Ishikawa F, Greiner DL (2007) Humanized mice in translational biomedical research. Nat Rev Immunol 7(2):118–130
Simon R (2010) Clinical trials for predictive medicine: new challenges and paradigms. Clin Trials 7(5):516–524
Singh M, Johnson L (2006) Using genetically engineered mouse models of cancer to aid drug development: an industry perspective. Clin Cancer Res 12(18):5312–5328
Singh M, Lima A, Molina R et al (2010) Assessing therapeutic responses in Kras mutant cancers using genetically engineered mouse models. Nat Biotechnol 28(6):585–593
Stockwell BR (2004) Exploring biology with small organic molecules. Nature 432(7019):846–854
Storrs EE (1971) The nine-banded armadillo: a model for leprosy and other biomedical research. Int J Lepr Other Mycobact Dis 39(3):703–714
Su LK, Kinzler KW, Vogelstein B et al (1992) Multiple intestinal neoplasia caused by a mutation in the murine homolog of the APC gene. Science 256(5057):668–670
Suggitt M, Bibby MC (2005) 50 Years of preclinical anticancer drug screening: empirical to target-driven approaches. Clin Cancer Res 11(3):971–981
Talmadge JE, Singh RK, Fidler IJ, Raz A (2007) Murine models to evaluate novel and conventional therapeutic strategies for cancer. Am J Pathol 170(3):793–804
Tan HT, Low J, Lim SG, Chung MCM (2009) Serum autoantibodies as biomarkers for early cancer detection. FEBS J 276(23):6880–6904
Traggiai E, Chicha L, Mazzucchelli L et al (2004) Development of a human adaptive immune system in cord blood cell-transplanted mice. Science 304(5667):104–107
Uronis JM, Threadgill DW (2009) Murine models of colorectal cancer. Mamm Genome 20(5):261–268
Van Dyke T, Jacks T (2002) Cancer modeling in the modern era: progress and challenges. Cell 108(2):135–144
Venter JC, Adams MD, Myers EW et al (2001) The sequence of the human genome. Science 291(5507):1304–1351
Voskoglou-Nomikos T, Pater JL, Seymour L (2003) Clinical predictive value of the in vitro cell line, human xenograft, and mouse allograft preclinical cancer models. Clin Cancer Res 9(11):4227–4239
Walrath JC, Hawes JJ, Van Dyke T, Reilly KM (2010) Genetically engineered mouse models in cancer research. Adv Cancer Res 106(10):113–164
Waterston RH, Lindblad-Toh K, Birney E et al (2002) Initial sequencing and comparative analysis of the mouse genome. Nature 420(6915):520–562

Watters JW, Cheng C, Majumder PK et al (2009) De novo discovery of a gamma-secretase inhibitor response signature using a novel in vivo breast tumor model. Cancer Res 69(23):8949–8957
Weiss B (2003) Mouse cancer models as a platform for performing preclinical therapeutic trials. Curr Opin Genet Dev 13(1):84–89
Weiss B, Shannon K (2003) Mouse cancer models as a platform for performing preclinical therapeutic trials. Curr Opin Genet Dev 13(1):84–89
Workman P, Aboagye EO, Balkwill F et al (2010) Guidelines for the welfare and use of animals in cancer research. Br J Cancer 102(11):1555–1577
Wu M, Jung L, Cooper AB et al (2009) Dissecting genetic requirements of human breast tumorigenesis in a tissue transgenic model of human breast cancer in mice. Proc Natl Acad Sci USA 106(17):7022–7027
Zambrowicz B (2003) Predicting drug efficacy: knockouts model pipeline drugs of the pharmaceutical industry. Curr Opin Pharmacol 3(5):563–570
Zambrowicz BP, Sands AT (2003) Knockouts model the 100 best-selling drugs – will they model the next 100? Nat Rev Drug Discov 2(1):38–51
Zan Y, Haag JD, Chen K-S et al (2003) Production of knockout rats using ENU mutagenesis and a yeast-based screening assay. Nat Biotechnol 21(6):645–651
Zhou Y, Rideout WM, Zi T et al (2010a) Chimeric mouse tumor models reveal differences in pathway activation between ERBB family- and KRAS-dependent lung adenocarcinomas. Nat Biotechnol 28(1):71–78
Zhou Y, Rideout WM, Zi T et al (2010b) Chimeric mouse tumor models reveal differences in pathway activation between ERBB family- and KRAS-dependent lung adenocarcinomas. Nat Biotechnol 28(1):71–78
Zielinska AE, Walker EA, Stewart PM, Lavery GG (2011) Biochemistry and physiology of hexose-6-phosphate knockout mice. Mol Cell Endocrinol 336(1–2):213–218
Zuber J, Radtke I, Pardee TS et al (2009) Mouse models of human AML accurately predict chemotherapy response. Genes Dev 23(7):877–889

Index

J.E. Green and T. Ried (eds.), *Genetically Engineered Mice for Cancer Research: Design, Analysis, Pathways, Validation and Pre-Clinical Testing*,
DOI 10.1007/978-0-387-69805-2, © Springer Science+Business Media, LLC 2012

H

MIX
Papier aus verantwortungsvollen Quellen
Paper from responsible sources
FSC® C105338

If you have any concerns about our products,
you can contact us on
ProductSafety@springernature.com

In case Publisher is established outside the EU,
the EU authorized representative is:
Springer Nature Customer Service Center GmbH
Europaplatz 3, 69115 Heidelberg, Germany

Printed by Libri Plureos GmbH
in Hamburg, Germany